Ninth Edition

Basic College Mathematics

Margaret L. Lial
American River College

Stanley A. Salzman
American River College

Diana L. Hestwood
Minneapolis Community and Technical College

With the assistance of
Jill K. Owens, Diablo Valley College
and
Linda C. Russell, Minneapolis Community and Technical College

PEARSON

Boston   Columbus   Indianapolis   New York   San Francisco   Upper Saddle River
Amsterdam   Cape Town   Dubai   London   Madrid   Milan   Munich   Paris   Montréal   Toronto
Delhi   Mexico City   São Paulo   Sydney   Hong Kong   Seoul   Singapore   Taipei   Tokyo

Editorial Director	Christine Hoag
Editor in Chief	Maureen O'Connor
Executive Content Editor	Kari Heen
Content Editor	Christine Whitlock
Senior Content Editor	Lauren Morse
Assistant Editor	Rachel Haskell
Senior Managing Editor	Karen Wernholm
Senior Production Project Manager	Kathleen A. Manley
Digital Assets Manager	Marianne Groth
Supplements Production Coordinator	Kerri Consalvo
Media Producer	Stephanie Green
Software Development	Rebecca Williams, MathXL; Mary Durnwald, TestGen
Marketing Manager	Rachel Ross
Senior Author Support/Technology Specialist	Joe Vetere
Rights and Permissions Advisor	Cheryl Besenjak
Image Manager	Rachel Youdelman
Procurement Manager	Evelyn Beaton
Procurement Specialist	Debbie Rossi
Media Procurement Specialist	Ginny Michaud
Associate Director of Design	Andrea Nix
Senior Designer	Barbara Atkinson
Text Design, Production Coordination, Composition, and Illustrations	Cenveo Publisher Services/ Nesbitt Graphics, Inc.
Cover Image	*Between Seasons* © Lorraine Cota Manley

For permission to use copyrighted material, grateful acknowledgment is made to the copyright holders on page P-1, which is hereby made part of this copyright page.

Many of the designations used by manufacturers and sellers to distinguish their products are claimed as trademarks. Where those designations appear in this book, and Pearson Education was aware of a trademark claim, the designations have been printed in initial caps or all caps.

Library of Congress Cataloging-in-Publication Data
Lial, Margaret L.
  Basic college mathematics. / Margaret L. Lial, Stanley A. Salzman,
Diana L. Hestwood.—9th ed.
    p. cm.
Includes bibliographical references and index.
ISBN: 978-0-321-82553-7
1. Mathematics—Textbooks. I. Salzman, Stanley A. II. Hestwood, Diana. III. Title.
QA37.3.L48 2013
513' .1—dc23
2012013816

2 3 4 5 6 7 8 9 10—V011—16 15 14 13

www.pearsonhighered.com

ISBN 13: 978-0-321-82553-7
ISBN 10: 0-321-82553-5

This book is dedicated to Margaret L. Lial

Always passionate about mathematics and teaching,

Always a valued colleague, a mentor, and a friend,

Always in our memory.

In appreciation of your lasting support and never-ending enthusiasm: family, colleagues, and more than a generation of motivated students.

Stan Salzman

This book is dedicated to my dad, who always told me when I was young that girls could learn math, and to my students at Minneapolis Community and Technical College, who keep me in touch with the real world.

Diana Hestwood

Contents

Preface

The ninth edition of *Basic College Mathematics* continues our ongoing commitment to provide the best possible text and supplements package that will help instructors teach and students succeed. To that end, we have addressed the diverse needs of today's students by creating a tightly coordinated text and technology package that includes integrated activities to help students improve their study skills, an attractive design, updated applications and graphs, helpful features, and careful explanations of concepts. We've also expanded the supplements and study aids. We've revamped the video series into a complete Lial Video Library with expanded video coverage and new, easier navigation. And we've added the new Lial MyWorkBook. We have also responded to the suggestions of users and reviewers and have added many new examples and exercises based on their feedback.

The program is designed to help students achieve success in a developmental mathematics program. It provides the necessary review and coverage of whole numbers, fractions, decimals, ratio and proportion, percent, and measurement, as well as an introduction to algebra and geometry and a preview of statistics. It is part of a series that also includes the following books:

- *Essential Mathematics*, Fourth Edition, by Lial and Salzman

- *Prealgebra,* Fifth Edition, by Lial and Hestwood

- *Introductory Algebra*, Tenth Edition, by Lial, Hornsby, and McGinnis

- *Intermediate Algebra,* Tenth Edition, by Lial, Hornsby, and McGinnis

- *Introductory and Intermediate Algebra,* Fifth Edition, by Lial, Hornsby, and McGinnis

- *Prealgebra and Introductory Algebra,* Fourth Edition, by Lial, Hestwood, Hornsby, and McGinnis

- *Developmental Mathematics: Basic Mathematics and Algebra,* Third Edition, by Lial, Hornsby, McGinnis, Salzman, and Hestwood

WHAT'S NEW IN THIS EDITION

The scope and sequence of topics in *Basic College Mathematics* has stood the test of time and rates highly with our reviewers. Therefore, you will find the table of contents intact, making the transition to the new edition easier.

▶ *Examples and Exercises*   Throughout the text, examples and exercises have been adjusted or replaced to reflect current data and practices. Applications have been updated and cover a wider variety of topics, such as the fields of technology, ecology, and health sciences.

▶ *Guided Solutions*   Selected exercises in the margins and in the exercise sets, marked with a **GS** icon, now show the first few solution steps. This provides guidance to students as they start learning a new concept or procedure and gets them off to a successful start. (See p. 301 margin, and p. 356, Exercises 3-4.)

▶ *Concept Checks*   The Concept Checks at the beginning of each exercise set assure students that they have necessary background skills or information to proceed. Concept Checks later in the exercise sets help students summarize and consolidate their learning by writing explanations, correcting common errors, and practicing mathematical processes. (See pp. 285-286.)

▶ *Vocabulary Tips*   Many students at this level of mathematics do not have strong reading skills. The vocabulary tips included in the margins throughout the book help them to learn the meaning of root words and prefixes commonly used in mathematics vocabulary (for example, *equ-*, *centi*, *tri-*), distinguish the mathematical meaning from the common usage of particular words (such as *volume, average*), and provide tips for remembering the difference between often-confused terms (such as LCM and LCD on p. 215).

▶ *Teaching Tips*   Although the mathematical content in this text is familiar to instructors, they may not have experience in teaching the material to adult students. The Teaching Tips, printed in the margins of the Annotated Instructor's Edition, provide helpful comments from colleagues with successful experience at this level. Common trouble spots are noted, with suggestions for improving student understanding. (See p. 276.) Simple, but effective, hands-on activities are also included. (See p. 548.)

▶ *Lial Video Library*   The Lial Video Library, available in MyMathLab and on the Video Resources DVD, provides students with a wealth of video resources to help them navigate the road to success. All video resources in the library include optional subtitles in English. The Lial Video Library includes Section Lecture Videos, Solutions Clips, Quick Review Lectures, and Chapter Test Prep Videos. The Chapter Test Prep Videos are also available on YouTube (searchable using author name and book title), or by scanning the QR code on the inside back cover for easy access.

▶ *MyWorkBook*   This new workbook provides Guided Examples and corresponding Now Try Exercises for each text objective. The extra practice exercises for every section of the text, with ample space for students to show their work, are correlated to Examples, Lecture Videos, and Exercise Solution Clips, to give students the help they need to successfully complete problems. Additionally, MyWorkBook lists the learning objectives and key vocabulary terms for every text section, along with vocabulary practice problems.

▶ *Math in the Media*   Each one-page activity presents a relevant look at how mathematics is used in the media. Designed to help instructors answer the often-asked question, "When will I ever use this stuff?," these activities ask students to read and interpret data from newspaper articles, the Internet, and other familiar, real-world sources. (See p. 196.) The activities are well-suited to collaborative work or they can be completed by individuals or used for open-ended class discussions.

▶ *Study Skills*   Thirteen carefully designed study skills activities provide opportunities for students to practice proven strategies for learning mathematics. Poor study skills and behaviors are major factors in low success rates in mathematics courses. Research shows that a few generic tips sprinkled here and there are not enough to help students change their study behaviors. Because students need specific instruction in study skills, we have contextualized them, integrating them into the text material. Topics include note taking, homework, study cards, math anxiety, test preparation, test taking, preparing for a final exam, and more. (See pp. 138–139, 180–181, and 182–183.) Most are located within the first few chapters so that students can use the skills throughout the course. (See the Contents for titles and locations.) The first activity, "Your Brain *Can* Learn Mathematics," explains how the brain actually learns and remembers so that students understand why the study skills will help them succeed in the course.

HALLMARK FEATURES

We believe students and instructors will welcome these familiar hallmark features.

▶ *Chapter Openers*   The new and engaging Chapter Openers portray real life situations making math relevant for students. (See Chapter 2, p. 113.)

▶ *Real-Life Applications*   We are always on the lookout for interesting data to use in real-life applications. As a result, we have included many new or updated examples and exercises throughout the text that focus on real-life applications of mathematics. Students are often asked to find data in a table, chart, graph, or advertisement. (See pp. 156 and 157.) These applied problems provide an up-to-date flavor that will appeal to and motivate students.

▶ *Figures and Photos*   Today's students are more visually oriented than ever. Thus, we have made a concerted effort to include mathematical figures, diagrams, tables, and graphs whenever possible. (See p. 157.) Many of the graphs use a style similar to that seen by students in today's print and electronic media. Photos have been incorporated to enhance applications in examples and exercises. (See p. 158.)

▶ *Emphasis on Problem Solving*   Introduced at the end of Chapter 1, our six-step process for solving application problems is integrated throughout the text. The six steps, *Read, Plan, Estimate, Solve, State the Answer,* and *Check,* are emphasized in boldface type and repeated in specific problem-solving examples in Chapters 1, 2, 3, 5, 6, 7, and 9. (See p. 162.)

▶ *Learning Objectives*   Each section begins with clearly stated, numbered objectives, and the material within sections is keyed to these objectives so that students know exactly what concepts are covered. (See p. 114.)

▶ *Pointers*   More pointers have been added to examples to provide students with important on-the-spot reminders and warnings about common pitfalls. (See pp. 132 and 162.)

▶ *Cautions and Notes*   These color-coded and boxed comments, one of the most popular features of previous editions, warn students about common errors and emphasize important ideas throughout the exposition. (See pp. 144–145.) Cautions are highlighted in yellow and Notes are highlighted with blue tabs.

▶ *Calculator Tips*   These optional tips, marked with a red calculator icon, offer helpful information and instruction for students using calculators in the course. (See p. 268.)

▶ *Margin Problems*   Margin problems, with answers immediately available on the bottom of the page, are found in every section of the text. (See pp. 121–122.) This key feature allows students to immediately practice the material covered in the examples in preparation for the exercise sets.

▶ *Ample and Varied Exercise Sets*   The text contains a wealth of exercises to provide students with opportunities to practice, apply, connect, and extend the skills they are learning. Numerous illustrations, tables, graphs, and photos help students visualize the problems they are solving. Problem types include skill building, writing, estimation, and calculator exercises, as well as applications and correct-the-error problems. In the Annotated Instructor's Edition of the text, the writing exercises are marked with an icon so that instructors may assign these problems at their discretion. Exercises suitable for calculator work are marked in both the student and instructor editions with a calculator icon . (See pp. 175–179.) Students can watch an instructor work through the complete solution for all exercises marked with a Play Button icon ▶ on the Videos on DVD or in MyMathLab.

▶ *Relating Concepts Exercises*   These help students tie concepts together and develop higher level problem-solving skills as they compare and contrast ideas, identify and describe patterns, and extend concepts to new situations. (See pp. 151, 294, and 306.) These exercises make great collaborative activities for pairs or small groups of students.

▶ *Solutions*   Solutions to selected section exercises are included in the back of the book (following the Answers section). This provides students with easily accessible step-by-step help in solving the exercises that are most commonly missed. Solutions are provided for the exercises marked with a square of blue color around the exercise number, for example, **15.**

▶ *Summary Exercises*   All chapters now include this helpful mid-chapter review. These exercises provide students with the all-important *mixed* practice they need at these critical points in their skill development. (See pp. 140–141.)

▶ *Ample Opportunity for Review*   Each chapter ends with a Chapter Summary featuring: Key Terms with definitions and helpful graphics, New Formulas, New Symbols, Test Your Word Power, and a Quick Review of each section's content with additional examples. Also included is a comprehensive set of Chapter Review Exercises keyed to individual sections, a set of Mixed Review Exercises, and a Chapter Test. Students can watch an instructor work out the full solutions to the Chapter Test problems in the Chapter Test Prep Videos. Beginning with Chapter 2, each chapter concludes with a set of Cumulative Review Exercises. (See pp. 184–195.)

▶ *Test Your Word Power*   This feature, incorporated into each Chapter Summary, helps students understand and master mathematical vocabulary. Key terms from the chapter are presented along with three possible definitions in a multiple-choice format. Answers and examples illustrating each term are provided. (See p. 185.)

STUDENT SUPPLEMENTS

Student's Solutions Manual

- By Jeffery A. Cole, Anoka-Ramsey Community College
- Provides detailed solutions to the odd-numbered section-level exercises and to all margin, Relating Concepts, Summary, Chapter Review, Chapter Test, and Cumulative Review Exercises
  ISBNs: 0-321-83660-X, 978-0-321-83660-1

NEW MyWorkBook

- Provides Guided Examples and corresponding Now Try Exercises for each text objective
- Refers students to correlated Examples, Lecture Videos, and Exercise Solution Clips
- Includes extra practice exercises for every section of the text with ample space for students to show their work
- Lists the learning objectives and key vocabulary terms for every text section, along with vocabulary practice problems
  ISBNs: 0-321-83682-0, 978-0-321-83682-3

NEW Lial Video Library

The Lial Video Library, available in MyMathLab and on the Video Resources DVD, provides students with a wealth of video resources to help them navigate the road to success! All video resources in the library include optional subtitles in English. The Lial Video Library includes the following resources:

- **Section Lecture Videos** offer a new navigation menu that allows students to easily focus on the key examples and exercises that they need to review in each section. Optional Spanish subtitles are available.
- **Solutions Clips** show an instructor working through the complete solutions to selected exercises from the text. Exercises with a solution clip are marked in the text and e-book with a Play Button icon ▶.
- **Quick Review Lectures** provide a short summary lecture of each key concept from the Quick Reviews at the end of every chapter in the text.
- **The Chapter Test Prep Videos** provide step-by-step solutions to all exercises from the Chapter Tests. These videos provide guidance and support when students need it the most: the night before an exam. The Chapter Test Prep Videos are also available on YouTube (searchable using author name and book title), or by scanning the QR code on the inside back cover for easy access.

INSTRUCTOR SUPPLEMENTS

Annotated Instructor's Edition

- Provides answers to all text exercises in color next to the corresponding problems
- Icons identify writing ✍ and calculator ▦ exercises
  ISBNs: 0-321-82631-0, 978-0-321-82631-2

Instructor's Solutions Manual (Download only)

- By Jeffery A. Cole, Anoka-Ramsey Community College
- Provides complete solutions to all exercises in the text
- Available for download at www.pearsonhighered.com
  ISBNs: 0-321-82568-3, 978-0-321-82568-1

Instructor's Resource Manual with Tests and Mini-Lectures (Download only)

- Contains a test bank with two diagnostic pretests, six free-response and two multiple-choice test forms per chapter, and two final exams
- Contains a mini-lecture for each section of the text with objectives, key examples, and teaching tips
- Includes a correlation guide from the eighth to the ninth edition and phonetic spellings for all key terms in the text
- Includes resources to help both new and adjunct faculty with course preparation and classroom management, by offering helpful teaching tips correlated to the sections of the text
- Available for download at www.pearsonhighered.com
  ISBNs: 0-321-57461-3, 978-0-321-57461-9

ADDITIONAL MEDIA SUPPLEMENTS

MyMathLab® MyMathLab® Online Course (access code required)

MyMathLab delivers **proven results** in helping individual students succeed. It provides **engaging experiences** that personalize, stimulate, and measure learning for each student. And, it comes from a **trusted partner** with educational expertise and an eye on the future.

To learn more about how MyMathLab combines proven learning applications with powerful assessment, visit **www.mymathlab.com** or contact your Pearson representative.

MyMathLab® Ready to Go Course (access code required)

These new Ready to Go courses provide students with all the same great MyMathLab features that they are used to but make it easier for instructors to get started. Each course includes pre-assigned homework and quizzes to make creating a course even simpler.

Ask your Pearson representative about the details for this particular course or to see a copy of this course.

MyMathLab® Plus/MyStatLab™ Plus

MyLabsPlus combines proven results and engaging experiences from MyMathLab® and MyStatLab™ with convenient management tools and a dedicated services team. Designed to support growing math and statistics programs, it includes additional features such as:

- **Batch Enrollment:** Schools can create the login name and password for every student and instructor, so everyone can be ready to start class on the first day. Automation of this process is also possible through integration with the school's Student Information System.
- **Login from your campus portal:** Instructors and their students can link directly from their campus portal into the MyLabsPlus courses. A Pearson service team works with the institution to create a single sign-on experience for instructors and students.
- **Advanced Reporting:** MyLabsPlus's advanced reporting enables instructors to review and analyze students' strengths and weaknesses by tracking their performance on tests, assignments, and tutorials. Administrators can review grades and assignments across all courses on a MyLabsPlus campus for a broad overview of program performance.
- **24/7 Support:** Students and instructors receive 24/7 support, 365 days a year, by email or online chat.

MyLabsPlus is available to qualified adopters. For more information, visit our website at *www.mylabsplus.com* or contact your Pearson representative.

MathXL® MathXL® Online Course (access code required)

MathXL® is the homework and assessment engine that runs MyMathLab. (MyMathLab is MathXL plus a learning management system.)

With MathXL, instructors can:

- Create, edit, and assign online homework and tests using algorithmically-generated exercises correlated at the objective level to the textbook.
- Create and assign their own online exercises and import TestGen tests for added flexibility.
- Maintain records of all student work tracked in MathXL's online gradebook.

With MathXL, students can:

- Take chapter tests in MathXL and receive personalized study plans and/or personalized homework assignments based on their test results.
- Use the study plan and/or the homework to link directly to tutorial exercises for the objectives they need to study.
- Access supplemental animations and video clips directly from selected exercises.

MathXL is available to qualified adopters. For more information, visit our website at *www.mathxl.com*, or contact your Pearson representative.

TestGen®

TestGen® (*www.pearsoned.com/testgen*) enables instructors to build, edit, print, and administer tests using a computerized bank of questions developed to cover all the objectives of the text. TestGen is algorithmically based, allowing instructors to create multiple but equivalent versions of the same question or test with the click of a button. Instructors can also modify test bank questions or add new questions. The software and testbank are available for download from Pearson Education's online catalog.

PowerPoint® Lecture Slides

- Present key concepts and definitions from the text
- Available for download at *www.pearsonhighered.com* or in MyMathLab

ACKNOWLEDGMENTS

The comments, criticisms, and suggestions of users, nonusers, instructors, and students have positively shaped this textbook over the years, and we are most grateful for the many responses we have received. The feedback gathered for this revision of the text was particularly helpful, and we especially wish to thank the following individuals who provided invaluable suggestions for this and the previous edition:

George Alexander, *University of Wisconsin*
Sonya Armstrong, *West Virginia State University*
Vernon Bridges , *Durham Technical Community College*
Solveig R. Bender, *William Rainey Harper College*
Barbara Brown, *Anoka-Ramsey Community College*
Ernie Chavez, *Gateway Community College*
Terry Joe Collins, *Hinds Community College*
Martha Daniels, *Central Oregon Community College*
Cheryl Eichenseer, *St. Charles Community College*
Matthew Flacche, *Camden Community College*
James V. Figliolia, *Cape Fear Community College*
Donna Foster, *Piedmont Technical College*
Lindsay A. Gold, *Sinclair Community College*
Mark Gollwitzer, *Greenville Technical College*
Lourdes Gonzalez, *Miami-Dade Community College*
Kimberly Gregor, *Delaware Technical & Community College*
Lance Hemlow, *Raritan Valley Community College*
Joe Howe, *St. Charles County Community College*
Matthew Hudock, *St Philip's College*
Rose Kaniper, *Burlington County College*

Cameron S. Kishel, *Columbus State Community College*
Douglas Lewis, *Yakima Valley Community College*
Yixia Lu, *South Suburban College*
Valerie Maley, *Cape Fear Community College*
Connie L. McLean, *Black Hawk College*
Judy Mee, *Oklahoma City Community College*
Wayne Miller, *Lee College*
John Notini, *William Rainey Harper College*
Kathy Peay, *Sampson Community College*
Thea Philliou, *College of Santa Fe*
Jane Roads, *Moberly Area Community College*
Richard D. Rupp, *Del Mar College*
Ellen Sawyer, *College of DuPage*
Lois Schuppig, *College of Mount St. Joseph*
Mary Lee Seitz, *Erie Community College—City Campus*
Kathryn Taylor, *Santa Ana College*
Mike Tieleman-Ward, *Anoka Technical College*
Sven Trenholm, *North Country Community College*
Bettie A. Truitt, *Black Hawk College*
Jackie Wing, *Angelina College*

Our sincere thanks go to the dedicated individuals at Pearson who have worked hard to make this revision a success: Maureen O'Connor, Kathy Manley, Barbara Atkinson, Michelle Renda, Rachel Ross, Kari Heen, Christine Whitlock, Lauren Morse, Stephanie Green, and Rachel Haskell. We are also grateful to Marilyn Dwyer and Carol Merrigan of Cenveo/Nesbitt Graphics for their excellent production work; Bonnie Boehme, for supplying her copyediting expertise; Beth Anderson, for her fine photo research; Lucie Haskins, for producing a useful Index; Jeff Cole, for writing the Solutions Manuals; Perian Herring, for writing many of the Vocabulary Tips and accuracy checking the manuscript; Sam Tinsley, for his accuracy checking; and Judy Martinez, for her excellent typing skills.

Special thanks go to Jill Owens, an instructor at Diablo Valley College, who assisted with manuscript development and provided accuracy checking throughout this edition, and to Janis M. Cimperman, Associate Professor of Mathematics at St. Cloud State University, for her careful review of the new material in selected chapters.

Linda Russell, who developed and wrote the Study Skills activities that appear throughout this text, also worked to make the text more readable for developmental-level students, wrote many of the vocabulary tips, and provided much-needed help during the production phase. Her many years of teaching and working with students at this level was invaluable.

The ultimate measure of this textbook's success is whether it helps students master basic skills, develop problem-solving techniques, and increase their confidence in learning and using mathematics. In order for us, as authors, to know what to keep and what to improve for the next edition, we need to hear from you, the instructor, and you, the student. Please tell us what you like and where you need additional help by sending an e-mail to *math@pearson.com*. We appreciate your feedback!

Stanley A. Salzman
Diana L. Hestwood

1 Whole Numbers

The Golden Gate Bridge, connecting San Francisco and Marin Counties, was completed in 1937 and is 8981 feet long. The bridge is supported by two main towers which are constructed with 1,200,000 rivets, rises 746 feet above the water, and has had more than 2,153,896,448 cars and trucks cross over it. In this chapter we discuss whole numbers, which are used daily in our lives.

Study Skills
YOUR BRAIN *CAN* LEARN MATHEMATICS

OBJECTIVES

1 **Describe how practice fosters dendrite growth.**

2 **Explain the effect of anxiety on the brain.**

Your brain knows how to learn, just as your lungs know how to breathe; however, there are important things you can do to maximize your brain's ability to do its work. This short introduction will help you choose effective strategies for learning mathematics. This is a simplified explanation of a complex process.

Your brain's outer layer is called the **neocortex**, which is where higher level thinking, language, reasoning, and purposeful behavior occur. The neocortex has about 100 billion (100,000,000,000) brain cells called **neurons.**

Learning Something New

▶ As you learn something new, threadlike branches grow out of each neuron. These branches are called **dendrites.**

▶ When the dendrite from one neuron grows close enough to the dendrite from another neuron, a connection is made. There is a small gap at the connection point called a **synapse.** One dendrite sends an electrical signal across the gap to another dendrite.

▶ *Learning = growth and connecting of dendrites.*

Remembering New Skills

▶ When you practice a skill just once or twice, the connections between neurons are very weak. If you do not practice the skill again, the dendrites at the connection points wither and die back. You have forgotten the new skill!

A neuron with several dendrites: one dendrite has developed a myelin coating through repeated practice.

A close-up view of the connection (synapse) between two dendrites.

▶ If you practice a new skill many times, the dendrites for that skill become coated with a fatty protein called **myelin.** Each time one dendrite sends a signal to another dendrite, the myelin coating becomes thicker and smoother, allowing the signals to move faster and with less interference. Thinking can now occur more quickly and easily, and *you will remember the skill for a long time* because the dendrite connections are very strong.

Become An Effective Student

▶ You grow dendrites specifically for the thing you are studying. If you practice dividing fractions, you will grow specialized dendrites just for dividing fractions. If you *watch other people* solve fraction problems, *you will grow dendrites for watching, not for solving.* So, be sure you are actively learning and practicing.

▶ If you practice something the *wrong* way, you will develop strong dendrite connections for doing it the wrong way! So, as you study, check frequently that you are getting correct answers.

▶ As you study a new topic that is related to things you already know, you will grow new dendrites, but your brain will also send signals throughout the network of dendrites for the related topics. In this way, you build a complex **neural network** that allows you to apply concepts, see differences and similarities between ideas, and understand relationships between concepts.

In the first few chapters of this textbook you will find "brain friendly" activities that are designed to help you grow and develop your own reliable neural networks for mathematics. Since you must grow your own dendrites (no one can grow them for you), these activities show you how to

▶ develop new dendrites,

▶ strengthen existing ones, and

▶ encourage the myelin coating to become thicker so signals are sent with less effort.

When you incorporate the activities into your regular study routine, you will discover that you understand better, remember longer, and forget less.

Also remember that *it does take time for dendrites to grow.* Trying to cram in several new concepts and skills at the last minute is not possible. Your dendrites simply can't grow that quickly. You can't expect to develop huge muscles by lifting weights for just one evening before a body building competition! In the same way, practice the study techniques *throughout the course* to facilitate strong growth of dendrites.

When Anxiety Strikes

If you are under stress or feeling anxious, such as during a test, your body secretes **adrenaline** into your system. Adrenaline in the brain blocks connections between neurons. In other words, you can't think! If you've ever experienced "blanking out" on a test, you know what adrenaline does. You'll learn several solutions to that problem in later activities.

Start Your Course Right!

▶ Attend all class sessions (especially the first one).

▶ Gather the necessary supplies.

▶ Carefully read the syllabus for the course, and ask questions if you don't understand.

1.1 Reading and Writing Whole Numbers

OBJECTIVES

1. Identify whole numbers.
2. Give the place value of a digit.
3. Write a number in words or digits.
4. Read a table.

VOCABULARY TIP

Place value In our decimal number system, each place has a value of 10 times the place to its right.

1 Identify the place value of the 4 in each whole number.

GS (a) 342

(b) 714

(c) 479

Knowing how to read and write numbers is important in mathematics.

OBJECTIVE ① Identify whole numbers. The **decimal system** of writing numbers uses the ten digits

$$0, 1, 2, 3, 4, 5, 6, 7, 8, 9$$

to write any number. These digits can be used to write **the whole numbers:**

$$0, 1, 2, 3, 4, 5, 6, 7, 8, 9, 10, 11, 12, 13 \ldots$$

The three dots indicate that the list goes on forever.

OBJECTIVE ② Give the place value of a digit. Each digit in a whole number has a **place value,** depending on its position in the whole number. The following place value chart shows the names of the different places used most often and has the whole number 402,759,780 entered.

The United States is the leading consumer of coffee in the world. Each day we drink 402,759,780 cups of coffee. Each of the 7s in 402,759,780 represents a different amount because of its position, or *place value*, within the number. The *place value* of the 7 on the left is 7 hundred-thousands (700,000). The *place value* of the 7 on the right is 7 hundreds (700).

EXAMPLE 1 Identifying Place Values

Identify the place value of 8 in each whole number.

> Each "8" has a different value.

(a) 28 — 8 ones

(b) 85 — 8 tens

(c) 869 — 8 hundreds

Notice that the value of 8 in each number is different, depending on its location (place) in the number.

◀ **Work Problem ① at the Side.**

EXAMPLE 2 Identifying Place Values

Identify the place value of each digit in the number 725,283.

7 2 5, 2 8 3
- 3 ones
- 8 tens
- 2 hundreds
- 5 thousands
- 2 ten-thousands
- 7 hundred-thousands

Continued on Next Page

Answers

1. (a) tens (b) ones (c) hundreds

Notice the comma between the hundreds and thousands position in the number 725,283 in **Example 2.**

·· **Work Problem 2 at the Side.** ▶

Using Commas

Commas are used to separate each group of three digits, starting from the right. This makes numbers easier to read. (An exception: Commas are frequently omitted in four-digit numbers such as 9748 or 1329.) Each three-digit group is called a **period.** Some instructors prefer to just call them **groups.**

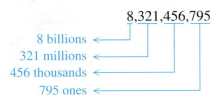

··· **EXAMPLE 3**   Knowing the Period or Group Names

Write the digits in each period of 8,321,456,795.

$$8,321,456,795$$

8 billions ←
321 millions ←
456 thousands ←
795 ones ←

·· **Work Problem 3 at the Side.** ▶

Use the following rule to read a number with more than three digits.

Writing Numbers in Words

Start at the left when writing a number in words or saying it aloud. Write or say the digit names in each period (group), followed by the name of the period, except for the period name "ones," which is *not* used.

OBJECTIVE ▶ 3 **Write a number in words or digits.** The following examples show how to write names for whole numbers.

··· **EXAMPLE 4**   Writing Numbers in Words

Write each number in words.

(a) 57
     This number means 5 tens and 7 ones, or 50 ones and 7 ones. Write the number as

fifty-seven.

··· **Continued on Next Page**

---

2 Identify the place value of each digit.

GS (a) 14,218
                            8 ones
                        _____ tens
                        _____ hundreds
                        _____ thousands
                        _____ ten-thousands

(b) 460,329

3 In the number 3,251,609,328 identify the digits in each period (group).

(a) billions period

(b) millions period

(c) thousands period

(d) ones period

Answers

2. (a) 1 : ten-thousands
         4 : thousands
         2 : hundreds
         1 : tens
         8 : ones
    (b) 4 : hundred-thousands
         6 : ten-thousands
         0 : thousands
         3 : hundreds
         2 : tens
         9 : ones

3. (a) 3 **(b)** 251 **(c)** 609 **(d)** 328

4 Write each number in words.

(a) 18

(b) 36

(c) 418

(d) 902

5 Write each number in words.

GS (a) 3104

three _____ ,

one _____ four

GS (b) 95,372

ninety-five _____ ,
three hundred seventy-two

(c) 100,075,002

(d) 11,022,040,000

6 Rewrite each number using digits.

(a) one thousand, four hundred thirty-seven

(b) nine hundred seventy-one thousand, six

(c) eighty-two million, three hundred twenty-five

Answers

4. (a) eighteen
(b) thirty-six
(c) four hundred eighteen
(d) nine hundred two
5. (a) three <u>thousand</u>, one <u>hundred</u> four
(b) ninety-five <u>thousand</u>, three hundred seventy-two
(c) one hundred million, seventy-five thousand, two
(d) eleven billion, twenty-two million, forty thousand
6. (a) 1437 (b) 971,006 (c) 82,000,325

(b) 94

ninety-four

Remember: Start at the *left* to read a number.

(c) 874

eight hundred seventy-four

(d) 601

six hundred one

◀ **Work Problem 4 at the Side.**

CAUTION

The word *and* should never be used when writing whole numbers. You will often hear someone say "five hundred *and* twenty-two," but the use of "and" is not correct since "522" is a whole number. When you work with decimal numbers, the word *and* is used to show the position of the decimal point. For example, 98.6 is read as "ninety-eight *and* six tenths." Practice with decimal numbers is a topic in **Chapter 4**.

EXAMPLE 5 Writing Numbers in Words by Using Period Names

Write each number in words.

(a) 725,283

seven hundred twenty-five **thousand,** two hundred eighty-three

Number in period   Name of period   Number in period (not necessary to write "ones")

(b) 7252

Careful: *Do not* use "and" when reading a whole number.

seven **thousand,** two hundred fifty-two

Name of period   No period name needed

(c) 111,356,075

one hundred eleven **million,** three hundred fifty-six **thousand,** seventy-five

The period name is not used for the ones period.

(d) 17,000,017,000

seventeen **billion,** seventeen **thousand**

◀ **Work Problem 5 at the Side.**

EXAMPLE 6 Writing Numbers in Digits

Rewrite each number using digits.

(a) six **thousand,** twenty-two

6022

With 4 digits or fewer, *no comma* is needed.

(b) two hundred fifty-six **thousand,** six hundred twelve

256,612

(c) nine **million,** five hundred fifty-nine

9,000,559

Zeros indicate there are no thousands.

◀ **Work Problem 6 at the Side.**

Calculator Tip

Does your calculator show a comma between each group of three digits? Probably not, but try entering a long number such as 34,629,075. Notice that there is no key with a comma on it, so you do not enter commas. A few calculators may show the position of the commas *above* the digits, like this

$$34'629'075$$

Most of the time you will have to write in the commas where needed.

OBJECTIVE ▶ **4** **Read a table.** A common way of showing number values is by using a **table**. Tables organize and display facts so that they are more easily understood. The following table shows some past facts and future predictions for the United States. These numbers give us a glimpse of what we can expect in the 21st century.

NUMBERS FOR THE 21ST CENTURY

Year	1990	2010	2020*
U.S. population	261 million	309 million	338 million
Household income	$42,936	$46,326	$55,735
Average yearly salary	$21,129	$28,834	$32,080

*Estimated figures

Source: *Family Circle* magazine; U.S. Census Bureau.

If you read from left to right along the row labeled "U.S. population," you find that the population in 1990 was 261 million, then the population in 2010 was 309 million, and the estimated population for 2020 is 338 million.

EXAMPLE 7   **Reading a Table**

Use the table to find each number, and write the number in words.

(a) The estimated household income in the year 2020
Read from left to right along the row labeled "Household income" until you reach the 2020 column and find $55,735.

Fifty-five thousand, seven hundred thirty-five dollars

(b) The average yearly salary in 1990
Read from left to right along the row labeled "Average yearly salary." In the 1990 column you find $21,129.

Remember: Use hyphens when necessary.

Twenty-one thousand, one hundred twenty-nine dollars

························ **Work Problem 7 at the Side.** ▶

Note

Notice in **Example 7** that hyphens are used when writing numbers in words. A hyphen is used when writing the numbers 21 through 99 (twenty-one through ninety-nine), except for numbers ending in zero (20, 30, 40, ... 90).

---

7 Use the table to find each number, and write the number in digits when given in words, or write the number in words when given in digits.

GS (a) The population in 2010

The U.S. population in the 2010 column is 309 million and is written in digits as 3__ __ ,000,000

(b) The estimated population in 2020

(c) Household income in 1990

(d) The estimated average yearly salary in 2020

Answers

7. **(a)** 0; 9; 309,000,000
(b) 338,000,000
(c) forty-two thousand, nine hundred thirty-six dollars
(d) thirty-two thousand, eighty dollars

1.1 Exercises

FOR EXTRA HELP

 Download the MyDashBoard App

 MyMathLab®

CONCEPT CHECK   *Choose the letter of the correct response.*

1. The digit in the hundreds place in the whole number 3065 is

   (a) 5   **(b)** 3   **(c)** 0   **(d)** 6

2. The digit in the ten-thousands place in the whole number 134,681 is

   (a) 6   **(b)** 3   **(c)** 8   **(d)** 1

*Write the digit for the given **place value** in each whole number. See Examples 1 and 2.*

3. 18,015
   ten-thousands
   hundreds

4. 86,332
   ten-thousands
   ones

5. 7,628,592,183
   millions
   thousands

6. 1,700,225,016
   billions
   millions

CONCEPT CHECK   *Identify the correct period.*

7. Write the digits in the thousands period in the whole number 552,687,318.

8. Write the digits in the millions period in the whole number 947,321,876,528.

*Write the digits for the given **period** (group) in each whole number. See Example 3.*

9. 3,561,435
   millions
   thousands
   ones

10. 100,258,100,006
   billions
   millions
   thousands
   ones

11. Do you think the fact that humans have four fingers and a thumb on each hand explains why we use a number system based on ten digits? Explain.

12. The decimal system uses ten digits. Fingers and toes are often referred to as digits. In your opinion, is there a relationship here? Explain.

CONCEPT CHECK   *Answer* true *or* false *for each statement.*

13. The number 23,115 is written in words as twenty-three thousand and one hundred and fifteen.

14. The number 37,886 is written in words as thirty-seven thousand, eight hundred eighty-six.

Write each number in words. See Examples 4 and 5.

15. 346,009
   three hundred forty six _____, _____

16. 218,033
   two hundred eighteen _____, thirty- _____

17. 25,756,665

18. 999,993,000

Write each number using digits. See Example 6.

19. sixty-three thousand, one hundred sixty-three

   6 __, 1 __ __

20. ninety-five thousand, one hundred eleven

   9 __, __ __ 1

21. ten million, two hundred twenty-three

22. one hundred million, two hundred

*Write the numbers from each sentence using digits. **See Example 6.***

23. There are three million, two hundred thousand parachute jumps in the United States each year. (*Source:* History Channel.)

24. A full-grown caterpillar is 27,000 times its birth size. A 9-pound human baby growing at the same rate would weigh two hundred forty-three thousand pounds by college graduation. (*Source: Spirit Magazine.*)

25. The number of cans of Pepsi Cola sold each day is fifty million, fifty-one thousand, five hundred seven. (*Source:* Andy Rooney, *60 Minutes.*)

26. In the United States we drink one hundred forty-six billion, three hundred eighty-five million cups of coffee every year. (*Source: Spirit Magazine.*)

27. There are fifty-four million, seven hundred fifty thousand Hot Wheels sold each year. (*Source:* Andy Rooney, *60 Minutes.*)

28. A middle-income family will typically spend two hundred twenty-one thousand dollars to raise a child to the age of eighteen. (*Source: Los Angeles Times.*)

29. Rewrite eight hundred trillion, six hundred twenty-one million, twenty thousand, two hundred fifteen by using digits.

30. Rewrite 2,153,896,448, the number of vehicles that have crossed the Gloden Gate Bridge, in words.

*The table at the right shows various ways people get to work. Use the table to answer Exercises 31–34. **See Example 7.***

31. Which method of transportation is least used? Write the number in words.

32. Which method of transportation is most used? Write the number in words.

33. Find the number of people who walk to work or work at home, and write it in words.

34. Find the number of people who carpool, and write it in words.

GETTING TO WORK

How workers 16 and over get to work:

Drive alone	**84,215,298**
Carpool	**15,377,634**
Walk or work at home	**7,894,911**
Use public transportation	**6,069,589**

Source: U.S. Census Bureau.

Study Skills
USING YOUR TEXTBOOK

Be sure to read Your Brain Can Learn Mathematics before this activity. It is at the beginning of this chapter. You'll find out how your brain learns and remembers.

Your textbook can be very helpful. Find out what it has to offer. First, let's look at some general features that will help in all chapters.

Table of Contents

Look at page v in the very front for the Table of Contents. Before you start Chapter 1, you should look at the Preface, which begins on page vii. On page x is a list of Supplementary Resources for students. If you are interested in any of these, ask your instructor if they are available.

Section Numbering

Each chapter is divided into sections, and each section has a number, such as 1.3 or 3.5. Your instructor will use these numbers to assign readings and homework.

Chapter 3 → 3.5 ← Section 5 within Chapter 3

Chapter Features

There are four features to pay special attention to as you work in your book.

▶ **Objectives.** Each section lists the objectives in the upper corner of the first page. The objectives are listed again as each one is introduced. An objective tells you *what you will be able to do after you complete the section*. An excellent way to check your learning is to go back to the list of objectives when you are finished with a section and ask yourself if you can do them all.

▶ **Margin Exercises.** The exercises in the blue shaded margins of the pages in your textbook give you immediate practice. **This is a perfect way to get your dendrites growing right away!** The answers are given at the bottom, so you can check yourself easily.

▶ **Cautions, Pointers, Notes, and Calculator Tips.**

- **Caution!** A yellow box is a comment about a common error that students make or a common trouble spot you will want to avoid.

- **Pointers** are little "clouds" next to worked examples. They point to specific places where common mistakes are made and give on-the-spot reminders.

- Look for the specially marked **Note** boxes. They contain hints, explanations, or interesting side comments about a topic.

- A small picture of a red calculator ▦ appears several places. In the main part of the chapter, the icon means that there is a **Calculator Tip,** which helps you learn more about using your calculator. A calculator in an Exercise section is a recommendation to use your calculator to work that exercise.

- **Vocabulary Tips** appear in the margins to help you learn the language of mathematics. Use them to create study cards for terminology that is new to you.

List a page number from Chapter 1 for each of these features:

A *Caution* appears on page _____.

A *Pointer* appears on page _____.

A *Note* appears on page _____.

A *Calculator Tip* appears on page _____.

End of Chapter Features

Go back to the Table of Contents again. What is listed at the end of each chapter?

▶ **Chapter Summary.** Turn to page 98 to find the Summary for Chapter 1. It lists the chapter's **Key Terms** (arranged in the order that they appear in the chapter) and **New Symbols** and/or **New Formulas**. Then, **Test Your Word Power** checks your understanding of the math vocabulary. Next is a **Quick Review section.** It lists each topic in the chapter and shows a worked example, with tips.

▶ **Review Exercises** Use these exercises as a way to check your understanding of all the concepts in the chapter. You can practice every type of problem. If you get stuck, the numbers inside the dark blue rectangles tell you which section of the chapter to go back to for more explanations. Make sure you do the **Mixed Review Exercises** to practice for tests.

▶ **Chapter Test.** Plan to take the test as a practice exam. That way you can be sure you really know how to work all types of problems without looking back at the chapter.

▶ **Cumulative Review (starting with Chapter 2)** These exercises help you maintain the skills you've learned in all previous chapters. Working on previous skills throughout the course will be a big help on the final exam.

> How will you make good use of the features at the end of each chapter?
> _____
> _____
> _____
> _____

Answers

How do you find out if you've worked the exercises correctly? Your textbook provides many of the answers. Throughout each chapter you should work the sample problems in the **margins.** The answers for those are at the **bottom of each page** in the margin area.

For homework, you can find the answers to all of the **odd-numbered section exercises** in the **Answers to Selected Exercises** section near the end of your textbook. Also, *all* of the answers are given for the Chapter Review Exercises, Chapter Tests, and Cumulative Reviews. Check your textbook now, and find the page on which the Answers section begins.

> *Flag the Answers section with a sticky note or other device, so that you can turn to it quickly.*

Solutions

The **Solutions** section near the end of the book shows how to solve some of the harder odd-numbered exercises step by step. In the exercise sets, look for the exercise numbers with a square of blue shading around them. These are the ones that have a solution in the back of the book.

Index

The **Index** is the last thing in your book. It lists all the topics, vocabulary, and concepts in alphabetical order. For example, look up the words below. There may be several subheadings listed under the main word, or several page numbers listed. Notice that the page number printed in bold type is where the word is introduced and defined. Write down the boldface page number for each word. Then go to that page and find the word.

> *Commutative property of multiplication* is defined on page _____ .
> *Factors* of numbers are defined on page _____ .
> *Rounding of mixed numbers* is explained on page _____ .

Why Are These Features Brain Friendly?

The textbook authors included text features that make it easier for you to understand the mathematics. **Your brain naturally seeks organization and predictability.** When you pay attention to the regular features of your textbook, you are allowing your brain to get familiar with all of the helpful tips, suggestions, and explanations that your book has to offer. You will make the best possible use of your textbook.

1.2 Adding Whole Numbers

OBJECTIVES

1. Add two single-digit numbers.
2. Add more than two numbers.
3. Add when regrouping (carrying) is not required.
4. Add with regrouping (carrying).
5. Use addition to solve application problems.
6. Check the answer in addition.

VOCABULARY TIP

Commutative property of addition You can remember the commutative property by thinking of the numbers "commuting," or changing places.

1 Add, and then change the order of numbers to write another addition problem.

(GS) (a) $2 + 6 \quad = \underline{\qquad}$

$6 + \underline{\qquad} = \underline{\qquad}$

(b) $9 + 5$

(c) $4 + 7$

(d) $6 + 9$

VOCABULARY TIP

Associative property of addition You can remember the associative property by thinking of two numbers associating with each other, and then one leaves to associate with another number.

Answers

1. **(a)** $8; 6 + 2 = 8$ **(b)** $14; 5 + 9 = 14$
   (c) $11; 7 + 4 = 11$ **(d)** $15; 9 + 6 = 15$

There are four baseballs at the left and two at the right. In all, there are six.

The process of finding the total is called **addition.** Here 4 and 2 were added to get 6. Addition is written with a + sign, so that

$$4 + 2 = 6.$$

OBJECTIVE ▶ **1** **Add two single-digit numbers.** In addition, the numbers being added are called **addends,** and the resulting answer is called the **sum** or **total.**

$$
\begin{array}{r}
4 \leftarrow \text{Addend} \\
+\ 2 \leftarrow \text{Addend} \\
\hline
6 \leftarrow \text{Sum (total)}
\end{array}
$$

Addition problems can also be written horizontally, as follows.

$$
\underset{\underset{\text{Addend}}{\uparrow}}{4} \ + \ \underset{\underset{\text{Addend}}{\uparrow}}{2} \ = \ \underset{\underset{\text{Sum}}{\uparrow}}{6}
$$

Commutative Property of Addition

By the **commutative property of addition,** changing the order of the addends in an addition problem does not change the sum.

For example, the sum of $4 + 2$ is the same as the sum of $2 + 4$. This allows the addition of the same numbers in a different order.

EXAMPLE 1 Adding Two Single-Digit Numbers

Add, and then change the order of numbers to write another addition problem.

(a) $5 + 3 = 8$ and $3 + 5 = 8$

(b) $7 + 8 = 15$ and $8 + 7 = 15$

(c) $8 + 3 = 11$ and $3 + 8 = 11$

(d) $8 + 8 = 16$

> Changing the order in addition does not change the sum.

◀ **Work Problem 1 at the Side.**

Associative Property of Addition

By the **associative property of addition,** changing the grouping of the addends in an addition problem does not change the sum.

For example, the sum of $3 + 5 + 6$ may be found as follows.

$$(3 + 5) + 6 = 8 + 6 = 14 \qquad \text{Parentheses tell us to add } 3 + 5 \text{ first.}$$

Another way to add the same numbers is

> Changing the grouping of addends does not change the sum.

$$3 + (5 + 6) = 3 + 11 = 14. \qquad \text{Parentheses tell us to add } 5 + 6 \text{ first.}$$

Either grouping gives a sum of 14 because of the associative property of addition.

OBJECTIVE **2** **Add more than two numbers.** To add several numbers, first write them in a column. Add the first number to the second. Add this sum to the third number. Continue until all the numbers are used.

EXAMPLE 2 **Adding More Than Two Numbers**

Add 2, 5, 6, 1, and 4.

$$\begin{array}{r} 2 \\ 5 \\ 6 \\ 1 \\ + 4 \\ \hline 18 \end{array}$$

2 + 5 = 7
7 + 6 = 13
13 + 1 = 14
14 + 4 = 18

····· **Work Problem 2 at the Side.** ▶

Note

By the commutative and associative properties of addition, numbers may also be added starting at the bottom of a column. Adding from the top or adding from the bottom will give the same answer.

OBJECTIVE **3** **Add when regrouping (carrying) is not required.** If numbers have two or more digits, you must arrange the numbers in columns so that the ones digits are in the same column, tens are in the same column, hundreds are in the same column, and so on. Next, you add column by column starting at the right.

EXAMPLE 3 **Adding without Regrouping**

Add 511 + 23 + 154 + 10.

First line up the numbers in columns, with the ones column at the right.

Hundreds in a column
Tens in a column
Ones in a column

$$\begin{array}{r} 511 \\ 23 \\ 154 \\ + 10 \end{array}$$

Ones digits at the right

Now start at the right and add the ones digits. Add the tens digits next, and finally, the hundreds digits.

$$\begin{array}{r} 511 \\ 23 \\ 154 \\ + 10 \\ \hline 698 \end{array}$$

Always begin addition in the ones column.

Sum of ones
Sum of tens
Sum of hundreds

The sum of the four numbers is 698.

2 Add each column of numbers.

(a) $\begin{array}{r} 3 \\ 8 \\ 5 \\ 4 \\ + 6 \end{array}$

(b) $\begin{array}{r} 5 \\ 6 \\ 3 \\ 2 \\ + 4 \end{array}$

(c) $\begin{array}{r} 9 \\ 6 \\ 8 \\ 7 \\ + 3 \end{array}$

(d) $\begin{array}{r} 3 \\ 8 \\ 6 \\ 4 \\ + 8 \end{array}$

Answers

2. (a) 26 (b) 20 (c) 33 (d) 29

3 Add.

(a)
```
  26
+ 73
```

(b)
```
  534
+ 265
```

(c)
```
  42,305
+ 11,563
```

4 Add with regrouping

GS (a)
```
   1
  66
+ 27
  _3
```

(b)
```
  58
+ 33
```

(c)
```
  56
+ 37
```

(d)
```
  34
+ 49
```

◄ **Work Problem 3** at the Side.

OBJECTIVE ▶ 4 Add with regrouping (carrying). If the sum of the digits in any column is more than 9, use **regrouping** (sometimes called **carrying**).

EXAMPLE 4 Adding with Regrouping

Add 47 and 29.
Add ones.

```
  47
+ 29
```
7 ones and 9 ones = 16 ones

Regroup 16 ones as 1 ten and 6 ones. Write 6 ones in the ones column and write 1 ten in the tens column.

```
  1
  47
+ 29
   6
```
Write 1 ten in the tens column.
7 ones and 9 ones = 16 ones
Write 6 ones in the ones column.

Add the digits in the tens column, including the regrouped 1.

```
  1
  47
+ 29
  76
```
1 ten + 4 tens + 2 tens = 7 tens

◄ **Work Problem 4** at the Side.

EXAMPLE 5 Adding with Regrouping

Add 324 + 7855 + 23 + 7 + 86.

Step 1 Add the digits in the ones column.

```
    2
  324
 7855
   23
    7
+  86
    5
```
Write 2 tens in the tens column.
Sum of the ones column is 25 ones.
Notice that 25 ones are regrouped as 2 tens and 5 ones.
Write 5 ones in the ones column.

Step 2 Add the digits in the tens column, including the regrouped 2.

```
   1 2
  324
 7855
   23
    7
+  86
   95
```
Write 1 hundred in the hundreds column.
Sum of the tens column is 19 tens.
Notice that 19 tens are regrouped as 1 hundred and 9 tens.
Write 9 in the tens column.

Continued on Next Page

Answers

3. **(a)** 99 **(b)** 799 **(c)** 53,868
4. **(a)** 9; 93 **(b)** 91 **(c)** 93 **(d)** 83

Step 3 Add the hundreds column, including the regrouped 1.

$$
\begin{array}{r}
{}^{1\,1\,2} \\
3 2 4 \\
7 8 5 5 \\
2 3 \\
7 \\
+ \quad 8 6 \\
\hline
2 9 5
\end{array}
$$

Write 1 thousand in the thousands column.

Notice that 12 hundreds are regrouped as 1 thousand and 2 hundreds.

Sum of the hundreds column is 12 hundreds.

Write 2 hundreds in the hundreds column.

Step 4 Add the thousands column, including the regrouped 1.

$$
\begin{array}{r}
{}^{1\,1\,2} \\
3 2 4 \\
7 8 5 5 \\
2 3 \\
7 \\
+ \quad 8 6 \\
\hline
8 2 9 5
\end{array}
$$

Sum of the thousands column is 8.

Finally, $324 + 7855 + 23 + 7 + 86 = 8295$.

·························· **Work Problem ❺ at the Side.** ▶

Note

For additional speed, you can try to regroup mentally. Do not write the regrouped number, but just remember it as you move to the top of the next column. Try this method, and use it if it works for you.

Work Problem ❻ at the Side. ▶

OBJECTIVE ▶ ❺ **Use addition to solve application problems.** In **Section 1.10** we will describe how to solve application problems in more detail. The next two examples are application problems that require adding.

EXAMPLE 6 **Applying Addition Skills**

On this map of the Walt Disney World area in Florida, the distance in miles from one location to another is written alongside the road. Find the shortest route from Altamonte Springs to Clear Lake.

·················· **Continued on Next Page**

---

❺ Add by regrouping as necessary.

GS **(a)**
$$
\begin{array}{r}
{}^{1}\quad {}^{2\,1} \\
1 6 2 \\
4 2 7 1 \\
3 7 2 \\
+ 8 9 7 6 \\
\hline
1 3, \_\_\, 1
\end{array}
$$

(b)
$$
\begin{array}{r}
7 8 2 1 \\
4 3 5 \\
7 2 \\
3 0 5 \\
+ 1 6 9 3 \\
\hline
\end{array}
$$

❻ Add by regrouping mentally.

(a)
$$
\begin{array}{r}
8 1 6 \\
3 6 3 \\
1 7 \\
2 \\
5 \\
+ 7 6 5 4 \\
\hline
\end{array}
$$

(b)
$$
\begin{array}{r}
1 5, 8 2 9 \\
7 6 5 \\
7 8 \\
1 5 \\
9 \\
7 \\
+ 1 3, 1 7 9 \\
\hline
\end{array}
$$

Answers

5. (a) 7; 8; 13,781 **(b)** 10,326

6. (a) 8857 **(b)** 29,882

7 Use the map in **Example 6** to find **GS** the shortest route from Lake Buena Vista to Conway.

$$
\begin{array}{r}
4 \\
6 \\
3 \\
+\underline{} \\
\hline
\underline{}
\end{array}
$$

Approach Add the mileage along various routes to determine the distances from Altamonte Springs to Clear Lake. Then select the shortest route.

Solution One way from Altamonte Springs to Clear Lake is through Orlando. Add the mileage numbers along this route.

$$
\begin{array}{rl}
8 & \text{Altamonte Springs to Pine Hills} \\
5 & \text{Pine Hills to Orlando} \\
+\ 8 & \text{Orlando to Clear Lake} \\
\hline
21 \rightarrow & \text{miles from Altamonte Springs to} \\
& \text{Clear Lake, going through Orlando}
\end{array}
$$

> Remember: Shortest distance is the fewest total miles.

Another way is through Casselberry, Bertha, and Winter Park. Add the mileage numbers along this route.

$$
\begin{array}{rl}
5 & \text{Altamonte Springs to Casselberry} \\
6 & \text{Casselberry to Bertha} \\
7 & \text{Bertha to Winter Park} \\
+\ 7 & \text{Winter Park to Clear Lake} \\
\hline
25 \rightarrow & \text{miles from Altamonte Springs to Clear Lake through} \\
& \text{Bertha and Winter Park}
\end{array}
$$

The shortest route from Altamonte Springs to Clear Lake is 21 miles through Orlando.

◀ **Work Problem 7 at the Side.**

8 The road is closed between Orlando and Clear Lake, so this route cannot be used. Use the map in **Example 6** to find the next shortest route from Orlando to Clear Lake.

EXAMPLE 7   **Finding a Total Distance**

Use the map in **Example 6** to find the total distance from Shadow Hills to Casselberry to Orlando and back to Shadow Hills.

Approach Add the mileage from Shadow Hills to Casselberry to Orlando and back to Shadow Hills to find the total distance.

Solution Use the numbers from the map.

$$
\begin{array}{rl}
9 & \text{Shadow Hills to Bertha} \\
6 & \text{Bertha to Casselberry} \\
5 & \text{Casselberry to Altamonte Springs} \\
8 & \text{Altamonte Springs to Pine Hills} \\
5 & \text{Pine Hills to Orlando} \\
8 & \text{Orlando to Clear Lake} \\
+\ 11 & \text{Clear Lake to Shadow Hills} \\
\hline
52 \rightarrow & \text{miles from Shadow Hills to Casselberry} \\
& \text{to Orlando and back to Shadow Hills}
\end{array}
$$

◀ **Work Problem 8 at the Side.**

EXAMPLE 8   **Finding a Perimeter**

Find the number of feet of hedges needed to enclose the Civil War Veterans Park shown.

1516 ft

385 ft   385 ft

> The short way to write feet is ft.

1516 ft

Approach Find the **perimeter,** or total distance around the park, by adding the lengths of all the sides.

········· **Continued on Next Page**

Answers

7. 6; 19 miles

8. 5 Orlando to Pine Hills
   8 Pine Hills to Altamonte Springs
   5 Altamonte Springs to Casselberry
   6 Casselberry to Bertha
   7 Bertha to Winter Park
+ 7 Winter Park to Clear Lake
 38 miles

Solution Use the lengths shown.

$$
\begin{array}{r}
1516 \\
385 \\
1516 \\
+\ 385 \\
\hline
3802 \text{ ft}
\end{array}
$$

The amount of hedge needed is 3802 ft, which is the perimeter of (distance around) the park.

···················· **Work Problem 9 at the Side.** ▶

OBJECTIVE ▶ 6 Check the answer in addition. Checking the answer is an important part of problem solving. A common method for checking addition is to re-add from bottom to top. This is an application of the commutative and associative properties of addition.

EXAMPLE 9   Checking Addition

Check the following addition.

Add down.
$$
\begin{array}{r}
\mathbf{1428} \\
738 \\
63 \\
125 \\
17 \\
+\ 485 \\
\hline
\mathbf{1428}
\end{array}
$$

Adding down and adding up should give the same answer.

Add from bottom to top to check addition.

To check, add up.

Here the answers agree, so the sum is probably correct.

EXAMPLE 10   Checking Addition

Check the following additions. Are they correct?

(a)
$$
\begin{array}{r}
785 \\
63 \\
+\ 185 \\
\hline
1033
\end{array}
\qquad
\begin{array}{r}
\mathbf{1033} \\
785 \\
63 \\
+\ 185 \\
\hline
\mathbf{1033}
\end{array}
$$

Correct, because both answers are the same.

To check, add up.

(b)
$$
\begin{array}{r}
635 \\
73 \\
831 \\
+\ 915 \\
\hline
2444
\end{array}
\qquad
\begin{array}{r}
\mathbf{2454} \\
635 \\
73 \\
831 \\
+\ 915 \\
\hline
\mathbf{2444}
\end{array}
$$

Error, because the answers are different.

To check, add up.

Avoid wrong answers by checking your work.

Re-add to find that the correct sum is 2454.

···················· **Work Problem 10 at the Side.** ▶

9 Solve the problem. Find the number of feet of fencing needed to enclose the solar electricity generating project shown.

526 ft

297 ft   297 ft

526 ft

10 Check the following additions. If an answer is incorrect, find the correct answer.

(a)
$$
\begin{array}{r}
63 \\
4 \\
9 \\
+\ 28 \\
\hline
104
\end{array}
$$

(b)
$$
\begin{array}{r}
927 \\
395 \\
64 \\
+\ 251 \\
\hline
1637
\end{array}
$$

(c)
$$
\begin{array}{r}
79 \\
218 \\
7 \\
+\ 639 \\
\hline
953
\end{array}
$$

(d)
$$
\begin{array}{r}
21,892 \\
11,746 \\
+\ 43,925 \\
\hline
79,563
\end{array}
$$

Answers

9. 1646 ft
10. (a) correct **(b)** correct
   (c) incorrect; should be 943
   (d) incorrect; should be 77,563

1.2 Exercises

Add. See Examples 1–3.

1. $\begin{array}{r} 43 \\ + 54 \\ \hline \end{array}$

2. $\begin{array}{r} 18 \\ + 11 \\ \hline \end{array}$

3. $\begin{array}{r} 56 \\ + 33 \\ \hline \end{array}$

4. $\begin{array}{r} 83 \\ + 15 \\ \hline \end{array}$

5. $\begin{array}{r} 317 \\ + 572 \\ \hline \end{array}$

6. $\begin{array}{r} 574 \\ + 325 \\ \hline \end{array}$

7. $\begin{array}{r} 318 \\ 151 \\ + 420 \\ \hline \end{array}$

8. $\begin{array}{r} 135 \\ 253 \\ + 410 \\ \hline \end{array}$

9. $\begin{array}{r} 6310 \\ 252 \\ + 1223 \\ \hline \end{array}$

10. $\begin{array}{r} 121 \\ 5705 \\ + 3163 \\ \hline \end{array}$

CONCEPT CHECK *Determine whether the following additions are* correct *or* incorrect.

11. $932 + 44 + 613 = 1589$

12. $517 + 131 + 250 = 1098$

13. $1251 + 4311 + 2114 = 7686$

14. $3241 + 1513 + 2014 = 6768$

Add. See Examples 1–3.

15. $12{,}142 + 43{,}201 + 23{,}103$

16. $41{,}124 + 12{,}302 + 23{,}500$

17. $3213 + 5715$

18. $6344 + 1655$

19. $38{,}204 + 21{,}020$ $\begin{array}{r} 38{,}2\,0\,4 \\ + 21{,}0\,2\,0 \\ \hline 59{,}\_\_4 \end{array}$

20. $63{,}251 + 36{,}305$ $\begin{array}{r} 6\,3{,}2\,5\,1 \\ + 3\,6{,}3\,0\,5 \\ \hline 9\,\_{,}\_5\_ \end{array}$

CONCEPT CHECK *Determine which answers are* correct *or* incorrect.

21. $\begin{array}{r} 87 \\ + 63 \\ \hline 150 \end{array}$

22. $\begin{array}{r} 19 \\ + 92 \\ \hline 101 \end{array}$

23. $\begin{array}{r} 86 \\ + 69 \\ \hline 155 \end{array}$

24. $\begin{array}{r} 37 \\ + 85 \\ \hline 132 \end{array}$

25. $\begin{array}{r} 47 \\ + 74 \\ \hline 111 \end{array}$

Add, regrouping as necessary. See Examples 4 and 5.

26. $\begin{array}{r} 97 \\ + 79 \\ \hline \end{array}$

27. $\begin{array}{r} 67 \\ + 78 \\ \hline \end{array}$

28. $\begin{array}{r} 96 \\ + 47 \\ \hline \end{array}$

29. $\begin{array}{r} 73 \\ + 29 \\ \hline \end{array}$

30. $\begin{array}{r} 68 \\ + 37 \\ \hline \end{array}$

31. $\begin{array}{r} 746 \\ + 905 \\ \hline \end{array}$

32. $\begin{array}{r} 621 \\ + 359 \\ \hline \end{array}$

33. $\begin{array}{r} 306 \\ + 848 \\ \hline \end{array}$

34. $\begin{array}{r} 798 \\ + 206 \\ \hline \end{array}$

35. $\begin{array}{r} 278 \\ + 135 \\ \hline \end{array}$

36. $\begin{array}{r} 172 \\ + 156 \\ \hline \end{array}$

37. $\begin{array}{r} 928 \\ + 843 \\ \hline \end{array}$

38. $\begin{array}{r} 686 \\ + 726 \\ \hline \end{array}$

39. $\begin{array}{r} 526 \\ + 884 \\ \hline \end{array}$

40. $\begin{array}{r} 116 \\ + 897 \\ \hline \end{array}$

41.   3574
    + 2817

42.   6871
    + 7528

43.   7896
    + 3728

44.   9382
    + 7586

45.   9625
    + 7986

46.   5718
      5623
    + 7436

47.   9056
        78
      6089
    + 731

48.   4022
      709
    8621
    + 37

49.   18
     708
    9286
    + 636

50.   1708
     321
      61
    + 8926

51.   422
    6074
    435
    + 8663

52.   6505
     173
    7044
    + 168

53.   321
    9603
      8
     21
    + 1604

54.   7631
    5983
      7
     36
    + 505

55.   2109
      63
      16
       3
    + 9887

56.   322
    6508
     93
    745
     18
    + 2005

57.   553
     97
    2772
    437
     63
    + 328

58.   3187
     810
     527
      76
    2665
    + 317

59.   413
      85
    9919
     602
      31
    + 1218

60.   576
    7934
      60
     781
    5968
    + 371

Check each addition. If an answer is incorrect, find the correct answer.
See Examples 9 and 10.

61. GS
    2 1
    832
    468
    + 791
    2091

62.
    326
    852
    + 679
    1857

63. GS
    7 9
    179
    214
    + 376
    759

64.
     17
    296
    713
    + 94
    1220

65.
    4713
      28
    615
    + 64
    5420

66.
    3 628
      72
    564
    + 7 319
    11,583

67.
    678
    7 952
     56
    718
    + 2 173
    11,377

68.
    516
    8 760
     24
    189
    + 1 723
    11,212

69.
    4 714
     27
     77
    8 878
    + 636
    14,332

70.
    6 715
     283
    9 617
     13
    + 81
    16,719

71. Explain the commutative property of addition in your own words. How is this used when checking an addition problem?

72. Explain the associative property of addition. How can this be used when adding columns of numbers?

For Exercises 73–76, use the map to find the shortest route between each pair of cities.
See Examples 6 and 7.

73. Southtown and Rena

Southtown to Thomasville   21 miles
Thomasville to Rena   + 12 miles

74. Elk Hill and Oakton

Elk Hill to Thomasville   18 miles
Thomasville to Oakton   + 17 miles

75. Thomasville and Murphy

76. Murphy and Thomasville

Solve each application problem.

77. The Twin Lakes Food Bank raised $3482 at a flea market and $12,860 at their annual auction. Find the total amount raised at these two events.

78. A ballpark vendor sold 185 hot dogs and 129 hamburgers. What was the total number of items sold?

79. There are 413 women and 286 men on the sales staff. How many people are on the sales staff?

80. One department in an office building has 283 employees while another department has 218 employees. How many employees are in the two departments?

81. This semester there are 13,786 students enrolled in on-campus day classes, 3497 students enrolled in night classes, and 2874 student's enrolled in on-line classes. Find the total number of students enrolled.

82. The number of tornadoes in each of the last 7 years has been 887, 223, 465, 683, 597, 214 and 1817. Find the total number of tornadoes in this 7-year period. (*Source:* Storm Prediction Center.)

Solve each problem involving perimeter. **See Example 8.**

83. Find the total distance around a lot that has been
developed as a go-cart track.

84. Because of heavy snowfall this winter, Maria needs
to put new rain gutters around her entire roof. How
many feet of gutters will she need?

85. Martin plans to frame his back patio with redwood
lumber. How many feet of lumber will he need?

86. The university agriculture department is expanding
the size of its experimental farm and wants to fence
the area around the farm. How many meters of
fencing will be needed?

Relating Concepts (Exercises 87–94)   For Individual or Group Work

Recall the place values of digits discussed in **Section 1.1** *and* **work
Exercises 87–94 in order.**

87. Write the largest four-digit number possible using
the digits 4, 1, 9, and 2. Use each digit once.

88. Using the digits 4, 1, 9, and 2, write the smallest
four-digit number possible. Use each digit once.

89. Write the largest five-digit number possible using
the digits 6, 2, and 7. Use each digit at least once.

90. Using the digits 6, 2, and 7, write the smallest five-
digit number possible. Use each digit at least once.

91. Write the largest seven-digit number possible using
the digits 4, 3, and 9. Use each digit at least twice.

92. Using the digits 4, 3, and 9, write the smallest seven-
digit number possible. Use each digit at least twice.

93. Explain your rule or procedure for writing the
largest number in **Exercise 91.**

94. Explain your rule or procedure for writing the
smallest number in **Exercise 92.**

1.3 Subtracting Whole Numbers

OBJECTIVES

1. Change addition problems to subtraction and subtraction problems to addition.
2. Identify the minuend, subtrahend, and difference.
3. Subtract when no regrouping (borrowing) is needed.
4. Check subtraction answers by adding.
5. Subtract with regrouping (borrowing).
6. Solve application problems with subtraction.

Suppose you have $9, and you spend $2 for parking. You then have $7 left. There are two different ways of looking at these numbers.

> **As an addition problem:**
>
> $$\$2 \quad + \quad \$7 \quad = \quad \$9$$
>
> Amount spent   Amount left   Original amount
>
> **As a subtraction problem:**
>
> $$\$9 \quad - \quad \$2 \quad = \quad \$7$$
>
> Original amount   Subtraction symbol   Amount spent   Amount left

OBJECTIVE ▶ 1 Change addition problems to subtraction and subtraction problems to addition. As shown in the box above, an addition problem can be changed to a subtraction problem and a subtraction problem can be changed to an addition problem.

---

1 Write two subtraction problems for each addition problem.

GS **(a)** $8 + 2 = 10$

$$10 - \underline{\quad} = 8$$
$$10 - \underline{\quad} = \underline{\quad}$$

(b) $7 + 4 = 11$

(c) $15 + 22 = 37$

(d) $23 + 55 = 78$

EXAMPLE 1 Changing Addition Problems to Subtraction

Change each addition problem to a subtraction problem.

(a) $4 + 1 = 5$

Two subtraction problems are possible:

$$5 - 1 = 4 \quad \text{or} \quad 5 - 4 = 1$$

These figures show each subtraction problem.

$$5 - 1 = 4 \qquad\qquad 5 - 4 = 1$$

(b) $8 + 7 = 15$

$$15 - 7 = 8 \quad \text{or} \quad 15 - 8 = 7$$

◀ Work Problem **1** at the Side.

EXAMPLE 2 Changing Subtraction Problems to Addition

Change each subtraction problem to an addition problem.

(a) $8 - 3 = 5$

$$8 = 3 + 5$$

It is also correct to write $8 = 5 + 3$.

> Recall that changing the order of addends does not change the sum.

Answers

1. (a) 2; $10 - 2 = 8$ or   8; 2; $10 - 8 = 2$
(b) $11 - 4 = 7$ or $11 - 7 = 4$
(c) $37 - 22 = 15$ or $37 - 15 = 22$
(d) $78 - 55 = 23$ or $78 - 23 = 55$

Continued on Next Page

(b) $18 - 13 = 5$

$18 = 13 + 5$ or $18 = 5 + 13$

(c) $29 - 13 = 16$

$29 = 13 + 16$ or $29 = 16 + 13$

·················· **Work Problem ❷ at the Side.** ▶

OBJECTIVE ▶ ❷ Identify the minuend, subtrahend, and difference. In subtraction, as in addition, the numbers in a problem have names. For example, in the problem $8 - 5 = 3$, the number 8 is the **minuend,** 5 is the **subtrahend,** and 3 is the **difference** or answer.

$$8 \quad - \quad 5 \quad = \quad 3 \leftarrow \text{Difference}$$

Minuend Subtrahend

The answer in subtraction is the difference.

$$\begin{array}{r} 8 \leftarrow \text{Minuend} \\ - 5 \leftarrow \text{Subtrahend} \\ \hline 3 \leftarrow \text{Difference} \end{array}$$

OBJECTIVE ▶ ❸ Subtract when no regrouping (borrowing) is needed. Subtract two numbers by lining up the numbers in columns so the digits in the ones place are in the same column, the tens digits are in the same column, the hundreds digits are in the same column, and so on. Next, subtract by columns, starting at the right with the ones column.

EXAMPLE 3 Subtracting Two Numbers

Subtract.

Tens digits are lined up in the same column.
Ones digits are lined up in the same column.

(a)
$$\begin{array}{r} 53 \\ - 21 \\ \hline 32 \end{array}$$
← 3 ones − 1 one = 2 ones
5 tens − 2 tens = 3 tens

Ones digits are lined up.

(b)
$$\begin{array}{r} 385 \\ - 165 \\ \hline 220 \end{array}$$
Subtract from right to left.
← 5 ones − 5 ones = 0 ones
8 tens − 6 tens = 2 tens
3 hundreds − 1 hundred = 2 hundreds

(c)
$$\begin{array}{r} 9437 \\ - 210 \\ \hline 9227 \end{array}$$
← 7 ones − 0 ones = 7 ones
3 tens − 1 ten = 2 tens
4 hundreds − 2 hundreds = 2 hundreds
9 thousands − 0 thousands = 9 thousands

·················· **Work Problem ❸ at the Side.** ▶

OBJECTIVE ▶ ❹ Check subtraction answers by adding. Use addition to check your answer to a subtraction problem. For example, check $8 - 3 = 5$ by *adding* 3 and 5.

$$3 + 5 = 8, \quad \text{so} \quad 8 - 3 = 5 \quad \text{is correct.}$$

❷ Write an addition problem for each subtraction problem.

GS **(a)** $7 - 5 = 2$

$7 = 5 +$ ____

(b) $9 - 4 = 5$

(c) $21 - 15 = 6$

(d) $58 - 42 = 16$

VOCABULARY TIP

Difference suggests comparing two things. In mathematics, this comparison is done by subtracting two numbers, the answer being the difference between them.

❸ Subtract.

(a)
$$\begin{array}{r} 74 \\ - 43 \end{array}$$

(b)
$$\begin{array}{r} 68 \\ - 24 \end{array}$$

(c)
$$\begin{array}{r} 429 \\ - 318 \end{array}$$

(d)
$$\begin{array}{r} 3927 \\ - 2614 \end{array}$$

Answers

2. (a) 2; $7 = 5 + 2$ or $7 = 2 + 5$
(b) $9 = 4 + 5$ or $9 = 5 + 4$
(c) $21 = 15 + 6$ or $21 = 6 + 15$
(d) $58 = 42 + 16$ or $58 = 16 + 42$

3. (a) 31 **(b)** 44 **(c)** 111 **(d)** 1313

4 Use addition to determine whether each answer is correct. If incorrect, what should it be?

GS (a)
$$\begin{array}{r} 76 \\ -\ 45 \\ \hline 31 \end{array} \qquad \begin{array}{r} 45 \\ +\ \underline{} \\ \hline \underline{} \end{array}$$

(b)
$$\begin{array}{r} 53 \\ -\ 22 \\ \hline 21 \end{array}$$

(c)
$$\begin{array}{r} 374 \\ -\ 251 \\ \hline 113 \end{array}$$

(d)
$$\begin{array}{r} 7531 \\ -\ 4301 \\ \hline 3230 \end{array}$$

VOCABULARY TIP

Regrouping refers to both "carrying" in addition and "borrowing" in subtraction.

---

5 Subtract.

(a)
$$\begin{array}{r} 58 \\ -\ 19 \\ \hline \end{array}$$

(b)
$$\begin{array}{r} 86 \\ -\ 38 \\ \hline \end{array}$$

(c)
$$\begin{array}{r} 41 \\ -\ 27 \\ \hline \end{array}$$

(d)
$$\begin{array}{r} 863 \\ -\ 47 \\ \hline \end{array}$$

(e)
$$\begin{array}{r} 762 \\ -\ 157 \\ \hline \end{array}$$

Answers

4. (a) 31; 76; correct (b) incorrect; should be 31
   (c) incorrect; should be 123 (d) correct

5. (a) 39 (b) 48 (c) 14 (d) 816
   (e) 605

EXAMPLE 4   **Checking Subtraction by Using Addition**

Use addition to check each answer. If the answer is incorrect, find the correct answer.

(a)
$$\begin{array}{r} 89 \\ -\ 47 \\ \hline 42 \end{array}$$

Rewrite as an addition problem, as shown in **Example 2**.

Subtraction problem $\begin{cases} \begin{array}{r} 89 \\ -\ 47 \\ \hline 42 \end{array} \end{cases}$ Addition problem $\qquad \begin{array}{r} \mathbf{47} \\ +\ \mathbf{42} \\ \hline \mathbf{89} \end{array}$

Because $47 + 42 = 89$, the subtraction was done correctly. ◁ Avoid errors by checking answers.

(b) $72 - 41 = 21$
   Rewrite as an addition problem.

$$72 = 41 + 21$$

But, $41 + 21 = 62$, **not** 72, so the subtraction was done **incorrectly**. Rework the original subtraction to get the correct answer, 31. Then, $41 + 31 = 72$.

(c)
$$\begin{array}{r} 374 \\ -\ 141 \\ \hline 233 \end{array} \longleftarrow \text{Match}$$
$141 + 233 = 374$

The answer checks.

◀ **Work Problem 4** at the Side.

OBJECTIVE ▶ **5** **Subtract with regrouping (borrowing).** When a digit in the minuend is less than the one directly below it, **regrouping** is necessary (also called **borrowing**).

EXAMPLE 5   **Subtracting with Regrouping**

Subtract 19 from 57.
Write the problem vertically.

$$\begin{array}{r} 57 \\ -\ 19 \\ \hline \end{array}$$

In the ones column, 7 is **less** than 9, so in order to subtract, we must regroup 1 ten as 10 ones.

5 tens − 1 ten = 4 tens → $\begin{array}{r} 4\ 17 \\ \cancel{5}\,\cancel{7} \\ -\ 1\ 9 \\ \hline \end{array}$ ← 1 ten = 10 ones, and 10 ones + 7 ones = 17 ones

Now subtract 9 ones from 17 ones in the ones column. Then subtract 1 ten from 4 tens in the tens column.

$$\begin{array}{r} {\scriptstyle 4\ 17} \\ \cancel{5}\,\cancel{7} \\ -\ 1\ 9 \\ \hline 3\ 8 \end{array} \quad \text{Difference}$$

Finally, $57 - 19 = 38$. Check by adding 19 and 38; you should get 57.

◀ **Work Problem 5** at the Side.

EXAMPLE 6   **Subtracting with Regrouping**

Subtract by regrouping when necessary.

(a)   7856
       − 137

Regroup 1 ten as 10 ones. ⟶   10 ones + 6 ones = 16 ones

```
        4  16
   7 8  5̶  6̶
 −   1  3  7
   7 7  1  9    Difference
```

(b)   635
       − 546

Regroup 1 ten   10 ones + 5 ones
as 10 ones.   = 15 ones

```
      2  15
   6  3̶  5̶     Need to regroup
 − 5  4  6     further because 2 is
         9     less than 4 in the
               tens column.
```

Regroup   10 tens + 2 tens
1 hundred   = 12 tens
as 10 tens

```
   5  12  15
   6̶  3̶   5̶
 − 5  4   6
      8   9    Difference
```

(c)   647
       − 489

```
      3  17
   6̶  4̶  7     Need to regroup further
 − 4  8  9     because 3 is less than
         8     8 in the tens column.
```

```
   5  13  17
   6̶  4̶   7
 − 4  8   9
   1  5   8    Difference
```

Work Problem 6 at the Side. ▶

Sometimes a minuend has zeros in some of the positions. In such cases, regrouping may be a little more complicated than what we have shown so far.

EXAMPLE 7   **Regrouping with Zeros**

Subtract.

```
     4607
   − 3168
```

There are no tens that can be regrouped into ones.
So you must first regroup 1 hundred as 10 tens.

Regroup 1 hundred ⟶   Write 10 tens.
as 10 tens.

```
      5  10
   4  6̶  0̶  7
 − 3  1  6  8
```

Now we may regroup from the tens position.

```
          9 ←       Regroup 1 ten as 10 ones.
      5 10 17 ←     10 tens − 1 ten = 9 tens.
   4  6̶ 0̶ 7̶
 − 3  1 6 8         10 ones + 7 ones = 17 ones
            9
```

Continued on Next Page

---

6 Subtract.

GS (a)
```
        8  12
     9  2̶  7
   −    4  3
   −  −     4
```

(b)   675
       − 86

(c)   477
       − 389

(d)   1417
       − 988

(e)   8739
       − 3892

7 Subtract.

(a)   206
       − 177

(b)   703
       − 415

(c)   7024
      − 2632

8 Subtract.

          9
        2 10 18
GS (a)   3̸ 0̸ 8̸
       − 1 5 9
       ─────
       − − 9

(b)   570
       − 368

(c)   1570
       − 983

(d)   7001
      − 5193

(e)   4000
      − 1782

Answers

7. (a) 29   (b) 288   (c) 4392

8. (a) 1; 4; 149   (b) 202   (c) 587
   (d) 1808   (e) 2218

Complete the problem.

             9
           5 10 17
         4 6̸ 0̸ 7̸
       − 3 1 6 8
       ─────────
         1 4 3 9   Difference

Check by adding 1439 and 3168; you should get 4607.

◄ **Work Problem 7 at the Side.**

EXAMPLE 8   Regrouping with Zeros

Subtract.

(a)   708
       − 149

Write 10 tens. ——— Regroup 1 ten as 10 ones.

Regroup 1 hundred as → 10 ones + 8 ones = 18 ones
10 tens.
              9
           6 10 18
          7̸ 0̸ 8̸   Remember to work
       − 1 4 9   from right to left.
       ───────
         5 5 9

(b)   380
       − 276

Regroup 1 ten as —— Write 10 ones.
10 ones.
           7 10
          3 8̸ 0̸
       − 2 7 6
       ───────
         1 0 4

(c)   9000
      − 6999

          9 9
        8 10 10 10
        9̸ 0̸ 0̸ 0̸   Be extra careful when
      − 6 9 9 9   zeros are involved.
      ─────────
        2 0 0 1

◄ **Work Problem 8 at the Side.**

As we have seen, an answer to a subtraction problem can be checked by adding.

EXAMPLE 9   Checking Subtraction by Using Addition

Use addition to check each answer.

                 CHECK
         613   275
(a)   − 275   + 338
       ───── Match ─────
         338   **613** ✓   Correct

········· **Continued on Next Page**

(b) $\begin{array}{r} 1915 \\ -\ 1635 \\ \hline 280 \end{array}$ *Match*   CHECK
$\begin{array}{r} 1635 \\ +\ 280 \\ \hline 1915 \end{array}$ ✓ Correct

(c) $\begin{array}{r} 15{,}803 \\ -\ 7\ 325 \\ \hline 8\ 578 \end{array}$ *No Match*   CHECK
$\begin{array}{r} 7\ 325 \\ +\ 8\ 578 \\ \hline 15{,}903 \end{array}$ Error

> It's always a good idea to check your work.

Rework the original problem to get the correct answer, 8478. Then, 7325 + 8478 **does** give 15,803.

.... **Work Problem ❾ at the Side.** ▶

OBJECTIVE ▶ ❻ Solve application problems with subtraction. As shown in the next example, subtraction can be used to solve an application problem.

| EXAMPLE 10 | Applying Subtraction Skills |

Use the table to find how much more, on average, a person with an Associate of Arts degree earns each year than a high school graduate.

EDUCATION PAYS

The more education adults get, the higher their annual earnings.

Education Level	Average Earnings
Not a high school graduate	$33,435
High school graduate	$43,165
Some college, no degree	$50,359
Associate of Arts degree	$54,861
Bachelor's degree	$82,197
Master's degree	$99,516
Doctoral degree	$129,773
Professional degree	$166,065

Note: Average annual earnings for workers between ages 25 and 64.

Source: U.S. Census Bureau and Pearson Education, Inc.

Approach The average earnings for a person with an Associate of Arts degree is $54,861 each year. The average for a high school graduate is $43,165. Find how much more a college graduate earns by subtracting $43,165 from $54,861.

Solution
$54,861 ⟵ Associate of Arts degree
$- \$43,165$ ⟵ High school graduate
$11,696 ⟵ More earnings

> Education pays.

On average, a person with an Associate of Arts degree earns $11,696 more each year than a high school graduate.

.... **Work Problem ❿ at the Side.** ▶

---

❾ Use addition to check each answer. If the answer is incorrect, find the correct answer.

GS (a) $\begin{array}{r} 357 \\ -\ 168 \\ \hline 189 \end{array}$ Check $\begin{array}{r} 1\ 6\ 8 \\ +\ \_\ \_\ \_ \\ \hline 3\ 5\ 7 \end{array}$

(b) $\begin{array}{r} 570 \\ -\ 328 \\ \hline 252 \end{array}$

(c) $\begin{array}{r} 14{,}726 \\ -\ 8\ 839 \\ \hline 5\ 887 \end{array}$

❿ Use the table from **Example 10** to find, on average,

(a) how much more a person with an Associate of Arts degree earns each year than a person who is not a high school graduate.

(b) how much more a person with a Bachelor's degree earns each year than a person with an Associate of Arts degree.

Answers
9. (a) 1; 8; 9; correct
  (b) incorrect; should be 242
  (c) correct
10. (a) $21,426 **(b)** $27,336

1.3 Exercises

 Download the MyDashBoard App

 MyMathLab®

CONCEPT CHECK   *Write in the number needed to complete the check.*

1. $\begin{array}{r} 48 \\ -32 \\ \hline 16 \end{array}$ $\begin{array}{r} 16 \\ +\underline{} \\ \hline 48 \end{array}$

2. $\begin{array}{r} 17 \\ -13 \\ \hline 4 \end{array}$ $\begin{array}{r} \underline{} \\ +13 \\ \hline 17 \end{array}$

3. $\begin{array}{r} 86 \\ -53 \\ \hline 33 \end{array}$ $\begin{array}{r} 53 \\ +33 \\ \hline \underline{} \end{array}$

4. $\begin{array}{r} 78 \\ -35 \\ \hline 43 \end{array}$ $\begin{array}{r} \underline{} \\ +43 \\ \hline 78 \end{array}$

5. $\begin{array}{r} 77 \\ -60 \\ \hline 17 \end{array}$ $\begin{array}{r} 60 \\ +\underline{} \\ \hline 77 \end{array}$

Work each subtraction problem. Use addition to check each answer. **See Examples 3 and 4.**

6. $\begin{array}{r} 87 \\ -63 \\ \hline \end{array}$

7. $\begin{array}{r} 335 \\ -122 \\ \hline \end{array}$

8. $\begin{array}{r} 602 \\ -301 \\ \hline \end{array}$

9. $\begin{array}{r} 552 \\ -451 \\ \hline \end{array}$

10. $\begin{array}{r} 888 \\ -215 \\ \hline \end{array}$

11. $\begin{array}{r} 7352 \\ -241 \\ \hline \end{array}$

12. $\begin{array}{r} 4420 \\ -310 \\ \hline \end{array}$

13. $\begin{array}{r} 5546 \\ -2134 \\ \hline \end{array}$

14. $\begin{array}{r} 1875 \\ -1362 \\ \hline \end{array}$

15. $\begin{array}{r} 6259 \\ -4148 \\ \hline \end{array}$

16. $\begin{array}{r} 9654 \\ -4323 \\ \hline \end{array}$

17. $\begin{array}{r} 24,392 \\ -11,232 \\ \hline \end{array}$

18. $\begin{array}{r} 57,921 \\ -34,801 \\ \hline \end{array}$

19. $\begin{array}{r} 46,253 \\ -5\,143 \\ \hline \end{array}$

20. $\begin{array}{r} 75,904 \\ -3\,702 \\ \hline \end{array}$

Use addition to check each subtraction problem. If an answer is not correct, find the correct answer. **See Example 4.**

21. $\begin{array}{r} 54 \\ -42 \\ \hline 12 \end{array}$ $\begin{array}{r} 42 \\ +\underline{} \\ \hline 54 \end{array}$

22. $\begin{array}{r} 87 \\ -43 \\ \hline 44 \end{array}$ $\begin{array}{r} 43 \\ +\underline{} \\ \hline 87 \end{array}$

23. $\begin{array}{r} 89 \\ -27 \\ \hline 63 \end{array}$

24. $\begin{array}{r} 47 \\ -35 \\ \hline 13 \end{array}$

25. $\begin{array}{r} 382 \\ -261 \\ \hline 131 \end{array}$

26. $\begin{array}{r} 754 \\ -342 \\ \hline 412 \end{array}$

27. $\begin{array}{r} 4683 \\ -3542 \\ \hline 1141 \end{array}$

28. $\begin{array}{r} 5217 \\ -4105 \\ \hline 1132 \end{array}$

29. $\begin{array}{r} 8643 \\ -1421 \\ \hline 7212 \end{array}$

30. $\begin{array}{r} 9428 \\ -3124 \\ \hline 6324 \end{array}$

31. **CONCEPT CHECK**   *Underline the correct answer.*

In subtraction, regrouping is necessary when the digit in the (*minuend/subtrahend*) is less value than the digit in the subtrahend which is directly (*above/below*) it.

32. **CONCEPT CHECK**   *Which of the subtraction problems will require regrouping?*

(a) $\begin{array}{r} 64 \\ -51 \\ \hline \end{array}$

(b) $\begin{array}{r} 763 \\ -473 \\ \hline \end{array}$

(c) $\begin{array}{r} 43,708 \\ -22,607 \\ \hline \end{array}$

(d) $\begin{array}{r} 6208 \\ -5126 \\ \hline \end{array}$

Subtract, regrouping when necessary. **See Examples 5–8.**

33. $\begin{array}{r} 75 \\ -37 \\ \hline \end{array}$

34. $\begin{array}{r} 86 \\ -28 \\ \hline \end{array}$

35. $\begin{array}{r} 94 \\ -49 \\ \hline \end{array}$

36. $\begin{array}{r} 68 \\ -39 \\ \hline \end{array}$

37. $\begin{array}{r} 57 \\ -38 \\ \hline \end{array}$

38. $\begin{array}{r} 47 \\ -29 \\ \hline \end{array}$

39. $\begin{array}{r} 828 \\ -547 \\ \hline \end{array}$

40. $\begin{array}{r} 916 \\ -618 \\ \hline \end{array}$

41. $\begin{array}{r} 771 \\ -252 \\ \hline \end{array}$

42. $\begin{array}{r} 973 \\ -788 \\ \hline \end{array}$

43. 7538
  − 479

44. 5863
  − 1295

45. 9988
  − 2399

46. 3576
  − 1658

47. 80
  − 73

48. 60
  − 37

49. 308
  − 289

50. 600
  − 599

51. 4041
  − 1208

52. 4602
  − 2063

53. 9305
  − 1530

54. 7120
  − 6033

55. 1580
  − 1077

56. 3068
  − 2105

57. 2006
  − 1850

58. 8203
  − 5365

59. 8240
  − 6056

60. 7050
  − 6045

61. 8503
  − 2816

62. 16,004
  − 5 087

63. 80,705
  − 61,667

64. 81,000
  − 55,456

65. 66,000
  − 34,444

66. 77,000
  − 65,308

67. 20,080
  − 13,496

CONCEPT CHECK *Fill in each blank with the correct response.*

68. To avoid errors when solving math problems, it's a good idea to _____ your work.

69. When checking the accuracy of an answer to an addition problem, you can use _____.

70. An answer to a subtraction problem may be checked using _____.

Use addition to check each subtraction problem. If an answer is incorrect, find the correct answer. **See Example 9.**

71. 9428
  − 4509
   4919

72. 1671
  − 1325
   1346

73. 2548
  − 2278
    270

74. 5274
  − 1130
   4144

75. 93,758
  − 52,869
   40,889

76. 82,357
  − 14,396
   68,961

77. 36,778
  − 17,405
   19,373

78. 34,821
  − 17,735
   17,735

79. An addition problem can be changed to a subtraction problem and a subtraction problem can be changed to an addition problem. Give two examples of each to demonstrate this.

80. Can you use the commutative and the associative properties in subtraction? Explain.

Solve each application problem. See Example 10.

81. A man burns 187 calories during 60 minutes of sitting at a computer while a woman burns 140 calories at the same activity. How many fewer calories does a woman burn than a man in 60 minutes? (*Source:* www.cookinglight.com)

82. A woman burns 302 calories during an hour of walking, while a man burns 403 calories doing the same activity. How many more calories does a man burn than a woman during an hour of walking? (*Source:* www.cookinglight.com)

83. In April 2011, there were 612 tornadoes, shattering the old record of 543. How many more tornadoes were there than the old record number? (*Source:* National Oceanic and Atmospheres Administration.)

84. With an estimated 327 deaths, the tornado outbreak in April 2011 was the third deadliest on record, behind 747 deaths in April 1925 and 332 deaths in April 1932. How many more deaths were there in the deadliest tornado outbreak than in the tornado outbreak of April 2011? (*Source:* Accu Weather.)

85. The top of each main tower of the Golden Gate Bridge is 746 feet above the water and 500 feet above the roadway. How far above the water is the roadway? (*Source:* gocalifornia.about.com)

$$\begin{array}{r} 746 \text{ feet above water} \\ -500 \text{ feet above roadway} \\ \hline \end{array}$$

86. In a recent three-month period there were 81,465 Ford Explorers and 70,449 Jeep Grand Cherokees sold. Which vehicle had greater sales? By how much? (*Source:* J. D. Power and Associates.)

87. Six years ago there were 6970 bridge and lock-tender jobs across the United States. Today there are 3700 that remain. How many of these jobs have been eliminated? (*Source:* Bureau of Labor Statistics.)

88. In 1964, its first year on the market, the Ford Mustang sold for $2500. In 2013, the Ford Mustang sold for $28,065. Find the increase in price. (*Source:* eBay.)

89. Patriot Flag Company manufactured 14,608 U.S. flags and sold 5069. How many flags remain unsold?

90. Eye exams have been given to 14,679 children in the school district. If there are 23,156 students in the school district, how many have not received eye exams?

91. The Jordanos now pay rent of $650 per month. If they buy a house, their housing expense will be $913 per month. How much more will they pay per month if they buy a house?

92. A retired couple who used to receive a Social Security payment of $1479 per month now receives $1568 per month. Find the amount of the monthly increase.

93. The distance from New York City to Buenos Aires, Argentina is 5299 miles, while the distance from Los Angeles to Dublin, Ireland is 5158 miles. How much further is one trip than the other? (*Source:* Map Crow Travel Calculator, mapcrow info)

94. In the year 2020 it is predicted that we will need 2,820,000 nurses in the United States, while only 1,810,000 nurses will be available. Find the shortage in the number of nurses. (*Source:* American Hospital Association.)

Solve each application problem. Add or subtract as necessary.

95. This year there were 264,311 hip replacement procedures in the United States. If 125,423 of the patients were 65 years of age or older, how many patients were under age 65? (*Source:* Federal Agency for Healthcare Research and Quality.)

96. This year there were 555,800 knee surgeries performed in the United States. The number of knee surgeries performed six years ago was 328,900. How many more of these surgeries were performed this year than six years ago? (*Source:* Agency for Healthcare Research and Quality.)

SUBWAY promotes healthy food choices by offering eight sandwiches that are low in fat. The nutritional information, printed on every SUBWAY napkin, appears below and includes information to answer Exercises 97–100. (Source: SUBWAY.)

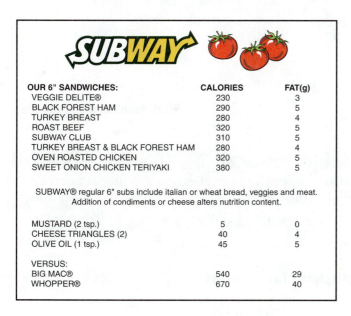

OUR 6" SANDWICHES:	CALORIES	FAT(g)
VEGGIE DELITE®	230	3
BLACK FOREST HAM	290	5
TURKEY BREAST	280	4
ROAST BEEF	320	5
SUBWAY CLUB	310	5
TURKEY BREAST & BLACK FOREST HAM	280	4
OVEN ROASTED CHICKEN	320	5
SWEET ONION CHICKEN TERIYAKI	380	5

SUBWAY® regular 6" subs include italian or wheat bread, veggies and meat. Addition of condiments or cheese alters nutrition content.

MUSTARD (2 tsp.)	5	0
CHEESE TRIANGLES (2)	40	4
OLIVE OIL (1 tsp.)	45	5
VERSUS:		
BIG MAC®	540	29
WHOPPER®	670	40

97. How many fewer calories and grams of fat are in a 6-inch Veggie Delite sandwich than a Big Mac?

98. How many fewer calories and grams of fat are in a 6-inch Turkey Breast and Black Forest Ham sandwich than a Whopper?

99. Find the total number of calories and grams of fat in an Oven Roasted Chicken sandwich with mustard and olive oil.

100. A customer ate two sandwiches, one with the least calories and one with the most calories. Find the total number of calories and grams of fat in the two sandwiches.

1.4 Multiplying Whole Numbers

OBJECTIVES

1. Identify the parts of a multiplication problem.
2. Do chain multiplication.
3. Multiply by single-digit numbers.
4. Use multiplication shortcuts for numbers ending in zeros.
5. Multiply by numbers having more than one digit.
6. Solve application problems with multiplication.

Suppose we want to know the total number of exercise bicycles available at the gym. The bicycles are arranged in four columns with three stations in each column. Adding the number 3 a total of 4 times gives 12.

$$3 + 3 + 3 + 3 = 12$$

This result can also be shown with a figure.

3 + 3 + 3 + 3

3 bicycles in each column

4 columns

OBJECTIVE ➊ **Identify the parts of a multiplication problem.** Multiplication is a shortcut for repeated addition. In the exercise bicycle example, instead of *adding* $3 + 3 + 3 + 3$ to get 12, we can *multiply* 3 by 4 to get 12. The numbers being multiplied are called **factors.** The answer is called the **product.** For example, the product of 3 and 4 can be written with the symbol $\times$, a raised dot, or parentheses, as follows.

$$
\begin{array}{r}
3 \\
\times\ 4 \\
\hline
12
\end{array}
$$

⟵ Factor (also called *multiplicand*)
⟵ Factor (also called *multiplier*)
⟵ Product (answer)

$3 \times 4 = 12$ *or* $3 \cdot 4 = 12$ *or* $(3)(4) = 12$ *or* $3(4) = 12$

◀ Work Problem ➊ at the Side.

➊ Identify the factors and the product in each multiplication problem.

(a) $8 \times 5 = 40$

(b) $6(4) = 24$

(c) $7 \cdot 6 = 42$

(d) $(3)(9) = 27$

Commutative Property of Multiplication

By the **commutative property of multiplication,** the product (answer) remains the same when the order of the factors is changed. For example,

$$3 \times 5 = 15 \quad \text{and} \quad 5 \times 3 = 15.$$

Multiply numbers in any order.

EXAMPLE 1   Multiplying Two Numbers

Multiply.

(a) $3 \times 4 = 12$

(b) $6 \cdot 0 = 0$

Multiply any number by zero and the answer is always zero.

(c) $4(8) = 32$

Continued on Next Page

Answers

1. **(a)** factors: 8, 5; product: 40
   (b) factors: 6, 4; product: 24
   (c) factors: 7, 6; product: 42
   (d) factors: 3, 9; product: 27

Learning the multiplication table will help you in later chapters.

Multiplication Table

×	1	2	3	4	5	6	7	8	9
1	1	2	3	4	5	6	7	8	9
2	2	4	6	8	10	12	14	16	18
3	3	6	9	12	15	18	21	24	27
4	4	8	12	16	20	24	28	32	36
5	5	10	15	20	25	30	35	40	45
6	6	12	18	24	30	36	42	48	54
7	7	14	21	28	35	42	49	56	63
8	8	16	24	32	40	48	56	64	72
9	9	18	27	36	45	54	63	72	81

Recall that any number multiplied by 1 is always the number itself.

Work Problem ② at the Side. ▶

OBJECTIVE ▶ ② Do chain multiplication. Some multiplications involve more than two factors.

Associative Property of Multiplication

By the **associative property of multiplication,** grouping the factors differently does not change the product.

EXAMPLE 2   Multiplying Three Numbers

Multiply $2 \times 3 \times 5$.

$$(2 \times 3) \times 5$$

Parentheses show what to do first.

$$6 \quad \times 5 = 30$$

Also,

$$2 \times (3 \times 5)$$

$$2 \times \quad 15 = 30$$

Either grouping results in the same product.

Work Problem ③ at the Side. ▶

🖩 **Calculator Tip**

The calculator approach to **Example 2** uses chain calculations.

$$2 \; ⊗ \; 3 \; ⊗ \; 5 \; ⊜ \; \mathbf{30}$$

A problem with more than two factors, such as the one in **Example 2**, is called a **chain multiplication** problem.

VOCABULARY TIP

Chain multiplication Just like a chain is a connected series of objects—a "chain of events," for example—chain multiplication is a series of numbers connected by the operation of multiplication.

② Multiply.

(a) 7×4

(b) 0×9

(c) $8(5)$

(d) $6 \cdot 5$

(e) $(1)(8)$

③ Multiply.

(a) $3 \times 2 \times 5$

(b) $4 \cdot 7 \cdot 1$

(c) $(8)(3)(0)$

Answers

2. (a) 28 **(b)** 0 **(c)** 40 **(d)** 30 **(e)** 8
3. (a) 30 **(b)** 28 **(c)** 0

4 Multiply.

GS **(a)** $\overset{1}{5}3$
      $\times\ \ 5$
      $\overline{\_\ \_\ 5}$

(b) 79
      $\times\ \ 0$

(c) 758
      $\times\ \ \ 8$

(d) 2831
      $\times\ \ \ \ 7$

(e) 4714
      $\times\ \ \ \ 8$

OBJECTIVE ▶ **3** **Multiply by single-digit numbers.** Regrouping may be needed in multiplication problems with larger factors.

EXAMPLE 3   **Multiplying with Regrouping**

Multiply.

(a) 53
      $\times\ 4$

Start by multiplying in the ones column.

$$\overset{1}{5}3$$
$$\underline{\times\ \ 4}$$
$$2$$

Write 1 ten in the tens column.
$4 \times 3 = \mathbf{12}$ ones
Write 2 ones in the ones column.

Next, multiply 4 times 5 tens.

$$\overset{1}{5}3$$
$$\underline{\times\ \ 4}$$
$$2$$

4×5 tens $=\ \mathbf{20}$ tens

Add the 1 ten that was written at the top of the tens column.

$$\overset{1}{5}3$$
$$\underline{\times\ \ 4}$$
$$\mathbf{212}$$

20 tens + 1 ten = 21 tens

(b) 724
      $\times\ \ \ 5$

Work as shown.

$$\overset{1\ 2}{724}$$
$$\underline{\times\ \ \ 5}$$
$$3620$$

Use regrouping here.

$5 \times 4 = \mathbf{20}$ ones; write 0 ones; write 2 tens in the tens column.

$5 \times 2 = \mathbf{10}$ tens; add the 2 regrouped tens to get 12 tens; write 2 tens; write 1 hundred in the hundreds column.

5×7 hundreds $= \mathbf{35}$ hundreds; add the 1 regrouped hundred to get 36 hundreds.

◀ **Work Problem** **4** **at the Side.**

OBJECTIVE ▶ **4** **Use multiplication shortcuts for numbers ending in zeros.** The product of two whole number factors is also called a **multiple** of either factor. For example, since 4 • 2 = 8, the whole number 8 is a multiple of both 4 and 2. *Multiples of 10* are very useful when multiplying. A **multiple of 10** is a whole number that ends in 0, such as 10, 20, or 30; 100, 200, or 300; 1000, 2000, or 3000; and so on. There is a short way to multiply by these multiples of 10. Look at the following examples.

$$26 \times 1 = 26$$

$$26 \times 10 = 260$$

$$26 \times 100 = 2600$$

$$26 \times 1000 = 26,000$$

Do you see a pattern? These examples suggest the rule that follows.

Multiplying by Multiples of 10

To multiply a whole number by 10, 100, or 1000, attach one, two, or three zeros, respectively, to the right of the whole number.

EXAMPLE 4   Using Multiples of 10 to Multiply

Multiply.

(a) $59 \times 10 = 590$
Attach 0.

(b) $74 \times 100 = 7400$
Attach 00.

(c) $803 \times 1000 = 803,000$ ← Attach 000.

·········· **Work Problem ⑤ at the Side. ▶**

You can also find the product of other multiples of 10 by attaching zeros.

EXAMPLE 5   Using Multiples of 10 to Multiply

Multiply.

(a) 75×3000
Multiply 75 by 3, and then attach three zeros.

$$75 \times 3000 = 225,000$$

$$\begin{array}{r} 75 \\ \times\ 3 \\ \hline 225 \end{array}$$
Attach 000.

Use useful shortcuts.

(b) 150×70
Multiply 15 by 7, and then attach two zeros.

$$150 \times 70 = 10,500 \leftarrow \text{Attach 00.}$$

$$\begin{array}{r} 15 \\ \times\ 7 \\ \hline 105 \end{array}$$

·········· **Work Problem ⑥ at the Side. ▶**

OBJECTIVE ⑤ Multiply by numbers having more than one digit. The next example shows multiplication when both factors have more than one digit.

EXAMPLE 6   Multiplying with More Than One Digit

Multiply 46 and 23.

First multiply 46 by 3.

$$\begin{array}{r} \overset{1}{4}6 \\ \times\ 3 \\ \hline 138 \end{array} \leftarrow 46 \times 3 = 138$$

Regrouping is needed here.

·········· **Continued on Next Page**

⑤ Multiply.

(a) $63 \times 10 = 63$ ____

(b) 305×100

(c) 714×1000

⑥ Multiply.

(a) 17×50

$$\begin{array}{r} 17 \\ \times\ 5 \\ \hline 85 \end{array} \_ \left\{ \begin{array}{l} \text{Attach} \\ \text{one zero.} \end{array} \right.$$

(b) 73×400

(c) $\begin{array}{r} 180 \\ \times\ 30 \end{array}$

(d) $\begin{array}{r} 4200 \\ \times\ 80 \end{array}$

(e) $\begin{array}{r} 800 \\ \times\ 600 \end{array}$

Answers

5. (a) 0; 630 **(b)** 30,500 **(c)** 714,000
6. (a) 0; 850 **(b)** 29,200 **(c)** 5400
(d) 336,000 **(e)** 480,000

7 Complete each multiplication.

GS **(a)**
$$\begin{array}{r} \overset{2}{\overset{2}{}} \\ 3\,5 \\ \times\ 5\,4 \\ \hline 1\,4\,0 \\ 1\,7\,5 \\ \hline \underline{\ \ }\ \underline{\ \ }\,9\,0 \end{array}$$

GS **(b)**
$$\begin{array}{r} \overset{2}{\overset{5}{}} \\ 7\,6 \\ \times\ 4\,9 \\ \hline 6\,8\,4 \\ 3\,0\,4 \\ \hline \underline{\ \ }\,\underline{\ \ }\,\underline{\ \ }\,\underline{\ \ } \end{array}$$

8 Multiply.

(a)
$$\begin{array}{r} 52 \\ \times\ 16 \end{array}$$

(b)
$$\begin{array}{r} 81 \\ \times\ 49 \end{array}$$

(c)
$$\begin{array}{r} 234 \\ \times\ 73 \end{array}$$

(d)
$$\begin{array}{r} 835 \\ \times\ 189 \end{array}$$

Now multiply 46 by 20.
$$\begin{array}{r} \overset{1}{} \\ 46 \\ \times\ 20 \\ \hline 920 \end{array} \leftarrow\ 46 \times 20 = 920$$

Add the results.
$$\begin{array}{r} 46 \\ \times\ 23 \\ \hline 138 \\ +\ 920 \\ \hline 1058 \end{array}$$
$138 \leftarrow 46 \times 3$
$920 \leftarrow 46 \times 20$
Add.

Both 138 and 920 are called **partial products.** As a common practice and to save time, the 0 in 920 is usually not written.

$$\begin{array}{r} 46 \\ \times\ 23 \\ \hline 138 \\ 92 \\ \hline 1058 \end{array}$$
{ 0 not written. Be very careful to place the 2 in the tens column.

◀ **Work Problem 7 at the Side.**

EXAMPLE 7 **Using Partial Products**

Multiply.

(a)
$$\begin{array}{r} 2\,3\,3 \\ \times\ 1\,3\,2 \\ \hline 4\,6\,6 \\ 6\,9\,9 \\ 2\,3\,3\ \ \\ \hline 3\,0,7\,5\,6 \end{array}$$
(Tens lined up)
(Hundreds lined up)
← Product

Be certain to align numbers in columns.

(b)
$$\begin{array}{r} 538 \\ \times\ 46 \end{array}$$

First multiply by 6.
$$\begin{array}{r} \overset{2\,4}{538} \\ \times\ 46 \\ \hline 3228 \end{array}$$
← Regrouping is needed here.

Now multiply by 4, being careful to line up the tens.
$$\begin{array}{r} \overset{1\ 3}{\overset{2\ 4}{5\,3\,8}} \\ \times\ 4\,6 \\ \hline 3\,2\,2\,8 \\ 2\,1\,5\,2\ \ \\ \hline 2\,4,7\,4\,8 \end{array}$$
Finally, add the partial products.

◀ **Work Problem 8 at the Side.**

Answers

7. **(a)** 1; 8; 1890 **(b)** 3724
8. **(a)** 832 **(b)** 3969
  (c) 17,082 **(d)** 157,815

When 0 appears in the multiplier, be sure to move the partial products to the left to account for the position held by the 0.

> **EXAMPLE 8**   **Multiplying with Zeros**

Multiply.

(a)
$$
\begin{array}{r}
1\ 3\ 7 \\
\times 3\ 0\ 6 \\
\hline
8\ 2\ 2 \\
0\ 0\ 0 \quad \text{(Tens lined up)} \\
4\ 1\ 1 \quad \text{(Hundreds lined up)} \\
\hline
4\ 1,9\ 2\ 2
\end{array}
$$

(b)

$$
\begin{array}{r}
1\ 4\ 0\ 6 \\
\times 2\ 0\ 0\ 1 \\
\hline
1\ 4\ 0\ 6 \\
0\ 0\ 0\ 0 \quad \leftarrow \text{(Zeros to line up tens)} \\
0\ 0\ 0\ 0 \quad \leftarrow \text{(Zeros to line up hundreds)} \\
2\ 8\ 1\ 2 \\
\hline
2,8\ 1\ 3,4\ 0\ 6
\end{array}
$$

Use extra caution when working with 0s.

$$
\begin{array}{r}
1\ 4\ 0\ 6 \\
\times 2\ 0\ 0\ 1 \\
\hline
1\ 4\ 0\ 6 \\
2\ 8\ 1\ 2\ 0\ 0 \quad \leftarrow \text{Zeros are written so this partial product starts in the thousands column.} \\
\hline
2,8\ 1\ 3,4\ 0\ 6
\end{array}
$$

> **Note**
>
> In **Example 8(b)** in the alternative method on the right, zeros were inserted so that thousands were placed in the thousands column. This is a commonly used shortcut.

········ **Work Problem 9 at the Side.** ▶

OBJECTIVE ▶ 6 Solve application problems with multiplication. The next example shows how multiplication can be used to solve an application problem.

> **EXAMPLE 9**   **Applying Multiplication Skills**

Find the total cost of 75 video games priced at $38 each.

Approach   To find the cost of all the video games multiply the number of games (75) by the cost of one video game ($38).

Solution   Multiply 75 by 38.

$$
\begin{array}{r}
7\ 5 \\
\times 3\ 8 \\
\hline
6\ 0\ 0 \\
2\ 2\ 5 \\
\hline
\$2\ 8\ 5\ 0
\end{array}
$$

The total cost of the video games is $2850.

> 🖩 **Calculator Tip**
>
> If you are using a calculator for **Example 9**, you will do this calculation.
>
> 75 ⊗ 38 ⊜ 2850

Work Problem 10 at the Side. ▶

---

9 Multiply.

GS (a)

$$
\begin{array}{r}
2\ 8 \\
\times\ 6\ 0 \\
\hline
1\ 6\ 8 \\
\hline
\end{array}
$$

(b)
$$
\begin{array}{r}
7\ 2\ 8 \\
\times\ 5\ 0 \\
\end{array}
$$

(c)
$$
\begin{array}{r}
5\ 6\ 2 \\
\times\ 1\ 0\ 9 \\
\end{array}
$$

(d)
$$
\begin{array}{r}
3\ 5\ 2\ 6 \\
\times\ 6\ 0\ 0\ 2 \\
\end{array}
$$

10 Find the total cost of the following items.

GS (a) 314 garden sprayers at $14 per sprayer

(b) 64 tires priced at $139 each

(c) 12 delivery vans at $28,300 per van

Answers

9. **(a)** 0; 0; 1680   **(b)** 36,400
     (c) 61,258   **(d)** 21,163,052
10. **(a)** $4396   **(b)** $8896   **(c)** $339,600

1.4 Exercises

FOR EXTRA HELP

Download the MyDashBoard App

MyMathLab®

CONCEPT CHECK   *Fill in each blank with the correct response.*

1. In a chain multiplication, if you multiply by the largest number first, rather than the smallest number first, the product will always be _____.

2. The property described in **Exercise 1** is the _____ property of multiplication.

3. When you multiply any number by zero, the answer is always _____.

4. When you multiply a whole number by 10, by 100 or by 1000, you can get the answer by attaching one, two, or three _____ to the _____ of the whole number.

Work each chain multiplication. ***See Example 2.***

5. $2 \times 6 \times 2$

6. $8 \times 6 \times 1$

7. $7 \cdot 8 \cdot 0$

8. $9 \cdot 0 \cdot 5$

9. $4 \cdot 1 \cdot 6$

10. $1 \cdot 5 \cdot 7$

11. $(4)(5)(2)$

12. $(4)(1)(9)$

13. Explain in your own words the commutative property of multiplication. How do the commutative properties of addition and multiplication compare to each other?

14. Explain in your own words the associative property of multiplication. How do the associative properties of addition and multiplication compare to each other?

Multiply. ***See Example 3.***

15. $\begin{array}{r} 35 \\ \times\ 6 \\ \hline \end{array}$

16. $\begin{array}{r} 53 \\ \times\ 7 \\ \hline \end{array}$

17. $\begin{array}{r} 34 \\ \times\ 7 \\ \hline \end{array}$

18. $\begin{array}{r} 76 \\ \times\ 5 \\ \hline \end{array}$

19. $\begin{array}{r} 642 \\ \times\ \ \ 5 \\ \hline \end{array}$

20. $\begin{array}{r} 472 \\ \times\ \ \ 4 \\ \hline \end{array}$

21. $\begin{array}{r} 624 \\ \times\ \ \ 3 \\ \hline \end{array}$

22. $\begin{array}{r} 852 \\ \times\ \ \ 7 \\ \hline \end{array}$

23. $\begin{array}{r} \overset{2\ 1}{2\,1\,5\,3} \\ \times\ \ \ \ \ \ 4 \\ \hline \_\ \_1\,2 \end{array}$

24. $\begin{array}{r} \overset{1\ 2}{1\,1\,3\,7} \\ \times\ \ \ \ \ \ 3 \\ \hline \_\ \_1\,1 \end{array}$

25. $\begin{array}{r} 2521 \\ \times\ \ \ \ \ 4 \\ \hline \end{array}$

26. $\begin{array}{r} 2544 \\ \times\ \ \ \ \ 3 \\ \hline \end{array}$

27. $\begin{array}{r} 2561 \\ \times\ \ \ \ \ 8 \\ \hline \end{array}$

28. $\begin{array}{r} 7326 \\ \times\ \ \ \ \ 5 \\ \hline \end{array}$

29. $\begin{array}{r} 36{,}921 \\ \times\ \ \ \ \ \ \ 7 \\ \hline \end{array}$

30. $\begin{array}{r} 28{,}116 \\ \times\ \ \ \ \ \ \ 4 \\ \hline \end{array}$

CONCEPT CHECK   *Fill in each blank with the correct response.*

31. You can use multiples of 10 to multiply 86×200. First, multiply _____ $\times$ 2 to get _____. Then, attach _____ zeros to the right of this number for a final answer of _____.

32. You can use multiples of 10 to multiply 7800×450. First, multiply 78 $\times$ _____ to get _____. Then, attach _____ zeros to the right of this number for a final answer of _____.

Multiply. ***See Examples 4 and 5.***

33. $\begin{array}{r} 80 \\ \times\ 6 \\ \hline \end{array}$

34. $\begin{array}{r} 70 \\ \times\ 5 \\ \hline \end{array}$

35. $\begin{array}{r} 740 \\ \times\ \ \ 3 \\ \hline \end{array}$

36. $\begin{array}{r} 400 \\ \times\ \ \ 8 \\ \hline \end{array}$

37. 600
× 6

38. 860
× 7

39. 125
× 30

40. 246
× 50

41. 1635
× 40

42. 7311
× 50

43. ▶ 900
× 300

44. 400
× 700

45. GS 43,000
× 2 000

46. GS 11,000
× 9 000

47. ▶ 970 • 50

48. 730 • 40

43
× 2
86   Attach 000,000.

11
× 9
99   Attach 000,000.

49. 800 • 900

50. 850 • 700

51. 9700 • 200

52. 10,050 • 300

Multiply. See Examples 6–8.

53. 28
× 17

54. 16
× 34

55. ▶ 75
× 32

56. 82
× 32

57. 83
× 45

58. (75)(21)

59. (58)(41)

60. (82)(67)

61. (67)(92)

62. (26)(33)

63. (28)(564)

64. (58)(312)

65. (619)(35)

66. (681)(47)

67. (55)(286)

68. 286
× 574

69. 735
× 112

70. 621
× 415

71. 538
× 342

72. 3228
× 751

73. 9352
× 264

74. 528
× 106

75. ▶ 215
× 307

76. 218
× 106

77. 428
× 201

78. 3706
× 208

79. 6310
× 3078

80. 3533
× 5001

81. 2195
× 1038

82. 1502
× 2009

83. A classmate of yours is not clear on how to use a shortcut to multiply a whole number by 10, by 100, or by 1000. Write a short note explaining how this can be done.

84. Show two ways to multiply when a 0 is in the multiplier. Use the problem 291 × 307 to show this.

Solve each application problem. **See Example 9.**

85. Carepanian Company, a health care supplier, purchased 300 cartons of Thera Bond Gym Balls. If there are 10 balls in each carton, find the total number of balls purchased.

86. A medical supply house has 30 bottles of vitamin C tablets, with each bottle containing 500 tablets. Find the total number of vitamin C tablets in the supply house.

87. The most expensive U.S. City for a hotel room is New York City, with an average cost of $194 a night. Find the cost of a 12-night stay. (*Source:* hotels.com)

88. Judge Judith Sheindlin, known as Judge Judy on court television, recently signed a four-year contract paying her $45 million each year. Find the total amount of her earnings on this contract. (*Source:* KSTE Radio News.)

89. The average amount of water used per person each day in the United States is 66 gallons. How much water does the average person use in one year? (1 year = 365 days). (*Source:* Oxfam.)

$$365 \times 66 = \underline{}\,\underline{},090$$

90. Squid are being hauled out of the Santa Barbara Channel by the ton. They are then processed, renamed calamari, and exported. Last night 27 fishing boats each hauled out 40 tons of squid. What was the total catch for the night? (*Source: Santa Barbara News Press.*)

27 × 40
27 × 4 = 108 Attach _____.

Find the total cost of the following items. **See Examples 7–9.**

91. 75 first-aid kits at $8 per kit

92. 27 days of child care at $82 per day

93. 65 rebuilt alternators at $24 per alternator

94. 62 wheelchair cushions at $44 per cushion

95. 206 laptop computers at $548 per computer

96. 520 printers at $219 per printer

Multiply.

97. 21 • 43 • 56

98. (600)(8)(75)(40)

Use addition, subtraction, or multiplication to solve each application problem.

99. In a forest-planting project, 450 trees are planted on each acre. Find the number of trees needed to plant 85 acres.

100. The largest living land mammal is the African elephant, and the largest mammal of all time is the blue whale. An African elephant weighs 15,225 pounds and a blue whale weighs 28 times that amount. Find the weight of the blue whale.

101. New York City has a population of 8,391,881, the largest in the country. Boston, in twenty-second place, has a population of 645,169. How many more people live in New York City than in Boston? (*Source:* U.S. Census Bureau.)

102. Los Angeles, the second largest city in the country, has a population of 3,849,378. Dallas, at ninth largest, has a population of 1,232,940. Find the difference in the population of these two cities. (*Source:* U.S. Census Bureau.)

103. A medical center purchased 12 laptop computers at $970 each and 8 printers at $315 each. Find the total cost of this equipment.

104. In the first 17 years of food drives, the Postal Workers have collected 954 million pounds of food. This year they collected 77 million pounds of food. Find the total food collection in 18 years. (*Source:* Stamp Out Hunger, U.S. Postal Service.)

Relating Concepts (Exercises 105–114)   For Individual or Group Work

Work Exercises 105–114 in order.

105. Add.
   (a) $189 + 263$
   (b) $263 + 189$

106. Your answers to **Exercise 105(a) and (b)** should be the same. This shows that the order of numbers in an addition problem does not change the sum. This is known as the _____ property of addition.

107. Add. Recall that parentheses show you what to do first.
   (a) $(65 + 81) + 135$
   (b) $65 + (81 + 135)$

108. Since the answers to **Exercise 107(a) and (b)** are the same, we see that grouping the numbers differently when adding does not change the sum. This is known as the _____ property of addition.

109. Multiply.
   (a) 220×72
   (b) 72×220

110. Since the answers to **Exercise 109(a) and (b)** are the same, we see that the product remains the same when the order of the factors is changed. This is known as the _____ property of multiplication.

111. Multiply. Recall that parentheses tell you what to do first.
   (a) $(26 \times 18) \times 14$
   (b) $26(18 \times 14)$

112. Since the answers to **Exercise 111(a) and (b)** are the same, we see that grouping the numbers differently when multiplying does not change the product. This is known as the _____ property of multiplication.

113. Do the commutative and associative properties apply to subtraction? Explain your answer using several examples.

114. Do you think that the commutative and associative properties will apply to division? Explain your answer using several examples.

1.5 Dividing Whole Numbers

OBJECTIVES

1. Write division problems in three ways.
2. Identify the parts of a division problem.
3. Divide 0 by a number.
4. Recognize that a number cannot be divided by 0.
5. Divide a number by itself.
6. Divide a number by 1.
7. Use short division.
8. Use multiplication to check the answer to a division problem.
9. Use tests for divisibility.

---

1 Write each division problem using two other symbols.

GS **(a)** $24 \div 6 = 4$

$$6\overline{)24}^{\,4} \qquad \frac{24}{\rule{1cm}{0.4pt}} = 4$$

(b) $9\overline{)36}^{\,4}$

(c) $48 \div 6 = 8$

(d) $\dfrac{42}{6} = 7$

Answers

1. **(a)** $6\overline{)24}^{\,4}$ and $\dfrac{24}{6} = 4$

   (b) $36 \div 9 = 4$ and $\dfrac{36}{9} = 4$

   (c) $6\overline{)48}^{\,8}$ and $\dfrac{48}{6} = 8$

   (d) $6\overline{)42}^{\,7}$ and $42 \div 6 = 7$

Suppose the total cost of lunch at a SUBWAY is $18 and is to be divided equally by three friends. Each person would pay $6, as shown here.

$18 total

$6 $6 $6

3 equal parts

OBJECTIVE 1 Write division problems in three ways. Just as $3 \cdot 6$, 3×6, and $(3)(6)$ are different ways of indicating the multiplication of 3 and 6, there are several ways to write 18 divided by 3.

Being divided   Divided by   Being divided

$$18 \div 3 = 6 \qquad 3\overline{)18}^{\,6} \qquad \frac{18}{3} = 6$$

Divided by   Being divided   Divided by

We will use all three division symbols, $\div$, $\overline{)}$, and $-$. In courses such as algebra, a slash symbol, $/$, or a fraction bar, $-$, is most often used.

EXAMPLE 1 Using Division Symbols

Write each division problem using two other symbols.

(a) $18 \div 6 = 3$

This division can also be written as shown below.

$$6\overline{)18}^{\,3} \quad \text{or} \quad \frac{18}{6} = 3 \quad \text{◄ Remember the three division symbols.}$$

(b) $\dfrac{15}{5} = 3 \qquad\qquad 15 \div 5 = 3 \quad \text{or} \quad 5\overline{)15}^{\,3}$

(c) $5\overline{)20}^{\,4} \qquad\qquad 20 \div 5 = 4 \quad \text{or} \quad \dfrac{20}{5} = 4$

◄ **Work Problem 1 at the Side.**

OBJECTIVE 2 Identify the parts of a division problem. In division, the number being divided is the **dividend**, the number divided by is the **divisor**, and the answer is the **quotient**.

$$\text{dividend} \div \text{divisor} = \text{quotient}$$

$$\text{divisor}\overline{)\text{dividend}}^{\,\text{quotient}} \qquad \frac{\text{dividend}}{\text{divisor}} = \text{quotient}$$

EXAMPLE 2   Identifying the Parts of a Division Problem

Identify the dividend, divisor, and quotient.

(a) $35 \div 7 = 5$

$$35 \div 7 = 5 \leftarrow \text{Quotient}$$

Dividend   Divisor

(b) $\dfrac{100}{20} = 5$

Dividend

$$\dfrac{\mathbf{100}}{\mathbf{20}} = \mathbf{5} \leftarrow \text{Quotient}$$

Divisor

> The answer in a division problem is the quotient.

(c) $8\overline{)72}$ with quotient 9

$$\mathbf{9} \leftarrow \text{Quotient}$$
$$8\overline{)72} \leftarrow \text{Dividend}$$

Divisor

···· **Work Problem ❷ at the Side.** ▶

OBJECTIVE ▶ ❸ Divide 0 by a number. If no money, or \$0, is divided equally among five people, each person gets \$0. The general rule for dividing 0 follows.

> **Dividing 0 by a Number**
>
> The number **0** divided by any nonzero number is **0**.

EXAMPLE 3   Dividing 0 by a Number

Divide.

(a) $0 \div 12 = 0$

(b) $0 \div 1728 = 0$

(c) $\dfrac{0}{375} = 0$

> Zero divided by any nonzero number is zero.

(d) $129\overline{)0}$ with quotient 0

···· **Work Problem ❸ at the Side.** ▶

Just as a subtraction such as $8 - 3 = 5$ can be written as the addition $8 = 3 + 5$, any division can be written as a multiplication. For example, $12 \div 3 = 4$ can be written as

$$3 \times 4 = 12 \quad \text{or} \quad 4 \times 3 = 12.$$

---

❷ Identify the dividend, divisor, and quotient.

(a) $15 \div 3 = 5$

(b) $18 \div 6 = 3$

(c) $\dfrac{28}{7} = 4$

(d) $9\overline{)27}$ with quotient 3

❸ Divide.

(a) $0 \div 5$

(b) $\dfrac{0}{9}$

(c) $\dfrac{0}{24}$

(d) $37\overline{)0}$

Answers

2. (a) dividend: 15; divisor: 3; quotient: 5
  (b) dividend: 18; divisor: 6; quotient: 3
  (c) dividend: 28; divisor: 7; quotient: 4
  (d) dividend: 27; divisor: 9; quotient: 3
3. all 0

4 Write each division problem as a
multiplication problem.

(a) $5\overline{)15}$ (with 3 above)

$5 \cdot 3 = \underline{\hspace{1cm}}$

$\underline{\hspace{1cm}} \cdot 5 = \underline{\hspace{1cm}}$

(b) $\dfrac{32}{4} = 8$

(c) $45 \div 9 = 5$

> **EXAMPLE 4**   **Changing Division Problems to Multiplication**

Change each division problem to a multiplication problem.

(a) $\dfrac{20}{4} = 5$ becomes $4 \cdot 5 = 20$ or $5 \cdot 4 = 20$.

(b) $8\overline{)48}$ (with 6 above)   becomes $8 \cdot 6 = 48$ or $6 \cdot 8 = 48$.

(c) $72 \div 9 = 8$ becomes $9 \cdot 8 = 72$ or $8 \cdot 9 = 72$.

◀ **Work Problem 4 at the Side.**

OBJECTIVE 4 Recognize that a number cannot be divided by 0.
Division of any number by 0 cannot be done. To see why, try to find

$$9 \div 0 = ?$$

As we have just seen, any division problem can be converted to a multiplication problem so that

divisor • quotient = dividend.

If you convert the preceding problem to its multiplication counterpart, it reads as follows.

$$0 \cdot ? = 9$$

You already know that 0 times any number must always be 0. Try any number you like to replace the "**?**" and you'll aways get 0 instead of 9. Therefore, the division problem $9 \div 0$ cannot be done. Mathematicians say it is **undefined** and have agreed never to divide by 0. However, $0 \div 9$ *can* be done. Check by rewriting it as a multiplication problem.

$$0 \div 9 = 0 \quad \text{because} \quad 9 \cdot 0 = 0 \text{ is true.}$$

> **Dividing a Number by 0**
>
> Since dividing any number by 0 cannot be done, we say that division by
> **0 is undefined.** It is impossible to compute an answer.

> **EXAMPLE 5**   **Dividing Numbers by 0**

All the following divisions are undefined.

(a) $\dfrac{6}{0}$ is undefined.

(b) $0\overline{)8}$ is undefined.

(c) $18 \div 0$ is undefined.   You **cannot** divide a number by zero.

(d) $\dfrac{3}{0}$ is undefined.

Answers

4. (a) $15; 5 \cdot 3 = 15$ or $3; 15; 3 \cdot 5 = 15$
   (b) $4 \cdot 8 = 32$ or $8 \cdot 4 = 32$
   (c) $9 \cdot 5 = 45$ or $5 \cdot 9 = 45$

Division Involving 0

$$0 \div \text{nonzero number} = 0 \quad \text{and} \quad \frac{0}{\text{nonzero number}} = 0$$

but

$$\text{nonzero number} \div 0 \quad \text{and} \quad \frac{\text{nonzero number}}{0} \quad \text{are } \textbf{undefined.}$$

CAUTION

When 0 is the divisor in a problem, you write "undefined" as the answer. Never divide by 0.

Work Problem ⑤ at the Side. ▶

▦ Calculator Tip

Try these two problems on your calculator. Jot down your answers.

9 ⊡ 0 ⊜ _____   0 ⊡ 9 ⊜ _____

When you try to divide by 0, the calculator cannot do it, so it shows the word "Error" or the letter "E" (for error) in the display. But, when you divide 0 by 9 the calculator displays 0, which is the correct answer.

OBJECTIVE ▶ ⑤ Divide a number by itself. What happens when a number is divided by itself? For example, what is $4 \div 4$ or $97 \div 97$?

Dividing a Number by Itself

Any *nonzero* number divided by itself is **1.**

EXAMPLE 6 **Dividing a Nonzero Number by Itself**

Divide.

(a) $16 \div 16 = 1$

(b) $32\overline{)32}^{\,1}$ ◀ *A nonzero number divided by itself is 1.*

(c) $\dfrac{57}{57} = 1$

·········· Work Problem ⑥ at the Side. ▶

OBJECTIVE ▶ ⑥ Divide a number by 1. What happens when a number is divided by 1? For example, what is $5 \div 1$ or $86 \div 1$?

Dividing a Number by 1

Any number divided by 1 is itself.

---

⑤ Divide. If the division is not possible, write "undefined."

(a) $\dfrac{4}{0}$

(b) $\dfrac{0}{4}$

(c) $0\overline{)36}$

(d) $36\overline{)0}$

(e) $100 \div 0$

(f) $0 \div 100$

⑥ Divide.

(a) $8 \div 8$

(b) $15\overline{)15}$

(c) $\dfrac{37}{37}$

Answers

5. (a) undefined **(b)** 0 **(c)** undefined
(d) 0 **(e)** undefined **(f)** 0
6. all 1

7 Divide.

(a) $9 \div 1$

(b) $1\overline{)18}$

(c) $\dfrac{43}{1}$

8 Divide using short division.

(a) $2\overline{)24}$

(b) $3\overline{)93}$

(c) $4\overline{)88}$

(d) $2\overline{)624}$

VOCABULARY TIP

Remainder In division, when the answer (quotient) is not a whole number, the portion left over (remains) is called the *remainder*.

9 Divide using short division.

(a) $2\overline{)125}$

(b) $3\overline{)215}$

(c) $4\overline{)538}$

(d) $\dfrac{819}{5}$

Answers

7. (a) 9 (b) 18 (c) 43
8. (a) 12 (b) 31 (c) 22 (d) 312
9. (a) 62 **R1** (b) 71 **R2** (c) 134 **R2**
   (d) 163 **R4**

EXAMPLE 7   **Dividing Numbers by 1**

Divide.

(a) $5 \div 1 = 5$

(b) $1\overline{)26}$ A number divided by 1 is itself.

(c) $\dfrac{41}{1} = 41$

◀ **Work Problem 7 at the Side.**

OBJECTIVE 7 Use short division. **Short division** is a method of dividing a number by a one-digit divisor.

EXAMPLE 8   **Using Short Division**

Divide using short division. $3\overline{)96}$

First, divide 9 by 3.

$$\begin{array}{c} 3 \\ 3\overline{)96} \end{array} \leftarrow \dfrac{9}{3} = 3$$

Next, divide 6 by 3.

$$\begin{array}{c} 32 \\ 3\overline{)96} \end{array} \leftarrow \dfrac{6}{3} = 2$$

◀ **Work Problem 8 at the Side.**

When two numbers do not divide exactly, the leftover portion is called the **remainder.** The remainder must always be less than the divisor.

EXAMPLE 9   **Using Short Division with a Remainder**

Divide 147 by 4 using short division.
Rewrite the problem.

$$4\overline{)147}$$

Because 1 cannot be divided by 4, divide 14 by 4. Notice that the 3 is placed over the 4 in 14.

Since 14 is being divided by 4, the answer (3) is placed over the 4.

$$\begin{array}{c} 3 \\ 4\overline{)14^27} \end{array} \qquad \dfrac{14}{4} = 3 \text{ with 2 left over}$$

Next, divide 27 by 4. The final number left over is the remainder. Use **R** to indicate the remainder, and write the remainder to the side.

$$\begin{array}{c} 3\ 6\ \textbf{R3} \\ 4\overline{)14^27} \end{array} \qquad \dfrac{27}{4} = 6 \text{ with 3 left over}$$

◀ **Work Problem 9 at the Side.**

EXAMPLE 10 Dividing with a Remainder

Divide 1809 by 7.

Divide 18 by 7.

$$\frac{2}{7)18^409} \qquad \frac{18}{7} = 2 \text{ with 4 left over}$$

Divide 40 by 7.

$$\frac{2\;5}{7)18^40^59} \qquad \frac{40}{7} = 5 \text{ with 5 left over}$$

Divide 59 by 7.

$$\frac{2\;5\;8\;\textbf{R}3}{7)18^40^59} \qquad \frac{59}{7} = 8 \text{ with 3 left over}$$

The remainder must be less than the divisor.

Work Problem 10 at the Side. ▶

Note

Short division takes practice but is useful when the divisor is a one-digit number.

OBJECTIVE 8 Use multiplication to check the answer to a division problem. Check the answer to a division problem as follows.

Checking Division

$$(\text{divisor} \times \text{quotient}) + \text{remainder} = \text{dividend}$$

Parentheses tell you what to do first: Multiply the divisor by the quotient, then add the remainder.

EXAMPLE 11 Checking Division by Using Multiplication

Check each answer.

(a) $5)\overline{458}$ 91 **R**3

$$(\text{divisor} \times \text{quotient}) + \text{remainder} = \text{dividend}$$

$$(5 \quad \times \quad 91) \quad + \quad 3$$

Be careful! Always add the remainder when checking division.

$$455 \quad + \quad 3 \quad = \quad \textbf{458}$$

Matches original dividend, so the division was done correctly.

Continued on Next Page

---

10 Divide.

(a) $4)\overline{5^13^10}$ 1 3 _ **R** _

(b) $\dfrac{515}{7}$

(c) $3)\overline{1885}$

(d) $6)\overline{1415}$

11 Use multiplication to check each division. If an answer is incorrect, give the correct answer.

GS (a) $\dfrac{32\ \text{R1}}{2)\overline{65}}$

$2 \times 32 = 64$

____ + ____ = 65

(b) $\dfrac{83\ \text{R4}}{7)\overline{586}}$

(c) $\dfrac{407\ \text{R2}}{3)\overline{1223}}$

(d) $\dfrac{476\ \text{R3}}{5)\overline{2383}}$

(b) $\dfrac{239\ \text{R4}}{6)\overline{1437}}$

(divisor × quotient) + remainder = dividend

$(6 \quad \times \quad 239) \quad + \quad 4$

$1434 \quad + \quad 4 \quad = \mathbf{1438}$

Does not match original dividend.

The answer does **not** check. Rework the original problem to get the correct answer, 239 **R3**. Then, $(6 \times 239) + 3$ **does** give 1437.

CAUTION

A common error when checking division is to forget to add the remainder. Be sure to add any remainder when checking a division problem.

◄ **Work Problem 11 at the Side.**

OBJECTIVE 9 Use tests for divisibility. It is often important to know whether a number is *divisible* by another number. You will find this useful in **Chapter 2** when writing fractions in lowest terms.

Divisibility

One whole number is **divisible** by another if the remainder is 0.

Use the following tests to decide whether one number is divisible by another number.

Tests for Divisibility

A number is divisible by

2	if it ends in 0, 2, 4, 6, or 8. These are the even numbers.
3	if the sum of its digits is divisible by 3.
4	if the last two digits make a number that is divisible by 4.
5	if it ends in 0 or 5.
6	if it is divisible by both 2 and 3.
7	has no simple test.
8	if the last three digits make a number that is divisible by 8.
9	if the sum of its digits is divisible by 9.
10	if it ends in 0.

The most commonly used tests are those for 2, 3, 5, and 10.

Answers

11. (a) 64; 1; correct **(b)** incorrect; should be 83 **R5**
  (c) correct **(d)** correct

Divisibility by 2

A number is divisible by **2** if the number ends in 0, 2, 4, 6, or 8. All even numbers are divisible by 2.

EXAMPLE 12 Testing for Divisibility by 2

Are the following numbers divisible by 2?

(a) 986

└─ Ends in 6

> All even numbers are divisible by 2.

Because the number ends in 6, which is an *even number,* the number 986 is divisible by 2.

(b) 3255 is not divisible by 2.

└─ Ends in 5, and not in 0, 2, 4, 6, or 8

⋯⋯⋯⋯⋯⋯⋯⋯⋯⋯⋯⋯⋯ **Work Problem 12 at the Side.** ▶

Divisibility by 3

A number is divisible by **3** if the sum of its digits is divisible by **3.**

EXAMPLE 13 Testing for Divisibility by 3

Are the following numbers divisible by 3?

(a) 4251

Add the digits.

$$4 \ 2 \ 5 \ 1$$

> If the sum of the digits is divisible by 3, the original number is divisible by 3.

$$4 + 2 + 5 + 1 = 12$$

Because 12 is divisible by 3, the number 4251 is also divisible by 3.

(b) 29,806

Add the digits.

$$2 \ 9 \ 8 \ 0 \ 6$$

$$2 + 9 + 8 + 0 + 6 = 25$$

Because 25 is *not* divisible by 3, the number 29,806 is *not* divisible by 3.

CAUTION

Be careful when testing for divisibility by adding the digits. This method works only for the numbers 3 and 9.

⋯⋯⋯⋯⋯⋯⋯⋯⋯⋯⋯⋯⋯ **Work Problem 13 at the Side.** ▶

---

12 Which numbers are divisible by 2?

(a) 258

(b) 307

(c) 4216

(d) 73,000

13 Which numbers are divisible by 3?

GS (a) 743

$7 + 4 + 3 = 14$

14 is *not* divisible by 3.

So, 743 is *not* divisible by 3.

GS (b) 5325

$5 + 3 + 2 + 5 = $ ____

____ is divisible by 3.

So, 5325 is divisible by ____.

(c) 374,214

(d) 205,633

Answers

12. all but b

13. 15; 15; 3; b and c

⑭ Which numbers are divisible by 5?

(a) 180

(b) 635

(c) 8364

(d) 206,105

Divisibility by 5 and by 10

A number is divisible by **5** if it ends in 0 or 5.
A number is divisible by **10** if it ends in 0.

EXAMPLE 14 **Testing for Divisibility by 5**

Are the following numbers divisible by 5?

(a) 12,900 ends in 0 and is divisible by 5.

If the number ends in 0 or 5, it's divisible by 5.

(b) 4325 ends in 5 and is divisible by 5.

(c) 392 ends in 2 and is *not* divisible by 5.

◀ Work Problem ⑭ at the Side.

EXAMPLE 15 **Testing for Divisibility by 10**

Are the following numbers divisible by 10?

(a) 700 and 9140 both end in 0 and are divisible by 10.

If the number ends in 0, it's divisible by 10.

(b) 355 and 18,743 do not end in 0 and are *not* divisible by 10.

◀ Work Problem ⑮ at the Side.

⑮ Which numbers are divisible by 10?

(a) 270

(b) 495

(c) 5030

(d) 14,380

Answers

14. all but (c)
15. all but (b)

1. CONCEPT CHECK   Write the three common symbols used to show multiplication.

2. CONCEPT CHECK   Write the three common symbols used to show division.

Write each division problem using two other symbols. **See Example 1.**

3. $24 \div 4 = 6$

4. $36 \div 3 = 12$

5. $\dfrac{45}{9} = 5$

6. $\dfrac{56}{8} = 7$

7. $2\overline{)16}$ with quotient 8

8. $8\overline{)48}$ with quotient 6

9. CONCEPT CHECK   When a number is divided by 1, the answer is always the _____ itself.

10. CONCEPT CHECK   When zero is divided by a number, the answer is always _____.

Divide. If the division is not possible, write "undefined." **See Examples 3–7.**

11. $9 \div 9$

12. $36 \div 9$

13. $\dfrac{14}{2}$

14. $\dfrac{10}{0}$

15. $22 \div 0$
GS   When 0 is the divisor, write _____ as the answer.

16. $6 \div 6$
GS   When a number is divided by itself, write _____ as the answer.

17. $\dfrac{24}{1}$

18. $\dfrac{12}{1}$

19. $15\overline{)0}$
GS   When dividing 0 by a nonzero number, the answer is _____.

20. $\dfrac{0}{12}$

21. $0\overline{)43}$
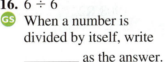

22. $\dfrac{8}{0}$

CONCEPT CHECK   *Use the tests for divisibility by 2, 3, 5, and 10. Circle the numbers that will divide evenly into the given number.*

23. 8670
   2   3   5   10

24. 13,785
   2   3   5   10

25. 9,221,784
   2   3   5   10

26. 5,409,720
   2   3   5   10

Divide by using short division. Use multiplication to check each answer. **See Examples 8–10.**

27. $3\overline{)75}$

28. $5\overline{)85}$

29. $7\overline{)126}$

30. $6\overline{)168}$

31. $4\overline{)1216}$

32. $5\overline{)2305}$

33. $4\overline{)2509}$

34. $8\overline{)1335}$

35. $6\overline{)9137}$

36. $9\overline{)8371}$

37. $6\overline{)1854}$

38. $8\overline{)856}$

39. 12,020 ÷ 4　　**40.** 8012 ÷ 4　　**41.** 30,036 ÷ 6　　**42.** 32,008 ÷ 8

43. 2434 ÷ 3　　**44.** 5993 ÷ 7　　**45.** 12,947 ÷ 5　　**46.** 33,285 ÷ 9

47. $\dfrac{21,040}{8}$　　**48.** $\dfrac{8199}{9}$　　**49.** $\dfrac{74,751}{6}$　　**50.** $\dfrac{72,543}{5}$

51. $\dfrac{71,776}{7}$　　**52.** $\dfrac{77,621}{3}$　　**53.** $\dfrac{128,645}{7}$　　**54.** $\dfrac{172,255}{4}$

Use multiplication to check each answer. If an answer is incorrect, find the correct answer.
See Example 11.

55. $5\overline{)1877}$ 375 R2
GS CHECK
$5 \cdot 375 + 2 =$ ___
correct

56. $3\overline{)1282}$ 427 R1
GS CHECK
$3 \cdot 427 + 1 =$ ___
correct

57. $3\overline{)5725}$ 1908 R2

58. $5\overline{)2158}$ 432 R3

59. $7\overline{)4692}$ 650 R2

60. $9\overline{)5974}$ 663 R5

61. $6\overline{)21,409}$ 3 568 R2

62. $6\overline{)3192}$ 532

63. $8\overline{)16,019}$ 2 002 R3

64. $8\overline{)33,664}$ 4 208

65. $6\overline{)69,140}$ 11,523 R2

66. $3\overline{)82,598}$ 27,532 R1

67. $9\overline{)86,655}$ 9 628 R7

68. $7\overline{)50,809}$ 7 258 R4

69. $8\overline{)222,576}$ 27,822

70. $4\overline{)311,216}$ 77,804

71. Explain in your own words how to check a division problem using multiplication. Be sure to tell what must be done if the quotient includes a remainder.

72. Describe the three divisibility rules that you think will be most useful and tell why.

Solve each application problem.

73. The Carnival Cruise Line has 2624 linen napkins. If it takes eight napkins to set each table, find the number of tables that can be set. (*Source: USA Today.*)

74. A school district will distribute 1620 new science books equally among 12 schools. How many books will each school receive?

75. In one 8-hour day Dreyer's Edy's can produce 76,800 ice cream drumsticks. How many are produced each hour? (*Source:* History Channel, *Modern Marvels: Snack Food Tech.*)

76. Tootsie Roll Industries produces 415,000,000 Tootsie Rolls in a 5-day week. Find the number produced each day. (*Source:* History Channel, *Modern Marvels: Snack Food Tech.*)

77. Lottery winnings of $436,500 are divided equally among nine Starbucks employees. Find the amount received by each employee.

78. How many 5-pound bags of organic whole wheat flour can be filled from a 17,175-pound bin of flour?

79. McDonald's Restaurants is hiring 660 new employees, from crew workers to managers, in one area. If 4 new employees are hired at each location, find the number of locations in this area. (*Source: Sacramento Bee.*)

80. The Wii remains the top-selling video game system. Nintendo sold 85 million Wiis in the first 5 years. Find the average number sold each year. (*Source: USA Today.*)

81. A class-action lawsuit settlement of $6,825,000 is divided evenly among six injured people. Find the amount received by each person.

82. A 12,000-square foot condominium at the edge of Central Park in Manhattan sold for a record $45,000,000. The buyer paid for the condominium in eight equal payments. Find the amount of each payment. (*Source: USA Today.*)

83. The record for picking blueberries in one day was set in the state of Maine and was 6900 pounds. Since there are 1000 blueberries in a pound, this amounted to 6,900,000 blueberries. If these berries were picked in 8 hours, find the average number of berries picked each hour. (*Source:* Discovery Channel, *Dirty Jobs.*)

84. A professional basketball player signed a 4-year contract for $21,937,500. How much is this each year?

Put a ✓ mark in the blank if the number at the left is divisible by the number at the top. Put an X in the blank if the number is not divisible by the number at the top. See Examples 12–15.

	2	3	5	10
85. 60	____	____	____	____
87. 92	____	____	____	____
89. 445	____	____	____	____
91. 903	____	____	____	____
93. 5166	____	____	____	____
95. 21,763	____	____	____	____

	2	3	5	10
86. 35	____	____	____	____
88. 96	____	____	____	____
90. 897	____	____	____	____
92. 500	____	____	____	____
94. 8302	____	____	____	____
96. 32,472	____	____	____	____

1.6 Long Division

If the total cost of 42 Sony iPod Docking Systems is $3066, we can find the cost of each docking system using **long division**. Long division is used to divide by a number with more than one digit.

OBJECTIVES

1. Do long division.
2. Divide numbers ending in 0 by numbers ending in 0.
3. Use multiplication to check division answers.

OBJECTIVE ▶ **1** **Do long division.** In long division, estimate the various numbers by using a **trial divisor** to get a **trial quotient**.

EXAMPLE 1   **Using a Trial Divisor and a Trial Quotient**

Divide. $42\overline{)3066}$

Because 42 is closer to 40 than to 50, use the first digit of the divisor as a trial divisor.

$$42 \quad \xleftarrow{\text{Using a trial divisor is a helpful tool.}}$$

↑
Trial divisor ──────┘

Try to divide the first digit of the dividend by 4. Since 3 cannot be divided by 4, use the first *two* digits, 30.

$$\frac{30}{4} = 7 \text{ with remainder 2}$$

↓
7 ← Trial quotient
$42\overline{)3066}$

↑
────── 7 goes over the 6, because $\frac{306}{42}$ is about 7.

Multiply 7 and 42 to get 294; next, subtract 294 from 306.

$$\begin{array}{r} 7 \\ 42\overline{)3066} \\ \underline{294} \leftarrow 7 \times 42 \\ 12 \leftarrow 306 - 294 \end{array}$$

This number (12) must be smaller than 42, the divisor.

Bring down the 6 at the right.

$$\begin{array}{r} 7 \\ 42\overline{)3066} \\ \underline{294}\downarrow \\ 126 \leftarrow 6 \text{ brought down} \end{array}$$

Use the trial divisor, 4.

First two digits of 126 → $\frac{12}{4} = 3$

$$\begin{array}{r} 73 \\ 42\overline{)3066} \\ \underline{294} \\ 126 \\ \underline{126} \leftarrow 3 \times 42 = 126 \\ 0 \end{array}$$

The cost of each docking system is $73.
Check the answer by multiplying 42 and 73. The product should be 3066.

VOCABULARY TIP

Trial divisor A method of determining how many times the true divisor goes into the dividend by estimating, using the leading digit of the divisor.

VOCABULARY TIP

Trial quotient The quotient that results from dividing the dividend by the trial divisor.

❶ Divide.

㊐ (a) 28)2296

$$
\begin{array}{r}
8 \\
28\overline{)2296} \\
224\downarrow \\
\hline
56 \\
56 \\
\hline
0
\end{array}
$$

(b) 16)1024

(c) 61)8784

(d) $\dfrac{2697}{93}$

❷ Divide.

(a) 24)1344

(b) 72)4472

(c) 65)5416

(d) 89)6649

CAUTION

The *first digit* of the quotient in long division must be placed in the proper position over the dividend.

◀ **Work Problem ❶ at the Side.**

EXAMPLE 2 **Dividing to Find a Trial Quotient**

Divide. 58)2730

Use 6 as a trial divisor, since 58 is closer to 60 than to 50.

First two digits of dividend ⟶ $\dfrac{27}{6}$ = 4 with 3 left over

4 ← Trial quotient
$$
\begin{array}{r}
4 \\
58\overline{)2730} \\
232 \leftarrow 4 \times 58 = 232 \\
\hline
41 \leftarrow 273 - 232 = 41 \text{ (smaller than 58,} \\
\text{the divisor)}
\end{array}
$$

Bring down the 0.

$$
\begin{array}{r}
4 \\
58\overline{)2730} \\
232\downarrow \\
\hline
410 \leftarrow 0 \text{ brought down}
\end{array}
$$

First two digits of 410 ⟶ $\dfrac{41}{6}$ = 6 with 5 left over

46 ← Trial quotient
$$
\begin{array}{r}
46 \\
58\overline{)2730} \\
232 \\
\hline
410 \\
348 \leftarrow 6 \times 58 = 348 \\
\hline
62 \leftarrow \text{Greater than 58}
\end{array}
$$

Do not leave a remainder that is **greater** than the divisor.

The remainder, 62, is greater than the divisor, 58, so 7 should be used instead of 6.

47 R4 ←
$$
\begin{array}{r}
47 \\
58\overline{)2730} \\
232 \\
\hline
410 \\
406 \leftarrow 7 \times 58 = 406 \\
\hline
4 \leftarrow 410 - 406
\end{array}
$$

Now the remainder, 4, is *less* than the divisor, 58.

◀ **Work Problem ❷ at the Side.**

Sometimes it is necessary to write a 0 in the quotient.

EXAMPLE 3 Writing Zeros in the Quotient

Divide: $34\overline{)7068}$

Start as in **Examples 1 and 2.**

$$
\begin{array}{r}
2 \\
34\overline{)7068} \\
68 \leftarrow 2 \times 34 = 68 \\
\overline{2} \leftarrow 70 - 68 = 2
\end{array}
$$

Bring down the 6.

$$
\begin{array}{r}
2 \\
34\overline{)7068} \\
68\downarrow \\
\overline{26} \leftarrow \text{6 brought down}
\end{array}
$$

Since 26 cannot be divided by 34, write a 0 in the quotient as a placeholder.

$$
\begin{array}{r}
2\mathbf{0} \leftarrow \text{0 in quotient} \\
34\overline{)7068} \\
68 \\
\overline{26}
\end{array}
$$

> Use a zero to hold a place in the quotient.

Bring down the final digit, the 8.

$$
\begin{array}{r}
20 \\
34\overline{)7068} \\
68\ \downarrow \\
\overline{268} \leftarrow \text{8 brought down}
\end{array}
$$

Complete the problem.

$$
\begin{array}{r}
207\ \ \mathbf{R}30 \\
34\overline{)7068} \\
68 \\
\overline{268} \\
238 \\
\overline{30}
\end{array}
$$

The quotient is 207 **R**30.

CAUTION

There **must be a digit** in the quotient (answer) above every digit in the dividend once the answer has begun. Notice in **Example 3** that a **0** was used to ensure a digit in the quotient above every digit in the dividend.

···· Work Problem ❸ at the Side. ▶

OBJECTIVE ❷ Divide numbers ending in 0 by numbers ending in 0.
When the divisor and dividend both contain zeros at the far right, recall that these numbers are multiples of 10. As with multiplication, there is a short way to divide these multiples of 10. Look at the following examples.

$$26{,}000 \div 1 = 26{,}000$$
$$26{,}000 \div 10 = 2600$$
$$26{,}000 \div 100 = 260$$
$$26{,}000 \div 1000 = 26$$

Do you see a pattern? These examples suggest the following rule.

---

❸ Divide.

GS (a) $17\overline{)1823}$

$$
\begin{array}{r}
1\_7\ \mathbf{R}\_ \\
17\overline{)1\,8\,2\,3} \\
1\,7\downarrow \\
\overline{1\,2} \\
0\downarrow \\
\overline{1\,2\,3} \\
1\,1\,9 \\
\overline{-}
\end{array}
$$

(b) $23\overline{)4791}$

(c) $39\overline{)15{,}933}$

(d) $78\overline{)23{,}462}$

Answers

3. (a) 0; 4; 4; 107 **R**4   (b) 208 **R**7
   (c) 408 **R**21   (d) 300 **R**62

4 Divide.

(a) $70 \div 10$

(b) $2600 \div 100$

(c) $505,000 \div 1000$

> **Dividing a Whole Number by 10, by 100, or by 1000**
>
> To divide a whole number by 10, by 100, or by 1000, drop the appropriate number of zeros from the whole number.

EXAMPLE 4   Dividing by Multiples of 10

Divide.

— One 0 in divisor

(a) $60 \div 10 = 6$

— 0 dropped

— Two zeros in divisor

(b) $3500 \div 100 = 35$

— 00 dropped

> The same number of zeros must be dropped from the divisor and quotient.

— Three zeros in divisor

(c) $915,000 \div 1000 = 915$

— 000 dropped

◀ **Work Problem 4 at the Side.**

---

5 Divide using the shortcut of dropping zeros.

(a) $50\overline{)6250}$ Drop one zero

Drop one zero $5\overline{)6\_\_}$
$\underline{5}$
12
$\underline{10}$
25
$\underline{25}$
0

(b) $130\overline{)131,040}$

(c) $3400\overline{)190,400}$

Now we'll find the quotients for other multiples of 10 by dropping zeros.

EXAMPLE 5   Dividing by Multiples of 10

Divide:

(a) $40\overline{)11,000}$ Drop one zero from the divisor and the dividend.

275
$4\overline{)1100}$
$\underline{8}$
30
$\underline{28}$
20
$\underline{20}$
0

Since $1100 \div 4$ is 275, then $11,000 \div 40$ is also 275.

(b) $3500\overline{)31,500}$ Drop two zeros from the divisor and the dividend.

9
$35\overline{)315}$
$\underline{315}$
0

Since $315 \div 35$ is 9, then $31,500 \div 3500$ is also 9.

> **Note**
>
> Dropping zeros when dividing by multiples of 10 **does not** change the quotient (answer).

◀ **Work Problem 5 at the Side.**

Answers

4. (a) 7   (b) 26   (c) 505
5. (a) 1; 2; 5; 125   (b) 1008   (c) 56

OBJECTIVE ❸ **Use multiplication to check division answers.** Answers in long division can be checked just as answers in short division were checked.

EXAMPLE 6 Checking Division by Using Multiplication

Check each answer.

(a)

$$48)\overline{5324} \quad 114 \text{ R}43$$

114
× 48   Multiply the quotient and the divisor.
912
456
5472
+ 43 ←Add the remainder.

> Remember to add the remainder when checking division.

5515 ←Result does *not* match dividend.

The answer does **not** check. Rework the original problem to get 110 **R**44. Then $(110 \times 48) + 44$ **does** give 5324.

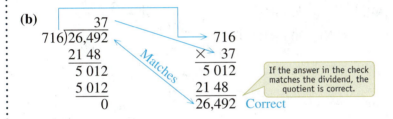

(b)

$$716)\overline{26,492} \quad 37$$

21 48
5 012
5 012
0

716
× 37
5 012
21 48
26,492 Correct

Matches

> If the answer in the check matches the dividend, the quotient is correct.

🖩 **Calculator Tip**

To check the answer to **Example 6(a)**, don't forget to add the remainder.

48 ⊗ 110 ⊕ 44 ⊜ **5324**

Add the remainder.

CAUTION

When checking a division problem, first multiply the quotient and the divisor. Then be sure to **add any remainder** before checking it against the original dividend.

················· Work Problem ❻ at the Side. ▶

---

❻ Decide whether each answer is correct. If the answer is incorrect, find the correct answer.

(a)

$$16)\overline{608} \quad 38$$

48
128
128
0

(b)

$$426)\overline{19,170} \quad 42 \text{ R}178$$

17 040
1 130
952
178

(c)

$$514)\overline{29,316} \quad 57 \text{ R}18$$

25 700
3 616
3 598
18

Answers

6. **(a)** correct **(b)** incorrect; should be 45
  (c) correct

1.6 Exercises   MyMathLab®

CONCEPT CHECK *Decide where the first digit in the quotient would be located. Then, without finishing the division, you can tell which of the three choices is the correct answer. Circle your choice.*

1. $50\overline{)2650}$

5   53   530

2. $14\overline{)476}$

3   34   304

3. $18\overline{)4500}$

2   25   250

4. $35\overline{)5600}$

16   160   1600

5. $86\overline{)10,327}$

12   120 **R7**   1200

6. $46\overline{)24,026}$

5   52   522 **R14**

7. $26\overline{)28,735}$

11   110   1105 **R5**

8. $12\overline{)116,953}$

974 **R2**   9746 **R1**   97,460

9. $21\overline{)149,826}$

71   713   7134 **R12**

10. $64\overline{)208,138}$

325 **R2**   3252 **R10**   32,521

11. $523\overline{)470,800}$

9 **R100**   90 **R100**   900 **R100**

12. $230\overline{)253,230}$

11   110   1101

Divide by using long division. Use multiplication to check each answer. See Examples 1–3, 5, and 6.

13. $18\overline{)1319}$ →
```
    73 R _
18)1319
   126
    59
    54
     5
```

GS **CHECK**

73 • 18 = 1314

1314 + ____ = 1319

14. $58\overline{)3654}$ →
```
      6 _
58)3654
   348
   174
   174
     0
```

GS **CHECK**

____ • 58 = 3654

15. $23\overline{)10,963}$

16. $83\overline{)39,692}$

17. $26\overline{)62,583}$

18. $28\overline{)84,249}$

19. $74\overline{)84,819}$

20. $238\overline{)186,948}$

21. $153\overline{)509,725}$

22. $308\overline{)26,796}$

23. $420\overline{)357,000}$

24. $900\overline{)153,000}$

Use multiplication to check each answer. If an answer is incorrect, find the correct answer. **See Example 6.**

$$\begin{array}{r}101\textbf{R}4\\\textbf{25. }35\overline{)3549}\end{array}$$

$$\begin{array}{r}42\textbf{R}26\\\textbf{26. }64\overline{)2712}\end{array}$$

$$\begin{array}{r}658\textbf{R}9\\\textbf{27. }28\overline{)18,424}\end{array}$$

$$\begin{array}{r}239\textbf{R}121\\\textbf{28. }145\overline{)34,776}\end{array}$$

$$\begin{array}{r}62\textbf{R}3\\\textbf{29. }614\overline{)38,068}\end{array}$$

$$\begin{array}{r}174\textbf{R}368\\\textbf{30. }557\overline{)97,286}\end{array}$$

Solve each application problem by using addition, subtraction, multiplication, or division as needed. **See Examples 3–5.**

31. The first female co-host of a game show, Vanna White, is "Television's Most Frequent Clapper." On the *Wheel of Fortune* she claps 720 times per episode, or 84,240 claps each season. Find the number of episodes filmed each season. (*Source: Guinness Book of World Records* and *Spirit Magazine.*)

32. The two main towers of the Golden Gate Bridge each used 600,000 rivets in their construction. To meet today's earthquake standards, each of these rivets is being replaced with high-strength bolts. If the bolts are shipped in containers holding 160 bolts, how many containers are used to complete the work on both towers? (*Source:* Consolidated Engineering Laboratories.*)

33. Don Gracey, the Mountain Timesmith, has serviced and repaired 636 clocks this year. He has worked on 272 wall clocks and 308 table clocks. The rest were standing floor clocks. Find the number of floor clocks he worked on this year.

total clocks − wall clocks − table clocks = floor clocks

  636   −   272   − 308   = _____

34. There are 24,000,000 business enterprises in the United States. If 7000 of these are larger businesses (over 500 employees), find the number of businesses that are small to mid-size. (*Source:* U.S. Census Bureau.)

total business − larger   = small to
enterprises   businesses   mid-size businesses

24,000,000 − 7000   = _____

35. To complete her college education, Judy Martinez received education loans of $34,080 including interest. Find her monthly payment if the loan is to be paid off in 96 months (8 years).

36. A consultant charged $19,800 for evaluating a school's compliance with the Americans with Disabilities Act. If the consultant worked 225 hours, find the rate charged per hour.

37. Each minute there is one diamond ring sold on eBay's U.S. site. Find the number of diamond rings sold in 30 days. (*Source: Time Style and Design.*)

60 minutes • 24 hours • 30 days = _____

38. A retired milkman in Indianapolis has eaten a Twinkie every day for the last 60 years. How many Twinkies has he eaten over this time period?
Hint: 1 year = 365 days. (*Source:* History Channel, *Modern Marvels: Snack Food Tech.*)

60 years • 365 days = _____

39. Don Gorske of Fond du Lac, Wisconsin has eaten 25,272 Big Macs over the last 39 years. (He is slim and claims a low cholestrol rate.) Find

(a) the average number of Big Macs he has eaten each year.

(b) whether Gorske has eaten more or less than 2 Big Macs each day.

Hint: 1 year = 365 days
(*Source: Guinness Book of World Records.*)

40. Major League baseball teams use more than 220,020 baseballs each season. If each of the 30 major league teams uses the same number of balls in a season, how many are used by each team? (*Source: Parade, The Sunday Newspaper Magazine.*)

Relating Concepts (Exercises 41–48) For Individual or Group Work

Knowing and using the rules of divisibility is necessary in problem solving.
Work Exercises 43–50 in order.

41. If you have $0 and you divide this amount among three people, how much will each receive?

42. When 0 is divided by any nonzero number, the result is ____.

43. Divide.
8 ÷ 0

44. We say that division by 0 is *undefined* because it is (*possible/impossible*) to compute the answer. Give an example involving cookies that will support your answer.

45. Divide.
(a) 14 ÷ 1   **(b)** 1)‾17‾

(c) $\dfrac{38}{1}$

46. Any number divided by 1 is the number itself. Is this also true when multiplying by 1? Give three examples that support your answer.

47. Divide.
(a) 32,000 ÷ 10
(b) 32,000 ÷ 100
(c) 32,000 ÷ 1000

48. Write a rule that explains the shortcut for doing divisions like the ones in **Exercise 47.**

Summary Exercises *Whole Numbers Computation*

CONCEPT CHECK *Write the digit for the given **place value** in each whole number.*

1. 631,548

ten-thousands _____

tens _____

2. 76,047,309

millions _____

hundred-thousands _____

3. 9,181,576,423

hundred-millions _____

thousands _____

Write each number in words.

4. 86,002

5. 425,208,733

Add or subtract as indicated.

6. 46
    + 51

7. 166 + 739

8. 82
    − 61

9. 798
    − 389

10. 6382 + 4062 + 7129

11. 75 + 81,579 + 506 + 4

12. 1704
     − 1027

13. 55,000
     − 17,326

14. 70,552
     − 34,663

CONCEPT CHECK *Multiply using the shortcut for multiples of 10. Circle the correct answer.*

15. 56 × 10

56   560   5600

16. 140 × 40

5600   560   56,000

17. 500
     × 700

3,500,000   35,000   350,000

18. 3600
     × 70

25,200   252,000   2,520,000

Write the numbers in each sentence using digits.

19. There are one million, two hundred thirty-eight thousand, two hundred one nonprofit charitable organizations in the United States. (*Source:* The Giving USA Foundation.)

20. More than five million, five hundred forty-nine thousand, three hundred seventy-five households depend on the National Flood Insurance Program as their main source of protection against flooding, the most common natural disaster in the United States. (*Source:* National Association of Realators.)

Multiply or divide as indicated. If the division cannot be done, write "undefined."

21. $8 \div 8$

22. $0 \div 9$

23. $\dfrac{12}{0}$

24. $\dfrac{15}{1}$

25. 7×8

26. $(6)(0)(5)$

27. $8 \cdot 4 \cdot 3$

28. $\dfrac{608}{2}$

29. $3\overline{)8252}$

30. $4569 \div 6$

31. $\begin{array}{r} 65 \\ \times\ 52 \\ \hline \end{array}$

32. $\begin{array}{r} 507 \\ \times\ 435 \\ \hline \end{array}$

33. $(28)(72)$

34. $(41)(36)$

35. $25\overline{)1950}$

36. $18\overline{)3780}$

37. $\begin{array}{r} 3602 \\ \times\ 5008 \\ \hline \end{array}$

38. $62\overline{)31,400}$

39. $630\overline{)32,760}$

40. $351\overline{)424,011}$

41. $4587 \div 8$

42. $72\overline{)2952}$

43. $\begin{array}{r} 662 \\ \times\ 315 \\ \hline \end{array}$

44. $\begin{array}{r} 2186 \\ \times\ 504 \\ \hline \end{array}$

Built in the 1930s, the San Francisco Golden Gate Bridge is one of the most recognizable landmarks in the world. Some Golden Gate Bridge facts are listed in the table below. Use them to answer Exercises 45–50.

GOLDEN GATE BRIDGE FACTS

Description	Number
Total length, including approaches	8981 feet
Length of middle span	4200 feet
Clearance above water	220 feet
Total weight when built	894,500 tons
Total weight today (lighter road surface)	887,000 tons
Height of two main towers	746 feet
Number of rivets used in each tower	600,000 rivets
Number of strands of wire to make each of the two suspension cables	27,572 strands
Wire needed to make the two cables	80,000 miles
Average number of bridge crossings each year	40 million crossings
Total bridge crossings since construction	2154 million crossings
Number of full-time touch-up painters	38 painters

Source: gocalifornia.about.com

45. Write the total number of cars that have crossed the bridge, using digits (no words).

46. Write the average number of bridge crossings, each year, using digits.

47. If each touch-up painter works 8 hours a day, 5 days a week, how many total hours are spent on painting each week?

48. Find the difference in the weight of the bridge when it was built and the weight of the bridge today.

49. How many feet of wire are there in the two main bridge cables? Write your answer entirely in words. *Hint:* 1 mile = 5280 feet.

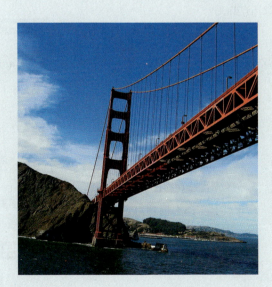

50. (a) Is the length of the bridge, including the approaches, more or less than a mile? How much more or less?

(b) Is the length of the middle span of the bridge more or less than a mile? How much more or less? *Hint:* 1 mile = 5280 feet.

1.7 Rounding Whole Numbers

OBJECTIVES

1. Locate the place to which a number is to be rounded.
2. Round numbers.
3. Round numbers to estimate an answer.
4. Use front end rounding to estimate an answer.

One way to get a quick check on an answer is to *round* the numbers in the problem. **Rounding** a number means finding a number that is close to the original number, but easier to work with.

For example, the county planning commissioner might be discussing the need for more affordable housing. To demonstrate this, she probably would not need to say that the county is in need of 8235 more affordable housing units—she probably could say that the county needs 8200 or even 8000 housing units.

OBJECTIVE ➊ Locate the place to which a number is to be rounded.
The first step in rounding a number is to locate the *place to be rounded.*

EXAMPLE 1   Finding the Place to Which a Number Is to Be Rounded

Locate and draw a line under the place to which each number is to be rounded.

(a) Round 83 to the nearest ten. Is 83 closer to 8̲0 or to 9̲0?

83 is closer to 80.   [83 is closer to 80 than to 90.]

Tens place

(b) Round 54,702 to the nearest thousand. Is it closer to 5̲4,000 or to 5̲5,000?

54,702 is closer to 55,000.

Thousands place

(c) Round 2,806,124 to the nearest hundred-thousand. Is it closer to 2,8̲00,000 or to 2,9̲00,000?

2,806,124 is closer to 2,800,000.

Hundred-thousands place

◀ **Work Problem ➊ at the Side.**

OBJECTIVE ➋ Round numbers. Use the rules for rounding whole numbers.

Rounding Whole Numbers

Step 1   Locate the *place* to which the number is to be rounded. Draw a line under that place.

Step 2(a) Look only at the next digit to the right of the one you underlined. If it is *5 or more, increase* the underlined digit by 1.

Step 2(b) If the next digit to the right is *4 or less, do not change* the digit in the underlined place.

Step 3   *Change* all digits to the right of the underlined place to zeros.

EXAMPLE 2   Using Rounding Rules for 4 or Less

Round 349 to the nearest hundred.

Step 1   Locate the place to which the number is being rounded. Draw a line under that place.

3̲49

Hundreds place

Continued on Next Page

VOCABULARY TIP

Rounding Rounding is a useful tool for estimating the answer to a problem.

➊ Locate and draw a line under the place to which each number is to be rounded. Then answer the question.

(a) 3̲73 (nearest ten)

    ⬆ Tens place

Is 373 closer to 370 or to 380?

_____

(b) 1482 (nearest thousand)

Is 1482 closer to 1000 or to 2000?

_____

(c) 89,512 (nearest hundred)

Is it closer to 89,500 or to 89,600?

_____

(d) 546,325 (nearest ten-thousand)

Is it closer to 540,000 or to 550,000?

_____

Answers

1. (a) 3̲73 is closer to 3̲70.
   (b) 1̲482 is closer to 1̲000.
   (c) 89,5̲12 is closer to 89,5̲00.
   (d) 54̲6,325 is closer to 55̲0,000.

Step 2 Because the next digit to the right of the underlined place is 4, which is 4 or less, do *not* change the digit in the underlined place.

Next digit is 4 or less.

349

3 remains 3.

> 4 or less, do not change underlined digit.

Step 3 Change all digits to the right of the underlined place to zeros.

3̲49 rounded to the nearest hundred is 300.

In other words, 349 is closer to 300 than to 400.

·············· **Work Problem ② at the Side.** ▶

> **EXAMPLE 3** **Using Rounding Rules for 5 or More**

Round 36,833 to the nearest thousand.

Step 1 Find the place to which the number is to be rounded and draw a line under that place.

36,833

Thousands

Step 2 Because the next digit to the right of the underlined place is 8, which is 5 or more, add 1 to the underlined place.

Next digit is 5 or more.

36,833

> 5 or more, add 1 to underlined digit.

Change 6 to 7.

Step 3 Change all digits to the right of the underlined place to zeros.

Change to 0.

36,833 rounded to the nearest thousand is 37,000.

Change 6 to 7.

In other words, 36,833 is closer to 37,000 than to 36,000.

·············· **Work Problem ③ at the Side.** ▶

> **EXAMPLE 4** **Using Rounding Rules**

(a) Round 2382 to the nearest ten.

Step 1 2382

Tens place

Step 2 The next digit to the right is 2, which is 4 or less.

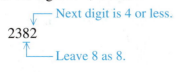

Next digit is 4 or less.

2382

Leave 8 as 8.

Step 3 23**8**2   Change to 0.

2382 rounded to the nearest ten is 2380.

In other words, 2382 is closer to 2380 than to 2390.

··········· **Continued on Next Page**

---

② Round to the nearest ten.

GS **(a)** 62

Next digit is ___ or less.

62

tens place

62 ← change to 0.

leave 6 as ___.

62 rounds to ___.

(b) 94

(c) 134

(d) 7543

③ Round to the nearest thousand.

(a) 3683

(b) 6502

(c) 84,621

(d) 55,960

Answers

2. **(a)** 4; 6; 60 **(b)** 90 **(c)** 130
   (d) 7540

3. **(a)** 4000 **(b)** 7000 **(c)** 85,000
   (d) 56,000

4 Round each number as indicated.

(a) 3458 to the nearest ten

(b) 6448 to the nearest hundred

GS (c) 73,077 to the nearest hundred

        Next digit
73,077 is _____ or more.

Hundreds place
rounds to 73, 1__ __.

(d) 85,972 to the nearest hundred

5 Round each number as indicated.

(a) 14,598 to the nearest
ten-thousand

(b) 724,518,715 to the nearest
million

Answers

4. (a) 3460   (b) 6400
  (c) 5; 0; 0; 73,100   (d) 86,000

5. (a) 10,000   (b) 725,000,000

(b) Round 13,961 to the nearest hundred.

Step 1   13,961
         └── Hundreds place

Step 2   The next digit to the right is 6.

        ┌── Next digit is 5 or more.
   13,961
       └── Change 9 to 10; write 0 and regroup 1 into thousands place.
      └── 3 + regrouped 1 = 4
      ┌── Change to 0.

Step 3   14,061

13,961 rounded to the nearest hundred is 14,000.
In other words, 13,961 is closer to 14,000 than to 13,900.

> **Note**
>
> In **Step 2** of **Example 4(b),** notice that the first three digits increased from 139 to 140 when we added 1 to the hundreds place.
>
>   (13,9)61 rounded to (14,0)00

◄ **Work Problem** **4** **at the Side.**

EXAMPLE 5   **Rounding Large Numbers**

(a) Round 37,892 to the nearest ten-thousand.

Step 1   37,892   *Remember to underline the place to which you are rounding.*
    └── Ten-thousands place

Step 2   The next digit to the right is 7.

        ┌── Next digit is 5 or more.
    37892
      └── Change 3 to 4.
   ┌── Change to 0.

Step 3   47,892

37,892 rounded to the nearest ten-thousand is 40,000.

(b) Round 528,498,675 to the nearest million.

Step 1   528,498,675
      └── Millions place

      ┌── Next digit is 4 or less.
Step 2   528,498,675
      └── Leave 8 as 8.

      ┌── Change to 0.   *Remember to change everything to the right of the place you have rounded to 0.*
Step 3   528,498,675

528,498,675 rounded to the nearest million is 528,000,000.

◄ **Work Problem** **5** **at the Side.**

Sometimes a number must be rounded to different places.

| EXAMPLE 6 | **Rounding to Different Places** |

Round 648 **(a)** to the nearest ten and **(b)** to the nearest hundred.

(a) to the nearest ten

648 rounded to the nearest ten is 650.

(b) to the nearest hundred

648 rounded to the nearest hundred is 600.

　　Notice that if 648 is rounded to the nearest ten (650), and then 650 is rounded to the nearest hundred, the result is 700. If, however, 648 is rounded directly to the nearest hundred, the result is 600 (not 700).

　　　　　　　　　　　　　　　　　　Work Problem 6 **at the Side.** ▶

> **CAUTION**
>
> Before rounding to a different place, always go back to the *original,* unrounded number.

| EXAMPLE 7 | **Applying Rounding Rules** |

Round each number to the nearest ten, nearest hundred, and nearest thousand.

(a) 4358
　　First round 4358 to the nearest ten.

```
              ┌──── Next digit is 5 or more.
              ↓
       4358
         ↑
         └──── Tens place (5 + 1 = 6)
```

4358 rounded to the nearest ten is 4360.
　　Now go back to 4358, the *original* number, before rounding to the nearest hundred.

```
            ┌──── Next digit is 5 or more.
            ↓
     4358                           ┌── Go back to the
       ↑                            │   original number.
       └──── Hundreds place (3 + 1 = 4)
```

4358 rounded to the nearest hundred is 4400.
　　Again, go back to the *original* number before rounding to the nearest thousand.

```
           ┌──── Next digit is 4 or less.
           ↓
    4358
      ↑
      └──── Thousands place stays the same.
```

4358 rounded to the nearest thousand is 4000.

　　　　　　　　　　　　　　　　　　Continued on Next Page

---

6 Round each number to the nearest ten and to the nearest hundred.

(a) 549

　　nearest ten: 5 __ 0

　　nearest hundred: __ __ 0

(b) 458

(c) 9308

Answers

6. **(a)** 5; 550; 5; 0; 500   **(b)** 460; 500
　　(c) 9310; 9300

7 Round each number to the nearest ten, nearest hundred, and nearest thousand.

(a) 4078

(b) 46,364

(c) 268,328

8 Estimate the answers by rounding each number to the nearest ten.

(a)   16
        74
        58
      + 31

GS (b) $53 \longrightarrow 50$
       $- 19 \longrightarrow \underline{-0}$
       30

GS (c) $46 \longrightarrow \underline{0}$
       $\times 74 \longrightarrow \underline{\times 7 }$
       $35\_\_$

Answers

7. (a) 4080; 4100; 4000
   (b) 46,360; 46,400; 46,000
   (c) 268,330; 268,300; 268,000

8. (a) 20 + 70 + 60 + 30 = 180
   (b) 2; 50 − 20 = 30
   (c) 5; 0; 0; 0; 70 × 50 = 3500

(b) 680,914
First, round to the nearest ten.

680,914 rounded to the nearest ten is 680,910.
Go back to 680,914, the *original* number, to round to the nearest hundred.

680,914 rounded to the nearest hundred is 680,900.
Go back to the *original* number to round to the nearest thousand.

680,914 rounded to the nearest thousand is 681,000.

◀ **Work Problem 7 at the Side.**

OBJECTIVE 3   Round numbers to estimate an answer. Numbers may be rounded to **estimate** an answer. An estimated answer is one that is close to the exact answer and may be used as a check when the exact answer is found. The "≈" sign is often used to show that an answer has been rounded or estimated and is almost equal to the exact answer; ≈ means "approximately equal to."

EXAMPLE 8   Using Rounding to Estimate an Answer

Estimate each answer by rounding to the nearest ten.

(a) $\begin{array}{r} 76 \longrightarrow 80 \\ 53 \longrightarrow 50 \\ 38 \longrightarrow 40 \\ + 91 \longrightarrow + 90 \\ \hline 260 \end{array}$ } Rounded to the nearest ten

260   Estimated answer

(b) $\begin{array}{r} 27 \\ - 14 \\ \hline \end{array}$ $\begin{array}{r} 30 \\ -10 \\ \hline 20 \end{array}$ } Rounded to the nearest ten

20   Estimated answer

(c) $\begin{array}{r} 16 \\ \times 21 \\ \hline \end{array}$ $\begin{array}{r} 20 \\ \times 20 \\ \hline 400 \end{array}$ } Rounded to the nearest ten

400   Estimated answer

◀ **Work Problem 8 at the Side.**

EXAMPLE 9   **Using Rounding to Estimate an Answer**

Estimate each answer by rounding to the nearest hundred.

(a)
$$
\begin{array}{r}
252 \longrightarrow 300 \\
749 \longrightarrow 700 \\
576 \longrightarrow 600 \\
+\,819 \longrightarrow +\,800 \\
\hline
2400
\end{array}
$$
Rounded to the nearest hundred

2400   Estimated answer

> The hundreds position is 3 places to the left.

(b)
$$
\begin{array}{r}
780 \quad\quad 800 \\
-\,536 \quad -\,500 \\
\hline
300
\end{array}
$$
Rounded to the nearest hundred

300   Estimated answer

(c)
$$
\begin{array}{r}
664 \quad\quad 700 \\
\times\,834 \quad \times\,800 \\
\hline
560{,}000
\end{array}
$$
Rounded to the nearest hundred

560,000   Estimated answer

Work Problem ⑨ at the Side. ▶

OBJECTIVE ▶ ④ Use front end rounding to estimate an answer. A convenient way to estimate an answer is to use *front end rounding*. With **front end rounding,** we round to the highest possible place so that all the digits become 0 except the first one. For example, suppose you want to buy a big flat-screen television for $2449, a home theater system for $1759, and a reclining chair for $525. Using front end rounding, you can estimate the total cost of these purchases.

Television	$2449 $\longrightarrow$	2000
Home theater system	$1759 $\longrightarrow$	2000
Reclining chair	$525 $\longrightarrow$	+ 500
		$4500 ← Estimated total cost

> Do not round the estimated answer.

EXAMPLE 10   **Using Front End Rounding to Estimate an Answer**

Estimate each answer using front end rounding.

(a)
$$
\begin{array}{r}
3825 \quad\quad 4000 \\
72 \quad\quad 70 \\
565 \quad\quad 600 \\
+\,2389 \quad +\,2000 \\
\hline
6670
\end{array}
$$
All digits changed to 0 except first digit, which is rounded

6670   Estimated answer

(b)
$$
\begin{array}{r}
6712 \quad\quad 7000 \\
-\,825 \quad -\,800 \\
\hline
6200
\end{array}
$$
First digit rounded and all others changed to 0

6200   Estimated answer

> Notice: Front end rounding leaves *only* one nonzero digit.

(c)
$$
\begin{array}{r}
725 \quad\quad 700 \\
\times\,86 \quad \times\,90 \\
\hline
63{,}000
\end{array}
$$

63,000   Estimated answer

Work Problem ⑩ at the Side. ▶

---

⑨ Estimate the answers by rounding each number to the nearest hundred.

(a)
$$
\begin{array}{r}
358 \\
743 \\
822 \\
+\,978 \\
\hline
\end{array}
$$

(b)
$$
\begin{array}{r}
842 \\
-\,475 \\
\hline
\end{array}
$$

(c)
$$
\begin{array}{r}
723 \\
\times\,478 \\
\hline
\end{array}
$$

VOCABULARY TIP

Front end rounding An estimation strategy that uses only the left-most digit of the number being rounded. This makes the number(s) the least accurate, but the easiest to work with.

⑩ Use front end rounding to estimate each answer.

GS (a)
$$
\begin{array}{r}
36 \longrightarrow 40 \\
3852 \longrightarrow 4000 \\
749 \longrightarrow 7\_\_ \\
+\,5474 \longrightarrow -\_000 \\
\hline
\end{array}
$$

(b)
$$
\begin{array}{r}
2583 \\
-\,765 \\
\hline
\end{array}
$$

(c)
$$
\begin{array}{r}
648 \\
\times\,67 \\
\hline
\end{array}
$$

Answers

9. (a) $400 + 700 + 800 + 1000 = 2900$
   (b) $800 - 500 = 300$
   (c) $500 \times 700 = 350{,}000$
10. (a) 0; 0; 5; $40 + 4000 + 700 + 5000 = 9740$
    (b) $3000 - 800 = 2200$
    (c) $70 \times 600 = 42{,}000$

1.7 Exercises

CONCEPT CHECK *The following numbers have been rounded. Decide whether the number has been rounded to the nearest ten, nearest hundred, nearest thousand, or nearest ten-thousand, and fill in the blank.*

1. 624 to the nearest _____ is 620.

2. 509 to the nearest _____ is 510.

3. 86,813 to the nearest _____ is 86,800.

4. 17,211 to the nearest _____ is 17,200.

5. 78,499 to the nearest _____ is 78,000.

6. 14,314 to the nearest _____ is 14,000.

7. 12,987 to the nearest _____ is 10,000.

8. 6599 to the nearest _____ is 10,000.

Round each number as indicated. See Examples 1–5.

9. 855 to the nearest ten

10. 946 to the nearest ten

11. 6771 to the nearest hundred

12. 5847 to the nearest hundred

13. 28,472 to the nearest hundred

GS 28,472

5 or more
hundreds place

14. 18,249 to the nearest hundred

GS 18,249

4 or less
hundreds place

15. 5996 to the nearest hundred

16. 4452 to the nearest hundred

17. 15,758 to the nearest thousand

18. 28,465 to the nearest thousand

19. 7,760,058,721 to the nearest billion

20. 4,468,523,628 to the nearest billion

21. 595,008 to the nearest ten-thousand

22. 725,182 to the nearest ten-thousand

23. 4,860,220 to the nearest million

24. 13,713,409 to the nearest million

Round each number to the nearest ten, nearest hundred, and nearest thousand. See Examples 6 and 7.

	Ten	Hundred	Thousand			Ten	Hundred	Thousand
25. 4476	_____	_____	_____		**26.** 6483	_____	_____	_____
27. 3374	_____	_____	_____		**28.** 7632	_____	_____	_____
29. 6048	_____	_____	_____		**30.** 7065	_____	_____	_____

	Ten	Hundred	Thousand		Ten	Hundred	Thousand
31. 5343	____	____	____	**32.** 7456	____	____	____
33. 19,539	____	____	____	**34.** 59,806	____	____	____
35. 26,292	____	____	____	**36.** 78,519	____	____	____
37. 93,706	____	____	____	**38.** 84,639	____	____	____

39. Write in your own words the three steps that you would use to round a number when the digit to the right of the place to which you are rounding is *5 or more.*

40. Write in your own words the three steps that you would use to round a number when the digit to the right of the place to which you are rounding is *4 or less.*

Estimate the answer by rounding each number to the nearest ten. Then find the exact answer. See Example 8.

41. *Estimate:*   *Exact:*

30	←Rounds to—	25
60	←	63
50	←	47
+ 80	←	+ 84

42. *Estimate:*   *Exact:*

60	←Rounds to—	56
20	←	24
90	←	85
+ 70	←	+ 71

43. *Estimate:*   *Exact:*

| | 78 |
| − ___ | − 43 |

44. *Estimate:*   *Exact:*

| | 57 |
| − ___ | − 24 |

45. *Estimate:*   *Exact:*

| | 67 |
| × ___ | × 34 |

46. *Estimate:*   *Exact:*

| | 53 |
| × ___ | × 75 |

Estimate the answer by rounding each number to the nearest hundred. Then find the exact answer. See Example 9.

47. *Estimate:*   *Exact:*

←Rounds to—	863
←	735
←	438
+ ___ ←	+ 792

48. *Estimate:*   *Exact:*

	623
	362
	189
× ___	+ 736

49. *Estimate:*   *Exact:*

$$\begin{array}{r} 883 \\ -\ 448 \\ \hline \end{array}$$

$$-\ \underline{\hspace{2em}}$$

50. *Estimate:*   *Exact:*

$$\begin{array}{r} 614 \\ -\ 276 \\ \hline \end{array}$$

$$-\ \underline{\hspace{2em}}$$

51. *Estimate:*   *Exact:*

$$\begin{array}{r} 752 \\ \times\ 375 \\ \hline \end{array}$$

$$\times\ \underline{\hspace{2em}}$$

52. *Estimate:*   *Exact:*

$$\begin{array}{r} 845 \\ \times\ 396 \\ \hline \end{array}$$

$$\times\ \underline{\hspace{2em}}$$

Estimate each answer using front end rounding. Then find the exact answer. **See Example 10.**

53. *Estimate:*   *Exact:*

$$\begin{array}{r} 8000 \\ 60 \\ 700 \\ +\ 4000 \\ \hline \end{array} \quad \xleftarrow{\text{Rounds to}} \quad \begin{array}{r} 8215 \\ 56 \\ 729 \\ +\ 3605 \\ \hline \end{array}$$

54. *Estimate:*   *Exact:*

$$\begin{array}{r} 3000 \\ 70 \\ 600 \\ +\ 7000 \\ \hline \end{array} \quad \xleftarrow{\text{Rounds to}} \quad \begin{array}{r} 2685 \\ 73 \\ 592 \\ +\ 7183 \\ \hline \end{array}$$

55. *Estimate:*   *Exact:*

$$\begin{array}{r} 687 \\ -\ 529 \\ \hline \end{array}$$

$$-\ \underline{\hspace{2em}}$$

56. *Estimate:*   *Exact:*

$$\begin{array}{r} 543 \\ -\ 174 \\ \hline \end{array}$$

$$-\ \underline{\hspace{2em}}$$

57. *Estimate:*   *Exact:*

$$\begin{array}{r} 939 \\ \times\ 29 \\ \hline \end{array}$$

$$\times\ \underline{\hspace{2em}}$$

58. *Estimate:*   *Exact:*

$$\begin{array}{r} 864 \\ \times\ 74 \\ \hline \end{array}$$

$$\times\ \underline{\hspace{2em}}$$

59. The number 3492 rounded to the nearest hundred is 3500, and 3500 rounded to the nearest thousand is 4000. But when 3492 is rounded directly to the nearest thousand it becomes 3000. Why is this true? Explain.

60. The use of rounding is helpful when estimating the answer to a problem. Why is this true? Give an example using either addition, subtraction, multiplication, or division to show how this works.

61. In 1900, the population of the United States was 76 million. Today it's 311 million. Round each of these numbers to the nearest ten-million. (*Source:* Reiman Publications and U.S. Census Bureau.)

62. Americans will eat more than 22,362,180 hot dogs in Major League ballparks this season. Round the number to the nearest ten-thousand and the nearest hundred-thousand. (*Source:* National Hot Dog and Sausage Council.)

63. There are 348,900 streets named Elm Street in the United States. Round this number to the nearest thousand and nearest ten-thousand. (*Source:* Expo Design Center.)

64. Of all the streets in the United States, the two most common candy-flavored street names are Peppermint and Chocolate. Ninety-five streets are named Peppermint and 27 are named Chocolate. Round each of these numbers to the nearest ten. (*Source:* Tel Atlas digital map database.)

65. There were 39,836,000 speeding tickets given in the United States last year. Round this number to the nearest ten-thousand, nearest hundred-thousand, and nearest million. (*Source:* Clark Howard Radio Show.)

66. Americans spend $57,463,625,000 on lottery tickets each year, or nine times as much as they spend on movie tickets. Round this number to the nearest ten-million, nearest hundred-million, and nearest billion. (*Source:* TLC-W, A Discovery Company.)

67. Ping-Pong is the fifth fastest growing sport in America. Today there are 19,265,780 players. Round this number to the nearest thousand, nearest ten-thousand, and the nearst hundred-thousand. (*Source: Parade, The Sunday Newspaper Magazine.*)

68. The highest annual production of Mitsubishi automobiles in Normal, Illinois was 221,543. Round this number to the nearest ten, nearest hundred, and nearest thousand. (*Source:* Newsweek.)

Relating Concepts (Exercises 69–75) For Individual or Group Work

To see how both rounding and front end rounding are used in solving problems,
work Exercises 69–75 in order.

69. A number rounded to the nearest thousand is 72,000. What is the *smallest* whole number this could have been before rounding?

70. A number rounded to the nearest thousand is 72,000. What is the *largest* whole number this could have been before rounding?

71. When front end rounding is used, a whole number rounds to 8000. What is the *smallest* possible original number?

72. When front end rounding is used, a whole number rounds to 8000. What is the *largest* possible original number?

The graph below shows the number of personal injuries in the United States each year for people participating in common activities.

WATCH YOUR STEP

Accidents while participating in common activities

2,788,000

534,883

3925 11,243 15,974 17,916

Brushing Teeth Reading Stapling Paper Bowling Bicycling Driving

Source: AARP Magazine.

73. Round the number of accidents occurring in each activity to the nearest ten.

74. Use front end rounding to round the number of accidents in each activity.

75. (a) What is one advantage of using front end rounding instead of rounding to the nearest ten?

(b) What is one disadvantage?

1.8 Exponents, Roots, and Order of Operations

OBJECTIVES

1. Identify an exponent and a base.
2. Find the square root of a number.
3. Use the order of operations.

OBJECTIVE **1** **Identify an exponent and a base.** The product $3 \cdot 3$ can be written as 3^2 (read as "3 squared"). The small raised number 2, called an **exponent,** says to use 2 factors of 3. The number 3 is called the **base.** Writing 3^2 as 9 is called *simplifying the expression.*

EXAMPLE 1   Simplifying Expressions

Identify the exponent and the base, and then simplify each expression.

(a) 4^3 Base $\longrightarrow 4^3 \longleftarrow$ Exponent $4^3 = 4 \times 4 \times 4 = 64$

> The small raised number is the exponent.

(b) $2^5 = 2 \times 2 \times 2 \times 2 \times 2 = 32$
The base is 2 and the exponent is 5.

◀ **Work Problem 1 at the Side.**

OBJECTIVE **2** **Find the square root of a number.** Because $3^2 = 9$, the number 3 is called the **square root** of 9. The square root of a number is one of two identical factors of that number. Square roots of numbers are written with the symbol $\sqrt{\ }$.

Square Root

$$\sqrt{\text{number} \cdot \text{number}} = \sqrt{\text{number}^2} = \text{number}$$

For example: $\sqrt{36} = \sqrt{6 \cdot 6} = \sqrt{6^2} = 6$ The square root of 36 is **6**.

To find the square root of 64 ask, "What number can be multiplied by itself (that is, *squared*) to give 64?" The answer is 8, so

$$\sqrt{64} = \sqrt{8 \cdot 8} = \sqrt{8^2} = 8.$$

A **perfect square** is a number that is the square of a *whole number.* The first few perfect squares are listed here.

Perfect Squares Table

$0 = 0^2$	$16 = 4^2$	$64 = 8^2$	$144 = 12^2$
$1 = 1^2$	$25 = 5^2$	$81 = 9^2$	$169 = 13^2$
$4 = 2^2$	$36 = 6^2$	$100 = 10^2$	$196 = 14^2$
$9 = 3^2$	$49 = 7^2$	$121 = 11^2$	$225 = 15^2$

EXAMPLE 2   Using Perfect Squares

Find each square root.

> Note that $\sqrt{16}$ is 4, not 4^2.

(a) $\sqrt{16}$ Because $4^2 = 16$, $\sqrt{16} = 4$.   (b) $\sqrt{49} = 7$

(c) $\sqrt{0} = 0$ (d) $\sqrt{169} = 13$

◀ **Work Problem 2 at the Side.**

OBJECTIVE **3** **Use the order of operations.** Frequently problems may have parentheses, exponents, and square roots, and may involve more than one operation. Work these problems by following the **order of operations.**

---

1 Identify the exponent and the base, and then simplify each expression.

(a) 4^2

(b) 5^3

(c) 3^4

(d) 2^6

VOCABULARY TIP

Perfect square It is helpful to have the first 12 or so perfect squares memorized so that they are easily recognizable.

2 Find each square root.

(a) $\sqrt{4}$

(b) $\sqrt{25}$

(c) $\sqrt{36}$

(d) $\sqrt{225}$

(e) $\sqrt{1}$

Answers

1. (a) 2; 4; 16 (b) 3; 5; 125
   (c) 4; 3; 81 (d) 6; 2; 64
2. (a) 2 (b) 5 (c) 6 (d) 15 (e) 1

Order of Operations

1. Do all operations inside *parentheses* or *other grouping symbols.*
2. Simplify any expressions with *exponents* and find any *square roots.*
3. *Multiply* or *divide,* proceeding from left to right.
4. *Add* or *subtract,* proceeding from left to right.

EXAMPLE 3 Understanding the Order of Operations

Use the order of operations to simplify each expression.

(a) $8^2 + 5 + 2$

$8^2 + 5 + 2$

$8 \cdot 8 + 5 + 2$ Evaluate exponent first; 8^2 is $8 \cdot 8$.

$64 + 5 + 2$ Add from left to right.

$69 + 2 = 71$

(b) $35 \div 5 \cdot 6$ Divide first (start at left).

$7 \cdot 6 = 42$ Multiply.

(c) $9 + (20 - 4) \cdot 3$ Work inside parentheses first.

$9 + 16 \cdot 3$ Multiply.

$9 + 48 = 57$ Add last.

(d) $12 \cdot \sqrt{16} - 8(4)$ Find the square root first.

$12 \cdot 4 - 8(4)$ Multiply from left to right.

$48 - 32 = 16$ Subtract last.

Work Problem ❸ at the Side. ▶

EXAMPLE 4 Using the Order of Operations

Use the order of operations to simplify each expression.

(a) $15 - 4 + 2$ Subtract first (start at left).

$11 + 2 = 13$ Add.

(b) $8 + (7 - 3) \div 2$ Work inside parentheses first.

$8 + 4 \div 2$ Divide.   *Add or subtract last.*

$8 + 2 = 10$ Add last.

(c) $4^2 \cdot 2^2 + (7 + 3) \cdot 2$ Work inside parentheses first.

$4^2 \cdot 2^2 + 10 \cdot 2$ Evaluate exponents.

$16 \cdot 4 + 10 \cdot 2$ Multiply from left to right.

$64 + 20 = 84$ Add last.

(d) $4 \cdot \sqrt{25} - 7 \cdot 2 + \dfrac{0}{5}$ Find the square root first.

Zero divided by any nonzero number is zero.

$4 \cdot 5 - 7 \cdot 2 + \dfrac{0}{5}$ Multiply or divide from left to right.

$20 - 14 + 0 = 6$ Add or subtract last.

Work Problem ❹ at the Side. ▶

❸ Simplify each expression.

(a) $4 + 5 + 2^2$ Evaluate exponents.

$4 + 5 + 2 \cdot 2$

$4 + 5 + \underline{\quad}$ Add left to right.

$9 + \underline{\quad} = \underline{\quad}$

(b) $3^2 + 2^3$

(c) $60 \div \sqrt{36} \div 2$

(d) $8 + 6(14 \div 2)$

❹ Simplify each expression.

(a) $12 - 6 + 4^2$ Evaluate exponents.

$12 - 6 + \underline{\quad} \cdot \underline{\quad}$

$12 - 6 + 16$ Add/subtract left to right.

$6 + 16 = \underline{\quad}$

(b) $2^3 + 3^2 - (5 \cdot 3)$

(c) $20 \div 2 + (7 - 5)$

(d) $15 \cdot \sqrt{9} - 8 \cdot \sqrt{4}$

Answers

3. **(a)** 4; 4; 13   **(b)** 17   **(c)** 5   **(d)** 50
4. **(a)** 4; 4; 22   **(b)** 2   **(c)** 12   **(d)** 29

1.8 Exercises

CONCEPT CHECK *Identify the exponent and the base.*

1. 3^2
exponent ____
base ____

2. 2^3
exponent ____
base ____

3. 5^2
exponent ____
base ____

4. 4^2
exponent ____
base ____

Identify the exponent and the base and then simplify each expression. **See Example 1.**

5. 8^2

6. 10^3

7. 15^2

8. 11^3

Use the Perfect Squares Table on the first page of this section to find each square root. **See Example 2.**

9. $\sqrt{16}$

10 $\sqrt{25}$

11. $\sqrt{64}$

12. $\sqrt{36}$

13. $\sqrt{100}$

14. $\sqrt{49}$

15. $\sqrt{144}$

16. $\sqrt{225}$

CONCEPT CHECK *Decide whether each statement is* true *or* false. *If it is* false, *explain why.*

17. The expression 5^2 means that 2 is used as a factor 5 times.

18. $4^2 = 8$

19. $1^3 = 3$

20. $6^1 = 1$

Fill in each blank. **See Example 2.**

21. $6^2 =$ ____ so $\sqrt{} = 6$
GS $6 \cdot 6 =$ ____ so $\sqrt{} = 6$

22. $9^2 =$ ____ so $\sqrt{} = 9$
GS $9 \cdot 9 =$ ____ so $\sqrt{} = 9$

23. $25^2 =$ ____ so $\sqrt{} = 25$

24. $50^2 =$ ____ so $\sqrt{} = 50$

25. $100^2 =$ ____ so $\sqrt{} = 100$

26. $60^2 =$ ____ so $\sqrt{} = 60$

27. Describe in your own words a perfect square. Of the two numbers 25 and 50, identify which is a perfect square and explain why.

28. Use the following list of words and phrases to write the four steps in the order of operations.

add	square root
exponents	subtract
multiply	divide

parentheses or other grouping symbols

CONCEPT CHECK   *Decide whether each statement is* true *or* false. *If it is* false, *explain why.*

29. $3^2 + 8 - 5 = 12$

30. $5^2 + 5 - 6 = 24$

31. $6 + 8 \div 2 = 7$

32. $4 + 5(6 - 4) = 18$

Simplify each expression by using the order of operations. **See Examples 3 and 4.**

33. $3^2 + 8 - 5$

$\underbrace{3^2} + 8 - 5$

$\underbrace{3 \cdot 3} + \underbrace{8 - 5}$

$9 \quad + \quad 3 =$

34. $5^2 + 5 - 6$

$\underbrace{5^2} + 5 - 6$

$\underbrace{5 \cdot 5} + 5 - 6$

$25 \quad + 5 - 6 =$

35. $25 \div 5(8 - 4)$

36. $36 \div 18(7 - 3)$

37. $5 \cdot 3^2 + \dfrac{0}{8}$

38. $8 \cdot 3^2 - \dfrac{10}{2}$

39. $4 \cdot 1 + 8(9 - 2) + 3$

40. $3 \cdot 2 + 7(3 + 1) + 5$

41. $2^2 \cdot 3^3 + (20 - 15) \cdot 2$

42. $4^2 \cdot 5^2 + (20 - 9) \cdot 3$

43. $5\sqrt{36} - 2(4)$

44. $2 \cdot \sqrt{100} - 3(4)$

45. $8(2) + 3 \cdot 7 - 7 =$

46. $10(3) + 6 \cdot 5 - 20$

47. $2^3 \cdot 3^2 + 3(14 - 4)$

48. $3^2 \cdot 4^2 + 2(15 - 6)$

49. $7 + 8 \div 4 + \dfrac{0}{7}$

50. $6 + 8 \div 2 + \dfrac{0}{8}$

51. $3^2 + 6^2 + (30 - 21) \cdot 2$

52. $4^2 + 5^2 + (25 - 9) \cdot 3$

53. GS $7 \cdot \sqrt{81} - 5 \cdot 6$

$7 \cdot \sqrt{81} - 5 \cdot 6$

$\quad 7 \cdot 9 - \quad 30$

$\qquad 63 - \quad 30 =$

54. GS $6 \cdot \sqrt{64} - 6 \cdot 5$

$6 \cdot \sqrt{64} - 6 \cdot 5$

$\quad 6 \cdot 8 - \quad 30$

$\qquad 48 - \quad 30 =$

55. $8 \cdot 2 + 5(3 \cdot 4) - 6$

56. $5 \cdot 2 + 3(5 + 3) - 6$

57. $4 \cdot \sqrt{49} - 7(5 - 2)$

58. $3 \cdot \sqrt{25} - 6(3 - 1)$

59. $7(4 - 2) + \sqrt{9}$

60. $5(4 - 3) + \sqrt{9}$

61. $7^2 + 3^2 - 8 + 5$

62. $3^2 - 2^2 + 3 - 2$

63. $5^2 \cdot 2^2 + (8 - 4) \cdot 2$

64. $5^2 \cdot 3^2 + (30 - 20) \cdot 2$

65. $5 + 9 \div 3 + 6 \cdot 3$

66. $8 + 3 \div 3 + 6 \cdot 3$

67. $8 \cdot \sqrt{49} - 6(9 - 4)$

68. $8 \cdot \sqrt{49} - 6(5 + 3)$

69. $5^2 - 4^2 + 3 \cdot 6$

70. $3^2 + 6^2 - 5 \cdot 8$

71. $8 + 8 \div 8 + 6 + \dfrac{5}{5}$

GS

$8 + \underline{8 \div 8} + 6 + \dfrac{5}{5}$

$8 + \quad 1 \quad + 6 + 1 =$

72. $3 + 14 \div 2 + 7 + \dfrac{8}{8}$

GS

$3 + \underline{14 \div 2} + 7 + \dfrac{8}{8}$

$3 + \quad 7 \quad + 7 + 1 =$

73. $6 \cdot \sqrt{25} - 7(2)$

74. $8 \cdot \sqrt{36} - 4(6)$

75. $9 \cdot \sqrt{16} - 3 \cdot \sqrt{25}$

76. $6 \cdot \sqrt{81} - 3 \cdot \sqrt{49}$

77. $7 \div 1 \cdot 8 \cdot 2 \div (21 - 5)$

78. $12 \div 4 \cdot 5 \cdot 4 \div (15 - 13)$

79. $15 \div 3 \cdot 2 \cdot 6 \div (14 - 11)$

80. $9 \div 1 \cdot 4 \cdot 2 \div (11 - 5)$

81. $6 \cdot \sqrt{25} - 4 \cdot \sqrt{16}$

82. $10 \cdot \sqrt{49} - 4 \cdot \sqrt{64}$

83. $5 \div 1 \cdot 10 \cdot 4 \div (17 - 9)$

84. $15 \div 3 \cdot 8 \cdot 9 \div (12 - 8)$

85. $8 \cdot 9 \div \sqrt{36} - 4 \div 2 + (14 - 8)$

86. $3 - 2 + 5 \cdot 4 \cdot \sqrt{144} \div \sqrt{36}$

87. $2 + 1 - 2 \cdot \sqrt{1} + 4 \cdot \sqrt{81} - 7 \cdot 2$

88. $6 - 4 + 2 \cdot 9 - 3 \cdot \sqrt{225} \div \sqrt{25}$

89. $5 \cdot \sqrt{36} \cdot \sqrt{100} \div 4 \cdot \sqrt{9} + 8$

90. $9 \cdot \sqrt{36} \cdot \sqrt{81} \div 2 + 6 - 3 - 5$

Study Skills
TAKING LECTURE NOTES

Study the set of sample math notes in this section, and read the comments about them. Then try to incorporate the techniques into your own math note taking in class.

January 2   <u>Exponents</u>

Exponents used to show repeated multiplication.

$$3 \cdot 3 \cdot 3 \cdot 3 \text{ can be written } 3^4 \quad \swarrow \text{exponent}$$
(how many times it's multiplied)

base
(the number being multiplied)

Read 3^2 as 3 to the 2nd power or 3 squared

        3^3 as 3 to the 3rd power or 3 cubed

        3^4 as 3 to the 4th power

          etc.

<u>Simplifying an expression with exponents</u>

    $\searrow$ actually do the repeated multiplication

$$2^3 \text{ means } 2 \cdot 2 \cdot 2 \text{ and } 2 \cdot 2 \cdot 2 = 8$$

☆ Careful! $\Big[$ 5^2 means $5 \cdot 5$ <u>NOT</u> $5 \cdot 2$

         so $5^2 = 5 \cdot 5 = 25$ BUT $5^2 \neq 10$

<u>Example</u>   <u>Explanation</u>

Simplify $\widehat{2^4} \cdot \widehat{3^2}$ Exponents mean <u>multiplication</u>.

$2 \cdot 2 \cdot 2 \cdot 2 \cdot 3 \cdot 3$ Use 2 as a factor 4 times.
                Use 3 as a factor 2 times.
                $2 \cdot 2 \cdot 2 \cdot 2$ is 16
    16   ·   9 $3 \cdot 3$ is 9 $16 \cdot 9$ is 144

      144   simplified result is 144
                (no exponents left)

▶ The **date and title** of the day's lecture topic are always at the top of every page. **Always begin a new day with a new page.**

▶ Note the **definitions** of base and exponent are written in parentheses— don't trust your memory!

▶ **Skipping lines** makes the notes easier to read.

▶ See how the **direction word** (*simplify*) is emphasized and explained.

▶ A **star marks an important concept.** This is a warning to avoid future mistakes. **Note the underlining**, too, which highlights the importance.

▶ Notice the two columns, which allow for the example and its explanation to be close together. **Whenever you know you'll be given a series of steps to follow, try the two-column method.**

▶ Note the **brackets and arrows**, which clearly show how the problem is set up to be simplified.

Now Try This

Find one or two people in your math class to work with. Compare each other's lecture notes over a period of a week or so. Ask yourself the following questions as you examine the notes.

1 What are you doing in your notes to show the **main points** or larger concepts? (Such as underlining, boxing, using stars, capital letters, etc.)

2 In what ways do you **set off the explanations** for worked problems, examples, or smaller ideas (subpoints)? (Such as indenting, using arrows, circling or boxing)

3 What does **your instructor do** to show that he or she is moving from one idea to the next? (Such as saying "Next" or "Any questions," "Now," or erasing the board, etc.)

4 **How do you mark** that in your notes? (Such as skipping lines, using dashes or numbers, etc.)

5 What **explanations (in words) do you give yourself** in your notes, so when those new dendrites that you grew during lecture are fading, you can read your notes and still remember the new concepts later when you try to do your homework?

6 What **did you learn** by examining your classmates' notes?

- _____
- _____
- _____

7 What **will you try** in your own note taking? List **four** techniques that you will use next time you take notes in math class.

- _____
- _____
- _____
- _____

Why Are These Notes Brain Friendly?

The notes are **easy to look at,** and you know that the brain responds to things that are visually pleasing. Other techniques that are visually memorable are the use of spacing (the two columns), stars, underlining, and circling. All of these methods **allow your brain to take note of important concepts and steps.**

The notes are also **systematic,** which means that they use certain techniques regularly. This way, your brain easily recognizes the topic of the day, the signals that show an important point, and the steps to follow for procedures. When you develop a system that you always use in your notes, your notes are easy to understand later when you are reviewing for a test.

1.9 Reading Pictographs, Bar Graphs, and Line Graphs

OBJECTIVES

1. Read and understand a pictograph.
2. Read and understand a bar graph.
3. Read and understand a line graph.

VOCABULARY TIP

Pictograph Every pictograph has a key. The key in the graph tells the number that each picture or symbol represents.

---

1 Use the pictograph to answer each question.

GS **(a)** Which of the U.S. stamps had the second greatest number of sales?

Each stamp symbol represents 20 million stamps. The greatest number of symbols (6) is for the Elvis stamp. The second greatest number of symbols is for the _____ stamp.

(b) Approximately how many more Rock and Roll/Rhythm and Blues stamps were sold than Art of Disney Romance stamps?

VOCABULARY TIP

Bar graphs are typically used to display data that fit into different categories.

Answers

1. **(a)** 4; Rock and Roll/Rhythm & Blues
   (b) about 20 million (or 20,000,000) more stamps

We have all heard the saying "A picture is worth a thousand words," and there may be some truth in this. Today, so much information and data are being presented in the form of pictographs, circle graphs, bar graphs, and line graphs that it is important to be able to read and understand these tools.

OBJECTIVE 1 **Read and understand a pictograph.** A **pictograph** is a graph that uses pictures or symbols. It displays information that can be compared easily. However, since a symbol is used to represent a certain quantity, it can be difficult to determine the amount represented by a fraction of a symbol.

The pictograph below compares the number of U.S. postage stamps sold in the five top releases. In this pictograph, it is difficult to determine what fractional amount is represented by the partial stamps.

ELVIS IS STILL KING

Five of the most popular U.S. postage stamps of all time.

Elvis
DC Comics Super Heroes
Legends of Baseball
Art of Disney Romance
Rock & Roll/ Rhythm & Blues

= 20 million stamps

Source: United States Postal Service.

EXAMPLE 1 **Using a Pictograph**

Use the pictograph to answer each question. *(Each stamp symbol represents 20 million stamps.)*

(a) Which of the U.S. postage stamps shown has the lowest number sold?

The row representing Legends of Baseball has the fewest symbols. This means that the lowest number of U.S. postage stamps sold was Legends of Baseball.

(b) Approximately how many more Elvis stamps were sold than the Art of Disney Romance stamps?

The row representing the Elvis stamps has three more symbols than that for the Art of Disney stamps. This means that 3 • 20 million or 60,000,000 more Elvis stamps were sold than the Art of Disney Romance stamps.

◀ Work Problem **1** at the Side.

OBJECTIVE 2 **Read and understand a bar graph. Bar graphs** are useful for showing comparisons. For example, the following bar graph shows how many adults out of every 100 surveyed chose each money secret that they kept from their partner.

MONEY SECRETS OF COUPLES

Number of Adults (for every 100 adults surveyed)

Secret Kept

Source: American Express.

Use a ruler to help read a bar graph.

Realize that the value of each bar is an estimation.

EXAMPLE 2 Using a Bar Graph

Use the bar graph to find the number of adults who picked "Bought item partner didn't agree with" as the secret they kept from their partner.

Use a ruler or straightedge to line up the top of the bar labeled "Bought item partner didn't agree with" with the numbers on the left edge of the graph, labeled "Number of Adults." We see that 57 out of 100 adults picked "Bought item partner didn't agree with."

············· **Work Problem ② at the Side.** ▶

OBJECTIVE ▶ ③ Read and understand a line graph. A **line graph** is often used for showing a trend. The following line graph shows the U.S. Census Bureau predictions for U.S. population growth to the year 2100.

ANOTHER CENTURY OF GROWTH

Population Estimates (in hundred-millions)

2000 2025 2050 2075 2100

Year

Source: U.S. Census Bureau.

Line graphs can show what is happening over time.

EXAMPLE 3 Using a Line Graph

Use the line graph to answer each question.

(a) What trend or pattern is shown in the graph?
The population will continue to increase.

(b) What is the estimated population for 2025?
Use a ruler or straightedge to line up the dot above the year labeled 2025 on the horizontal line with the numbers along the left edge of the graph. Notice that the label on the left side says "in hundred-millions." Since the 2025 dot is halfway between 3 and 4, the population in 2025 is halfway between 3 • 100,000,000 and 4 • 100,000,000 or 300,000,000 and 400,000,000. That means that the predicted population in 2025 is about 350,000,000 people.

············· **Work Problem ③ at the Side.** ▶

---

② Use the bar graph to find the approximate number of adults who picked each money secret they kept from their partner.

(a) Secret account or card

(b) Misrepresented purchase price

(c) Hid purchases

(d) Hid debt

(e) Did not tell credit score

VOCABULARY TIP

Line graphs can be used to show how something changes over time. These graphs often have peaks and valleys to indicate upward and downward trends.

③ Use the line graph to find the predicted population of the United States for each year.

GS **(a)** 2050

The dot above 2050 lines up with ____. This means that the predicted population will be __ __ __ , 000,000.

(b) 2075

(c) 2100

Answers
2. **(a)** 27 out of 100 **(b)** 39 out of 100
  (c) 42 out of 100   **(d)** 26 out of 100
  (e) 12 out of 100
3. **(a)** 4; 4; 0; 0; 400,000,000
  (b) 475,000,000   **(c)** 575,000,000

1.9 Exercises

The following pictograph shows the number of retail stores for the seven companies with the greatest number of outlets. Use the pictograph to answer Exercises 1–8.
See Example 1.

SOMETHING IN STORE
While Walmart has the greatest amount of sales, it trails other chains in number of stores.

Dollar General
7-Eleven
Family Dollar
CVS
Walgreens
Rite-Aid
Walmart

= 500 stores

Source: T. D. Linx.

1. **CONCEPT CHECK** *Fill in the blanks.*
   The number of pictures or symbols for the Walgreens retail stores is _____. Since each symbol represents _____ stores, Walgreens has
   _____ • _____ = _____ stores.

2. **CONCEPT CHECK** *Fill in the blanks.*
   The number of pictures or symbols for the CVS retail stores is _____. Since each symbol represents _____ stores, CVS has
   _____ • _____ = _____ stores.

3. Find the number of Family Dollar retail stores.
   GS Family Dollar has 9 symbols.
   Each symbol is 500 stores
   $9 \cdot 500 =$ _____

4. Approximately how many retail stores does
   GS 7-Eleven have?
   7-Eleven has $10\frac{1}{2}$ symbols.
   Each symbol is 500 stores.
   $(10 \cdot 500) + \frac{1}{2} \cdot 500 =$ _____

5. Which company has the greatest number of retail stores? How many is that?

6. Which companies have the least number of retail stores? How many does each one have?

7. How many fewer stores does Family Dollar have than Dollar General?

8. How many more retail stores does Walgreens have than Walmart?

The following bar graph shows the results of a survey that was taken of 100 working adults to determine how they chose their careers. Use the bar graph to answer Exercises 11–16.
See Example 2.

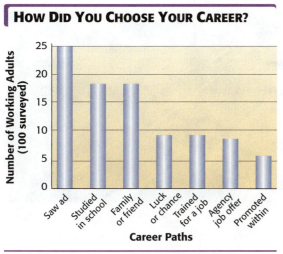

HOW DID YOU CHOOSE YOUR CAREER?

Number of Working Adults (100 surveyed)

Career Paths

Saw ad / Studied in school / Family or friend / Luck or chance / Trained for a job / Agency job offer / Promoted within

Source: Market Facts/TeleNation for Career Education Corporation.

9. CONCEPT CHECK What is the number of working adults who were surveyed?

10. CONCEPT CHECK How many career paths were included in the survey?

11. How many people found their careers as a result of training for a job?

12. How many people found their careers because they studied for the career in school?

13. (a) Which career path was taken by the greatest number of people?

The bar graph shows that the greatest number of people _____.

(b) How many people used this path?

From the bar graph _____ people used this path.

14. (a) Which career path was taken by the least number of people?

The bar graph shows that the least number of people were _____.

(b) How many people used this path?

From the bar graph, _____ people used this path.

15. How many more people found their careers as a result of "Studied in school" than "Luck or chance"?

16. Find the total number of people who found their careers as a result of either "Studied in school" or "Trained for a job."

Sun Solar Products collected installation data and prepared the following line graph. Remembering that the solar installation data are shown in thousands of installations, use the line graph to answer Exercises 17–22. **See Example 3.**

17. Which year had the greatest number of installations? How many installations were there?

The line graph shows that _____ had the greatest number.

There were 7 • 1000 = _____ .

18. Which two years had the least number of installations? How many installations were there in each of those years?

The line graph shows that _____ and _____ had the least number.

The dots for both years are halfway between 2000 and 3000, or _____ installations.

19. Find the increase in the number of installations from 2013 to 2014.

20. Find the decrease in the number of installations from 2012 to 2013.

21. Give three possible explanations for the decrease in solar installations from 2011 to 2013.

22. Give three possible explanations for the increase in solar installations from 2013 to 2014.

Getting a correct answer in mathematics always depends on following the order of operations. Insert grouping symbols (parentheses) so that each expression will result in the given number when simplified. **Work Exercises 23–27 in order.**

23. $7 - 2 \cdot 3 - 6$; simplifies to 9

24. $4 + 2 \cdot 5 + 1$; simplifies to 36

25. $36 \div 3 \cdot 3 \cdot 4$; simplifies to 16

26. $56 \div 2 \cdot 2 \cdot 2 + \dfrac{0}{6}$; simplifies to 7

27. The Good Shepherd Ranch owns one section of land (one mile by one mile) shown below as Parcel 1. Parcels 2 and 3 are leased from neighbors.

   (a) Use the order of operations to write an expression for the distance around the combined parcels.

   (b) How many feet of barbed wire are needed for a three-strand barbed wire fence around the parcels?

   (c) How many miles of barbed wire is this? (*Hint:* 1 mile = 5280 feet.)

1.10 Solving Application Problems

OBJECTIVES

1. Find indicator words in application problems.
2. Solve application problems.
3. Estimate an answer.

VOCABULARY TIP

Indicator words These key words are useful in translating **word** problems from English into **math,** a necessary skill in becoming a good problem solver.

Most problems involving applications of mathematics are written in sentence form. You need to read the problem carefully to decide how to solve it.

OBJECTIVE ▶ 1 Find indicator words in application problems. As you read an application problem, look for **indicator words** that help you determine whether to use addition, subtraction, multiplication, or division. Some of these indicator words are shown here.

Addition	Subtraction	Multiplication	Division	Equals
plus	less	product	divided by	is
more	subtract	double	divided into	the same as
more than	subtracted from	triple	quotient	equals
added to	difference	times	goes into	equal to
increased by	less than	of	divide	yields
sum	fewer	twice	divided equally	results in
total	decreased by	twice as much	per	are
sum of	loss of			
increase of	minus			
gain of	take away			

CAUTION

The word *and* does not always indicate addition, so it does not appear as an indicator word in the table above. Notice how the "and" shows the location of different operation signs in the examples below.

The sum of 6 *and* 2 is 6 + 2.
The difference of 6 *and* 2 is 6 − 2.
The product of 6 *and* 2 is 6 • 2.
The quotient of 6 *and* 2 is 6 ÷ 2.

OBJECTIVE ▶ 2 Solve application problems. Solve application problems by using the following six steps.

Solving an Application Problem

Step 1 **Read** the problem carefully and be certain you *understand* what the problem is asking. It may be necessary to read the problem several times.

Step 2 Before doing any calculations, **work out a plan** and try to visualize the problem. Draw a sketch if possible. Know which facts are given and which must be found. Use *indicator words* to help decide on the *plan* (whether you will need to add, subtract, multiply, or divide). *Always* **estimate** *the final answer.*

Step 3 **Estimate** a *reasonable answer* by using rounding.

Step 4 **Solve** the problem by using the facts given and your plan.

Step 5 **State the answer.**

Step 6 **Check** your work. If the answer does not seem reasonable, begin again by reading the problem. *Check your answer to see if it is reasonable.*

CAUTION

Do **NOT** make the mistake of trying the solve the problem before you know what is being asked.

OBJECTIVE ▶ **3** **Estimate an answer.** The six problem-solving steps give a systematic approach for solving word problems. Each of the steps is important, but special emphasis should be placed on Step 3, estimating a *reasonable answer*. Many times an "answer" just does not fit the problem.

What is a reasonable answer? Read the problem and try to determine the approximate size of the answer. Should the answer be part of a dollar, a few dollars, hundreds, thousands, or even millions of dollars? For example, if a problem asks for the cost of a man's shirt, would an answer of $20 be reasonable? $2000? $2? $200?

CAUTION

Always estimate the answer, then look at your final result to be sure it fits your estimate and is reasonable. This step will give greater success in problem solving.

Work Problem **1** at the Side. ▶

EXAMPLE 1   **Applying Division**

A community group has raised $8260 for charity. Equal amounts are given to the Food Bank, Children's Center, Boy Scouts of America, and the Women's Shelter. How much did each group receive?

Step 1 **Read.** A reading of the problem shows that the four charities divided $8260 equally.

> Read and **understand** a problem before you begin.

Step 2 **Work out a plan.** The indicator words, ***divided equally,*** show that the amount each received can be found by dividing $8260 by 4.

Step 3 **Estimate.** Round $8260 to $8000. Then $8000 ÷ 4 = $2000, so a reasonable answer would be a little greater than $2000 each.

Step 4 **Solve.** Find the actual answer by dividing $8260 by 4.

$$\frac{2065}{4\overline{)8260}}$$

Step 5 **State the answer.** Each charity received $2065.

Step 6 **Check.** The exact answer of $2065 is reasonable, as $2065 is close to the estimated answer of $2000. Is the exact answer of $2065 correct? Check by multiplying.

> Remember: Check your work.

$2065 ← Amount received by each charity

$\times$ 4 ← Number of charities

$8260 ← Total raised; matches number given in problem

Work Problem **2** at the Side. ▶

1 Pick the most reasonable answer for each problem.

GS **(a)** A grocery clerk's hourly wage: $1.40; $14; $140

_____ is eliminated (low)

_____ is eliminated (high)

_____ is most reasonable

(b) The total length of five sport-utility vehicles: 8 ft; 18 ft; 80 ft; 800 ft

(c) The cost of heart bypass surgery: $1000; $100,000; $10,000,000

2 Solve each problem.

(a) On a recent geology field trip, 84 fossils were collected. If the fossils are divided equally among John, Sean, Jenn, and Kara, how many fossils will each receive?

(b) This week there are 408 children attending a winter sports camp. If 12 children are assigned to each camp counselor, how many counselors are needed?

Answers

1. (a) $1.40; $140; $14 **(b)** 80 ft **(c)** $100,000
2. (a) 21 fossils **(b)** 34 counselors

③ Solve each problem.

(a) During the semester, Cindy Fong received the following points on examinations and quizzes: 92, 81, 83, 98, 15, 14, 15, and 12. Find her total points for the semester.

(b) Stephanie Dixon works at the telephone order desk of a catalog sales company. One week she had the following number of customer contacts: Monday, 78; Tuesday, 64; Wednesday, 118; Thursday, 102; and Friday, 196. How many customer contacts did she have that week?

EXAMPLE 2 Applying Addition

One week, Andrea Abriani, operations manager, decided to total the stroller production at Safe T First Strollers. The daily figures were 7642 strollers on Monday, 8150 strollers on Tuesday, 7916 strollers on Wednesday, 8419 strollers on Thursday, and 7704 strollers on Friday. Find the total number of strollers for the week.

Step 1 **Read.** In this problem, the number of strollers for each day is given and the total strollers for the week must be found.

Step 2 **Work out a plan.** Add the daily stroller figures to arrive at the weekly total.

Step 3 **Estimate.** Because there were about 8000 strollers per day for a week of five days, a reasonable estimate would be 5 • 8000 = 40,000 strollers.

Step 4 **Solve.** Find the exact answer by adding the stroller numbers for the 5 days.

$$
\begin{array}{r}
3\,9,8\,3\,1 \leftarrow \text{Check by adding up.}\\
7\ 6\,4\,2\\
8\ 1\,5\,0\\
7\ 9\,1\,6\\
8\ 4\,1\,9\\
+\ 7\ 7\,0\,4\\
\hline
3\,9,8\,3\,1 \leftarrow \text{Number of strollers for the week}
\end{array}
$$

Add from bottom to top to check addition.

Step 5 **State the answer.** Abriani's total figure for the week was 39,831 strollers.

Step 6 **Check.** The exact answer of 39,831 strollers is close to the estimate of 40,000 strollers, so it is reasonable. Add up the columns to check the exact answer.

▦ **Calculator Tip**

The calculator solution to **Example 2** uses chain calculations.

7642 ⊕ 8150 ⊕ 7916 ⊕ 8419 ⊕ 7704 ⊜ **39,831**

◀ Work Problem **③** at the Side.

EXAMPLE 3 Determining Whether Subtraction Is Necessary

The number of miles driven this year is 3028 fewer than the number driven last year. The miles driven last year was 16,735. Find the number of miles driven this year.

Step 1 **Read.** In this problem, the miles driven decreased from last year to this year. The miles driven last year and the decrease in miles driven are given. This year's miles driven must be found.

Step 2 **Work out a plan.** The indicator word, *fewer,* shows that subtraction must be used to find the number of miles driven this year.

Step 3 **Estimate.** Because the driving last year was about 17,000 miles, and the decrease in driving is about 3000 miles, a reasonable estimate would be 17,000 − 3000 = 14,000 miles.

Answers

3. (a) 410 points (b) 558 customer contacts

Continued on Next Page

Step 4 **Solve.** Find the exact answer by subtracting 3028 from 16,735.

$$
\begin{array}{r}
16{,}735 \\
-\ 3\ 028 \\
\hline
13{,}707
\end{array}
$$

Step 5 **State the answer.** The driving this year is 13,707 miles.

Step 6 **Check.** The exact answer of 13,707 is reasonable, as it is close to the estimate of 14,000. Check by adding.

$$
\begin{array}{r}
13{,}707 \\
+\ 3\ 028 \\
\hline
16{,}735
\end{array}
$$

13,707 ← miles driven this year
+ 3 028 ← decrease in miles driven
16,735 ← miles driven last year; matches number given in problem

> Remember: Check your work.

Work Problem ❹ at the Side. ▶

| **EXAMPLE 4** | **Solving a Two-Step Problem** |

In May, a landlord received $720 from each of eight tenants. After paying $2180 in expenses, how much rent money did the landlord have left?

Step 1 **Read.** The problem asks for the amount of rent remaining after expenses have been paid.

Step 2 **Work out a plan.** The wording *from each of eight tenants* indicates that the eight rents must be totaled. Since the rents are all the same, use multiplication to find the total rent received. Then, subtract expenses.

Step 3 **Estimate.** The amount of rent is about $700, making the total rent received about $700 • 8 = $5600. The expenses are about $2000. A reasonable estimate of the amount remaining is $5600 − $2000 = $3600.

Step 4 **Solve.** Find the exact amount by first multiplying $720 by 8 (the number of tenants).

$$
\begin{array}{r}
\$\ 720 \\
\times\quad 8 \\
\hline
\$5760
\end{array}
$$
← Total rent

Then subtract the $2180 in expenses from $5760.

$$
\begin{array}{r}
\$5760 \\
-\ \$2180 \\
\hline
\$3580
\end{array}
$$

> The exact answer is close to the estimate and reasonable.

Step 5 **State the answer.** The amount remaining is $3580.

Step 6 **Check.** The exact answer of $3580 is reasonable, since it is close to the estimated answer of $3600. Check the answer by adding the expenses to the amount remaining and then dividing by 8.

$$\$3580 + \$2180 = \$5760$$

$$
\begin{array}{r}
\$720 \\
8\overline{)5760}
\end{array}
$$

> Always check your work.

Matches the rent amount given in the problem

Work Problem ❺ at the Side. ▶

---

❹ Solve each problem.

(a) Alaska is our largest state, with an area of 663,267 square miles. The second largest state is Texas, with an area of 268,580 square miles. How much larger is Alaska than Texas? (*Source:* U.S. Census Bureau.)

(b) The Antique Military Vehicle Collectors (AMVC) had $14,863 in their club treasury bank account. After writing a check for $1180 to rent a display hall, find the amount remaining in the club account.

❺ Solve each problem.

GS (a) An automobile insurance company purchased 318 Samsung digital cameras at a cost of $129 each. After a rebate of $2470, find the final cost of the cameras.

318 • $_____ each = $41,022

$_____ − $2470 = $_____

(b) An Internet book company had sales of 12,628 books with a profit of $6 for each book sold. If 863 books were returned, how much profit remains?

Answers

4. (a) 394,687 square miles   (b) $13,683
5. (a) $129; $41,022; $38,552   (b) $70,590

1.10 Exercises

     MyMathLab®

1. **CONCEPT CHECK**   A problem asks for the number of hours a person worked each day. Which would *not* be a reasonable answer?

    (a) 4 hours   **(b)** 8 hours   **(c)** 25 hours

2. **CONCEPT CHECK**   A problem asks for the hourly earnings of a part-time student employee. Which would *not* be a reasonable answer?

    (a) $9/hour   **(b)** $100/hour   **(c)** $12/hour

3. **CONCEPT CHECK**   A problem asks for the cost of lunch at a fast food restaurant. Which answer is reasonable?

    (a) $5   **(b)** $50   **(c)** $500

4. **CONCEPT CHECK**   You calculate the gas mileage for your car. Which answer is reasonable? (mpg = miles per gallon)

    (a) 5 mpg   **(b)** 25 mpg   **(c)** 125 mpg

Solve each application problem. First use front end rounding to estimate the answer. Then find the exact answer. **See Examples 1–4.**

5. Last week, SUBWAY sold 602 Veggie Delite sandwiches, 935 ham sandwiches, 1328 turkey breast sandwiches, 757 roast beef sandwiches, and 1586 SUBWAY club sandwiches. Find the total number of sandwiches sold.

    Estimate: 600 + 900 + _____ + 800 + _____

    = _____

    Exact:

6. During a recent week, Radio Flyer, Inc. manufactured 32,815 Model #18 wagons, 4875 steel miniwagons, 1975 wood 40-inch wagons, 15,308 scooters, and 9815 new-design plastic wagons. Find the total number of items manufactured.

    Estimate: _____ + 5000 + 2000 + 20,000

    + _____ = _____

    Exact:

7. Paying ahead for a rental car can save 35% at Budget.com. They pay-at-the-counter base rate for a seven-day, full-size car rental is $296. The rate for the same car when paying ahead using the Pay Now rate at Budget.com is $192. How much is saved? (*Source:* Budget Car Rental.)

    Estimate: $300 − $200 = _____

    Exact:

8. The U.S. population will rise from 311 million today to 478 million by 2100. Find the expected increase in population. (*Source:* United Nations population division.)

    Estimate:

    Exact:

9. A packing machine can package 236 first-aid kits each hour. At this rate, find the number of first-aid kits packaged in 24 hours.

    Estimate:

    Exact:

10. If 450 admission tickets to a classic car show are sold each day, how many tickets are sold in a 12-day period?

    Estimate: 500 × 10

    Exact:

11. Clarence Hanks, coordinator of Toys for Tots, has collected 2628 toys. If his group can give the same number of toys to each of 657 children, how many toys will each child receive?

Estimate:

Exact:

12. If profits of $680,000 are divided evenly among a firm's 1000 employees, how much money will each employee receive?

Estimate:

Exact:

13. Turn down the thermostat in the winter and you can save money and energy. In the upper Midwest, setting back the thermostat from 68° to 55° at night can save $34 per month on fuel. Find the amount of money saved in five months.

Estimate:

Exact:

14. The cost of tuition and fees at a community college is $785 per quarter. If Gale Klein has five quarters remaining, find the total amount that she will need for tuition and fees.

Estimate:

Exact:

The sesquicentennial anniversary (150 years) of the beginning of the Civil War occurred in 2011. The loss of life at the conclusion of the war still stands as the highest war casualty count in American history. The table below shows the number of deaths and cause of death for both the Southern and the Northern American States. Use these data to answer Exercises 15–20.

Union (Northern) Deaths	Confederate (Southern) Deaths
Battle 110,070	Battle 94,120
Disease 250,152	Disease 164,300

Source: *The Civil War, Strange and Fascinating Facts* by Burke Davis.

15. Find the total number of Union deaths in the Civil War.

Estimate: 100,000 + 300,000 = _____

Exact:

16. Find the total number of Confederate deaths in the Civil War.

Estimate:

Exact:

17. How many more Union deaths resulted from disease than battle?

Estimate:

Exact:

18. How many more Confederate deaths were caused by disease than battle?

Estimate:

Exact:

19. Find the total loss of life (North and South) in the Civil War.

Estimate:

Exact:

20. How many more Union deaths than Confederate deaths were there in the Civil War?

Estimate:

Exact:

21. Ronda Biondi decides to establish a monthly budget. She will spend $695 for rent, $340 for food, $435 for child care, $240 for transportation, $180 for other expenses, and she will put the remainder in savings. If her monthly take-home pay is $2240, find her monthly savings.

Estimate:

Exact:

22. Robert Heisner had $2874 in his bank account. He paid $308 for auto repairs, $580 for a dishwasher, and $778 for an insurance payment. Find the amount remaining in his account.

Estimate:

Exact:

23. There are 43,560 square feet in one acre. How many square feet are there in 138 acres?

Estimate:

Exact:

24. The number of gallons of water polluted each day in an industrial area is 209,670. How many gallons of water are polluted each year? (Use a 365-day year.)

Estimate:

Exact:

The Internet was used to find the following minivan optional features and the price of each feature. Use this information to answer Exercises 25–28.

Safety and Exterior Options		Interior Options	
Option	Cost	Option	Cost
VIP Plus Security System	$299	Carpet floor mats	$321
Roof rack crossbars	$185	Cargo nets	$51
Mudguards	$99	Interface kit for iPod	$299
Alloy wheel locks	$67	Dual screen entertainment system	$1799
Paint protection	$395	Wireless headphones	$82
Lower body moulding	$209	XM Satellite Radio	$449

Source: www.edmunds.com

25. Find the total cost of all Safety and Exterior Options listed.

Estimate:

Exact:

26. Find the total cost of all Interior Options listed.

Estimate:

Exact:

27. A new-car dealer offers an option value package that includes VIP Plus Security System, alloy wheel locks, paint protection, and lower body moulding for $785. If Jill buys the value package instead of paying for each option separately, how much will she save?

Estimate:

Exact:

28. A new-car dealer offers an option package that includes VIP Plus Security System, roof rack cross-bars, paint protection, carpet floor mats, cargo nets, and XM Satellite Radio for a total of $1495. How much will Samuel save if he buys the option package instead of paying for each option separately?

Estimate:

Exact:

29. The Enabling Supply House purchased 6 wheelchairs at $1256 each and 15 speech compression recorder-players at $895 each. Find the total cost.

Estimate:

Exact:

30. A college bookstore buys 17 laptop computers at $506 each and 13 printers at $482 each. Find the total cost.

Estimate:

Exact:

31. Being able to identify indicator words is helpful in determining how to solve an application problem. Write three indicator words for each of these operations: add, subtract, multiply, and divide. Write two indicator words that mean equals.

32. Identify and explain the six steps used to solve an application problem. You may refer to the text if you need help, but use your own words.

Solve each application problem. See Examples 1–4.

33. Steve Edwards, manager, decided to total his sales at SUBWAY. The daily sales figures were $2358 on Monday, $3056 on Tuesday, $2515 on Wednesday, $1875 on Thursday, $3978 on Friday, $3219 on Saturday, and $3008 on Sunday. Find his total sales for the week.

34. The numbers of visitors at a war veterans' memorial during one week are 5318; 2865; 4786; 1998; 3899; 2343; and 7221. Find the total attendance for the week.

35. A car weighs 2425 pounds. If its 582-pound engine is removed and replaced with a 634-pound engine, what will the car weigh?

36. Estelle Alan has $2324 in her preschool operating account. She spends $734 from this account, and then the class parents raise $568 in a rummage sale. Find the balance in the account after she deposits the money from the rummage sale.

37. In a recent survey of Reno/Lake Tahoe hotels, the cost per night at Harrah's in Reno was $45, while the cost at Harrah's in Lake Tahoe was $99 per night. Find the amount saved on a 7-night stay at Harrah's in Reno instead of staying at Harrah's in Lake Tahoe. (*Source:* Harrah's Casinos and Hotels.)

38. The most expensive hotel room in a recent study was the Ritz-Carlton at $645 per night, while the least expensive was Motel 6 at $74 per night. Find the amount saved in a 4-night stay at Motel 6 instead of staying at the Ritz-Carlton. (*Source:* Ritz-Carlton/Motel 6.)

39. A youth soccer association raised $7588 through fund-raising projects. After expenses of $838 were paid, the balance of the money was divided evenly among the 18 teams. How much did each team receive?

40. Feather Farms Egg Ranch collected 3545 eggs in the morning and 2575 eggs in the afternoon. If the eggs are packed in flats containing 30 eggs each, find the number of flats needed for packing.

41. A theater owner wants to provide enough seating for 1250 people. The main floor has 30 rows of 25 seats in each row. If the balcony has 25 rows, how many seats must be in each balcony row to satisfy the owner's seating requirements?

42. Jennie makes 24 grapevine wreaths per week to sell to gift shops. She works 40 weeks a year and packages six wreaths per box. If she ships equal quantities to each of five shops, find the number of boxes each store will receive.

Chapter 1 *Summary*

Key Terms

1.1

whole numbers The whole numbers are 0, 1, 2, 3, 4, 5, 6, 7, 8, and so on.

place value The place value of each digit in a whole number is determined by its position in the whole number.

table A table is a display of facts in rows and columns.

1.2

addition The process of finding the total is addition.

addends The numbers being added in an addition problem are addends.

sum (total) The answer in an addition problem is called the sum.

commutative property of addition The commutative property of addition states that the order of numbers in an addition problem can be changed without changing the sum.

associative property of addition The associative property of addition states that grouping the addition of numbers differently does not change the sum.

regrouping The process of regrouping is used in an addition problem when the sum of the digits in a column is greater than 9.

perimeter The perimeter is the distance around the outside edges of a flat figure.

1.3

minuend The number from which another number is being subtracted is the minuend.

subtrahend The subtrahend is the number being subtracted in a subtraction problem.

difference The answer in a subtraction problem is called the difference.

regrouping The process of regrouping is used in subtraction if a digit is less than the one directly below it.

1.4

factors The numbers being multiplied are called factors. For example, in $3 \times 4 = 12$, both 3 and 4 are factors.

product The answer in a multiplication problem is called the product.

commutative property of multiplication The commutative property of multiplication states that changing the order of the factors in a multiplication problem does not change the product.

associative property of multiplication The associative property of multiplication states that grouping the factors differently does not change the product.

chain multiplication problem A multiplication problem having more than two factors is a chain multiplication problem.

multiple The product of two whole number factors is a multiple of those numbers.

multiple of 10 A whole number that ends in 0, such as 10, 20, or 30; 100, 200, or 300.

partial products The products found when multiplying numbers having two or more digits.

1.5

dividend The number being divided by another number in a division problem is the dividend.

divisor The divisor is the number by which you are dividing in a division problem.

quotient The answer in a division problem is called the quotient.

undefined Dividing any number by 0 is *undefined*. Dividing by zero cannot be done.

short division A method of dividing a number by a one-digit divisor is short division.

remainder The remainder is the number left over when two numbers do not divide evenly.

1.6

long division The process of long division is used to divide by a number with more than one digit.

trial divisors A method of determining how many times the true divisor goes into the dividend by estimating.

trial quotient The quotient that results from dividing the dividend by the trial divisor.

1.7

rounding Rounding is used to find a number that is close to the original number, but easier to work with. Use the $\approx$ sign, which means "approximately equal to."

estimate An estimated answer is one that is close to the exact answer.

front end rounding Rounding to the highest possible place so that all the digits become zeros except the first one is front end rounding.

exponent The exponent is the small raised number (2) in the expression 3^2.

base The base is the number 3 in the expression 3^2.

1.8

square root The square root of a whole number is the number that can be multiplied by itself to produce the given number.

perfect square A number that is the square of a whole number is a perfect square.

order of operations For problems or expressions with more than one operation, the order of operations tells what to do first, second, and so on to get the correct answer.

1.9

pictograph A graph that uses pictures or symbols to show data is a pictograph.

bar graph A graph that uses bars of various heights to show quantity is a bar graph.

line graph A graph that uses dots connected by lines to show trends is a line graph.

1.10

indicator words Words in a problem that indicate the necessary operations—addition, subtraction, multiplication, or division—are indicator words.

New Symbols

≈ This sign is used to show that an answer has been estimated. It means "is approximately equal to."

√ The symbol for square root.

5^2 The small raised 2 is an exponent; it tells how many times to use 5 (the base) as a factor in multiplication.

Test Your Word Power

See how well you have learned the vocabulary in this chapter.

1 When using **addends** you are performing
   A. division
   B. subtraction
   C. addition.

2 The subtrahend is the
   A. number being multiplied
   B. number being subtracted
   C. number being added.

3 A **factor** is
   A. one of two or more numbers being added
   B. one of two or more numbers being multiplied
   C. one of two or more numbers being divided.

4 The **divisor** is
   A. the number being multiplied
   B. always the largest number
   C. the number doing the dividing.

5 We use **rounding** to
   A. avoid solving a problem
   B. help estimate a reasonable answer
   C. find the remainder.

6 A **perfect square** is
   A. the square of a whole number
   B. the same as a square root
   C. similar to a perfect triangle.

Answers to Test Your Word Power

1. C; *Example:* In $2 + 3 = 5$, the 2 and the 3 are addends.

2. B; *Example:* In $5 - 4 = 1$, the 4 is the subtrahend.

3. B; *Example:* In $3 \times 5 = 15$, the numbers 3 and 5 are factors.

4. C; *Example:* In $8 \div 4 = 2, \frac{8}{4} = 2$, and $4\overline{)8}$, the 4 is the divisor.

5. B; *Example:* We can use rounding to estimate our answer and then determine whether the exact answer is reasonable.

6. A; *Example:* 25 is a perfect square because $5^2 = 25$ and 5 is a whole number.

Quick Review

Concepts	Examples

1.1 **Reading and Writing Whole Numbers**

Do not use the word *and* when writing a whole number. Commas help divide the periods or groups for ones, thousands, millions, and billions. A comma is not needed when a number has four digits or fewer.

795 is written *seven hundred ninety-five.*

9,768,002 is written *nine million, seven hundred sixty-eight thousand, two*

Concepts	**Examples**

1.2 Adding Whole Numbers

Add from top to bottom, starting with the ones column and working left. To check, add from bottom to top.

$$
\begin{array}{r}
1140 \\
687 \\
26 \\
9 \\
+\ 418 \\
\hline
1140
\end{array}
$$

(Add up to check.)   } Addends   Sum

1.2 Commutative Property of Addition

Changing the order of the addends in an addition problem does not change the sum.

$$2 + 4 = 6$$
$$4 + 2 = 6$$

By the commutative property, the sum is the same.

1.2 Associative Property of Addition

Grouping the addends differently when adding does not change the sum.

$$(2 + 3) + 4 = 5 + 4 = 9$$
$$2 + (3 + 4) = 2 + 7 = 9$$

By the associative property, the sum is the same.

1.3 Subtracting Whole Numbers

Subtract the subtrahend from the minuend to get the difference, using regrouping when necessary. To check, add the difference to the subtrahend to get the minuend.

Problem   *Check*

$$
\begin{array}{r}
\overset{6\ 12\ 18}{4\,7\,3\,8} \\
-\ 649 \\
\hline
4089
\end{array}
$$
← Minuend
Subtrahend
Difference

$$
\begin{array}{r}
4089 \\
+\ 649 \\
\hline
4738
\end{array}
$$

1.4 Multiplying Whole Numbers

Use ×, • (a raised dot), or parentheses to indicate multiplication.

The numbers being multiplied are called *factors*. The multiplicand is being multiplied by the multiplier, giving the product. When the multiplier has more than one digit, partial products must be used and added to find the product.

$$3 \times 4 \quad \text{or} \quad 3 \cdot 4 \quad \text{or} \quad (3)(4) \quad \text{or} \quad 3(4)$$

$$
\begin{array}{r}
78 \\
\times\ 24 \\
\hline
312 \\
156 \\
\hline
1872
\end{array}
$$

Multiplicand } Factors
Multiplier
Partial product
Partial product (move one position left)
Product

1.4 Commutative Property of Multiplication

The product in a multiplication problem remains the same when the order of the factors is changed.

$$3 \times 4 = 12$$
$$4 \times 3 = 12$$

By the commutative property, the product is the same.

1.4 Associative Property of Multiplication

Grouping the factors differently when multiplying does not change the product.

$$(2 \times 3) \times 4 = 6 \times 4 = 24$$
$$2 \times (3 \times 4) = 2 \times 12 = 24$$

By the associative property, the product is the same.

1.5 Dividing Whole Numbers

$\div$ and $\overline{)}$ mean divide.

Also a ——, as in $\frac{25}{5}$, means to divide the top number (dividend) by the bottom number (divisor).

$$
\begin{array}{r}
22 \\
4\overline{)88} \\
88 \\
\hline
0
\end{array}
$$

Divisor →   ← Quotient   ← Dividend

$$88 \div 4 = 22$$

Dividend | Quotient
Divisor

$$\frac{88}{4} = 22 \leftarrow \text{Quotient}$$

Dividend

Concepts	Examples

1.7 Rounding Whole Numbers

Rules for Rounding

Step 1 Locate the place to be rounded, and draw a line under it.

Step 2 If the next digit to the right is 5 or more, increase the underlined digit by 1. If the next digit is 4 or less, do not change the underlined digit.

Step 3 Change all digits to the right of the underlined place to zeros.

Round 726 to the nearest ten.

Next digit is 5 or more.

726

Tens place increases from 2 to 3.

726 rounds to 730.

Round 1,498,586 to the nearest million.

Next digit is 4 or less.

1,498,586

Millions place does not change.

1,498,586 rounds to 1,000,000.

1.7 Front End Rounding

Front end rounding is rounding to the highest possible place so that all the digits become 0 except the first digit.

Round each number using front end rounding.

76 rounds to 80.

348 rounds to 300.

6512 rounds to 7000.

23,751 rounds to 20,000.

652,179 rounds to 700,000.

1.8 Order of Operations

Problems may have several operations. Work these problems using the order of operations.

1. Do all operations inside parentheses or other grouping symbols.
2. Simplify any expressions with exponents and find any square roots $\left(\sqrt{\ }\right)$.
3. Multiply or divide proceeding from left to right.
4. Add or subtract proceeding from left to right.

Simplify, using the order of operations.

$7 \cdot \sqrt{9} - 4 \cdot 5$ Find the square root.

$7 \cdot 3 \ - 4 \cdot 5$ Multiply from left to right.

$21 \ - \ 20 = 1$ Subtract.

1.9 Reading Pictographs, Bar Graphs, and Line Graphs

A *pictograph* uses pictures or symbols to show data.

A *bar graph* uses bars of various heights to show quantity.

A *line graph* uses dots connected by lines to show trends.

When reading a pictograph, be certain that you determine the quantity represented by each picture or symbol.

When reading a bar graph, use a straightedge to line up the top of the bar with the numbers along the left edge of the graph.

When reading a line graph, use a straightedge to line up the dot with the numbers along the left edge of the graph.

Concepts	Examples

1.10 Application Problems

Steps for Solving an Application Problem

Step 1 **Read** the problem carefully, perhaps several times.

Step 2 **Work out a plan** before starting. Draw a sketch if possible.

Step 3 **Estimate** a reasonable answer.

Step 4 **Solve** the problem.

Step 5 **State the answer.**

Step 6 **Check** your work. If the answer is not reasonable, start over.

Manuel earns $118 on Sunday, $87 on Monday, and $63 on Tuesday. Find his total earnings for the 3 days.

Step 1 The earnings for each day are given, and the total for the 3 days must be found.

Step 2 Add the daily earnings to find the total.

Step 3 Since the earnings were about $100 + $90 + $60 = $250, a reasonable estimate would be approximately $250.

Step 4   **$268**   Check by adding up

$$\begin{array}{r} \$118 \\ 87 \\ +\ \ 63 \\ \hline \$268 \end{array}$$ Total earnings

Step 5 Manuel's total earnings are $268.

Step 6 The exact answer is reasonable, because it is close to the estimate of $250.

Chapter 1   *Review Exercises*

If you need help with any of these Review Exercises, look in the section indicated inside the dark blue rectangles.

1.1   *Write the digits for the given period or group in each number.*

1. 6573

thousands

ones

2. 36,215

thousands

ones

3. 105,724

thousands

ones

4. 1,768,710,618

billions

millions

thousands

ones

Rewrite each number in words.

5. 728

6. 15,310

7. 319,215

8. 62,500,005

Rewrite each number in digits.

9. ten thousand, eight

10. two hundred million, four hundred fifty-five

1.2   *Add.*

11.
```
   72
+ 38
```

12.
```
   54
+ 67
```

13.
```
   807
  4606
+   51
```

14.
```
  8215
     9
+ 7433
```

15.
```
  2130
   453
  8107
+  296
```

16.
```
  5684
   218
  2960
+  983
```

17.
```
   5 732
  11,069
      37
   1 595
+ 22,169
```

18.
```
   3 451
  12,286
      43
   1 291
+ 32,784
```

1.3 *Subtract.*

19. $\begin{array}{r} 64 \\ -\ 28 \\ \hline \end{array}$ **20.** $\begin{array}{r} 46 \\ -\ 19 \\ \hline \end{array}$ **21.** $\begin{array}{r} 375 \\ -\ 186 \\ \hline \end{array}$ **22.** $\begin{array}{r} 573 \\ -\ 389 \\ \hline \end{array}$

23. $\begin{array}{r} 7416 \\ -\ 567 \\ \hline \end{array}$ **24.** $\begin{array}{r} 5210 \\ -\ 883 \\ \hline \end{array}$ **25.** $\begin{array}{r} 2210 \\ -\ 1986 \\ \hline \end{array}$ **26.** $\begin{array}{r} 99{,}704 \\ -\ 73{,}838 \\ \hline \end{array}$

1.4 *Multiply.*

27. $\begin{array}{r} 7 \\ \times\ 7 \\ \hline \end{array}$ **28.** $\begin{array}{r} 8 \\ \times\ 0 \\ \hline \end{array}$ **29.** 8(4)   **30.** 8(8)

31. (5)(9)   **32.** (6)(7)   **33.** $7 \cdot 8$ **34.** $9 \cdot 9$

Work each chain multiplication.

35. $5 \times 4 \times 2$ **36.** $9 \times 1 \times 5$ **37.** $4 \times 4 \times 3$ **38.** $2 \times 2 \times 2$

39. (6)(0)(8)   **40.** (7)(1)(6)   **41.** $6 \cdot 1 \cdot 8$ **42.** $7 \cdot 7 \cdot 0$

Multiply.

43. $\begin{array}{r} 28 \\ \times\ 3 \\ \hline \end{array}$ **44.** $\begin{array}{r} 46 \\ \times\ 8 \\ \hline \end{array}$ **45.** $\begin{array}{r} 58 \\ \times\ 9 \\ \hline \end{array}$ **46.** $\begin{array}{r} 98 \\ \times\ 1 \\ \hline \end{array}$

47. $\begin{array}{r} 625 \\ \times\ 8 \\ \hline \end{array}$ **48.** $\begin{array}{r} 374 \\ \times\ 8 \\ \hline \end{array}$ **49.** $\begin{array}{r} 1349 \\ \times\ 4 \\ \hline \end{array}$ **50.** $\begin{array}{r} 9163 \\ \times\ 5 \\ \hline \end{array}$

51. $\begin{array}{r} 7456 \\ \times\ 2 \\ \hline \end{array}$ **52.** $\begin{array}{r} 2880 \\ \times\ 7 \\ \hline \end{array}$ **53.** $\begin{array}{r} 93{,}105 \\ \times\ 5 \\ \hline \end{array}$ **54.** $\begin{array}{r} 21{,}873 \\ \times\ 8 \\ \hline \end{array}$

55.
$$\begin{array}{r} 35 \\ \times\ 25 \\ \hline \end{array}$$

56.
$$\begin{array}{r} 74 \\ \times\ 32 \\ \hline \end{array}$$

57.
$$\begin{array}{r} 98 \\ \times\ 12 \\ \hline \end{array}$$

58.
$$\begin{array}{r} 68 \\ \times\ 75 \\ \hline \end{array}$$

59.
$$\begin{array}{r} 472 \\ \times\ 33 \\ \hline \end{array}$$

60.
$$\begin{array}{r} 392 \\ \times\ 77 \\ \hline \end{array}$$

61.
$$\begin{array}{r} 4051 \\ \times\ 219 \\ \hline \end{array}$$

62.
$$\begin{array}{r} 1527 \\ \times\ 328 \\ \hline \end{array}$$

Find each total cost.

63. 30 scientific calculators at $12 per calculator

64. 76 subscribers at $14 per subscription

65. 318 drill bit sets at $64 per set

66. 114 earplugs at $6 per earplug

Multiply by using the shortcut for multiples of 10.

67.
$$\begin{array}{r} 280 \\ \times\ 50 \\ \hline \end{array}$$

68.
$$\begin{array}{r} 340 \\ \times\ 70 \\ \hline \end{array}$$

69.
$$\begin{array}{r} 517 \\ \times\ 400 \\ \hline \end{array}$$

70.
$$\begin{array}{r} 637 \\ \times\ 500 \\ \hline \end{array}$$

71.
$$\begin{array}{r} 16,000 \\ \times\ 8\ 000 \\ \hline \end{array}$$

72.
$$\begin{array}{r} 43,000 \\ \times\ 2\ 100 \\ \hline \end{array}$$

1.5 *Divide. If the division is not possible, write "undefined."*

73. $20 \div 4$

74. $35 \div 5$

75. $42 \div 7$

76. $18 \div 9$

77. $\dfrac{54}{9}$

78. $\dfrac{36}{9}$

79. $\dfrac{49}{7}$

80. $\dfrac{0}{6}$

81. $\dfrac{148}{0}$

82. $\dfrac{0}{23}$

83. $\dfrac{64}{8}$

84. $\dfrac{81}{9}$

1.5-1.6 *Divide.*

85. $4\overline{)328}$

86. $3\overline{)294}$

87. $6\overline{)26{,}532}$

88. $76\overline{)26{,}752}$

89. $2704 \div 18$

90. $15{,}525 \div 125$

1.7 *Round as indicated.*

91. 817 to the nearest ten

92. 15,208 to the nearest hundred

93. 20,643 to the nearest thousand

94. 67,485 to the nearest ten-thousand

Round each number to the nearest ten, nearest hundred, and nearest thousand.
Remember to round from the original number.

	Ten	Hundred	Thousand
95. 3487	_____	_____	_____
96. 20,065	_____	_____	_____
97. 98,201	_____	_____	_____
98. 352,118	_____	_____	_____

1.8 *Find each square root by using the Perfect Squares Table from this chapter.*

99. $\sqrt{16}$
100. $\sqrt{49}$
101. $\sqrt{144}$
102. $\sqrt{196}$

Identify the exponent and the base, and then simplify each expression.

103. 7^3
104. 3^6
105. 5^3
106. 4^5

Simplify each expression by using the order of operations.

107. $7^2 - 15$
108. $6^2 - 10$
109. $2 \cdot 3^2 \div 2$

110. $9 \div 1 \cdot 2 \cdot 2 \div (11 - 2)$
111. $\sqrt{9} + 2(3)$
112. $6 \cdot \sqrt{16} - 6 \cdot \sqrt{9}$

1.9 *The bar graph shows the number of parents out of 100 surveyed who nag their children about performing certain household chores.*

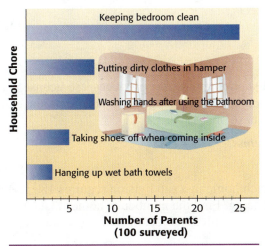

CLEAN UP YOUR ROOM

Source: Opinion Research Corporation for the Soap and Detergent Association.

113. How many parents nagged their children about washing hands after using the bathroom?

114. Find the number of parents who nagged their children about taking shoes off when coming inside.

115. Which household chore was nagged about by the greatest number of parents? How many were there?

116. Which household chore was nagged about by the least number of parents? How many did this?

1.10 *Solve each application problem. First use front end rounding to estimate the answer. Then find the exact answer.*

117. Bank of America processes 40 million checks each day. Find the number of checks processed by the bank in a year. Use a 365-day year. (*Source:* Bank of America.)

Estimate:

Exact:

118. A pulley on an evaporative cooler turns 1400 revolutions per minute. How many revolutions will the pulley turn in 60 minutes?

Estimate:

Exact:

119. The two most populated states are California, with 37,341,989 people, and Texas, with 25,268,418. Find the difference in population. (*Source:* U.S. Census Bureau.)

Estimate:

Exact:

120. The two least populated states are Alaska, with a population of 721,523, and Wyoming, with a population of 568,300. How many more people does Alaska have than Wyoming? (*Source:* U.S. Census Bureau.)

Estimate:

Exact:

121. The mechanic tells Kara Jantzi that the transmission on her Ford Escape has blown up. The cost of a new transaxle is $2633, labor is 8 hours at $90 per hour, and the sales tax is $230. Find the total cost to replace her transmission. (*Source:* Undisclosed auto repair shop.)

Estimate:

Exact:

122. Moving to new home, Scott Samon rented a U-haul truck for $55 plus $2 per mile. After completing the move, the odometer shows that he has driven the truck 89 miles. Find the cost of the truck rental.

Estimate:

Exact:

123. Find the total cost if SUBWAY buys 32 baking ovens at $1538 each and 28 warming ovens at $887 each.

Estimate:

Exact:

124. A newspaper carrier has 62 customers who take the paper daily and 21 customers who take the paper on weekends only. A daily customer pays $16 per month and a weekend-only customer pays $7 per month. Find the total monthly collections.

Estimate:

Exact:

125. This holiday season, the average amount consumers plan to spend on holiday shopping for others is $620. If they plan to spend $107 on holiday shopping for themselves, how much more do they plan to spend on others than themselves? (*Source:* National Retail Federation.)

Estimate:

Exact:

126. Rachel Leach pays $520 for rent and $385 for her car payment. If she started with $1924 in her bank account, how much remains in her account?

Estimate:

Exact:

127. A food canner uses 1 pound of pork for every 175 cans of pork and beans. How many pounds of pork are needed for 8750 cans?

Estimate:

Exact:

128. A stamping machine produces 986 license plates each hour. How long will it take to produce 32,538 license plates?

Estimate:

Exact:

129. Nitrogen sulfate is used in farming to enrich nitrogen-poor soil. If 625 pounds of nitrogen sulfate are spread per acre, how many acres can be spread with 32,500 pounds of nitrogen sulfate?

Estimate:

Exact:

130. Each home in a subdivision requires 180 feet of fencing. Find the number of homes that can be fenced with 5760 feet of fencing material.

Estimate:

Exact:

Mixed Review Exercises*

Perform the indicated operations.

131. 4(83)

132. 7(64)

133.
```
  309
-  56
```

134.
```
  835
- 247
```

135.
```
  662
+ 379
```

136.
```
  789
+ 872
```

137.
```
 38,140
-  6 078
```

138.
```
 29,156
-  4 209
```

139. 21 ÷ 7

140. $\dfrac{42}{6}$

141.
```
   7 218
       3
      18
   1 791
  82,623
+  1 982
```

142.
```
   3 812
       5
      22
   1 836
  75,134
+  2 369
```

143. $\dfrac{9}{0}$

144. $\dfrac{7}{1}$

145. 27,600 ÷ 4

*The order of exercises in this final group does not correspond to the order in which topics occur in the chapter. This random ordering should help you prepare for the chapter test in yet another way.

146. 18,480 ÷ 8

147. 8430
× 128

148. 21,702
× 6

149. 34)3672

150. 68)14,076

151. Rewrite 376,853 in words.

152. Rewrite 408,610 in words.

153. Round 8749 to the nearest hundred.

154. Round 400,503 to the nearest thousand.

Find each square root.

155. $\sqrt{64}$

156. $\sqrt{81}$

Find each total cost.

157. 308 pairs of knee guards at $18 per pair

158. 84 dishwashers at $370 per dishwasher

159. 208 baseball hats at $11 per hat

160. 607 boxes of avocados at $26 per box

Solve each application problem.

161. There are 52 playing cards in a deck. How many cards are there in nine decks?

162. A group of neighbors in Salem, Oregon founded an organization known as Salem Harvest. Picking fruits and vegetables to be distributed to those in need, 1700 volunteers picked 31 pounds of produce each. How many pounds did the volunteers pick? (*Source: AARP Magazine.*)

163. Push-type gasoline-powered lawn mowers cost $100 less than self-propelled mowers that you walk behind. If a self-propelled mower costs $380, find the cost of a push-type mower.

164. The Country Day School wants to raise $218,450 to construct and equip a computer lab. If $103,815 has already been raised, how much more is needed?

American River Raft Rentals lists the following daily raft rental fees. Notice that there is an additional $2 launch fee payable to the park system for each raft rented. Use this information to solve Exercises 165 and 166.

AMERICAN RIVER RAFT RENTALS

Size	Rental Fee	Launch Fee
4-person	$28	$2
6-person	$38	$2
10-person	$70	$2
12-person	$75	$2
16-person	$85	$2

Source: American River Raft Rentals.

165. On a recent Tuesday the following rafts were rented: 6 4-person; 15 6-person; 10 10-person; 3 12-person; and 2 16-person. Find the total receipts, including the $2 per-raft launch fee.

166. On the 4th of July the following rafts were rented: 38 4-person; 73 6-person; 58 10-person; 34 12-person; and 18 16-person. Find the total receipts, including the $2 per-raft launch fee.

1 World Trade Center will be the tallest building in the United States and one of the world's giants. The world's tallest building is the Burj Dubai in the United Arab Emirates. The pictogram below shows the height, in feet, of some of the tallest buildings in the world. Use the information to solve Exercises 167–170.

167. How much taller is the Burj Dubai than the Empire State Building?

168. How much taller will 1 WTC in New York be than the Petronas Tower in Malaysia?

169. (a) What is the combined height of all the buildings shown?

(b) Is the combined height of these six buildings greater or less than two miles? How much greater or less than two miles? (*Hint:* 1 mile = 5280 feet.)

170. The height of the Willis Tower in Chicago is equivalent to the length of how many football fields? (*Hint:* A football field is 100 yards long and 1 yard = 3 feet.)

The World of Giants
(Heights given in feet)

Burj Dubai Dubai, UAE 2717 ft; 1 WTC New York 1776 ft; Taipei 101 Taiwan 1667 ft; World Financial Center Shanghai 1614 ft; Petronas Towers Kuala Lampur, Malaysia 1483 ft; Willis Tower Chicago 1451 ft; Empire State Building, New York 1250 ft

Source: Council on Tall Buildings and Urban Habitat; Silverstein Properties; The Port Authority of New York and New Jersey Associated Press.

Chapter 1 Test

The Chapter Test Prep Videos with test solutions are available on DVD, in **MyMathLab**, and on **YouTube**—search "LialBasicCollegeMath" and click on "Channels."

Write each number in words.

1. 9205

2. 25,065

3. Use digits to write four hundred twenty-six thousand, five.

Add.

4.
```
    853
     66
   4022
 + 3589
```

5.
```
  17,063
       7
      12
   1 505
  93,710
 +   333
```

Subtract.

6.
```
   9009
 − 7964
```

7.
```
   9075
 − 2869
```

Multiply

8. $7 \times 6 \times 4$

9. $57 \cdot 3000$

10. $85(19)$

11.
```
   7381
 × 603
```

Divide. If the division is not possible, write "undefined."

12. $16\overline{)112{,}752}$

13. $\dfrac{835}{0}$

14. $19{,}241 \div 42$

15. $280\overline{)44{,}800}$

Round as indicated.

16. 6347 to the nearest ten

17. 76,489 to the nearest thousand

Simplify each expression.

18. $5^2 + 8(2)$

19. $7 \cdot \sqrt{64} - 14 \cdot 2$

Solve each application problem. First use front end rounding to estimate the answer. Then find the exact answer.

20. Judy Martinez collects the following monthly rents from the tenants in her fourplex: $485, $500, $515, and $425. After she pays expenses of $785, how much does she have left?

Estimate:

Exact:

21. The major producer of ethanol made from corn is the United States. If 374 gallons of ethanol can be produced from the corn grown on one acre of land, how many acres are needed to produce 86,394 gallons. (*Source:* Earth Policy Institute.)

Estimate:

Exact:

22. Sadie Simms paid $528 for tires, $195 for brakes, and $235 for a timing belt. If this money was withdrawn from her bank account, which had a beginning balance of $1906, find her new balance.

Estimate:

Exact:

23. The technicians at a chicken ranch identify the sex of 48 baby chicks each minute for 4 hours in the morning and 36 baby chicks each minute for 3 hours in the afternoon. Find the total number of baby chicks identified. (*Source:* Discovery Channel, *Dirty Jobs.*)

Estimate:

Exact:

24. Explain in your own words the rules for rounding numbers. Give an example of rounding a number to the nearest ten-thousand.

25. List the six steps for solving application problems.

2

Multiplying and Dividing Fractions

The recipe shown below uses Jelly Belly jelly beans and will make $2\frac{1}{2}$ dozen (30) cookies. But suppose you wanted to make 3 dozen, 10 dozen, or even $3\frac{1}{2}$ dozen cookies? In this chapter we discuss multiplication and division of fractions, which you need to know when cooking or baking.

NEST COOKIES

$1\frac{1}{2}$ cups all purpose flour	$\frac{1}{2}$ cup sugar	2 cups shredded coconut
1 tsp. baking powder	1 egg white	5 oz. Jelly Belly
$\frac{1}{2}$ tsp. salt	1 tsp. vanilla	jelly beans
$\frac{1}{2}$ cup shortening	2 Tbs. milk	assorted flavors

Heat oven to 375 F. Sift together flour, baking powder, and salt and set aside. In large bowl beat shortening, sugar, egg white, and vanilla until well blended. Add flour mixture and milk until blended. Stir in coconut.

Roll dough into a ball, divide in half. Roll 15 one-inch balls from each half and place on ungreased baking sheet. Make thumb print depression in center of each ball to form nest. Bake 6 minutes. Remove from oven and place 4 Jelly Belly beans in center of each cookie. Return to oven and bake 5 more minutes. Transfer cookies to wire rack to cool. Makes 30 cookies.

2.1 Basics of Fractions

OBJECTIVES

1. Use a fraction to show how many parts of a whole are shaded.

2. Identify the numerator and denominator.

3. Identify proper and improper fractions.

In **Chapter 1** we discussed whole numbers. Many times, however, we find that parts of whole numbers are considered. One way to write parts of a whole is with **fractions.** Another way is with decimals, which is discussed in **Chapter 4.**

OBJECTIVE ▶ 1 Use a fraction to show how many parts of a whole are shaded. The number $\frac{1}{8}$ is a fraction that represents 1 of 8 equal parts. Read $\frac{1}{8}$ as "one eighth."

> A fraction may be used to represent *part* of a whole.

EXAMPLE 1 Identifying Fractions

Use fractions to represent the shaded portions and the unshaded portions of each figure.

(a) The figure on the left has 6 equal parts. The 1 shaded part is represented by the fraction $\frac{1}{6}$. The *un*shaded part is $\frac{5}{6}$.

$\frac{1}{6}$ shaded

$\frac{5}{6}$ unshaded

$\frac{7}{10}$ unshaded $\frac{3}{10}$ shaded

(b) The 3 shaded parts of the 10-part figure on the right are represented by the fraction $\frac{3}{10}$. The *un*shaded part is $\frac{7}{10}$.

◀ **Work Problem 1 at the Side.**

A fraction can also be used to represent more than one whole object.

EXAMPLE 2 Representing Fractions Greater Than 1

Use a fraction to represent the shaded part of each figure.

(a)

$\frac{1}{4}$

Whole object

An area equal to 5 of the $\frac{1}{4}$ parts is shaded, so $\frac{5}{4}$ is shaded.

(b)

> A fraction may be used to represent more than 1 whole.

Whole object

$\frac{1}{3}$

An area equal to 4 of the $\frac{1}{3}$ parts is shaded, so $\frac{4}{3}$ is shaded.

◀ **Work Problem 2 at the Side.**

OBJECTIVE ▶ 2 Identify the numerator and denominator. In the fraction $\frac{2}{3}$, the number 2 is the **numerator** and 3 is the **denominator.** The line (bar) between the numerator and the denominator is the *fraction bar.*

Fraction bar → $\frac{2 ← \text{Numerator}}{3 ← \text{Denominator}}$

---

1 Write fractions for the shaded portions and the unshaded portions of each figure.

(a)

(b)

VOCABULARY TIP

Fraction The word *fraction* actually comes from the Latin *fractio*, which means "to break." Mathematically speaking, the bottom number of a fraction tells us how many parts a whole is broken into.

2 Write fractions for the shaded portions of each figure.

(a)
$\frac{1}{7}$

(b)

Answers

1. (a) $\frac{3}{4}$; $\frac{1}{4}$ (b) $\frac{1}{6}$; $\frac{5}{6}$

2. (a) $\frac{8}{7}$ (b) $\frac{7}{4}$

Numerators and Denominators

The **denominator** of a fraction shows the number of equivalent parts in the whole, and the **numerator** shows how many parts are being considered.

Note

A fraction bar, —, is one of the division symbols. Because division by 0 is undefined, a fraction with a denominator of 0 is also undefined.

EXAMPLE 3   Identifying Numerators and Denominators

Identify the numerator and denominator in each fraction.

(a) $\dfrac{3}{4}$ (b) $\dfrac{8}{5}$

$\dfrac{3 \leftarrow \text{Numerator}}{4 \leftarrow \text{Denominator}}$ $\dfrac{8 \leftarrow \text{Numerator}}{5 \leftarrow \text{Denominator}}$

················· Work Problem ❸ at the Side. ▶

OBJECTIVE ❸ **Identify proper and improper fractions.** Fractions can be identified as *proper* or *improper* fractions.

Proper and Improper Fractions

If the numerator of a fraction is *less* than the denominator, the fraction is a **proper fraction.** A proper fraction is less than 1 whole.

If the numerator is *greater than or equal to* the denominator, the fraction is an **improper fraction.** An improper fraction is greater than or equal to 1 whole.

Proper Fractions			**Improper Fractions**		
$\dfrac{5}{8}$	$\dfrac{3}{5}$	$\dfrac{23}{24}$	$\dfrac{6}{5}$	$\dfrac{10}{10}$	$\dfrac{115}{112}$

EXAMPLE 4   Classifying Types of Fractions

(a) Identify all proper fractions in this list.

$$\dfrac{3}{4} \quad \dfrac{5}{9} \quad \dfrac{17}{5} \quad \dfrac{9}{7} \quad \dfrac{3}{3} \quad \dfrac{12}{25} \quad \dfrac{1}{9} \quad \dfrac{5}{3}$$

Proper fractions have a numerator that is less than the denominator. The proper fractions are shown below.

$\dfrac{3}{4} \leftarrow$ 3 is less than 4. $\dfrac{5}{9} \quad \dfrac{12}{25} \quad \dfrac{1}{9}$ — A proper fraction is less than 1.

(b) Identify all improper fractions in the list in part (a).

Improper fractions have a numerator that is equal to or greater than the denominator. The improper fractions are shown below.

$\dfrac{17}{5} \leftarrow$ 17 is greater than 5. $\dfrac{9}{7} \quad \dfrac{3}{3} \quad \dfrac{5}{3}$ — An improper fraction is equal to or greater than 1.

················· Work Problem ❹ at the Side. ▶

---

❸ Identify the numerator and the denominator. Draw a picture with shaded parts to show each fraction. Drawings may vary, but should have the correct number of parts.

(a) $\dfrac{2}{3}$ (b) $\dfrac{1}{4}$

(c) $\dfrac{8}{5}$ (d) $\dfrac{5}{2}$

❹ From the following group of fractions:

$$\dfrac{2}{3} \quad \dfrac{4}{3} \quad \dfrac{3}{4} \quad \dfrac{8}{8} \quad \dfrac{3}{1} \quad \dfrac{1}{3}$$

GS (a) list all proper fractions;

$$\underline{\quad\quad} \quad \underline{\quad\quad} \quad \dfrac{1}{\underline{\quad}}$$
$$3 \quad 4$$

GS (b) list all improper fractions.

$$\dfrac{4}{\underline{\quad}} \quad \underline{\quad\quad} \quad \dfrac{3}{\underline{\quad}}$$
$$8$$

Answers

3. (a) N: 2; D: 3 (b) N: 1; D: 4

(c) N: 8; D: 5

(d) N: 5; D: 2

4. (a) $\dfrac{2}{3}$; $\dfrac{3}{4}$; $\dfrac{1}{3}$ (b) $\dfrac{4}{3}$; $\dfrac{8}{8}$; $\dfrac{3}{1}$

2.1 Exercises

FOR EXTRA HELP

 MyMathLab®

Download the MyDashBoard App

CONCEPT CHECK *Identify the numerator and denominator.*

Numerator   Denominator   Numerator   Denominator

1. $\frac{4}{5}$ _____   _____   **2.** $\frac{5}{6}$ _____   _____

3. $\frac{9}{8}$ _____   _____   **4.** $\frac{7}{5}$ _____   _____

CONCEPT CHECK *Fill in the blanks to complete each sentence.*

5. The fraction $\frac{3}{8}$ represents _____ of the _____ equal parts into which a whole is divided.

6. The fraction $\frac{7}{16}$ represents _____ of the _____ equal parts into which a whole is divided.

7. The fraction $\frac{5}{24}$ represents _____ of the _____ equal parts into which a whole is divided.

8. The fraction $\frac{24}{32}$ represents _____ of the _____ equal parts into which a whole is divided.

Write fractions to represent the shaded and unshaded portions of each figure.
See **Examples 1 and 2.**

9.

10.

11.

12.

13.

14.

15. What fraction of these 6 bills has a life span of 2 years or greater? What fraction has a life of 4 years or less? What fraction has a life of 9 years?

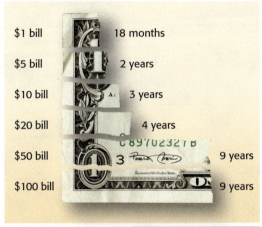

A Bill's Life

A $1 bill lasts about 18 months as compared with the average lifespan of other denominations:

$1 bill	18 months
$5 bill	2 years
$10 bill	3 years
$20 bill	4 years
$50 bill	9 years
$100 bill	9 years

Source: Federal Reserve System; Bureau of Engraving and Printing.

16. What fraction of the 9 coins shown are pennies? What fraction are nickels? What fraction of the coins are dimes?

17. In an American Sign Language (ASL) class of 25 students, 8 are hearing impaired. What fraction of the students are hearing impaired?

$$\frac{8}{\rule{1cm}{0.4pt}}$$ ← hearing impaired students (numerator)
← total students (denominator)

18. A supermarket has 215 shopping carts. If 76 of the shopping carts are in the parking lot and the rest are in the store, what fraction are in the store?

215 total shopping carts
− 76 in parking lot
139 shopping carts in store

Fraction of carts in store: $\dfrac{139}{\rule{1cm}{0.4pt}}$

19. There are 520 rooms in a hotel. If 217 of the rooms are reserved for nonsmokers, what fraction of the rooms are for smokers?

20. Of the 46 employees at the college bookstore, 15 are full-time while the rest are part-time student help. What fraction of the bookstore employees work part-time?

List the proper and improper fractions in each group. **See Example 4.**

	Proper	**Improper**
21. $\dfrac{8}{5}$ $\dfrac{1}{3}$ $\dfrac{5}{8}$ $\dfrac{6}{6}$ $\dfrac{12}{2}$ $\dfrac{7}{16}$ ▶	_____	_____
22. $\dfrac{1}{3}$ $\dfrac{3}{8}$ $\dfrac{16}{12}$ $\dfrac{10}{8}$ $\dfrac{6}{6}$ $\dfrac{3}{4}$	_____	_____
23. $\dfrac{3}{4}$ $\dfrac{3}{2}$ $\dfrac{5}{5}$ $\dfrac{9}{11}$ $\dfrac{7}{15}$ $\dfrac{19}{18}$ ▶	_____	_____
24. $\dfrac{12}{12}$ $\dfrac{15}{11}$ $\dfrac{13}{12}$ $\dfrac{11}{8}$ $\dfrac{17}{17}$ $\dfrac{19}{12}$	_____	_____

25. Write a fraction of your own choice. Label the parts of the fraction and write a sentence describing what each part represents. Draw a picture with shaded parts showing your fraction.

26. Give one example of a proper fraction and one example of an improper fraction. What determines whether a fraction is proper or improper? Draw pictures with shaded parts showing these fractions.

Study Skills

HOMEWORK: HOW, WHY, AND WHEN

OBJECTIVES

1. Select an appropriate strategy for homework.
2. Use textbook features effectively.

It is best for your brain if you keep up with the reading and homework in your math class. Remember that the more times you work with the information, the more dendrites you grow! So, give yourself every opportunity to read, work problems, and review your mathematics.

You have two choices for reading your math textbook. Read the short descriptions below and decide which will be best for you.

Preview before Class; Read Carefully after Class

Maddy learns best by listening to her instructor explain things. She "gets it" when she sees the instructor work problems on the board. She likes to ask questions in class and put the information in her notes. She has learned that it helps if she has *previewed* the section before the lecture, so she knows generally what to expect in class. *But after the class instruction*, when Maddy gets home, she finds that she can understand the math textbook easily. She remembers what her instructor said, and she can double-check her notes if she gets confused. So, Maddy does her **careful** reading of the section in her text **after** hearing the classroom lecture on the topic.

Read Carefully before Class

De'Lore, on the other hand, feels he learns well by reading on his own. He prefers to read the section and try working the example problems before coming to class. That way, he already knows what the instructor is going to talk about. Then, he can follow the instructor's examples more easily. It is also easier for him to take notes in class. De'Lore likes to have his questions answered right away, which he can do if he has already read the chapter section. So, De'Lore **carefully** reads the section in his text **before** he hears the classroom lecture on the topic.

Notice that there is **no one right way** to work with your textbook. You always must figure out what works best for you. Note also that both Maddy and De'Lore work with one section at a time. **The key is that you read the textbook regularly!** The rest of this activity will give you some ideas of how to make the most of your reading.

Try the following steps as you **read** your math textbook.

▶ Read slowly. Read only one section—or even part of a section—at a time.

▶ Do the sample problems in the margins **as you go.** Check them right away. The answers are at the bottom of the page.

▶ If your mind wanders, work problems on separate paper and write explanations in your own words.

▶ Make study cards as you read each section. Pay special attention to the yellow and blue boxes in the book. Make cards for new vocabulary, rules, procedures, formulas, and sample problems.

▶ **NOW,** you are ready to do your homework assignment!

Why Are These Reading Techniques Brain Friendly?

The steps at the left encourage you to be **actively working with the material** in your text. Your brain grows dendrites when it is doing something.

These methods require you to **try several different techniques,** not just the same thing over and over. Your brain loves variety!

Also, the techniques allow you to **take small breaks** in your learning. Those rest periods are crucial for good dendrite growth.

Study Skills

Continued from page 119

Why Are These Homework Suggestions Brain Friendly?

Your brain will grow dendrites as you study the worked examples in the text and **try doing them yourself** on separate paper. So, when you see similar problems in the homework, you will already have dendrites to work from.

Giving yourself a practice test by trying to remember the steps (without looking at your card) is an excellent way to reinforce what you are learning.

Correcting errors right away is how you learn and reinforce the correct procedures. It is hard to unlearn a mistake, so always check to see that you are on the right track.

Now Try This

Which steps for reading this book will be most helpful for you?

1 _____

2 _____

3 _____

Homework

Instructors assign homework so you can grow your own dendrites (learn the material) and then coat the dendrites with myelin through practice (remember the material). Really! In learning, you get good at what you practice. So, completing homework every day will strengthen your neural network and prepare you for exams.

If you have read each section in your textbook according to the steps above, you will probably encounter few difficulties with the exercises in the homework. Here are some additional suggestions that will help you succeed with the homework.

▶ If you **have trouble with a problem,** find a similar worked example in the section. Pay attention to *every line* of the worked example to see how to get from step to step. Work it yourself too, on separate paper; don't just look at it.

▶ If it is **hard to remember the steps** to follow for certain procedures, write the steps on a separate card. Then write a short explanation of each step. Keep the card nearby while you do the exercises, but try *not* to look at it.

▶ If you **aren't sure you are working the assigned exercises correctly,** choose two or three odd-numbered problems that are a similar type and work them. Then check the answers in the Answers section of your book and see if you are doing them correctly. If you aren't, go back to the section in the text and review the examples and find out how to correct your errors. Finally, when you are sure you understand, try the assigned problems again.

▶ **Make sure you do some homework every day,** even if the math class does not meet each day!

Now Try This

What are your biggest homework concerns?
List your two main concerns and a **brain friendly solution** for each one.

1 Concern: _____

   Solution: _____

2 Concern: _____

   Solution: _____

2.2 Mixed Numbers

Suppose you had three whole trays of muffins and half of another tray. You would state this as a whole number and a fraction.

OBJECTIVE 1 Identify mixed numbers. When a whole number and a fraction are written together, the result is a **mixed number.** For example, the mixed number

$$3\frac{1}{2} \quad \text{represents} \quad 3 + \frac{1}{2}$$

or 3 wholes and $\frac{1}{2}$ of a whole. Read $3\frac{1}{2}$ as "three and one-half." As this figure shows, the mixed number $3\frac{1}{2}$ is equal to the improper fraction $\frac{7}{2}$.

Notice that $\frac{7}{2}$ is equal to $3\frac{1}{2}$.

$$\frac{2}{2} + \frac{2}{2} + \frac{2}{2} + \frac{1}{2} = \frac{7}{2}$$

Work Problem ❶ at the Side. ▶

OBJECTIVE 2 Write mixed numbers as improper fractions. Use the following steps to write $3\frac{1}{2}$ as an improper fraction without drawing a figure.

Step 1 Multiply 3 and 2.

$$3\frac{1}{2} \qquad 3 \cdot 2 = 6$$

Step 2 Add 1 to the product.

$$3\frac{1}{2} \qquad 6 + 1 = 7$$

Step 3 Use 7, from Step 2, as the numerator of the improper fraction and 2 as the denominator.

$$3\frac{1}{2} = \frac{7}{2}$$

Same denominator

In summary, use the following steps to *write a mixed number as an improper fraction.*

Writing a Mixed Number as an Improper Fraction

Step 1 *Multiply* the denominator of the fraction and the whole number.

Step 2 *Add* to this product the numerator of the fraction.

Step 3 Write the result of Step 2 as the *numerator* of the improper fraction and the original denominator as the *denominator.*

---

❶ (a) Use these diagrams to write $1\frac{2}{3}$ as an improper fraction.

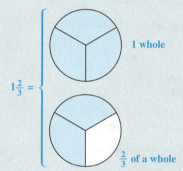

$1\frac{2}{3} =$

1 whole

$\frac{2}{3}$ of a whole

GS (b) Use these diagrams to write $2\frac{1}{4}$ as an improper fraction.

$2\frac{1}{4} =$

2 wholes

$\frac{1}{4}$ of a whole

Since each of these diagrams is divided into ____ pieces, the denominator will be ____. The number of pieces shaded is ____.

VOCABULARY TIP

Mixed number A memory tip is to think "whole number times the bottom plus the top."

Answers

1. (a) $\frac{5}{3}$ **(b)** 4; 4; 9; $\frac{9}{4}$

2 Write as improper fractions.

(GS) (a) $6\frac{1}{2}$

$6 \cdot \underline{\quad} = \underline{\quad}$

$12 + 1 = 13$

$6\frac{1}{2} = \dfrac{\underline{\quad}}{\underline{\quad}}$

(b) $7\frac{3}{4}$

(c) $4\frac{7}{8}$

(d) $8\frac{5}{6}$

3 Write as whole or mixed numbers.

(GS) (a) $\dfrac{6}{5}$

Divide ____ by ____

$\begin{array}{r} 1 \leftarrow \text{Whole number part} \\ 5\overline{)6} \\ \underline{5} \\ 1 \leftarrow \text{Remainder} \end{array}$

So $\dfrac{6}{5} = 1\dfrac{1}{\underline{\quad}}$

(b) $\dfrac{9}{4}$

(c) $\dfrac{35}{5}$

(d) $\dfrac{78}{7}$

Answers

2. (a) $2; 12; \dfrac{13}{2}$ **(b)** $\dfrac{31}{4}$ **(c)** $\dfrac{39}{8}$ **(d)** $\dfrac{53}{6}$

3. (a) $6; 5; 1\dfrac{1}{5}$ **(b)** $2\dfrac{1}{4}$ **(c)** 7 **(d)** $11\dfrac{1}{7}$

EXAMPLE 1 **Writing a Mixed Number as an Improper Fraction**

Write $7\frac{2}{3}$ as an improper fraction (numerator greater than denominator).

Step 1 $7\frac{2}{3}$ $\quad 7 \cdot 3 = 21 \quad$ Multiply 7 and 3.

Step 2 $7\frac{2}{3}$ $\quad 21 + 2 = 23 \quad$ Add 2. The numerator of the improper fractions is 23.

Step 3 $7\frac{2}{3} = \dfrac{23}{3} \quad$ Use the same denominator.

Always use the same denominator.

◀ **Work Problem 2 at the Side.**

OBJECTIVE 3 Write improper fractions as mixed numbers. Write an improper fraction as a mixed number as follows.

Writing an Improper Fraction as a Mixed Number

Write an **improper fraction** as a mixed number by dividing the numerator by the denominator. The quotient is the whole number (of the mixed number), the remainder is the numerator of the fraction part, and the denominator stays the same.

EXAMPLE 2 **Writing Improper Fractions as Mixed Numbers**

Write each improper fraction as a mixed number.

Divide numerator by denominator.

(a) $\dfrac{17}{5}$ $\quad$ Divide 17 by 5. $\longrightarrow$ $\begin{array}{r} 3 \leftarrow \textbf{Whole number part} \\ 5\overline{)17} \\ \underline{15} \\ 2 \leftarrow \text{Remainder} \end{array}$

The quotient **3** is the whole number part of the mixed number. The remainder **2** is the numerator of the fraction, and the denominator stays as **5**.

$\dfrac{17}{5} = 3\dfrac{2}{5} \leftarrow \text{Remainder}$

The denominator stays the same.

We can check this by using a diagram in which $\frac{17}{5}$ is shaded.

$\frac{1}{5}$

$\frac{5}{5} = 1$ (whole) $\quad$ $\frac{5}{5} = 1$ (whole) $\quad$ $\frac{5}{5} = 1$ (whole) $\quad$ $\frac{2}{5}$

3 wholes

(b) $\dfrac{24}{4}$ $\quad$ Divide 24 by 4. $\longrightarrow$ $\begin{array}{r} 6 \\ 4\overline{)24} \\ \underline{24} \\ 0 \leftarrow \text{No remainder} \end{array}$ $\quad$ so $\dfrac{24}{4} = 6$

No remainder, so no fraction part; just a whole number.

◀ **Work Problem 3 at the Side.**

2.2 Exercises

 FOR EXTRA HELP

 Download the MyDashBoard App

 MyMathLab®

CONCEPT CHECK *Decide whether each statement is* true *or* false. *If it is* false, *explain why.*

1. The fraction $\frac{12}{12}$ is an improper fraction.

2. The fraction $\frac{2}{3}$ is a proper fraction.

3. The mixed number $7\frac{2}{5}$ can be changed to the improper fraction $\frac{14}{5}$.

4. Some mixed numbers cannot be changed to an improper fraction.

5. The mixed number $6\frac{1}{2}$ written as an improper fraction is $\frac{12}{2}$.

6. The mixed number $5\frac{5}{6}$ written as an improper fraction is $\frac{35}{6}$.

Write each mixed number as an improper fraction. **See Example 1.**

7. $1\frac{1}{4}$

8. $2\frac{1}{2}$

9. $4\frac{3}{5}$

10. $8\frac{1}{4}$

11. $8\frac{1}{2}$

12. $1\frac{7}{11}$

13. $10\frac{1}{8}$

Find the new numerator:

$10 \cdot 8 = 80$ Multiply.

$80 + 1 = 81$ Add.

$10\frac{1}{8} =$

14. $12\frac{2}{3}$

Find the new numerator:

$12 \cdot 3 = 36$ Multiply.

$36 + 2 = 38$ Add.

$12\frac{2}{3} =$

15. $10\frac{3}{4}$

16. $3\frac{3}{8}$

17. $5\frac{4}{5}$

18. $2\frac{8}{9}$

19. $8\frac{3}{5}$

20. $3\frac{4}{7}$

21. $4\frac{10}{11}$

22. $11\frac{5}{8}$

23. $32\frac{3}{4}$

24. $15\frac{3}{10}$

25. $18\frac{5}{12}$

26. $19\frac{8}{11}$

27. $17\frac{14}{15}$

28. $9\frac{5}{16}$

29. $7\frac{19}{24}$

30. $9\frac{7}{12}$

CONCEPT CHECK *Decide whether each statement is* true *or* false. *If it is false,* explain why.

31. The improper fraction $\frac{4}{3}$ written as a mixed number is $1\frac{1}{4}$.

32. An improper fraction cannot always be written as a whole number or mixed number.

33. Some improper fractions can be written as a whole number with no fraction part.

34. The improper fraction $\frac{48}{6}$ can be written as the whole number 8.

Write each improper fraction as a whole or mixed number. **See Example 2.**

35. $\frac{4}{3}$

$$\begin{array}{r} 1 \leftarrow \text{Whole number} \\ 3\overline{)4} \quad \text{part} \\ \underline{3} \\ 1 \leftarrow \text{Remainder} \end{array}$$

36. $\frac{11}{9}$

$$\begin{array}{r} 1 \leftarrow \text{Whole number} \\ 9\overline{)11} \quad \text{part} \\ \underline{9} \\ 2 \leftarrow \text{Remainder} \end{array}$$

37. $\frac{9}{4}$

38. $\frac{7}{2}$

39. $\frac{54}{6}$

40. $\frac{63}{9}$

41. $\frac{38}{5}$

42. $\frac{33}{7}$

43. $\frac{63}{4}$

44. $\frac{19}{5}$

45. $\frac{47}{9}$

46. $\frac{65}{9}$

47. $\frac{65}{8}$

48. $\frac{37}{6}$

49. $\frac{84}{5}$

50. $\frac{92}{3}$

51. $\frac{112}{4}$

52. $\frac{117}{9}$

53. $\frac{183}{7}$

54. $\frac{212}{11}$

55. Your classmate asks you how to change a mixed number to an improper fraction. Write a couple of sentences and give an example to show her how this is done.

56. Explain in a sentence or two how to change an improper fraction to a mixed number. Give an example to show how this is done.

Write each mixed number as an improper fraction.

57. $250\frac{1}{2}$

58. $185\frac{3}{4}$

59. $333\frac{1}{3}$

60. $138\frac{4}{5}$

61. $522\frac{3}{8}$

62. $622\frac{1}{4}$

Write each improper fraction as a whole or mixed number.

63. $\dfrac{617}{4}$

64. $\dfrac{760}{8}$

65. $\dfrac{2565}{15}$

66. $\dfrac{2915}{16}$

67. $\dfrac{3917}{32}$

68. $\dfrac{5632}{64}$

Relating Concepts (Exercises 69–74) For Individual or Group Work

Knowing the basics of fractions is necessary in problem solving.
Work Exercises 69–74 in order.

69. Which of these fractions are proper fractions?

$$\frac{2}{3} \quad \frac{4}{5} \quad \frac{8}{5} \quad \frac{3}{4} \quad \frac{6}{6} \quad \frac{7}{10}$$

70. (a) The proper fractions in **Exercise 69** are the ones where the _____ is less than the _____.

(b) Draw a picture with shaded parts to show each proper fraction in **Exercise 69.**

(c) The proper fractions in **Exercise 69** are all (*less/greater*) than 1.

71. Which of these fractions are improper fractions?

$$\frac{5}{5} \quad \frac{3}{4} \quad \frac{10}{3} \quad \frac{2}{3} \quad \frac{5}{6} \quad \frac{6}{5}$$

72. (a) The improper fractions in **Exercise 71** are the ones where the _____ is equal to or greater than the _____.

(b) Draw a picture with shaded parts to show each improper fraction in **Exercise 71.**

(c) The improper fractions in **Exercise 71** are all equal to or (*less/greater*) than 1.

73. Identify which of these fractions can be written as whole or mixed numbers, and then write them as whole or mixed numbers.

$$\frac{5}{3} \quad \frac{7}{8} \quad \frac{7}{7} \quad \frac{11}{6} \quad \frac{4}{5} \quad \frac{15}{16}$$

74. (a) The fractions that can be written as whole or mixed numbers in **Exercise 73** are (*proper/improper*) fractions, and their value is always (*less than/greater than or equal to*) 1.

(b) Draw a picture with shaded parts to show each whole or mixed number in **Exercise 73.**

2.3 Factors

OBJECTIVES

1. Find factors of a number.
2. Identify prime numbers and composite numbers.
3. Find prime factorizations.

VOCABULARY TIP

Factors of a number A *factor* is a whole number that divides *exactly* into a whole number, leaving *no* remainder.

1 Find all the whole number factors of each number.

(GS) **(a)** 18

1, 2, _____ , 6, _____ , 18

(b) 16

(c) 36

(d) 80

VOCABULARY TIP

Prime numbers Because a prime number can be divided evenly only by itself and 1, it is the basic building block of all numbers.

2 Which of the following are prime?

4, 7, 9, 13, 17, 19, 29, 33

Answers

1. **(a)** 3; 9; 1, 2, 3, 6, 9, 18 **(b)** 1, 2, 4, 8, 16
   (c) 1, 2, 3, 4, 6, 9, 12, 18, 36
   (d) 1, 2, 4, 5, 8, 10, 16, 20, 40, 80
2. 7, 13, 17, 19, 29

OBJECTIVE 1 Find factors of a number. You will recall that numbers multiplied to give a product are called **factors**. Because 2 • 5 = 10, both 2 and 5 are factors of 10. The numbers 1 and 10 are also factors of 10, because 1 • 10 = 10. The various tests for divisibility show that 1, 2, 5, and 10 are the only whole number factors of 10. The products 2 • 5 and 1 • 10 are called **factorizations** of 10.

Note

The tests to decide whether one number is divisible by another number were shown in **Chapter 1.** You might want to review these. The tests that you will use most often are those for 2, 3, 5, and 10.

EXAMPLE 1   Using Factors

Find all possible two-number factorizations of each number.

(a) 12

$$1 \cdot 12 = 12 \qquad 2 \cdot 6 = 12 \qquad 3 \cdot 4 = 12$$

The factors of 12 are 1, 2, 3, 4, 6, and 12.

(b) 60

$$\begin{array}{ll} 1 \cdot 60 = 60 & 2 \cdot 30 = 60 \\ 3 \cdot 20 = 60 & 4 \cdot 15 = 60 \\ 5 \cdot 12 = 60 & 6 \cdot 10 = 60 \end{array}$$

> The factors of a number all divide evenly into that number.

The factors of 60 are 1, 2, 3, 4, 5, 6, 10, 12, 15, 20, 30, and 60.

◀ **Work Problem 1** at the Side.

OBJECTIVE 2 Identify prime numbers and composite numbers. Whole numbers that have only two factors are called **prime numbers**. They can only be divided evenly by themselves and 1.

Prime Numbers

A **prime number** is a whole number that has exactly *two different* factors, *itself* and *1.*

The number 3 is a prime number, since it can be divided evenly only by itself and 1. The number 8 is **not** a prime number (it is composite), since 8 can be divided evenly by 2 and 4, as well as by itself and 1.

CAUTION

A prime number has **only two** different factors, itself and 1. The number 1 is not a prime number because it does not have *two different* factors; the only factor of 1 is 1.

EXAMPLE 2   Finding Prime Numbers

Which of the following numbers are prime?

> A prime number can be divided evenly only by the number itself and by 1.

2   5   11   15   27

The number 15 can be divided by 3 and 5, so it is **not** prime. Also, because 27 can be divided by 3 and 9, then 27 is **not** prime. The other numbers in the list, 2, 5, and 11, are divisible only by themselves and 1, so they are prime.

◀ **Work Problem 2** at the Side.

Composite Numbers

A number that is not prime is called a **composite number.** The numbers 0 and 1 are neither prime nor composite.

VOCABULARY TIP

Composite number A number that is not a prime number is a composite number.

EXAMPLE 3 Identifying Composite Numbers

Which of the following numbers are composite?

(a) 6

Because 6 has factors of **2** and **3**, as well as 6 and 1, the number 6 is composite.

(b) 11

The number 11 has only two factors, 11 and 1. It is *not* composite. (It is a prime number.)

(c) 25

Because 25 has a factor of **5**, as well as 25 and 1, 25 is composite.

·········· Work Problem ❸ at the Side. ▶

❸ Which of these numbers are composite?

2, 4, 5, 6, 8, 10, 11, 13, 19, 21, 27, 28, 33, 36, 42

OBJECTIVE ❸ **Find prime factorizations.** For reference, here are the prime numbers less than 50.

2	3	5	7	11
13	17	19	23	29
31	37	41	43	47

These are the prime numbers less than 50.

The **prime factorization** of a number can be especially useful when we are adding or subtracting fractions and need to find a common denominator or write a fraction in lowest terms.

VOCABULARY TIP

Prime factorization Every whole number has its own unique prime factorization. The order of the factors may differ, but the factors themselves will be unique to that number.

Prime Factorization

A **prime factorization** of a number is a factorization in which every factor is a *prime number.*

❹ Find the prime factorization of each number.

GS (a) 8

$2 \cdot \underline{} \cdot \underline{}$

GS (b) 28

$2 \cdot \underline{} \cdot \underline{}$

(c) 18

EXAMPLE 4 Determining the Prime Factorization

Find the prime factorization of 12.

Try to divide 12 by the first prime, 2.

$$12 \div 2 = 6,$$

└─ First prime

so

$$12 = 2 \cdot 6.$$

Try to divide 6 by the prime, 2.

$$6 \div 2 = 3,$$

so

$$12 = 2 \cdot 2 \cdot 3.$$

└─ Factorization of 6

Because all factors are prime, the prime factorization of 12 is

$$2 \cdot 2 \cdot 3.$$ All these factors are prime numbers.

·········· Work Problem ❹ at the Side. ▶

(d) 40

Answers

3. 4, 6, 8, 10, 21, 27, 28, 33, 36, 42
4. (a) 2; 2; 2 • 2 • 2 **(b)** 2; 7; 2 • 2 • 7
 (c) 2 • 3 • 3 **(d)** 2 • 2 • 2 • 5

⑤ Find the prime factorization of each number. Write the factorization with exponents.

GS (a) 36

$$2 \cdot 2 \cdot 3 \cdot \underline{\quad}$$

$$2^2 \cdot \underline{\quad}$$

(b) 54

(c) 60

(d) 81

⑥ Write the prime factorization of each number using exponents.

(a) 48

(b) 44

(c) 90

(d) 120

(e) 180

Answers

5. **(a)** 3; 3^2; $2^2 \cdot 3^2$ **(b)** $2 \cdot 3^3$
   (c) $2^2 \cdot 3 \cdot 5$ **(d)** 3^4
6. **(a)** $2^4 \cdot 3$ **(b)** $2^2 \cdot 11$
   (c) $2 \cdot 3^2 \cdot 5$ **(d)** $2^3 \cdot 3 \cdot 5$
   (e) $2^2 \cdot 3^2 \cdot 5$

CAUTION

All prime numbers are odd numbers except the number 2. Be careful, though, because *all odd numbers are not prime numbers*. For example, 9, 15, and 21 are odd numbers, but are *not* prime numbers.

EXAMPLE 5   Factoring by Using the Division Method

Find the prime factorization of 48.

1	Continue to divide until the quotient is 1.
3)$\overline{3}$	Divide 3 by 3 (second prime).
2)$\overline{6}$	Divide 6 by 2.
2)$\overline{12}$	Divide 12 by 2.
2)$\overline{24}$	Divide 24 by 2.
2)$\overline{48}$	Divide 48 by 2. (first prime)

The divisors are all prime factors.

Because all factors (divisors) are prime, the prime factorization of 48 is

$$2 \cdot 2 \cdot 2 \cdot 2 \cdot 3.$$

In **Chapter 1,** we wrote $2 \cdot 2 \cdot 2 \cdot 2$ as 2^4, so the prime factorization of 48 can be written, using exponents, as

$$48 = 2 \cdot 2 \cdot 2 \cdot 2 \cdot 3 = 2^4 \cdot 3.$$

◀ **Work Problem ⑤ at the Side.**

Note

When using the division method of factoring, the last quotient found is 1. The "1" is never used as a prime factor because 1 is neither prime nor composite. Besides, 1 times any number is the number itself.

EXAMPLE 6   Using Exponents with Prime Factorization

Find the prime factorization of 225.

1 is not a prime factor.

1	Continue to divide until the quotient is 1.
5)$\overline{5}$	Divide 5 by 5.
5)$\overline{25}$	25 is not divisible by 3; use 5.
3)$\overline{75}$	Divide 75 by 3.
3)$\overline{225}$	225 is not divisible by 2; use 3.

All the divisors are prime factors.

Write the prime factorization.

$$225 = 3 \cdot 3 \cdot 5 \cdot 5$$

Or, using exponents,

$$225 = 3^2 \cdot 5^2$$

◀ **Work Problem ⑥ at the Side.**

Another method of factoring is a *factor tree*.

EXAMPLE 7   Factoring by Using a Factor Tree

Find the prime factorization of each number using a factor tree.

(a) 30

Try to divide by the first prime, 2. Write the factors under the 30. Circle the 2, since it is a prime.

Since 15 cannot be divided evenly by 2, try the next prime, 3.

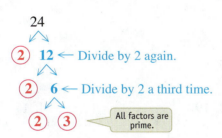

Divide until all factors are prime.

③ ⑤ ← Circle, because they are primes.

No uncircled factors remain, so the prime factorization (the circled factors) has been found.

$$30 = 2 \cdot 3 \cdot 5$$

(b) 24

Divide by 2.

24

② 12 ← Divide by 2 again.

② 6 ← Divide by 2 a third time.

② ③

All factors are prime.

$24 = 2 \cdot 2 \cdot 2 \cdot 3$ or, using exponents, $24 = 2^3 \cdot 3$

(c) 45

Because 45 cannot be divided by 2, try 3.

45

③ 15 ← Divide by 3 again.

③ ⑤

$45 = 3 \cdot 3 \cdot 5$ or, using exponents, $45 = 3^2 \cdot 5$

Note

The diagrams used in **Example 7** look like tree branches, and that is why this method is referred to as using a factor tree.

· **Work Problem 7 at the Side. ▶**

---

7 Complete each factor tree and give the prime factorization.

(a) 28

② 14

◯ ◯

(b) 35

⑤

(c) 78

Answers

7. (a)   28

② 14

② ⑦

$28 = 2 \cdot 2 \cdot 7 = 2^2 \cdot 7$

(b)   35

⑤ ⑦

$35 = 5 \cdot 7$

(c) 78

② 39

③ ⑬

$78 = 2 \cdot 3 \cdot 13$

2.3 Exercises

   MyMathLab®

CONCEPT CHECK   *Decide whether the the following are* true *or* false.

1. All the factors for the number 8 are 2, 4, 6, and 8.

2. All the factors for the number 12 are 1, 2, 3, 4, 6, and 12.

3. All the factors for the number 15 are 1, 3, 5, and 15.

4. All the factors for the number 28 are 2, 4, 7, and 14.

Find all the factors of each number. **See Example 1.**

5. 48

6. 30

7. 56   Factors are: 1, 2, 4, 7, 8, 14, 28, _____

8. 72   Factors are: 1, 2, 3, 4, 6, 8, 9, 12, 18, 24, 36, _____

9. 36

10. 20

11. 40

12. 60

13. 64

14. 84

15. 82

16. 39

CONCEPT CHECK   *Decide whether each number is* prime *or* composite.

17. 6

18. 9

19. 5

20. 16

21. 10

22. 13

23. 19

24. 17

25. 25

26. 48

27. 47

28. 45

CONCEPT CHECK   *Find the prime factorization of each number. Circle the letter of the correct answer.*

29. 40
   (a) $2 \cdot 4 \cdot 6$
   (b) $2^3 \cdot 5$
   (c) $2 \cdot 4 \cdot 5$

30. 36
   (a) $3 \cdot 12$
   (b) $2^2 \cdot 6$
   (c) $2^2 \cdot 3^2$

31. 100
   (a) $2^2 \cdot 5^2$
   (b) $5^2 \cdot 4$
   (c) $2 \cdot 5 \cdot 10$

Find the prime factorization of each number. Write answers with exponents when repeated factors appear. **See Examples 4–7.**

32. 8

33. 6

34. 20

35. 25
   $5 \cdot \underline{} = 25$

36. 56
   $\underline{} \cdot \underline{} \cdot 2 \cdot 7 = 56$

37. 68

38. 70

39. 72

40. 64

41. 44   **42.** 104   **43.** 100

44. 112   **45.** 125   **46.** 135

47. 180   **48.** 300   **49.** 320

50. 480   **51.** 360   **52.** 400

53. Give a definition in your own words of both a prime number and a composite number. Give three examples of each. Which whole numbers are neither prime nor composite?

54. With the exception of the number 2, all prime numbers are odd numbers. Nevertheless, all odd numbers are not prime numbers. Explain why these statements are true.

55. Explain the difference between finding all possible factors of 24 and finding the prime factorization of 24.

56. Use the division method to find the prime factorization of 36. Can you divide by 3s before you divide by 2s? Does the order of division change the answers?

Find the prime factorization of each number. Write answers using exponents.

57. 350   **58.** 640   **59.** 960   **60.** 1000

61. 1560   **62.** 2000   **63.** 1260   **64.** 2200

Relating Concepts (Exercises 65–70) For Individual or Group Work

An understanding of factors and factorization will be needed to solve fraction problems.
Work Exercises 65–70 in order.

65. A prime number is a whole number that has exactly two different factors, itself and 1. List all prime numbers less than 50.

66. Explain what it is about the numbers in **Exercise 65** that makes them prime.

67. The number 2 is an even number and a prime number. Can any other even numbers be prime numbers? Explain.

68. Can a multiple of a prime number be prime (for example, 6, 9, 12, and 15 are multiples of 3)? Explain.

69. Find the prime factorization of 2100. Do not use exponents in your answer.

70. Write the answer to **Exercise 69** using exponents for repeated factors.

2.4 Writing a Fraction in Lowest Terms

VOCABULARY TIP

Equivalent fractions The word *equivalent* means being of equal value. So, equivalent fractions have the same value, even though they may look different.

VOCABULARY TIP

Common factor A number also called the common divisor.

1 Decide whether the number in blue is a common factor of the other two numbers.

(a) 6, 12; **2**   (b) 32, 64; **8**

(c) 32, 56; **16**   (d) 75, 81; **1**

VOCABULARY TIP

Lowest terms The process of reducing a fraction to lowest terms is also called *simplifying* the fraction.

2 Are the following fractions in lowest terms?

(a) $\frac{4}{5}$ (b) $\frac{6}{18}$

(c) $\frac{9}{15}$ (d) $\frac{17}{46}$

Answers

1. (a) yes (b) yes (c) no (d) yes
2. (a) yes (b) no (c) no (d) yes

When working problems involving fractions, we must often compare two fractions to determine whether they represent the same portion of a whole. Look at the two cases of soda.

$\frac{3}{4}$ full $\frac{18}{24}$ full

The cases of soda show areas that are $\frac{3}{4}$ full and $\frac{18}{24}$ full. Because the full areas are equivalent (the same portion), the fractions $\frac{3}{4}$ and $\frac{18}{24}$ are **equivalent fractions.** They each represent the same portion of the whole.

$$\frac{3}{4} = \frac{18}{24}$$

Because the numbers 18 and 24 both have 6 as a factor, 6 is called a **common factor** of the numbers. Other common factors of 18 and 24 are 1, 2, and 3.

◀ Work Problem **1** at the Side.

OBJECTIVE ▶ 1️⃣ **Tell whether a fraction is written in lowest terms.** The fraction $\frac{3}{4}$ is written in *lowest terms* because the numerator and denominator have no common factor other than 1. However, the fraction $\frac{18}{24}$ is **not** in lowest terms because its numerator and denominator have common factors of 6, 3, 2, and 1.

> **Writing a Fraction in Lowest Terms**
>
> A fraction is written in **lowest terms** when the numerator and denominator have no common factor other than 1.

EXAMPLE 1   Understanding Lowest Terms

Are the following fractions in lowest terms?

(a) $\frac{3}{8}$ — No common factors other than 1.

The numerator and denominator have no common factor other than 1, so the fraction is in lowest terms.

(b) $\frac{21}{36}$ — 3 is a common factor of 21 and 36.

The numerator and denominator have a common factor of 3, so the fraction is **not in lowest terms.**

◀ Work Problem **2** at the Side.

OBJECTIVE ▶ 2️⃣ **Write a fraction in lowest terms using common factors.** There are two common methods for writing a fraction in lowest terms. These methods are shown in the next examples. The first method works best when the numerator and denominator are small numbers.

EXAMPLE 2   Writing Fractions in Lowest Terms

Write each fraction in lowest terms.

(a) $\dfrac{18}{24}$

The greatest common factor of 18 and 24 is 6. Divide both numerator and denominator by **6**.

$$\frac{18}{24} = \frac{18 \div 6}{24 \div 6} = \frac{3}{4}$$

> Divide numerator and denominator by the greatest common factor, 6.

(b) $\dfrac{30}{50} = \dfrac{30 \div 10}{50 \div 10} = \dfrac{3}{5}$ Divide both numerator and denominator by 10.

(c) $\dfrac{24}{42} = \dfrac{24 \div 6}{42 \div 6} = \dfrac{4}{7}$ Divide both numerator and denominator by 6.

(d) $\dfrac{60}{72}$

Suppose we thought that 4 was the greatest common factor of 60 and 72. Dividing by 4 would give

$$\frac{60}{72} = \frac{60 \div 4}{72 \div 4} = \frac{15}{18}. \;\leftarrow \text{Not in lowest terms}$$

But $\frac{15}{18}$ is **not** in lowest terms, because 15 and 18 have a common factor of 3. So we divide by 3.

$$\frac{15}{18} = \frac{15 \div 3}{18 \div 3} = \frac{5}{6}$$

> *Continue* dividing until there is no common factor other than 1.

The fraction $\frac{60}{72}$ could have been written in lowest terms in one step by dividing by 12, the greatest common factor of 60 and 72.

$$\frac{60}{72} = \frac{60 \div 12}{72 \div 12} = \frac{5}{6} \;\leftarrow \text{Same answer as above}$$

Continue dividing until the fraction is in lowest terms.

Note

Dividing the numerator and denominator by the same number results in an equivalent fraction.

In **Example 2,** we wrote fractions in lowest terms by dividing by a common factor. This method is summarized in the following steps.

The Method of Dividing by a Common Factor

Step 1   Find a number that will divide evenly into both the numerator and denominator. This number is a *common factor.*

Step 2   *Divide* both numerator and denominator by the common factor.

Step 3   *Check* to see whether the new fraction has any common factors (besides 1). If it does, repeat *Steps 2 and 3.* If the only common factor is 1, the fraction is in lowest terms.

········· **Work Problem ❸ at the Side.** ▶

---

❸ Write in lowest terms.

GS (a) $\dfrac{8}{16}$

$$\frac{8}{16} = \frac{8 \div 8}{16 \div \underline{\quad}} = \frac{1}{\underline{\quad}}$$

GS (b) $\dfrac{9}{12}$

$$\frac{9}{12} = \frac{9 \div \underline{\quad}}{12 \div 3} = \frac{\underline{\quad}}{4}$$

(c) $\dfrac{28}{42}$

(d) $\dfrac{30}{80}$

(e) $\dfrac{16}{40}$

Answers

3. (a) $8; 2; \dfrac{1}{2}$ **(b)** $3; 3; \dfrac{3}{4}$ **(c)** $\dfrac{2}{3}$
   (d) $\dfrac{3}{8}$ **(e)** $\dfrac{2}{5}$

OBJECTIVE **3** **Write a fraction in lowest terms using prime factors.**
The method of writing a fraction in lowest terms by division works well for fractions with small numerators and denominators. For larger numbers, when common factors are not obvious, use the method of *prime factors,* which is shown in the next example.

EXAMPLE 3   **Using Prime Factors**

Write each fraction in lowest terms.

(a) $\dfrac{24}{42}$

Write the prime factorization of both numerator and denominator. See the previous section for help.

$$\frac{24}{42} = \frac{2 \cdot 2 \cdot 2 \cdot 3}{2 \cdot 3 \cdot 7}$$

Just as with the method used in **Example 2,** divide both numerator and denominator by any common factors. Write a **1** by each factor that has been divided.

$$2 \div 2 \text{ is } 1 \qquad \frac{24}{42} = \frac{\overset{1}{2} \cdot 2 \cdot 2 \cdot \overset{1}{3}}{\underset{1}{2} \cdot \underset{1}{3} \cdot 7} \qquad 3 \div 3 \text{ is } 1$$

Multiply the remaining factors in both numerator and denominator.

$$\frac{24}{42} = \frac{1 \cdot 2 \cdot 2 \cdot 1}{1 \cdot 1 \cdot 7} = \frac{4}{7}$$

> $\frac{4}{7}$ is equivalent to $\frac{24}{42}$ but is written in lowest terms.

Finally, $\frac{24}{42}$ written in lowest terms is $\frac{4}{7}$.

(b) $\dfrac{162}{54}$

Write the prime factorization of both numerator and denominator.

$$\frac{162}{54} = \frac{2 \cdot 3 \cdot 3 \cdot 3 \cdot 3}{2 \cdot 3 \cdot 3 \cdot 3}$$

Now divide by the common factors. *Do not forget to write the 1s.*

$$\frac{162}{54} = \frac{\overset{1}{2} \cdot \overset{1}{3} \cdot \overset{1}{3} \cdot \overset{1}{3} \cdot 3}{\underset{1}{2} \cdot \underset{1}{3} \cdot \underset{1}{3} \cdot \underset{1}{3}}$$

> Remember to write in the 1s when dividing by a common factor.

$$= \frac{1 \cdot 1 \cdot 1 \cdot 1 \cdot 3}{1 \cdot 1 \cdot 1 \cdot 1} = \frac{3}{1} = 3$$

(c) $\dfrac{18}{90}$

$$\frac{18}{90} = \frac{\overset{1}{2} \cdot \overset{1}{3} \cdot \overset{1}{3}}{\underset{1}{2} \cdot \underset{1}{3} \cdot \underset{1}{3} \cdot 5} = \frac{1 \cdot 1 \cdot 1}{1 \cdot 1 \cdot 1 \cdot 5} = \frac{1}{5}$$

> All factors of the numerator were divided, and $1 \cdot 1 \cdot 1 = 1$.

Continued on Next Page

In **Example 3,** we wrote fractions in lowest terms using prime factors. This method is summarized as follows.

> **The Method of Prime Factors**
>
> **Step 1** Write the *prime factorization* of both numerator and denominator.
>
> **Step 2** Use slashes to show you are *dividing* both numerator and denominator by common factors.
>
> **Step 3** *Multiply* the remaining factors in the numerator and denominator.

·········· **Work Problem ❹ at the Side.** ▶

OBJECTIVE ▶ ❹ Determine whether two fractions are equivalent. The next example shows how to decide whether two fractions are equivalent.

EXAMPLE 4 Determining Whether Two Fractions Are Equivalent

Determine whether each pair of fractions is equivalent. In other words, do both fractions represent the same part of a whole?

(a) $\dfrac{16}{48}$ and $\dfrac{24}{72}$

Use the method of prime factors to write each fraction in lowest terms.

$$\frac{16}{48} = \frac{\cancel{2} \cdot \cancel{2} \cdot \cancel{2} \cdot \cancel{2}}{\cancel{2} \cdot \cancel{2} \cdot \cancel{2} \cdot \cancel{2} \cdot 3} = \frac{1 \cdot 1 \cdot 1 \cdot 1}{1 \cdot 1 \cdot 1 \cdot 1 \cdot 3} = \frac{1}{3}$$ ← Equivalent

$$\frac{24}{72} = \frac{\cancel{2} \cdot \cancel{2} \cdot \cancel{2} \cdot \cancel{3}}{\cancel{2} \cdot \cancel{2} \cdot \cancel{2} \cdot \cancel{3} \cdot 3} = \frac{1 \cdot 1 \cdot 1 \cdot 1}{1 \cdot 1 \cdot 1 \cdot 1 \cdot 3} = \frac{1}{3}$$ ←

$\left(\dfrac{1}{3} = \dfrac{1}{3}\right)$

(b) $\dfrac{32}{52}$ and $\dfrac{64}{112}$

$$\frac{32}{52} = \frac{\cancel{2} \cdot \cancel{2} \cdot 2 \cdot 2 \cdot 2}{\cancel{2} \cdot \cancel{2} \cdot 13} = \frac{2 \cdot 2 \cdot 2}{1 \cdot 1 \cdot 13} = \frac{8}{13}$$ ← Not equivalent

$$\frac{64}{112} = \frac{\cancel{2} \cdot \cancel{2} \cdot \cancel{2} \cdot \cancel{2} \cdot 2 \cdot 2}{\cancel{2} \cdot \cancel{2} \cdot \cancel{2} \cdot \cancel{2} \cdot 7} = \frac{1 \cdot 1 \cdot 1 \cdot 1 \cdot 2 \cdot 2}{1 \cdot 1 \cdot 1 \cdot 1 \cdot 7} = \frac{4}{7}$$ ←

$\left(\dfrac{8}{13} \neq \dfrac{4}{7}\right)$

(c) $\dfrac{75}{15}$ and $\dfrac{60}{12}$

$$\frac{75}{15} = \frac{\cancel{3} \cdot \cancel{5} \cdot 5}{\cancel{3} \cdot \cancel{5}} = \frac{1 \cdot 1 \cdot 5}{1 \cdot 1} = 5$$ ←

Equivalent (5 = 5)

$$\frac{60}{12} = \frac{\cancel{2} \cdot \cancel{2} \cdot \cancel{3} \cdot 5}{\cancel{2} \cdot \cancel{2} \cdot \cancel{3}} = \frac{1 \cdot 1 \cdot 1 \cdot 5}{1 \cdot 1 \cdot 1} = 5$$ ←

·········· **Work Problem ❺ at the Side.** ▶

---

❹ Use the method of prime factors to write each fraction in lowest terms.

GS (a) $\dfrac{12}{36}$

$$\frac{12}{36} = \frac{2 \cdot 2 \cdot 3}{2 \cdot \underline{\quad} \cdot \underline{\quad} \cdot 3}$$

$$\frac{\cancel{2} \cdot \cancel{2} \cdot \cancel{3}}{\cancel{2} \cdot \cancel{2} \cdot \cancel{3} \cdot 3} = \frac{}{3}$$

GS (b) $\dfrac{32}{56} = \dfrac{2 \cdot 2 \cdot 2 \cdot 2 \cdot \underline{\quad}}{2 \cdot 2 \cdot \underline{\quad} \cdot 7}$

$$\frac{\cancel{2} \cdot \cancel{2} \cdot \cancel{2} \cdot 2 \cdot 2}{\cancel{2} \cdot \cancel{2} \cdot \cancel{2} \cdot 7} = \frac{4}{\underline{\quad}}$$

(c) $\dfrac{74}{111}$ **(d)** $\dfrac{124}{340}$

❺ Is each pair of fractions equivalent?

(a) $\dfrac{24}{48}$ and $\dfrac{36}{72}$

(b) $\dfrac{45}{60}$ and $\dfrac{50}{75}$

(c) $\dfrac{20}{4}$ and $\dfrac{110}{22}$

(d) $\dfrac{120}{220}$ and $\dfrac{180}{320}$

Answers

4. (a) $2; 3; \dfrac{1}{3}$ **(b)** $2; 2; 7; \dfrac{4}{7}$

(c) $\dfrac{2}{3}$ **(d)** $\dfrac{31}{85}$

5. (a) equivalent **(b)** not equivalent
  (c) equivalent **(d)** not equivalent

2.4 Exercises

FOR EXTRA HELP

  MyMathLab®

Download the MyDashBoard App

CONCEPT CHECK *Underline the correct answer or fill in the blank in each of the following.*

1. A number can be divided by 2 if the number is an (*odd/even*) number.

2. A number can be divided by 5 if the number ends in _____ or _____.

3. Any number can be divided by 10 if the number ends in _____.

4. If the sum of a number's digits is divisible by _____, the number is divisible by 3.

Put a ✓ mark in the blank if the number at the left is divisible by the number at the top.
Put an ✗ in the blank if the number is not divisible by the number at the top.

	2	3	5	10			2	3	5	10
5. 60	___	___	___	___		**6.** 90	___	___	___	___
7. 48	___	___	___	___		**8.** 36	___	___	___	___
9. 160	___	___	___	___		**10.** 175	___	___	___	___
11. 138	___	___	___	___		**12.** 150	___	___	___	___

CONCEPT CHECK *Decide* true *or* false *whether each pair of fractions is equivalent.*

13. $\dfrac{6}{8} = \dfrac{3}{4}$

14. $\dfrac{7}{12} = \dfrac{1}{2}$

15. $\dfrac{3}{8} = \dfrac{5}{16}$

16. $\dfrac{4}{12} = \dfrac{1}{3}$

Write each fraction in lowest terms. **See Example 2.**

17. GS $\dfrac{15}{25}$

▶ $\dfrac{15}{25} = \dfrac{15 \div 5}{25 \div 5} =$

18. GS $\dfrac{32}{48}$

$\dfrac{32}{48} = \dfrac{32 \div 16}{48 \div 16} =$

19. $\dfrac{36}{42}$

20. $\dfrac{22}{33}$

21. $\dfrac{56}{64}$

22. $\dfrac{21}{35}$

23. $\dfrac{180}{210}$

24. $\dfrac{72}{80}$

25. ▶ $\dfrac{72}{126}$

26. $\dfrac{73}{146}$

27. $\dfrac{12}{600}$

28. $\dfrac{8}{400}$

29. $\dfrac{96}{132}$

30. $\dfrac{165}{180}$

31. $\dfrac{60}{108}$

32. $\dfrac{112}{128}$

Write the numerator and denominator of each fraction as a product of prime factors and divide by the common factors. Then write the fraction in lowest terms. **See Example 3.**

33. $\dfrac{18}{24}$

34. $\dfrac{16}{64}$

35. ▶ $\dfrac{35}{40}$

36. $\dfrac{20}{32}$

37. ▶ $\dfrac{90}{180}$

38. $\dfrac{36}{48}$

39. $\dfrac{36}{12}$

40. $\dfrac{192}{48}$

41. $\dfrac{72}{225}$

42. $\dfrac{65}{234}$

Write each fraction in lowest terms. Then state whether the fractions are equivalent or not equivalent. **See Example 4.**

43. **GS** ▶ $\dfrac{3}{6}$ and $\dfrac{18}{36}$ $\quad \dfrac{3 \div 3}{6 \div 3} = \dfrac{\rule{1cm}{0.4pt}}{\rule{1cm}{0.4pt}}$

$\quad$ and $\dfrac{18 \div \rule{0.6cm}{0.4pt}}{36 \div \rule{0.6cm}{0.4pt}} = \dfrac{\rule{0.6cm}{0.4pt}}{\rule{0.6cm}{0.4pt}}$

44. **GS** $\dfrac{3}{8}$ and $\dfrac{27}{72}$ $\quad \dfrac{3 \div \rule{0.6cm}{0.4pt}}{8 \div \rule{0.6cm}{0.4pt}} = \dfrac{\rule{0.6cm}{0.4pt}}{\rule{0.6cm}{0.4pt}}$

$\quad$ and $\dfrac{27 \div 9}{72 \div 9} = \dfrac{\rule{0.6cm}{0.4pt}}{\rule{0.6cm}{0.4pt}}$

45. $\dfrac{10}{24}$ and $\dfrac{12}{30}$

46. $\dfrac{15}{35}$ and $\dfrac{18}{40}$

47. $\dfrac{15}{24}$ and $\dfrac{35}{52}$

48. $\dfrac{21}{33}$ and $\dfrac{9}{12}$

49. $\dfrac{14}{16}$ and $\dfrac{35}{40}$

50. $\dfrac{27}{90}$ and $\dfrac{24}{80}$

51. $\dfrac{48}{6}$ and $\dfrac{72}{8}$

52. $\dfrac{33}{11}$ and $\dfrac{72}{24}$

53. $\dfrac{25}{30}$ and $\dfrac{65}{78}$

54. $\dfrac{24}{72}$ and $\dfrac{30}{90}$

55. What does it mean when a fraction is written in lowest terms? Give three examples.

56. Explain what equivalent fractions are, and give an example of a pair of equivalent fractions. Show that they are equivalent.

Write each fraction in lowest terms.

57. $\dfrac{160}{256}$

58. $\dfrac{363}{528}$

59. $\dfrac{238}{119}$

60. $\dfrac{570}{95}$

Study Skills
USING STUDY CARDS

OBJECTIVES

1. **Create study cards for all new terms.**
2. **Create study cards for new procedures.**

You may have used "flash cards" in other classes before. In math, study cards can be helpful, too. However, they are different because the main things to remember in math are *not* necessarily terms and definitions; they are *sets of steps to follow* to solve problems (and how to know which set of steps to follow) and *concepts about how math works* (principles). So, the cards will look different but will be just as useful.

In this two-part activity, you will find four types of study cards to use in math. Look carefully at what kinds of information to put on them and where to put it. Then use them the way you would any flash card:

▶ to quickly review when you have a few minutes,

▶ to do daily reviews,

▶ to review before a test.

Remember, the most helpful thing about study cards is making them. While you are making them, you have to do the kind of thinking that is most brain friendly which improves your neural network of dendrites. After each card description you will find an assignment to try. It is marked **NOW TRY THIS.**

New Vocabulary Cards

For **new vocabulary cards,** put the word (spelled correctly) and the page number where it is found on the front of the card. On the back, write:

▶ the definition (in your own words if possible),

▶ an example, an exception (if there are any),

▶ any related words, and

▶ a sample problem (if appropriate).

Why Are Study Cards Brain Friendly?

- Making cards is **active.**
- Cards are **visually** appealing.
- **Repetition** is good for your brain.

For details see Using Study Cards Revisited following **Section 2.7.**

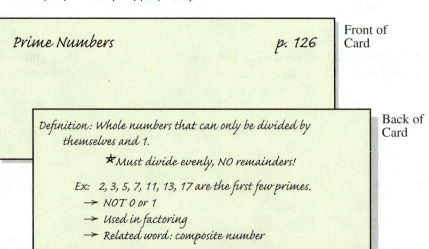

Front of Card

Prime Numbers   *p. 126*

Back of Card

Definition: Whole numbers that can only be divided by themselves and 1.
★*Must divide evenly, NO remainders!*
Ex: 2, 3, 5, 7, 11, 13, 17 are the first few primes.
→ *NOT 0 or 1*
→ *Used in factoring*
→ *Related word: composite number*

Now Try This

List four new vocabulary words/concepts you need to learn right now. Make a card for each one.

_____   _____   _____   _____

Procedure ("Steps") Cards

For **procedure cards,** write the name of the procedure at the top on the front of the card. Then write each step _in words_. If you need to know abbreviations for some words, include them along with the whole words written out. On the back, put an example of the procedure, showing each step you need to take. You can review by looking at the front and practicing a new worked example, or by looking at the back and remembering what the procedure is called and what the steps are.

Front of Card

Writing a fraction in lowest terms using prime factors

Use this method with larger denominators.

Step 1:   Write prime factorization of numerator and denominator.

Step 2:   Divide out all common factors.

Step 3:   Multiply remaining factors.

Back of Card

Example: Write this fraction in lowest terms.

$$\frac{64}{112} = \frac{2 \cdot 2 \cdot 2 \cdot 2 \cdot 2 \cdot 2}{2 \cdot 2 \cdot 2 \cdot 2 \cdot 7}$$

← Prime factors of 64.

← Prime factors of 112.

Divide out common factors of 2

$$= \frac{\cancel{2}^1 \cdot \cancel{2}^1 \cdot \cancel{2}^1 \cdot \cancel{2}^1 \cdot 2 \cdot 2}{\cancel{2}_1 \cdot \cancel{2}_1 \cdot \cancel{2}_1 \cdot \cancel{2}_1 \cdot 7} = \frac{1 \cdot 1 \cdot 1 \cdot 1 \cdot 2 \cdot 2}{1 \cdot 1 \cdot 1 \cdot 1 \cdot 7} = \frac{4}{7}$$

multiply remaining factors

lowest terms

Now Try This

What procedure are you learning right now? Make a "steps" card for it.

Procedure: _____

Summary Exercises   *Fraction Basics*

CONCEPT CHECK   *Write fractions to represent the shaded and unshaded portions of each figure.*

1.

2.

3.

CONCEPT CHECK   *Identify the numerator and denominator.*

	Numerator	Denominator		Numerator	Denominator

4. $\dfrac{3}{4}$ _____   _____

5. $\dfrac{8}{5}$ _____   _____

CONCEPT CHECK   *List the proper and improper fractions.*

Proper   Improper

6. $\dfrac{8}{2}$ $\dfrac{3}{5}$ $\dfrac{16}{7}$ $\dfrac{8}{8}$ $\dfrac{4}{25}$ $\dfrac{1}{32}$ _____   _____

The bar graph below shows the countries that have won the Boston Marathon in the Men's Wheelchair Division. Use the graph to answer Exercises 7–10. Write fractions in lowest terms.

Boston Marathon—Men's Wheelchair Division

In the past, Ernst Van Dyk of South Africa has won the Boston Marathon in the Men's Wheelchair Division 9 times, while Franz Nietispach of Switzerland and Jim Knaub of the United States have each won 5 times. The most winning countries are:

USA — 12
South Africa — 9
Switzerland — 7
Canada — 3
France — 3
Japan — 2

Source: Wikipedia.

7. What fraction of the winners were from Switzerland?

8. What fraction of the winners were not from either France or South Africa?

9. What fraction of the winners were from either Japan or the United States?

10. What fraction of the winners were not from Canada?

Write each improper fraction as a mixed number or whole number.

11. $\dfrac{5}{2}$ **12.** $\dfrac{11}{8}$ **13.** $\dfrac{9}{7}$ **14.** $\dfrac{8}{3}$

15. $\dfrac{64}{8}$ **16.** $\dfrac{45}{9}$ **17.** $\dfrac{36}{5}$ **18.** $\dfrac{47}{10}$

Write each mixed number as an improper fraction.

19. $3\dfrac{1}{3}$

20. $5\dfrac{3}{8}$

21. $6\dfrac{4}{5}$

22. $10\dfrac{3}{5}$

23. $12\dfrac{3}{4}$

24. $4\dfrac{10}{13}$

25. $11\dfrac{5}{6}$

26. $23\dfrac{5}{8}$

Write the prime factorization of each number using exponents.

27. 10

28. 55

29. 36

30. 81

31. 280

32. 360

Write each fraction in lowest terms.

33. $\dfrac{4}{12}$

34. $\dfrac{7}{14}$

35. $\dfrac{4}{16}$

36. $\dfrac{15}{25}$

37. $\dfrac{18}{24}$

38. $\dfrac{30}{36}$

39. $\dfrac{56}{64}$

40. $\dfrac{6}{300}$

41. $\dfrac{25}{200}$

42. $\dfrac{125}{225}$

43. $\dfrac{88}{154}$

44. $\dfrac{70}{126}$

Write the numerator and denominator of each fraction as a product of prime factors and divide by the common factors. Then write the fraction in lowest terms.

45. $\dfrac{24}{36}$

46. $\dfrac{80}{160}$

47. $\dfrac{126}{42}$

48. $\dfrac{96}{112}$

2.5 Multiplying Fractions

OBJECTIVES

1. Multiply fractions.
2. Use a multiplication shortcut.
3. Multiply a fraction and a whole number.
4. Find the area of a rectangle.

1 Use these figures to find $\frac{1}{4}$ of $\frac{1}{2}$.

$\frac{1}{4}$ of $\frac{1}{2}$

$\frac{1}{4}$ of $\frac{1}{2}$ is $\underline{\quad\quad}$.

OBJECTIVE **1** **Multiply fractions.** Suppose that you give $\frac{1}{2}$ of your Energy Bar to your kickboxing partner Ally. Then Ally gives $\frac{1}{2}$ of her share to Jake. How much of the Energy Bar does Jake get to eat?

Start with a sketch showing the Energy Bar cut in half (2 equal pieces).

$\frac{1}{2}$ to Ally

Next, take $\frac{1}{2}$ of the shaded area. (Here we are dividing $\frac{1}{2}$ into 2 equal parts and shading one darker than the other.)

The sketch shows that Jake gets $\frac{1}{4}$ of the Energy Bar.

$\frac{1}{2}$ of $\frac{1}{2}$ to Jake

Jake gets $\frac{1}{2}$ of $\frac{1}{2}$ of the Energy Bar. When used between two fractions, the word **of** tells us to multiply.

$$\frac{1}{2} \text{ of } \frac{1}{2} \quad \text{means} \quad \frac{1}{2} \cdot \frac{1}{2}$$

Jake's share of the Energy Bar is

$$\frac{1}{2} \cdot \frac{1}{2} = \frac{1}{4}.$$

◀ **Work Problem 1 at the Side.**

The rule for multiplying fractions follows.

> **Multiplying Fractions**
>
> Multiply two fractions by multiplying the numerators and multiplying the denominators.

Answer

1. $\frac{1}{8}$

Use this rule to find the product of $\frac{2}{3}$ and $\frac{1}{3}$, that is, to multiply $\frac{2}{3}$ by $\frac{1}{3}$.

$$\frac{2}{3} \cdot \frac{1}{3} = \frac{2 \cdot 1}{3 \cdot 3} \leftarrow \text{Multiply numerators.}$$
$$\phantom{\frac{2}{3} \cdot \frac{1}{3} = }\leftarrow \text{Multiply denominators.}$$

Finish multiplying.

$$\frac{2}{3} \cdot \frac{1}{3} = \frac{2 \cdot 1}{3 \cdot 3} = \frac{2}{9} \quad \begin{matrix} \leftarrow 2 \cdot 1 = 2 \\ \leftarrow 3 \cdot 3 = 9 \end{matrix}$$

> Multiply numerators.

> Multiply denominators.

Check that the final result is in lowest terms. $\frac{2}{9}$ is in lowest terms because 2 and 9 have no common factor other than 1.

EXAMPLE 1 Multiplying Fractions

Multiply. Write answers in lowest terms.

(a) $\dfrac{5}{8} \cdot \dfrac{3}{4}$

Multiply the numerators and multiply the denominators.

$$\frac{5}{8} \cdot \frac{3}{4} = \frac{5 \cdot 3}{8 \cdot 4} = \frac{15}{32} \quad \left[\text{Already in lowest terms.} \right]$$

Notice that 15 and 32 have no common factors other than 1, so the answer is in lowest terms.

(b) $\dfrac{4}{7} \cdot \dfrac{2}{5}$

$$\frac{4}{7} \cdot \frac{2}{5} = \frac{4 \cdot 2}{7 \cdot 5} = \frac{8}{35} \leftarrow \text{Lowest terms}$$

(c) $\dfrac{5}{8} \cdot \dfrac{3}{4} \cdot \dfrac{1}{2}$

$$\frac{5}{8} \cdot \frac{3}{4} \cdot \frac{1}{2} = \frac{5 \cdot 3 \cdot 1}{8 \cdot 4 \cdot 2} = \frac{15}{64} \leftarrow \text{Lowest terms}$$

· **Work Problem ❷ at the Side. ▶**

OBJECTIVE ▶ ❷ Use a multiplication shortcut. A **multiplication shortcut** that can be used with fractions is shown in **Example 2.**

EXAMPLE 2 Using the Multiplication Shortcut

Multiply $\frac{5}{6}$ and $\frac{9}{10}$. Write the answer in lowest terms.

$$\frac{5}{6} \cdot \frac{9}{10} = \frac{5 \cdot 9}{6 \cdot 10} = \frac{45}{60} \leftarrow \text{Not in lowest terms}$$

The numerator and denominator have a common factor other than 1, so write the prime factorization of each number.

$$\frac{5}{6} \cdot \frac{9}{10} = \frac{5 \cdot 9}{6 \cdot 10} = \frac{5 \cdot 3 \cdot 3}{2 \cdot 3 \cdot 2 \cdot 5} \quad \left[\text{Write the prime factorization of each number.} \right]$$

· **Continued on Next Page**

---

❷ Multiply. Write answers in lowest terms.

GS (a) $\dfrac{1}{2} \cdot \dfrac{3}{4}$

$$\frac{1}{2} \cdot \frac{3}{4} = \frac{3}{\underline{}} \quad \begin{matrix} \leftarrow 1 \cdot 3 \\ \leftarrow 2 \cdot 4 \end{matrix}$$

(b) $\dfrac{3}{5} \cdot \dfrac{1}{3}$

(c) $\dfrac{5}{6} \cdot \dfrac{1}{2} \cdot \dfrac{1}{8}$

(d) $\dfrac{1}{2} \cdot \dfrac{3}{4} \cdot \dfrac{3}{8}$

Answers

2. **(a)** $8; \dfrac{3}{8}$ **(b)** $\dfrac{1}{5}$ **(c)** $\dfrac{5}{96}$ **(d)** $\dfrac{9}{64}$

Next, divide by the common factors of 5 and 3.

$$\frac{5}{6} \cdot \frac{9}{10} = \frac{5 \cdot 9}{6 \cdot 10} = \frac{\overset{1}{\cancel{5}} \cdot \overset{1}{\cancel{3}} \cdot 3}{2 \cdot \underset{1}{\cancel{3}} \cdot 2 \cdot \underset{1}{\cancel{5}}}$$

Finally, multiply the remaining factors in the numerator and in the denominator.

$$\frac{5}{6} \cdot \frac{9}{10} = \frac{1 \cdot 1 \cdot 3}{2 \cdot 1 \cdot 2 \cdot 1} = \frac{3}{4} \quad \leftarrow \text{Lowest terms}$$

As a shortcut, instead of writing the prime factorization of each number, find the product of $\frac{5}{6}$ and $\frac{9}{10}$ as follows.

First, divide by 5, a common factor of both 5 and 10.

> Divide 5 by 5 to get 1.

$$\frac{\overset{1}{\cancel{5}}}{6} \cdot \frac{9}{\underset{2}{\cancel{10}}}$$

> Divide 10 by 5 to get 2.

Next, divide by 3, a common factor of both 6 and 9.

> Divide 9 by 3 to get 3.

$$\frac{\overset{1}{\cancel{5}}}{\underset{2}{\cancel{6}}} \cdot \frac{\overset{3}{\cancel{9}}}{\underset{2}{\cancel{10}}}$$

> Divide 6 by 3 to get 2.

Finally, multiply numerators and multiply denominators.

$$\frac{1 \cdot 3}{2 \cdot 2} = \frac{3}{4}$$

CAUTION

When using the multiplication shortcut, you are dividing a numerator and a denominator by a common factor. Be certain that you divide a numerator and a denominator **by the same number.** If you do all possible divisions, your answer will be in lowest terms.

EXAMPLE 3 **Using the Multiplication Shortcut**

Use the multiplication shortcut to find each product. Write the answers in lowest terms and as mixed numbers where possible.

(a) $\dfrac{6}{11} \cdot \dfrac{7}{8}$

Divide both 6 and 8 by their common factor of 2. Notice that 7 and 11 have no common factor. Then multiply.

$$\frac{\overset{3}{\cancel{6}}}{11} \cdot \frac{7}{\underset{4}{\cancel{8}}} = \frac{3 \cdot 7}{11 \cdot 4} = \frac{21}{44} \quad \leftarrow \text{Lowest terms}$$

(b) $\dfrac{7}{10} \cdot \dfrac{20}{21}$

> Divide a numerator and a denominator by the same number.

Divide 7 and 21 by 7, then divide 10 and 20 by 10.

$$\frac{\overset{1}{\cancel{7}}}{\underset{1}{\cancel{10}}} \cdot \frac{\overset{2}{\cancel{20}}}{\underset{3}{\cancel{21}}} = \frac{1 \cdot 2}{1 \cdot 3} = \frac{2}{3} \quad \leftarrow \text{Lowest terms}$$

Continued on Next Page

(c) $\dfrac{35}{12} \cdot \dfrac{32}{25}$

$$\dfrac{\overset{7}{\cancel{35}}}{\underset{3}{\cancel{12}}} \cdot \dfrac{\overset{8}{\cancel{32}}}{\underset{5}{\cancel{25}}} = \dfrac{7 \cdot 8}{3 \cdot 5} = \dfrac{56}{15} \quad \text{or} \quad 3\dfrac{11}{15} \leftarrow \text{Mixed number}$$

(d) $\dfrac{2}{3} \cdot \dfrac{8}{15} \cdot \dfrac{3}{4}$

$$\dfrac{\overset{1}{\cancel{2}}}{\underset{1}{\cancel{3}}} \cdot \dfrac{\overset{4}{\cancel{8}}}{15} \cdot \dfrac{\overset{1}{\cancel{3}}}{\underset{1}{\cancel{4}} \, \underset{2}{}} = \dfrac{1 \cdot 4 \cdot 1}{1 \cdot 15 \cdot 1} = \dfrac{4}{15} \leftarrow \text{Lowest terms}$$

This shortcut is especially helpful when the fractions involve large numbers.

> **Note**
>
> There is no specific order that must be used when dividing numerators and denominators, as long as both a numerator and a denominator are divided by the *same* number each time.

· **Work Problem ❸ at the Side.** ▶

OBJECTIVE ❸ **Multiply a fraction and a whole number.** The rule for multiplying a fraction and a whole number follows.

> **Multiplying a Whole Number and a Fraction**
>
> Multiply a whole number and a fraction by writing the whole number as a fraction with a denominator of 1.

For example, write the whole numbers 8, 10, and 25 as follows.

$$8 = \dfrac{8}{1} \qquad 10 = \dfrac{10}{1} \qquad 25 = \dfrac{25}{1} \quad \boxed{\text{Write the whole number over 1.}}$$

EXAMPLE 4 **Multiplying by Whole Numbers**

Multiply. Write answers in lowest terms and as whole numbers where possible.

(a) $8 \cdot \dfrac{3}{4}$

Write 8 as $\frac{8}{1}$ and multiply.

$$8 \cdot \dfrac{3}{4} = \dfrac{\overset{2}{\cancel{8}}}{1} \cdot \dfrac{3}{\underset{1}{\cancel{4}}} = \dfrac{2 \cdot 3}{1 \cdot 1} = \dfrac{6}{1} = 6 \quad \boxed{\dfrac{6}{1} \text{ is the same as } 6 \div 1, \text{ which equals } 6.}$$

· · · · · · · · · · · · · · · · · · · **Continued on Next Page**

❸ Use the multiplication shortcut to find each product.

GS (a) $\dfrac{3}{4} \cdot \dfrac{2}{5}$

$$\dfrac{3}{\cancel{4}} \cdot \dfrac{\overline{2}}{5} = \underline{}$$

(b) $\dfrac{6}{11} \cdot \dfrac{33}{21}$

(c) $\dfrac{20}{4} \cdot \dfrac{3}{40} \cdot \dfrac{1}{3}$

(d) $\dfrac{18}{17} \cdot \dfrac{1}{36} \cdot \dfrac{2}{3}$

Answers

3. (a) $2; 1; \dfrac{3}{\underset{2}{\cancel{4}}} \cdot \dfrac{\overset{1}{\cancel{2}}}{5} = \dfrac{3}{10}$

(b) $\dfrac{\overset{2}{\cancel{6}}}{\underset{1}{\cancel{11}}} \cdot \dfrac{\overset{3}{\cancel{33}}}{\underset{7}{\cancel{21}}} = \dfrac{6}{7}$

(c) $\dfrac{\overset{1}{\cancel{20}}}{4} \cdot \dfrac{\overset{1}{\cancel{3}}}{\underset{2}{\cancel{40}}} \cdot \dfrac{1}{\underset{1}{\cancel{3}}} = \dfrac{1}{8}$

(d) $\dfrac{\overset{1}{\cancel{18}}}{17} \cdot \dfrac{1}{\underset{2}{\cancel{36}}} \cdot \dfrac{\overset{1}{\cancel{2}}}{3} = \dfrac{1}{51}$

④ Multiply. Write answers in lowest terms and as whole numbers or mixed numbers where possible.

GS (a) $8 \cdot \dfrac{1}{8}$

$$\dfrac{\overline{8}}{\underline{\quad}} \times \dfrac{1}{\underset{1}{8}} = \dfrac{1}{1} =$$

VOCABULARY TIP

Rectangle The formula for finding the area of a rectangle works for any figure with four straight sides and four right angles.

(b) $\dfrac{3}{4} \cdot 5 \cdot \dfrac{5}{3}$

(c) $\dfrac{3}{5} \cdot 40$

(d) $\dfrac{3}{25} \cdot \dfrac{5}{11} \cdot 99$

Answers

4. **(a)** $1; 1; 1$ **(b)** $6\dfrac{1}{4}$ **(c)** 24 **(d)** $5\dfrac{2}{5}$

(b) $15 \cdot \dfrac{5}{6}$

$$15 \cdot \dfrac{5}{6} = \dfrac{\overset{5}{\cancel{15}}}{1} \cdot \dfrac{5}{\underset{2}{\cancel{6}}} = \dfrac{5 \cdot 5}{1 \cdot 2} = \dfrac{25}{2} = 12\dfrac{1}{2}$$

◀ **Work Problem ④ at the Side.**

OBJECTIVE ▶ ④ Find the area of a rectangle. To find the area of a rectangle (the amount of surface inside the rectangle), use this formula.

Area of a Rectangle

The area of a rectangle is equal to the length multiplied by the width.

$$\textbf{Area} = \textbf{length} \cdot \textbf{width}$$

For example, the rectangle shown here has an area of 12 square feet (12 ft²).

1ft 1 square foot (1 ft²)
3ft
1ft
4ft

Area = length • width — Area is the amount of surface.
Area = 4 ft • 3 ft
Area = 12 ft²

Other units for measuring area are square inches (in.²), square yards (yd²), and square miles (mi²). (See **Chapter 8** for more information on area.)

EXAMPLE 5 Applying Fraction Skills

To find the area of each rectangle, multiply its length by its width.

(a) Find the area of each rectangular shower tile.

$\dfrac{3}{4}$ ft

$\left|\leftarrow \dfrac{11}{12} \text{ ft} \rightarrow\right|$

$$\text{Area} = \text{length} \cdot \text{width}$$

$$\text{Area} = \dfrac{11}{12} \cdot \dfrac{3}{4}$$

$$= \dfrac{11}{\underset{4}{\cancel{12}}} \cdot \dfrac{\overset{1}{\cancel{3}}}{4} \qquad \begin{array}{l}\text{Divide numerator and} \\ \text{denominator by 3, so} \\ 3 \div 3 \text{ is 1, and } 12 \div 3 \text{ is 4.}\end{array}$$

$$= \dfrac{11}{16} \text{ square foot (ft²)}$$

Continued on Next Page

(b) Find the area of this rectangular SUV running board.

$\frac{7}{10}$ yd

$\frac{5}{14}$ yd

Multiply the length by the width.

$$\text{Area} = \frac{7}{10} \cdot \frac{5}{14}$$

$$= \frac{\overset{1}{\cancel{7}}}{\underset{2}{\cancel{10}}} \cdot \frac{\overset{1}{\cancel{5}}}{\underset{2}{\cancel{14}}} \qquad \text{\color{blue}Divide 7 and 14 by 7.} \\ \text{\color{blue}Divide 10 and 5 by 5.}$$

$$= \frac{1}{4} \text{ square yard (yd}^2\text{)}$$

· **Work Problem ❺ at the Side.** ▶

---

❺ Find the area of each rectangle.

(a)

$\frac{1}{3}$ yd

$\frac{3}{4}$ yd

(b) a community college campus that is $\frac{3}{8}$ mile by $\frac{1}{3}$ mile

|← $\frac{1}{3}$ mile   →|

$\frac{3}{8}$ mile

(c) a parcel of land that is $\frac{9}{7}$ mile by $\frac{7}{12}$ mile

|← $\frac{9}{7}$ mile   →|

$\frac{7}{12}$ mile

Answers

5. (a) $\frac{1}{4}$ yd^2 **(b)** $\frac{1}{8}$ mi^2 **(c)** $\frac{3}{4}$ mi^2

2.5 Exercises

     MyMathLab®

CONCEPT CHECK   *Fill in the blanks with the correct response.*

1. To multiply two or more fractions, you _____ the numerators and you multiply the _____.

2. To write a fraction answer in lowest terms, you must divide both the _____ and _____ by a common factor.

3. A shortcut when multiplying fractions is to _____ both a numerator and a _____ by the same number.

4. Using the shortcut when multiplying fractions should result in an answer that is in _____.

Multiply. Write answers in lowest terms. **See Examples 1–3.**

5. $\dfrac{1}{3} \cdot \dfrac{3}{4}$

6. $\dfrac{2}{5} \cdot \dfrac{3}{4}$

7. $\dfrac{2}{7} \cdot \dfrac{1}{5}$

8. $\dfrac{2}{3} \cdot \dfrac{1}{2}$

9. $\dfrac{8}{5} \cdot \dfrac{15}{32}$

10. $\dfrac{5}{9} \cdot \dfrac{4}{3}$

11. $\dfrac{2}{3} \cdot \dfrac{7}{12} \cdot \dfrac{9}{14}$

$$\dfrac{\overset{1}{\cancel{2}}}{\underset{1}{\cancel{3}}} \cdot \dfrac{\overset{1}{\cancel{7}}}{\underset{4}{\cancel{12}}} \cdot \dfrac{\overset{\overset{3}{\cancel{9}}}{}}{\underset{\underset{1}{}}{\cancel{14}}} =$$

12. $\dfrac{7}{8} \cdot \dfrac{16}{21} \cdot \dfrac{1}{2}$

$$\dfrac{\overset{1}{\cancel{7}}}{\underset{1}{\cancel{8}}} \cdot \dfrac{\overset{\overset{1}{\cancel{2}}}{\cancel{16}}}{\underset{3}{\cancel{21}}} \cdot \dfrac{1}{\underset{1}{\cancel{2}}} =$$

13. $\dfrac{3}{4} \cdot \dfrac{5}{6} \cdot \dfrac{2}{3}$

14. $\dfrac{5}{8} \cdot \dfrac{16}{25}$

15. $\dfrac{6}{11} \cdot \dfrac{22}{15}$

16. $\dfrac{14}{25} \cdot \dfrac{65}{48} \cdot \dfrac{15}{28}$

17. $\dfrac{35}{64} \cdot \dfrac{32}{15} \cdot \dfrac{27}{72}$

18. $\dfrac{16}{25} \cdot \dfrac{35}{32} \cdot \dfrac{15}{64}$

19. $\dfrac{39}{42} \cdot \dfrac{7}{13} \cdot \dfrac{7}{24}$

CONCEPT CHECK   *Decide whether each of the following is* true *or* false. *If false, explain why.*

20. When multiplying a fraction by a whole number, the whole number should be rewritten as the number over 1.

21. $\dfrac{4}{5} \cdot 8 = \dfrac{\overset{1}{\cancel{4}}}{5} \cdot \dfrac{1}{\underset{2}{\cancel{8}}} = \dfrac{1}{10}$

Multiply. Write answers in lowest terms and as whole or mixed numbers where possible.
See Example 4.

22. $5 \cdot \dfrac{4}{5}$

23. $20 \cdot \dfrac{3}{4}$

$$\dfrac{\overset{5}{\cancel{20}}}{1} \cdot \dfrac{3}{\underset{1}{\cancel{4}}} = \dfrac{15}{1} =$$

24. $36 \cdot \dfrac{2}{3}$

$$\dfrac{\overset{12}{\cancel{36}}}{1} \cdot \dfrac{2}{\underset{1}{\cancel{3}}} = \dfrac{24}{1} =$$

25. $36 \cdot \dfrac{5}{8} \cdot \dfrac{9}{15}$

26. $30 \cdot \dfrac{3}{10}$

27. $100 \cdot \dfrac{21}{50} \cdot \dfrac{3}{4}$

28. $400 \cdot \dfrac{7}{8}$

29. $\dfrac{2}{5} \cdot 200$

30. $\dfrac{6}{7} \cdot 245$

31. $142 \cdot \dfrac{2}{3}$

32. $\dfrac{12}{25} \cdot 430$

33. $\dfrac{28}{21} \cdot 640 \cdot \dfrac{15}{32}$

34. $\dfrac{21}{13} \cdot 520 \cdot \dfrac{7}{20}$

35. $\dfrac{54}{38} \cdot 684 \cdot \dfrac{5}{6}$

36. $\dfrac{76}{43} \cdot 473 \cdot \dfrac{5}{19}$

Find the area of each rectangle. ***See Example 5.***

37.

$\frac{3}{4}$ mile

$\frac{1}{3}$ mile

38.

$\frac{1}{4}$ ft

$\frac{7}{8}$ ft

39.

$\frac{3}{4}$ meter

12 meters

40.

$\frac{3}{8}$ in.

8 in.

41.

$\frac{3}{10}$ mi

$\frac{5}{6}$ mi

42.
$\frac{3}{8}$ mi

$\frac{7}{5}$ mi

43. Write in your own words the rule for multiplying fractions. Make up an example problem to show how this works.

44. A useful shortcut when multiplying fractions is to divide a numerator and a denominator by the same number. Describe how this works and give an example.

Find the area of each rectangle in these application problems. Write answers in lowest terms and as whole or mixed numbers where possible. ***See Example 5.***

45. Find the area of a heating-duct grill having a length of 2 yd and a width of $\frac{3}{4}$ yd.

Hint: Area = length • width.

46. Find the area of a HD television having a width of 2 yd and a height of $\frac{15}{16}$ yd.

47. A wildfire is contained in a rectangular area measuring $\frac{7}{8}$ mile by 4 miles. Find the total area of the containment.

48. A rectangular flood plain is $\frac{3}{4}$ mile wide by 7 miles long. Find the area of the flood plain.

49. The Sunny Side Soccer Park is $\frac{1}{4}$ mile long and $\frac{3}{16}$ mile wide, while the Creek Side Soccer Park is $\frac{3}{8}$ mile long and $\frac{1}{8}$ mile wide. Which park has the larger area?

50. The Rocking Horse Ranch is $\frac{3}{4}$ mile long and $\frac{2}{3}$ mile wide. The Silver Spur Ranch is $\frac{5}{8}$ mile long and $\frac{4}{5}$ mile wide. Which ranch has the larger area?

Relating Concepts (Exercises 51–56) For Individual or Group Work

Front end rounding can be used to estimate an answer when multiplying fractions.
Work Exercises 51–56 in order. *Round exact answers to the nearest whole number.*

The bar graph shows the number of supermarkets in several states. The greatest
number are in California, while the least number are in Wyoming.

SUPERMARKETS BY THE NUMBER

State	Number
California	3753
New York	2202
North Carolina	1349
Washington	826
Iowa	441
New Hampshire	163
Alaska	98
Wyoming	80

Source: The Nielsen Company.

51. Use front end rounding to estimate the total number of supermarkets in these states.

$$4000 + 2000 + 1000 + 800 + 400 + 200 + 100 + 80 =$$

_____ .

52. Find the exact total number of supermarkets in these states.

$$3753 + 2202 + 1349 + 826 + 441 + 163 + 98 + 80 =$$

_____ .

53. If $\frac{4}{5}$ of the supermarkets in New York are in medium to large population areas, use front end rounding to estimate, and then find the exact number of stores in these population areas.

Estimate:

Exact:

54. If $\frac{3}{8}$ of the supermarkets in New Hampshire are in shopping centers, use front end rounding to estimate, and then find the exact number of stores in these locations.

Estimate:

Exact:

Compare the estimated answers and the exact answers in **Exercises 53 and 54.** *How can you get an estimated answer that is closer to the exact answer? Try rounding the number of stores to some multiple of the denominator in the fraction that has two nonzero digits.*

55. Refer to **Exercise 53.** Round the number of supermarkets in New York to some multiple of the denominator in $\frac{4}{5}$ that has *two* nonzero digits. Now estimate the answer, showing your work.

56. Refer to **Exercise 54.** Round the number of supermarkets in New Hampshire to some multiple of the denominator in $\frac{3}{8}$ that has *two* nonzero digits. Now estimate the answer, showing your work.

2.6 Applications of Multiplication

OBJECTIVE

1 Solve fraction application problems using multiplication.

1 Solve each problem. Use the six problem solving steps.

GS (a) Erich and Sabrina Means are saving $\frac{3}{8}$ of their income for the down payment on their first home. If they have a combined annual income of $81,576, how much can they save in a year?

Estimate:

Round 81,576 to 82,000.

$$\frac{1}{2} \cdot \frac{82,000}{1} = \underline{\qquad}$$

Solve: to find the exact answer.

$$\frac{3}{8} \cdot \frac{\overset{10,197}{81,576}}{1} = \underline{\qquad}$$

State the answer: $30,591 can be saved in a year.

Check: Exact answer is close to estimate.

(b) A retiring firefighter will receive $\frac{5}{8}$ of her highest annual salary as retirement income. If her highest annual salary is $62,504, how much will she receive as retirement income?

OBJECTIVE ▶ 1 **Solve fraction application problems using multiplication.**
Many application problems are solved by multiplying fractions. Use the following indicator words for multiplication.

product
double
triple
times
of (when "of" follows a fraction)
twice
twice as much

> Always look for indicator words.

Look for these indicator words in the following examples.

EXAMPLE 1 | **Applying Indicator Words**

Lois Stevens gives $\frac{1}{10}$ of her income to her church. One month she earned $2980. How much did she give to the church that month?

Step 1 **Read** the problem. The problem asks us to find the amount of money given to the church.

Step 2 **Work out a plan.** Stevens gave $\frac{1}{10}$ *of* her income. The indicator word is *of*. When it follows a fraction, the word *of* indicates multiplication, so find the amount given to the church by multiplying $\frac{1}{10}$ and $2980.

Step 3 **Estimate** a reasonable answer. Round the income of $2980 to $3000. Then divide $3000 by 10 to find $\frac{1}{10}$ of the income (one of 10 equal parts). Our estimate is $3000 ÷ 10 = $300. (Recall the shortcut for dividing by 10; drop one 0 from the dividend.)

Step 4 **Solve** the problem.

> Divide by the common factor of 10.

$$\text{amount} = \frac{1}{\underset{1}{\cancel{10}}} \cdot \frac{\overset{298}{\cancel{2980}}}{1} = \frac{298}{1} = 298$$

Step 5 **State the answer.** Stevens gave $298 to her church that month.

Step 6 **Check.** The exact answer, $298, is close to our estimate of $300.

◀ **Work Problem 1 at the Side.**

EXAMPLE 2 | **Solving a Fraction Application Problem**

Of the 39 students in Sharon Martin's high school economics class, $\frac{2}{3}$ plan to go to college. How many plan to go to college?

Step 1 **Read** the problem. The problem asks us to find the number of students who plan to go to college.

Continued on Next Page

Answers

1. (a) *Estimate:* $41,000; *Exact:* $30,591
   (b) $39,065

Step 2 **Work out a plan.** Reword the problem to read

$$\frac{2}{3} \text{ of the students plan to go to college.}$$

↑ Indicator word for multiplication
when it follows a fraction

Step 3 **Estimate** a reasonable answer. Round the number of students in the class from 39 to 40. Then, $\frac{1}{2}$ of 40 is 20. Since $\frac{2}{3}$ is more than $\frac{1}{2}$, our estimate is that "more than 20 students" plan to go to college.

Step 4 **Solve** the problem. Find the number who plan to go to college by multiplying $\frac{2}{3}$ and 39.

$$\text{number who plan to go} = \frac{2}{3} \cdot 39$$

$$= \frac{2}{\overset{}{\underset{1}{3}}} \cdot \frac{\overset{13}{39}}{1} = \frac{26}{1} = 26$$

Step 5 **State the answer.** 26 students plan to go to college.

Check that the answer is reasonable.

Step 6 **Check.** The exact answer, 26, fits our estimate of "more than 20."

·········· **Work Problem ② at the Side.** ▶

EXAMPLE 3 **Finding a Fractional Part of a Fraction**

In her will, a woman divides her estate into 6 equal parts. Five of the 6 parts are given to relatives. Of the sixth part, $\frac{1}{3}$ goes to the Salvation Army. What fraction of her total estate goes to the Salvation Army?

Step 1 **Read** the problem. The problem asks for the fraction of an estate that goes to the Salvation Army.

Step 2 **Work out a plan.** Reword the problem to read
the Salvation Army gets $\frac{1}{3}$ **of** $\frac{1}{6}$.

↑ Indicator word for multiplication when it follows a fraction

Step 3 Estimate a reasonable answer. If the estate is divided into 6 equal parts and each of these parts was divided into 3 equal parts, we would have 6 • 3 = 18 equal parts. Our estimate is $\frac{1}{18}$.

Step 4 **Solve** the problem. The Salvation Army gets $\frac{1}{3}$ **of** $\frac{1}{6}$. Indicator word
To find the fraction that the Salvation Army is to receive, multiply $\frac{1}{3}$ and $\frac{1}{6}$.

$$\text{fraction to Salvation Army} = \frac{1}{3} \cdot \frac{1}{6}$$

$$= \frac{1}{18}$$

Step 5 **State the answer.** The Salvation Army gets $\frac{1}{18}$ of the total estate.

Step 6 **Check.** The exact answer, $\frac{1}{18}$, matches our estimate. Remember to check your work.

·········· **Work Problem ③ at the Side.** ▶

② At Sid's Pharmacy, $\frac{5}{16}$ of the prescriptions are paid by a third party (insurance company). If 3696 prescriptions are filled, find the number paid by a third party. Use the six problem-solving steps.

③ At our college, $\frac{1}{3}$ of the students **GS** speak a foreign language. Of those speaking a foreign language, $\frac{3}{4}$ speak Spanish. What fraction of the students speak Spanish? Use the six problem-solving steps.

$$\frac{1}{3} \cdot \frac{3}{4}$$

$$\frac{1}{\overset{}{3}} \cdot \frac{\overset{1}{3}}{\underline{}} = \underline{}$$

Answers

2. 1155 prescriptions
3. 1; 4; $\frac{1}{4}$ speak Spanish

4 Solve each problem using the six problem-solving steps. Use the circle graph in **Example 4.**

(a) What fraction of the children buy food from vending machines?

GS (b) What number of children buy food from vending machines?

Estimate: $\dfrac{1}{5}$ of 2500 is 500

Solve: $\dfrac{1}{\overset{}{\underset{1}{\cancel{5}}}} \cdot \dfrac{\overset{500}{\cancel{2500}}}{1} =$ _____

State the answer: 500 children

Check: Answer is the same as the estimate.

(c) What fraction of the children buy food from a convenience store or street vendor?

(d) What number of children buy food from a convenience store or street vendor?

Answers

4. (a) $\dfrac{1}{5}$ (b) 5; 500 children

   (c) $\dfrac{1}{10}$ (d) 250 children

| EXAMPLE 4 | Using Fractions with a Circle Graph |

The circle graph, or pie chart, shows where children 8 to 17 years of age are most likely to make food purchases when away from home. If 2500 children were in the survey, find the number of children who buy food in the school cafeteria.

FINDING FOOD

Children ages 8 to 17 are most likely to make food purchases at the following locations:

The circle represents all of the 2500 children in the survey.

Source: Pursuant Inc. for American Dietetic Association Foundation.

Step 1 **Read** the problem. The problem asks for the number of children who buy food in the school cafeteria.

Step 2 **Work out a plan.** Reword the problem to read

$\dfrac{2}{5}$ **of** 2500 children buy food in the school cafeteria.

Indicator word for multiplication when it follows a fraction

Step 3 **Estimate** a reasonable answer. $\frac{1}{2}$ of 2500 people is 1250 people. $\frac{2}{5}$ is less than $\frac{1}{2}$, so our estimate is "less than 1250 people."

Step 4 **Solve** the problem. Find the number who buy food in the school cafeteria by multiplying $\frac{2}{5}$ and 2500.

$$\text{number in school cafeteria} = \dfrac{2}{5} \cdot 2500$$

$$= \dfrac{2}{5} \cdot \dfrac{2500}{1}$$

$$= \dfrac{2}{\underset{1}{\cancel{5}}} \cdot \dfrac{\overset{500}{\cancel{2500}}}{1} \qquad \text{Divide both numerator and denominator by 5.}$$

$$= \dfrac{1000}{1} = 1000$$

Step 5 **State the answer.** 1000 children buy food in the school cafeteria.

Step 6 **Check.** The exact answer, 1000 children, fits our estimate of "less than 1250 children."

◀ **Work Problem** **4** at the Side.

2.6 Exercises

  Download the MyDashBoard App   MyMathLab®

1. **CONCEPT CHECK** Circle the words that are indicator words for multiplication.

more than	of	sum
times	twice	difference
triple	greater than	product
less than	all together	quotient
divided evenly	after a decrease	twice as much

2. **CONCEPT CHECK** Underline the correct answer. The final step when solving an application problem is to (*solve the problem/check your work*),

3. **CONCEPT CHECK** Fill in the blank. When you multiply length by width you are finding the _____ of a rectangular surface.

4. **CONCEPT CHECK** Fill in the blank. When calculating area, the length and the width must be in the same units of measurement. If the measurements are both in miles, the answer will be in _____ miles and shown as mi².

Solve each application problem. Look for indicator words. **See Examples 1–4.**

5. A digital photo frame measures $\frac{3}{4}$ ft by $\frac{2}{3}$ ft. Find its area.

   Hint: Area = length • width

6. The rectangular floor of Darby's dog house measures $\frac{14}{15}$ yd by $\frac{3}{4}$ yd. Find its area.

7. A cookie sheet is $\frac{4}{3}$ ft by $\frac{2}{3}$ ft. Find its area.

8. Each day there are 16 million people who shop at flea markets. If $\frac{2}{5}$ of these people purchase produce at the flea market, how many purchase produce? (*Source:* National Flea Market Association.)

9. Pete is helping Collin make a rectangular mahogany lamp table for Carolyn's birthday. Find the area of the top of the table if it is $\frac{4}{5}$ yd long by $\frac{3}{8}$ yd wide.

10. The average person consumes 160 bowls of cereal each year. If $\frac{3}{10}$ of the cereal is eaten in the summer months, how many bowls of cereal are eaten in the summer months? (*Source:* Target Stores.)

11. Dan Crump had expenses of $6848 during one semester of college. His part-time job provided $\frac{3}{8}$ of the amount he needed. How much did he earn on his job?

12. The average household does 400 loads of wash each year. If $\frac{3}{8}$ of the wash loads are done in the winter months, how many wash loads are done in the winter months? (*Source:* Target Stores.)

13. The city with the most expensive daily parking fee is New York City (Midtown) at $40. The daily parking fee in Boston (third most expensive) is $\frac{7}{8}$ as much as New York City. Find the daily parking fee in Boston. (*Source:* Colliers International.)

14. The daily parking fee in Boston (third most expensive) is $35. In San Francisco (fourth most expensive), the daily parking fee is $\frac{4}{5}$ the cost of Boston. How much is the daily parking fee in San Francisco? (*Source:* Colliers International.)

15. At the Garlic Festival Fun Run, $\frac{7}{12}$ of the runners are women. If there are 1560 runners, how many of the runners are
   (a) women?
   (b) men?

16. A hotel has 408 rooms. Of these rooms, $\frac{9}{17}$ are nonsmoking rooms. How many rooms are
   (a) nonsmoking?
   (b) smoking?

Most electric cars can go 100 miles before they need to be recharged, which could take up to eight hours. The circle graph below shows how long people would be willing to wait to recharge an electric car. Use this information to work Exercises 17–20.

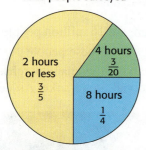

TIME TO RECHARGE

How long would you wait to recharge your electric car?

1020 people surveyed

4 hours $\frac{3}{20}$

2 hours or less $\frac{3}{5}$

8 hours $\frac{1}{4}$

Source: Deloitte Touch Tohmatsu Limited [DTTL].

17. Which response was given by the least number of people? How many people gave this response?

18. Which response was given by the greatest number of people? How many gave this response?

19. Find the fraction and the total number of people who would be willing to wait 4 hours or less to recharge the car.

20. Find the fraction and the total number of people who would be willing to wait 4 hours or more to recharge the car.

21. Without actually adding the fractions given for all the groups in the "Time to Recharge" circle graph, explain why their sum has to be 1.

22. Refer to **Exercise 21.** Suppose you added all the fractions for the groups and did not get 1 as an answer. List some possible explanations.

The table shows the earnings for the Owens family last year, and the circle graph shows how they spent their earnings. Use this information to answer Exercises 23–28.

Month	Earnings	Month	Earnings
January	$6075	July	$7040
February	$5812	August	$5232
March	$6488	September	$5670
April	$6030	October	$7012
May	$5820	November	$6465
June	$6398	December	$7958

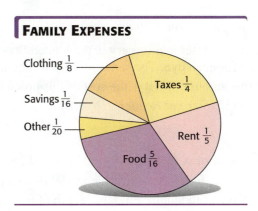

FAMILY EXPENSES

Clothing $\frac{1}{8}$

Savings $\frac{1}{16}$

Other $\frac{1}{20}$

Taxes $\frac{1}{4}$

Rent $\frac{1}{5}$

Food $\frac{5}{16}$

23. Find the Owens family's total income for the year.

24. How much of their annual earnings went to taxes?

25. Find the amount of their rent for the year. ▶

26. How much did they spend for food during the year?

27. Find their annual savings.

28. How much of their annual income was spent on clothing?

29. Here is how one student solved a multiplication problem. Find the error and solve the problem correctly.

$$\frac{9}{10} \times \frac{20}{21} = \frac{\overset{3}{\cancel{9}}}{\underset{1}{\cancel{10}}} \times \frac{\overset{2}{\cancel{20}}}{\underset{3}{\cancel{21}}} = \frac{6}{3} = 2$$

30. When two whole numbers are multiplied, the product is always larger than the numbers being multiplied. When two proper fractions are multiplied, the product is always smaller than the numbers being multiplied. Are these statements true? Why or why not?

Solve each application problem.

31. The cost of laser eye surgery in the United States is $2000 for each eye. The same surgery in Thailand is $\frac{3}{8}$ of this amount. What is the cost of this procedure in Thailand? (*Source: Reader's Digest.*)

32. A knee replacement in the United States costs $36,300, while in Mexico the same procedure costs $\frac{3}{50}$ of this amount. Find the cost of a knee replacement in Mexico. (*Source: Reader's Digest.*)

33. A collector of scale model World War II ships wants to know the length of a $\frac{1}{128}$ scale model of a ship that was 256 feet in length. Find the length of the scale model. (*Source:* Lilliput Motor Company, LTD.)

34. United Parcel Service (UPS) has trucks built to their specifications and they weigh 10,000 pounds. The company is testing a new truck that weighs $\frac{1}{10}$ less than the currently used trucks in order to save on fuel. How much does the test truck weigh?

35. LaDonna Washington is running for city council. She needs to get $\frac{2}{3}$ of her votes from senior citizens and 27,000 votes in all to win. How many votes does she need from voters other than the senior citizens?

36. The average debt of a graduate with a bachelor's degree from a four-year college is $24,400. If $\frac{5}{16}$ of the amount was borrowed in the last year of college, find the amount borrowed in the first years. (*Source:* U.S. Department of Education.)

37. A will states that $\frac{7}{8}$ of an estate is to be divided among relatives. Of the remaining estate, $\frac{1}{4}$ goes to the American Cancer Society. What fraction of the estate goes to the American Cancer Society?

38. A couple has $\frac{2}{5}$ of their total investments in real estate. Of the remaining investments, $\frac{1}{3}$ is invested in bonds. What fraction of the total investments is in bonds?

2.7 Dividing Fractions

OBJECTIVE **1** **Find the reciprocal of a fraction.** To divide fractions, we need to know how to find the **reciprocal** of a fraction.

> ### Reciprocal of a Fraction
>
> Two numbers are reciprocals of each other if their product is 1. To find the reciprocal of a fraction, switch the numerator and denominator.

For example, the reciprocal of $\frac{3}{4}$ is $\frac{4}{3}$.

$$\text{Fraction} \quad \frac{3}{4} \diagdown\!\!\!\!\times \frac{4}{3} \quad \text{Reciprocal}$$

> **Note**
>
> Notice that you invert, or "flip," a fraction to find its reciprocal.

EXAMPLE 1   **Finding Reciprocals**

Find the reciprocal of each fraction.

> Flip a fraction to find the reciprocal.

(a) The reciprocal of $\frac{1}{4}$ is $\frac{4}{1}$ because $\frac{1}{4} \cdot \frac{4}{1} = \frac{4}{4} = 1$

(b) The reciprocal of $\frac{2}{3}$ is $\frac{3}{2}$ because $\frac{2}{3} \cdot \frac{3}{2} = \frac{6}{6} = 1$

(c) The reciprocal of $\frac{3}{5}$ is $\frac{5}{3}$ because $\frac{3}{5} \cdot \frac{5}{3} = \frac{15}{15} = 1$

(d) The reciprocal of 8 is $\frac{1}{8}$ because $\frac{8}{1} \cdot \frac{1}{8} = \frac{8}{8} = 1$ Think of 8 as $\frac{8}{1}$.

·········· **Work Problem** **1** **at the Side.** ▶

> **Note**
>
> Every number has a reciprocal except 0. The number 0 has no reciprocal because there is no number that can be multiplied by 0 to get 1.
>
> $$0 \cdot (\text{reciprocal}) = 1$$
>
> ↑
> There is no number to use here that will give an answer of 1. When you multiply by 0, you always get 0.

OBJECTIVES

1 Find the reciprocal of a fraction.

2 Divide fractions.

3 Solve application problems in which fractions are divided.

VOCABULARY TIP

Reciprocal Taking the reciprocal of a number involves alternating, or switching the numerator and denominator.

1 Find the reciprocal of each fraction.

GS **(a)** $\frac{4}{5} \diagdown\!\!\!\!\times \dfrac{5}{\underline{}}$

GS **(b)** $\frac{3}{8} \diagdown\!\!\!\!\times \dfrac{\underline{}}{3}$

(c) $\frac{9}{4}$

(d) 16

Answers

1. **(a)** $4; \frac{5}{4}$ **(b)** $8; \frac{8}{3}$ **(c)** $\frac{4}{9}$ **(d)** $\frac{1}{16}$

In **Chapter 1,** we saw that the division problem $12 \div 3$ asks how many 3s are in 12. In the same way, the division problem $\frac{2}{3} \div \frac{1}{6}$ asks how many $\frac{1}{6}$s are in $\frac{2}{3}$. The figure illustrates $\frac{2}{3} \div \frac{1}{6}$.

Ask: How many $\frac{1}{6}$s are in $\frac{2}{3}$?

The figure shows that there are 4 of the $\frac{1}{6}$ pieces in $\frac{2}{3}$, or

$$\frac{2}{3} \div \frac{1}{6} = 4.$$

OBJECTIVE ▶ **2** **Divide fractions.** We will use reciprocals to divide fractions.

> ### Dividing Fractions
>
> To divide two fractions, multiply the first fraction by the reciprocal of the divisor (the second fraction).

EXAMPLE 2 **Dividing One Fraction by Another**

Divide. Write answers in lowest terms and as mixed numbers where possible.

(a) $\dfrac{7}{8} \div \dfrac{15}{16}$

The reciprocal of $\dfrac{15}{16}$ is $\dfrac{16}{15}$. | Be sure to find the reciprocal of the divisor (the second fraction).

Reciprocals

$$\frac{7}{8} \div \frac{15}{16} = \frac{7}{8} \cdot \frac{16}{15}$$

Use $\frac{16}{15}$, the reciprocal of $\frac{15}{16}$, and multiply.

Change division to multiplication.

$$= \frac{7}{\underset{1}{8}} \cdot \frac{\overset{2}{16}}{15} \qquad \text{Divide the numerator and denominator by the common factor of 8.}$$

$$= \frac{7 \cdot 2}{1 \cdot 15} \qquad \text{Multiply.}$$

$$= \frac{14}{15} \leftarrow \text{Lowest terms}$$

Continued on Next Page

(b) $\dfrac{\dfrac{4}{5}}{\dfrac{3}{10}}$

$\dfrac{\dfrac{4}{5}}{\dfrac{3}{10}} = \dfrac{4}{5} \div \dfrac{3}{10}$ Rewrite by using the $\div$ symbol for division.

$= \dfrac{4}{\overset{}{\underset{1}{5}}} \cdot \dfrac{\overset{2}{10}}{3}$ The reciprocal of $\frac{3}{10}$ is $\frac{10}{3}$. Change $\div$ to " $\bullet$ ". Divide the numerator and denominator by the common factor of 5.

$= \dfrac{4 \cdot 2}{1 \cdot 3}$ Multiply.

$= \dfrac{8}{3} = 2\dfrac{2}{3}$ Rewrite the answer as a mixed number.

CAUTION

Be certain that the divisor fraction is changed to its reciprocal *before* you divide numerators and denominators by common factors.

Work Problem ❷ at the Side. ▶

EXAMPLE 3 **Dividing with a Whole Number**

Divide. Write all answers in lowest terms and as whole or mixed numbers where possible.

(a) $5 \div \dfrac{1}{4}$ (*Hint:* How many quarters are in $5?)

Write 5 as $\frac{5}{1}$. Next, use the reciprocal of $\frac{1}{4}$, which is $\frac{4}{1}$.

$5 \div \dfrac{\mathbf{1}}{\mathbf{4}} = \dfrac{5}{1} \cdot \dfrac{\mathbf{4}}{\mathbf{1}}$ Reciprocal of $\frac{1}{4}$ is $\frac{4}{1}$.

Reciprocals

$= \dfrac{5 \cdot 4}{1 \cdot 1}$ Multiply.

$= \dfrac{20}{1} = 20$ Rewrite the answer as a whole number.

Continued on Next Page

❷ Divide. Write answers in lowest terms and as mixed numbers where possible.

GS (a) $\dfrac{1}{4} \div \dfrac{2}{3}$

$\dfrac{1}{4} \cdot \dfrac{3}{\rule{1cm}{0.4pt}} = \dfrac{\rule{1cm}{0.4pt}}{8}$

(b) $\dfrac{3}{8} \div \dfrac{5}{8}$

GS (c) $\dfrac{\dfrac{2}{3}}{\dfrac{4}{5}} = \dfrac{2}{3} \div \dfrac{4}{5}$

$= \dfrac{\overset{1}{2}}{3} \cdot \dfrac{\rule{1cm}{0.4pt}}{\underset{2}{4}}$

$= \dfrac{1 \cdot \rule{1cm}{0.4pt}}{3 \cdot 2} =$

(d) $\dfrac{\dfrac{5}{6}}{\dfrac{7}{12}}$

Answers

2. (a) $2; 3; \dfrac{3}{8}$ **(b)** $\dfrac{3}{5}$ **(c)** $5; 5; \dfrac{5}{6}$ **(d)** $1\dfrac{3}{7}$

3 Divide. Write answers in lowest terms and as whole or mixed numbers where possible.

GS (a) $10 \div \dfrac{1}{2}$

$$= \dfrac{\rule{1cm}{0.4pt}}{1} \cdot \dfrac{2}{1}$$

$$= \dfrac{10 \cdot 2}{1 \cdot 1} =$$

(b) $6 \div \dfrac{6}{7}$

(c) $\dfrac{4}{5} \div 6$

(d) $\dfrac{3}{8} \div 4$

4 Solve each problem using the six problem-solving steps.

(a) How many $\frac{5}{6}$-ounce dispensers can be filled with 40 ounces of eye drops?

(b) Find the number of $\frac{4}{5}$-quart bottles that can be filled from a 120-quart cask.

Answers

3. **(a)** 10; 20 **(b)** 7 **(c)** $\dfrac{2}{15}$ **(d)** $\dfrac{3}{32}$

4. **(a)** 48 dispensers **(b)** 150 bottles

(b) $\dfrac{2}{3} \div 6$

Write 6 as $\frac{6}{1}$. The reciprocal of $\frac{6}{1}$ is $\frac{1}{6}$.

$$\underset{\text{Reciprocals}}{\dfrac{2}{3} \div \dfrac{6}{1} = \dfrac{2}{3} \cdot \dfrac{1}{6}}$$

> Careful! Divide out common factors **after** changing to the reciprocal and to multiplication.

$$= \dfrac{\overset{1}{2}}{3} \cdot \dfrac{1}{\underset{3}{6}}$$

Divide the numerator and denominator by 2, then multiply.

$$= \dfrac{1 \cdot 1}{3 \cdot 3} = \dfrac{1}{9}$$

Lowest terms

◀ **Work Problem 3 at the Side.**

OBJECTIVE ▶ 3 **Solve application problems in which fractions are divided.** Many application problems require division of fractions. Recall from **Chapter 1** that typical indicator words for division are *goes into, per, divide, divided by, divided equally,* and *divided into.*

EXAMPLE 4 **Applying Fraction Skills**

Goldie, the manager of the Burnside Deli, must fill a 10-gallon kosher dill pickle crock with salt brine. She has only a $\frac{2}{3}$-gallon container to use. How many times must she fill the $\frac{2}{3}$-gallon container and empty it into the 10-gallon crock?

Step 1 **Read** the problem. We need to find the number of times Goldie needs to use a $\frac{2}{3}$-gallon container in order to fill a 10-gallon crock.

Step 2 **Work out a plan.** We can solve the problem by finding the number of times $\frac{2}{3}$ goes into 10.

Step 3 **Estimate** a reasonable answer. Round $\frac{2}{3}$ gallon to 1 gallon. In order to fill the 10-gallon container, she would have to use the 1-gallon container 10 times, so our estimate is 10.

Step 4 **Solve** the problem.

$$\underset{\text{Reciprocals}}{10 \div \dfrac{2}{3} = \dfrac{10}{1} \cdot \dfrac{3}{2}}$$

The reciprocal of $\frac{2}{3}$ is $\frac{3}{2}$.
Change " $\div$ " to " $\cdot$."

$$= \dfrac{\overset{5}{10}}{1} \cdot \dfrac{3}{\underset{1}{2}}$$

Divide the numerator and denominator by 2, and then multiply.

$$= \dfrac{15}{1} = 15$$

Step 5 **State the answer.** Goldie must fill the container 15 times.

Step 6 **Check.** The exact answer, 15 times, is reasonably close to our estimate of 10 times.

◀ **Work Problem 4 at the Side.**

Some content goes here.

EXAMPLE 5 **Applying Fraction Skills**

At the Happi-Time Day Care Center, $\frac{6}{7}$ of the total budget goes to classroom operation. If there are 18 classrooms and each one receives the same amount, what fraction of the operating amount does each classroom receive?

Step 1 **Read** the problem. Since $\frac{6}{7}$ of the total budget must be split into 18 parts, we must find the fraction of the classroom operating amount received by each classroom.

Step 2 **Work out a plan.** We must divide the fraction of the total budget going to classroom operation $\left(\frac{6}{7}\right)$ by the number of classrooms (18).

Step 3 **Estimate** a reasonable answer. Round $\frac{6}{7}$ to 1. If all of the operating expenses (1 whole) were divided between 18 classrooms, each classroom would receive $\frac{1}{18}$ of the operating expenses, our estimate.

Step 4 **Solve** the problem. We solve by dividing $\frac{6}{7}$ by 18.

$$\frac{6}{7} \div 18 = \frac{6}{7} \div \frac{18}{1}$$

Write 18 as $\frac{18}{1}$.

$$= \frac{\overset{1}{6}}{7} \cdot \frac{1}{\underset{3}{18}}$$

The reciprocal of $\frac{18}{1}$ is $\frac{1}{18}$. Change "÷" to "•".

Divide the numerator and denominator by 6.

$$= \frac{1}{21}$$

Multiply.

Step 5 **State the answer.** Each classroom receives $\frac{1}{21}$ of the total budget.

Step 6 **Check.** The exact answer, $\frac{1}{21}$, is close to our estimate of $\frac{1}{18}$.

··· **Work Problem 5 at the Side.** ▶

5 Solve each problem using the six problem-solving steps.

(a) The top 12 employees at Mayfield Manufacturing will divide $\frac{3}{4}$ of the annual bonus money. What fraction of the bonus money will each employee receive?

Estimate: $1 \div 12$ is $\dfrac{1}{12}$

Solve: $\dfrac{3}{4} \div 12 = \dfrac{3}{4} \div \dfrac{12}{1}$

$$= \frac{\overset{}{3}}{4} \cdot \frac{1}{\overset{}{12}}$$

——

State the answer: = ____ of the bonus money

Check: Answer is close to estimate.

(b) A winning lottery ticket was purchased by 8 employees of United States Marketing and Promotions (USMP). They will donate $\frac{1}{5}$ of the total winnings to pay the medical expenses of a fellow employee, and then divide the remaining winnings evenly. What fraction of the prize money will each receive?

Answers

5. (a) 1; 4; $\dfrac{1}{16}$ of the bonus money

(b) $\dfrac{1}{10}$ of the lottery prize money

2.7 Exercises

FOR EXTRA HELP

MyMathLab®

CONCEPT CHECK *Fill in the blank.*

1. When you invert or flip a fraction, you have the _____ of the fraction.

2. To find the reciprocal of a whole number, you must first write the whole number over _____, and then invert it.

3. To divide by a fraction, you must first _____ the divisor and then change division to _____.

4. After completing a fraction division problem, it is best to write the answer in _____ terms.

CONCEPT CHECK *Find the reciprocal of each number.*

5. $\dfrac{3}{8}$

6. $\dfrac{2}{5}$

7. $\dfrac{5}{6}$

8. $\dfrac{12}{7}$

9. $\dfrac{8}{5}$

10. $\dfrac{13}{20}$

11. 4

12. 10

Divide. Write answers in lowest terms and as whole or mixed numbers where possible.
*See **Examples 2 and 3**.*

13. $\dfrac{1}{2} \div \dfrac{3}{4}$

14. $\dfrac{5}{8} \div \dfrac{7}{8}$

15. $\dfrac{7}{8} \div \dfrac{1}{3}$

16. $\dfrac{7}{8} \div \dfrac{3}{4}$

17. $\dfrac{3}{4} \div \dfrac{5}{3}$

18. $\dfrac{4}{5} \div \dfrac{9}{4}$

19. $\dfrac{7}{9} \div \dfrac{7}{36}$

20. $\dfrac{5}{8} \div \dfrac{5}{16}$

21. $\dfrac{15}{32} \div \dfrac{5}{64}$

22. $\dfrac{7}{12} \div \dfrac{14}{15}$

23. $\dfrac{\frac{13}{20}}{\frac{4}{5}}$

24. $\dfrac{\frac{9}{10}}{\frac{3}{5}}$

25. $\dfrac{\frac{5}{6}}{\frac{25}{24}}$

$\dfrac{5}{6} \div \dfrac{25}{24}$

$\dfrac{\overset{1}{\cancel{5}}}{\underset{1}{\cancel{6}}} \cdot \dfrac{\overset{4}{\cancel{24}}}{\underset{5}{\cancel{25}}} =$

26. $\dfrac{\frac{28}{15}}{\frac{21}{5}}$

$\dfrac{28}{15} \div \dfrac{21}{5}$

$\dfrac{\overset{4}{\cancel{28}}}{\underset{3}{\cancel{15}}} \cdot \dfrac{\overset{1}{\cancel{5}}}{\underset{3}{\cancel{21}}} =$

27. $12 \div \dfrac{2}{3}$

28. $7 \div \dfrac{1}{4}$

29. $\dfrac{\frac{18}{3}}{\frac{3}{4}}$

30. $\dfrac{\frac{12}{3}}{\frac{3}{4}}$

31. $\dfrac{\frac{4}{7}}{8}$

32. $\dfrac{\frac{7}{10}}{3}$

Solve each application problem by using division. ***See Examples 4 and 5.***

33. Veterinarian Jasmine Cato has $\frac{8}{9}$ quart of medication. She prescribes this medication for 4 horses in her care. If she divides the medication evenly, how much will each horse receive?

$$\frac{8}{9} \div 4 = \frac{\overset{2}{\cancel{8}}}{9} \cdot \frac{1}{\underset{1}{\cancel{4}}} =$$

34. Harold Pishke, barber, has 15 quarts of conditioning shampoo. If he wants to put this shampoo into $\frac{3}{8}$-quart containers, how many containers can be filled?

$$\frac{15}{1} \div \frac{3}{8} = \frac{\overset{5}{\cancel{15}}}{1} \cdot \frac{8}{\underset{1}{\cancel{3}}} =$$

35. Some college roommates want to make pancakes for their neighbors. They need 5 cups of flour, but have only a $\frac{1}{3}$-cup measuring cup. How many times will they need to fill their measuring cup?

36. How many $\frac{3}{8}$ pound bags of Jelly Belly jelly beans can be filled with 408 pounds of jelly beans? (*Source:* Jelly Belly Candy Company.)

37. How many $\frac{1}{8}$-ounce eye drop dispensers can be filled with 11 ounces of eye drops?

38. It is estimated that each guest at a party will eat $\frac{5}{16}$ pound of peanuts. How many guests can be served with 10 pounds of peanuts?

39. Metal fasteners used in furniture assembly weigh $\frac{5}{32}$ pound and are packaged in 25-pound cartons. Find the number of fasteners in each carton.

40. The new residential subdivision features large $\frac{3}{4}$-acre lots. How many lots are there in the 210- acre-subdivision? (Do not consider streets, sidewalks, or other improvements.)

41. Your classmate is confused on how to divide by a fraction. Write a short note telling him how this should be done.

42. If you multiply positive proper fractions, the product is less than the fractions multiplied. When you divide by a proper fraction, is the quotient less than the fractions in the problem? Prove your answer with examples.

Solve each application problem using multiplication or division.

43. The recipe for a Jelly Belly Express loafcake calls for **GS** $\frac{3}{4}$ pound of Jelly Belly jelly beans in assorted colors. If you want to make 16 cakes, how many pounds of Jelly Belly jelly beans will you need?

$$\frac{3}{4} \cdot 16 = \frac{3}{\underset{1}{4}} \cdot \frac{\overset{4}{16}}{1} =$$

44. In a recent study, it was found that one month after leaving the hospital only $\frac{7}{8}$ of the 1520 heart attack patients were still taking the life-saving drugs prescribed for them. How many of these patients were still taking their drugs? (*Source:* Dr. Michael Ho, Denver Veterans Medical Center.)

45. Broadly Plumbing finds that $\frac{3}{4}$ can of pipe joint compound is needed for each new home. How many homes can be plumbed with 156 cans of compound?

46. The Auto Technology Center at American River Community College purchases differential fluid in drums. If each car serviced needs $\frac{2}{3}$ gallon of lubricant, how many cars can be serviced with a 50-gallon drum?

47. In recordings of 186 patient visits, doctors failed to mention a new drug's side effects or how long to take the drug in $\frac{2}{3}$ of the visits.

(a) How many visits did doctors fail to discuss these issues with patients?

(b) How many times did they discuss them? (*Source:* Dr. Deijieng Tam, UCLA.)

48. Hans Bueff has completed $\frac{4}{5}$ of his kayak excursion down the Colorado River. If the trip is a total of 285 miles,

(a) how many miles has he gone?

(b) how many miles remain?

49. A dish towel manufacturer requires $\frac{3}{8}$ yard of cotton fabric for each towel. Find the number of dish towels that can be made from 912 yards of fabric.

50. A local McDonald's expects 300 applicants but only has jobs for $\frac{1}{60}$ of those who apply. How many job openings are these?

Relating Concepts (Exercises 51–56) For Individual or Group Work

Many application problems are solved using multiplication and division of fractions.
Work Exercises 51–54 in order.

51. Perhaps the most common indicator word for multiplication is the word *of* (when it follows a fraction). Circle the words in the list below that are also indicator words for multiplication.

more than	per
double	twice
times	product
less than	difference
equals	twice as much

52. Circle the words in the list below that are indicator words for division.

fewer	sum of
goes into	divide
per	quotient
equals	double
loss of	divided by

53. To divide two fractions, multiply the first fraction by the _____ of the second fraction.

54. Find the reciprocals for each number.

$$\frac{3}{4} \quad \frac{7}{8} \quad 5 \quad \frac{12}{19}$$

The size of an antique U.S.A. postage stamp is shown here. Use this to answer Exercises 55 and 56.

$\frac{15}{16}$ in.

$\frac{15}{16}$ in.

55. (a) Explain how to find the perimeter (distance around the edges) of any flat, equal-sided, 3-, 4-, 5-, or 6-sided figure using multiplication.

(b) Find the perimeter of the stamp using multiplication.

56. Find the area of the postage stamp. Explain how to find the area of any rectangle.

Study Skills
USING STUDY CARDS REVISITED

OBJECTIVES

1 Create study cards for difficult problems.

2 Create study cards of quiz problems.

This is the second part of the Study Cards activity. As you get further into a chapter, you can choose particular problems that will serve as a good test review. Here are two more types of study cards that will help you.

Tough Problems Card

When you are doing your homework and find yourself saying, "This is really hard," or "I'm having trouble with this," make a **tough problem** study card! On the front, write out the procedure to work the type of problem *in words*. If there are special notes (like what *not* to do), include them. On the back, work at least one example; make sure you label what you are doing.

Front of Card

$\underline{Warning}$: Division is NOT commutative. The order in which you write the numbers DOES matter.

Example: $1\frac{1}{3} \div 4 = \frac{1\cancel{4}}{3} \cdot \frac{1}{\cancel{4}_1} = \frac{1}{3}$ ← Very different answers!

But $4 \div 1\frac{1}{3} = \frac{1\cancel{4}}{1} \cdot \frac{3}{\cancel{4}_1} = \frac{3}{1} = 3$ ←

In an application problem, do NOT assume the numbers are given in the correct order. Use estimation to check that the answer is $\underline{reasonable}$!

Back of Card

Maite painted 4 windows using $1\frac{1}{3}$ cans of paint. How much paint did she use on each window?

Try $4 \div 1\frac{1}{3}$

↓ ↓ (round)

$4 \div 1 = 4$ cans on each window ← Estimate

$\underline{Not}$ reasonable — she only used $1\frac{1}{3}$ cans in all!

Need to find: paint on each window

↓ ↓ ↓

$1\frac{1}{3}$ cans ÷ $4 = \frac{1\cancel{4}}{3} \cdot \frac{1}{\cancel{4}_1} = \frac{1}{3}$ can

Reasonable!

Now Try This

Choose three types of difficult problems, and work them out on *study cards*. Be sure to put the words for solving the problem on one side and the worked problem on the other side.

Practice Quiz Cards

Make up a few **quiz cards** for each type of problem you learn, and use them to prepare for a test. Choose two or three problems from the different sections of the chapter. Be sure you don't just choose the easiest problems! Put the problem **with the direction words** (like *solve, simplify, estimate*) on the front, and work the problem on the back. If you like, put the page number from the text there, too. When you review, you work the problem on a separate paper, and check it by looking at the back.

Front of Card

Solve this application problem.

Tiffany's monthly income is $1275. She spends $\frac{2}{5}$ of her income on rent and utilities. How much does she pay for rent and utilities?

Back of Card

Proper fraction followed by "of" indicates multiplication.

She spends $\frac{2}{5}$ of her income

$$\frac{2}{5} \cdot 1275 = \frac{2}{\overset{1}{\cancel{5}}} \cdot \frac{\overset{255}{\cancel{1275}}}{1} = \frac{510}{1} = 510$$

Tiffany spends $510 on rent and utilities.

First, making the study cards is an **active technique** that really gets your dendrites growing. You have to make decisions about what is most important and how to put it on the card. This kind of thinking is more in depth than just memorizing, and as a result, you will understand the concepts better and remember them longer.

Second, the cards are **visually appealing** (if you write neatly and try some color). Your brain responds to pleasant visual images, and again, you will remember longer and may even be able to "picture in your mind" how your cards look. This will help you during tests.

Third, because study cards are small and portable, you can review them easily whenever you have a few minutes. Even while you're waiting for a bus or have a few minutes between classes you can take out your cards and read them to yourself. Your **brain really benefits from repetition**; each time you review your cards your dendrites are growing thicker and stronger. After a while, the information will become automatic and you will remember it for a long time.

2.8 Multiplying and Dividing Mixed Numbers

OBJECTIVES

1. Estimate the answer and multiply mixed numbers.

2. Estimate the answer and divide mixed numbers.

3. Solve application problems with mixed numbers.

Earlier in this chapter we worked with mixed numbers—a whole number and a fraction written together. Many of the fraction problems you encounter in everyday life involve mixed numbers.

OBJECTIVE ▶ 1 Estimate the answer and multiply mixed numbers. When multiplying mixed numbers, it is a good idea to estimate the answer first.

To estimate the answer, round each mixed number to the nearest whole number. If the numerator is *half* of the denominator or *more*, round up the whole number part. If the numerator is *less* than half the denominator, leave the whole number as it is.

$$1\frac{5}{8} \quad \begin{matrix} \leftarrow \text{5 is more than 4.} \\ \leftarrow \text{Half of 8 is 4.} \end{matrix} \Bigg\} \quad 1\frac{5}{8} \text{ rounds up to 2.}$$

$$3\frac{2}{5} \quad \begin{matrix} \leftarrow \text{2 is less than } 2\frac{1}{2}. \\ \leftarrow \text{Half of 5 is } 2\frac{1}{2}. \end{matrix} \Bigg\} \quad 3\frac{2}{5} \text{ rounds to 3.}$$

> Round mixed numbers to the nearest whole number when estimating.

◀ **Work Problem ① at the Side.**

After estimating the answer, multiply the mixed numbers by using the following steps.

Multiplying Mixed Numbers
Step 1 *Change* each mixed number to an improper fraction.
Step 2 *Multiply* as fractions.
Step 3 *Simplify* the answer, which means to write it in *lowest terms,* and change it to a mixed number or whole number where possible.

① Round each mixed number to the nearest whole number.

GS (a) $4\frac{2}{3}$ ← Half of 3 is $1\frac{1}{2}$

2 is (*less than/more than*) half of 3.

$4\frac{2}{3}$ rounds to ____.

GS (b) $3\frac{2}{5}$ ← Half of 5 is $2\frac{1}{2}$

2 is (*less than/more than*) half of 5.

$3\frac{2}{5}$ rounds to ____.

(c) $5\frac{3}{4}$ **(d)** $4\frac{7}{12}$

(e) $1\frac{1}{2}$ **(f)** $8\frac{4}{9}$

EXAMPLE 1 Multiplying Mixed Numbers

First estimate the answer. Then multiply to get an exact answer. Simplify your answers.

(a) $2\frac{1}{2} \cdot 3\frac{1}{5}$

Estimate the answer by rounding the mixed numbers.

$$2\frac{1}{2} \text{ rounds to 3} \quad \text{and} \quad 3\frac{1}{5} \text{ rounds to 3}$$

$$3 \cdot 3 = 9 \quad \text{Estimated answer}$$

To find the exact answer, change each mixed number to an improper fraction.

Step 1

$$2\frac{1}{2} = \frac{5}{2} \quad \text{and} \quad 3\frac{1}{5} = \frac{16}{5}$$

Answers

1. (a) more than; 5 **(b)** less than; 3 **(c)** 6
(d) 5 **(e)** 2 **(f)** 8

Continued on Next Page

Next, multiply.

$$\underbrace{2\frac{1}{2} \cdot 3\frac{1}{5} = \frac{5}{2} \cdot \frac{16}{5}}_{\text{Step 1}} = \underbrace{\frac{\overset{1}{\cancel{5}}}{\underset{1}{\cancel{2}}} \cdot \frac{\overset{8}{\cancel{16}}}{\cancel{5}} = \frac{1 \cdot 8}{1 \cdot 1}}_{\text{Step 2}} = \underbrace{\frac{8}{1} = 8}_{\text{Step 3}}$$

> **Remember:** Change to improper fractions, then divide out common factors, and finally multiply.

The estimated answer is 9 and the exact answer is 8. The exact answer is reasonable.

(b) $3\frac{5}{8} \cdot 4\frac{4}{5}$

$3\frac{5}{8}$ rounds to 4   and $4\frac{4}{5}$ rounds to 5

$4 \cdot 5 = 20$ Estimated answer

Now find the exact answer.

$$\underbrace{3\frac{5}{8} \cdot 4\frac{4}{5} = \frac{29}{8} \cdot \frac{24}{5}}_{\text{Step 1}} = \underbrace{\frac{29}{\underset{1}{\cancel{8}}} \cdot \frac{\overset{3}{\cancel{24}}}{5} = \frac{29 \cdot 3}{1 \cdot 5} = \frac{87}{5}}_{\text{Step 2}}$$

$$\underbrace{\frac{87}{5} = 17\frac{2}{5}}_{\text{Step 3}}$$

> Simplify this answer by writing it as a mixed number.

The estimate was 20, so the exact answer of $17\frac{2}{5}$ is reasonable.

(c) $1\frac{3}{5} \cdot 3\frac{1}{3}$

$1\frac{3}{5}$ rounds to 2   and $3\frac{1}{3}$ rounds to 3

$2 \cdot 3 = 6$ Estimated answer

The exact answer is shown below.

$$1\frac{3}{5} \cdot 3\frac{1}{3} = \frac{8}{\underset{1}{\cancel{5}}} \cdot \frac{\overset{2}{\cancel{10}}}{3} = \frac{8 \cdot 2}{1 \cdot 3} = \frac{16}{3} = 5\frac{1}{3}$$

The estimate was 6, so the exact answer of $5\frac{1}{3}$ is reasonable.

·················· **Work Problem ② at the Side. ▶**

---

② First estimate the answer. Then multiply to find the exact answer. Simplify your answers.

GS **(a)** $3\frac{1}{4} \cdot 6\frac{2}{3}$

____ • ____

= ____ estimate

Exact: $3\frac{1}{4} \cdot 6\frac{2}{3}$

$\frac{13}{\underset{1}{\cancel{4}}} \cdot \frac{\overset{5}{\cancel{20}}}{3} = \frac{65}{3} =$ ____

(b) $4\frac{2}{3} \cdot 2\frac{3}{4}$

____ • ____

= ____ estimate

(c) $3\frac{3}{5} \cdot 4\frac{4}{9}$

____ • ____

= ____ estimate

(d) $5\frac{1}{4} \cdot 3\frac{3}{5}$

____ • ____

= ____ estimate

Answers

2. **(a)** *Estimate:* 3; 7; 21; *Exact:* $21\frac{2}{3}$

   (b) *Estimate:* $5 \cdot 3 = 15$; *Exact:* $12\frac{5}{6}$

   (c) *Estimate:* $4 \cdot 4 = 16$; *Exact:* 16

   (d) *Estimate:* $5 \cdot 4 = 20$; *Exact:* $18\frac{9}{10}$

OBJECTIVE ▶ **2** **Estimate the answer and divide mixed numbers.** Just as you did when multiplying mixed numbers, it is also a good idea to estimate the answer when dividing mixed numbers. To divide mixed numbers, use the following steps.

Dividing Mixed Numbers

Step 1 *Change* each mixed number to an improper fraction.

Step 2 Use the *reciprocal* of the second fraction (divisor).

Step 3 *Change* division to multiplication.

Step 4 *Simplify* the answer, which means to write it in *lowest terms,* and change it to a mixed number or whole number where possible.

Note

Recall that the reciprocal of a fraction is found by interchanging the numerator and the denominator.

EXAMPLE 2 **Dividing Mixed Numbers**

First estimate the answer. Then divide to find the exact answer. Simplify your exact answers.

(a) $2\dfrac{2}{5} \div 1\dfrac{1}{2}$

First estimate the answer by rounding each mixed number to the nearest whole number.

$$2\dfrac{2}{5} \quad \div \quad 1\dfrac{1}{2}$$

$$\downarrow \quad \text{Rounded} \quad \downarrow$$

$$2 \quad \div \quad 2 = 1 \quad \text{Estimated answer}$$

To find the exact answer, first change each mixed number to an improper fraction.

$$2\dfrac{2}{5} \div 1\dfrac{1}{2} = \overbrace{\dfrac{12}{5} \div \dfrac{3}{2}}^{\text{Step 1}}$$

> Change mixed numbers to improper fractions.

Next, use the reciprocal of the second fraction and change division to multiplication.

$$\dfrac{12}{5} \div \dfrac{3}{2} = \overset{\text{Step 2}}{\dfrac{\overset{4}{\cancel{12}}}{5}} \cdot \overset{\text{Step 3}}{\dfrac{2}{\underset{1}{\cancel{3}}}} = \dfrac{4 \cdot 2}{5 \cdot 1} = \overset{\text{Step 4}}{\dfrac{8}{5}} = 1\dfrac{3}{5} \quad \text{Exact answer simplified}$$

Reciprocals

> Remember: Use the reciprocal of the *divisor* (the *second* fraction).

The estimate was 1, so the exact answer of $1\dfrac{3}{5}$ is reasonable.

Continued on Next Page

(b) $8 \div 3\frac{3}{5}$

$$8 \quad \div \quad 3\frac{3}{5}$$

Rounded

$$8 \quad \div \quad 4 = 2 \quad \text{Estimate}$$

Now find the exact answer.

Reciprocals

$$8 \div 3\frac{3}{5} = \frac{8}{1} \div \frac{18}{5} = \frac{8}{1} \cdot \frac{5}{\underset{9}{\cancel{18}}} = \frac{20}{9} = 2\frac{2}{9}$$

Write 8 as $\frac{8}{1}$.

> Divide out common factors only *after* you have changed to the reciprocal and are multiplying.

The estimate was 2, so the exact answer of $2\frac{2}{9}$ is reasonable.

(c) $4\frac{3}{8} \div 5$

$$4\frac{3}{8} \quad \div \quad 5$$

Rounded   Reciprocals

$$4 \quad \div \quad 5 = \frac{4}{1} \div \frac{5}{1} = \frac{4}{1} \cdot \frac{1}{5} = \frac{4}{5} \quad \text{Estimate}$$

The exact answer is shown below.

Reciprocals

$$4\frac{3}{8} \div 5 = \frac{35}{8} \div \frac{5}{1} = \frac{35}{8} \cdot \frac{1}{\underset{1}{\cancel{5}}} = \frac{7}{8}$$

Write 5 as $\frac{5}{1}$.

The estimate was $\frac{4}{5}$, so the exact answer of $\frac{7}{8}$ is reasonable.

························· **Work Problem ❸ at the Side.** ▶

OBJECTIVE ❸ **Solve application problems with mixed numbers.** The next two examples show how to solve application problems involving mixed numbers.

EXAMPLE 3   **Applying Multiplication Skills**

The local Habitat for Humanity chapter is looking for 11 contractors who will each donate $3\frac{1}{4}$ days of labor to a community building project. How many days of labor will be donated in all?

Step 1   **Read** the problem. The problem asks for the total days of labor donated by the 11 contractors.

Step 2   **Work out a plan.** Multiply the number of contractors (11) and the amount of labor that each donates ($3\frac{1}{4}$ days).

··················· **Continued on Next Page**

---

❸ First estimate the answer. Then divide to find the exact answer. Simplify all answers.

GS **(a)** $\quad 3\frac{1}{8} \div 6\frac{1}{4}$

$$\downarrow \qquad \downarrow$$

$$\underline{\quad 3 \quad} \div \underline{\quad 6 \quad}$$

$$= \underline{\quad\quad} \textit{ estimate}$$

Exact: $3\frac{1}{8} \div 6\frac{1}{4} =$

$$\frac{25}{8} \div \frac{25}{4} = \frac{\overset{1}{\cancel{25}}}{\underset{2}{\cancel{8}}} \cdot \frac{\overset{1}{\cancel{4}}}{\underset{1}{\cancel{25}}} =$$

(b) $\quad 10\frac{1}{3} \div 2\frac{1}{2}$

$$\downarrow \qquad\qquad \downarrow$$

$$\underline{\quad\quad} \div \underline{\quad\quad}$$

$$= \underline{\quad\quad} \textit{ estimate}$$

(c) $\quad 8 \quad \div \quad 5\frac{1}{3}$

$$\downarrow \qquad\qquad \downarrow$$

$$\underline{\quad\quad} \div \underline{\quad\quad}$$

$$= \underline{\quad\quad} \textit{ estimate}$$

(d) $\quad 13\frac{1}{2} \div 18$

$$\downarrow \qquad\qquad \downarrow$$

$$\underline{\quad\quad} \div \underline{\quad\quad}$$

$$= \underline{\quad\quad} \textit{ estimate}$$

Answers

3. (a) *Estimate:* $3 \div 6 = \frac{1}{2}$; *Exact:* $\frac{1}{2}$

    (b) *Estimate:* $10 \div 3 = 3\frac{1}{3}$; *Exact:* $4\frac{2}{15}$

    (c) *Estimate:* $8 \div 5 = 1\frac{3}{5}$; *Exact:* $1\frac{1}{2}$

    (d) *Estimate:* $14 \div 18 = \frac{7}{9}$; *Exact:* $\frac{3}{4}$

4 Use the six problem-solving
GS steps. Simplify the answer.
If one car requires $2\frac{5}{8}$ quarts
of paint, find the number of
quarts needed to paint 16 cars.

Estimate: _____ • 16 = 48

Solve: $2\frac{5}{8} \cdot 16 = \frac{21}{8} \cdot \frac{16}{1}$

$$= \frac{21}{\underset{1}{8}} \cdot \frac{\overset{2}{16}}{1}$$

State the answer: _____ quarts of
paint are needed.

Check: Answer is close to
estimate.

5 Use the six problem-solving
steps. Simplify all answers.

(a) The manufacture of one out-
board engine propeller
requires $4\frac{3}{4}$ pounds of brass.
How many propellers can
be manufactured from
57 pounds of brass?

(b) Jack Armstrong Trucking
uses $21\frac{3}{4}$ quarts of motor oil
for each oil change on his
diesel engine truck. Find the
number of oil changes that
can be made with 609 quarts
of oil.

Step 3 **Estimate** a reasonable answer. Round $3\frac{1}{4}$ days to 3 days. Multiply
3 days by 11 contractors (3 • 11) to get an estimate of 33 days.

Step 4 **Solve** the problem. Find the exact answer.

$$11 \cdot 3\frac{1}{4} = 11 \cdot \frac{13}{4}$$

$$= \frac{11}{1} \cdot \frac{13}{4} = \frac{143}{4} = 35\frac{3}{4}$$

Step 5 **State the answer.** The community building
project will receive $35\frac{3}{4}$ days of donated labor.

> Always check to see
> if the answer is close
> to the estimate.

Step 6 **Check.** The exact answer, $35\frac{3}{4}$ days, is close to our estimate of
33 days.

◀ **Work Problem 4 at the Side.**

EXAMPLE 4 **Applying Division Skills**

A dome tent for backpacking requires $7\frac{1}{4}$ yards of nylon material. How many
tents can be made from $21\frac{3}{4}$ yards of material?

Step 1 **Read** the problem. The problem asks how many tents can be made
from $21\frac{3}{4}$ yards of material.

Step 2 **Work out a plan.** Divide the number of yards of cloth ($21\frac{3}{4}$ yd) by
the number of yards needed for one tent ($7\frac{1}{4}$ yd).

Step 3 **Estimate** a reasonable answer.

$$21\frac{3}{4} \qquad \div \qquad 7\frac{1}{4}$$

↓ Rounded ↓

> Recall ≈ means
> "approximately equal to."

$$22 \qquad \div \qquad 7 \approx 3 \text{ tents} \qquad \text{Estimate}$$

Step 4 **Solve** the problem.

$$21\frac{3}{4} \div 7\frac{1}{4} = \frac{87}{4} \div \frac{29}{4}$$

$$= \frac{\overset{3}{87}}{\underset{1}{4}} \cdot \frac{\overset{1}{4}}{\underset{1}{29}} = \frac{3}{1} = 3 \qquad \text{Matches estimate}$$

Step 5 **State the answer.** 3 tents can be made from $21\frac{3}{4}$ yards of cloth.

Step 6 **Check.** The exact answer, 3, matches our estimate.

◀ **Work Problem 5 at the Side.**

Note

When rounding mixed numbers to estimate the answer to a problem, the
estimated answer usually varies somewhat from the exact answer. However, the importance of the estimated answer is that it will show you
whether your exact answer is reasonable or not.

2.8 EXERCISES

FOR EXTRA HELP

  MyMathLab®

Download the MyDashBoard App

CONCEPT CHECK *Decide whether each statement is* true *or* false. *If it is* false, *say why.*

1. When multiplying two mixed numbers, the reciprocal of the second mixed number must be used.

2. If you were dividing a mixed number by the whole number 10, the reciprocal of 10 would be $\frac{10}{1}$.

3. To round mixed numbers before estimating the answer, decide whether the numerator of the fraction part is less than or more than half of the denominator.

4. When rounding mixed numbers to estimate the answer to a problem, the estimated answer can vary quite a bit from the exact answer. However, it can still show whether the exact answer is reasonable.

First estimate the answer. Then multiply to find the exact answer. Simplify all answers.
See Example 1.

5. $4\frac{1}{2} \cdot 1\frac{3}{4}$

Estimate:

___ • ___ = ___

Exact:

6. $2\frac{1}{2} \cdot 2\frac{1}{4}$

Estimate:

___ • ___ = ___

Exact:

7. $1\frac{2}{3} \cdot 2\frac{7}{10}$

Estimate:

___ • ___ = ___

Exact:

8. $4\frac{1}{2} \cdot 2\frac{1}{4}$

Estimate:

___ • ___ = ___

Exact:

9. $3\frac{1}{9} \cdot 1\frac{2}{7}$

Estimate:

___ • ___ = ___

Exact:

10. $6\frac{1}{4} \cdot 3\frac{1}{5}$

Estimate:

___ • ___ = ___

Exact:

11. $8 \cdot 6\frac{1}{4}$

Estimate:

___ • ___ = ___

Exact:

12. $6 \cdot 2\frac{1}{3}$

Estimate:

___ • ___ = ___

Exact:

13. $4\frac{1}{2} \cdot 2\frac{1}{5} \cdot 5$

Estimate:

___ • ___ • ___ = ___

Exact:

14. $5\frac{1}{2} \cdot 1\frac{1}{3} \cdot 2\frac{1}{4}$

Estimate:

___ • ___ • ___ = ___

Exact:

15. $3 \cdot 1\frac{1}{2} \cdot 2\frac{2}{3}$

Estimate:

___ • ___ • ___ = ___

Exact:

16. $\frac{2}{3} \cdot 3\frac{2}{3} \cdot \frac{6}{11}$

Estimate:

___ • ___ • ___ = ___

Exact:

CONCEPT CHECK *Circle the letter of the correct answer.*

17. $3\frac{1}{4} \cdot 7\frac{5}{8}$

     (a) 21   **(b)** 32   **(c)** 28   **(d)** 24

18. $5\frac{3}{4} \cdot 2\frac{2}{5}$

     (a) 12   **(b)** 18   **(c)** 10   **(d)** 15

19. $2\frac{1}{8} \div 1\frac{3}{4}$

     (a) 2   **(b)** 1   **(c)** $1\frac{1}{2}$ **(d)** 3

20. $8\frac{3}{5} \div 2\frac{3}{8}$

     (a) 4   **(b)** $3\frac{1}{2}$ **(c)** $4\frac{1}{2}$ **(d)** 3

First estimate the answer. Then divide to find the exact answer. Simplify all answers.
See Example 2.

21. $1\frac{1}{4} \div 3\frac{3}{4}$

Estimate:

____ ÷ ____ = ____

Exact:

22. $1\frac{1}{8} \div 2\frac{1}{4}$

Estimate:

____ ÷ ____ = ____

Exact:

23. $2\frac{1}{2} \div 3$

Estimate:

____ ÷ ____ = ____

Exact: $\dfrac{5}{2} \div \dfrac{3}{1}$

$\dfrac{5}{2} \cdot \dfrac{1}{3} =$

24. $2\frac{3}{4} \div 2$

Estimate:

____ ÷ ____ = ____

Exact: $2\frac{3}{4} \div \dfrac{2}{1}$

$\dfrac{11}{4} \cdot \dfrac{1}{2} = \dfrac{11}{8} =$

25. $9 \div 2\frac{1}{2}$

Estimate:

____ ÷ ____ = ____

Exact:

26. $5 \div 1\frac{7}{8}$

Estimate:

____ ÷ ____ = ____

Exact:

27. $\dfrac{5}{8} \div 1\frac{1}{2}$

Estimate:

____ ÷ ____ = ____

Exact:

28. $\dfrac{3}{4} \div 2\frac{1}{2}$

Estimate:

____ ÷ ____ = ____

Exact:

29. $1\frac{7}{8} \div 6\frac{1}{4}$

Estimate:

____ ÷ ____ = ____

Exact:

30. $8\frac{2}{5} \div 3\frac{1}{2}$

Estimate:

____ ÷ ____ = ____

Exact:

31. $5\frac{2}{3} \div 6$

Estimate:

____ ÷ ____ = ____

Exact:

32. $5\frac{3}{4} \div 2$

Estimate:

____ ÷ ____ = ____

Exact:

For Exercises 33–50, first estimate the answer. Then find each exact answer and simplify it if possible. **See Examples 3 and 4.** *Use the recipe for Carrot Cake Cupcakes to work Exercises 33–36.*

CARROT CAKE CUPCAKES

12 paper bake cups
$1\frac{3}{4}$ cups flour
1 cup packed brown sugar
1 tsp. baking powder
1 tsp. baking soda
1 tsp. ground cinnamon
$\frac{1}{2}$ tsp. salt
1 cup shredded carrots
$\frac{3}{4}$ cup applesauce

$\frac{1}{3}$ cup vegetable oil
1 large egg
$\frac{1}{2}$ tsp. vanilla extract
1 container (16 oz.) ready-to-
   spread cream cheese frosting
Shredded coconut,
   tinted green
3 oz. Jelly Belly jelly beans,
   Orange Sherbet flavor

Preheat oven to 350° F. Place 12 bake cups in a muffin pan; set aside. In a large bowl, using a wire whisk, stir together flour, brown sugar, baking powder, baking soda, cinnamon, and salt. In a medium bowl, combine carrots, applesauce, oil, egg, and vanilla until blended. Add carrot mixture to flour mixture, stir well. Spoon batter into bake cups, filling $\frac{2}{3}$ full. Bake until toothpick inserted in center comes out clean,

20–25 minutes. Cool cupcakes in pan 10 minutes; remove to wire rack and cool completely. Frost with cream cheese frosting. To make carrot design on cupcakes, place gourmet Jelly Belly jelly beans in a carrot shape on each cupcake, top with green coconut for carrot top. Makes 12 cupcakes.

33. If 30 cupcakes are baked ($2\frac{1}{2}$ times the recipe), find the amount of each ingredient.

  (a) Applesauce

   Estimate:

   Exact:

  (b) Salt

   Estimate:

   Exact:

  (c) Flour

   Estimate:

   Exact:

34. If 18 cupcakes are baked ($1\frac{1}{2}$ times the recipe), find the amount of each ingredient.

  (a) Flour

   Estimate:

   Exact:

  (b) Applesauce

   Estimate:

   Exact:

  (c) Vegetable oil

   Estimate:

   Exact:

35. How much of each ingredient is needed if you bake one-half of the recipe?

  (a) Vanilla extract

   Estimate:

   Exact:

  (b) Applesauce

   Estimate:

   Exact:

  (c) Flour

   Estimate:

   Exact:

36. How much of each ingredient is needed if you bake one-third of the recipe?

  (a) Flour

   Estimate:

   Exact:

  (b) Salt

   Estimate:

   Exact:

  (c) Applesauce

   Estimate:

   Exact:

37. A new condominium conversion project requires $11\frac{3}{4}$ gallons of paint for each unit. How many units can be painted with 329 gallons of paint?

Estimate:

Exact:

38. According to an old English system of time units, a moment is $1\frac{1}{2}$ minutes. How many moments are there in an 8-hour work day? (8 hours = 480 minutes). (*Source:* hightechscience.org/funfacts.htm)

Estimate:

Exact:

39. A manufacturer of floor jacks is ordering steel tubing to make the handles for the jack shown below. How much steel tubing is needed to make 45 of these jacks? The symbol for inches is ". For example, 5" means 5 inches. (*Source:* Harbor Freight Tools.)

Estimate:

Exact:

40. A wheelbarrow manufacturer uses handles made of hardwood. Find the amount of wood that is necessary to make 182 handles. The longest dimension shown in the advertisement below is the handle length. (*Source:* Harbor Freight Tools.)

Estimate:

Exact:

41. Write the three steps for multiplying mixed numbers. Use your own words.

42. Refer to **Exercise 41.** In your own words, write the additional step that must be added to the rule for multiplying mixed numbers to make it the rule for dividing mixed numbers.

43. The average cell phone contains $\frac{1}{1000}$ ounce of gold worth about $1\frac{2}{5}$. If you could extract the gold from all of the 130 million cell phones junked each year, how much money would you have from the sale of the gold? (*Source:* Discovery Channel, *Gold Rush Alaska.*)

Estimate:

Exact:

44. The manager of the flooring department at The Home Depot determines that each apartment unit requires $62\frac{1}{2}$ square yards of carpet. Find the number of apartment units that can be carpeted with 6750 square yards of carpet.

Estimate:

Exact:

45. Foxworthy Forest Products owns a truck that holds $1\frac{1}{4}$ cords of firewood. How many trips will he need to make to deliver 140 cords of firewood?

Estimate:

Exact:

46. BAE Roofing has completed an agreement to re-roof homes in a 20-year old residential development. Each home needs $31\frac{1}{2}$ squares of roofing material and the company has 1827 squares of material in inventory. How many homes will they be able to re-roof with this material?

Estimate:

Exact:

Use the information on bottle jacks in the advertisement to answer Exercises 47–48. The " symbol is for inches.

47. A mechanic needs a hydraulic lift that will raise a car 4 times as high as the standard jack shown.

(a) How high must it lift?

(b) Will a mechanic 6 feet tall be able to fit under the vehicle without bending down? *Hint:* 1 ft = 12 in.

(a) *Estimate:*

   Exact:

(b)

48. A race car driver needs a jack that will raise a vehicle only $\frac{1}{3}$ as high as the low-profile jack pictured.

(a) How high must it lift?

(b) Will a 6-inch part fit under the car?

(a) *Estimate:*

   Exact:

(b)

49. A shark can swim $5\frac{1}{2}$ times faster than a person. If a person can swim $6\frac{1}{8}$ miles per hour, how fast can a shark swim? (*Source:* Discovery Channel, *Animal Planet*.)

Estimate:

Exact:

50. A flooring contractor needs $24\frac{2}{7}$ boxes of tile to cover a kitchen floor. If there are 24 homes in a subdivision, how many boxes of tile are needed to cover all of the floors?

Estimate:

Exact:

Study Skills
REVIEWING A CHAPTER

OBJECTIVES

1. Use the Chapter Quick Review to see examples of every type of problem.
2. Create study cards for vocabulary.
3. Practice by doing review and mixed review exercises.
4. Take the Chapter Test as a practice test.

This activity is really about **preparing for tests.** Some of the suggestions are ideas that you will learn to use a little later in the term, but get started trying them out now. Often, the first chapters in your math textbook will be review, so it is good to practice some of the study techniques on material that is not too challenging.

Chapter Reviewing Techniques

Use these **chapter reviewing techniques**.

▶ **Make a study card for each vocabulary word and concept.** Include a definition, an example, a sketch, and a page reference. Include the symbol or formula if there is one. See the *Using Study Cards* activity for a quick look at some sample study cards.

▶ **Go back to the section** to find more explanations or information about any new vocabulary, formulas, or symbols.

▶ **Use the Chapter Summary** to see examples of each type of problem. Do not expect it to substitute for reading and working through the whole chapter! First, take the Test Your Word Power quiz to check your understanding of new vocabulary. The answers follow the quiz. Then read the Quick Review. **Pay special attention to the dark blue headings.** Check the explanations for the solutions to problems given. Try to think about how all the topics in the **whole chapter** are related.

▶ **Study your lecture notes** to see what your instructor has emphasized in class. Then review that material in your text.

▶ **Do the Review Exercises** to practice every type of problem.
  ✓ Check your answers **after** you're done with each **section of exercises.**
  ✓ If you get stuck on a problem, **first** check the Chapter Quick Review. If that doesn't clear up your confusion, then check the section and your lecture notes.
  ✓ Pay attention to **direction words** for the problems, such as *simplify, round, solve*, and *estimate*.
  ✓ Make **study cards for especially difficult problems.**

▶ **Do the Mixed Review exercises.** This is a good check to see if you can still do the problems when they are in mixed-up order. **Check your answers carefully** in the Answers section in the back of your book. Are your answers **exact** and **complete?** Make sure you are **labeling** answers correctly, using the right **units.** For example, does your answer need to include *$, cm^2, ft*, and so on?

▶ **Take the Chapter Test as if it is a real test.** If your instructor has skipped sections in the chapter, ask which problems to skip on the test before you start.

✓ **Time yourself** just as you would for a real test.

✓ **Use a calculator or notes** just as you would be permitted to (or not) on a real test.

✓ **Take the test in one sitting,** just like a real test is given in one sitting.

✓ **Show all your work.** Practice showing your work just the way your instructor has asked you to show it.

✓ **Practice neatness.** Can someone else follow your steps?

✓ **Check your answers** in the back of the book.

Notice that reviewing a chapter will take some time. Remember that it takes time for dendrites to grow! You cannot grow a good network of dendrites by rushing through a review in one night. But if you use the suggestions over a few days or evenings, you will notice that you understand the material more thoroughly and remember it longer.

Now Try This

Follow the reviewing techniques listed above for your next test. For each technique, write a comment about how it worked for you in the spaces below.

1 **Make a study card for each vocabulary word and concept.**

2 **Go back to the section** to find more explanations or information.

3 **Take the Test Your Word Power quiz and use the Quick Review** to review each concept in the chapter.

4 **Study your lecture notes** to see what your instructor has emphasized in class.

5 **Do the Review Exercises,** following the specific suggestions on the previous page.

6 **Do the Mixed Review exercises.**

7 **Take the Chapter Test** as if it is a real test.

Why Are These Review Activities Brain Friendly?

You have already become familiar with the features of your textbook. This activity requires you to make good use of them. Your **brain needs repetition** to strengthen dendrites and the connections between them. By following the steps outlined here, you will be reinforcing the concepts, procedures, and skills you need to use for tests (and for the next chapters).

The combination of techniques provides repetition in different ways. That **promotes good branching of dendrites** instead of just relying on one branch, or route, to connect to the other dendrites. A thorough review of each chapter will **solidify your dendrite connections.** It will help you be sure that you understand the concepts **completely and accurately.** Also, taking the Chapter Test will **simulate the testing situation,** which gives you practice in test taking conditions.

Study Skills

TIPS FOR TAKING MATH TESTS

OBJECTIVES

1. Apply suggestions to tests and quizzes.
2. Develop a set of "best practices" to apply while testing.

Improving Your Test Score

To Improve Your Test Score	Comments
Come prepared with a pencil, eraser, and calculator, if allowed. If you are easily distracted, sit in the corner farthest from the door.	*Working in pencil lets you erase,* keeping your work neat and readable.
Scan the entire test, note the point value of different problems, and plan your time accordingly. Allow at least five minutes to check your work at the end of the testing time.	If you have 50 minutes to do 20 problems, $50 \div 20 = 2.5$ minutes per problem. *Spend less time on easy ones,* more time on problems with higher point values.
Read directions carefully, and circle any significant words. When you finish a problem, read the directions again to make sure you did what was asked.	*Pay attention to announcements* written on the board or made by your instructor. Ask if you don't understand. You don't want to get problems wrong because you misread the directions!
Show your work. Most math teachers give partial credit if some of the steps in your work are correct, even if the final answer is wrong. *Write neatly.* If you like to scribble when first working or checking a problem, do it on scratch paper.	*If your instructor can't read your writing, you won't get credit for it.* If you need more space to work, ask if you can use extra pieces of paper that you hand in with your test paper.
Check that the *answer to an application problem is reasonable* and makes sense. Read the problem again to make sure you've answered the question.	*Use common sense.* Can the father really be seven years old? Would a month's rent be $32,140? Label your answer: $, years, inches, etc.
To check for careless errors, you need to *rework the problem again, without looking at your previous work.* Cover up your work with a piece of scratch paper, and pretend you are doing the problem for the first time. Then compare the two answers.	If you just "look over" your work, your mind can make the same mistake again without noticing it. Reworking the problem from the beginning *forces you to rethink it.* If possible, use a different method to solve the problem the second time.
Do not try to review up until the last minute before the test. Instead, go for a walk, do some deep breathing, and arrive just in time for the test. Ignore other students.	Listening to anxious classmates before the test *may cause you to panic.* Moderate exercise and deep breathing will calm your mind.

Reducing Anxiety

To Reduce Anxiety	Comments
Do a "knowledge dump" as soon as you get the test. Write important notes to yourself in a corner of the test paper: formulas, or common errors you want to watch out for.	Writing down tips and things that you've memorized *lets you relax;* you won't have to worry about forgetting those things and can refer to them as needed.
Do the easy problems first in order to build confidence. If you feel your anxiety starting to build, *immediately* stop for a minute, close your eyes, and take several slow, deep breaths.	Greater confidence helps you *get the easier problems correct.* Anxiety causes shallow breathing, which leads to confusion and reduced concentration. Deep breathing calms you.
As you work on more difficult problems, *notice your "inner voice."* You may have negative thoughts such as, "I can't do it," or "who cares about this test anyway." In your mind, yell, "STOP" and take several deep, slow breaths. Or, replace the negative thought with a positive one.	Here are *examples of positive statements.* Try writing one of them on the top of your test paper. • I know I can do it. • I can do this one step at a time. • I've studied hard, and I'll do the best I can.
If you still can't solve a difficult problem when you come back to it the second time, *make a guess and do not change it.* In this situation, your first guess is your best bet. Do not change an answer just because you're a little unsure. *Change it only if you find an obvious mistake.*	If you are thinking about changing an answer, be sure you have a good reason for changing it. If you cannot find a specific error, leave your first answer alone. *When the tests are returned, check to see if changing answers helped or hurt you.*
Read the harder problems twice. Write down *anything* that might help solve the problem: a formula, a picture, etc. If you still can't get it, circle the problem and *come back to it later.* Do *not* erase any of the things you wrote down.	If you know even a *little* bit about the problem, write it down. The *answer may come to you* as you work on it, or you may get partial credit. Don't spend too long on any one problem. Your subconscious mind will work on the tough problem while you go on with the test.
Ignore students who finish early. Use the entire test time. *You do not get extra credit for finishing early.* Use the extra time to rework problems and correct careless errors.	Students who leave early are often the ones who didn't study or who are too anxious to continue working. If they bother you, *sit as far from the door as possible.*

Chapter 2 Summary

Key Terms

Chapter 2 **Summary**

2.1

numerator The number above the fraction bar in a fraction is called the numerator. It shows how many of the equivalent parts are being considered.

denominator The number below the fraction bar in a fraction is called the denominator. It shows the number of equal parts in a whole.

proper fraction In a proper fraction, the numerator is smaller than the denominator. The fraction is less than 1.

improper fraction In an improper fraction, the numerator is greater than or equal to the denominator. The fraction is equal to or greater than 1.

2.2

mixed number A mixed number includes a fraction and a whole number written together.

2.3

factors Numbers that are multiplied to give a product are factors.

factorizations The numbers that can be multiplied to give a specific number (product) are factorizations of that number.

prime number A prime number is a whole number other than 0 and 1 that has exactly two factors, itself and 1.

composite number A composite number has at least one factor other than itself and 1.

prime factorization In a prime factorization, every factor is a prime number.

2.4

equivalent fractions Two fractions are equivalent when they represent the same portion of a whole.

common factor A common factor is a number that can be divided evenly into two or more whole numbers.

lowest terms A fraction is written in lowest terms when its numerator and denominator have no common factor other than 1.

2.5

multiplication shortcut When multiplying or dividing fractions, the process of dividing a numerator and denominator by a common factor can be used as a shortcut.

2.7

reciprocal Two numbers are reciprocals of each other if their product is 1. To find the reciprocal of a fraction, interchange the numerator and the denominator. Zero does not have a reciprocal.

New Formula

Area of a rectangle: Area = length • width

Test Your Word Power

See how well you have learned the vocabulary in this chapter.

1 A **numerator** is
A. a number greater than 5
B. the number above the fraction bar in a fraction
C. the number below the fraction bar in a fraction.

2 A **proper fraction**
A. has a value less than 1
B. has a whole number and a fraction
C. has a value greater than 1.

3 A **mixed number** is
A. less than 1
B. a whole number and a fraction written together
C. a number multiplied by another number.

4 A **factor** is
A. one of two or more numbers that are added to get another number
B. the answer in division
C. one of two or more numbers that are multiplied to get another number.

5 A whole number greater than 1 is **prime** if
A. it cannot be factored
B. it has just one factor
C. it has only itself and 1 as factors.

6 A **common factor** can
A. only be divided by itself and 1
B. be divided evenly into two or more whole numbers
C. never be divided by 2.

7 A fraction is in **lowest terms** when
A. it cannot be divided
B. its numerator and denominator have no common factor other than 1
C. it has a value less than 1.

8 To find the **reciprocal** of a fraction,
A. multiply it by itself
B. interchange the numerator and the denominator
C. change it to an improper fraction.

Answers to Test Your Word Power

1. B; *Example:* In $\frac{3}{8}$, the numerator is 3.

2. A; *Example:* $\frac{1}{2}, \frac{3}{4}$, and $\frac{7}{8}$ are all proper fractions with a value less than 1.

3. B; *Example:* $2\frac{3}{8}$ and $5\frac{3}{4}$ are mixed numbers.

4. C; *Example:* Since $3 \cdot 5 = 15$, the numbers 3 and 5 are factors of 15.

5. C; *Example:* 3, 5, and 11 are prime numbers; 4, 8, and 12 are composite numbers.

6. B; *Example:* 3 is a common factor of both 6 and 9 because it can be evenly divided into each of them.

7. B; *Example:* $\frac{3}{8}, \frac{4}{5}$, and $\frac{5}{6}$ are in lowest terms but $\frac{6}{8}, \frac{3}{6}$, and $\frac{2}{4}$ are not.

8. B; *Example:* The reciprocal of $\frac{3}{8}$ is $\frac{8}{3}$, the reciprocal of $\frac{25}{4}$ is $\frac{4}{25}$, and the reciprocal of 6 or $\frac{6}{1}$ is $\frac{1}{6}$.

Quick Review

Concepts	Examples

2.1 Types of Fractions

Proper Numerator smaller than denominator; a value less than 1

$$\frac{2}{3} \quad \frac{3}{4} \quad \frac{15}{16} \quad \frac{1}{8} \qquad \text{Proper fractions}$$

Improper Numerator equal to or greater than denominator; a value equal to or greater than 1

$$\frac{17}{8} \quad \frac{19}{12} \quad \frac{11}{2} \quad \frac{5}{3} \quad \frac{7}{7} \qquad \text{Improper fractions}$$

Concepts	Examples

2.2 Converting Fractions

Mixed to Improper Multiply denominator by whole number, add numerator, and place over denominator.

Improper to Mixed Divide numerator by denominator and place remainder over denominator.

$$7\frac{2}{3} = \frac{23}{3} \leftarrow (3 \cdot 7) + 2$$

Same denominator

$$\frac{17}{5} = 3\frac{2}{5} \leftarrow \text{Remainder}$$

Same denominator

$$\begin{array}{r} 3 \\ 5\overline{)17} \\ \underline{15} \\ 2 \end{array} \leftarrow \begin{array}{l}\text{Divide numerator}\\ \text{by denominator.}\end{array}$$

$\leftarrow$ Remainder

2.3 Prime Numbers

Determine whether a whole number is evenly divisible only by itself and 1. (By definition, 0 and 1 are not prime.)

The prime numbers less than 50 are 2, 3, 5, 7, 11, 13, 17, 19, 23, 29, 31, 37, 41, 43, 47.

2.3 Finding the Prime Factorization of a Number

Divide each factor by a prime number using a diagram that forms the shape of tree branches.

Find the prime factorization of 30. Use a factor tree.

Prime factors are circled.
$30 = \mathbf{2 \cdot 3 \cdot 5}$

2.4 Writing Fractions in Lowest Terms

Divide the numerator and denominator by the greatest common factor.

Write $\frac{30}{42}$ in lowest terms.

$$\frac{30}{42} = \frac{30 \div 6}{42 \div 6} = \frac{5}{7}$$

2.5 Multiplying Fractions

1. Multiply the numerators and multiply the denominators.

2. Write answers in lowest terms if the multiplication shortcut was not used.

Multiply.

$$\frac{6}{11} \cdot \frac{7}{8} = \frac{\overset{3}{6}}{11} \cdot \frac{7}{\underset{4}{8}} = \frac{3 \cdot 7}{11 \cdot 4} = \frac{21}{44}$$

2.7 Finding the Reciprocal

To find the reciprocal of a fraction, interchange the numerator and denominator.

Find the reciprocal of each fraction.

$\frac{3}{4}$ The reciprocal of $\frac{3}{4}$ is $\frac{4}{3}$.

$\frac{8}{5}$ The reciprocal of $\frac{8}{5}$ is $\frac{5}{8}$.

9   The reciprocal of **9** is $\frac{1}{9}$.

Concepts	Examples

2.7 Dividing Fractions

Use the reciprocal of the divisor (second fraction) and change division to multiplication.

Divide.

$$\frac{25}{36} \div \frac{15}{18} = \frac{25}{36} \cdot \frac{18}{15} = \frac{5 \cdot 1}{2 \cdot 3} = \frac{5}{6}$$

Reciprocals

2.8 Multiplying Mixed Numbers

First estimate the answer. Then follow these steps.

Step 1 *Change* each mixed number to an improper fraction.

Step 2 *Multiply*.

Step 3 *Simplify* the answer, which means to write it in *lowest terms,* and change it to a mixed number or whole number where possible.

First estimate the answer. Then multiply to get the exact answer.

Estimate: *Exact:*

$$1\frac{3}{5} \cdot 3\frac{1}{3} \qquad 1\frac{3}{5} \cdot 3\frac{1}{3} = \frac{8}{5} \cdot \frac{10}{3}$$

Rounded

$$2 \cdot 3 = 6 \qquad\qquad = \frac{8 \cdot 2}{1 \cdot 3}$$

$$= \frac{16}{3} = 5\frac{1}{3}$$

Close to estimate

2.8 Dividing Mixed Numbers

First estimate the answer. Then follow these steps.

Step 1 *Change* each mixed number to an improper fraction.

Step 2 Use the *reciprocal* of the second fraction (divisor).

Step 3 *Change* division to multiplication.

Step 4 *Simplify* the answer, which means to write it in *lowest terms,* and change it to a mixed number or whole number where possible.

First estimate the answer. Then divide to get the exact answer.

Estimate: *Exact:*

$$3\frac{5}{9} \div 2\frac{2}{5} \qquad 3\frac{5}{9} \div 2\frac{2}{5} = \frac{32}{9} \div \frac{12}{5}$$

Rounded

$$4 \div 2 = 2 \qquad\qquad = \frac{32}{9} \cdot \frac{5}{12} = \frac{40}{27}$$

$$= 1\frac{13}{27}$$

Close to estimate

Chapter 2 Review Exercises

2.1 *Write the fraction that represents each shaded portion.*

1.

2.

3.

List the proper and improper fractions in each group.

	Proper	**Improper**

4. $\dfrac{1}{8}$ $\dfrac{4}{3}$ $\dfrac{5}{5}$ $\dfrac{3}{4}$ $\dfrac{2}{3}$ _____   _____

5. $\dfrac{6}{5}$ $\dfrac{15}{16}$ $\dfrac{16}{13}$ $\dfrac{1}{8}$ $\dfrac{5}{3}$ _____   _____

2.2 *Write each mixed number as an improper fraction. Write each improper fraction as a mixed number.*

6. $4\dfrac{3}{4}$

7. $9\dfrac{5}{6}$

8. $\dfrac{27}{8}$

9. $\dfrac{63}{5}$

2.3 *Find all factors of each number.*

10. 6

11. 24

12. 55

13. 90

Write the prime factorization of each number by using exponents.

14. 27

15. 150

16. 420

Simplify each expression.

17. 5^2

18. $6^2 \cdot 2^3$

19. $8^2 \cdot 3^3$

20. $4^3 \cdot 2^5$

2.4 *The purity of gold is regulated by law and is the same in all parts of the world. Exercises 21–24 show the purity of 24-kt (karat), 18-kt, 14-kt, and 10-kt gold by comparing the parts of gold to the parts of alloy (metals other than gold). Write a fraction in lowest terms to show the portion that is gold. (Source: Costco Wholesale.)*

21. 24 kt. = (24 parts gold, 0 parts alloy)

22. 18 kt. = (18 parts gold, 6 parts alloy)

23. 14 kt. = (14 parts gold, 10 parts alloy)

24. 10 kt. = (10 parts gold, 14 parts alloy)

Write the numerator and denominator of each fraction as a product of prime factors. Then, write the fraction in lowest terms.

25. $\dfrac{25}{60}$

26. $\dfrac{384}{96}$

Decide whether each pair of fractions is equivalent or not equivalent, using the method of prime factors.

27. $\dfrac{3}{4}$ and $\dfrac{48}{64}$

28. $\dfrac{5}{8}$ and $\dfrac{70}{120}$

29. $\dfrac{2}{3}$ and $\dfrac{360}{540}$

2.5–2.8 *Multiply. Write answers in lowest terms, and as mixed numbers or whole numbers where possible.*

30. $\dfrac{4}{5} \cdot \dfrac{3}{4}$

31. $\dfrac{3}{10} \cdot \dfrac{5}{8}$

32. $\dfrac{70}{175} \cdot \dfrac{5}{14}$

33. $\dfrac{44}{63} \cdot \dfrac{3}{11}$

34. $\dfrac{5}{16} \cdot 48$

35. $\dfrac{5}{8} \cdot 1000$

Divide. Write answers in lowest terms, and as mixed numbers or whole numbers where possible.

36. $\dfrac{2}{3} \div \dfrac{1}{2}$

37. $\dfrac{5}{6} \div \dfrac{1}{2}$

38. $\dfrac{\frac{15}{18}}{\frac{10}{30}}$

39. $\dfrac{\frac{3}{4}}{\frac{3}{8}}$

40. $7 \div \dfrac{7}{8}$

41. $18 \div \dfrac{3}{4}$

42. $\dfrac{5}{8} \div 3$

43. $\dfrac{2}{3} \div 5$

44. $\dfrac{\frac{12}{13}}{3}$

Find the area of each rectangle.

45.

$2\frac{3}{4}$ ft

$\frac{1}{2}$ ft

46.

$4\frac{1}{2}$ yd

$\frac{7}{8}$ yd

47. Ceramic tile is being installed on the floor of a meeting hall that is 108 ft long and $72\frac{3}{4}$ ft wide. Find the area.

48. Find the area of a display shelf that is a rectangle measuring 6 ft long and $\frac{11}{12}$ ft wide.

First estimate the answer. Then multiply or divide to find the exact answer. Simplify all answers.

49. $5\frac{1}{2} \cdot 1\frac{1}{4}$

Estimate:

____ • ____ = ____

Exact:

50. $2\frac{1}{4} \cdot 7\frac{1}{8} \cdot 1\frac{1}{3}$

Estimate:

____ • ____ • ____ = ____

Exact:

51. $15\frac{1}{2} \div 3$

Estimate:

____ ÷ ____ = ____

Exact:

52. $4\frac{3}{4} \div 6\frac{1}{3}$

Estimate:

____ ÷ ____ = ____

Exact:

Solve each application problem by using the six problem-solving steps.

53. Blue Diamond Almonds has 320 tons of almonds. How many $\frac{5}{8}$ ton bins will be needed to store the almonds?

54. The founding partner of Professional Networking owns $\frac{2}{5}$ of the business, while 4 other equal partners own the rest. What fraction of the total business is owned by each of the other partners?

55. How many window-blind pull cords can be made from $157\frac{1}{2}$ yards of cord if $4\frac{3}{8}$ yards of cord are needed for each blind? First estimate, and then find the exact answer.

Estimate:

Exact:

56. A gallon of water weighs $8\frac{1}{3}$ pounds. Find the weight of the water in two 50-gallon aquariums. First estimate, and then find the exact answer.

Estimate:

Exact:

57. Ebony Wilson purchased 100 pounds of rice at the food co-op. After selling $\frac{1}{4}$ of this to her neighbor, she gave $\frac{2}{3}$ of the remaining rice to her parents. How many pounds of rice does she have left?

58. Sheila Spinney, a recent college graduate, receives a salary of $2976 each month. She pays $\frac{3}{8}$ of this amount in taxes, social security, and a retirement plan. Of the remainder, $\frac{9}{10}$ goes for basic living expenses. How much money remains?

59. The results of a back-to-school fundraiser to purchase school supplies showed that 6 schools will divide $\frac{7}{8}$ of the total amount raised. What fraction of the total amount raised will each school receive?

60. In a morning of deep-sea fishing, 5 fishermen catch $\frac{4}{5}$ ton of salmon. If they divide the fish evenly, how much will each receive?

Mixed Review Exercises

Multiply or divide as indicated. Simplify all answers.

61. $\dfrac{1}{2} \cdot \dfrac{3}{4}$

62. $\dfrac{2}{3} \cdot \dfrac{3}{5}$

63. $12\dfrac{1}{2} \cdot 2\dfrac{1}{2}$

64. $8\dfrac{1}{3} \cdot 3\dfrac{2}{5}$

65. $\dfrac{\frac{4}{5}}{8}$

66. $\dfrac{\frac{5}{8}}{4}$

67. $\dfrac{15}{31} \cdot 62$

68. $3\dfrac{1}{4} \div 1\dfrac{1}{4}$

Write each mixed number as an improper fraction. Write each improper fraction as a mixed number.

69. $\dfrac{8}{5}$

70. $\dfrac{153}{4}$

71. $5\dfrac{2}{3}$

72. $38\dfrac{3}{8}$

Write the numerator and denominator of each fraction as a product of prime factors; then write the fraction in lowest terms.

73. $\dfrac{8}{12}$

74. $\dfrac{108}{210}$

Write each fraction in lowest terms.

75. $\dfrac{75}{90}$

76. $\dfrac{48}{72}$

77. $\dfrac{44}{110}$

78. $\dfrac{87}{261}$

Solve each application problem.

79. The label of the Roundup Weed and Grass Killer says to mix $2\dfrac{1}{2}$ ounces of the product with 1 gallon of water. How many ounces are needed for 50 gallons of water? First estimate, then find the exact answer. (*Source:* Roundup Weed and Grass Killer.)

Estimate:

Exact:

80. Valley Farms purchased some diesel fuel additive. The instructions say to use $7\dfrac{1}{4}$ quarts of additive for each tank of fuel. How many quarts are needed for $9\dfrac{1}{3}$ tanks? First estimate, then find the exact answer.

Estimate:

Exact:

81. The antique U.S.A. stamp is $1\dfrac{3}{4}$ in. by $\dfrac{7}{8}$ in. Find its area.

82. A patio table top is $\dfrac{7}{8}$ yard by $2\dfrac{1}{4}$ yards. What is the area?

Chapter 2 Test The Chapter Test Prep Videos with test solutions are available on DVD, in MyMathLab, and on You[Tube]—*search* "LialBasicCollegeMath" and click on "Channels."

Write a fraction to represent each shaded portion.

1.

2.

3. Identify all the proper fractions in this list:

$$\frac{2}{3} \quad \frac{4}{4} \quad \frac{6}{7} \quad \frac{5}{2} \quad \frac{1}{4} \quad \frac{5}{8} \quad \frac{30}{18}$$

4. Write $3\frac{3}{8}$ as an improper fraction.

5. Write $\frac{123}{4}$ as a mixed number.

6. Find all factors of 18.

Find the prime factorization of each number. Write the answers using exponents.

7. 45

8. 144

9. 500

Write each fraction in lowest terms.

10. $\frac{36}{48}$

11. $\frac{60}{72}$

12. The method of prime factors is used to write a fraction in lowest terms. Briefly explain how this is done. Use the fraction $\frac{56}{84}$ to show how this works.

13. Explain how to multiply fractions. What additional steps must be taken when dividing fractions?

Multiply or divide. Write answers in lowest terms, and as mixed numbers or whole numbers where possible.

14. $\frac{3}{4} \cdot \frac{4}{9}$

15. $54 \cdot \frac{2}{3}$

16. A rectangular barbecue grill is $\frac{15}{16}$ yard by $\frac{4}{9}$ yard. Find the area of the grill.

17. The Sierra College Conservation Club planted 8760 Douglas Fir seedlings. If $\frac{3}{8}$ of these seedlings are not expected to survive, find the number of seedlings that do survive.

18. $\dfrac{3}{4} \div \dfrac{5}{6}$

19. $\dfrac{\frac{7}{4}}{9}$

20. To complete a custom-designed cabinet, oak trim pieces must be cut exactly $2\frac{1}{4}$ inches long so that they can be used as dividers in a spice rack. Find the number of pieces that can be cut from a piece of oak that is 54 inches in length.

First estimate the answer. Then find the exact answer. Simplify all answers.

21. $4\dfrac{1}{8} \cdot 3\dfrac{1}{2}$

Estimate:

Exact:

22. $1\dfrac{5}{6} \cdot 4\dfrac{1}{3}$

Estimate:

Exact:

23. $9\dfrac{3}{5} \div 2\dfrac{1}{4}$

Estimate:

Exact:

24. $\dfrac{8\frac{1}{2}}{1\frac{3}{4}}$

Estimate:

Exact:

25. A new vaccine is synthesized at the rate of $2\frac{1}{2}$ grams per day. How many grams can be synthesized in $12\frac{1}{4}$ days?

Estimate:

Exact:

Chapters 1–2   *Cumulative Review Exercises*

Name the digit that has the given place value in each number.

1. 783
hundreds
tens

2. 8,621,785
millions
ten-thousands

Add, subtract, multiply, or divide as indicated.

3.
$$\begin{array}{r} 71 \\ 23 \\ 47 \\ +\ 36 \end{array}$$

4.
$$\begin{array}{r} 82{,}121 \\ 5\ 468 \\ 316 \\ +\ 61{,}294 \end{array}$$

5.
$$\begin{array}{r} 6537 \\ -\ 2085 \end{array}$$

6.
$$\begin{array}{r} 4{,}819{,}604 \\ -\ 1{,}597{,}783 \end{array}$$

7.
$$\begin{array}{r} 83 \\ \times\ 9 \end{array}$$

8. $9 \cdot 4 \cdot 2$

9.
$$\begin{array}{r} 3784 \\ \times\ \ \ \ 573 \end{array}$$

10.
$$\begin{array}{r} 563 \\ \times\ \ \ 800 \end{array}$$

11. $\dfrac{63}{7}$

12. $18\overline{)136{,}458}$

13. $33{,}886 \div 4$

14. $492\overline{)10{,}850}$

Round each number to the nearest ten, nearest hundred, and nearest thousand.

	Ten	**Hundred**	**Thousand**
15. 6583	_____	_____	_____
16. 76,271	_____	_____	_____

Simplify.

17. $2^5 - 6(4)$

18. $\sqrt{36} - 2 \cdot 3 + 5$

Solve each application problem using the six problem-solving steps.

19. A Gettysburg tour van uses 9 gallons of fuel on a half-day tour and 17 gallons of fuel on a full-day tour. Find the total number of gallons of fuel used in 26 half-day tours and 18 full-day tours.

20. As a result of unvaccinated travelers and children, there have been outbreaks of several diseases. Last year there were 4519 cases of mumps and 21,291 cases of pertussis (whooping cough) in the United States. How many more cases of pertussis than mumps were there? (*Source:* Centers for Disease Control and Prevention.)

21. A typical adult loses 100 hairs a day out of approximately 120,000 hairs. If the lost hairs were not replaced, find the number of hairs remaining after two years. (1 year = 365 days.)

22. A group of 22 health care workers will divide 4136 work hours evenly this month. Find the number of hours each will work.

23. The standard for low-flow toilets in the United States is $1\frac{3}{5}$ gallons of water per flush. Find the number of gallons of water used in 160 flushes. (*Source:* Wikipedia, The Free Encyclopedia.)

24. A welder needs angle iron pieces that are $3\frac{1}{3}$ inches long. Find the number of pieces that can be cut from a piece of angle iron that is 70 inches in length.

Write proper or improper for each fraction.

25. $\dfrac{2}{3}$

26. $\dfrac{6}{6}$

27. $\dfrac{9}{18}$

Write each mixed number as an improper fraction. Write each improper fraction as a whole or mixed number.

28. $3\dfrac{3}{8}$

29. $6\dfrac{2}{5}$

30. $\dfrac{14}{7}$

31. $\dfrac{103}{8}$

Find the prime factorization of each number. Write answers using exponents.

32. 72

33. 126

34. 350

Simplify each expression.

35. $4^2 \cdot 2^2$

36. $2^3 \cdot 6^2$

37. $2^3 \cdot 4^2 \cdot 5$

Write each fraction in lowest terms.

38. $\dfrac{42}{48}$

39. $\dfrac{24}{36}$

40. $\dfrac{30}{54}$

Multiply or divide as indicated. Simplify all answers.

41. $\dfrac{1}{2} \cdot \dfrac{3}{4}$

42. $30 \cdot \dfrac{2}{3} \cdot \dfrac{3}{5}$

43. $7\dfrac{1}{2} \cdot 3\dfrac{1}{3}$

44. $\dfrac{3}{5} \div \dfrac{5}{8}$

45. $\dfrac{7}{8} \div 1\dfrac{1}{2}$

46. $3 \div 1\dfrac{1}{4}$

Math in the Media

RECIPES

Rachael Ray is a Food Network television host, bestselling cookbook author, and the editor of her own lifestyle magazine. Rachael Ray's recipes can be found on her Web site, www.rachaelray.com, on her television show, or in one of her many cookbooks. The recipe at the side is from her book *Rachael Ray: 30-Minute Meals*.

Alaska Burgers

1 pound 93% lean ground beef
$\frac{1}{2}$ medium Spanish onion, minced or processed
4 shakes Worcestershire sauce
$\frac{1}{4}$ teaspoon allspice (1 good pinch)
$\frac{1}{2}$ teaspoon ground cumin (two good pinches)
Cracked black pepper
$\frac{1}{3}$ pound brick of smoked cheddar cheese, cut into $\frac{1}{2}$ inch slices

4 fresh, crusty onion rolls
Thick-sliced tomato and lettuce to top

Mix beef, onion, Worcestershire, allspice, cumin, and black pepper in a bowl. Separate a quarter of the mixture. Take a slice of the smoked cheese and place it in the middle of the mixture. Form the pattie shape around the cheese filling. Patties should be no more than $\frac{3}{4}$ inch thick. Repeat with rest of mixture to have a total of 4 patties.

Heat a nonstick griddle or frying pan to medium hot. Cook burgers 5 to 6 minutes on each side. Meat should be cooked through and cheese melted. Check each burger with an instant-read thermometer for an internal temp of 170°F for well done if undercooking concerns you. Or cut into one and check the color of the meat.

Salt burgers after preparation to your taste. (Salting beef before cooking draws out juices and flavor.) Top with tomato slices and lettuce. Serves 4.

1. Following the recipe, **(a)** what is the weight of one $\frac{1}{2}$-inch slice of cheese, and **(b)** what is the thickness of a $\frac{1}{3}$-pound brick of smoked cheddar cheese?

2. According to the recipe, **(a)** how many teaspoons of allspice are in 8 good pinches, and **(b)** how many teaspoons of ground cumin are in 7 good pinches?

3. Suppose you are preparing Alaska Burgers for 18 guests. By what factor will you change the ingredient amounts?

4. You know that of the 15 guests at your next party, 5 large eaters will eat $1\frac{1}{2}$ burgers each, 5 children will eat $\frac{1}{2}$ burger each, and the rest of the guests will each eat 1 burger. **(a)** How many burgers will you need, and **(b)** by what factor will you change the ingredient amounts?

5. Fill in the blanks with the ingredient amounts needed to make 9 servings of Alaska Burgers

   Lean ground beef   _____

   Spanish onions   _____

   Worcestershire sauce _____

   Allspice   _____

   Ground cumin   _____

   Cheddar cheese   _____

6. If you have $5\frac{3}{4}$ pounds of beef, how many servings of Alaska Burgers can be prepared?

3

Adding and Subtracting Fractions

Bryan Berg has set eight Guinness world records for the tallest freestanding card structures—no glue, no tape, no hidden supports. At the African American Museum in Dallas, Texas, using 1060 decks of cards, he set a new world record of 25 feet, $9\frac{7}{16}$ inches. In this chapter we will work with the addition and subtraction of fractions and mixed numbers.

3.1 Adding and Subtracting Like Fractions

OBJECTIVES

1. Define like and unlike fractions.
2. Add like fractions.
3. Subtract like fractions.

In **Chapter 2** we looked at the basics of fractions and then practiced with multiplication and division of fractions and mixed numbers. In this chapter we will work with addition and subtraction of fractions and mixed numbers.

OBJECTIVE ▶ 1 **Define like and unlike fractions.** Fractions with the same denominators are **like fractions**. Fractions with different denominators are **unlike fractions**.

---

1 Next to each pair of fractions write *like* or *unlike*.

(a) $\dfrac{2}{5}$ $\dfrac{3}{5}$ _____

(b) $\dfrac{2}{3}$ $\dfrac{3}{4}$ _____

(c) $\dfrac{7}{12}$ $\dfrac{11}{12}$ _____

(d) $\dfrac{3}{8}$ $\dfrac{3}{16}$ _____

| EXAMPLE 1 | Identifying Like and Unlike Fractions |

(a) $\dfrac{3}{4}, \dfrac{1}{4}, \dfrac{5}{4}, \dfrac{6}{4},$ and $\dfrac{4}{4}$ are **like** fractions.

↑ ↑ ↑ ↑   ↑
            All denominators are the same.

(b) $\dfrac{7}{12}$ and $\dfrac{12}{7}$ are **unlike** fractions.

> Unlike fractions have different denominators.

↑   ↑
Denominators are different.

Note

Like fractions have the *same* denominator.

◀ **Work Problem ① at the Side.**

OBJECTIVE ▶ 2 **Add like fractions.** The figures below show you how to add the fractions $\frac{2}{7}$ and $\frac{4}{7}$.

As the figures show,

$$\frac{2}{7} + \frac{4}{7} = \frac{6}{7}.$$

Add like fractions as follows.

Adding Like Fractions

Step 1 Add the numerators to find the numerator of the sum.

Step 2 Write the denominator of the like fractions as the denominator of the sum.

Step 3 Write the sum in lowest terms.

Answers

1. (a) like **(b)** unlike **(c)** like
(d) unlike

EXAMPLE 2 **Adding Like Fractions**

Add and write the sum in lowest terms.

(a) $\dfrac{1}{5} + \dfrac{2}{5}$

Add numerators.

$\frac{3}{5}$ is already in lowest terms.

$$\frac{1}{5} + \frac{2}{5} = \frac{1 + 2}{5} = \frac{3}{5} \begin{array}{l} \leftarrow \text{Sum of numerators} \\ \leftarrow \text{Same denominator} \end{array}$$

(b) $\dfrac{1}{12} + \dfrac{7}{12} + \dfrac{1}{12}$ — Fractions are ready to add if they are *like* fractions.

Add numerators.

Step 1 $\dfrac{\overbrace{1 + 7 + 1}}{12}$

Step 2 $= \dfrac{9}{12} \begin{array}{l} \leftarrow \text{Sum of numerators} \\ \leftarrow \text{Same denominator} \end{array}$

Step 3 $= \dfrac{9 \div 3}{12 \div 3} = \dfrac{3}{4} \leftarrow \text{Lowest terms}$

CAUTION

Fractions may be added **only** if they have like denominators.

· · · · · · · · · · · · · · · · · · **Work Problem ② at the Side.** ▶

OBJECTIVE ▶ ③ **Subtract like fractions.** The figures below show $\frac{7}{8}$ broken into $\frac{4}{8}$ and $\frac{3}{8}$.

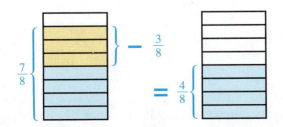

Subtracting $\frac{3}{8}$ from $\frac{7}{8}$ gives the answer $\frac{4}{8}$, or

$$\frac{7}{8} - \frac{3}{8} = \frac{4}{8}.$$

---

② Add and write the sums in lowest terms.

GS (a) $\dfrac{3}{8} + \dfrac{1}{8}$

$$\frac{3 + 1}{8} = \frac{\rule{1cm}{0.4pt}}{8} = \frac{\rule{1cm}{0.4pt}}{2}$$

(b) $\dfrac{2}{9}$
$+ \dfrac{5}{9}$

(c) $\dfrac{3}{16} + \dfrac{1}{16}$

(d) $\dfrac{3}{10} + \dfrac{1}{10} + \dfrac{4}{10}$

Answers

2. **(a)** $4; \dfrac{1}{2}$ **(b)** $\dfrac{7}{9}$ **(c)** $\dfrac{1}{4}$ **(d)** $\dfrac{4}{5}$

3 Find the difference and simplify.

GS **(a)** $\dfrac{5}{6} - \dfrac{1}{6}$

$\dfrac{5-1}{6} = \dfrac{}{6} = \dfrac{}{3}$

(b) $\dfrac{16}{10}$

$-\dfrac{7}{10}$

(c) $\dfrac{15}{3} - \dfrac{5}{3}$

(d) $\dfrac{25}{32}$

$-\dfrac{6}{32}$

Write $\frac{4}{8}$ in lowest terms.

$$\frac{7}{8} - \frac{3}{8} = \frac{4 \div 4}{8 \div 4} = \frac{1}{2}$$

The steps for subtracting like fractions are very similar to those for adding like fractions.

Subtracting Like Fractions

Step 1 Subtract the numerators to find the numerator of the difference.

Step 2 Write the denominator of the like fractions as the denominator of the difference.

Step 3 Write the answer in lowest terms.

EXAMPLE 3 **Subtracting Like Fractions**

Find the difference and simplify the answer.

(a) $\dfrac{15}{16} - \dfrac{3}{16}$

Subtract numerators.

Step 1 $\dfrac{15}{16} - \dfrac{3}{16} = \dfrac{15 - 3}{16}$

Step 2 $= \dfrac{12}{16}$ ← Difference of numerators
← Same denominator

Step 3 $= \dfrac{12 \div 4}{16 \div 4} = \dfrac{3}{4}$ ← Lowest terms

(b) $\dfrac{13}{4} - \dfrac{6}{4}$ | Fractions are ready to subtract if they are *like* fractions.

Subtract numerators.

$\dfrac{13}{4} - \dfrac{6}{4} = \dfrac{13 - 6}{4}$ ← Difference of numerators
← Same denominator

$= \dfrac{7}{4}$

To simplify the answer, write $\frac{7}{4}$ as a mixed number.

$\dfrac{7}{4} = 1\dfrac{3}{4}$ | Always simplify the answer.

CAUTION

Fractions may be subtracted **only** if they have like denominators.

Answers

3. **(a)** $4; \dfrac{2}{3}$ **(b)** $\dfrac{9}{10}$ **(c)** $3\dfrac{1}{3}$ **(d)** $\dfrac{19}{32}$

◀ **Work Problem** **3** at the Side.

3.1 Exercises

FOR EXTRA HELP

 Download the MyDashBoard App

MyMathLab®

CONCEPT CHECK *Next to each pair of fractions write* like *or* unlike.

1. $\dfrac{3}{8}$ $\dfrac{5}{8}$

2. $\dfrac{5}{16}$ $\dfrac{1}{4}$

3. $\dfrac{3}{5}$ $\dfrac{3}{4}$

4. $\dfrac{5}{12}$ $\dfrac{7}{12}$

CONCEPT CHECK *Fill in the blank.*

5. In order to add or subtract fractions, they must be _____ fractions.

6. After adding or subtracting like fractions, the answer should be written in _____.

Find the sum and simplify the answer. See Example 2.

7. $\dfrac{3}{8} + \dfrac{2}{8}$

8. $\dfrac{1}{5} + \dfrac{3}{5}$

9. $\dfrac{1}{4} + \dfrac{1}{4}$

10. $\begin{array}{r} \dfrac{9}{10} \\ + \dfrac{3}{10} \\ \hline \end{array}$

11. $\begin{array}{r} \dfrac{13}{12} \\ + \dfrac{5}{12} \\ \hline \end{array}$

12. $\begin{array}{r} \dfrac{2}{9} \\ + \dfrac{1}{9} \\ \hline \end{array}$

13. $\dfrac{7}{12} + \dfrac{3}{12} = \dfrac{7+3}{12} = \dfrac{10}{12} =$

14. $\dfrac{4}{15} + \dfrac{2}{15} + \dfrac{5}{15} = \dfrac{4+2+5}{15} =$

15. $\dfrac{3}{8} + \dfrac{7}{8} + \dfrac{2}{8}$

16. $\dfrac{4}{9} + \dfrac{1}{9} + \dfrac{7}{9}$

17. $\dfrac{2}{54} + \dfrac{8}{54} + \dfrac{12}{54}$

18. $\dfrac{7}{64} + \dfrac{15}{64} + \dfrac{20}{64}$

Find the difference and simplify the answer. See Example 3.

19. $\dfrac{7}{8} - \dfrac{4}{8}$

20. $\dfrac{2}{3} - \dfrac{1}{3}$

21. $\dfrac{10}{11} - \dfrac{4}{11}$

22. $\dfrac{4}{5} - \dfrac{3}{5}$

23. $\dfrac{9}{10} - \dfrac{3}{10} = \dfrac{9-3}{10} = \dfrac{6}{10} =$

24. $\dfrac{7}{14} - \dfrac{3}{14} = \dfrac{7-3}{14} = \dfrac{4}{14} =$

25. $\begin{array}{r} \dfrac{31}{21} \\ - \dfrac{7}{21} \\ \hline \end{array}$

26. $\begin{array}{r} \dfrac{43}{24} \\ - \dfrac{13}{24} \\ \hline \end{array}$

27. $\begin{array}{r} \dfrac{27}{40} \\ - \dfrac{19}{40} \\ \hline \end{array}$

28. $\begin{array}{r} \dfrac{38}{55} \\ - \dfrac{16}{55} \\ \hline \end{array}$

29. $\dfrac{47}{36} - \dfrac{5}{36}$

30. $\dfrac{76}{45} - \dfrac{21}{45}$

31. $\dfrac{73}{60} - \dfrac{7}{60}$

32. $\dfrac{181}{100} - \dfrac{31}{100}$

33. In your own words, write an explanation of how to add like fractions. Use three steps in your explanation.

34. Describe in your own words the difference between *like* fractions and *unlike* fractions. Give three examples of each type.

Solve each application problem. Write answers in lowest terms.

35. The Fair Oaks Save the Bluffs Committee raised $\frac{2}{9}$ of their target goal last year and another $\frac{5}{9}$ of the goal this year. What fraction of their goal has been raised?

36. After an initial payment to a winner at an arcade fundraiser, the organization still owed the winner $\frac{7}{10}$ of her total winnings. If the organization pays the winner another $\frac{3}{10}$ of the winnings, what fraction is still owed?

37. Julie Circle, a landscaper, is working on a commercial job and completed only $\frac{1}{8}$ of the irrigation system in the first week. If she completed $\frac{5}{8}$ of the system in the second week, what fraction of the irrigation system has she completed?

$$\frac{1}{8} + \frac{5}{8} = \frac{6}{8} =$$

38. On a wild shopping spree, Angie Gragg spent $\frac{3}{24}$ of the day shopping in the morning and another $\frac{5}{24}$ of the day shopping after lunch. What fraction of the day did she spend shopping?

$$\frac{3}{24} + \frac{5}{24} = \frac{8}{24} =$$

39. An organic farmer purchased $\frac{9}{10}$ acre of land one year and $\frac{3}{10}$ acre the next year. She then planted carrots on $\frac{7}{10}$ acre of the land and squash on the remainder. How much land is planted with squash?

40. A forester planted $\frac{5}{12}$ acre in seedlings in the morning and $\frac{11}{12}$ acre in the afternoon. That night, $\frac{7}{12}$ acre of seedlings were destroyed by frost. How many acres of seedlings remain?

Study Skills
MANAGING YOUR TIME

Many college students find themselves juggling a difficult schedule and multiple responsibilities. Perhaps you are going to school, working part time, and managing family demands. Here are some tips to help you develop good time management skills and habits.

OBJECTIVES

1 **Create a semester schedule.**

2 **Create a "to do" list.**

▶ **Read the syllabus for each class.** Check on class policies, such as attendance, late homework, and make-up tests. Find out how you are graded. Keep the syllabus in your notebook.

▶ **Make a semester or quarter calendar.** Put test dates and major due dates for *all* your classes on the same calendar. That way you will see which weeks are the really busy ones. Try using a different color pen for each class. Your brain responds well to the use of color. A semester calendar is on the next page.

▶ **Make a weekly schedule.** After you fill in your classes and other regular responsibilities (such as work, picking up kids from school, etc.), block off some study periods during the day that you can guarantee you will use for studying. Aim for 2 hours of study for each 1 hour you are in class.

▶ **Make "To Do" lists.** Then use them by crossing off the tasks as you complete them. You might even number them in the order they need to be done (most important ones first).

▶ **Break big assignments into smaller chunks.** They won't seem so big that way. Make deadlines for each small part so you stay on schedule.

▶ **Give yourself small breaks in your studying.** Do not try to study for hours at a time! Your brain needs rest between periods of learning. Try to give yourself a 10 minute break each hour or so. You will learn more and remember it longer.

▶ **If you get off schedule, just try to get back on schedule tomorrow.** We all slip from time to time. All is not lost! Make a new "to do" list and start doing the most important things first.

▶ **Get help when you need it.** Talk with your instructor during office hours. Also, most colleges have a Learning Center, tutoring center, or counseling office. If you feel lost and overwhelmed, ask for help. Someone can help you decide what to do first and what to spend your time on right away.

Which two or three of the suggestions above will you try this week? How do you think they will help you?

1. _____

2. _____

3. _____

Why Are These Techniques Brain Friendly?

Your brain appreciates some order. It enjoys a little routine, for example, choosing the same study time and place each day. You will find that you quickly settle in to your reading or homework.

Also, your brain **functions better when you are calm.** Too much rushing around at the last minute to get your homework and studying done sends hostile chemicals to your brain and makes it more difficult for you to learn and remember. So, a little planning can really pay off.

Building rest into your schedule is good for your brain. Remember, it takes time for dendrites to grow.

We've suggested using color on your calendars. This too, is brain friendly. Remember, your brain **likes pleasant colors and visual material** that are nice to look at. Messy and hard to read calendars will not be helpful, and you probably won't look at them often.

Study Skills

Continued from page 203

SEMESTER CALENDAR

WEEK	MON	TUES	WED	THUR	FRI	SAT	SUN
1							
2							
3							
4							
5							
6							
7							
8							
9							
10							
11							
12							
13							
14							
15							
16							

3.2 Least Common Multiples

Only *like* fractions can be added or subtracted. So, we must rewrite *unlike* fractions as *like* fractions before we can add or subtract them.

OBJECTIVE ▶ **1** **Find the least common multiple (LCM).** We can rewrite unlike fractions as like fractions by finding the *least common multiple* of the denominators.

> **Least Common Multiple (LCM)**
>
> The **least common multiple (LCM)** of two whole numbers is the smallest whole number divisible by both of those numbers.

EXAMPLE 1   **Finding the Least Common Multiple (LCM)**

Find the least common multiple of 6 and 9.

First, find the multiples of 6.

$$\underline{6 \cdot 1} \quad \underline{6 \cdot 2} \quad \underline{6 \cdot 3} \quad \underline{6 \cdot 4} \quad \underline{6 \cdot 5} \quad \underline{6 \cdot 6} \quad \underline{6 \cdot 7} \quad \underline{6 \cdot 8}$$
$$6, \quad 12, \quad 18, \quad 24, \quad 30, \quad 36, \quad 42, \quad 48, \ldots$$

(The three dots at the end of the list show that the list continues in the same pattern without stopping.) Now, find the multiples of 9.

$$\underline{9 \cdot 1} \quad \underline{9 \cdot 2} \quad \underline{9 \cdot 3} \quad \underline{9 \cdot 4} \quad \underline{9 \cdot 5} \quad \underline{9 \cdot 6} \quad \underline{9 \cdot 7} \quad \underline{9 \cdot 8}$$
$$9, \quad 18, \quad 27, \quad 36, \quad 45, \quad 54, \quad 63, \quad 72, \ldots$$

The smallest number found in *both* lists is 18, so 18 is the **least common multiple** of 6 and 9; the number 18 is the smallest whole number divisible by both 6 and 9.

Multiples of 6:   6, 12, **18**, 24, 30, 36, 42, 48, . . .

Multiples of 9:   9, **18**, 27, 36, 45, 54, 63, 72, . . .

> The *smallest* number in *both* lists is the LCM.

18 is the smallest number found in both lists. **18** is the least common multiple (LCM) of 6 and 9.

· Work Problem **1** at the Side. ▶

OBJECTIVE ▶ **2** **Find the least common multiple using multiples of the largest number.** There are several ways to find the least common multiple. If the numbers are small, the least common multiple can often be found by inspection. Can you think of a number that can be divided evenly by both 3 and 4? The number 12 will work; it is the least common multiple of 3 and 4. You can also find the least common multiple by writing multiples of the larger number.

In this case, 4 is larger than 3, so write the multiples of 4.

$$4, 8, 12, 16, 20, \ldots$$

Now, check each multiple of 4 to see if it is divisible by 3.

4 is *not* divisible by 3
8 is *not* divisible by 3
12 *is* divisible by 3

The first multiple of 4 that is divisible by 3 is 12, so 12 is the least common multiple of 3 and 4.

OBJECTIVES

1 Find the least common multiple (LCM).

2 Find the least common multiple using multiples of the largest number.

3 Find the least common multiple using prime factorization.

4 Find the least common multiple using an alternative method.

5 Write a fraction with an indicated denominator.

VOCABULARY TIP

Least common multiple (LCM) A common multiple can be found by simply multiplying together the numbers under consideration. To make simplifying fractions easier, it's helpful to find the **least common multiple (LCM).** This is the *smallest* number into which the original numbers divide.

1 **(a)** List the multiples of 5.
   GS   5, _10_, _15_, ____,
   ____, ____, _35_,
   ____, . . .

(b) List the multiples of 8.
   8, ____, ____, ____,
   ____, ____, ____, . . .

(c) Find the least common multiple of 5 and 8.

Answers

1. **(a)** 5, 10, 15, 20, 25, 30, 35, 40, . . .
   (b) 8, 16, 24, 32, 40, 48, 56, . . .
   (c) 40

2 Use multiples of the larger number to find the least common multiple in each set of numbers.

GS (a) 2 and 5

Multiples of 5 are:

5, 10, 15, 20, . . .

What is the first multiple of 5 that is divisible by 2? _____

So the LCM of 2 and 5 is _____.

(b) 3 and 9

(c) 6 and 8

(d) 4 and 7

3 Use prime factorization to find the LCM for each pair of numbers.

GS (a) 15 and 18

$15 = 3 \cdot 5$

$18 = $ _____ $\cdot$ _____ $\cdot$ _____

LCM = _____ $\cdot$ _____ $\cdot$ _____ $\cdot$

_____ = _____

(b) 12 and 20

Answers

2. (a) 10, 10   (b) 9   (c) 24   (d) 28

3. (a)

$$15 = 3 \cdot \circled{5}$$
$$18 = \circled{2} \cdot \circled{3} \cdot \circled{3}$$

$$\overset{15}{\wedge}$$

LCM = $2 \cdot 3 \cdot 3 \cdot 5 = 90$

$$\underset{18}{\vee\wedge}$$

(b)

$$12 = \circled{2} \cdot \circled{2} \cdot \circled{3}$$
$$20 = 2 \cdot 2 \cdot \circled{5}$$

$$\overset{12}{\wedge}$$

LCM = $2 \cdot 2 \cdot 3 \cdot 5 = 60$

$$\underset{20}{\vee\wedge}$$

EXAMPLE 2 Finding the Least Common Multiple (LCM)

Use multiples of the larger number to find the least common multiple of 6 and 9.

Start by writing the first few multiples of 9.

$$\overbrace{9, 18, 27, 36, 45, 54, \ldots}^{\text{Multiples of 9}}$$

Now check each multiple of 9 to see if it is divisible by 6. The first multiple of 9 that is divisible by 6 is 18.

$$9, \mathbf{18}, 27, 36, 45, 54, \ldots$$

First multiple divisible by 6, because $18 \div 6 = 3$

The least common multiple (LCM) of 6 and 9 is 18.

◀ **Work Problem 2 at the Side.**

OBJECTIVE ▶ 3 Find the least common multiple using prime factorization. **Example 2** shows how to find the least common multiple of two numbers by making a list of the multiples of the *larger* number. Although this method works well if both numbers are fairly small, it is usually easier to find the least common multiple for larger numbers by using *prime factorization,* as shown in the next example. (See **Chapter 2** for further review.)

EXAMPLE 3 Using Prime Factorization to Find the LCM

Use prime factorization to find the least common multiple of 9 and 12.

Start by finding the prime factorization of each number.

$$9 = 3 \cdot 3$$
$$12 = 2 \cdot 2 \cdot 3$$

Prime factorizations of 9 and 12.

Circle the factors where they appear the greatest number of times in either factorization.

$$9 = \circled{3} \cdot \circled{3}$$

3 appears most often in this factorization.

$$12 = \circled{2} \cdot \circled{2} \cdot 3$$

2 appears most often in this factorization.

The LCM is the product of the circled factors.

Factors of 9

$$\text{LCM} = 3 \cdot 3 \cdot 2 \cdot 2 = 36$$

The product of the circled factors is the LCM.

Factors of 12

The product of the prime factors, 36, is the least common multiple. Check to see that 36 is divisible by 9 (yes) and by 12 (yes). The smallest whole number divisible by both 9 and 12 is 36.

CAUTION

Notice that we did *not* repeat the factors that 9 and 12 have in common. In this case, the **3** in $2 \cdot 2 \cdot \mathbf{3} = 12$ was *not* used because 3 is already included in $3 \cdot 3 = 9$.

◀ **Work Problem 3 at the Side.**

EXAMPLE 4   **Using Prime Factorization**

Find the least common multiple of 12, 18, and 20.

Find the prime factorization of each number. Then use the prime factors to build the LCM.

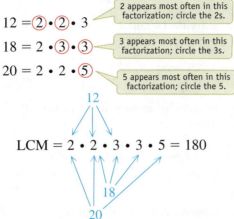

$12 = ②•②• 3$ 2 appears most often in this factorization; circle the 2s.

$18 = 2 •③•③$ 3 appears most often in this factorization; circle the 3s.

$20 = 2 • 2 •⑤$ 5 appears most often in this factorization; circle the 5.

$LCM = 2 • 2 • 3 • 3 • 5 = 180$

Check to see that 180 is divisible by 12 (yes) and by 18 (yes) and by 20 (yes). This smallest whole number divisible by 12, 18, and 20 is 180. The LCM is 180.

························ **Work Problem ④ at the Side.** ▶

EXAMPLE 5   **Finding the Least Common Multiple**

Find the least common multiple for each set of numbers.

(a) 5, 6, 35

Find the prime factorization for each number.

$5 =⑤$ Circle either 5, but not both.

$6 =②•③$

$35 = 5 •⑦$

$LCM = 2 • 3 • 5 • 7 = 210$

The least common multiple of 5, 6, and 35 is 210.

(b) 10, 20, 24

Find the prime factorization for each number.

$10 = 2 • 5$

$20 = 2 • 2 •⑤$

$24 =②•②•②•③$

$LCM = 2 • 2 • 2 • 3 • 5 = 120$

The least common multiple of 10, 20, and 24 is 120.

···················· **Work Problem ⑤ at the Side.** ▶

OBJECTIVE ▶ ④ **Find the least common multiple using an alternative method.** Some people like the following *alternative method* for finding the least common multiple for larger numbers. Try both methods, and *use the one you prefer.* As a review, a list of the first few prime numbers follows.

First few prime numbers ⟶ 2, 3, 5, 7, 11, 13, 17

④ Find the least common multiple of the denominators in each set of fractions.

(a) $\dfrac{3}{8}$ and $\dfrac{6}{5}$

(b) $\dfrac{5}{6}$ and $\dfrac{1}{14}$

(c) $\dfrac{4}{9}, \dfrac{5}{18},$ and $\dfrac{7}{24}$

⑤ Find the least common multiple for each set of numbers.

GS **(a)** 4, 8, 9

$4 = 2 • 2$

$8 = 2 • 2 • 2$

$9 = 3 • 3$

$LCM = \underline{\ \ } • \underline{\ \ } • \underline{\ \ } • \underline{\ \ } • \underline{\ \ } • \underline{\ \ } = \underline{\ \ }$

(b) 3, 6, 8

(c) 15, 20, 30, 40

Answers

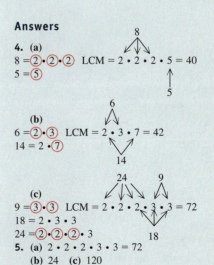

4. (a)
$8 =②•②•②$ $LCM = 2 • 2 • 2 • 5 = 40$
$5 =⑤$

(b)
$6 =②•③$ $LCM = 2 • 3 • 7 = 42$
$14 = 2 •⑦$

(c)
$9 =③•③$ $LCM = 2 • 2 • 2 • 3 • 3 = 72$
$18 = 2 • 3 • 3$
$24 =②•②•②• 3$

5. (a) $2 • 2 • 2 • 3 • 3 = 72$
(b) 24   **(c)** 120

6 In the following problems, the divisions have already been worked out. Multiply the prime numbers on the left to find the least common multiple.

GS (a) 2 | 6 15
    3 | 3 15
    5 | 1 5
       1 1

2 • ____ • ____ = ____

(b) 2 | 20 36
    2 | 10 18
    3 | 5 9
    3 | 5 3
    5 | 5 1
      1 1

EXAMPLE 6   **Alternative Method for Finding the Least Common Multiple**

Find the least common multiple for each set of numbers.

(a) 14 and 21

Start by trying to divide 14 and 21 by the first prime number, which is 2. Use the following shortcut.

14 divided by 2 is 7.　21 cannot be divided evenly by 2, so bring it down.

$$2 \underline{|14 \quad 21}$$
$$7 \quad 21$$

Because 21 cannot be divided evenly by 2, cross out 21 and bring it down. Divide by 3, the second prime.

7 cannot be divided evenly by 3, so bring it down.　21 divided by 3 is 7.

$$2 \underline{|14 \quad 21}$$
$$3 \underline{|7 \quad 21}$$
$$7 \quad 7$$

Since 7 cannot be divided evenly by the third prime, 5, skip 5. Divide by 7, the fourth prime.

Multiply these prime numbers to get the LCM.

$$2 \underline{|14 \quad 21}$$
$$3 \underline{|7 \quad 21}$$
$$7 \underline{|7 \quad 7}$$
$$1 \quad 1 \quad \text{All quotients are 1.}$$

When all quotients are 1, multiply the prime numbers on the left side.

$$\text{least common multiple} = 2 \cdot 3 \cdot 7 = 42$$

The least common multiple of 14 and 21 is 42.

(b) 6, 15, 18

Divide by 2.

$$2 \underline{|6 \quad 15 \quad 18} \quad \text{Cross out 15 and bring}$$
$$3 \quad 15 \quad 9 \quad\quad \text{it down.}$$

Divide by 3.

$$2 \underline{|6 \quad 15 \quad 18}$$
$$3 \underline{|3 \quad 15 \quad 9}$$
$$1 \quad 5 \quad 3$$

Divide by 3 again, since the remaining 3 can be divided.

$$2 \underline{|6 \quad 15 \quad 18}$$
$$3 \underline{|3 \quad 15 \quad 9}$$
$$3 \underline{|1 \quad 5 \quad 3}$$
$$1 \quad 5 \quad 1$$

Finally, divide by 5.

$$2 \underline{|6 \quad 15 \quad 18}$$
$$3 \underline{|3 \quad 15 \quad 9}$$
$$3 \underline{|1 \quad 5 \quad 3}$$
$$5 \underline{|1 \quad 5 \quad 1}$$
$$1 \quad 1 \quad 1 \quad \text{All quotients are 1.}$$

Multiply the prime numbers on the left side.

$$2 \cdot 3 \cdot 3 \cdot 5 = 90 \leftarrow \text{Least common multiple}$$

◀ **Work Problem 6 at the Side.**

Answers

6. (a) 3; 5; 30 **(b)** 180

EXAMPLE 7   Find the Least Common Multiple Using Either Method

Find the least common multiple of 12, 21, and 24. For **(a)** use the prime factorization method. Then, for **(b)** use the alternative method.

(a) Find the prime factorization for each number.

$$12 = 2 \cdot 2 \cdot ③$$
$$21 = 3 \cdot ⑦$$
$$24 = ② \cdot ② \cdot ② \cdot 3$$

The product of the circled factors is the LCM.

$$\text{LCM} = 2 \cdot 2 \cdot 2 \cdot 3 \cdot 7 = 168$$

(b) Use the alternative method to find the LCM.

$$
\begin{array}{r|ccc}
2 & 12 & 21 & 24 \\
2 & 6 & 21 & 12 \\
2 & 3 & 21 & 6 \\
3 & 3 & 21 & 3 \\
7 & 1 & 7 & 1 \\
\hline
  & 1 & 1 & 1
\end{array}
$$

All quotients are 1.

The product of the prime numbers on the left is the LCM.

$$\text{LCM} = \mathbf{2 \cdot 2 \cdot 2 \cdot 3 \cdot 7} = 168$$

················· **Work Problem 7 at the Side.** ▶

OBJECTIVE ▶ 5 Write a fraction with an indicated denominator. Before adding or subtracting *unlike* fractions, you must find the least common multiple, which is then used as the denominator of the fractions.

EXAMPLE 8   Writing a Fraction with an Indicated Denominator

Write the fraction $\frac{2}{3}$ with a denominator of 15.

Find a numerator, so that these fractions are equivalent.

$$\frac{2}{3} = \frac{?}{15}$$

To find the new numerator, first divide **15** by **3.**

$$\frac{2}{3} = \frac{?}{15} \qquad 15 \div 3 = 5$$

Multiply both numerator and denominator of the fraction $\frac{2}{3}$ by 5.

$$\frac{2}{3} = \frac{2 \cdot 5}{3 \cdot 5} = \frac{10}{15}$$

Multiplying by $\frac{5}{5}$ is the same as multiplying by 1, because $\frac{5}{5} = 1$.

This process is just the opposite of writing a fraction in lowest terms. Check the answer by writing $\frac{10}{15}$ in lowest terms; you should get $\frac{2}{3}$ again.

---

7 Find the least common multiple of each set of numbers. Use whichever method you prefer.

GS (a) 3, 6, 10

$$
\begin{array}{r|ccc}
2 & 3 & 6 & 10 \\
3 & 3 & 3 & 5 \\
5 & & 1 & \\
\hline
  & 1 & 1 & 1
\end{array}
$$

$$2 \cdot \underline{} \cdot 5 = \underline{}$$

(b) 15, 40

(c) 9, 24

(d) 8, 21, 24

Answers

7. (a) 1; 5; 3; 30   **(b)** 120   **(c)** 72   **(d)** 168

8 Rewrite each fraction with the indicated denominator.

(a) $\dfrac{1}{4} = \dfrac{?}{16}$

(b) $\dfrac{5}{3} = \dfrac{?}{15}$

(c) $\dfrac{7}{16} = \dfrac{?}{32}$

(d) $\dfrac{6}{11} = \dfrac{?}{33}$

EXAMPLE 9 **Writing Fractions with a New Denominator**

Rewrite each fraction with the indicated denominator.

(a) $\dfrac{3}{8} = \dfrac{?}{48}$

Divide 48 by 8, to get 6. Now multiply both the numerator and the denominator of $\frac{3}{8}$ by 6.

$$\frac{3}{8} = \frac{3 \cdot 6}{8 \cdot 6} = \frac{18}{48} \qquad \text{Multiply numerator and denominator by 6.}$$

> Multiplying a number by 1 does *not* change the number, and $\frac{6}{6} = 1$.

That is, $\frac{3}{8} = \frac{18}{48}$. As a check, write $\frac{18}{48}$ in lowest terms; you should get $\frac{3}{8}$ again.

(b) $\dfrac{7}{6} = \dfrac{?}{42}$

Divide 42 by 6, to get 7. Next, multiply both the numerator and the denominator of $\frac{7}{6}$ by 7.

$$\frac{7}{6} = \frac{7 \cdot 7}{6 \cdot 7} = \frac{49}{42} \qquad \text{Multiply numerator and denominator by 7.}$$

> Multiplying by $\frac{7}{7}$ is the same as multiplying by 1.

This shows that $\frac{7}{6} = \frac{49}{42}$. As a check, write $\frac{49}{42}$ in lowest terms. Did you get $\frac{7}{6}$ again?

Note

In **Example 8,** on the previous page, the fraction $\frac{2}{3}$ was multiplied by $\frac{5}{5}$. In **Example 9,** the fraction $\frac{3}{8}$ was multiplied by $\frac{6}{6}$ and the fraction $\frac{7}{6}$ was multiplied by $\frac{7}{7}$. The fractions, $\frac{5}{5}$, $\frac{6}{6}$, and $\frac{7}{7}$ are all equal to 1.

$$\frac{5}{5} = 1 \qquad \frac{6}{6} = 1 \qquad \frac{7}{7} = 1$$

Recall that any number multiplied by 1 is the number itself.

◀ **Work Problem** **8** at the Side.

Answers

8. (a) $\dfrac{4}{16}$ (b) $\dfrac{25}{15}$ (c) $\dfrac{14}{32}$ (d) $\dfrac{18}{33}$

3.2 Exercises

 FOR EXTRA HELP   Download the MyDashBoard App   MyMathLab®

CONCEPT CHECK *Write either* true *or* false *for each statement. If* false, *explain why.*

1. The least common multiple (LCM) of 4 and 8 is 8.

2. The least common multiple (LCM) of 6 and 5 is 25.

3. The least common multiple (LCM) of 9 and 4 is 36.

4. The least common multiple (LCM) of 3 and 7 is 28.

Use multiples of the larger number to find the least common multiple in each set of numbers. **See Examples 1 and 2.**

5. 3 and 6

6. 2 and 4

7. 3 and 5

8. 3 and 7

9. 4 and 9

10. 6 and 8

GS

$\overbrace{\text{Multiples of 8}}$

8 16 24 32 40

What is the first multiple divisible by 6? _____

11. 12 and 16

GS

$\overbrace{\text{Multiples of 16}}$

16 32 48 64 80

What is the first multiple divisible by 16? _____

12. 25 and 75

13. 20 and 50

Find the least common multiple of each set of numbers. Use any method. **See Examples 3–7.**

14. 4, 10

GS Prime factorizations

4 = ②·②

10 = 2·⑤

LCM = ____ · ____ · ____ = ____

15. 8, 10

GS Prime factorizations

8 = ②·②·②

10 = 2·⑤

LCM = ____ · ____ · ____ · ____ = ____

16. 12, 20

17. 9 and 15

18. 6, 9, 12

19. 20, 24, 30

20. 8, 9, 12, 18

21. 4, 6, 8, 10

22. 12, 15, 18, 20

23. 6, 8, 9, 27, 36 **24.** 8, 10, 12, 16, 36 **25.** 5, 6, 8, 25, 30

Rewrite each fraction with a denominator of 24. **See Examples 8 and 9.**

26. $\dfrac{2}{3} =$ **27.** $\dfrac{3}{8} =$ **28.** $\dfrac{3}{4} =$

29. $\dfrac{5}{12} =$ **30.** $\dfrac{5}{6} =$ **31.** $\dfrac{7}{8} =$

32. CONCEPT CHECK Circle the fraction that is equivalent to $\frac{2}{3}$.

$$\frac{7}{8} \quad \frac{3}{4} \quad \frac{12}{16} \quad \frac{8}{12}$$

33. CONCEPT CHECK Circle the fraction that is equivalent to $\frac{3}{4}$.

$$\frac{4}{5} \quad \frac{18}{22} \quad \frac{21}{28} \quad \frac{9}{15}$$

34. CONCEPT CHECK Circle the fraction that is equivalent to $\frac{9}{4}$.

$$\frac{36}{16} \quad \frac{35}{20} \quad \frac{21}{8} \quad \frac{15}{6}$$

35. CONCEPT CHECK Circle the fraction that is equivalent to $\frac{7}{8}$.

$$\frac{14}{15} \quad \frac{21}{24} \quad \frac{8}{7} \quad \frac{35}{48}$$

Rewrite each fraction with the indicated denominator.

36. $\dfrac{1}{2} = \dfrac{}{6}$ **37.** $\dfrac{2}{3} = \dfrac{}{9}$ **38.** $\dfrac{3}{4} = \dfrac{}{16}$

39. $\dfrac{7}{8} = \dfrac{}{32}$ **40.** $\dfrac{8}{5} = \dfrac{}{20}$ **41.** $\dfrac{3}{16} = \dfrac{}{64}$

42. $\dfrac{5}{8} = \dfrac{}{40}$ **43.** $\dfrac{9}{7} = \dfrac{}{56}$ **44.** $\dfrac{3}{2} = \dfrac{}{64}$

45. $\dfrac{7}{4} = \dfrac{}{48}$

$\dfrac{7}{4} = \dfrac{?}{48}$

Divide 48 by 4 to get _____.

Now multiply 7 by _____ to get the new numerator, which is _____.

46. $\dfrac{5}{6} = \dfrac{}{120}$

$\dfrac{5}{6} = \dfrac{?}{120}$

Divide 120 by 6 to get _____.

Now multiply 5 by _____ to get the new numerator, which is _____.

47. $\dfrac{8}{11} = \dfrac{}{132}$

48. $\dfrac{4}{15} = \dfrac{}{165}$

49. $\dfrac{3}{16} = \dfrac{}{144}$

50. $\dfrac{7}{16} = \dfrac{}{112}$

51. There are several methods for finding the least common multiple (LCM). Do you prefer the method using multiples of the largest number, the method using prime factorizations, or the alternative method for finding the least common multiple? Why? Would you ever use the other methods?

52. Explain in your own words how to write a fraction with an indicated denominator. As part of your explanation, show how to change $\frac{3}{4}$ to a fraction having 12 as a denominator. Also explain how you could check your answer.

Find the least common multiple of the denominators of each pair of fractions.

53. $\dfrac{25}{400}, \dfrac{38}{1800}$

54. $\dfrac{53}{600}, \dfrac{115}{4000}$

55. $\dfrac{109}{1512}, \dfrac{23}{392}$

56. $\dfrac{61}{810}, \dfrac{37}{1170}$

Relating Concepts (Exercises 57–64) For Individual or Group Work

Most people think that addition and subtraction of fractions are more difficult than multiplication and division of fractions. This is probably because a common denominator must be used. **Work Exercises 57–64 in order.**

57. Fractions with the same denominators are _____ fractions and fractions with different denominators are _____ fractions.

58. To subtract like fractions, first subtract the _____ to find the numerator of the difference. Write the denominator of the like fractions as the _____ of the difference. Finally, write the answer in _____ terms.

59. The _____ common multiple (LCM) of two numbers is the (*smallest/largest*) whole number divisible by both those numbers.

60. The following shows the common multiples for both 8 and 10. What is the least common multiple for these two numbers?

Multiples of 8:   8, 16, 24, 32, 40, 48, 56, 64, 72, 80, 88, . . .

Multiples of 10:   10, 20, 30, 40, 50, 60, 70, 80, 90, . . .

Find the least common multiple for each set of numbers. Use whichever method you like best.

61. 5, 7, 14, 10

62. 25, 18, 30, 5

63. Explain why the least common multiple for 8, 3, 5, 4, and 10 is not 240. Find the least common multiple.

64. Explain why the least common multiple of 55 and 1760 is 1760.

3.3 Adding and Subtracting Unlike Fractions

OBJECTIVES

1 Add unlike fractions.

2 Add unlike fractions vertically.

3 Subtract unlike fractions.

4 Subtract unlike fractions vertically.

OBJECTIVE 1 Add unlike fractions. In this section, we add and subtract unlike fractions. To add unlike fractions, we must first change them to like fractions (fractions with the same denominator). For example, the figures below show $\frac{3}{8}$ and $\frac{1}{4}$.

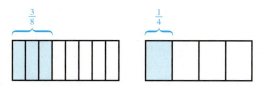

These fractions can be added by changing them to like fractions. Make like fractions by changing $\frac{1}{4}$ to the equivalent fraction $\frac{2}{8}$.

Now you can add the fractions.

$$\frac{3}{8} + \frac{1}{4} = \frac{3}{8} + \frac{2}{8} = \frac{5}{8}$$

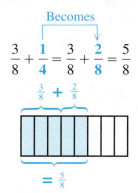

VOCABULARY TIP

LCM and LCD The least common multiple **(LCM)** and the least common denominator **(LCD)** are similar. When the LCM of two or more numbers is found and then used as a new denominator, it is called the least common denominator (LCD). You must find the LCD in order to add or subtract unlike fractions.

Use the following steps to add or subtract unlike fractions.

Adding or Subtracting Unlike Fractions

Step 1 Rewrite the *unlike fractions* as *like fractions* with the least common multiple as their new denominator. This new denominator is called the **least common denominator (LCD).**

Step 2 Add or subtract as with like fractions.

Step 3 Simplify the answer by writing it in lowest terms and as a whole or mixed number where possible.

EXAMPLE 1 Adding Unlike Fractions

Add $\frac{2}{3}$ and $\frac{1}{9}$.

The least common multiple of 3 and 9 is 9, so first rewrite the fractions as like fractions with a denominator of 9. This is the *least common denominator (LCD)* of 3 and 9.

Continued on Next Page

1 Add.

(GS) (a) $\dfrac{1}{2} + \dfrac{3}{8}$

$$\dfrac{}{8} + \dfrac{3}{8} = \dfrac{}{8}$$

(b) $\dfrac{3}{4} + \dfrac{1}{8}$

(c) $\dfrac{3}{5} + \dfrac{3}{10}$

(d) $\dfrac{1}{12} + \dfrac{5}{6}$

2 Add. Simplify all answers.

(GS) (a) $\dfrac{3}{10} + \dfrac{1}{5}$

$$\dfrac{3}{10} + \dfrac{}{10} = \dfrac{}{} = \dfrac{}{}$$

(b) $\dfrac{5}{8} + \dfrac{1}{3}$

(c) $\dfrac{1}{10} + \dfrac{1}{3} + \dfrac{1}{6}$

Answers

1. (a) $4; \dfrac{7}{8}$ **(b)** $\dfrac{7}{8}$ **(c)** $\dfrac{9}{10}$ **(d)** $\dfrac{11}{12}$

2. (a) $2; \dfrac{5}{10}; \dfrac{1}{2}$ **(b)** $\dfrac{23}{24}$ **(c)** $\dfrac{3}{5}$

Step 1
$$\dfrac{2}{3} = \dfrac{?}{9}$$

Divide 9 by 3, getting 3. Next, multiply numerator and denominator by 3.

$$\dfrac{2}{3} = \dfrac{2 \cdot 3}{3 \cdot 3} = \dfrac{6}{9} \qquad \text{⟨ } \tfrac{6}{9} \text{ is equivalent to } \tfrac{2}{3}.$$

Now, add the like fractions $\dfrac{6}{9}$ and $\dfrac{1}{9}$.

Becomes

> Both fractions must have the *same* denominator **before** you add them.

Step 2
$$\dfrac{2}{3} + \dfrac{1}{9} = \dfrac{6}{9} + \dfrac{1}{9} = \dfrac{6+1}{9} = \dfrac{7}{9}$$

Step 3 Step 3 is not needed because $\dfrac{7}{9}$ is already in lowest terms.

◀ **Work Problem** **1** **at the Side.**

EXAMPLE 2 **Adding Fractions**

Add each pair of fractions using the three steps. Simplify all answers.

(a) $\dfrac{1}{3} + \dfrac{1}{6}$

The least common multiple of 3 and 6 is 6. Rewrite both fractions as fractions with a least common denominator of 6.

Rewritten as like fractions

Step 1 $\dfrac{1}{3} + \dfrac{1}{6} = \dfrac{2}{6} + \dfrac{1}{6}$ ⟨ 6 is the LCD (least common denominator).

Add numerators.

Step 2 $\dfrac{2}{6} + \dfrac{1}{6} = \dfrac{2+1}{6} = \dfrac{3}{6}$ ← Sum of numerators
← Least common denominator

Step 3 $\dfrac{3}{6} = \dfrac{1}{2}$ ← Lowest terms

(b) $\dfrac{6}{15} + \dfrac{3}{10}$

The least common multiple of 15 and 10 is 30, so rewrite both fractions with a least common denominator of 30.

Rewritten as like fractions

Step 1 $\dfrac{6}{15} + \dfrac{3}{10} = \dfrac{12}{30} + \dfrac{9}{30}$ ⟨ The least common multiple of the denominators, 30, is also the LCD.

Add numerators.

Step 2 $\dfrac{12}{30} + \dfrac{9}{30} = \dfrac{12+9}{30} = \dfrac{21}{30}$

Step 3 $\dfrac{21}{30} = \dfrac{7}{10}$ ← Lowest terms

◀ **Work Problem** **2** **at the Side.**

OBJECTIVE **2** **Add unlike fractions vertically.** Fractions can also be added vertically (one fraction written below the other).

EXAMPLE 3 Vertical Addition of Fractions

Add the following fractions vertically.

(a)

$$\frac{3}{8} = \frac{3 \cdot 3}{8 \cdot 3} = \frac{9}{24} \quad \text{← 24 is the LCD.}$$

Rewritten as like fractions

$$+\frac{7}{12} = \frac{7 \cdot 2}{12 \cdot 2} = \frac{14}{24}$$

$$\frac{23}{24} \quad \begin{array}{l} \text{← Add the numerators.} \\ \text{← Denominator is 24, the LCD.} \end{array}$$

$\frac{23}{24}$ is in simplest form.

(b)

$$\frac{2}{9} = \frac{2 \cdot 4}{9 \cdot 4} = \frac{8}{36} \quad \text{← 36 is the LCD.}$$

Rewritten as like fractions

$$+\frac{1}{4} = \frac{1 \cdot 9}{4 \cdot 9} = \frac{9}{36}$$

$$\frac{17}{36} \quad \begin{array}{l} \text{← Add the numerators.} \\ \text{← Denominator is 36, the LCD.} \end{array}$$

$\frac{17}{36}$ is in simplest form.

············· **Work Problem** **3** **at the Side.** ▶

OBJECTIVE **3** **Subtract unlike fractions.** The next example shows subtraction of unlike fractions.

EXAMPLE 4 Subtracting Unlike Fractions

Subtract. Simplify all answers.

As with addition, rewrite unlike fractions with a least common denominator.

(a) $\dfrac{3}{4} - \dfrac{3}{8}$

Rewritten as like fractions

Step 1 $\dfrac{3}{4} - \dfrac{3}{8} = \dfrac{6}{8} - \dfrac{3}{8}$ ← The LCD is 8.

Subtract numerators.

Step 2 $\dfrac{6}{8} - \dfrac{3}{8} = \dfrac{6-3}{8} = \dfrac{3}{8}$ ← Difference of numerators ← Least common denominator

Step 3 Not needed because $\frac{3}{8}$ is in lowest terms.

············· **Continued on Next Page**

---

3 Add the following fractions vertically.

GS **(a)**
$$\frac{5}{8} = \frac{5 \cdot 3}{8 \cdot \underline{\ }} = \frac{15}{\underline{\ }}$$
$$+\frac{1}{12} = \frac{1 \cdot 2}{12 \cdot 2} = \frac{2}{24}$$
$$\overline{\overline{\underline{\ }}}$$

(b)
$$\frac{7}{16}$$
$$+\frac{1}{4}$$
$$\overline{}$$

(c)
$$\frac{1}{8}$$
$$\frac{5}{24}$$
$$+\frac{7}{16}$$
$$\overline{}$$

Answers

3. (a) $3; 24; \frac{17}{24}$ (b) $\frac{11}{16}$ (c) $\frac{37}{48}$

4 Subtract. Simplify all answers.

GS **(a)** $\dfrac{5}{8} - \dfrac{1}{4}$

$\dfrac{5}{8} - \dfrac{}{8} =$

(b) $\dfrac{4}{5} - \dfrac{3}{4}$

(b) $\dfrac{3}{4} - \dfrac{7}{12}$

Rewritten as like fractions

Step 1 $\quad \dfrac{3}{4} - \dfrac{7}{12} = \dfrac{9}{12} - \dfrac{7}{12}$

Subtract numerators.

Step 2 $\quad \dfrac{9}{12} - \dfrac{7}{12} = \dfrac{9-7}{12} = \dfrac{2}{12}$ ← Subtract the numerators.
← Denominator is 12, the LCD.

Step 3 $\quad \dfrac{2}{12} = \dfrac{1}{6}$ ← Lowest terms   Always simplify the final answer.

◄ **Work Problem** **4** **at the Side.**

OBJECTIVE **4** Subtract unlike fractions vertically.

EXAMPLE 5 **Vertical Subtraction of Fractions**

Subtract the following fractions vertically. Simplify all answers.

(a)
$$\begin{array}{r} \dfrac{4}{5} = \dfrac{4 \cdot 8}{5 \cdot 8} = \dfrac{32}{40} \\[2mm] -\dfrac{3}{8} = \dfrac{3 \cdot 5}{8 \cdot 5} = \dfrac{15}{40} \\[2mm] \hline \dfrac{17}{40} \end{array}$$

Rewritten as like fractions

$\dfrac{17}{40}$ ← Subtract numerators.
$\phantom{\dfrac{17}{40}}$ ← Denominator is 40, the LCD.

$\frac{17}{40}$ is in simplest form.

(b)
$$\begin{array}{r} \dfrac{3}{7} = \dfrac{3 \cdot 12}{7 \cdot 12} = \dfrac{36}{84} \\[2mm] -\dfrac{5}{12} = \dfrac{5 \cdot 7}{12 \cdot 7} = \dfrac{35}{84} \\[2mm] \hline \dfrac{1}{84} \end{array}$$

Rewritten as like fractions

$\dfrac{1}{84}$ ← Subtract numerators.
$\phantom{\dfrac{1}{84}}$ ← Denominator is 84, the LCD.

$\frac{1}{84}$ is in simplest form.

◄ **Work Problem** **5** **at the Side.**

---

5 Subtract vertically. Simplify all answers.

GS **(a)** $\dfrac{7}{8} = \dfrac{7 \cdot }{8 \cdot 3} = \dfrac{21}{}$

$\quad -\dfrac{2}{3} = \dfrac{2 \cdot 8}{3 \cdot 8} = \dfrac{16}{24}$

(b) $\dfrac{5}{6}$

$\quad -\dfrac{1}{12}$

Answers

4. (a) $2; \dfrac{3}{8}$ (b) $\dfrac{1}{20}$

5. (a) $3; 24; \dfrac{5}{24}$ (b) $\dfrac{3}{4}$

3.3 Exercises

 FOR EXTRA HELP

 Download the MyDashBoard App

MyMathLab®

CONCEPT CHECK *Fill in the blank with the correct response.*

1. To add or subtract unlike fractions, the first step is to rewrite the fractions as _____ fractions.

2. To rewrite unlike fractions as like fractions, you must find the _____ (LCD).

Add the following fractions. Simplify all answers. **See Examples 1–3.**

3. $\dfrac{3}{4} + \dfrac{1}{8}$

4. $\dfrac{1}{6} + \dfrac{2}{3}$

5. $\dfrac{2}{3} + \dfrac{2}{9}$

6. GS $\dfrac{3}{7} + \dfrac{1}{14}$ LCD is 14.

$\dfrac{6}{14} + \dfrac{1}{14}$

$= \dfrac{7}{14} =$ simplest form

7. GS $\dfrac{9}{20} + \dfrac{3}{10}$ LCD is 20.

$\dfrac{9}{20} + \dfrac{6}{20}$

$= \dfrac{15}{20} =$ simplest form

8. $\dfrac{5}{8} + \dfrac{1}{4}$

9. $\dfrac{3}{5} + \dfrac{3}{8}$

10. $\dfrac{5}{7} + \dfrac{3}{14}$

11. $\dfrac{2}{9} + \dfrac{5}{12}$

12. GS $\dfrac{1}{4} + \dfrac{2}{9} + \dfrac{1}{3}$ LCD is 36.

$\dfrac{9}{36} + \dfrac{8}{36} + \dfrac{12}{36} = \dfrac{}{36}$ Already in simplest form

13. GS $\dfrac{3}{7} + \dfrac{2}{5} + \dfrac{1}{10}$ LCD is 70.

$\dfrac{30}{70} + \dfrac{28}{70} + \dfrac{7}{70}$

$= \dfrac{65}{70} =$ simplest form

14. $\dfrac{3}{10} + \dfrac{2}{5} + \dfrac{3}{20}$

15. $\dfrac{1}{3} + \dfrac{3}{8} + \dfrac{1}{4}$

16. $\dfrac{4}{15} + \dfrac{1}{6} + \dfrac{1}{3}$

17. $\dfrac{5}{12} + \dfrac{2}{9} + \dfrac{1}{6}$

18. $\begin{array}{r} \dfrac{2}{3} \\ + \dfrac{1}{6} \\ \hline \end{array}$

19. $\begin{array}{r} \dfrac{1}{4} \\ + \dfrac{1}{8} \\ \hline \end{array}$

20. $\begin{array}{r} \dfrac{7}{12} \\ + \dfrac{1}{8} \\ \hline \end{array}$

21. $\begin{array}{r} \dfrac{5}{12} \\ + \dfrac{1}{16} \\ \hline \end{array}$

22. $\begin{array}{r} \dfrac{3}{7} \\ + \dfrac{1}{3} \\ \hline \end{array}$

Subtract the following fractions. Simplify all answers. **See Example 4.**

23. $\dfrac{5}{6} - \dfrac{1}{3}$ LCD is 6.

$\dfrac{5}{6} - \dfrac{2}{6} = \dfrac{3}{6} =$ simplest form

24. $\dfrac{3}{4} - \dfrac{5}{8}$ LCD is 8.

$\dfrac{6}{8} - \dfrac{5}{8} =$ simplest form

25. $\dfrac{2}{3} - \dfrac{1}{6}$

26. $\dfrac{5}{8} - \dfrac{1}{4}$

27. $\dfrac{2}{3} - \dfrac{1}{5}$

28. $\dfrac{5}{6} - \dfrac{7}{9}$

29. $\dfrac{5}{12} - \dfrac{1}{4}$

30. $\dfrac{5}{7} - \dfrac{1}{3}$

31. $\dfrac{8}{9} - \dfrac{7}{15}$

32.
$$\begin{array}{r} \frac{4}{5} \\ -\frac{1}{3} \\ \hline \end{array}$$

33.
$$\begin{array}{r} \frac{7}{8} \\ -\frac{4}{5} \\ \hline \end{array}$$

34.
$$\begin{array}{r} \frac{5}{8} \\ -\frac{1}{3} \\ \hline \end{array}$$

35.
$$\begin{array}{r} \frac{5}{12} \\ -\frac{1}{16} \\ \hline \end{array}$$

36.
$$\begin{array}{r} \frac{7}{12} \\ -\frac{1}{3} \\ \hline \end{array}$$

Solve each application problem.

4-Piece Chisel Set

LOT NO. 42429

- Cutting-edge widths of $\frac{1}{4}$", $\frac{1}{2}$", $\frac{3}{4}$", and 1"
- Heat-treated, high-carbon steel
- Straight bevel

SALE! $4⁹⁷

REGULAR PRICE $7.99

Use the newspaper advertisement for this 4-piece chisel set to answer Exercises 37–38. (*Source:* Harbor Freight Tools.)

37. Find the difference in the cutting-edge width of the two chisels with the widest blades. The symbol " is for inches.

38. Find the difference in the cutting-edge width of the two chisels with the narrowest blades. The " symbol is for inches.

39. A sports and entertainment center has $\frac{4}{5}$ of its total area devoted to seating of fans and guests. If $\frac{3}{8}$ of the seating area is used for general admission seating and the rest for reserved seating, find the fraction of the total area used for reserved seating.

40. A dairy farmer must vaccinate $\frac{5}{8}$ of her cows this week. If she vaccinates $\frac{3}{16}$ of the herd on Monday, and $\frac{1}{4}$ of the herd on Wednesday, what fraction of the herd remains to be vaccinated?

41. When installing cabinets for The Home Depot, Sarah Bryn must be certain that the proper type and size of mounting screw is used. Find the total length of the screw shown.

$\frac{1}{8}$ in. $\frac{1}{4}$ in. $\frac{2}{5}$ in.

42. When installing a computer chassis, Bonnie Bottorff must be certain that the proper type and size of bolt is used. Find the total length of the bolt shown.

$\frac{1}{5}$ in. $\frac{1}{3}$ in. $\frac{1}{4}$ in.

43. Bill Newton is a general contractor. He began a job with $\frac{3}{4}$ of a tank of fuel in his backhoe. He used $\frac{1}{3}$ of the tank in the morning and $\frac{3}{8}$ of the tank in the afternoon. What fraction of the tank of fuel remains?

44. Cliff Dinsmore is coordinating the refurbishing of the community swimming pool. The pool was $\frac{7}{8}$ full when workers began draining it. By noon, another $\frac{3}{16}$ of the pool had been drained. An additional $\frac{1}{3}$ of the pool was drained in the afternoon. Find the fraction of the pool water remaining.

45. Step 1 in adding or subtracting unlike fractions is to rewrite the fractions so they have the least common multiple as a denominator. Explain in your own words why this is necessary.

46. Briefly list the three steps used for addition and subtraction of unlike fractions.

A survey of 1200 users of social networking sites showed that honesty was not always practiced. Refer to the circle graph to answer Exercises 47–50.

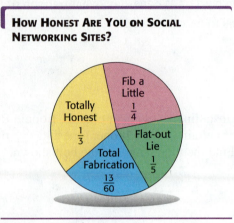

HOW HONEST ARE YOU ON SOCIAL NETWORKING SITES?

Fib a Little $\frac{1}{4}$

Totally Honest $\frac{1}{3}$

Flat-out Lie $\frac{1}{5}$

Total Fabrication $\frac{13}{60}$

Source: *USA Today.*

47. What fraction of those surveyed are totally honest?

48. What fraction of those surveyed fib a little?

49. Which response was given most often? How many people gave this response? What fraction of the users gave this response and the "fib a little" response?

50. Which response was given least often? How many people gave this response? What fraction of the users gave this response and the "total fabrication" response?

51. Find the diameter of the hole in the mounting bracket shown. (The diameter is the distance across the center of the hole.)

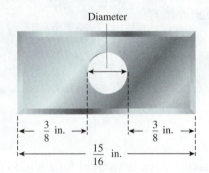

Diameter

$\frac{3}{8}$ in. $\frac{3}{8}$ in.

$\frac{15}{16}$ in.

52. Chakotay is fitting a turquoise stone into a bear claw pendant. Find the diameter of the hole in the pendant. (The diameter is the distance across the center of the hole.)

$\frac{3}{16}$ in. $\frac{3}{16}$ in.

$\frac{7}{8}$ in.

3.4 Adding and Subtracting Mixed Numbers

Recall that a mixed number is the sum of a whole number and a fraction. For example,

$$3\frac{2}{5} \quad \text{means} \quad 3 + \frac{2}{5}.$$

OBJECTIVE ▶ 1 Estimate an answer, then add or subtract mixed numbers. Add or subtract mixed numbers by adding or subtracting the fraction parts and then the whole number parts. It is a good idea to estimate the answer first, as we did when multiplying and dividing mixed numbers in **Chapter 2.**

Work Problem **1** at the Side. ▶

1 As a review of mixed numbers, write each mixed number as an improper fraction and each improper fraction as a mixed number.

(a) $\dfrac{9}{2}$

(b) $\dfrac{8}{3}$

(c) $4\dfrac{3}{4}$

(d) $3\dfrac{7}{8}$

EXAMPLE 1 Adding and Subtracting Mixed Numbers

First estimate the answer. Then add or subtract to find the exact answer.

(a) $16\dfrac{1}{8} + 5\dfrac{5}{8}$

Estimate: *Exact:*

$$16 \xleftarrow{\text{Rounds to}} \left\{ \quad 16\frac{1}{8} \right.$$

$$+ \ \ 6 \xleftarrow{\text{Rounds to}} \left\{ + \ 5\frac{5}{8} \right.$$

> The exact answer is close to the estimate.

First, estimate the answer. $\quad 22 \qquad 21\dfrac{6}{8} = 21\dfrac{3}{4} \leftarrow$ Lowest terms

Sum of whole numbers ⟶   ⟵ Sum of fractions

In lowest terms $\frac{6}{8}$ is $\frac{3}{4}$, so the exact answer of $21\frac{3}{4}$ is in lowest terms. The exact answer is *reasonable* because it is close to the estimate of 22.

(b) $8\dfrac{5}{8} - 3\dfrac{1}{12}$

Estimate: *Exact:*

$$9 \xleftarrow{\text{Rounds to}} \left\{ \quad 8\frac{5}{8} = \ \ 8\frac{15}{24} \right.$$

$$- \ 3 \xleftarrow{\text{Rounds to}} \left\{ - 3\frac{1}{12} = - 3\frac{2}{24} \right.$$

24 is the least common denominator.

$$6 \qquad\qquad\qquad 5\frac{13}{24} \leftarrow \text{Lowest terms}$$

Subtract whole numbers. ⟶   ⟵ Subtract fractions.

The exact answer of $5\frac{13}{24}$ is *reasonable* because it is close to the estimated answer of 6. Check by adding $5\frac{13}{24}$ and $3\frac{1}{12}$; the sum should be $8\frac{5}{8}$.

Continued on Next Page

Answers

1. (a) $4\dfrac{1}{2}$ (b) $2\dfrac{2}{3}$ (c) $\dfrac{19}{4}$ (d) $\dfrac{31}{8}$

2 First estimate, and then add or subtract to find the exact answer.

GS **(a)** *Estimate:*   *Exact:*

$$7 \xleftarrow{\text{Rounds to}} \left\{ \quad 6\frac{7}{8} = 6\frac{7}{8} \right.$$

$$+2 \xleftarrow{\text{Rounds to}} \left\{ +2\frac{1}{4} = 2\frac{2}{8} \right.$$

(b) *Estimate:*   *Exact:*

$$\xleftarrow{\text{Rounds to}} \left\{ \quad 4\frac{7}{9} \right.$$

$$- \xleftarrow{\text{Rounds to}} \left\{ -2\frac{2}{3} \right.$$

3 First estimate, and then add to find the exact answer.

GS **(a)** *Estimate:*   *Exact:*

$$10 \xleftarrow{\text{Rounds to}} \left\{ \quad 9\frac{3}{4} \right.$$

$$+\ 8 \xleftarrow{\text{Rounds to}} \left\{ +7\frac{1}{2} \right.$$

(b) *Estimate:*   *Exact:*

$$\xleftarrow{\text{Rounds to}} \left\{ \quad 15\frac{4}{5} \right.$$

$$+ \xleftarrow{\text{Rounds to}} \left\{ +12\frac{2}{3} \right.$$

Answers

2. **(a)** $7 + 2 = 9; 9\frac{1}{8}$

    (b) $5 - 3 = 2; 2\frac{1}{9}$

3. **(a)** $10 + 8 = 18; 17\frac{1}{4}$

    (b) $16 + 13 = 29; 28\frac{7}{15}$

Note

When estimating, if the numerator is *half* of the denominator or *more*, round up the whole number part. If the numerator is *less* than *half* the denominator, leave the whole number part as it is.

◀ **Work Problem 2 at the Side.**

When you add the fraction parts of mixed numbers, the sum may be greater than 1. If this happens, simplify the fraction and regroup in the whole number column.

EXAMPLE 2   **Simplify and Regroup When Adding Mixed Numbers**

First estimate, and then add $9\frac{5}{8} + 13\frac{7}{8}$.

Estimate: *Exact:*

$$10 \xleftarrow{\text{Rounds to}} \left\{ \quad 9\frac{5}{8} \right.$$

First, add the fractions, then the whole numbers.

$$+14 \xleftarrow{\text{Rounds to}} \left\{ +13\frac{7}{8} \right.$$

$$24 \qquad\qquad 22\frac{12}{8}$$

Sum of whole numbers ⟶   ⟵ Sum of fractions

The improper fraction $\frac{12}{8}$ can be written in lowest terms as $\frac{3}{2}$. Then $\frac{3}{2} = 1\frac{1}{2}$, so the simplified sum is

Becomes   *Becomes*

$$22\frac{12}{8} = 22 + \frac{12}{8} = 22 + \frac{3}{2} = 22 + 1\frac{1}{2} = 23\frac{1}{2}.$$

The estimate was 24, so the exact answer of $23\frac{1}{2}$ is reasonable.

Note

When adding mixed numbers, first add the fraction parts, then add the whole number parts. Finally, combine the two answers and simplify.

◀ **Work Problem 3 at the Side.**

OBJECTIVE ▶ 2 **Estimate an answer, then subtract mixed numbers by regrouping.** When subtracting mixed numbers, **regrouping** is necessary when the fraction part of the first number is less than the fraction part of the second number.

EXAMPLE 3   **Regroup When Subtracting Mixed Numbers**

First estimate, and then subtract to find the exact answer.

(a) $7 - 2\frac{5}{6}$

Continued on Next Page

Estimate: *Exact:*

$$7 \xleftarrow{\text{Rounds to}} \left\{ \begin{array}{c} 7 \end{array} \right.$$

> There is no fraction here from which to subtract $\frac{5}{6}$.

$$\frac{-\,3}{4} \xleftarrow{\text{Rounds to}} \left\{ \begin{array}{c} -\,2\dfrac{5}{6} \\ \hline \end{array} \right.$$

It is **not** possible to subtract $\frac{5}{6}$ without regrouping the whole number **7** first.

Regroup 7 as 6 + 1.

$$7 = \overbrace{6 + \mathbf{1}}$$

$$1 = \frac{6}{6}$$

$$= 6 + \frac{\mathbf{6}}{\mathbf{6}}$$

$$= 6\frac{6}{6}$$

Now you can subtract.

$$7 = \qquad 6\frac{6}{6}$$

> 7 was rewritten as $6\frac{6}{6}$.

$$-\,2\frac{5}{6} = -\,2\frac{5}{6}$$

$$\overline{\qquad\qquad\qquad 4\frac{1}{6}}$$

> Exact answer is close to the estimate.

The estimate was 4, so the exact answer of $4\frac{1}{6}$ is reasonable.

(b) $8\dfrac{1}{3} - 4\dfrac{3}{5}$

Estimate: *Exact:*

$$8 \xleftarrow{\text{Rounds to}} \left\{ \begin{array}{c} 8\dfrac{1}{3} = \quad 8\dfrac{5}{15} \end{array} \right.$$

$$\frac{-\,5}{3} \xleftarrow{\text{Rounds to}} \left\{ \begin{array}{c} -\,4\dfrac{3}{5} = -\,4\dfrac{9}{15} \\ \hline \end{array} \right.$$

15 is the least common denominator.

It is **not** possible to subtract $\frac{9}{15}$ from $\frac{5}{15}$, so regroup the whole number **8.**

Regroup 8 as 7 + 1.

$$8\frac{5}{15} = 8 + \frac{5}{15} = \overbrace{7 + \mathbf{1}} + \frac{5}{15}$$

$$1 = \frac{15}{15}$$

$$= 7 + \frac{\mathbf{15}}{\mathbf{15}} + \frac{5}{15}$$

$$= 7 + \frac{\mathbf{20}}{\mathbf{15}} \quad \leftarrow \frac{15}{15} + \frac{5}{15}$$

$$= 7\frac{20}{15}$$

Continued on Next Page

4 First estimate and then subtract to find the exact answer.

(a) *Estimate:*   *Exact:*

$$7 \xleftarrow{\text{Rounds to}} \begin{cases} 7\frac{1}{3} = 7\frac{2}{6} \end{cases}$$

$$-5 \xleftarrow{\text{Rounds to}} \begin{cases} -4\frac{5}{6} = 4\frac{5}{6} \end{cases}$$

(b) *Estimate:*   *Exact:*

$$\xleftarrow{\text{Rounds to}} \begin{cases} 4\frac{5}{8} \end{cases}$$

$$\xleftarrow{\text{Rounds to}} \begin{cases} -2\frac{15}{16} \end{cases}$$

(c) *Estimate:*   *Exact:*

$$\xleftarrow{\text{Rounds to}} \begin{cases} 15 \end{cases}$$

$$\xleftarrow{\text{Rounds to}} \begin{cases} -6\frac{4}{9} \end{cases}$$

5 Add or subtract by changing mixed numbers to improper fractions. Simplify answers.

(a) $3\frac{3}{8} = \frac{27}{8} = \frac{27}{8}$

$+2\frac{1}{2} = \frac{5}{2} = \frac{}{8}$ $\left. \begin{array}{c} \\ \\ \end{array} \right\}$ 8 is LCD.

(b) $6\frac{3}{4}$

$-4\frac{2}{3}$

Answers

4. (a) $7 - 5 = 2$; $2\frac{1}{2}$

   (b) $5 - 3 = 2$; $1\frac{11}{16}$

   (c) $15 - 6 = 9$; $8\frac{5}{9}$

5. (a) $\frac{27}{8} + \frac{20}{8} = \frac{47}{8} = 5\frac{7}{8}$

   (b) $\frac{25}{12} = 2\frac{1}{12}$

Now you can subtract.

$$8\frac{1}{3} = 8\frac{5}{15} = 7\frac{20}{15}$$

$$-4\frac{3}{5} = 4\frac{9}{15} = 4\frac{9}{15}$$

$$\overline{3\frac{11}{15}}$$

The exact answer is $3\frac{11}{15}$ (lowest terms), which is reasonable because it is close to the estimate of 3.

◀ **Work Problem 4 at the Side.**

OBJECTIVE 3 Add or subtract mixed numbers using an alternative method. An alternative method for adding or subtracting mixed numbers is to first change the mixed numbers to improper fractions. Then rewrite the unlike fractions as like fractions. Finally, add or subtract the numerators and write the answer in lowest terms.

EXAMPLE 4   Adding or Subtracting Mixed Numbers

Add or subtract.

(a) $2\frac{3}{8} = \frac{19}{8} = \frac{19}{8}$

$+3\frac{3}{4} = \frac{15}{4} = \frac{30}{8}$ ⟵ 8 is the least common denominator.

Rewrite $2\frac{3}{8}$ as $\frac{19}{8}$ and $3\frac{3}{4}$ as $\frac{15}{4}$.

$\frac{49}{8} = 6\frac{1}{8}$ ⟵ Answer as mixed number

(b) $4\frac{2}{3} = \frac{14}{3} = \frac{70}{15}$

$-2\frac{1}{5} = \frac{11}{5} = \frac{33}{15}$ ⟵ 15 is the least common denominator.

$\frac{37}{15} = 2\frac{7}{15}$ Simplify the answer by writing it as a mixed number.

Improper fractions

◀ **Work Problem 5 at the Side.**

Note

The advantage of this alternative method of adding or subtracting mixed numbers is that it eliminates the need to regroup. It is also the most useful method for working with algebraic fractions. (See **Chapter 9.**) However, if the mixed numbers are large, then the numerators of the improper fractions may become so large that they are difficult to work with. In such cases, you may want to keep the numbers as mixed numbers.

3.4 Exercises

 FOR EXTRA HELP

 MyMathLab®

Download the MyDashBoard App

CONCEPT CHECK *Round each of the following mixed numbers to the nearest whole number.*

1. $5\frac{1}{3}$

2. $6\frac{3}{10}$

3. $8\frac{4}{5}$

4. $12\frac{1}{2}$

5. $15\frac{7}{15}$

6. $20\frac{5}{8}$

7. $16\frac{2}{3}$

8. $3\frac{5}{12}$

First estimate the answer. Then add to find the exact answer. Write answers as mixed numbers. **See Examples 1 and 2.**

9. *Estimate:* GS

$$6 \xleftarrow{\text{Rounds to}} \{ \quad 5\frac{1}{2} = 5\frac{3}{6}$$

$$+ 3 \xleftarrow{\text{Rounds to}} \{ + 3\frac{1}{3} = 3\frac{2}{6}$$

$$\overline{\quad 9 \quad}$$

Exact:

10. *Estimate:* GS

$$7 \xleftarrow{\hspace{1cm}} \{ \quad 6\frac{3}{5} = 6\frac{6}{10}$$

$$+ 7 \xleftarrow{\hspace{1cm}} \{ + 7\frac{1}{10} = 7\frac{1}{10}$$

$$\overline{\quad 14 \quad}$$

Exact:

11. *Estimate:*

$$7\frac{1}{3}$$

$$+ 4\frac{1}{6}$$

Exact:

12. *Estimate:*

$$+$$

Exact:

$$10\frac{1}{4}$$

$$+ 5\frac{5}{8}$$

13. *Estimate:*

$$+$$

Exact:

$$\frac{5}{8}$$

$$+ 3\frac{7}{12}$$

14. *Estimate:*

$$+$$

Exact:

$$12\frac{4}{5}$$

$$+ \frac{7}{10}$$

15. *Estimate:*

$$+$$

Exact:

$$24\frac{5}{6}$$

$$+ 18\frac{5}{6}$$

16. *Estimate:*

$$+$$

Exact:

$$14\frac{6}{7}$$

$$+ 15\frac{1}{2}$$

17. *Estimate:*

$$+$$

Exact:

$$33\frac{3}{5}$$

$$+ 18\frac{1}{2}$$

18. *Estimate:*

$$+$$

Exact:

$$18\frac{5}{8}$$

$$+ 6\frac{2}{3}$$

19. *Estimate:*

$$+$$

Exact:

$$22\frac{3}{4}$$

$$+ 15\frac{3}{7}$$

20. *Estimate:*

$$+$$

Exact:

$$7\frac{1}{4}$$

$$+ 25\frac{7}{8}$$

21. *Estimate:*   *Exact:*

$$12\frac{8}{15}$$

$$18\frac{3}{5}$$

$+$ _____ $+\ 14\frac{7}{10}$

_____

22. *Estimate:*   *Exact:*

$$14\frac{9}{10}$$

$$8\frac{1}{4}$$

$+$ _____ $+\ 13\frac{3}{5}$

_____

First estimate the answer. Then subtract to find the exact answer. Simplify all answers.
See Examples 1 and 3.

23. *Estimate:*   *Exact:*

$15 \longleftarrow$ _____ $14\frac{7}{8} = 14\frac{7}{8}$

$-\ 12 \longleftarrow$ _____ $-\ 12\frac{1}{4} = 12\frac{2}{8}$

_____   _____   _____

3

24. *Estimate:*   *Exact:*

$15 \longleftarrow$ _____ $14\frac{3}{4} = 14\frac{6}{8}$

$-\ 11 \longleftarrow$ _____ $-\ 11\frac{3}{8} = 11\frac{3}{8}$

_____   _____   _____

4

25. *Estimate:*   *Exact:*

$$12\frac{2}{3}$$

$-$ $-\ 1\frac{1}{5}$

_____   _____

26. *Estimate:*   *Exact:*

$$11\frac{9}{20}$$

$-$ $-\ 4\frac{3}{5}$

_____   _____

27. *Estimate:*   *Exact:*

$$28\frac{3}{10}$$

$-$ $-\ 6\frac{1}{15}$

_____   _____

28. *Estimate:*   *Exact:*

$$15\frac{7}{20}$$

$-$ $-\ 6\frac{1}{8}$

_____   _____

29. *Estimate:*   *Exact:*

$17 = 16\frac{8}{8}$

$-$ $-\ 6\frac{5}{8} = 6\frac{5}{8}$

_____   _____   _____

30. *Estimate:*   *Exact:*

$22 = 21\frac{6}{6}$

$-$ $-\ 4\frac{5}{6} = 4\frac{5}{6}$

_____   _____   _____

31. *Estimate:*   *Exact:*

$$18\frac{3}{4}$$

$-$ $-\ 5\frac{4}{5}$

_____   _____

32. *Estimate:*   *Exact:*

$$14\frac{5}{8}$$
$$- \underline{\hspace{1.5em}} \qquad - 3\frac{2}{3}$$

33. *Estimate:*   *Exact:*

$$19\frac{2}{3}$$
$$- \underline{\hspace{1.5em}} \qquad - 11\frac{3}{4}$$

34. *Estimate:*   *Exact:*

$$20\frac{3}{5}$$
$$- \underline{\hspace{1.5em}} \qquad - 12\frac{7}{15}$$

CONCEPT CHECK *Write each mixed number as an improper fraction and each improper fraction as a mixed number.*

35. $3\frac{3}{4}$

36. $7\frac{7}{8}$

37. $\frac{12}{5}$

38. $\frac{24}{7}$

39. $5\frac{3}{8}$

40. $6\frac{2}{15}$

41. $\frac{56}{3}$

42. $\frac{29}{8}$

Add or subtract by changing mixed numbers to improper fractions. Write answers as mixed numbers when possible. **See Example 4.**

43.
$$7\frac{5}{8} = \frac{61}{8} = \frac{61}{8}$$
$$+ 1\frac{1}{2} = \frac{3}{2} = \frac{12}{8}$$
$$\overline{} \quad \frac{73}{8} =$$

44.
$$8\frac{3}{4} = \frac{35}{4} = \frac{70}{8}$$
$$+ 1\frac{5}{8} = \frac{13}{8} = \frac{13}{8}$$
$$\overline{} \quad \frac{83}{8} =$$

45.
$$4\frac{2}{3}$$
$$+ 6\frac{5}{6}$$

46.
$$3\frac{3}{5}$$
$$+ 6\frac{1}{2}$$

47.
$$2\frac{2}{3}$$
$$+ 1\frac{1}{6}$$

48.
$$4\frac{1}{2}$$
$$+ 2\frac{3}{4}$$

49.
$$3\frac{1}{4}$$
$$+ 3\frac{2}{3}$$

50.
$$2\frac{4}{5}$$
$$+ 5\frac{1}{3}$$

51.
$$1\frac{3}{8}$$
$$+ 6\frac{3}{4}$$

52.
$$1\frac{5}{12}$$
$$+ 1\frac{7}{8}$$

53.
$$3\frac{1}{2}$$
$$- 2\frac{2}{3}$$

54.
$$4\frac{1}{4}$$
$$- 3\frac{7}{12}$$

55. $8\dfrac{3}{4}$

$-5\dfrac{7}{8}$

56. 5

$-4\dfrac{7}{8}$

57. $7\dfrac{1}{4}$

$-4\dfrac{2}{3}$

58. $4\dfrac{1}{10}$

$-3\dfrac{7}{8}$

59. $9\dfrac{1}{5}$

$-3\dfrac{3}{4}$

60. 9

$-7\dfrac{5}{6}$

61. $6\dfrac{3}{7}$

$-2\dfrac{2}{3}$

62. $8\dfrac{2}{15}$

$-6\dfrac{1}{2}$

63. In your own words, explain the steps you would take to add two large mixed numbers.

64. When subtracting mixed numbers, explain when you need to regroup. Explain how to regroup using your own example.

First estimate the answer. Then solve each application problem.

65. At the beginning of this chapter you read about Bryan Berg, who builds houses of cards. While in high school, Bryan built a house of cards $14\dfrac{1}{2}$ ft tall, setting his first world record. Today his current world record is $25\dfrac{3}{4}$ ft tall. How much taller is his current world record than his first world record? (*Source: Guinness World Records.*)

Estimate:

Exact:

66. The average age of a fast-food worker has increased over the past 12 years from $21\dfrac{3}{4}$ years of age to $29\dfrac{1}{2}$. How much older is the average fast-food worker today than 12 years ago? (*Source:* U.S. Census Bureau.)

Estimate:

Exact:

Use the newspaper advertisement for these industrial quality professional pliers to answer Exercises 67–70. The " symbol is for inches and 2-15/16" means $2\frac{15}{16}$ inches. (Source: Harbor Freight Tools.)

	DESCRIPTION	MAX. JAW CAP.	LOT NO.	PRICE
A	6" SLIP JOINT	2-15/16"	94380	$2.99
B	6" LONG NOSE	1-1/2"	94378	$4.29
C	8" SLIP JOINT	3-3/8"	94381	$4.29
D	7" LINEMAN'S	1-1/2"	94382	$5.29
E	7" DIAGONAL	7/8"	94383	$5.29

	DESCRIPTION	MAX. JAW CAP.	LOT NO.	PRICE
F	8" LINEMAN'S	1-7/16"	94385	$6.29
G	5" CURVED JAW LOCKING	1-3/8"	94289	$3.29
H	12" LONG NOSE LOCKING	2-1/4"	94285	$6.29
I	10" CURVED JAW LOCKING	2-1/4"	94286	$7.29
J	12" GROOVE JOINT	2"	94288	$8.29

67. How much larger is the maximum jaw capacity of the 8″ slip joint pliers than of the 6″ slip joint pliers?

Estimate:

Exact:

68. How much larger is the maximum jaw capacity of the 8″ lineman's pliers than of the 5″ curver jaw locking pliers?

Estimate:

Exact:

69. What is the difference in the maximum jaw capacity of the smallest pliers and that of the largest pliers?

Estimate:

Exact:

70. What is the difference in the maximum jaw capacity of the smallest pliers and that of the third smallest pliers?

Estimate:

Exact:

Storehouse
34 PC. GEAR HOSE CLAMP ASSORTMENT
LOT NO. 1420

Includes: two 2-3/4", two 2-1/2", four 2-1/4", four 2", six 1-3/4", six 1-1/2", six 1-1/4", four 9/16" through 1-1/16"

SALE!
$5⁹⁷

SAVE 40%

REGULAR PRICE $9.99

A mechanic buys the 34-piece hose clamp assortment shown at the left. Use the advertisement to answer Exercises 71–72. The " symbol is for inches and 2-3/4" means $2\frac{3}{4}$ inches. (Source: Harbor Freight Tools.)

71. Find the difference in size between the largest hose clamp and the smallest hose clamp.

Estimate:

Exact:

72. Find the difference in size between the second to largest hose clamp and the smallest hose clamp.

Estimate:

Exact:

73. The four sides of Pam Prentiss' vegetable garden are $15\frac{1}{2}$ feet, $18\frac{3}{4}$ feet, $24\frac{1}{4}$ feet, and $30\frac{1}{2}$ feet. How many feet of fencing are needed to go around the garden?

Estimate:

Exact:

74. On a recent vacation to Canada, Janeen Cartmill drove for $7\frac{3}{4}$ hours on the first day, $5\frac{1}{4}$ hours on the second day, $6\frac{1}{2}$ hours on the third day, and 9 hours on the fourth day. How many hours did she drive altogether?

Estimate:

Exact:

75. A craftsperson must attach a lead strip around all four sides of a stained glass window before it is installed. Find the length of lead stripping needed.

$23\frac{3}{4}$ in.

$34\frac{1}{2}$ in.

Estimate:

Exact:

76. To complete a custom order, Zak Morten of Home Depot must find the number of inches of brass trim needed to go around the four sides of the lamp base plate shown. Find the length of brass trim needed.

$5\frac{1}{8}$ in.

$9\frac{7}{8}$ in.

Estimate:

Exact:

77. A museum humidifier contains 100 gallons of water. The system uses $10\frac{1}{4}$ gallons of water on Monday, $13\frac{1}{2}$ gallons on Tuesday, $8\frac{7}{8}$ gallons on Wednesday, $18\frac{3}{4}$ gallons on Thursday, $12\frac{3}{8}$ gallons on Friday, $9\frac{1}{2}$ gallons on Saturday, and $14\frac{1}{8}$ gallons on Sunday. Find the total number of gallons of water remaining.

Estimate:

Exact:

78. Scott Alamo had 16 cubic yards of mulch delivered to the home he recently purchased. Using his wheelbarrow, he spread $3\frac{3}{4}$ yards of the mulch in his front yard, $4\frac{7}{8}$ yards in his back yard, $2\frac{1}{2}$ yards in the side yard, and used the remainder in the children's outdoor play area. Find the number of cubic yards of mulch remaining for the play area.

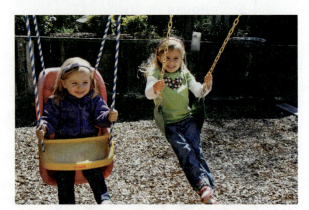

Estimate:

Exact:

79. The exercise yard at the correction center has four sides and is surrounded by $527\frac{1}{24}$ ft of security fencing. Three sides of the yard measure $107\frac{2}{3}$ ft, $150\frac{3}{4}$ ft, and $138\frac{5}{8}$ ft. Find the length of the fourth side.

$138\frac{5}{8}$ ft

? ft

$150\frac{3}{4}$ ft

$107\frac{2}{3}$ ft

Estimate:

Exact:

80. Three sides of a parking lot are $108\frac{1}{4}$ ft, $162\frac{3}{8}$ ft, and $143\frac{1}{2}$ ft. The total distance around the lot is $518\frac{3}{4}$ ft. What is the length of the fourth side?

$108\frac{1}{4}$ ft

$162\frac{3}{8}$ ft

? ft

$143\frac{1}{2}$ ft

Estimate:

Exact:

81. A freight car is loaded with Morton Salt products consisting of $58\frac{1}{2}$ tons of coarse rock salt, $23\frac{5}{8}$ tons of medium rock salt, $16\frac{5}{6}$ tons of table salt, and $29\frac{1}{4}$ tons of animal salt lick blocks. The weight of the unloaded freight car is $58\frac{1}{3}$ tons. Find the weight of the loaded freight car.

Estimate:

Exact:

82. Bryan Berg, from the opening page of this chapter, built houses of cards reaching heights of $14\frac{1}{2}$ ft, $19\frac{3}{8}$ ft, $23\frac{5}{12}$ ft, and $25\frac{3}{4}$ ft (his current world record). Find the total height of these four houses of cards. (*Source: Reader's Digest.*)

Estimate:

Exact:

Find the unknown length, labeled with a question mark, in each figure.

83.

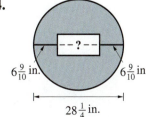

?

$2\frac{3}{8}$ in. $2\frac{3}{8}$ in.

$9\frac{7}{16}$ in.

84.

- - ? - -

$6\frac{9}{10}$ in. $6\frac{9}{10}$ in.

$28\frac{1}{4}$ in.

85.

86.

Relating Concepts (Exercises 87–92) For Individual or Group Work

Most fraction problems include fractions with different denominators.
Work Exercises 87–92 in order.

87. To add or subtract fractions, we must first rewrite them as like fractions. Rewrite each fraction with the indicated denominator.

(a) $\dfrac{5}{9} = \dfrac{}{54}$ (b) $\dfrac{7}{12} = \dfrac{}{48}$

(c) $\dfrac{5}{8} = \dfrac{}{40}$ (d) $\dfrac{11}{5} = \dfrac{}{120}$

88. When rewriting unlike fractions as like fractions with the least common multiple as a denominator, the new denominator is called the _____ _____ _____, or LCD.

89. Add or subtract as indicated. Write answers in lowest terms.

(a) $\dfrac{5}{8} + \dfrac{1}{3}$ (b) $\dfrac{19}{20} - \dfrac{5}{12}$

(c) $\begin{array}{r} \dfrac{7}{12} \\ \dfrac{3}{16} \\ + \dfrac{3}{24} \\ \hline \end{array}$ (d) $\begin{array}{r} \dfrac{6}{7} \\ - \dfrac{2}{3} \\ \hline \end{array}$

90. A common method for adding or subtracting mixed numbers is to add or subtract the _____ _____ and then add or subtract the whole number parts.

91. Another method for adding or subtracting mixed numbers is to first change the mixed numbers to _____ fractions. After adding or subtracting, write the answer in lowest terms and as a mixed number when possible. This method is difficult to use if the mixed numbers are (*large/small*).

92. Add or subtract these fractions as indicated. First use the method where you add or subtract fraction parts and then whole number parts. Then use the method where you change each mixed number to an improper fraction before adding or subtracting. Do you get the same answer using both methods? Which method do you prefer?

(a) $\begin{array}{r} 4\dfrac{5}{8} \\ + 3\dfrac{3}{4} \\ \hline \end{array}$ (b) $\begin{array}{r} 12\dfrac{2}{5} \\ - 8\dfrac{7}{8} \\ \hline \end{array}$

Summary Exercises *Operations with Fractions*

CONCEPT CHECK   *Write* proper *or* improper *for each fraction.*

1. $\dfrac{3}{4}$
2. $\dfrac{4}{3}$
3. $\dfrac{10}{10}$
4. $\dfrac{11}{12}$

Write each fraction in lowest terms.

5. $\dfrac{30}{36}$
6. $\dfrac{175}{200}$
7. $\dfrac{15}{35}$
8. $\dfrac{115}{235}$

Add, subtract, multiply, or divide as indicated. Simplify all answers.

9. $\dfrac{3}{4} \cdot \dfrac{2}{3}$
10. $\dfrac{7}{12} \cdot \dfrac{9}{14}$
11. $56 \cdot \dfrac{5}{8}$

12. $\dfrac{5}{8} \div \dfrac{3}{4}$
13. $\dfrac{35}{45} \div \dfrac{10}{15}$
14. $21 \div \dfrac{3}{8}$

15. $\dfrac{7}{8} + \dfrac{2}{3}$
16. $\dfrac{5}{8} + \dfrac{3}{4} + \dfrac{7}{16}$
17. $\dfrac{7}{12} + \dfrac{5}{6} + \dfrac{2}{3}$

18. $\dfrac{5}{6} - \dfrac{3}{4}$
19. $\dfrac{7}{8} - \dfrac{5}{12}$
20. $\dfrac{4}{5} - \dfrac{2}{3}$

First estimate the answer. Then add, subtract, multiply, or divide to find the exact answer.

21. $3\dfrac{1}{2} \cdot 2\dfrac{1}{4}$

   Estimate:

   ____ • ____ = ____

   Exact:

22. $5\dfrac{3}{8} \cdot 3\dfrac{1}{4}$

   Estimate:

   ____ • ____ = ____

   Exact:

23. $8 \cdot 5\dfrac{2}{3} \cdot 2\dfrac{3}{8}$

   Estimate:

   ____ • ____ • ____ = ____

   Exact:

24. $4\dfrac{3}{8} \div 3\dfrac{3}{4}$

   Estimate:

   ____ ÷ ____ = ____

   Exact:

25. $6\dfrac{7}{8} \div 2$

   Estimate:

   ____ ÷ ____ = ____

   Exact:

26. $4\dfrac{5}{8} \div \dfrac{3}{4}$

   Estimate:

   ____ ÷ ____ = ____

   Exact:

27. *Estimate:*   *Exact:*

$\xleftarrow{\text{Rounds to}}\Big\{ \quad 5\dfrac{2}{3}$

$+ \quad \xleftarrow{\text{Rounds to}}\Big\{ \quad + 4\dfrac{1}{4}$

_____   _____

28. *Estimate:*   *Exact:*

$18\dfrac{5}{12}$

$+ \qquad\qquad + \ 9\dfrac{3}{4}$

_____   _____

29. *Estimate:*   *Exact:*

$14\dfrac{3}{5}$

$+ \qquad\qquad + 10\dfrac{2}{3}$

_____   _____

30. *Estimate:*   *Exact:*

$8\dfrac{3}{4}$

$- \qquad\qquad - 3\dfrac{4}{5}$

_____   _____

31. *Estimate:*   *Exact:*

14

$- \qquad\qquad - 7\dfrac{3}{8}$

_____   _____

32. *Estimate:*   *Exact:*

$31\dfrac{5}{6}$

$- \qquad\qquad - 22\dfrac{7}{12}$

_____   _____

Find the least common multiple of each set of numbers.

33. 8, 10

34. 4, 15

35. 3, 5, 10

36. 6, 8, 16

37. 9, 18, 24

38. 4, 12, 21

Rewrite each fraction with the indicated denominator.

39. $\dfrac{5}{6} = \dfrac{}{42}$

40. $\dfrac{3}{4} = \dfrac{}{16}$

41. $\dfrac{3}{7} = \dfrac{}{28}$

42. $\dfrac{5}{8} = \dfrac{}{40}$

43. $\dfrac{3}{9} = \dfrac{}{45}$

44. $\dfrac{11}{12} = \dfrac{}{60}$

Study Skills
MAKING A MIND MAP

M ind mapping is a visual way to show information that you have learned. It is an excellent way to review. Mapping is flexible and can be personalized, which is helpful for your memory. Your brain likes to see things that are **pleasing** to look at, **colorful**, and that **show connections** between ideas. Take advantage of that by creating maps that

▶ are easy to read,

▶ use color in a systematic way, and

▶ clearly show you how different concepts are related (using arrows or dotted lines, for example).

Directions for Making a Mind Map

Here are some general directions for making a map. After you read them, work on completing the map that has been started for you on the next page. It is from **Chapters 2 and 3: Fractions.**

▶ To begin a mind map, write the concept in the center of a piece of paper and either circle it or draw a box around it.

▶ Make a line out from the center concept, and draw a box large enough to write the definition of the concept.

▶ Think of the other aspects (subpoints) of the concept that you have learned, such as procedures to follow or formulas. Make a separate line and box connecting each subpoint to the center.

▶ From each of the new boxes, add the information you've learned. You can continue making new lines and boxes or circles, or you can list items below the new information.

▶ Use color to highlight the major points. For example, everything related to one subpoint might be the same color. That way you can easily see related ideas.

▶ You may also use arrows, underlining, or small drawings to help yourself remember.

OBJECTIVES

1 Create mind maps for appropriate concepts.

2 Visually show how concepts relate to each other using arrows or lines.

Why Is Mapping Brain Friendly?

Remember that your brain grows dendrites when you are **actively thinking** about and working with information. Making a map requires you to think hard about *how to place the information*, *how to show connections* between parts of the map, and *how color will be useful*. It also takes a lot of thinking to fill in all related details and **show how those details connect to the larger concept.** All that thinking will let your brain grow a complex, many-branched neural network of interconnected dendrites. It is time well spent.

Study Skills

Continued from page 237

Try This Fractions Mind Map

On a separate paper, make a map that summarizes Computations with Fractions. Follow the directions and use the starter map below.

▶ The longest rectangles are *instructions* for all four operations. (The first one starts "Rewrite all numbers as fractions. . ." and the second one is at the bottom of the map.)

▶ Notice the wavy dividing lines that separate the map into two sides.

▶ Your job is to complete the map by writing the steps used in multiplying and dividing fractions and the steps used in adding and subtracting fractions.

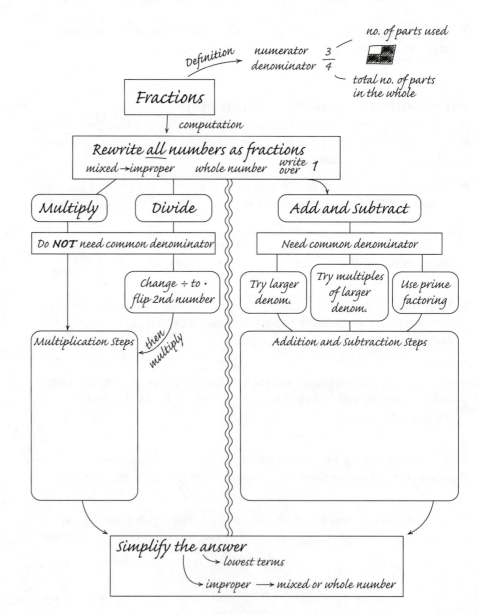

3.5 Order Relations and the Order of Operations

There are times when we want to compare the size of two numbers. For example, we might want to know which is the greater amount, the larger size, or the longer distance.

Fractions, like whole numbers, can be graphed on a number line. Fractions divide the space between whole numbers into equal parts.

OBJECTIVES

1. Identify the greater or lesser of two fractions.

2. Use exponents with fractions.

3. Use the order of operations with fractions.

OBJECTIVE 1 **Identify the greater or lesser of two fractions.** To compare the size of two numbers, place the two numbers on a number line and use the following rule.

> **Comparing the Size of Two Numbers**
>
> The number farther to the *left* on the number line is always *less*, and the number farther to the *right* on the number line is always *greater*.

For example, on the number line above, $\frac{1}{2}$ is to the *left* of $\frac{4}{3}$, so $\frac{1}{2}$ is *less than* $\frac{4}{3}$.

> **Work Problem** 1 **at the Side.** ▶

Write *order relations* using the symbols shown below.

> **Symbols Used to Show Order Relations**
>
> <   is less than   >   is greater than

EXAMPLE 1   **Using Less-Than and Greater-Than Symbols**

Rewrite the following using < and > symbols.

(a) $\frac{1}{2}$ is less than $\frac{4}{3}$.

$\frac{1}{2}$ **is less than** $\frac{4}{3}$ is written as $\frac{1}{2} < \frac{4}{3}$.

> $\frac{1}{2}$ is farther to the *left* on the number line, so it is *less than* $\frac{4}{3}$.

(b) $\frac{9}{4}$ is greater than 1.

$\frac{9}{4}$ **is greater than** 1 is written as $\frac{9}{4} > 1$.

> $\frac{9}{4}$ is farther to the *right* on the number line, so it is *greater than* 1.

(c) $\frac{5}{3}$ is less than $\frac{11}{4}$.

$\frac{5}{3}$ **is less than** $\frac{11}{4}$ is written as $\frac{5}{3} < \frac{11}{4}$.

.... **Continued on Next Page**

1 Locate each fraction on the number line.

GS (a) $\frac{2}{3}$ Put a dot on the number line between 0 and 1.

(b) $1\frac{1}{2}$

(c) $2\frac{3}{4}$

Answer

1.

2 Use the number line on the previous page to help you write $<$ or $>$ in each blank to make a true statement.

GS **(a)** 1 ____ $\dfrac{5}{4}$ The point on the symbol points to the lesser value.

(b) $\dfrac{8}{3}$ ____ $\dfrac{3}{2}$

(c) 0 ____ 1

(d) $\dfrac{17}{8}$ ____ $\dfrac{8}{4}$

3 Write $<$ or $>$ in each blank to make a true statement.

GS **(a)** $\dfrac{7}{8}$ ____ $\dfrac{3}{4}$ The point on the symbol points to the lesser value.

(b) $\dfrac{13}{8}$ ____ $\dfrac{15}{9}$

(c) $\dfrac{9}{4}$ ____ $\dfrac{7}{3}$

(d) $\dfrac{9}{10}$ ____ $\dfrac{14}{15}$

Answers
2. (a) $<$ (b) $>$ (c) $<$ (d) $>$
3. (a) $>$ (b) $<$ (c) $<$ (d) $<$

Note

A number line is a very useful tool when working with order relations.

◀ **Work Problem 2 at the Side.**

The fraction $\frac{7}{8}$ represents 7 of 8 equal parts, while $\frac{3}{8}$ means 3 of 8 equal parts. Because $\frac{7}{8}$ represents more of the equal parts, $\frac{7}{8}$ is greater than $\frac{3}{8}$.

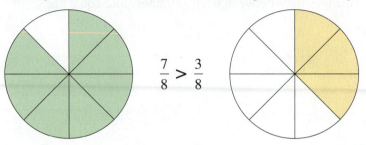

$$\frac{7}{8} > \frac{3}{8}$$

To identify the greater fraction, use the following steps.

Identifying the Greater Fraction

Step 1 Write the fractions as like fractions (same denominators).

Step 2 Compare the numerators. The fraction with the greater numerator is the greater fraction.

EXAMPLE 2 **Identifying the Greater Fraction**

Determine which fraction in each pair is greater.

(a) $\dfrac{7}{8}, \dfrac{9}{10}$

First, write the fractions as like fractions. The least common multiple for 8 and 10 is 40.

$$\frac{7}{8} = \frac{7 \cdot 5}{8 \cdot 5} = \frac{35}{40} \quad \text{and} \quad \frac{9}{10} = \frac{9 \cdot 4}{10 \cdot 4} = \frac{36}{40}$$

Rewrite both fractions with 40 as the denominator.

Look at the numerators. Because 36 is greater than 35, $\frac{36}{40}$ is greater than $\frac{35}{40}$. Then, because $\frac{36}{40}$ is equivalent to $\frac{9}{10}$,

$$\frac{9}{10} > \frac{7}{8} \quad \text{or} \quad \frac{7}{8} < \frac{9}{10}.$$

The greater fraction is $\frac{9}{10}$.

(b) $\dfrac{8}{5}, \dfrac{23}{15}$

The least common multiple of 5 and 15 is 15.

$$\frac{8}{5} = \frac{8 \cdot 3}{5 \cdot 3} = \frac{24}{15} \quad \text{and} \quad \frac{23}{15} = \frac{23}{15}$$

This shows that $\frac{8}{5}$ is greater than $\frac{23}{15}$, or

$$\frac{8}{5} > \frac{23}{15}.$$

◀ **Work Problem 3 at the Side.**

OBJECTIVE ▶ **2** **Use exponents with fractions.** Exponents were used in **Chapter 1** to write repeated multiplication.

$$\overbrace{3^2}^{\text{Exponent}} = \underbrace{3 \cdot 3}_{\substack{\text{Two} \\ \text{factors of 3}}} = 9 \quad \text{and} \quad \overbrace{5^3}^{\text{Exponent}} = \underbrace{5 \cdot 5 \cdot 5}_{\substack{\text{Three} \\ \text{factors of 5}}} = 125$$

The next example shows exponents used with fractions.

EXAMPLE 3 Using Exponents with Fractions

Simplify.

(a) $\left(\dfrac{1}{2}\right)^3$

$$\left(\dfrac{1}{2}\right)^3 = \overbrace{\dfrac{1}{2} \cdot \dfrac{1}{2} \cdot \dfrac{1}{2}}^{\text{Three factors of } \frac{1}{2}} = \dfrac{1}{8}$$

> $\frac{1}{2}$ is multiplied by itself three times.

(b) $\left(\dfrac{5}{8}\right)^2$

$$\left(\dfrac{5}{8}\right)^2 = \overbrace{\dfrac{5}{8} \cdot \dfrac{5}{8}}^{\text{Two factors of } \frac{5}{8}} = \dfrac{25}{64}$$

(c) $\left(\dfrac{3}{4}\right)^2 \cdot \left(\dfrac{2}{3}\right)^3$

$$\left(\dfrac{3}{4}\right)^2 \cdot \left(\dfrac{2}{3}\right)^3 = \overbrace{\left(\dfrac{3}{4} \cdot \dfrac{3}{4}\right)}^{\substack{\text{Two factors} \\ \text{of } \frac{3}{4}}} \cdot \overbrace{\left(\dfrac{2}{3} \cdot \dfrac{2}{3} \cdot \dfrac{2}{3}\right)}^{\substack{\text{Three factors} \\ \text{of } \frac{2}{3}}}$$

$$= \dfrac{\overset{1}{\cancel{3}} \cdot \overset{1}{\cancel{3}} \cdot \overset{1}{\cancel{2}} \cdot \overset{1}{\cancel{2}} \cdot \overset{1}{\cancel{2}}}{\underset{2}{\cancel{4}} \cdot \underset{2}{\cancel{4}} \cdot \underset{1}{\cancel{3}} \cdot \underset{1}{\cancel{3}} \cdot 3} \quad \text{Divide out all the common factors.}$$

$$= \dfrac{1}{6} \quad \text{The fraction is in lowest terms after all common factors are divided out.}$$

·············· **Work Problem** **4** **at the Side.** ▶

OBJECTIVE ▶ **3** **Use the order of operations with fractions.** Recall the *order of operations* from **Chapter 1.**

> **Order of Operations**
>
> **1.** Do all operations inside *parentheses or other grouping symbols.*
> **2.** Simplify any expressions with *exponents* and find any *square roots.*
> **3.** *Multiply* or *divide,* proceeding from left to right.
> **4.** *Add* or *subtract,* proceeding from left to right.

4 Simplify.

GS **(a)** $\left(\dfrac{1}{2}\right)^4$

$$\overbrace{\dfrac{1}{2} \cdot \dfrac{1}{2} \cdot \dfrac{1}{2} \cdot \dfrac{\underline{\quad}}{\underline{\quad}}}^{\text{Four factors of } \frac{1}{2}} =$$

(b) $\left(\dfrac{3}{4}\right)^2$

(c) $\left(\dfrac{1}{2}\right)^3 \cdot \left(\dfrac{2}{3}\right)^2$

(d) $\left(\dfrac{1}{5}\right)^2 \cdot \left(\dfrac{5}{3}\right)^2$

Answers

4. **(a)** $1; 2; \dfrac{1}{16}$ **(b)** $\dfrac{9}{16}$ **(c)** $\dfrac{1}{18}$ **(d)** $\dfrac{1}{9}$

5 Simplify by using the order of operations.

GS **(a)** $\dfrac{5}{9} - \dfrac{3}{4}\left(\dfrac{2}{3}\right)$

$\dfrac{5}{9} - \dfrac{\overset{1}{\cancel{3}}}{\underset{2}{\cancel{4}}}\left(\dfrac{2}{\underset{1}{\cancel{3}}}\right)$ Multiply first.

$\dfrac{5}{9} - \dfrac{1}{2} = \dfrac{10}{18} - \dfrac{9}{\underline{}} =$

(b) $\dfrac{3}{4}\left(\dfrac{2}{3} \cdot \dfrac{3}{5}\right)$

(c) $\dfrac{7}{8}\left(\dfrac{2}{3}\right) - \left(\dfrac{1}{2}\right)^2$

(d) $\dfrac{\left(\dfrac{5}{6}\right)^2}{\dfrac{4}{3}}$

The next example shows how to apply the order of operations with fractions.

EXAMPLE 4   **Using the Order of Operations with Fractions**

Simplify by using the order of operations.

(a) $\dfrac{1}{3} + \dfrac{1}{2}\left(\dfrac{4}{5}\right)$

Multiply $\frac{1}{2}\left(\frac{4}{5}\right)$ first because multiplication and division are done *before* adding.

> Do **not** add $\frac{1}{3} + \frac{1}{2}$ as the first step.

$$\dfrac{1}{3} + \dfrac{1}{\underset{1}{\cancel{2}}}\left(\dfrac{\overset{2}{\cancel{4}}}{5}\right) = \dfrac{1}{3} + \dfrac{2}{5}$$

Now add. The least common denominator of 3 and 5 is 15.

$$\dfrac{1}{3} + \dfrac{2}{5} = \dfrac{5}{15} + \dfrac{6}{15} = \dfrac{11}{15} \leftarrow \text{Lowest terms}$$

(b) $\dfrac{3}{8}\left(\dfrac{1}{2} + \dfrac{1}{3}\right)$

$$\dfrac{3}{8}\left(\dfrac{1}{2} + \dfrac{1}{3}\right) = \dfrac{3}{8}\left(\dfrac{3}{6} + \dfrac{2}{6}\right) \quad \text{Work inside parentheses first.}$$

$$= \dfrac{3}{8}\left(\dfrac{5}{6}\right)$$

$$= \dfrac{\overset{1}{\cancel{3}}}{8}\left(\dfrac{5}{\underset{2}{\cancel{6}}}\right) \quad \begin{array}{l}\text{Divide numerator and}\\ \text{denominator by 3.}\\ \text{Then multiply.}\end{array}$$

$$= \dfrac{5}{16} \leftarrow \text{Lowest terms}$$

(c) $\left(\dfrac{2}{3}\right)^2 - \dfrac{4}{5}\left(\dfrac{1}{2}\right)$

$$\left(\dfrac{2}{3}\right)^2 - \dfrac{4}{5}\left(\dfrac{1}{2}\right) = \dfrac{4}{9} - \dfrac{4}{5}\left(\dfrac{1}{2}\right) \quad \begin{array}{l}\text{Simplify the}\\ \text{expression with}\\ \text{the exponent.}\\ \frac{2}{3} \cdot \frac{2}{3} \text{ is } \frac{4}{9}.\end{array}$$

> No work inside the parentheses, so applying the exponent is next.

$$= \dfrac{4}{9} - \dfrac{\overset{2}{\cancel{4}}}{5}\left(\dfrac{1}{\underset{1}{\cancel{2}}}\right) \quad \text{Multiply next.}$$

$$= \dfrac{4}{9} - \dfrac{2}{5}$$

$$= \dfrac{20}{45} - \dfrac{18}{45} \quad \begin{array}{l}\text{Subtract last. (Least com-}\\ \text{mon denominator is 45.)}\end{array}$$

$$= \dfrac{2}{45} \leftarrow \text{Lowest terms}$$

◄ **Work Problem** **5** **at the Side.**

Answers

5. (a) $18; \dfrac{1}{18}$ **(b)** $\dfrac{3}{10}$ **(c)** $\dfrac{1}{3}$ **(d)** $\dfrac{25}{48}$

3.5 Exercises

 Download the MyDashBoard App

 MyMathLab®

CONCEPT CHECK *Locate each fraction in Exercises 1–12 on the following number line.*

1. $\dfrac{1}{2}$

2. $\dfrac{1}{4}$

3. $\dfrac{3}{2}$

4. $\dfrac{5}{4}$

5. $\dfrac{7}{3}$

6. $\dfrac{11}{4}$

7. $2\dfrac{1}{6}$

8. $3\dfrac{4}{5}$

9. $\dfrac{7}{2}$

10. $\dfrac{7}{8}$

11. $3\dfrac{1}{4}$

12. $1\dfrac{7}{8}$

Write < or > to make a true statement. **See Examples 1 and 2.**

13. $\dfrac{1}{2}$ ____ $\dfrac{3}{8}$

14. $\dfrac{5}{8}$ ____ $\dfrac{3}{4}$

15. $\dfrac{5}{6}$ ____ $\dfrac{11}{12}$

16. $\dfrac{13}{18}$ ____ $\dfrac{5}{6}$

17. $\dfrac{5}{12}$ ____ $\dfrac{3}{8}$

18. $\dfrac{17}{24}$ ____ $\dfrac{5}{6}$

19. $\dfrac{11}{18}$ ____ $\dfrac{5}{9}$

20. $\dfrac{7}{12}$ ____ $\dfrac{11}{20}$

CONCEPT CHECK *Write* true *or* false *for each statement, Also show how to complete each solution.*

21. $\left(\dfrac{1}{2}\right)^2 = \dfrac{1}{4}$

$\left(\dfrac{1}{2}\right)^2 = \dfrac{1}{2} \cdot \dfrac{1}{\rule{1cm}{0.4pt}} = \dfrac{1}{\rule{1cm}{0.4pt}}$

22. $\left(\dfrac{3}{8}\right)^2 = \dfrac{6}{16}$

$\left(\dfrac{3}{8}\right)^2 = \dfrac{3}{8} \cdot \dfrac{\rule{1cm}{0.4pt}}{8} = \dfrac{\rule{1cm}{0.4pt}}{64}$

23. $\left(\dfrac{2}{5}\right)^3 = \dfrac{6}{15}$

24. $\left(\dfrac{5}{6}\right)^3 = \dfrac{125}{216}$

Simplify. **See Example 3.**

25. $\left(\dfrac{1}{3}\right)^2$

26. $\left(\dfrac{2}{3}\right)^2$

27. $\left(\dfrac{5}{8}\right)^2$

28. $\left(\dfrac{7}{8}\right)^2$

29. $\left(\dfrac{3}{4}\right)^2$

30. $\left(\dfrac{3}{5}\right)^3$

31. $\left(\dfrac{4}{5}\right)^3$

32. $\left(\dfrac{4}{7}\right)^3$

33. $\left(\dfrac{3}{2}\right)^4 = \dfrac{3}{2} \cdot \dfrac{3}{2} \cdot \dfrac{3}{2} \cdot \dfrac{3}{2} =$

34. $\left(\dfrac{4}{3}\right)^4 = \dfrac{4}{3} \cdot \dfrac{4}{3} \cdot \dfrac{4}{3} \cdot \dfrac{4}{3} =$

35. $\left(\dfrac{3}{4}\right)^4$

36. $\left(\dfrac{2}{3}\right)^5$

37. Describe in your own words what a number line is, and draw a picture of one. Be sure to include how it works and how it can be used.

38. You have used the order of operations with whole numbers and again with fractions. List from memory the steps in the order of operations.

CONCEPT CHECK *Use the order of operations to simplify each expression. Circle the correct answer.*

39. $2^4 - 4(3)$

  30 4 12 2

40. $3^2 + 4(1)$

  9 13 11 10

41. $3 \cdot 2^2 - \dfrac{6}{3}$

  9 4 2 10

42. $5 \cdot 2^3 - \dfrac{6}{2}$

  37 12 27 22

Use the order of operations to simplify each expression. **See Example 4.**

43. $\left(\dfrac{1}{2}\right)^2 \cdot 4$

44. $\left(\dfrac{1}{4}\right)^2 \cdot 4$

45. $\left(\dfrac{3}{4}\right)^2 \cdot \left(\dfrac{1}{3}\right)$

46. $\left(\dfrac{2}{3}\right)^3 \cdot \left(\dfrac{1}{2}\right)$

47. $\left(\dfrac{4}{5}\right)^2 \cdot \left(\dfrac{5}{6}\right)^2$

48. $\left(\dfrac{5}{8}\right)^2 \cdot \left(\dfrac{4}{25}\right)^2$

49. $6\left(\dfrac{2}{3}\right)^2 \left(\dfrac{1}{2}\right)^3$

50. $9\left(\dfrac{1}{3}\right)^3 \left(\dfrac{4}{3}\right)^2$

51. $\dfrac{3}{5}\left(\dfrac{1}{3}\right) + \dfrac{2}{5}\left(\dfrac{3}{4}\right)$

$\underbrace{\dfrac{3}{5} \cdot \dfrac{1}{3}}_{} + \underbrace{\dfrac{2}{5} \cdot \dfrac{3}{4}}_{}$

$\dfrac{3}{15} + \dfrac{6}{20}$

$\dfrac{12}{60} + \dfrac{18}{60} = \dfrac{30}{60} =$

52. $\dfrac{1}{4}\left(\dfrac{3}{4}\right) + \dfrac{3}{8}\left(\dfrac{4}{3}\right)$

$\underbrace{\dfrac{1}{4} \cdot \dfrac{3}{4}}_{} + \underbrace{\dfrac{3}{8} \cdot \dfrac{4}{3}}_{}$

$\dfrac{3}{16} + \dfrac{12}{24}$

$\dfrac{9}{48} + \dfrac{24}{48} = \dfrac{33}{48} =$

53. $\dfrac{1}{2} + \left(\dfrac{1}{2}\right)^2 - \dfrac{3}{8}$

54. $\dfrac{2}{3} + \left(\dfrac{1}{3}\right)^2 - \dfrac{5}{9}$

55. $\left(\dfrac{1}{3} + \dfrac{1}{6}\right) \cdot \dfrac{1}{2}$

56. $\left(\dfrac{3}{5} - \dfrac{3}{20}\right) \cdot \dfrac{4}{3}$

57. $\dfrac{9}{8} \div \left(\dfrac{2}{3} + \dfrac{1}{12}\right)$

58. $\dfrac{6}{5} \div \left(\dfrac{3}{5} - \dfrac{3}{10}\right)$

59. GS $\underbrace{\left(\dfrac{7}{8} - \dfrac{3}{4}\right) \div \dfrac{3}{2}}$

$\left(\dfrac{7}{8} - \dfrac{6}{8}\right) \div \dfrac{3}{2}$

$\dfrac{1}{8} \div \dfrac{3}{2}$

$\dfrac{1}{8} \cdot \dfrac{2}{3} = \dfrac{2}{24} =$

60. GS $\underbrace{\left(\dfrac{4}{5} - \dfrac{3}{10}\right) \div \dfrac{4}{5}}$

$\left(\dfrac{8}{10} - \dfrac{3}{10}\right) \div \dfrac{4}{5}$

$\dfrac{5}{10} \div \dfrac{4}{5}$

$\dfrac{5}{10} \cdot \dfrac{5}{4} = \dfrac{25}{40} =$

61. $\dfrac{3}{8}\left(\dfrac{1}{4} + \dfrac{1}{2}\right) \cdot \dfrac{32}{3}$

62. $\dfrac{1}{3}\left(\dfrac{4}{5} - \dfrac{3}{10}\right) \cdot \dfrac{4}{2}$

63. ▶ $\left(\dfrac{3}{4}\right)^2 - \left(\dfrac{1}{2} - \dfrac{1}{6}\right) \div \dfrac{4}{3}$

64. $\left(\dfrac{2}{3}\right)^2 - \left(\dfrac{5}{8} - \dfrac{1}{2}\right) \div \dfrac{3}{2}$.

65. $\left(\dfrac{7}{8} - \dfrac{1}{4}\right) - \dfrac{2}{3}\left(\dfrac{3}{4}\right)^2$

66. $\left(\dfrac{5}{6} - \dfrac{7}{12}\right) - \dfrac{3}{4}\left(\dfrac{1}{3}\right)^2$

67. $\left(\dfrac{3}{4}\right)^2\left(\dfrac{2}{3} - \dfrac{5}{9}\right) - \dfrac{1}{4}\left(\dfrac{1}{8}\right)$

68. $\left(\dfrac{2}{3}\right)^2\left(\dfrac{1}{2} - \dfrac{1}{8}\right) - \dfrac{2}{3}\left(\dfrac{1}{8}\right)$

Solve each application problem.

69. The population of Las Vegas, Nevada has had an increase of $\dfrac{11}{50}$ since the turn of the century. During this same period, the population in Atlanta, Georgia has had an increase of $\dfrac{5}{30}$. Which city has had a higher rate of population growth? (*Source:* Census Bureau estimates.)

70. Jacob Evan weighs $22\dfrac{11}{16}$ pounds and his stroller weighs $22\dfrac{5}{8}$ pounds. Which weighs more, Jacob or the stroller?

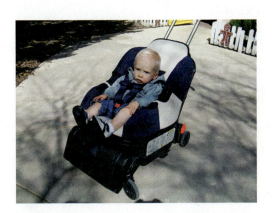

Relating Concepts (Exercises 71–80) For Individual or Group Work

You often need to use order relations and the order of operations when solving problems. **Work Exercises 71–80 in order.**

71. When comparing the size of two numbers, the symbol _____ means **is less than** and the symbol _____ means **is greater than.**

72. (a) To identify the greater of two fractions, we must first write the fractions as _____ fractions and then compare the _____. The fraction with the greater _____ is the greater fraction.

(b) Write four pairs of fractions, all with different denominators. Write the symbol for **less than** or for **greater than** between each pair.

Use the order of operations to simplify each expression.

73. $\left(\dfrac{2}{3}\right)^2 - \left(\dfrac{4}{5} - \dfrac{3}{10}\right) \div \dfrac{5}{4}$

74. $\left(\dfrac{1}{2}\right)^2 \left(\dfrac{3}{8} - \dfrac{1}{4}\right) + \dfrac{2}{3} \div \left(\dfrac{1}{3}\right)$

Simplify, then place the results on the number line.

75. $\left(\dfrac{2}{3}\right)^2$

76. $\left(\dfrac{3}{2}\right)^2$

77. $\left(\dfrac{1}{2}\right)^3$

78. $\left(\dfrac{5}{3}\right)^2$

79. $4 + 2 - 2^2$

80. $\left(\dfrac{3}{4}\right)^2 + \left(\dfrac{5}{8} - \dfrac{1}{4}\right) \div \dfrac{2}{3}$

Study Skills
PREPARING FOR TESTS

OBJECTIVES

1 Restate the importance of sleep and good nutrition as it affects learning.

2 Explain the effect of anxiety and stress on learning.

Many things besides studying can improve your test scores. You may not realize that eating the right foods, and getting enough exercise and sleep, can also improve your scores. Your brain (and therefore your ability to think) is affected by the condition of your whole body. So, part of your preparation for tests includes keeping yourself in good physical shape as well as spending time on the actual course material. Try these suggestions and see the difference.

Performance Health Tips

To Improve Your Test Score	Explanation
Get *seven to eight hours of sleep* the night before the exam. (It's helpful to get that much sleep *every* night.)	*Fatigue and exhaustion* reduce efficiency. They also cause poor memory and recall. If you didn't sleep much the night before a test, 20 minutes of relaxation or meditation can help.
Eat a *small, high-energy meal* about two hours before the test. Start the meal with a small amount of protein such as fish, chicken, or nonfat yogurt. Include carbohydrates if you like, but no high-fat foods.	Just 3 to 4 ounces of protein increases the amount of a chemical in the brain called tyrosine, which *improves your alertness, accuracy, and motivation.* High-fat foods dull your mind and slow down your brain.
Drink plenty of water. Don't wait until you feel thirsty; your body is already dehydrated by the time you feel it.	Research suggests that staying well hydrated improves the electrochemical communications in your brain.
Give your brain the time it needs to grow dendrites!	*Cramming doesn't work;* your brain cannot grow dendrites that quickly. *Studying every day* is the way to give your brain the time it needs.

Anxiety Prevention Tips

To Prevent Anxiety	Explanation
Practice slow, deep breathing for five minutes each day. Then do a minute or two of deep breathing right before the test. Also, if you feel your anxiety building during the test, stop for a minute, close your eyes, and do some deep breathing.	When *test anxiety* hits, you breathe more quickly and shallowly, which causes hyperventilation. Symptoms may be confusion, inability to concentrate, shaking, dizziness, and more. Slow, deep breathing will *calm you and prevent panic.*

Study Skills

Continued from page 247

Anxiety Prevention Tips *(Continued)*

To Prevent Anxiety	Explanation
Do 15 to 20 minutes of **moderate exercise** (like walking) shortly before the test. Daily exercise is even better!	**Exercise reduces stress** and will help prevent "blanking out" on a test. Exercise also increases your alertness, clear thinking, and energy.
To help you sleep the night before the test, or any time you need to calm down, **eat high carbohydrate foods** such as popcorn, bread, rice, crackers, muffins, bagels, pasta, corn, baked potatoes (not fries or chips), and cereals.	Carbohydrates increase the level of a chemical in the brain called serotonin, which has a *calming effect on the mind*. It reduces feelings of tension and stress and improves your ability to concentrate. You only need to eat a small amount, like half a bagel, to get this effect.
Before the test, **go easy on caffeinated beverages** such as coffee, tea, and soft drinks. Do not eat candy bars or other sugary snacks.	Extra caffeine can **make you jittery,** "hyper," and shaky for the test. It can increase the tendency to panic. Too much sugar causes negative emotional reactions in some people.

Now Try This

What will you do to improve your next test score?
List the three or four tips you think will help you the most.

1 _____

2 _____

3 _____

4 _____

What changes will you have to make in order to try the tips you chose?

_____

_____

_____

_____

See *Tips for Taking Math Tests* and *Preparing for Your Final Exam* for more ideas about managing anxiety. (Check the Table of Contents to find their locations.)

Chapter 3   *Summary*

Key Terms

3.1

like fractions Fractions with the same denominator are called *like fractions*.

unlike fractions Fractions with different denominators are called *unlike fractions*.

3.2

least common multiple Given two or more whole numbers, the least common multiple is the smallest whole number that is divisible by all the numbers.

LCM The abbreviation for *least common multiple* is LCM.

3.3

least common denominator When unlike fractions are rewritten as like fractions with the least common multiple as their denominator, the new denominator is the least common denominator.

LCD The abbreviation for *least common denominator* is LCD.

3.4

regrouping when adding fractions Regrouping is used in the addition of mixed numbers when the sum of the fractions is greater than 1.

regrouping when subtracting fractions Regrouping is used in the subtraction of mixed numbers when the fraction part of the first number is less than the fraction part of the second number.

New Symbols

$<$ is less than ($2 < 5$)

$>$ is greater than ($4 > 2$)

Test Your Word Power

See how well you have learned the vocabulary in this chapter.

1 **Like fractions** are
  A. fractions that are equivalent
  B. fractions that have the same numerator
  C. fractions that have the same denominator.

2 Two or more fractions are **unlike fractions** if
  A. they are not equivalent
  B. they have different numerators
  C. they have different denominators.

3 The abbreviation **LCM** stands for
  A. the largest common multiple
  B. the longest common multiple
  C. the least common multiple.

4 The **least common multiple** is
  A. the smallest whole number that is divisible by each of two or more numbers
  B. the smallest numerator
  C. the smallest denominator.

5 The abbreviation **LCD** stands for
  A. the largest common denominator
  B. the least common denominator
  C. the least common divisor.

6 The **least common denominator** is
  A. needed when multiplying fractions
  B. needed when dividing fractions
  C. the least common multiple of the denominators in a fraction problem.

Answers to Test Your Word Power

1. C; *Example:* Because the fractions $\frac{3}{8}$ and $\frac{10}{8}$ both have 8 as a denominator, they are like fractions.

2. C; *Example:* The fractions $\frac{2}{3}$ and $\frac{3}{4}$ are unlike fractions because they have different denominators.

3. C; *Example:* LCM is the abbreviation for least common multiple.

4. A; *Example:* The least common multiple of 4 and 5 is 20 because 20 is the smallest number into which both 4 and 5 will divide evenly.

5. B; *Example:* LCD is the abbreviation for least common denominator.

6. C; *Example:* The least common denominator of the fractions $\frac{2}{3}$ and $\frac{1}{2}$ is 6 because 6 is the least common multiple of 3 and 2. When written using the least common denominator, $\frac{2}{3}$ and $\frac{1}{2}$ become $\frac{4}{6}$ and $\frac{3}{6}$, respectively.

Quick Review

Concepts	**Examples**
3.1 **Adding Like Fractions**	
Add numerators and keep the same denominator. Simplify the answer.	$\dfrac{3}{4} + \dfrac{1}{4} + \dfrac{5}{4} = \dfrac{3+1+5}{4} = \dfrac{9}{4} = 2\dfrac{1}{4}$
3.1 **Subtracting Like Fractions**	
Subtract numerators and keep the same denominator. Simplify the answer.	$\dfrac{7}{8} - \dfrac{5}{8} = \dfrac{7-5}{8} = \dfrac{2}{8} = \dfrac{2 \div 2}{8 \div 2} = \dfrac{1}{4}$
3.2 **Finding the Least Common Multiple (LCM)**	
Method of using multiples of the larger number: List the first few multiples of the larger number. Then find the first multiple that is divisible by the smaller number.	$\dfrac{1}{3} + \dfrac{1}{4}$ 4, 8, 12, 16, … ← Multiples of 4 First multiple divisible by 3 ($12 \div 3 = 4$) The least common multiple (LCM) of 3 and 4 is 12.
3.2 **Finding the Least Common Multiple (LCM)**	
Method of prime numbers: First find the prime factorization of each number. Then use the prime factors to build the least common multiple.	Factors of 9 $9 = 3 \cdot 3$ $\text{LCM} = 3 \cdot 3 \cdot 5 = 45$ $15 = 3 \cdot 5$ Factors of 15 The least common multiple (LCM) of 9 and 15 is 45.
3.2 **Finding the Least Common Multiple (LCM)**	
Alternative method: Start by trying to divide the numbers by the first prime number. Continue dividing by prime numbers until all quotients are 1. The product of the prime numbers is the least common multiple.	4 and 6 $\quad \begin{array}{c|cc} 2 & 4 & 6 \\ 2 & 2 & 3 \\ 3 & 1 & 3 \\ \hline & 1 & 1 \end{array}$ The least common multiple (LCM) of 4 and 6 = **2 · 2 · 3 = 12**.
3.3 **Adding Unlike Fractions**	
Step 1 Find the least common multiple (LCM).	$\dfrac{1}{3} + \dfrac{1}{4} + \dfrac{1}{10}$ $\qquad$ LCM of 3, 4, and 10 = 60
Step 2 Rewrite the fractions with the least common multiple as the denominator.	$\dfrac{1}{3} = \dfrac{20}{60} \quad \dfrac{1}{4} = \dfrac{15}{60} \quad \dfrac{1}{10} = \dfrac{6}{60}$
Step 3 Add the numerators, placing the sum over the common denominator, and simplify the answer.	$\dfrac{20}{60} + \dfrac{15}{60} + \dfrac{6}{60} = \dfrac{41}{60}$ $\qquad$ Lowest terms

Concepts	Examples

3.3 Subtracting Unlike Fractions

Step 1 Find the least common multiple (LCM).

Step 2 Rewrite the fractions with the least common multiple as the denominator.

Step 3 Subtract the numerators, place the difference over the common denominator, and simplify the answer.

$$\frac{5}{8} - \frac{1}{3} \qquad \text{LCM of 8 and 3} = 24$$

$$\frac{5}{8} = \frac{15}{24} \quad \frac{1}{3} = \frac{8}{24}$$

$$\frac{15}{24} - \frac{8}{24} = \frac{7}{24} \qquad \text{Lowest terms}$$

3.4 Adding Mixed Numbers

Round the numbers and estimate the answer. Then find the exact answer using these steps.

Step 1 Add the fractions, using a common denominator.

Step 2 Add the whole numbers.

Step 3 Combine the sums of the whole numbers and the fractions. Simplify the fraction part when necessary.

Compare the exact answer to the estimate to see if it is reasonable.

Estimate: *Exact:*

$$10 \xleftarrow{\text{Rounds to}} \left\{ \quad 9\frac{2}{3} = 9\frac{8}{12} \right.$$

$$+7 \xleftarrow{\text{Rounds to}} \left\{ +6\frac{3}{4} = 6\frac{9}{12} \right.$$

$$17 \qquad\qquad 15\frac{17}{12} = 16\frac{5}{12}$$

The exact answer of $16\frac{5}{12}$ is reasonable because it is close to the estimate of 17.

3.4 Subtracting Mixed Numbers

Round the numbers and estimate the answer. Then find the exact answer using these steps.

Step 1 Subtract the fractions, regrouping if necessary.

Step 2 Subtract the whole numbers.

Step 3 Combine the differences of the whole numbers and the fractions. Simplify the fraction part when necessary.

Compare the exact answer to the estimate to see if it is reasonable.

Estimate: *Exact:*

$$9 \xleftarrow{\text{Rounds to}} \left\{ \quad 8\frac{5}{8} = 8\frac{15}{24} = 7\frac{39}{24} \right.$$

$$-4 \xleftarrow{\text{Rounds to}} \left\{ -3\frac{11}{12} = 3\frac{22}{24} = 3\frac{22}{24} \right.$$

$$5 \qquad\qquad 4\frac{17}{24}$$

The exact answer of $4\frac{17}{24}$ is reasonable because it is close to the estimate of 5.

Concepts	Examples

3.4 Adding or Subtracting Mixed Numbers Using an Alternative Method

Step 1 Change the mixed numbers to improper fractions.

Step 2 Rewrite the unlike fractions as like fractions.

Step 3 Add or subtract the numerators and simplify the answer.

Add.

$$2\frac{2}{3} = \frac{8}{3} = \frac{64}{24}$$

$$+\,1\frac{3}{8} = \frac{11}{8} = +\frac{33}{24}$$

24 is the least common denominator.

$$\frac{97}{24} = 4\frac{1}{24}$$ Answer as mixed number

Improper fractions

Subtract.

$$8\frac{2}{3} = \frac{26}{3} = \frac{104}{12}$$

$$-\,5\frac{3}{4} = \frac{23}{4} = -\frac{69}{12}$$

12 is the least common denominator.

$$\frac{35}{12} = 2\frac{11}{12}$$ Answer as mixed number

Improper fractions

3.5 Identifying the Greater of Two Fractions

With unlike fractions, change to like fractions first. The fraction with the greater numerator is the greater fraction. Use these symbols:

<   is less than

>   is greater than

Identify the greater fraction.

$$\frac{7}{8} \qquad \frac{9}{10}$$

$$\frac{7}{8} = \frac{7 \cdot \mathbf{5}}{8 \cdot \mathbf{5}} = \frac{35}{40}$$

$$\frac{9}{10} = \frac{9 \cdot \mathbf{4}}{10 \cdot \mathbf{4}} = \frac{36}{40}$$

$\frac{36}{40}$ is greater than $\frac{35}{40}$, so $\frac{9}{10} > \frac{7}{8}$.

$\frac{9}{10}$ is greater.

3.5 Using the Order of Operations with Fractions

Follow the order of operations.

1. Do all operations inside parentheses or other grouping symbols.

2. Simplify any expressions with exponents and find any square roots.

3. Multiply or divide, proceeding from left to right.

4. Add or subtract, proceeding from left to right.

Simplify by using the order of operations.

$$\frac{1}{2}\left(\frac{2}{3}\right) - \left(\frac{1}{4}\right)^2$$ Apply the exponent first.

$$= \frac{1}{2}\left(\frac{\overset{1}{2}}{3}\right) - \frac{1}{16}$$ Multiply next.

$$= \frac{1}{3} - \frac{1}{16}$$ Subtract last.

$$= \frac{16}{48} - \frac{3}{48}$$ Rewrite fractions so they have a common denominator. Subtract numerators.

$$= \frac{13}{48}$$ Simplest form

Chapter 3 Review Exercises

3.1 *Add or subtract. Write answers in lowest terms.*

1. $\dfrac{5}{7} + \dfrac{1}{7}$

2. $\dfrac{4}{9} + \dfrac{3}{9}$

3. $\dfrac{1}{8} + \dfrac{3}{8} + \dfrac{2}{8}$

4. $\dfrac{5}{16} - \dfrac{3}{16}$

5. $\dfrac{5}{10} + \dfrac{3}{10}$

6. $\dfrac{5}{12} - \dfrac{3}{12}$

7. $\dfrac{36}{62} - \dfrac{10}{62}$

8. $\dfrac{68}{75} - \dfrac{43}{75}$

Solve each application problem. Write answers in lowest terms.

9. Jaime Villagranna earns $\frac{7}{12}$ of his income installing kitchen cabinets for Home Depot and $\frac{4}{12}$ of his income by operating his own cabinet business. What fraction of his total income comes from the two jobs?

10. At the annual Cub Scout Pinewood Derby competition, $\frac{3}{8}$ of the events were completed before lunch and $\frac{5}{8}$ after lunch. How much more was completed after lunch than before lunch?

3.2 *Find the least common multiple of each set of numbers.*

11. 5, 2

12. 3, 4

13. 10, 12, 20

14. 3, 8, 4

15. 6, 8, 5, 15

16. 15, 9, 20

Rewrite each fraction using the indicated denominator.

17. $\dfrac{2}{3} = \dfrac{}{12}$

18. $\dfrac{3}{8} = \dfrac{}{56}$

19. $\dfrac{2}{5} = \dfrac{}{25}$

20. $\dfrac{5}{9} = \dfrac{}{81}$

21. $\dfrac{4}{5} = \dfrac{}{40}$

22. $\dfrac{5}{16} = \dfrac{}{64}$

3.1–3.3   *Add or subtract. Write answers in lowest terms.*

23. $\dfrac{1}{2} + \dfrac{1}{3}$

24. $\dfrac{1}{5} + \dfrac{3}{10} + \dfrac{3}{8}$

25. $\begin{aligned}\dfrac{5}{12}\\ +\dfrac{5}{24}\\ \hline\end{aligned}$

26. $\dfrac{2}{3} - \dfrac{1}{4}$

27. $\begin{aligned}\dfrac{7}{8}\\ -\dfrac{1}{3}\\ \hline\end{aligned}$

28. $\begin{aligned}\dfrac{11}{12}\\ -\dfrac{4}{9}\\ \hline\end{aligned}$

Solve each application problem.

29. The San Juan School District operates an after school program for students. This year $\frac{2}{5}$ of the students played after school sports, $\frac{1}{6}$ participated in arts and crafts, and $\frac{1}{3}$ spent their time in tutoring and study hall. What fraction of the total students participated in these activities?

30. When budgeting for her wedding, Kara Salzie plans to spend $\frac{1}{3}$ of her total budget on the wedding site, $\frac{3}{8}$ on food and beverages, $\frac{3}{16}$ on photos and video, and $\frac{1}{16}$ on entertainment. What portion of her budget will be spent on these four categories?

3.4 *First estimate the answer. Then add or subtract to find the exact answer. Simplify all exact answers.*

31. *Estimate:*   *Exact:*

Rounds to → $18\dfrac{5}{8}$

$+$ Rounds to → $+\,13\dfrac{3}{4}$

---

32. *Estimate:*   *Exact:*

$22\dfrac{2}{3}$

$+$ $+\,15\dfrac{4}{9}$

---

33. *Estimate:*   *Exact:*

$12\dfrac{3}{5}$

$8\dfrac{5}{8}$

$+$ $+\,10\dfrac{5}{16}$

---

34. *Estimate:*   *Exact:*

$31\dfrac{3}{4}$

$-$ $-\,14\dfrac{2}{3}$

---

35. *Estimate:*   *Exact:*

34

$-$ $-\,15\dfrac{2}{3}$

---

36. *Estimate:*   *Exact:*

$215\dfrac{7}{16}$

$-$ $-\,136$

Add or subtract by changing mixed numbers to improper fractions. Simplify all answers.

37. $5\dfrac{2}{5}$

$+\,3\dfrac{7}{10}$

---

38. $4\dfrac{3}{4}$

$+\,5\dfrac{2}{3}$

---

39. 5

$-\,1\dfrac{3}{4}$

---

40. $6\dfrac{1}{2}$

$-\,4\dfrac{5}{6}$

---

41. $8\dfrac{1}{3}$

$-\,2\dfrac{5}{6}$

---

42. $5\dfrac{5}{12}$

$-\,2\dfrac{5}{8}$

First estimate the answer and then find the exact answer for each application problem. Simplify all exact answers.

43. Two long-distance runners began an $18\frac{3}{4}$ mile run. They ran uphill $5\frac{5}{8}$ miles, downhill $7\frac{1}{3}$ miles, and the rest of the course was level. Find the distance of the level portion of the course.

Estimate:

Exact:

44. The Boys Scouts had a paper drive. They collected $28\frac{2}{3}$ tons of newspapers on Saturday and $24\frac{3}{4}$ tons on Sunday. Find the total weight of the newspapers collected.

Estimate:

Exact:

45. On a recent fishing trip to Kemmerer, Wyoming, Roy Abriani caught four fish. One fish was a German brown trout weighing $7\frac{1}{2}$ pounds, while the other three were rainbow trout weighing $2\frac{3}{4}$ pounds, $4\frac{7}{8}$ pounds, and $3\frac{3}{8}$ pounds. Find the total weight of the four fish.

Estimate:

Exact:

46. The Safeway World Championship Pumpkin Weigh-off awarded first place to the grower of a pumpkin weighing $1535\frac{3}{8}$ pounds. The second place pumpkin weighed $1475\frac{11}{16}$ pounds. How much more did the first place pumpkin weigh? (*Source: Sacramento Bee.*)

Estimate:

Exact:

3.5   *Locate each fraction in Exercises 47–50 on the number line.*

47. $\frac{3}{8}$

48. $\frac{7}{4}$

49. $\frac{8}{3}$

50. $3\frac{1}{5}$

Write < or > in each blank to make a true statement.

51. $\frac{2}{3}$ ____ $\frac{3}{4}$

52. $\frac{3}{4}$ ____ $\frac{7}{8}$

53. $\frac{1}{2}$ ____ $\frac{7}{15}$

54. $\frac{7}{10}$ ____ $\frac{8}{15}$

55. $\frac{9}{16}$ ____ $\frac{5}{8}$

56. $\frac{7}{20}$ ____ $\frac{8}{25}$

57. $\frac{19}{36}$ ____ $\frac{29}{54}$

58. $\frac{19}{132}$ ____ $\frac{7}{55}$

Simplify each expression.

59. $\left(\dfrac{1}{2}\right)^2$

60. $\left(\dfrac{2}{3}\right)^2$

61. $\left(\dfrac{3}{10}\right)^3$

62. $\left(\dfrac{3}{8}\right)^4$

Simplify by using the order of operations.

63. $8\left(\dfrac{1}{4}\right)^2$

64. $12\left(\dfrac{3}{4}\right)^2$

65. $\left(\dfrac{2}{3}\right)^2 \cdot \left(\dfrac{3}{8}\right)^2$

66. $\dfrac{7}{8} \div \left(\dfrac{1}{8} + \dfrac{3}{4}\right)$

67. $\left(\dfrac{1}{2}\right)^2 \cdot \left(\dfrac{1}{4} + \dfrac{1}{2}\right)$

68. $\left(\dfrac{1}{4}\right)^3 + \left(\dfrac{5}{8} + \dfrac{3}{4}\right)$

Mixed Review Exercises

Simplify. Use the order of operations as necessary.

69. $\dfrac{7}{8} - \dfrac{1}{8}$

70. $\dfrac{7}{10} - \dfrac{3}{10}$

71. $\dfrac{29}{32} - \dfrac{5}{16}$

72. $\dfrac{1}{4} + \dfrac{1}{8} + \dfrac{5}{16}$

73. $\begin{aligned} 6\tfrac{2}{3} \\ -\,4\tfrac{1}{2} \\ \hline \end{aligned}$

74. $\begin{aligned} 9\tfrac{1}{2} \\ +\,16\tfrac{3}{4} \\ \hline \end{aligned}$

75. $\begin{aligned} 7 \\ -\,1\tfrac{5}{8} \\ \hline \end{aligned}$

76. $\begin{aligned} 2\tfrac{3}{5} \\ 8\tfrac{5}{8} \\ +\,\tfrac{5}{16} \\ \hline \end{aligned}$

77. $\begin{aligned} 32\tfrac{5}{12} \\ -\,17 \\ \hline \end{aligned}$

78. $\dfrac{7}{22} + \dfrac{3}{22} + \dfrac{3}{11}$

79. $\left(\dfrac{1}{4}\right)^2 \cdot \left(\dfrac{2}{5}\right)^3$

80. $\dfrac{3}{8} \div \left(\dfrac{1}{2} + \dfrac{1}{4}\right)$

81. $\left(\dfrac{2}{3}\right)^2 \cdot \left(\dfrac{1}{3} + \dfrac{1}{6}\right)$

82. $\left(\dfrac{2}{3}\right)^3 + \left(\dfrac{2}{3} - \dfrac{5}{9}\right)$

Write < or > in each blank to make a true statement.

83. $\dfrac{2}{3}$ _____ $\dfrac{7}{12}$

84. $\dfrac{8}{9}$ _____ $\dfrac{15}{8}$

85. $\dfrac{17}{30}$ _____ $\dfrac{36}{60}$

86. $\dfrac{5}{8}$ _____ $\dfrac{17}{30}$

Find the least common multiple of each set of numbers.

87. 12, 18

88. 6, 8, 10, 12

89. 9, 14, 21

Rewrite each fraction using the indicated denominator.

90. $\dfrac{2}{3} = \dfrac{}{27}$

91. $\dfrac{9}{12} = \dfrac{}{144}$

92. $\dfrac{4}{5} = \dfrac{}{75}$

First estimate the answer and then find the exact answer for each application problem.
Simplify exact answers.

93. A cement contractor needs $13\frac{1}{2}$ ft of wire mesh for a concrete walkway and $22\frac{3}{8}$ ft of wire mesh for a driveway. If the contractor starts with a roll of wire that is $92\frac{3}{4}$ ft long, find the number of feet remaining after the two jobs have been completed.

94. A baker had four 50-pound bags of sugar. She used $68\frac{1}{2}$ pounds of sugar to bake cakes, $76\frac{5}{8}$ pounds for baking pies, and $33\frac{1}{4}$ pounds for baking cookies. How many pounds of sugar remain?

Estimate:

Exact:

Estimate:

Exact:

Chapter 3 Test

Add or subtract. Write answers in lowest terms.

1. $\dfrac{5}{8} + \dfrac{1}{8}$

2. $\dfrac{1}{16} + \dfrac{7}{16}$

3. $\dfrac{7}{10} - \dfrac{3}{10}$

4. $\dfrac{7}{12} - \dfrac{5}{12}$

Find the least common multiple of each set of numbers.

5. 2, 3, 4

6. 6, 3, 5, 15

7. 6, 9, 27, 36

Add or subtract. Write answers in lowest terms.

8. $\dfrac{3}{8} + \dfrac{1}{4}$

9. $\dfrac{2}{9} + \dfrac{5}{12}$

10. $\dfrac{7}{8} - \dfrac{2}{3}$

11. $\dfrac{2}{5} - \dfrac{3}{8}$

First estimate the answer. Then add or subtract to find the exact answer. Simplify exact answers.

12. $7\dfrac{2}{3} + 4\dfrac{5}{6}$

Estimate:

Exact:

13. $16\dfrac{2}{5} - 11\dfrac{2}{3}$

Estimate:

Exact:

14. $18\dfrac{3}{4} + 9\dfrac{2}{5} + 12\dfrac{1}{3}$

Estimate:

Exact:

15. $24 - 18\dfrac{3}{8}$

Estimate:

Exact:

16. Most students say that "addition and subtraction of fractions are more difficult than multiplication and division of fractions." Why do you think they say this? Do you agree with these students?

17. Explain a method of estimating an answer to addition and subtraction problems involving mixed numbers. Could your estimated answer vary from the exact answer? If it did, what would the estimation accomplish?

First estimate the answer and then find the exact answer for each application problem. Simplify exact answers.

18. In one week, a kennel owner used $10\frac{3}{8}$ pounds of puppy chow, $84\frac{1}{2}$ pounds of dry kibble, $36\frac{5}{6}$ pounds of high-protein mature dog mix, and $8\frac{1}{3}$ pounds of fresh ground meat products. Find the total number of pounds used.

   Estimate:

   Exact:

19. A painting contractor arrived at a 6-unit apartment complex with $147\frac{1}{2}$ gallons of exterior paint. If his crew sprayed $68\frac{1}{2}$ gallons on the wood siding and rolled $37\frac{3}{8}$ gallons on the masonry exterior, find the number of gallons of paint remaining.

   Estimate:

   Exact:

Write $<$ or $>$ between each pair of fractions to make a true statement.

20. $\dfrac{3}{4} \underline{\hspace{1cm}} \dfrac{17}{24}$

21. $\dfrac{19}{24} \underline{\hspace{1cm}} \dfrac{17}{36}$

Simplify. Use the order of operations as needed.

22. $\left(\dfrac{1}{3}\right)^3 \cdot 54$

23. $\left(\dfrac{3}{4}\right)^2 - \left(\dfrac{7}{8} \cdot \dfrac{1}{3}\right)$

24. $4\left(\dfrac{7}{8} - \dfrac{7}{16}\right)$

25. $\dfrac{5}{6} + \dfrac{4}{3}\left(\dfrac{3}{8}\right)$

Chapters 1–3 *Cumulative Review Exercises*

1. In this number, name the digit that has the given place value.   5,639,428

   millions   ten-thousands

   thousands   hundreds

2. Round 59,803 to the nearest ten, then to the nearest hundred, and finally to the nearest thousand.

Use front end rounding to estimate each answer. Then find the exact answer.

3. *Estimate:*   *Exact:*

   $\underline{}$ $\begin{array}{r} 24{,}276 \\ -\ 9\,887 \\ \hline \end{array}$

4. *Estimate:*   *Exact:*

   $)\overline{}$ $35)\overline{112{,}385}$

Add, subtract, multiply, or divide as indicated.

5. $\begin{array}{r} 375{,}899 \\ 521{,}742 \\ +\ 357{,}968 \\ \hline \end{array}$

6. $\begin{array}{r} 3{,}896{,}502 \\ -\ 1{,}094{,}807 \\ \hline \end{array}$

7. $5(8)(4)$

8. $\begin{array}{r} 962 \\ \times\ 384 \\ \hline \end{array}$

9. $8)\overline{1080}$

10. $13{,}467 \div 5$

Use front end rounding to estimate the answer to each application problem. Then find the exact answer.

Handicap Parking Sign

Accessible Aisle

White on Blue Background

18 ft

9 ft   5 ft

11. The Americans with Disabilities Act provides the single parking space design shown at the left. Find the perimeter of (distance around) this parking space, including the accessible aisle.

    Estimate:

    Exact:

12. The single parking space design in **Exercise 11** measures 18 ft by 14 ft. Find its area.

    Estimate:

    Exact:

Round the mixed numbers in each problem to the nearest whole number and estimate the answer. Then find the exact answer. Simplify exact answers.

13. The first car to be marketed specifically to women was the 1953 Nash Metropolitan. Ads said, "Metropolitan is a good thing in a small package." The car could get $47\frac{3}{10}$ miles per gallon and the gas tank capacity was $8\frac{3}{4}$ gallons. How many miles could the car go on a full tank of gasoline? (*Source:* uniquecarsandparts.com and Wikipedia.)

Estimate:

Exact:

14. Fleas can jump 130 times higher than their own height. In human terms, how far could a $70\frac{1}{4}$ inch person jump? (*Source:* High Tech Productions Science and Technology Center.)

Estimate:

Exact:

Simplify.

15. $2^4 \cdot 3^2$

16. $5^2 - 2(8)$

17. $\sqrt{25} + 5 \cdot 9 - 6$

18. $\frac{2}{3}\left(\frac{4}{5} - \frac{2}{3}\right)$

19. $\frac{3}{4} \div \left(\frac{1}{3} + \frac{1}{2}\right)$

20. $\frac{7}{8} + \left(\frac{3}{4}\right)^2 - \frac{3}{8}$

21. $\frac{3}{4} \cdot \frac{2}{3}$

22. $42 \cdot \frac{7}{8}$

23. $\frac{25}{40} \div \frac{10}{35}$

24. $9 \div \frac{2}{3}$

First estimate the answer and then add or subtract to find the exact answer. Write exact answers as mixed numbers.

25. Estimate: Exact:

Rounds to $\left\{ \begin{array}{l} 3\frac{3}{8} \end{array} \right.$

$+ \underline{\hspace{1cm}}$ Rounds to $\left\{ + 4\frac{1}{2} \right.$

26. Estimate: Exact:

$21\frac{7}{8}$

$+ \underline{\hspace{1cm}}$ $+ 4\frac{5}{12}$

27. Estimate: Exact:

5

$- \underline{\hspace{1cm}}$ $- 2\frac{3}{8}$

Locate each fraction in Exercises 28–31 on the number line at the right.

28. $2\frac{3}{4}$

29. $\frac{1}{9}$

30. $\frac{5}{3}$

31. $\frac{10}{3}$

```
+----+----+----+----+----+
0    1    2    3    4
```

Write < or > in each blank to make a true statement.

32. $\frac{3}{5}$ ___ $\frac{5}{8}$

33. $\frac{17}{20}$ ___ $\frac{3}{4}$

34. $\frac{7}{12}$ ___ $\frac{11}{18}$

4

Decimals

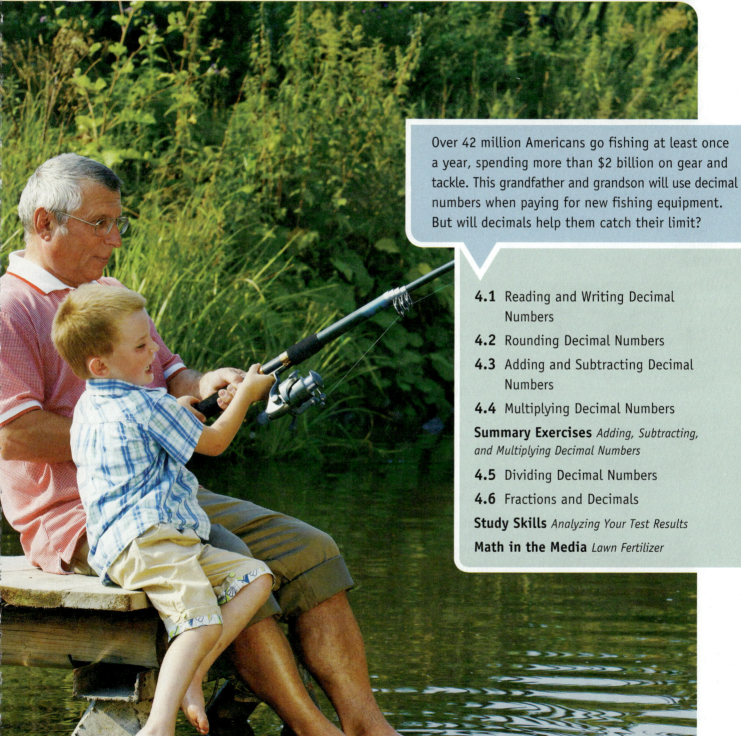

Over 42 million Americans go fishing at least once a year, spending more than $2 billion on gear and tackle. This grandfather and grandson will use decimal numbers when paying for new fishing equipment. But will decimals help them catch their limit?

263

4.1 Reading and Writing Decimal Numbers

OBJECTIVES

1. Write parts of a whole using decimals.
2. Identify the place value of a digit.
3. Read and write decimals in words.
4. Write decimals as fractions or mixed numbers.

VOCABULARY TIP

Deci The prefix **deci** means **tenth**. For example, a **deci**meter is one **tenth** of a meter.

---

1 There are 10 dimes in one dollar. Each dime is $\frac{1}{10}$ of a dollar. Label the yellow shaded portion of each dollar as a fraction, as a decimal, and in words.

(a)

(b)

(c)

Answers

1. **(a)** $\frac{1}{10}$; 0.1; one tenth

   (b) $\frac{3}{10}$; 0.3; three tenths

   (c) $\frac{9}{10}$; 0.9; nine tenths

Fractions are used to represent parts of a whole. In this chapter, **decimals** are used as another way to show parts of a whole. For example, our money system is based on decimals. One dollar is divided into 100 equal parts. One cent ($0.01) is one of the parts, and a dime ($0.10) is 10 of the parts. Metric measurement (see **Chapter 7**) is also based on decimals.

OBJECTIVE ▶ 1 Write parts of a whole using decimals. Decimals are used when a whole is divided into 10 equal parts, or into 100 or 1000 or 10,000 equal parts. In other words, decimals are fractions with denominators that are a power of 10. For example, the square below is cut into 10 equal parts. Written as a fraction, each part is $\frac{1}{10}$ of the whole. Written as a decimal, each part is 0.1. Both $\frac{1}{10}$ and 0.1 are read as "*one tenth.*"

$\frac{1}{10}$ 0.1

One-tenth of the square is shaded.

The dot in 0.1 is called the **decimal point.**

$$0.1$$
Decimal point ⎯⎯↑

The square at the right has **7** of its 10 parts shaded.

Written as a *fraction,* $\dfrac{7}{10}$ of the square is shaded.

Written as a *decimal,* 0.**7** of the square is shaded.

Both $\frac{7}{10}$ and 0.7 are read as "*seven tenths.*"

$\frac{7}{10}$

0.7

Seven-tenths of the square is shaded.

◀ **Work Problem 1 at the Side.**

The square below is cut into 100 equal parts.

Written as a *fraction,* each part is $\dfrac{1}{100}$ of the whole.

Written as a decimal, each part is **0.01** of the whole.

Both $\frac{1}{100}$ and 0.01 are read as "*one hundredth.*"

$\frac{1}{100}$ 0.01

This square has 87 parts shaded.

Written as a fraction, $\dfrac{87}{100}$ of the total area is shaded.

Written as a decimal, **0.87** of the total area is shaded.

Both $\frac{87}{100}$ and 0.87 are read as "*eighty-seven hundredths.*"

Work Problem ② at the Side. ▶

Example 1 below shows several numbers written as fractions, as decimals, and in words.

| EXAMPLE 1 | Using the Decimal Forms of Fractions |

Fraction	Decimal	Read As
(a) $\dfrac{4}{10}$	0.4	four tenths
(b) $\dfrac{9}{100}$	0.09	nine hundredths
(c) $\dfrac{71}{100}$	0.71	seventy-one hundredths
(d) $\dfrac{8}{1000}$	0.008	eight thousandths
(e) $\dfrac{45}{1000}$	0.045	forty-five thousandths
(f) $\dfrac{832}{1000}$	0.832	eight hundred thirty-two thousandths

· Work Problem ③ at the Side. ▶

OBJECTIVE ▶ ② **Identify the place value of a digit.** The decimal point separates the *whole number part* from the *fractional part* in a decimal number. In the chart below, you see that the **place value** names for fractional parts are similar to those on the whole number side, but end in "*ths*."

Decimal Place Value Chart

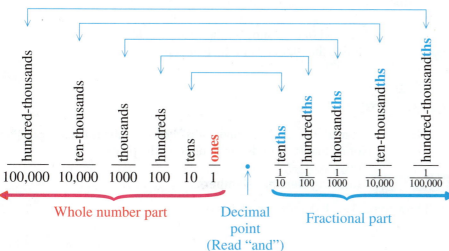

hundred-thousands	ten-thousands	thousands	hundreds	tens	ones		tenths	hundredths	thousandths	ten-thousandths	hundred-thousandths
100,000	10,000	1000	100	10	1		$\frac{1}{10}$	$\frac{1}{100}$	$\frac{1}{1000}$	$\frac{1}{10,000}$	$\frac{1}{100,000}$

Whole number part ◀——   Decimal point (Read "and")   Fractional part ——▶

Note

Notice that the **ones** place is at the center of the place value chart. There is no "oneths" place.

Also notice that each place is 10 times the value of the place to its right.

Finally, be sure to write a **hyphen** (dash) in ten–thousand**ths** and hundred–thousand**ths**.

---

② Write the portion of each square that is shaded as a fraction, as a decimal, and in words.

(a)

$\dfrac{3}{10} = 0.\underline{\quad} = \underline{\qquad}$ tenths

(b)

③ Write each decimal as a fraction.

(a) $0.7 = \dfrac{7}{\underline{\quad}}$

(b) 0.2

(c) 0.03

(d) 0.69

(e) 0.047

(f) 0.351

Answers

2. (a) 0.3; three tenths

(b) $\dfrac{41}{100}$; 0.41; forty-one hundredths

3. (a) 10; $\dfrac{7}{10}$ (b) $\dfrac{2}{10}$ (c) $\dfrac{3}{100}$

(d) $\dfrac{69}{100}$ (e) $\dfrac{47}{1000}$ (f) $\dfrac{351}{1000}$

4 Identify the place value of each digit.

(a) 971.54

(b) 0.4

(c) 5.60

(d) 0.0835

5 Tell how to read each decimal in words.

GS (a) 0.6 six _____

GS (b) 0.46 forty-six _____

(c) 0.05   (d) 0.409

(e) 0.0003   (f) 0.2703

CAUTION

If a number does *not* have a decimal point, it is a *whole number*. A whole number has no fractional part. If you want to show the decimal point in a whole number, it is just to the **right** of the digit in the ones place. Here are two examples.

$$8 = 8. \qquad\qquad 306 = 306.$$

Decimal point   Decimal point

EXAMPLE 2 **Identifying the Place Value of a Digit**

Identify the place value of each digit.

Notice the hyphen (dash) in ten-thousandths and in hundred-thousandths.

(a) 178.36

(b) 0.00935

```
        hundreds                              ones
          tens                               tenths
          ones                             hundredths
        tenths                             thousandths
      hundredths                         ten-thousandths
                                       hundred-thousandths
   1  7  8 . 3  6                    0 . 0  0  9  3  5
```

Notice in **Example 2(b)** that we do *not* use commas on the right side of the decimal point.

◀ **Work Problem 4 at the Side.**

OBJECTIVE 3 **Read and write decimals in words.** A decimal is read according to its form as a fraction.

```
ones
tenths
0 . 9
```
We read 0.9 as "nine tenths" because 0.9 is the same as $\frac{9}{10}$. Notice that 0.9 ends in the tenths place.

```
ones
tenths
hundredths
0 . 0  2
```
We read 0.02 as "two hundredths" because 0.02 is the same as $\frac{2}{100}$. Notice that 0.02 ends in the hundredths place.

EXAMPLE 3 **Reading Decimal Numbers**

Tell how to read each decimal in words.

(a) 0.3

Because $0.3 = \frac{3}{10}$, read the decimal as three <u>ten</u>ths.

(b) 0.49   Read it as: forty-nine <u>hundred</u>ths.

(c) 0.08   Read it as: eight <u>hundred</u>ths.

Think: $0.08 = \frac{8}{100}$ so write *hundredths*.

(d) 0.918   Read it as: nine hundred eighteen <u>thousand</u>ths.

(e) 0.0106   Read it as: one hundred six <u>ten-thousand</u>ths.

Think: $0.0106 = \frac{106}{10,000}$

◀ **Work Problem 5 at the Side.**

Answers

```
                hundreds              ones
                  tens              tenths
                  ones
                tenths
              hundredths
```
4. (a) 9 7 1 . 5 4   (b) 0 . 4

```
        ones                    ones
      tenths                  tenths
    hundredths              hundredths
                           thousandths
                         ten-thousandths
```
(c) 5 . 6 0   (d) 0 . 0 8 3 5

5. (a) six tenths
   (b) forty-six hundredths
   (c) five hundredths
   (d) four hundred nine thousandths
   (e) three ten-thousandths
   (f) two thousand seven hundred three ten-thousandths

Reading Decimal Numbers

Step 1 Read any whole number part to the *left* of the decimal point as you normally would.

Step 2 Read the decimal point as "*and.*"

Step 3 Read the part of the number to the *right* of the decimal point as if it were an ordinary whole number.

Step 4 Finish with the place value name of the rightmost digit; these names all end in "*ths.*"

Note

If there is *no whole number part,* you will use only Steps 3 and 4.

EXAMPLE 4 **Reading Decimals**

Read each decimal.

(a)

9 is in tenths place.

16.9

sixteen **and** nine **tenths**

Remember to say or write "and" *only* when you see a decimal point.

16.9 is read "sixteen **and** nine tenths."

(b)

5 is in hundredths place.

482.35

four hundred eighty-two **and** thirty-five **hundredths**

482.35 is read "four hundred eighty-two and thirty-five hundredths."

3 is in thousandths place.

(c) 0.063 is "sixty-three **thousandths**." (no whole number part)

(d) 11.1085 is "eleven **and** one thousand eighty-five **ten-thousandths**."

CAUTION

Use "and" when reading a decimal point. A common mistake is to read the whole number 405 as "four hundred *and* five." But there is **no decimal point** shown in 405, so it is read "four hundred five."

························ Work Problem **6** at the Side. ▶

OBJECTIVE **4** **Write decimals as fractions or mixed numbers.** Knowing how to read decimals will help you when writing decimals as fractions.

Writing Decimals as Fractions or Mixed Numbers

Step 1 The digits to the right of the decimal point are the numerator of the fraction.

Step 2 The denominator is 10 for tenths, 100 for hundredths, 1000 for thousandths, 10,000 for ten-thousandths, and so on.

Step 3 If the decimal has a whole number part, it will be written as a mixed number with the same whole number part.

6 Tell how to read each decimal in words.

(a) 3.8 is read

three and eight _____.

(b) 15.001

(c) 0.0073

(d) 64.309

Answers

6. (a) three and eight tenths
(b) fifteen and one thousandth
(c) seventy-three ten-thousandths
(d) sixty-four and three hundred nine thousandths

7 Write each decimal as a fraction or mixed number.

(a) 0.7

(b) 12.21

(c) 0.101

(d) 0.007

(e) 1.3717

8 Write each decimal as a fraction or mixed number in lowest terms.

GS (a) 0.5 Write in lowest terms.

$$\frac{5}{10} = \frac{1}{\underline{\hspace{1cm}}}$$

GS (b) 12.6 Write in lowest terms.

$$12.6 = 12\frac{6}{10} = 12\frac{\underline{\hspace{0.8cm}}}{\underline{\hspace{0.8cm}}}$$

(c) 0.85

(d) 3.05

(e) 0.225

(f) 420.0802

Answers

7. (a) $\frac{7}{10}$ (b) $12\frac{21}{100}$ (c) $\frac{101}{1000}$

   (d) $\frac{7}{1000}$ (e) $1\frac{3717}{10,000}$

8. (a) $\frac{1}{2}$ (b) $12\frac{3}{5}$ (c) $\frac{17}{20}$ (d) $3\frac{1}{20}$

   (e) $\frac{9}{40}$ (f) $420\frac{401}{5000}$

EXAMPLE 5 **Writing Decimals as Fractions or Mixed Numbers**

Write each decimal as a fraction or mixed number.

(a) 0.19

The digits to the right of the decimal point, 19, are the numerator of the fraction. The denominator is 100 for hundredths because the rightmost digit is in the hundredths place.

$$0.19 = \frac{19}{100} \leftarrow 100 \text{ for hundredths}$$

Hundredths place ⤴

(b) 0.863

$$0.863 = \frac{863}{1000} \leftarrow 1000 \text{ for thousandths}$$

Thousandths place ⤴

(c) 4.0099

The whole number part stays the same.

$$4.0099 = 4\frac{99}{10,000} \leftarrow 10,000 \text{ for ten-thousandths}$$

Ten-thousandths place ⤴

◄ **Work Problem 7 at the Side.**

CAUTION

After you write a decimal as a fraction or a mixed number, make sure the fraction is in lowest terms.

EXAMPLE 6 **Writing Decimals as Fractions or Mixed Numbers in Lowest Terms**

Write each decimal as a fraction or mixed number in lowest terms.

(a) $0.4 = \frac{4}{10} \leftarrow 10$ for tenths   Write $\frac{4}{10}$ in lowest terms.

$$\frac{4}{10} = \frac{4 \div 2}{10 \div 2} = \frac{2}{5} \leftarrow \text{Lowest terms}$$

(b) $0.75 = \frac{75}{100} = \frac{75 \div 25}{100 \div 25} = \frac{3}{4} \leftarrow \text{Lowest terms}$

The whole number part stays the same.

(c) $18.105 = 18\frac{105}{1000} = 18\frac{105 \div 5}{1000 \div 5} = 18\frac{21}{200} \leftarrow \text{Lowest terms}$

(d) $42.8085 = 42\frac{8085}{10,000} = 42\frac{8085 \div 5}{10,000 \div 5} = 42\frac{1617}{2000} \leftarrow \text{Lowest terms}$

◄ **Work Problem 8 at the Side.**

Calculator Tip

In this book we will write a 0 in the ones place for decimal fractions. We write **0**.45 instead of just .45, to emphasize that there is no whole number. Many calculators show these zeros also. Try entering ● 4 5 ; the display probably shows 0.45 even though you did not press 0.

4.1 Exercises

   FOR EXTRA HELP Download the MyDashBoard App **MyMathLab**®

CONCEPT CHECK *Circle the correct answer in Exercises 1–4.*

1. The number 11.084 has how many decimal places?

2 3 5

2. The number 0.7185 has how many decimal places?

4 5 1

3. In 22.85, the 8 means what?

8 tenths

8 hundredths

8 ones

4. In 57.213, the 3 means what?

3 thousandths

3 tenths

3 hundredths

Identify the digit that has the given place value. **See Example 2.**

5. 70.489

tens

ones

tenths

6. 135.296

ones

tenths

tens

7. 0.83472

thousandths

ten-thousandths

tenths

8. 0.51968

tenths

ten-thousandths

hundredths

9. 149.0832

hundreds

hundredths

ones

10. 3458.712

hundreds

hundredths

tenths

11. 6285.7125

thousands

thousandths

hundredths

12. 5417.6832

thousands

thousandths

ones

Write the decimal number that has the specified place values. **See Example 2.**

13. Fill in the blanks for this decimal number:

GS 0 ones, 5 hundredths, 1 ten, 4 hundreds, 2 tenths

410.____ ____

14. Fill in the blanks for this decimal number:

GS 7 tens, 9 tenths, 3 ones, 6 hundredths, 8 hundreds

____ ____ ____.96

15. 3 thousandths, 4 hundredths, 6 ones, 2 ten-thousandths, 5 tenths

16. 8 ten-thousandths, 4 hundredths, 0 ones, 2 tenths, 6 thousandths

17. 4 hundredths, 4 hundreds, 0 tens, 0 tenths, 5 thousandths, 5 thousands, 6 ones

18. 7 tens, 7 tenths, 6 thousands, 6 thousandths, 3 hundreds, 3 hundredths, 2 ones

Write each decimal as a fraction or mixed number in lowest terms.
See Examples 5 and 6.

19. 0.7   **20.** 0.1   **21.** 13.4   **22.** 9.8

23. 0.25   **24.** 0.55   **25.** 0.66   **26.** 0.33

27. 10.17   **28.** 31.99   **29.** 0.06   **30.** 0.08

31. 0.205   **32.** 0.805   **33.** 5.002   **34.** 4.008

35. 0.686   **36.** 0.492

*Tell how to read each decimal in words. **See Examples 3 and 4.***

37. 0.5

38. 0.2

39. 0.78

40. 0.55

41. 0.105

42. 0.609

43. 12.04

44. 86.09

45. 1.075

46. 4.025

*For each exercise, rewrite the words as a decimal number. **See Examples 3 and 4.***

47. Fill in the blanks to write six and seven tenths as a decimal number.

_____ • _____

48. Fill in the blanks to write eight and twelve hundredths as a decimal number.

_____ • _____ _____

49. thirty-two hundredths

50. one hundred eleven thousandths

51. four hundred twenty and eight thousandths

52. two hundred and twenty-four thousandths

53. seven hundred three ten-thousandths

54. eight hundred and six hundredths

55. seventy-five and thirty thousandths

56. sixty and fifty hundredths

57. CONCEPT CHECK Anne read the number 4302 as "four thousand three hundred and two." Explain what is wrong with the way Anne read the number.

58. CONCEPT CHECK Jerry read the number 9.0106 as "nine and one hundred and six ten-thousandths." Explain the error he made.

The grandfather on the first page of this chapter needs to select the correct fishing line for his grandson's reel. Fishing line is sold according to how many pounds of "pull" the line can withstand before breaking. Use the table to answer Exercises 59–62. Write all fractions in lowest terms. (Note: The diameter of the fishing line is its thickness.)

RELATING FISHING LINE DIAMETER TO TEST STRENGTH

Test Strength (pounds)	Average Diameter (inches)
4	0.008
8	0.010
12	0.013
14	0.014
17	0.015
20	0.016

Source: Berkley Outdoor Technologies Group.

The diameter is the distance across the end of the line (or its thickness).

59. Write the diameter of 8-pound test line in words and as a fraction.

60. Write the diameter of 17-pound test line in words and as a fraction.

61. What is the test strength of the line with a diameter of $\frac{13}{1000}$ inch?

62. What is the test strength of the line with a diameter of sixteen thousandths inch?

CONCEPT CHECK   *Suppose your job is to take phone orders for precision parts. Use the table, and in Exercises 63–66, write the correct part number that matches what you hear the customer say over the phone. In Exercises 67–68, write the words you would say to the customer.*

Part Number	Size in Centimeters
3-A	0.06
3-B	0.26
3-C	0.6
3-D	0.86
4-A	1.006
4-B	1.026
4-C	1.06
4-D	1.6
4-E	1.602

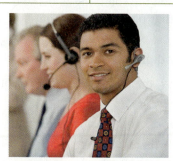

63. "Please send the six-tenths centimeter bolt."

Part number _____.

64. "The part missing from our order was the one and six hundredths size."

Part number _____.

65. "The size we need is one and six thousandths centimeters."

Part number _____.

66. "Do you still stock the twenty-six hundredths centimeter bolt?"

Part number _____.

67. "What size is part number 4-E?" Write your answer in words.

68. "What size is part number 4-B?" Write your answer in words.

Relating Concepts (Exercises 69–76)   For Individual or Group Work

Use your knowledge of place value to **work Exercises 69–76 in order.**

69. Look back at the decimal place value chart early in this section. What do you think would be the names of the next four places to the *right* of hundred-thousandths? What information did you use to come up with these names?

70. A common mistake is to think that the first place to the right of the decimal point is "oneths" and the second place is "tenths." Why might someone make that mistake? How would you explain why there is no "oneths" place?

71. Use your answer to **Exercise 69** to write 0.72436955 in words.

72. Use your answer to **Exercise 69** to write 0.000678554 in words.

73. Write 8006.500001 in words.

74. Write 20,060.000505 in words.

75. Write this decimal in numbers.

three hundred two thousand forty ten-millionths

76. Write this decimal in numbers.

nine billion, eight hundred seventy-six million, five hundred forty-three thousand, two hundred ten and one hundred million two hundred thousand three hundred billionths

4.2 Rounding Decimal Numbers

OBJECTIVES

1. Learn the rules for rounding decimals.
2. Round decimals to any given place.
3. Round money amounts to the nearest cent or nearest dollar.

In **Chapter 1,** you learned how to round whole numbers. For example, 89 rounded to the nearest ten is 90, and 8512 rounded to the nearest hundred is 8500.

OBJECTIVE ① **Learn the rules for rounding decimals.** It is also important to be able to **round** decimals. For example, a store is selling 2 candy mints for $0.75 but you want only one mint. The price of each mint is $0.75 ÷ 2, which is $0.375, but you cannot pay part of a cent. Is $0.375 closer to $0.37 or to $0.38? Actually, it's exactly halfway between. When this happens in everyday situations, the rule is to round *up*. The store will charge you $0.38 for the mint.

Rounding Decimals

Step 1   Find the place to which the rounding is being done. Draw a "cut-off" line *after* that place to show that you are cutting off and dropping the rest of the digits.

Step 2   Look *only* at the *first* digit you are cutting off.

Step 3(a) If this digit is *4 or less,* the part of the number you are keeping *stays the same.*

Step 3(b) If this digit is *5 or more,* you must *round up* the part of the number you are keeping.

Step 4   You can use the ≈ symbol or the $\doteq$ symbol to indicate that the rounded number is now an approximation (close, but not exact). Both symbols mean "is approximately equal to." In this book we will use the ≈ symbol.

CAUTION

Do ***not*** move the decimal point when rounding.

OBJECTIVE ② **Round decimals to any given place.** The following examples show you how to round decimals.

EXAMPLE 1   **Rounding a Decimal Number**

Round 14.39652 to the nearest thousandth.

Step 1   Draw a "cut-off" line after the thousandths place.

$$1\ 4\ .\ 3\ 9\ 6\ |\ 5\ 2$$

You are cutting off the 5 and 2. They will be dropped.

Thousandths

Step 2   Look *only* at the *first* digit you are cutting off. Ignore the other digits you are cutting off.

$$1\ 4\ .\ 3\ 9\ 6\ |\ 5\ 2$$

Look *only* at the 5. Ignore the 2.

Continued on Next Page

Step 3 If the first digit you are cutting off is *5 or more*, round up the part of the number you are keeping.

$$14.396\,|\!5\,2$$
$$+\ \ 0.001$$
$$\overline{14.397}$$

First digit cut is *5 or more*, so round up by adding 1 thousandth to the part you are keeping.

Think: Rounding to *thousandths* means *three* decimal places.

So, 14.39652 rounded to the nearest thousandth is 14.397
We can write 14.39652 ≈ 14.397

CAUTION

When rounding whole numbers in **Chapter 1,** you kept all the digits but changed some to zeros. With decimals, you cut off and *drop the extra digits.* In **Example 1** above, 14.39652 rounds to 14.397, **not** 14.39700.

Work Problem **1** at the Side. ▶

In **Example 1**, the rounded number 14.397 had *three decimal places.* **Decimal places** are the number of digits to the *right* of the decimal point. The first decimal place is tenths, the second is hundredths, the third is thousandths, and so on.

EXAMPLE 2   **Rounding Decimals to Different Places**

Round to the place indicated.

(a) 5.3496 to the nearest tenth

Tenths is one decimal place.

Step 1 Draw a cut-off line after the tenths place.

$$5.3\,|\!4\,9\,6$$

Tenths

You are cutting off the 4, 9, and 6. They will be dropped.

Step 2 $$5.3\,|\!\mathbf{4}\,9\,6$$

Look *only* at the 4.

Ignore these digits.

Step 3 $$5.3\,|\!\mathbf{4}\,9\,6$$
$$5.3 \leftarrow \text{Stays the same}$$

First digit cut is *4 or less,* so the part you are keeping stays the same.

5.3496 rounded to the nearest tenth is 5.3 (*one decimal place for tenths*).
We can write 5.3496 ≈ 5.3
Notice: 5.3496 does *not* round to 5.3000, which would be ten-thousandths.

(b) 0.69738 to the nearest hundredth

Step 1 $$0.69\,|\,7\,3\,8$$

Hundredths

Draw a cut-off line after the hundredths place.

Look *only* at the 7.

Step 2 $$0.69\,|\,\mathbf{7}\,3\,8$$

Continued on Next Page

1 Round to the nearest thousandth.

(a) 0.33492

$$0.334\,|\,9\,2$$

Thousandths

First digit cut is *5 or more.*

(b) 8.00851

(c) 265.42038

(d) 10.70180

2 Round to the place indicated.

GS **(a)** 0.8988 to the nearest hundredth

First digit cut is *5 or more*, so round up.

0 . 8 9 | 88

0 . 8 9 ← Keep this part.
+ 0 . 0 1 ← To round up, add 1 hundredth.
0 . _ _

(b) 5.8903 to the nearest hundredth

(c) 11.0299 to the nearest thousandth

(d) 0.545 to the nearest tenth

First digit cut is *5 or more*, so round up by adding 1 hundredth to the part you are keeping.

Step 3   0 . 6 9 | 7 3 8

0 . 6 9   ← Keep this part.
+ 0 . 0 1   ← To round up, add 1 hundredth.
0 . 7 0   ← 9 + 1 is 10; write 0 and regroup 1 to the tenths place.

0.69738 rounded to the nearest hundredth is 0.70. Hundredths is *two* decimal places so you *must* write the 0 in the hundredths place. We can write 0.69738 ≈ 0.70

Think: Rounding to *hundredths* means *two* decimal places.

CAUTION

If a *rounded* number has a 0 in the rightmost place, you *must* keep the 0. As shown above, 0.69738 rounded to the nearest hundredth is 0.7**0**. Do *not* write 0.7, which is rounded to tenths instead of hundredths.

(c) 0.01806 to the nearest thousandth

First digit cut is *4 or less*, so the part you are keeping stays the same.

0 . 0 1 8 | 0 6

0 . 0 1 8 ← Stays the same

Rounding to *thousandths* means *three* decimal places.

0.01806 rounded to the nearest thousandth is 0.018
We can write 0.01806 ≈ 0.018

(d) 57.976 to the nearest tenth

First digit cut is *5 or more*, so round up by adding 1 tenth to the part you are keeping.

57.9 | 76

57.9   ← Keep this part.
+ 0.1   ← To round up, add 1 tenth.
58.0   ← 9 + 1 is 10; write the 0 and regroup the 1 to the ones place.

Be sure to write the 0 in the tenths place.

57.976 rounded to the nearest tenth is 58.0. We can write 57.976 ≈ 58.**0** You *must* write the 0 in the tenths place to show that the number was rounded to the nearest tenth.

CAUTION

Check that your rounded answer shows *exactly* the number of decimal places asked for in the problem. Be sure your answer shows *one* decimal place if you rounded to *tenths*, *two* decimal places for *hundredths*, *three* decimal places for *thousandths*, and so on.

◀ **Work Problem** **2** **at the Side.**

OBJECTIVE **3** **Round money amounts to the nearest cent or nearest dollar.** In many everyday situations, such as shopping in a store, money amounts are rounded to the nearest cent. There are 100 cents in a dollar.

Each cent is $\dfrac{1}{100}$ of a dollar.

Another way to write $\frac{1}{100}$ is 0.01. So rounding to the *nearest cent* is the same as rounding to the *nearest hundredth of a dollar*.

Answers

2. **(a)** 0.90 **(b)** 5.89 **(c)** 11.030 **(d)** 0.5

EXAMPLE 3   **Rounding to the Nearest Cent**

How much will you pay in each shopping situation? Round each money amount to the nearest cent.

(a) $2.4238 (Is it closer to $2.42 or to $2.43?)

$2.42 | 38   First digit cut is *4 or less,* so the part you are keeping stays the same.

$2.42 ⟵ You pay.   *Rounding to the nearest cent is rounding to hundredths.*

You pay $2.42 because $2.4238 is closer to $2.42 than to $2.43.

(b) $0.695 (Is it closer to $0.69 or to $0.70?)

$0.69 | 5   *5 or more;* round up.

$0.69
+ $0.01 ⟵ To round up, add 1 hundredth (1 cent).
$0.70 ⟵ You pay.

···· **Work Problem ❸ at the Side.** ▶

Note

Some stores round *all* money amounts up to the next higher cent, even if the next digit is *4 or less.* In **Example 3(a)** above, some stores would round $2.4238 *up* to $2.43, even though it is closer to $2.42.

It is also common to round money amounts to the nearest dollar. For example, you can do that on your federal and state income tax returns to make the calculations easier. Rounding to the nearest dollar is rounding to the ones place, that is, to the nearest whole number.

EXAMPLE 4   **Rounding to the Nearest Dollar**

Round to the nearest dollar.

(a) $48.69 (Is it closer to $48 or to $49?)

Draw a cut-off line after the ones place.   First digit cut is *5 or more,* so round up by adding $1.

$48 | . 6 9

$48
+ 1
$49

Write $49 not $49.00

$48.69 is closer to $49 than to $48.
So $48.69 rounded to the nearest dollar is $49

CAUTION

$48.69 rounded to the nearest dollar is $49. Write the answer as **$49** to show that the rounding is to the *nearest dollar*. Writing $49.00 would show rounding to the *nearest cent*.

···· **Continued on Next Page**

---

❸ Round each money amount to the nearest cent.

GS **(a)** $14.595

$14 . 5 9 | 5   First digit cut is *5 or more,* so round up.

$14 . 5 9
+ 0 . 0 1 ⟵ To round up, add 1 hundredth. (1 cent).
$14 . _ _ ⟵ You pay.

(b) $578.0663

(c) $0.849

(d) $0.0548

Answers

3. **(a)** $14.60   **(b)** $578.07   **(c)** $0.85
    (d) $0.05

4 Round to the nearest dollar.

GS **(a)** $29.10

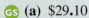

First digit cut is *4 or less,* so the part you keep stays the same.

$29|.10

(b) $136.49

(c) $990.91

(d) $5949.88

(e) $49.60

(f) $0.55

(g) $1.08

(b) $594.36 (Is it closer to $594 or $595?)

Draw a cut-off line after the ones place.

$594|.36

$594

First digit cut is *4 or less,* so the part you keep stays the same.

Careful! Write $594 (*not* $594.00)

$594.36 rounded to the nearest dollar is $594

(c) $399.88 (Is it closer to $399 or to $400?)

5 *or more,* so round up by adding $1.

$399|.88

$399
+   1
$400

You are rounding to the nearest whole number, so do *not* show any decimal places.

$399.88 rounded to the nearest dollar is $400

(d) $2689.50 (Is it closer to $2689 or $2690?)

5 *or more,* so round up by adding $1.

$2689|.50

$2689
+   1
$2690

Careful! Write $2690 (*not* $2690.00)

$2689.50 rounded to the nearest dollar is $2690

Note

When rounding $2689.50 to the nearest dollar, above, notice that it is exactly halfway between $2689 and $2690. When this happens in everyday situations, the rule is to round *up.* (Scientists working with technical data may use a more complicated rule when rounding numbers that are exactly in the middle.)

(e) $0.61 (Is it closer to $0 or to $1?)

5 *or more,* so round up.

$0|.61

Write the rounded amount as $1, *not* $1.00

$0.61 rounded to the nearest dollar is $1

Calculator Tip

Accountants and other people who work with money amounts often set their calculators to automatically round to two decimal places (nearest cent) or to round to zero decimal places (nearest dollar). Your calculator may have this feature.

◀ **Work Problem** **4** at the Side.

4.2 Exercises

   Download the MyDashBoard App **MyMathLab®**

1. **CONCEPT CHECK** Which digit would you look at when deciding how to round 4.8073 to the nearest tenth?

   _____

2. **CONCEPT CHECK** Which digit would you look at when deciding how to round 875.639 to the nearest hundredth?

   _____

3. **CONCEPT CHECK** Explain how to round 5.70961 to the nearest thousandth.

4. **CONCEPT CHECK** Explain how to round 10.028 to the nearest tenth.

Round each number to the place indicated. **See Examples 1 and 2.**

First digit cut is *5 or more.*

5. 16.8|974 to the nearest tenth

First digit cut is *4 or less.*

7. 0.956|47 to the nearest thousandth

9. 0.799 to the nearest hundredth

11. 3.66062 to the nearest thousandth

13. 793.988 to the nearest tenth

15. 0.09804 to the nearest ten-thousandth

17. 9.0906 to the nearest hundredth

19. 82.000151 to the nearest ten-thousandth

First digit cut is *5 or more.*

6. 193.84|5 to the nearest hundredth

First digit cut is *4 or less.*

8. 96.8158|4 to the nearest ten-thousandth

10. 0.952 to the nearest tenth

12. 1.5074 to the nearest hundredth

14. 476.1196 to the nearest thousandth

16. 176.004 to the nearest tenth

18. 30.1290 to the nearest thousandth

20. 0.400594 to the nearest ten-thousandth

Nardos is grocery shopping. The store will round the amount she pays for each item to the nearest cent. Write the rounded amounts. **See Example 3.**

21. Soup is three cans for $2.45, so one can is $0.81666. Nardos pays _____.

22. Orange juice is two cartons for $3.89, so one carton is $1.945. Nardos pays _____.

23. Facial tissue is four boxes for $4.89, so one box is $1.2225. Nardos pays _____.

24. Muffin mix is three packages for $1.99, so one package is $0.66333. Nardos pays _____.

25. Candy bars are six for $4.19, so one bar is $0.6983. Nardos pays _____.

26. Spaghetti is four boxes for $4.39, so one box is $1.0975. Nardos pays _____.

As she gets ready to do her income tax return, Ms. Chen rounds each amount to the nearest dollar. Write the rounded amounts. **See Example 4.**

27. Income from job, $48,649.60

28. Income from interest on bank account, $69.58

29. Donations to charity, $840.08

30. Federal withholding, $6064.49

Round each money amount as indicated. **See Examples 3 and 4.**

31. $499.98 to the nearest dollar

32. $9899.59 to the nearest dollar

33. $0.996 to the nearest cent

34. $0.09929 to the nearest cent

35. $999.73 to the nearest dollar

36. $9999.80 to the nearest dollar

The table lists speed records for various types of transportation. Use the table to answer Exercises 37–40.

Record	Speed (miles per hour)
Land speed record (specially built car)	763.04
Motorcycle speed record (conventional motorcycle)	252.662
Fastest roller coaster	134.8
Fastest military jet	2193.167
Boeing 757-300 airplane (regular passenger service)	509
Indianapolis 500 auto race (fastest average winning speed)	185.981
Daytona 500 auto race (fastest average winning speed)	177.602

Source: *Guinness World Records* and *World Almanac and Book of Facts.*

37. Round these speed records to the nearest whole number.

   (a) Motorcycle

   (b) Roller coaster

38. Round these speed records to the nearest hundredth.

   (a) Daytona 500 average winning speed

   (b) Indianapolis 500 average winning speed

39. Round these speed records to the nearest tenth.

   (a) Indianapolis 500 average winning speed

   (b) Land speed record

40. Round these speed records to the nearest hundred.

   (a) military jet

   (b) Boeing 757-300 airplane

Relating Concepts (Exercises 41–44) For Individual or Group Work

Use your knowledge about rounding money amounts to **work Exercises 41–44 in order.**

41. Explain what happens when you round $0.499 to the nearest dollar. Why does this happen?

42. Look again at **Exercise 41.** How else could you round $0.499 that would be more helpful? What kind of guideline does this suggest about rounding to the nearest dollar?

43. Explain what happens when you round $0.0015 to the nearest cent. Why does this happen?

44. Suppose you want to know which of these amounts is less, so you round them both to the nearest cent.

   $0.5968 $0.6014

   Explain what happens. Describe what you could do instead of rounding to the nearest cent.

4.3 Adding and Subtracting Decimal Numbers

OBJECTIVES

1. Add decimals.
2. Subtract decimals.
3. Estimate the answer when adding or subtracting decimals.

OBJECTIVE **1** **Add decimals.** When adding or subtracting *whole* numbers in **Chapter 1,** you lined up the numbers in columns so that you were adding ones to ones, tens to tens, and so on. A similar idea applies to adding or subtracting *decimal* numbers. With decimals, you line up the decimal points to make sure you are adding tenths to tenths, hundredths to hundredths, and so on.

Adding and Subtracting Decimals

Step 1 Write the numbers in columns with the decimal points lined up.

Step 2 If necessary, write in zeros so both numbers have the same number of decimal places. Then add or subtract as if they were whole numbers.

Step 3 Line up the decimal point in the answer directly below the decimal points in the problem.

EXAMPLE 1 **Adding Decimal Numbers**

Find each sum.

(a) 16.92 and 48.34

Step 1 Write the numbers in columns with the decimal points lined up.

$$
\begin{array}{r}
\text{tens}\ \text{ones}\ .\ \text{tenths}\ \text{hundredths}\\
1\ 6\ .\ 9\ 2\\
+\ 4\ 8\ .\ 3\ 4\\
\end{array}
$$

Decimal points are lined up.

Step 2 Add as if these were whole numbers.

$$
\begin{array}{r}
\overset{1\ 1}{16}.92\\
+\ 48.34\\
\hline
65.26\\
\end{array}
$$

Step 3 Decimal point in answer is lined up under decimal points in problem.

(b) 5.897 + 4.632 + 12.174

Write the numbers vertically with decimal points lined up. Then add.

$$
\begin{array}{r}
\overset{1\ 1\ \ 2\ 1}{5}.897\\
4.632\\
+\ 12.174\\
\hline
22.703\\
\end{array}
$$

When you rewrite the numbers in columns, and then add, be careful to line up the decimal points.

Work Problem 1 at the Side. ▶

In **Example 1(a)** above, both numbers had *two decimal places* (two digits to the right of the decimal point). In **Example 1(b),** all the numbers had *three decimal places* (three digits to the right of the decimal point). That made it easy to add tenths to tenths, hundredths to hundredths, and so on.

1 Find each sum.

GS **(a)** 2.86 + 7.09

$$
\begin{array}{r}
2.86\\
+\ 7.09\\
\hline
.\\
\end{array}
$$

Decimal points are lined up.

(b) 13.761 + 8.325

(c) 0.319 + 56.007 + 8.252

(d) 39.4 + 0.4 + 177.2

Answers

1. **(a)** 9.95 **(b)** 22.086 **(c)** 64.578
(d) 217.0

2 Find each sum.

GS **(a)** $6.54 + 9.8$

$$
\begin{array}{r}
6 \, . \, 5 \, 4 \\
+ \ 9 \, . \, 8 \, \textcolor{blue}{0} \\
\hline
\_ \_ \, . \, \_ \_
\end{array}
$$

(b) $0.831 + 222.2 + 10$

(c) $8.64 + 39.115 + 3.0076$

(d) $5 + 429.823 + 0.76$

If the number of decimal places does *not* match, you can write in zeros as placeholders to make them match. This is shown in **Example 2**.

EXAMPLE 2   Writing Zeros as Placeholders before Adding

Find each sum.

(a) $7.3 + 0.85$

There are two decimal places in 0.85 (tenths and hundredths), so write zero in the hundredths place in 7.3 so that it has two decimal places also.

$$
\begin{array}{r}
7.3\textcolor{blue}{0} \leftarrow \text{One 0 is} \\
+ \ 0.85 \quad \text{written in.} \\
\hline
8.15
\end{array}
$$

$\textbf{7.30}$ is equivalent to 7.3 because

$7\dfrac{\textbf{30}}{\textbf{100}}$ in lowest terms is $7\dfrac{3}{10}$

(b) $6.42 + 9 + 2.576$

Write in zeros so that all the addends have three decimal places. Notice how the whole number 9 is written with the decimal point at the *far right* side. (If you put the decimal point on the *left* side of the 9, you would turn it into the decimal fraction 0.9.)

Write the decimal point in 9 on the *right* side.

$$
\begin{array}{r}
6 \, . \, 4 \, 2 \, \textcolor{blue}{0} \leftarrow \text{One 0 is written in.} \\
9 \, . \, \textcolor{blue}{0 \, 0 \, 0} \leftarrow \left\{ \begin{array}{l}\text{9 is a whole number; decimal point} \\ \text{and three zeros are written in.}\end{array}\right. \\
+ \ 2 \, . \, 5 \, 7 \, 6 \leftarrow \text{No zeros are needed.} \\
\hline
1 \, 7 \, . \, 9 \, 9 \, 6
\end{array}
$$

Decimal points are lined up.

Note

Writing zeros to the right of a *decimal* number does *not* change the value of the number, as shown in **Example 2(a)** above.

◀ **Work Problem 2 at the Side.**

OBJECTIVE 2 Subtract decimals. Subtraction of decimals is done in much the same way as addition of decimals. You can check the answers to subtraction problems using addition, as you did with whole numbers.

EXAMPLE 3   Subtracting Decimal Numbers

Find each difference. Check your answers using addition.

(a) 15.82 from 28.93

Watch the order of the numbers when you see "from." Subtraction is **not** commutative like addition. So the number you are subtracting "from" goes first.

Step 1

$$
\begin{array}{r}
28 \, . \, 93 \\
- \ 15 \, . \, 82
\end{array}
$$

Line up decimal points. Then you will be subtracting hundredths from hundredths and tenths from tenths.

·········· **Continued on Next Page**

Answers

2. (a) 16.34   **(b)** 233.031   **(c)** 50.7626
   (d) 435.583

Step 2

$$\begin{array}{r} 28.93 \\ -\ 15.82 \\ \hline 13.11 \end{array}$$ ← Subtract as if they were whole numbers.

Both numbers have two decimal places: no need to write in zeros.

Decimal point in answer is lined up.

Step 3

Check the answer by adding 13.11 and 15.82. If the subtraction is done correctly, the sum will be 28.93.

(b) 146.35 minus 58.98
Regrouping is needed here.

$$\begin{array}{r} \overset{0\ 13\ 15}{\cancel{1}\cancel{4}\cancel{6}} . \overset{12\ 15}{\cancel{3}\cancel{5}} \\ -\ 58.98 \\ \hline 87.37 \end{array}$$

Line up decimal points.

Check the answer by adding 87.37 and 58.98. If you did the subtraction correctly, the sum will be 146.35. (If it *isn't*, rework the problem.)

························ **Work Problem** ❸ **at the Side.** ▶

EXAMPLE 4 Writing Zeros as Placeholders before Subtracting

Find each difference.

(a) 16.5 from 28.362
Use the same steps as in **Example 3** above. Remember to write in zeros so both numbers have three decimal places.

Line up decimal points.

16.500 is equivalent to 16.5

$$\begin{array}{r} 28.362 \\ -\ 16.500 \\ \hline 11.862 \end{array}$$ ← Write two zeros.
← Subtract as usual.

Check the answer by adding.

$$\begin{array}{r} 16.500 \\ +\ 11.862 \\ \hline 28.362 \end{array}$$ ← Matches minuend in original problem.

(b) 59.7 − 38.914

$$\begin{array}{r} 59.700 \\ -\ 38.914 \\ \hline 20.786 \end{array}$$ ← Write two zeros.
← Subtract as usual.

(c) 12 less 5.83

12.00 is equivalent to 12

$$\begin{array}{r} 12.00 \\ -\ 5.83 \\ \hline 6.17 \end{array}$$ ← Write a decimal point and two zeros.
← Subtract as usual.

························ **Work Problem** ❹ **at the Side.** ▶

OBJECTIVE ❸ **Estimate the answer when adding or subtracting decimals.**
A common error in working decimal problems by hand is to misplace the decimal point in the answer. Or, when using a calculator, you may accidentally press the wrong key. **Estimating** the answer will help you avoid these mistakes. Start by using *front end rounding* on each number (as you did with whole numbers). Here are several examples. Notice that in the rounded numbers, only the leftmost digit is something other than 0.

3.25	rounds to	3		0.812	rounds to	1
532.6	rounds to	500		26.397	rounds to	30
7094.2	rounds to	7000		351.24	rounds to	400

❸ Find each difference. Check your answers using addition.

GS **(a)** 22.7 from 72.9

$$\begin{array}{r} 72.9 \\ -\ 22.7 \\ \hline \_\_.\_ \end{array}$$

(b) 6.425 from 11.813

(c) 20.15 − 19.67

❹ Find each difference. Check your answers using addition.

GS **(a)** 18.651 from 25.3

$$\begin{array}{r} 25.300 \\ -\ 18.651 \end{array}$$

(b) 5.816 − 4.98

(c) 40 less 3.66

(d) 1 − 0.325

Answers

3. **(a)** 50.2; 50.2 + 22.7 = 72.9
   (b) 5.388; 5.388 + 6.425 = 11.813
   (c) 0.48; 0.48 + 19.67 = 20.15
4. **(a)** 6.649; 6.649 + 18.651 = 25.3
   (b) 0.836; 0.836 + 4.98 = 5.816
   (c) 36.34; 36.34 + 3.66 = 40
   (d) 0.675; 0.675 + 0.325 = 1

5 First, use front end rounding and estimate each answer. Then add or subtract to find the exact answer.

GS **(a)** $2.83 + 5.009 + 76.1$

Estimate:		Exact:
3	⟵	2.83**0**
5	⟵	5.009
+ 80	⟵	76.1**00**

(b) 11.365 from 58

Estimate: *Exact:*

(c) $398.81 + 47.658 + 4158.7$

Estimate: *Exact:*

(d) Find the difference between 12.837 meters and 46.091 meters.

Estimate: *Exact:*

(e) $19.28 plus $1.53

Estimate: *Exact:*

Answers

5. **(a)** *Estimate:* $3 + 5 + 80 = 88$;
     Exact: 83.939
   (b) *Estimate:* $60 - 10 = 50$;
     Exact: 46.635
   (c) *Estimate:* $400 + 50 + 4000 = 4450$;
     Exact: 4605.168
   (d) *Estimate:* $50 - 10 = 40$;
     Exact: 33.254 meters
   (e) *Estimate:* $20 + $2 = 22;
     Exact: $20.81

EXAMPLE 5   **Estimating Decimal Answers**

Use front end rounding to round each number. Then add or subtract the rounded numbers to get an estimated answer. Finally, find the exact answer.

(a) Find the sum of 194.2 and 6.825.

Estimate:	*Exact:*
200 ⟵ Rounds to	194.200
+ 7 ⟵ Rounds to	+ 6.825
207	201.025

The estimate goes out to the hundreds place (three places to the *left* of the decimal point), and so does the exact answer. Therefore, the decimal point is probably in the correct place in the exact answer.

(b) $69.42 + $13.78

Estimate:	*Exact:*
$70 ⟵ Rounds to	$69.42
+ 10 ⟵ Rounds to	+ 13.78
$80	$83.20 ⟵ Exact answer is close to estimate, so it is reasonable.

(c) Find the difference between 0.92 ft and 8 ft.
Use subtraction to find the difference between two numbers. The larger number, 8, is written on top.

Estimate:	*Exact:*
8 ⟵ Rounds to	8.**00** ⟵ Write a decimal point and two zeros.
− 1 ⟵ Rounds to	− 0.92
7	7.08 ft

0.92 has a digit in the ones place, so round to the ones place (nearest whole number).

(d) Subtract 1.8614 from 7.3.

Estimate:	*Exact:*
7 ⟵ Rounds to	7.3**000** ⟵ Write three zeros.
− 2 ⟵ Rounds to	− 1.8614
5	5.4386 ⟵ Exact answer is close to estimate.

◀ **Work Problem 5 at the Side.**

🖩 **Calculator Tip**

If you are *adding* numbers, you can enter them in any order on your calculator. Try these; jot down the answers.

9.82 ⊕ 1.86 ⊜ _____   1.86 ⊕ 9.82 ⊜ _____

The answers are the same because addition is *commutative.* (See **Chapter 1.**) But subtraction is *not* commutative. It *does* matter which number you enter first. Try these:

9.82 ⊖ 1.86 ⊜ _____   1.86 ⊖ 9.82 ⊜ _____

The second answer has a negative sign (−) next to it. A negative number is *less* than 0. If it was shown on your bank statement, you'd be "in the hole" by $7.96. (See **Chapter 9** for more about negative numbers.)

4.3 Exercises

FOR EXTRA HELP

 Download the MyDashBoard App

 MyMathLab®

CONCEPT CHECK *Rewrite each addition or subtraction in columns. Write in any decimal points or zeros, as needed. Do not complete the calculation.*

1. $6.42 + 10.163$

2. $7 + 9.204$

3. $20 - 9.1263$

4. $137.06 - 12$

Find each sum or difference. **See Examples 1–4.**

5. $5.69 + 11.79$

6. $0.7759 + 9.8883$

7. $8.263 - 0.5$

8. $47.658 - 20.9$

9. $76.5 + 0.506$

10. $1.87 + 9.749$

11. $21 - 0.896$

12. $9 - 1.183$

13. Subtract 0.291 from 0.4

14. Subtract 0.088 from 0.35

15. $39.76005 + 182 + 4.799 + 98.31 + 5.9999$

16. $489.76 + 0.9993 + 38 + 8.55087 + 80.697$

This drawing of a human skeleton shows the average length of the longest bones, in inches. Use the drawing to answer Exercises 17–20. (Source: The Human Body.)

17. (a) What is the combined length of the humerus and radius bones?

(b) What is the difference in the lengths of these two bones?

18. (a) What is the total length of the femur and tibia bones?

(b) How much longer is the femur than the tibia?

19. (a) Find the sum of the lengths of the humerus, ulna, femur, and tibia.

(b) How much shorter is the 8th rib than the 7th rib?

20. (a) What is the difference in the lengths of the two bones in the lower arm?

(b) What is the difference in the lengths of the two bones in the lower leg?

7th rib 9.45 in.

Humerus 14.35 in.

Radius 10.4 in.

8th rib 9.06 in.

Ulna 11.1 in.

Femur 19.88 in.

Tibia 16.94 in.

Fibula 15.94 in.

21. **CONCEPT CHECK**   Explain and correct the error that a student made when he added $0.72 + 6 + 39.5$ this way:

$$
\begin{array}{r}
0.72 \\
6 \\
+\ 39.50 \\
\hline
40.28
\end{array}
$$

22. **CONCEPT CHECK**   Explain the difference between saying "subtract 2.9 from 8" and saying "2.9 minus 8."

Use front end rounding to round each number. Then add or subtract the rounded numbers to get an estimated answer. Finally, find the exact answer. **See Example 5.**

23. The *Estimate* is already done. Now find the *Exact* answer.

Estimate:	*Exact:*
$20 ←	$19.74
− 7 ←	− 6.58
$13	

24. The *Estimate* is already done. Now find the *Exact* answer.

Estimate:	*Exact:*
$30 ←	$27.96
− 8 ←	− 8.39
$22	

25. *Estimate:*   *Exact:*

$$
\begin{array}{r}
392.7 \\
0.865 \\
+\ \ \ 21.08 \\
\hline
\end{array}
$$

$$+ \underline{\hspace{1cm}}$$

26. *Estimate:*   *Exact:*

$$
\begin{array}{r}
38.55 \\
7.716 \\
+\ \ \ 0.6 \\
\hline
\end{array}
$$

$$+ \underline{\hspace{1cm}}$$

27. What is 8.6 less 3.751?

Estimate: *Exact:*

28. What is 31.7 less 4.271?

Estimate: *Exact:*

29. *Estimate:*   *Exact:*

$$
\begin{array}{r}
62.8173 \\
539.99 \\
+\ \ \ 5.629 \\
\hline
\end{array}
$$

$$+ \underline{\hspace{1cm}}$$

30. *Estimate:*   *Exact:*

$$
\begin{array}{r}
332.607 \\
12.5 \\
+\ \ 823.3949 \\
\hline
\end{array}
$$

$$+ \underline{\hspace{1cm}}$$

*Use your estimation skills to pick the most reasonable answer for each example. Do **not** solve the problems. Circle your choice.*

31. $12 − 11.725$

    2.75   0.275   27.5

32. $20 − 1.37$

    0.1863   1.863   18. 63

33. $6.5 + 0.007$

    6.507   0.6507   65.07

34. $9.67 + 0.09$

    0.976   9.76   0.00976

35. $456.71 − 454.9$

    18.1   181   1.81

36. $803.25 − 0.6$

    802.65   0.80265   8.0265

37. $6004.003 + 52.7172$

    60.567202   605.67202   6056.7202

38. $128.35 + 97.0093$

    2253.593   225.3593   0.2253593

First use front end rounding to round each number and estimate the answer. Then find the exact answer. Use the information in the table below for Exercises 39–42.

INTERNET USERS IN SELECTED COUNTRIES

Country	Number of Users
China	420 million
United States	239.9 million
Japan	99.14 million
India	81 million
Nigeria	43.98 million
Mexico	30.6 million
Canada	26.2 million
World total	1966.5 million

Source: www.internetworldstats.com

39. How many fewer Internet users are there in Mexico than in India?

Estimate:

Exact:

40. How many more users are there in China than in Nigeria?

Estimate:

Exact:

41. How many Internet users are there in all the countries listed in the table?

Estimate:

Exact:

42. Using the exact answer from **Exercise 41,** calculate the number of worldwide Internet users in countries other than the ones in the table.

Estimate:

Exact:

43. The tallest known land mammal, a prehistoric ancestor of the rhino, was 6.4 meters tall. Compare the rhino's height to the combined heights of these three NBA basketball players: Dirk Nowitzki at 2.13 meters, Kobe Bryant at 1.98 meters, and Kevin Love at 2.08 meters. Is their combined height greater or less than the rhino's height? By how much?
(*Source:* www.NBA.com/players)

6.4 meters

Estimate:

Exact:

44. At a bakery, Sue Chee bought $7.42 worth of muffins and $10.09 worth of croissants for a staff party and a $0.69 cookie for herself. How much change did she receive from two $10 bills?

Estimate:

Exact:

45. Namiko is comparing two boxes of chicken nuggets. One box weighs 9.85 ounces and the other weighs 10.5 ounces. What is the difference in the weight of the two boxes?

Estimate:

Exact:

46. Sammy works in a veterinarian's office. He weighed two young kittens. One was 3.9 ounces and the other was 4.05 ounces. What was the difference in the weight of the two kittens?

Estimate:

Exact:

Find the perimeter of (distance around) each figure by adding the lengths of the sides.

47.

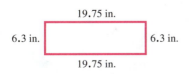

19.75 in.

6.3 in.   6.3 in.

19.75 in.

Estimate:

Exact:

48.

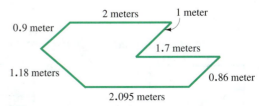

2 meters   1 meter

0.9 meter

1.7 meters

1.18 meters

0.86 meter

2.095 meters

Estimate:

Exact:

The grandfather and grandson on the first page of this chapter are buying fishing equipment. They brought along the store's sale insert from the Sunday paper. Use the information below on sale prices to answer Exercises 49–52. When estimating, round to the nearest whole number.

★ **Fishing Opener Sale** ★
Catch your limit of savings!

Bobbers 3 for 87¢

8-pound test fishing line
regular $4.84
invisible $7.47
fluorescent $5.14
No-See Line

Environmentally safe
tin split shot
$2.07

Tackle boxes
Two trays $7.96
Three trays $9.96

Leaded split shot
94¢

Spinning reels: $9.88, $12.54, $18.84, $24.96
Spinning rods: $9.97, $18.97, $22.96, $28.94

Source: Walmart.

49. What is the difference in price between the fluorescent and regular fishing line?

Estimate:

Exact:

50. How much more does the least expensive spinning rod cost than the least expensive spinning reel?

Estimate:

Exact:

51. Find the total cost of the second highest priced spinning reel, two packages of tin split shot, and a three-tray tackle box. Sales tax for all the items was $2.31.

Estimate:

Exact:

52. The grandfather bought three bobbers on sale. He also bought some SPF15 sunscreen for $7.53 and a flotation vest for $44.96. Sales tax was $3.74. How much did he spend in all?

Estimate:

Exact:

Olivia Sanchez kept track of her expenses for one month. Use her list to answer Exercises 53–58.

MONTHLY EXPENSES

Rent	$994
Car payment	$290.78
Car repairs, gas	$205
Cable TV	$49.95
Internet access	$29.95
Electricity	$40.80
Cell phone	$57.32
Groceries	$186.81
Entertainment	$97.75
Clothing, laundry	$107

53. What were Olivia's total expenses for the month?

54. How much did Olivia pay for cell phone, cable TV, and Internet access?

55. What was the difference in the amounts spent for groceries and for the car payment?

56. Compare the amount Olivia spent on entertainment to the amount spent on car repairs and gas. What is the difference?

57. How much more did Olivia spend on rent than on all her car expenses?

58. How much less did Olivia spend on clothing and laundry than on all her car expenses?

Find the length of the dashed line in each rectangle or circle.

59.

0.91 cm   0.7 cm   b

← 3 centimeters →

60.

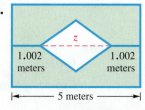

z

1.002 meters   1.002 meters

← 5 meters →

61.

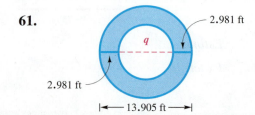

2.981 ft

q

2.981 ft

← 13.905 ft →

4.4 Multiplying Decimal Numbers

OBJECTIVE ▸ **1** **Multiply decimals.** The decimals 0.3 and 0.07 can be multiplied by writing them as fractions.

$$0.3 \times 0.07 = \frac{3}{10} \times \frac{7}{100} = \frac{3 \times 7}{10 \times 100} = \frac{21}{1000} = 0.021$$

1 decimal place + 2 decimal places → 3 decimal places

Can you see a way to multiply decimals without writing them as fractions? Use these steps. Remember that each number in a multiplication problem is called a *factor*, and the answer is called the *product*.

Multiplying Decimal Numbers

Step 1   Multiply the numbers (the factors) as if they were whole numbers.

Step 2   Find the *total* number of decimal places in *both* factors.

Step 3   Write the decimal point in the product (the answer) so it has the same number of decimal places as the total from Step 2. You may need to write in extra zeros on the left side of the product to get the correct number of decimal places.

Note

When multiplying decimals, you do **not** need to line up decimal points. (You **do** need to line up decimal points when adding or subtracting.)

EXAMPLE 1   **Multiplying Decimal Numbers**

Find the product of 8.34 and 4.2

Step 1   Multiply the numbers as if they were whole numbers.

```
    8.3 4
  ×   4.2
  ──────────
  1 6 6 8
  3 3 3 6
  ──────────
  3 5 0 2 8
```

> You do *not* have to line up decimal points when multiplying.

Step 2   Count the total number of decimal places in both factors.

```
    8.3 4  ← 2 decimal places
  ×   4.2  ← 1 decimal place
  ──────────
  1 6 6 8       3 total decimal places
  3 3 3 6
  ──────────
  3 5 0 2 8
```

Step 3   Count over 3 places in the product and write the decimal point.

```
    8.3 4  ← 2 decimal places
  ×   4.2  ← 1 decimal place
  ──────────
  1 6 6 8       3 total decimal places
  3 3 3 6
  ──────────
  3 5.0 2 8  ← 3 decimal places in product
```

Count over 3 places; count from *right* to *left*.

Work Problem ① at the Side. ▸

OBJECTIVES

1 Multiply decimals.

2 Estimate the answer when multiplying decimals.

① Find each product.

GS (a)
```
    2.6   ← 1 decimal place
  × 0.4   ← 1 decimal place
  ──────────
          ← 2 decimal places
            in the product
```

(b)
```
   45.2
  × 0.25
```

GS (c)
```
  0.104   ← 3 decimal places
  ×   7   ← 0 decimal places
  ──────────
          ← 3 decimal places
            in the product
```

(d)
```
   3.18
  × 2.23
```

(e)
```
   611
  × 3.7
```

Answers

1. (a) 1.04   (b) 11.300   (c) 0.728
(d) 7.0914   (e) 2260.7

❷ Find each product.

(a) 0.04×0.09

(b) $(0.2)(0.008)$

(c) $(0.003)^2$ *Hint:* Recall that the 2 is an exponent, so multiply $(0.003)(0.003)$.

(d) $(0.0081)(0.003)$

(e) $(0.11)(0.0005)$

❸ First use front end rounding and estimate the answer. Then find the exact answer.

(a) $(11.62)(4.01)$

(b) $(5.986)(33)$

(c) $8.31(4.2)$

(d) 58.6×17.4

EXAMPLE 2 Writing Zeros as Placeholders in the Product

Find the product: $(0.042)(0.03)$.
Start by multiplying, then count decimal places.

$$
\begin{array}{r}
0.0\,4\,2 \leftarrow \text{3 decimal places} \\
\times \quad 0.0\,3 \leftarrow \text{2 decimal places} \\
\hline
1\,2\,6 \leftarrow \text{5 decimal places needed in product}
\end{array}
$$

After multiplying, the answer has only three decimal places, but five are needed, so write two zeros on the *left* side of the answer.

$$
\begin{array}{r}
0.0\,4\,2 \\
\times \quad 0.0\,3 \\
\hline
0\,0\,1\,2\,6
\end{array}
\qquad
\begin{array}{r}
0.0\,4\,2 \leftarrow \text{3 decimal places} \\
\times \quad 0.0\,3 \leftarrow \text{2 decimal places} \\
\hline
.0\,0\,1\,2\,6 \leftarrow \text{5 decimal places}
\end{array}
$$

Write two zeros on *left* side of answer.　　Now count over 5 places and write in the decimal point.

The final product is 0.00126, which has five decimal places.

◀ **Work Problem ❷ at the Side.**

OBJECTIVE ❷ Estimate the answer when multiplying decimals. If you are doing multiplication problems by hand, estimating the answer helps you check that the decimal point is in the right place. When you are using a calculator, estimating helps you catch an error like pressing the ⊕ key instead of the ⊗ key.

EXAMPLE 3 Estimating before Multiplying

First estimate the answer to $(76.34)(12.5)$ using front end rounding. Then find the exact answer.

Estimate:　　　*Exact:*

$$
\begin{array}{r}
80 \\
\times\,10 \\
\hline
800
\end{array}
\qquad
\begin{array}{r}
7\,6.3\,4 \leftarrow \text{2 decimal places} \\
\times \quad 1\,2.5 \leftarrow \text{1 decimal place} \\
\hline
3\,8\,1\,7\,0 \quad \text{3 decimal places needed in product} \\
1\,5\,2\,6\,8 \\
7\,6\,3\,4 \\
\hline
9\,5\,4.2\,5\,0
\end{array}
$$

Both the estimate and the exact answer go out to the hundreds place, so the decimal point in 954.250 is probably in the correct place.

◀ **Work Problem ❸ at the Side.**

🖩 Calculator Tip

When working with money amounts, you may need to write a 0 in your answer. For example, try multiplying $\$3.54 \times 5$ on your calculator. Write down the result.

$$3.54 \,⊗\, 5 \,=\, _____$$

Notice that the result is 17.7, which is *not* the way to write a money amount. You have to write the 0 in the hundredths place: $\$17.70$ is correct. The calculator does not show the "extra" 0 because:

$$17.70 \text{ or } 17\frac{70}{100} \text{ simplifies to } 17\frac{7}{10} \text{ or } 17.7$$

So keep an eye on your calculator—it doesn't know when you're working with money amounts.

Answers

2. (a) 0.0036 (b) 0.0016 (c) 0.000009
   (d) 0.0000243 (e) 0.000055

3. (a) $(10)(4) = 40$; 46.5962
   (b) $(6)(30) = 180$; 197.538
   (c) $8(4) = 32$; 34.902
   (d) $60 \times 20 = 1200$; 1019.64

4.4 Exercises

 FOR EXTRA HELP   MyMathLab®

CONCEPT CHECK *Fill in the blanks for Exercises 1–4.*

1. In 4.2×3.46, how many decimal places will the product have? _____

2. In 8.071×2.79, how many decimal places will the product have? _____

3. In $(0.12)(0.03)$, how many zeros do you have to write in the product as placeholders? _____

4. In $(0.0006)(0.07)$, how many zeros do you have to write in the product as placeholders? _____

*Find each product. **See Example 1.***

5.
```
  0.042
× 3.2
```

6.
```
  0.571
× 2.9
```

7.
```
  21.5
× 7.4
```

8.
```
  85.4
× 3.5
```

9. $(0.666)(23.4)$

10. $(0.799)(0.896)$

11.
```
  $51.88
×    665
```

12.
```
  $736.75
×     118
```

CONCEPT CHECK *Use the fact that $72 \times 6 = 432$ to solve Exercises 13–18 by simply counting decimal places and writing the decimal point in the correct location.*

13. $72 \times 0.6 = \quad 4\ 3\ 2$

14. $7.2 \times 6 = \quad 4\ 3\ 2$

15. $(7.2)(0.06) = \quad 4\ 3\ 2$

16. $(0.72)(0.6) = \quad 4\ 3\ 2$

17. $0.72\,(0.06) = \quad 4\ 3\ 2$

18. $72\,(0.0006) = \quad 4\ 3\ 2$

*Find each product. **See Example 2.***

19. $(0.006)(0.0052)$

20. $(0.0052)(0.009)$

21. $(0.005)^2$

22. $(0.03)^2$

*First use front end rounding to round each number and estimate the answer. Then find the exact answer. **See Example 3.***

23. *Estimate:*
```
  40  ← Rounds to
×  5  ← Rounds to
─────
 200
```
Exact:
```
  39.6
× 4.8
```

24. *Estimate:*
```
  20  ← Rounds to
×  2  ← Rounds to
─────
  40
```
Exact:
```
  18.7
× 2.3
```

25. *Estimate:*
```
×
```
Exact:
```
  37.1
× 42
```

26. *Estimate:*
```
×
```
Exact:
```
  5.08
× 71
```

27. *Estimate:*
```
×
```
Exact:
```
  6.53
× 4.6
```

28. *Estimate:*
```
×
```
Exact:
```
  7.51
× 8.2
```

29. *Estimate:*
$(\_\_\_)(\_\_\_) = \_\_\_$
Exact:
$(2.809)(6.85) =$

30. *Estimate:*
$(\_\_\_)(\_\_\_) = \_\_\_$
Exact:
$(73.52)(22.34) =$

Even with most of the problem missing, you can tell whether or not these answers are reasonable. Circle reasonable *or* unreasonable. *If the answer is unreasonable, move the decimal point, or insert a decimal point, to make the answer reasonable.*

31. How much was his car payment? $28.90

reasonable

unreasonable, should be _____

32. How many hours did she work today? 25 hours

reasonable

unreasonable, should be _____

33. How tall is her son? 60.5 in.

reasonable

unreasonable, should be _____

34. How much does he pay for rent now? $6.92

reasonable

unreasonable, should be _____

35. What is the price of one gallon of milk? $419

reasonable

unreasonable, should be _____

36. How long is the living room? 16.8 feet

reasonable

unreasonable, should be _____

37. How much did Mrs. Brown's baby weigh? 0.095 pound

reasonable

unreasonable, should be _____

38. What was the sale price of the jacket? $1.49

reasonable

unreasonable, should be _____

Solve each application problem. Round money answers to the nearest cent when necessary.

39. LaTasha worked 50.5 hours over the last two weeks. She earns $18.73 per hour. How much did she make?

40. Michael's time card shows 42.2 hours at $10.03 per hour. What are his earnings?

41. Sid needs 0.6 meter of canvas material to make a carry-all bag that fits on his wheelchair. If canvas is $4.09 per meter, how much will Sid spend? (*Note:* $4.09 *per* meter means $4.09 for *one* meter.)

42. How much will Mrs. Nguyen pay for 3.5 yards of lace trim that costs $0.87 per yard?

43. Michelle filled the tank of her pickup truck with regular unleaded gas. Use the information shown on the pump to find how much she paid for gas.

GALLONS	PRICE PER GALLON	GALLONS	PRICE PER GALLON
10.329	$ 3.979	20.510	$ 3.759
SUPRA UNLEADED Minimum Octane Rating 90		UNLEADED REGULAR Minimum Octane Rating 87	

Source: Holiday.

44. Ground beef and chicken legs are on sale. Juma bought 1.7 pounds of legs. Use the information in the ad to find the amount she paid.

BIG ONE FOODS — Sale

Ground Beef $2.09 per pound — For juicy burgers

Chicken Legs $0.98 per pound

PRICES GOOD THROUGH SUNDAY!

45. Ms. Rolack is a real estate broker who helps people sell their homes. Her fee is 0.07 times the price of the home. What was her fee for selling a $289,500 home?

46. Josh Hamilton of the Texas Rangers baseball team had a batting average of 0.359 in the 2010 season. He went to bat 518 times. How many hits did he make? (*Hint:* Multiply the number of times at bat by his batting average.) Round to the nearest whole number. (*Source: World Almanac and Book of Facts.*)

Paper money in the United States has not always been the same size. Shown below are the measurements of bills printed before 1929 and the measurements from 1929 on. Use this information to answer Exercises 47–50. (Source: www.moneyfactory.com)

Before 1929

3.125 in.

7.4218 in.

From 1929 on

2.61 in.

6.14 in.

47. (a) Find the area of each bill, rounded to the nearest tenth. (*Hint:* Multiply to find area.)

(b) What is the difference in the rounded areas?

48. (a) Find the perimeter of each bill, to the nearest hundredth. (*Hint:* Add to find perimeter.)

(b) How much less is the perimeter of today's bills than the bills printed before 1929?

49. The thickness of one piece of today's paper money is 0.0043 inch.
(a) If you had a pile of 100 bills, how high would the pile be?

(b) How high would a pile of 1000 bills be?

50. (a) Use your answers from **Exercise 49** to find the number of bills in a pile that is 43 inches high.

(b) How much money would you have if the pile is all $20 bills?

51. Judy Lewis pays $48.96 per month for basic cable TV. The one-time installation fee was $89. How much will she pay for cable over two years? How much would she pay in two years for the deluxe cable package that costs $109.78 per month?

52. Chuck's car payment is $420.27 per month for four years. He also made a down payment of $5000 at the time he bought the car. How much will he pay altogether?

53. Barry bought 16.5 meters of rope at $0.47 per meter and three meters of wire at $1.05 per meter. How much change did he get from three $5 bills?

54. Susan bought a 46-inch LED Smart HDTV that cost $1249.97. She paid $66.59 per month for 24 months. How much could she have saved by paying for the HDTV when she bought it?

▦ *Use the information from the Look Smart online catalog to answer Exercises 55–56.*
Disregard sales tax.

Knit Shirt Ordering Information		
43-2A	Short-sleeved, solid colors	$14.75 each
43-2B	Short-sleeved, stripes	$16.75 each
43-3A	Long-sleeved, solid colors	$18.95 each
43-3B	Long-sleeved, stripes	$21.95 each
XXL size, add $2 per shirt.		
Monogram, $4.95 each. Gift box, $5 each.		

Total Price of All Items (excluding monograms and gift boxes)	Shipping, Packing, and Handling
$0–25.00	$5.50
$25.01–75.00	$7.95
$75.01–125.00	$9.95
$125.01 +	$11.95
Shipping to each additional address, add $4.25.	

55. **(a)** What is the total cost, including shipping, of sending three short-sleeved, solid-color shirts, size M, with monograms, in a gift box to your aunt for her birthday?

(b) How much did the monograms, gift box, and shipping add to the cost of your gift?

56. **(a)** Suppose you order one of each type of shirt for yourself, adding a monogram to each of the solid-color shirts. At the same time, you order three long-sleeved striped shirts, in the XXL size, shipped to your dad in a gift box. Find the total cost of your order.

(b) What is the difference in total cost (excluding shipping) between the shirts for yourself and the gift for your dad?

Relating Concepts (Exercises 57–58) For Individual or Group Work

*Look for patterns in the multiplications as you **work Exercises 57 and 58 in order**.*

57. Do these multiplications:

$(5.96)(10) =$ _____ $(3.2)(10) =$ _____

$(0.476)(10) =$ _____ $(80.35)(10) =$ _____

$(722.6)(10) =$ _____ $(0.9)(10) =$ _____

What pattern do you see? Write a "rule" for multiplying by 10. What do you think the rule is for multiplying by 100? by 1000? Write the rules and try them out on the numbers above.

58. Do these multiplications:

$(59.6)(0.1) =$ _____ $(3.2)(0.1) =$ _____

$(0.476)(0.1) =$ _____ $(80.35)(0.1) =$ _____

$(65)(0.1) =$ _____ $(523)(0.1) =$ _____

What pattern do you see? Write a "rule" for multiplying by 0.1. What do you think the rule is for multiplying by 0.01? by 0.001? Write the rules and try them out on the numbers above.

Summary Exercises *Adding, Subtracting, and Multiplying Decimal Numbers*

CONCEPT CHECK *Write each decimal as a fraction or mixed number in lowest terms.*

1. 0.8

2. 6.004

3. 0.35

Write each decimal in words.

4. 94.5

5. 2.0003

6. 0.706

Write each decimal in numbers.

7. five hundredths

8. three hundred nine ten-thousandths

9. ten and seven tenths

Round to the place indicated.

10. 6.1873 to the nearest hundredth

11. 0.95 to the nearest tenth

12. 0.42025 to the nearest thousandth

13. $0.893 to the nearest cent

14. $3.0017 to the nearest cent

15. $99.64 to the nearest dollar

Simplify.

16. 0.27(3.5)

17. 50 − 0.3801

18. $\begin{array}{r} 0.205 \\ \times\ \ \ \ 9 \\ \hline \end{array}$

19. 25($3.74)

20. (0.004)(1.22)

21. 3.7 − 1.55

22. 0.95 + 10.005

23. 3.6 + 0.718 + 9 + 5.0829

24. 32.305 − 28 − 0.0007

25. 8.9 − 0.4 − 0.03 − 7.6

26. $18 + $2.09 + $90 + $0.75

27. (0.99)(83.672)   Round your answer to the nearest hundredth.

28. Find the perimeter and area of this rectangular computer chip. Round your answers to the nearest tenth.

1.1 in.

0.45 in.   0.45 in.

1.1 in.

29. Find the perimeter and area of this rectangular watch face. Round your answers to the nearest hundredth.

0.875 in.

0.75 in.   0.75 in.

0.875 in.

Use the table on heaviest fruits and vegetables to answer Exercises 30–34.

HEAVIEST FRUITS AND VEGETABLES

Fruit/Vegetable	Weight (in pounds)	Location
Apple	4.0625	Japan
Grapefruit	6.75	Australia
Onion	16.52	England
Peach	1.6	Michigan, USA
Radish	68.563	Japan
Pumpkin	1810	Wisconsin, USA
Strawberry	0.5	England
Tomato	7.75	Oklahoma, USA

Source: Guinness World Records, www.funonthenet.inc, Associated Press.

30. (a) Identify the heaviest and lightest weight items in the table.

(b) Find the difference in weight between the heaviest and lightest items.

31. If you used the radish, tomato, and onion in a salad, what would be their combined weight to the nearest tenth of a pound?

32. (a) How much would six grapefruits weigh?

(b) Find the total weight of a dozen strawberries.

33. (a) How much heavier is the tomato than the apple?

(b) How much less does the peach weigh than the apple?

34. (a) Round the weight of each fruit or vegetable to the nearest whole number.

(b) Use the rounded numbers to find an estimated total weight if you had one of each item in the table. Then find the exact total weight.

35. A craft cooperative paid $60.32 for enough fabric to make eight baby blankets for their store. They sold the blankets for $18.95 each. How much profit was made on the blankets?

4.5 Dividing Decimal Numbers

There are two kinds of decimal division problems: those in which a decimal is divided by a whole number, and those in which a number is divided by a decimal. First recall the parts of a division problem from **Chapter 1.**

$$
\begin{array}{r}
8 \leftarrow \text{Quotient} \\
\text{Divisor} \rightarrow 4\overline{)33} \leftarrow \text{Dividend} \\
\underline{32} \\
1 \leftarrow \text{Remainder}
\end{array}
$$

OBJECTIVE 1 Divide a decimal by a whole number. When the divisor is a whole number, use these steps.

OBJECTIVES

1 Divide a decimal by a whole number.

2 Divide a number by a decimal.

3 Estimate the answer when dividing decimals.

4 Use the order of operations with decimals.

> **Dividing Decimals by Whole Numbers**
>
> **Step 1** Write the decimal point in the quotient (answer) directly above the decimal point in the dividend.
>
> **Step 2** Divide as if both numbers were whole numbers.

1 Find each quotient. Check the quotients by multiplying.

GS (a)
$$
\begin{array}{r}
2\,3\,.\,\rule{0.5em}{0.4pt} \\
4\overline{)9\,3\,.\,6} \\
\underline{8} \\
1\,3
\end{array}
$$
Finish the division. Then multiply the quotient by 4. You should get 93.6 if the quotient is correct.

EXAMPLE 1 Dividing Decimals by Whole Numbers

Find each quotient. Check the quotients by multiplying.

(a) 21.93 by 3

Dividend — Divisor

Rewrite the division problem. $3\overline{)21.93}$

Step 1 Write the decimal point in the quotient directly above the decimal point in the dividend.

Decimal points lined up

$3\overline{)21\,.\,93}$

Step 2 Divide as if the numbers were whole numbers.

$$
\begin{array}{r}
7.31 \\
3\overline{)21.93}
\end{array}
$$

Check by multiplying the quotient times the divisor.

The quotient (answer) is 7.31.

CHECK

Matches, so 7.31 is correct.

$$
\begin{array}{r}
7.31 \\
\times \quad 3 \\
\hline
21.93
\end{array}
$$

(b) $9\overline{)470.7}$

Divisor — Dividend

Write the decimal point in the quotient above the decimal point in the dividend. Then divide as if the numbers were whole numbers.

Decimal points lined up

$$
\begin{array}{r}
52.3 \\
9\overline{)470.7} \\
\underline{45} \\
20 \\
\underline{18} \\
2\,7 \\
\underline{2\,7} \\
0
\end{array}
$$

CHECK

$$
\begin{array}{r}
52.3 \\
\times \quad 9 \\
\hline
470.7
\end{array}
$$

Matches

Multiply the quotient by the divisor. The result should match the dividend.

The quotient is 52.3.

(b) $6\overline{)6.804}$

(c) $11\overline{)278.3}$

(d) $0.51835 \div 5$

(e) $213.45 \div 15$

Answers

1. (a) 23.4; (23.4)(4) = 93.6
 (b) 1.134; (1.134)(6) = 6.804
 (c) 25.3; (25.3)(11) = 278.3
 (d) 0.10367; (0.10367)(5) = 0.51835
 (e) 14.23; (14.23)(15) = 213.45

········· **Work Problem 1 at the Side.** ▶

2 Divide. Check each quotient by multiplying.

(GS) (a) $5\overline{)6.4}$ → $5\overline{)6\,.\,4\,0}$

Finish the division.

(b) $30.87 \div 14$

(c) $\dfrac{259.5}{30}$

(d) $0.3 \div 8$

| EXAMPLE 2 | Writing Extra Zeros to Complete a Division |

Divide 1.5 by 8. Check the quotient by multiplying.

Keep dividing until the remainder is 0, or until the digits in the quotient begin to repeat in a pattern. In **Example 1(b),** you ended up with a remainder of 0. But sometimes you run out of digits in the dividend before that happens. If so, write extra zeros on the right side of the dividend so you can continue dividing.

$$
\begin{array}{r}
0.1 \\
8\overline{)1.5} \\
\underline{8} \\
7
\end{array}
$$

← All digits have been used.

← Remainder is not yet 0.

Write a 0 after the 5 in the dividend so you can continue dividing. Keep writing more zeros in the dividend, if needed. Recall that writing zeros to the *right* of a decimal number does *not* change its value.

CHECK

Three zeros needed to complete the division.

Use multiplication to check the quotient.

Matches dividend, so 0.1875 is correct.

Stop dividing when the remainder is 0.

> **CAUTION**
>
> When dividing decimals, notice that the dividend might *not* be the greater number. In **Example 2** above, the dividend is 1.5, which is *less* than the divisor 8.

◀ **Work Problem 2 at the Side.**

🖩 **Calculator Tip**

In **Chapter 1,** you learned that when *multiplying* numbers, you can enter them in any order because multiplication is commutative. But division is *not* commutative. It *does* matter which number you enter first. Try **Example 2** both ways; jot down your answers.

$1.5 \;÷\; 8 \;=\;$ _____ $8 \;÷\; 1.5 \;=\;$ _____

Notice that the first answer, 0.1875, matches the result from **Example 2.** But the second answer is much different: 5.333333333. Be careful to enter the dividend first.

The next example shows a quotient (answer) that must be rounded because you will never get a remainder of 0.

EXAMPLE 3   **Rounding a Decimal Quotient**

Divide 4.7 by 3. Round the quotient to the nearest thousandth. Write extra zeros in the dividend so you can continue dividing.

$$
\begin{array}{r}
1.5\,6\,6\,6 \\
3\overline{)4.7\,0\,0\,0} \leftarrow \text{Three zeros added so far} \\
\underline{3} \\
1\,7 \\
\underline{1\,5} \\
2\,0 \\
\underline{1\,8} \\
2\,0 \\
\underline{1\,8} \\
2\,0 \\
\underline{1\,8} \\
2 \leftarrow \text{Remainder is still not 0.}
\end{array}
$$

Notice that the digit 6 in the answer is repeating. It will continue to do so. The remainder will *never be 0*. There are two ways to show that the answer is a **repeating decimal** that goes on forever. You can write three dots after the answer, or you can write a bar above the digits that repeat (in this case, the 6).

$$1.5\underbrace{666 \ldots}_{\text{Three dots}} \quad \text{or} \quad 1.5\overline{6} \quad \begin{array}{l}\leftarrow \text{Bar above} \\ \text{repeating digit}\end{array}$$

When repeating decimals occur, round the quotient according to the directions in the problem. In this example, to round to thousandths, divide out one *more* place, to ten-thousandths.

$$4.7 \div 3 = 1.5666\ldots \quad \text{rounds to} \quad 1.567$$

> Nearest thousandth is *three* decimal places.

Check the answer by multiplying 1.567 by 3. Because 1.567 is a rounded answer, the check will not give exactly 4.7, but it should be very close.

$$(1.567)(3) = 4.701 \begin{array}{l}\leftarrow \text{Does not equal exactly 4.7} \\ \text{because 1.567 was rounded.}\end{array}$$

CAUTION

When checking quotients that you've rounded, the check will *not* match the dividend exactly, but it should be very close.

················ Work Problem **3** at the Side. ▶

OBJECTIVE ▶ 2 Divide a number by a decimal. To divide by a *decimal* divisor, first change the divisor to a whole number. Then divide as before. To see how this is done, write the problem in fraction form. Here is an example.

$$1.2\overline{)6.36} \quad \text{can be written} \quad \frac{6.36}{1.2}$$

In **Chapter 3** you learned that multiplying the numerator and denominator by the same number gives an equivalent fraction. We want the divisor (1.2) to be a whole number. Multiplying by 10 will accomplish that.

$$\underset{\substack{\text{Decimal} \\ \text{divisor}}}{\Big\{} \frac{6.36}{1.2} = \frac{(6.36)(10)}{(1.2)(10)} = \frac{63.6}{12} \underset{\substack{\text{Whole number} \\ \text{divisor}}}{\Big\}}$$

---

3 Divide. Round quotients to the nearest thousandth. If it is a repeating decimal, also write the answer using a bar. Check your quotients by multiplying.

(a) $13\overline{)2\,6\,7\,.\,0\,1\,\overline{0}}$

Use your calculator.

$267.01 : 13 \approx$ _____

Is the quotient a repeating decimal? _____

So round your answer to the nearest thousandth.

_____

(b) $6\overline{)20.5}$

(c) $\dfrac{10.22}{9}$

(d) $16.15 \div 3$

(e) $116.3 \div 11$

Answers

3. (a) 20.53923; no repeating digits visible on calculator: 20.539 (rounded); (20.539)(13) = 267.007
   (b) 3.417 (rounded); 3.41$\overline{6}$; (3.417)(6) = 20.502
   (c) 1.136 (rounded); 1.13$\overline{5}$; (1.136)(9) = 10.224
   (d) 5.383 (rounded); 5.38$\overline{3}$; (5.383)(3) = 16.149
   (e) 10.573 (rounded); 10.5$\overline{72}$; (10.573)(11) = 116.303

4 Divide. If the quotient does not come out even, round to the nearest hundredth.

GS **(a)** $0.2\overline{)1.04}$

(b) $0.06\overline{)1.8072}$

(c) $0.005\overline{)32}$

(d) $8.1 \div 0.025$

(e) $\dfrac{7}{1.3}$

(f) $5.3091 \div 6.2$

The short way to multiply by 10 is to move the decimal point *one place* to the *right* in both the divisor and the dividend.

$$1.2\overline{)6.3\,6} \quad \text{is equivalent to} \quad 12\overline{)63.6}$$

> **Note**
>
> Moving the decimal points the **same** number of places in **both** the divisor and dividend will **not** change the answer.

Dividing by a Decimal Number

Step 1 Count the number of decimal places in the divisor and move the decimal point that many places to the *right*. (This changes the divisor to a whole number.)

Step 2 Move the decimal point in the dividend the *same* number of places to the *right*. (Write in extra zeros if needed.)

Step 3 Write the decimal point in the quotient directly above the decimal point in the dividend. Then divide as usual.

EXAMPLE 4 **Dividing by Decimal Numbers**

(a) $0.003\overline{)27.69}$

Move the decimal point in the divisor *three* places to the *right* so 0.003 becomes the whole number 3. To move the decimal point in the dividend the same number of places, write in an extra 0.

(b) Divide 5 by 4.2. Round to the nearest hundredth.

Move the decimal point in the divisor one place to the right so 4.2 becomes the whole number 42. The decimal point in the dividend starts on the right side of 5 and is also moved one place to the right.

$$
\begin{array}{r}
1.190 \\
4.2\overline{)5.0000} \\
\underline{42} \\
80 \\
\underline{42} \\
380 \\
\underline{378} \\
20
\end{array}
$$

← In order to round to hundredths, divide out one *more* place, to thousandths.

Move the decimal points the *same* number of places.

Round the quotient. It is 1.19 (rounded to the nearest hundredth).

◀ **Work Problem 4** at the Side.

OBJECTIVE **3** **Estimate the answer when dividing decimals.** Estimating answers helps you catch errors. Compare the estimate to your exact answer. If they are very different, work the problem again.

EXAMPLE 5 **Estimating before Dividing**

First use front end rounding to round each number and estimate the answer. Then divide to find the exact answer.

$$580.44 \div 2.8$$

Here is how one student solved this problem. She rounded 580.44 to 600 and rounded 2.8 to 3 to estimate the answer.

Estimate:

$$\begin{array}{r} \mathbf{200} \\ 3\overline{)600} \end{array}$$

Very different; need to rework the problem

Exact:

$$\begin{array}{r} 2\,7.3 \\ 2.8\overline{)5\,8\,0.4\,4} \\ \underline{5\,6} \\ 2\,0\,4 \\ \underline{1\,9\,6} \\ 8\,4 \\ \underline{8\,4} \\ 0 \end{array}$$

The estimate is too different. There is an error in the exact answer.

Notice that the estimate, which is in the hundreds, is very different from the exact answer, which is only in the tens. This tells the student that she needs to rework the problem. Can you find the error?
(The exact answer should be 207.3, which fits with the estimate of 200.)

············· **Work Problem** **5** **at the Side.** ▶

OBJECTIVE **4** **Use the order of operations with decimals.** Use the order of operations when a decimal problem involves more than one operation, as you did with whole numbers in **Chapter 1.**

Order of Operations

1. Do all operations inside *parentheses* or *other grouping symbols.*
2. Simplify any expressions with *exponents* and find any *square roots.*
3. *Multiply* or *divide,* proceeding from left to right.
4. *Add* or *subtract,* proceeding from left to right.

EXAMPLE 6 **Using the Order of Operations**

Use the order of operations to simplify each expression.

(a) $2.5 + \mathbf{6.3^2} + 9.62$ Apply the exponent: $(6.3)(6.3)$ is 39.69

   $\mathbf{2.5 + 39.69} + 9.62$ Add from left to right.

   $42.19 \quad + 9.62$

   51.81

(b) $1.82 + (\mathbf{6.7 - 5.2})(5.8)$ Work inside parentheses.

   $1.82 + \quad (\mathbf{1.5})(\mathbf{5.8})$ Multiply next.

   $1.82 + \quad 8.7$ Add last.

   10.52

········· **Continued on Next Page**

---

5 Decide whether each answer is reasonable by using front end rounding to estimate the answer. If the exact answer is *not* reasonable, find and correct the error.

GS **(a)** $42.75 \div 3.8 = 1.125$

   Estimate: $40 \div 4 = 10$

   Answer of 1.125 is **not** reasonable. Rework.

   $$3.8\overline{)42.7\,5\,0}$$

(b) $807.1 \div 1.76 = 458.580$
   to nearest thousandth

   Estimate:

(c) $48.63 \div 52 = 93.519$
   to nearest thousandth

   Estimate:

(d) $9.0584 \div 2.68 = 0.338$

   Estimate:

Answers

5. **(a)** Exact answer should be 11.25.
   (b) Estimate is $800 \div 2 = 400$; answer is reasonable.
   (c) Estimate is $50 \div 50 = 1$; answer is not reasonable, should be 0.935 (rounded).
   (d) Estimate is $9 \div 3 = 3$; answer is not reasonable, should be 3.38.

6 Use the order of operations to simplify each expression. The black brace shows you where to start.

(a) $4.6 - 0.79 + \underbrace{1.5^2}$

(b) $\underbrace{3.64 \div 1.3} \times 3.6$

(c) $0.08 + 0.6(\underbrace{3 - 2.99})$

(d) $10.85 - \underbrace{2.3(5.2)} \div 3.2$

(c) $\underline{3.7^2} - 1.8 \div 5\,(1.5)$ Apply the exponent.

> CAREFUL! Do **not** subtract yet.

$13.69 - \underline{1.8 \div 5}\,(1.5)$ Multiply and divide from left to right, so first divide 1.8 by 5 to get 0.36

$13.69 - \underline{0.36\,(1.5)}$ Then multiply 0.36 by 1.5

$\underbrace{13.69 - \quad 0.54}$ Subtract last.

13.15

◀ **Work Problem 6 at the Side.**

▦ **Calculator Tip**

You may want to use a calculator to check your work. Most scientific calculators that have parentheses keys ( ) can handle calculations like those in **Example 6** if you just enter the numbers in the order given. For example, the keystrokes for **Example 6(b)** on the previous page are:

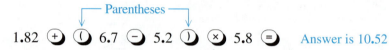

Parentheses

$1.82 \;\boxplus\; (\; 6.7 \;\boxminus\; 5.2 \;)\; \boxtimes\; 5.8 \;\boxminus\;$ Answer is 10.52

Standard, four-function calculators generally do not have parentheses keys and will *not* give the correct answer if you simply enter the numbers in the order given.

Check the instruction manual that came with your calculator for information on "order of calculations" to see if your model has the rules for order of operations built into it. For a quick check, try entering this problem.

$2 \;\boxplus\; 2 \;\boxtimes\; 2 \;\boxminus\;$

If the result is 6, the calculator follows the order of operations. If the result is 8, it does *not* have the rules built into it. To see why this test works, do the calculations by hand.

Follow the order of operations.	**Work from left to right.**
$2 + \underbrace{2 \times 2}$ Multiply before adding.	$\underbrace{2 + 2} \times 2$
$2 + \quad 4$	$4 \quad \times 2$
$6 \longleftarrow$ **Correct**	$8 \longleftarrow$ **Incorrect**

Answers

6. **(a)** 6.06   **(b)** 10.08   **(c)** 0.086
    (d) 7.1125

4.5 Exercises

   MyMathLab®

1. **CONCEPT CHECK** Which problem below has a whole number for the divisor? Find it and then rewrite the problem using the $\overline{)}$ symbol.

   (a) Divide 0.25 by 0.05   **(b)** 25 ÷ 0.5   **(c)** 25.5 ÷ 5

2. **CONCEPT CHECK** Which problem below has a whole number for the divisor? Find it and then rewrite the problem using the $\overline{)}$ symbol.

   (a) 0.423 ÷ 0.07   **(b)** 42.3 ÷ 7   **(c)** Divide 423 by 0.7

3. **CONCEPT CHECK** In $2.2\overline{)8.24}$ how do you make 2.2 a whole number, and how does that change 8.24? Use arrows to show how to move the decimal points.

4. **CONCEPT CHECK** In $5.1\overline{)10.5}$ how do you make 5.1 a whole number, and how does that change 10.5? Use arrows to show how to move the decimal points.

*Find each quotient. **See Examples 1 and 4.***

5. $7\overline{)27.3}$

6. $8\overline{)50.4}$

7. $\dfrac{4.23}{9}$

8. $\dfrac{1.62}{6}$

9. $0.05\overline{)20.01}$

10. $0.08\overline{)16.04}$

11. $1.5\overline{)54.0}$
   GS

12. $2.4\overline{)132.0}$
   GS

CONCEPT CHECK *Use the fact that* $108 ÷ 18 = 6$ *to work Exercises 13–16 simply by moving decimal points.*

13. $0.108 ÷ 1.8$

14. $10.8 ÷ 18$

15. $0.018\overline{)108}$

16. $0.18\overline{)1.08}$

*Divide. Round quotients to the nearest hundredth if necessary. **See Examples 3 and 4.***

17. $4.6\overline{)116.38}$

18. $2.6\overline{)4.992}$

19. $\dfrac{3.1}{0.006}$

20. $\dfrac{1.7}{0.09}$

Divide. Round quotients to the nearest thousandth.

21. $240.8 ÷ 9$

22. $76.43 ÷ 7$

23. $0.034\overline{)342.81}$

24. $0.043\overline{)1748.4}$

Decide whether each answer is reasonable *or* unreasonable *by rounding the numbers and estimating the answer. If the exact answer is not reasonable, find the correct answer. **See Example 5.***

25. $37.8 ÷ 8 = 47.25$

   Estimate:

26. $345.6 ÷ 3 = 11.52$

   Estimate:

27. $54.6 ÷ 48.1 = 1.135$

   Estimate:

28. $2428.8 ÷ 4.8 = 56$

   Estimate:

29. $307.02 ÷ 5.1 = 6.2$

   Estimate:

30. $395.415 ÷ 5.05 = 78.3$

   Estimate:

Solve each application problem. Round money answers to the nearest cent, if necessary.

31. Rob discovered that his daughter's favorite brand of tights is on sale. He decided to buy one pair as a surprise for her. How much did he pay?

Special Purchase!
Girls'
Tights
6 pairs for $23.98
Stock up now!

32. The bookstore has a special price on notepads. How much did Randall pay for one notepad?

Notepads 4 for $3.69

33. It will take 21 equal monthly payments for Aimee to pay off her credit card balance of $1408.68. How much is she paying each month?

34. Marcella Anderson bought 2.6 meters of microfiber woven suede fabric for $33.77. How much did she pay per meter? (*Hint:* Cost *per* meter means the cost for *one* meter.)

35. Darren Jackson earned $476.80 for 40 hours of work. Find his earnings per hour.

36. Adrian Webb bought 108 patio blocks to build a backyard patio. He paid $237.60. Find the cost per block. (*Hint:* Cost *per* block means the cost for *one* block.)

37. It took 12.3 gallons of gas to fill the gas tank of Kim's car. She had driven 344.1 miles since her last fill-up. How many miles per gallon did her car get? Round to the nearest tenth.

38. Mr. Rodriquez pays $53.19 each month to Household Finance. How many months will it take him to pay off $1436.13?

39. Soup is on sale at six cans for $3.25, or you may purchase individual cans for $0.57. How much will you save per can if you buy six cans? Round to the nearest cent.

40. Nadia's diet allows her to eat 3.5 ounces of chicken nuggets. The package weighs 10.5 ounces and contains 15 nuggets. How many nuggets can Nadia eat?

Use the table of world records for the women's long jump (through the year 2010) to answer Exercises 41–46. To find an average, add up the values you are interested in and then divide the sum by the number of values. Round your answers to the nearest hundredth.

Athlete	Country	Year	Length (meters)
Galina Christyakova	USSR	1988	7.52
Jackie Joyner-Kersee	U.S.	1994	7.49
Heike Drechsler	Germany	1992	7.48
Jackie Joyner-Kersee	U.S.	1987	7.45
Jackie Joyner-Kersee	U.S.	1988	7.40
Jackie Joyner-Kersee	U.S.	1991	7.32
Jackie Joyner-Kersee	U.S.	1996	7.20
Chioma Ajunwa	Nigeria	1996	7.12
Fiona May	Italy	2000	7.09
Tatyana Lebedeva	Russia	2004	7.07

Source: CNNSI.com

41. Find the average length of the long jumps made by Jackie Joyner-Kersee.

42. Find the average length of all the long jumps listed in the table.

43. How much longer was the fifth-longest jump than the sixth-longest jump?

44. If the first athlete in the table made five jumps of the same length, what would be the total distance jumped?

45. What was the total length jumped by the top three athletes in the table?

46. How much less was the shortest jump than the next-to-shortest jump?

47. For $2.2^2 + (9.5 - 3.1)$ list what you would do, in the correct order, to simplify the expression.
 (a) First

 (b) Second

 (c) Third

48. For $60.41 - (0.4 + 5.07)(3)$ list what you would do, in the correct order, to simplify the expression.
 (a) First

 (b) Second

 (c) Third

Use the order of operations to simplify each expression. **See Example 6.**

49. $7.2 - 5.2 + 3.5^2$

50. $6.2 + 4.3^2 - 9.72$

51. $38.6 + 11.6(13.4 - 10.4)$

52. $2.25 - 1.06(4.85 - 3.95)$

53. $8.68 - 4.6(10.4) \div 6.4$

54. $25.1 + 11.4 \div 7.5(3.75)$

55. $33 - 3.2(0.68 + 9) - 1.3^2$

56. $0.6 + (1.89 + 0.11) \div 0.004(0.5)$

57. In 2010, the U.S. Treasury printed about 26,000,000 pieces of paper money each day. The printing presses run 24 hours a day. How many pieces of money are printed, to the nearest whole number:
(a) each hour?

(b) each minute?

(c) each second?

26,000,000 pieces of paper money are printed each day.

(*Source:* www.moneyfactory.com)

58. Mach 1 is the speed of sound. Dividing a vehicle's speed by the speed of sound gives its speed on the Mach scale. In 1997, a specially built car with two 110,000-horsepower engines broke the world land speed record by traveling 763.035 miles per hour. The speed of sound that day was 748.11 miles per hour. What was the car's Mach speed, to the nearest hundredth? (*Source:* Associated Press.)

General Mills will give a school 10¢ for each box top logo from its cereals and other products. A school can earn up to $10,000 per year. Use this information to answer Exercises 59–62. Round your answers to the nearest whole number when necessary. (Source: General Mills.)

59. How many box tops would a school need to collect in one year to earn the maximum amount?

60. (Complete **Exercise 59** first.) If a school has 550 children, how many box tops would each child need to collect in one year to reach the maximum?

61. How many box tops would need to be collected during each of the 38 weeks in the school year to reach the maximum amount?

62. How many box tops would each of the 550 children need to collect during each of the 38 weeks of school to reach the maximum amount?

Relating Concepts (Exercises 63–64) For Individual or Group Work

Look for patterns as you work Exercises 63 and 64 in order.

63. Do these division problems:

$3.77 \div 10 =$ _____ $9.1 \div 10 =$ _____

$0.886 \div 10 =$ _____ $30.19 \div 10 =$ _____

$406.5 \div 10 =$ _____ $6625.7 \div 10 =$ _____

(a) What pattern do you see? Write a "rule" for dividing by 10. What do you think the rule is for dividing by 100? by 1000? Write the rules and try them out on the numbers above.

(b) Compare your rules to the ones you wrote in the previous section of this chapter for **Exercise 57**. How are they different?

64. Do these division problems:

$40.2 \div 0.1 =$ _____ $7.1 \div 0.1 =$ _____

$0.339 \div 0.1 =$ _____ $15.77 \div 0.1 =$ _____

$46 \div 0.1 =$ _____ $873 \div 0.1 =$ _____

(a) What pattern do you see? Write a "rule" for dividing by 0.1. What do you think the rule is for dividing by 0.01? by 0.001? Write the rules and try them out on the numbers above.

(b) Compare your rules to the ones you wrote in the previous section of this chapter for **Exercise 58**. How are they different?

4.6 Fractions and Decimals

Writing fractions as equivalent decimals can help you do calculations more easily or compare the size of two numbers.

OBJECTIVE ▶ 1 Write fractions as equivalent decimals. Recall from **Chapter 1** that a fraction is one way to show division. For example, $\frac{3}{4}$ means $3 \div 4$. If you are doing the division by hand, write it as $4\overline{)3}$. When you do the division, the result is 0.75, the decimal equivalent of $\frac{3}{4}$.

> **Writing a Fraction as an Equivalent Decimal**
>
> **Step 1** Divide the numerator of the fraction by the denominator.
>
> **Step 2** If necessary, round the answer to the place indicated.

Work Problem ❶ at the Side. ▶

EXAMPLE 1   **Writing Fractions or Mixed Numbers as Decimals**

(a) Write $\frac{1}{8}$ as a decimal.

$\frac{1}{8}$ means $1 \div 8$. Write it as $8\overline{)1}$. The decimal point in the dividend is on the *right* side of the 1. Write extra zeros in the dividend so you can continue dividing until the remainder is 0.

$$\frac{1}{8} \;\longrightarrow\; 1 \div 8 \;\longrightarrow\; 8\overline{)1} \;\longrightarrow\; \begin{array}{r} 0.125 \\ 8\overline{)1.000} \\ \underline{8} \\ 20 \\ \underline{16} \\ 40 \\ \underline{40} \\ 0 \end{array}$$

Decimal points lined up

← Three extra zeros needed

← Remainder is 0.

Therefore, $\frac{1}{8} = 0.125$.

To check this, write 0.125 as a fraction, then change it to lowest terms.

$$0.125 = \frac{125}{1000} \quad \text{In lowest terms} \quad \frac{125 \div 125}{1000 \div 125} = \frac{1}{8} \quad \leftarrow \text{Original fraction}$$

> **⌨ Calculator Tip**
>
> When using your calculator to write fractions as decimals, enter the numbers from the top down. Remember that the order in which you enter the numbers *does* matter in division. **Example 1(a)** above works like this:
>
> $\dfrac{1}{8} \Big\downarrow$ Top down   Enter 1 ÷ 8 =   Answer is 0.125
>
> What happens if you enter 8 ÷ 1 = ? Do you see why that cannot possibly be correct? (Answer: $8 \div 1 = 8$. A proper fraction like $\frac{1}{8}$ *cannot* be equivalent to a whole number.)

···················· **Continued on Next Page** ·············

OBJECTIVES

1 Write fractions as equivalent decimals.

2 Compare the size of fractions and decimals.

❶ Rewrite each fraction so you could do the division by hand. Do *not* complete the division.

GS (a) $\frac{1}{9}$ is written $9\overline{)}$

(b) $\frac{2}{3}$ is written $\overline{)}$

(c) $\frac{5}{4}$ is written $\overline{)}$

(d) $\frac{3}{10}$ is written $\overline{)}$

(e) $\frac{21}{16}$ is written $\overline{)}$

(f) $\frac{1}{50}$ is written $\overline{)}$

Answers

1. (a) $9\overline{)1}$ (b) $3\overline{)2}$ (c) $4\overline{)5}$
  (d) $10\overline{)3}$ (e) $16\overline{)21}$ (f) $50\overline{)1}$

2 Write each fraction or mixed number as a decimal.

(GS) **(a)** $\dfrac{1}{4}$ ➝ $4\overline{)1.\,0\,0}^{\,0.\,\_\,\_}$

(GS) **(b)** $2\dfrac{1}{2} = \dfrac{5}{2}$ ➝ $2\overline{)5.0}$

(c) $\dfrac{5}{8}$

(d) $4\dfrac{3}{5}$

(e) $\dfrac{7}{8}$

(b) Write $2\dfrac{3}{4}$ as a decimal.

One method is to divide 3 by 4 to get 0.75 for the fraction part. Then add the whole number part to 0.75.

$$\dfrac{3}{4} \quad\longrightarrow\quad 4\overline{)3.00}^{\,0.75}$$

$$\begin{array}{r} 2\,8 \\ \hline 20 \\ 20 \\ \hline 0 \end{array}$$

Fraction part ➝

$$\begin{array}{r} 2.00 \leftarrow \text{Whole number part} \\ +\,0.75 \\ \hline 2.75 \end{array}$$

So, $2\dfrac{3}{4} = 2.75$ **CHECK** $2.75 = 2\dfrac{75}{100} = 2\dfrac{3}{4}$ ← Lowest terms

Whole number parts match.

A second method is to first write $2\dfrac{3}{4}$ as an improper fraction and then divide numerator by denominator.

$$2\dfrac{3}{4} = \dfrac{11}{4}$$

$$\dfrac{11}{4} \quad\longrightarrow\quad 11 \div 4 \quad\longrightarrow\quad 4\overline{)11} \quad\longrightarrow\quad 4\overline{)11.00}^{\,2.75} \quad \leftarrow \text{Two extra zeros needed}$$

$$\begin{array}{r} 8 \\ \hline 3\,0 \\ 2\,8 \\ \hline 2\,0 \\ 2\,0 \\ \hline 0 \end{array}$$

Whole number parts match.

So, $2\dfrac{3}{4} = 2.75$

$\dfrac{3}{4}$ is equivalent to $\dfrac{75}{100}$ or 0.75

·········· ◀ **Work Problem 2 at the Side.**

EXAMPLE 2   **Writing a Fraction as a Decimal with Rounding**

Write $\dfrac{2}{3}$ as a decimal and round to the nearest thousandth.

$\dfrac{2}{3}$ means $2 \div 3$. To round to thousandths, divide out one *more* place, to ten-thousandths.

$$\dfrac{2}{3} \quad\longrightarrow\quad 2 \div 3 \quad\longrightarrow\quad 3\overline{)2} \quad\longrightarrow\quad 3\overline{)2.0000}^{\,0.6666} \quad \begin{array}{l} \leftarrow \text{Four zeros needed} \\ \text{for ten-thousandths} \end{array}$$

Be careful to divide in the correct order!

$$\begin{array}{r} 1\,8 \\ \hline 20 \\ 18 \\ \hline 20 \\ 18 \\ \hline 20 \\ 18 \\ \hline 2 \end{array}$$

Written as a repeating decimal, $\dfrac{2}{3} = 0.\overline{6}$ ← Bar above repeating digit

Rounded to the nearest thousandth, $\dfrac{2}{3} \approx 0.667$

·········· **Continued on Next Page**

Answers

2. (a) 0.25 **(b)** 2.5 **(c)** 0.625
   (d) 4.6 **(e)** 0.875

🖩 **Calculator Tip**

Try **Example 2** on your calculator. Enter 2 ÷ 3. Which answer do you get?

0.666666667   or   **0.6666666**

Many scientific calculators will show a 7 as the last digit. Because the sixes keep on repeating forever, the calculator automatically rounds in the last decimal place it has room to show. If you have a 10-digit display space, the calculator is rounding as shown below.

0.6666666666 (11 digits)   rounds to   0.666666667

Next digit is 5 or more, so 6 rounds to 7.

Other calculators, especially standard, four-function ones, may *not* round. They just cut off, or *truncate,* the extra digits. Such a calculator would show 0.6666666 in the display.

Would this difference in calculators show up when changing $\frac{1}{3}$ to a decimal? Why not? (Answer: The repeating digit is 3, which is *4 or less,* so it stays as 3 whether it's rounded or not.)

··· Work Problem ❸ at the Side. ▶

OBJECTIVE ❷ **Compare the size of fractions and decimals.** You can use a number line to compare fractions and decimals. For example, the number line below shows the space between 0 and 1. The locations of some commonly used fractions are marked, along with their decimal equivalents.

The next number line shows the locations of some commonly used fractions between 0 and 1 that are equivalent to *repeating* decimals. The decimal equivalents use a bar above repeating digits.

EXAMPLE 3 **Using a Number Line to Compare Numbers**

Use the number lines above to decide whether to write >, <, or = in the blank between each pair of numbers.

(a) 0.6875 _____ 0.625

You learned in **Chapter 3** that the number farther to the right on the number line is the greater number. Look at the first number line above. Because 0.6875 is to the *right* of 0.625, use the > symbol.

0.6875 is greater than 0.625   can be written as   0.6875 > 0.625

The larger end of the > symbol faces the greater number.

··· **Continued on Next Page**

❸ Write as decimals. Round to the
🖩 nearest thousandth.

(a) $\frac{1}{3}$

GS (b) $2\frac{7}{9} = \frac{25}{9} = 25 \div 9$

Now use your calculator to do the division and round the answer to the nearest thousandth.

(c) $\frac{10}{11}$

(d) $\frac{3}{7}$

(e) $3\frac{5}{6}$

Answers

3. All answers are rounded.
  (a) 0.333 **(b)** 2.778 **(c)** 0.909
  (d) 0.429 **(e)** 3.833

4 Use the number lines on the previous page to help you decide whether to write <, >, or = in each blank.

(a) 0.4375 _____ 0.5

(b) 0.75 _____ 0.6875

(c) 0.625 _____ 0.0625

(d) $\frac{2}{8}$ _____ 0.375

(e) $0.8\overline{3}$ _____ $\frac{5}{6}$

(f) $\frac{1}{2}$ _____ $0.\overline{5}$

(g) $0.\overline{1}$ _____ $0.1\overline{6}$

(h) $\frac{8}{9}$ _____ $0.\overline{8}$

(i) $0.\overline{7}$ _____ $\frac{4}{6}$

(j) $\frac{1}{4}$ _____ 0.25

5 Arrange each group in order from least to greatest.

(a) 0.7, 0.703, 0.7029

(b) 6.39, 6.309, 6.401, 6.4

(c) 1.085, $1\frac{3}{4}$, 0.9

(d) $\frac{1}{4}, \frac{2}{5}, \frac{3}{7}$, 0.428

Answers

4. (a) < (b) > (c) > (d) < (e) =
   (f) < (g) < (h) = (i) > (j) =
5. (a) 0.7, 0.7029, 0.703
   (b) 6.309, 6.39, 6.4, 6.401
   (c) 0.9, 1.085, $1\frac{3}{4}$ (d) $\frac{1}{4}, \frac{2}{5}$, 0.428, $\frac{3}{7}$

(b) $\frac{3}{4}$ _____ 0.75

On the first number line, $\frac{3}{4}$ and 0.75 are at the same point on the number line. They are equivalent.

$$\frac{3}{4} = 0.75$$

(c) 0.5 _____ $0.\overline{5}$

On the second number line, 0.5 is to the *left* of $0.\overline{5}$ (which is actually 0.555 . . .) so use the < symbol.

0.5 is less than $0.\overline{5}$ can be written as 0.5 < $0.\overline{5}$

> The smaller end of the < symbol points to the lesser number.

(d) $\frac{2}{6}$ _____ $0.\overline{3}$

Write $\frac{2}{6}$ in lowest terms as $\frac{1}{3}$.

On the second number line you can see that $\frac{1}{3} = 0.\overline{3}$.

◀ **Work Problem 4 at the Side.**

Fractions can also be compared by first writing each one as a decimal. The decimals can then be compared by writing each one with the same number of decimal places.

EXAMPLE 4   Arranging Numbers in Order

Write each group of numbers in order, from least to greatest.

(a) 0.49   0.487   0.4903

It is easier to compare decimals if they are all tenths, or all hundredths, and so on. Because 0.4903 has four decimal places (ten-thousandths), write zeros to the right of 0.49 and 0.487 so they also have four decimal places. Recall that writing zeros to the right of a decimal number does *not* change its value (see the section on adding decimals). Then find the least and greatest number of ten-thousandths.

0.49 = 0.4900 = **4900** ten-thousandths ← 4900 is in the middle.

0.487 = 0.4870 = **4870** ten-thousandths ← 4870 is the least.

0.4903 = **4903** ten-thousandths ← 4903 is the greatest.

From least to greatest, the correct order is: 0.487   0.49   0.4903

(b) $2\frac{5}{8}$ 2.63   2.6

Write $2\frac{5}{8}$ as $\frac{21}{8}$ and divide $8\overline{)21}$ to get the decimal form, 2.625. Then, because 2.625 has three decimal places, write zeros so all the numbers have three decimal places.

$2\frac{5}{8}$ = 2.625 = 2 and **625** thousandths ← 625 is in the middle.

2.63 = 2.630 = 2 and **630** thousandths ← 630 is the greatest.

2.6 = 2.600 = 2 and **600** thousandths ← 600 is the least.

From least to greatest, the correct order is: 2.6 $2\frac{5}{8}$ 2.63

◀ **Work Problem 5 at the Side.**

4.6 Exercises

MyMathLab®

CONCEPT CHECK   *In Exercises 1 and 2, circle all the divisions that are correctly set up to convert the given fraction to a decimal. If a division is **not** set up correctly, fix it.*

1. (a) $\frac{2}{5} \rightarrow 2\overline{)5}$ **(b)** $\frac{1}{7} \rightarrow 7\overline{)1}$ **(c)** $\frac{3}{8} \rightarrow 8\overline{)3}$ **2. (a)** $\frac{4}{9} \rightarrow 9\overline{)4}$ **(b)** $\frac{1}{3} \rightarrow 1\overline{)3}$ **(c)** $\frac{5}{6} \rightarrow 6\overline{)5}$

CONCEPT CHECK   *In Exercises 3 and 4, show the correct division set up for converting the given fraction to a decimal. Place the decimal points in the dividend and quotient. Then write the first digit in the quotient. Finally, explain what you will do next in order to continue dividing.*

3. $\frac{3}{4} \rightarrow$ **4.** $\frac{1}{8} \rightarrow$

CONCEPT CHECK   *In Exercises 5 and 6, write* greater than *or* less than *between each pair of numbers.*

5. (a) $\frac{2}{5}$ is _____ 0.5   **6. (a)** 0.1 is _____ $\frac{1}{5}$

  (b) $\frac{3}{4}$ is _____ 0.6   **(b)** $\frac{2}{3}$ is _____ 0.5

  (c) 0.2 is _____ $\frac{5}{8}$ **(c)** 0.7 is _____ $\frac{1}{2}$

*Write each fraction or mixed number as a decimal. Round to the nearest thousandth if necessary. **See Examples 1 and 2.***

7. Finish the **GS** **division.** $\frac{1}{2} \rightarrow 2\overline{)1.0}^{\;0.}$ **8. Finish the** **GS** **division.** $\frac{1}{4} \rightarrow 4\overline{)1.00}^{\;0.}$

9. $\frac{3}{4}$ **10.** $\frac{1}{10}$ **11.** $\frac{3}{10}$ **12.** $\frac{7}{10}$ **13.** $\frac{9}{10}$

14. $\frac{4}{5}$ **15.** $\frac{3}{5}$ **16.** $\frac{2}{5}$ **17.** $\frac{7}{8}$ **18.** $\frac{3}{8}$

19. $2\frac{1}{4}$ **20.** $1\frac{1}{2}$ **21.** $14\frac{7}{10}$ **22.** $23\frac{3}{5}$ **23.** $3\frac{5}{8}$

24. $2\frac{7}{8}$ **25.** $6\frac{1}{3}$ **26.** $5\frac{2}{3}$ **27.** $\frac{5}{6}$

28. $\frac{1}{6}$ **29.** $1\frac{8}{9}$ **30.** $5\frac{4}{7}$

Find each decimal or fraction equivalent. Write fractions in lowest terms.

Fraction	Decimal	Fraction	Decimal
31. _____	0.4	**32.** _____	0.75
33. _____	0.625	**34.** _____	0.111
35. _____	0.35	**36.** _____	0.9
37. $\dfrac{7}{20}$	_____	**38.** $\dfrac{1}{40}$	_____
39. _____	0.04	**40.** _____	0.52
41. $\dfrac{1}{5}$	_____	**42.** $\dfrac{1}{8}$	_____
43. _____	0.09	**44.** _____	0.02

Solve each application problem.

45. The average length of a newborn baby is 20.8 inches. Charlene's baby is 20.08 inches long. Is her baby longer or shorter than the average? By how much?

46. The patient in room 830 is supposed to get 8.3 milligrams of medicine. She was actually given 8.03 milligrams. Did she get too much or too little medicine? What was the difference?

47. Ginny Brown hoped her crops would get $3\frac{3}{4}$ inches of rain this month. The newspaper said the area received 3.8 inches of rain. Was that more or less than Ginny had hoped for? By how much?

48. The rats in a medical experiment gained $\frac{3}{8}$ ounce. They were expected to gain 0.3 ounce. Was their actual gain more or less than expected? By how much?

49. The label on the bottle of vitamins says that each capsule contains 0.5 gram of calcium. When checked, each capsule had 0.505 gram of calcium. Was there too much or too little calcium? What was the difference?

50. The glass mirror of the Hubble telescope had to be repaired in space because it would not focus properly. The problem was that the mirror's outer edge had a thickness of 0.6248 centimeter when it was supposed to be 0.625 centimeter. Was the edge too thick or too thin? By how much? (*Source:* NASA.)

51. CONCEPT CHECK Precision Medical Parts makes an artificial heart valve that must measure between 0.998 centimeter and 1.002 centimeters. Circle the lengths that are acceptable.

1.01 cm   0.9991 cm   1.0007 cm   0.99 cm

52. CONCEPT CHECK The white rats in a medical experiment must start out weighing between 2.95 ounces and 3.05 ounces. Circle the weights that can be used.

3.0 ounces   2.995 ounces   3.055 ounces

3.005 ounces

Arrange each group of numbers in order, from least to greatest. **See Example 4.**

53. 0.54, 0.5455, 0.5399

54. 0.76, 0.7, 0.7006

55. 5.8, 5.79, 5.0079, 5.804

56. 12.99, 12.5, 13.0001, 12.77

57. 0.628, 0.62812, 0.609, 0.6009

58. 0.27, 0.281, 0.296, 0.3

59. 5.8751, 4.876, 2.8902, 3.88

60. 0.98, 0.89, 0.904, 0.9

61. 0.043, 0.051, 0.006, $\frac{1}{20}$

62. 0.629, $\frac{5}{8}$, 0.65, $\frac{7}{10}$

63. $\frac{3}{8}$, $\frac{2}{5}$, 0.37, 0.4001

64. 0.1501, 0.25, $\frac{1}{10}$, $\frac{1}{5}$

Four boxes of fishing line are in the sale bin. The thicker the line, the stronger it is. The diameter of the fishing line is its thickness. Use the information on the boxes to answer Exercises 65–68.

65. Which color box has the strongest line?

66. Which color box has the line with the least strength?

67. Which color box has the line that is $\frac{1}{125}$ inch in diameter?

68. What is the difference in line diameter between the blue and purple boxes?

Some rulers for technical occupations show each inch divided into tenths. Use this scale drawing for Exercises 69–74. Change the measurements on the drawing to decimals and round them to the nearest tenth of an inch.

69. Length **(a)** is _____

70. Length **(b)** is _____

71. Length **(c)** is _____

72. Length **(d)** is _____

73. Length **(e)** is _____

74. Length **(f)** is _____

Relating Concepts (Exercises 75–78)   For Individual or Group Work

*Use your knowledge of fractions and decimals to **work Exercises 75–78 in order.***

75. (a) Explain how you can tell that Keith made an error *just by looking at his final answer.* Here is his work.

$$\frac{5}{9} = 5\overline{)9.0}^{\,1.8} \quad \text{so} \quad \frac{5}{9} = 1.8$$

(b) Show the correct way to change $\frac{5}{9}$ to a decimal. Explain why your answer makes sense.

76. (a) How can you prove to Sandra that $2\frac{7}{20}$ is *not* equivalent to 2.035? Here is her work.

$$2\frac{7}{20} = 20\overline{)7.00}^{\,0.35} \quad \text{so} \quad 2\frac{7}{20} = 2.035$$

(b) What is the correct answer? Show how to prove that it is correct.

77. Ving knows that $\frac{3}{8} = 0.375$. How can he write $1\frac{3}{8}$ as a decimal *without* having to do a division? How can he write $3\frac{3}{8}$ as a decimal? $295\frac{3}{8}$? Explain your answer.

78. Iris has found a shortcut for writing mixed numbers as decimals.

$$2\frac{7}{10} = 2.7 \qquad 1\frac{13}{100} = 1.13$$

Does her shortcut work for all mixed numbers? Explain when it works and why it works.

Study Skills
ANALYZING YOUR TEST RESULTS

After taking a test, many students heave a big sigh of relief and try to forget it ever happened. Don't fall into this trap! An exam is a learning opportunity. It gives you clues about *what your instructor thinks is important*, what *concepts and skills are valued* in mathematics, and *if you are on the right track*.

OBJECTIVES

1. Determine the reason for errors.
2. Develop a plan to avoid test-taking errors.
3. Review material to correct misunderstandings.

Immediately After the Test

Jot down problems that caused you trouble. Find out how to solve them by checking your textbook, looking at your notes, or asking your instructor or tutor (if available). You might see those same problems again on a final exam.

After the Test Is Returned

Find out what you got wrong and why you had points deducted. Write down the problem so you can learn how to do it correctly. Sometimes you only have a short time in class to review your test. *If you need more time*, ask your instructor if you can look at the test in his or her office.

Find Out Why You Made the Errors You Made

Here is a list of typical reasons for making errors on math tests.

1. You read the directions wrong.
2. You read the question wrong or skipped over something.
3. You made a computation error (maybe even an easy one).
4. Your answer is not accurate.
5. Your answer is not complete.
6. You labeled your answer wrong. For example, you labeled it "feet" and it should have been "feet^2."
7. You didn't show your work.
8. *You didn't understand the concept.
9. *You were unable to go from words (in a word problem) to setting up the problem.
10. *You were unable to apply a procedure to a new situation.
11. You were so anxious that you made errors even when you knew the material.

The first seven errors are **test-taking errors.** They are easy to correct if you decide to carefully read test questions and directions, proofread or rework your problems, show all your work, and double check units and labels every time.

The three starred errors (*) are **test preparation errors.** Remember that to grow a complex neural network, you need to practice the kinds of problems that you will see on the tests. So, for example, if application problems are difficult for you, you must *do more application problems!* If you have practiced the study skills techniques, however, you are less likely to make these kinds of errors on tests because you will have a deeper understanding of course concepts and you will be able to remember them better.

The last error isn't really an error. **Anxiety** can play a big part in your test results. Go back to the *Preparing for Tests* activity and read the suggestions about exercise and deep breathing. Recall from the *Your Brain* **Can** *Learn Mathematics* activity that when you are anxious, your body produces adrenaline. The presence of *adrenaline in the brain blocks connections* between dendrites. If you can *reduce the adrenaline* in your system, you will be able to *think more clearly* during your test. Just five minutes of brisk walking right before your test can help do that. Also, *practicing a relaxation technique while you do your homework* will make it more likely that you can benefit from using the technique during a test. *Deep breathing* is helpful because it gets *oxygen into your brain*. When you are anxious you tend to breathe more shallowly, which can make you feel confused and easily distracted.

Make a Plan for the Next Test

Make a plan for your next test based on your results from this test. You might review the Chapter Summary and work the problems in the Chapter Review Exercises or the Chapter Test. Ask your instructor or a tutor (if available) for more help if you are confused about any of the problems.

Now Try This

Below is a record sheet to track your progress in test-taking. Use it to find out if you make particular kinds of errors. Then you can work specifically on correcting them. Just place a check in the box when you made one of the errors. If you take more than four tests, make your own grid on separate paper.

Test-taking Errors

Test #	Read directions wrong	Read question wrong	Computation error	Not exact or accurate	Not complete	Labeled wrong	Didn't show work
1							
2							
3							
4							

Test Preparation Errors

Test #	Didn't understand concept	Didn't set up problem correctly	Couldn't apply concept to new situation
1			
2			
3			
4			

Anxiety

Test #	Felt anxious *before* the exam	Felt anxious *during* the exam	Blanked out on questions	Got questions wrong that I knew how to do
1				
2				
3				
4				

What will you do to avoid test-taking errors?

_____

_____

_____

What will you do to avoid test preparation errors?

_____

_____

_____

What will you do to reduce anxiety?

_____

_____

_____

Chapter 4 Summary

Key Terms

4.1

decimals Decimals, like fractions, are used to show parts of a whole.

decimal point A decimal point is the dot that is used to separate the whole number part from the fractional part of a decimal number.

place value A place value is assigned to each place to the right or left of the decimal point. Whole numbers, such as ones and tens, are to the *left* of the decimal point. Fractional parts, such as tenths and hundredths, are to the *right* of the decimal point.

4.2

rounding Rounding is "cutting off" a number after a certain place, such as rounding to the nearest hundredth. The rounded number is less accurate than the original number. You can use the symbol "≈" to mean "is approximately equal to."

decimal places Decimal places are the number of digits to the *right* of the decimal point. For example, 6.37 has two decimal places, and 4.706 has three decimal places.

4.3

estimating Estimating is the process of rounding the numbers in a problem and getting an approximate answer. This helps you check that the decimal point is in the correct place in the exact answer.

4.5

repeating decimal A repeating decimal is a decimal number with one or more digits that repeat forever; it never ends. For example, in 0.1666 . . . , the digit 6 continues to repeat. Use three dots to indicate that it is a repeating decimal. Or, write the number with a bar above the repeating digits, as in $0.1\overline{6}$. (Use the dots or the bar, but not both.)

New Symbols

$3.8\overline{6}$ ⟵——— Bar above repeating digit(s) in a decimal number

3.866 . . .   Three dots indicate a repeating decimal.

Test Your Word Power

See how well you have learned the vocabulary in this chapter.

① Decimal numbers are like fractions in that they both
  A. need common denominators
  B. have decimal points
  C. represent parts of a whole.

② Decimal places refer to
  A. the digits from 0 to 9
  B. digits to the left of the decimal point
  C. digits to the right of the decimal point.

③ When a decimal number is rounded, it
  A. always ends in 0
  B. is less accurate than the original number
  C. is less than one whole.

④ The decimal point
  A. separates the whole number part from the fractional part
  B. separates tenths from hundredths
  C. is at the far left side of a whole number.

⑤ The number $0.\overline{3}$ is an example of
  A. an estimate
  B. a repeating decimal
  C. a rounded number.

⑥ The place value names on the right side of the decimal point are
  A. ones, tens, hundreds, and so on
  B. ones, tenths, hundredths, and so on
  C. tenths, hundredths, thousandths, and so on.

Answers To Test Your Word Power

1. C; *Example:* For 0.7, the whole is cut into ten equal parts, and you are interested in 7 of the parts.

2. C; *Examples:* The number 6.87 has two decimal places; 0.309 has three decimal places.

3. B; *Example:* When 0.815 is rounded to 0.8 it is accurate only to the nearest tenth, while the original number was accurate to the nearest thousandth.

4. A; *Example:* In 5.42, the whole number part is 5 ones, and the decimal part is 42 hundredths.

5. B; *Example:* The bar above the 3 in $0.\overline{3}$ indicates that the 3 repeats forever.

6. C; *Example:* In 6.219, the 2 is in the tenths place, the 1 is in the hundredths place, and the 9 is in the thousandths place.

Quick Review

Concepts	Examples

4.1 Reading and Writing Decimal Numbers

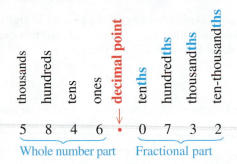

Write each decimal in words.

4.1 Writing Decimal Numbers as Fractions

The digits to the right of the decimal point are the numerator. The place value of the rightmost digit determines the denominator.

Always write the fractions in lowest terms.

Write 0.45 as a fraction in lowest terms.

The numerator is 45. The rightmost digit, 5, is in the hundredths place, so the denominator is 100. Then write the fraction in lowest terms.

$$\frac{45}{100} = \frac{45 \div 5}{100 \div 5} = \frac{9}{20} \quad \longleftarrow \text{Lowest terms}$$

4.2 Rounding Decimal Numbers

Find the place to which you are rounding. Draw a cut-off line to the right of that place; the rest of the digits will be dropped. Look *only* at the first digit being cut. If it is *4 or less,* the part you are keeping stays the same. If it is *5 or more,* the part you are keeping rounds up. Do not move the decimal point when rounding. Write "≈" to mean "is approximately equal to."

Round 0.17952 to the nearest thousandth.

First digit cut is *5 or more,* so round up.

$$0.179\,|\,52$$

$$
\begin{array}{r}
0.179 \quad \longleftarrow \text{Keep this part.} \\
+\ 0.001 \quad \longleftarrow \text{To round up, add 1 thousandth.} \\
\hline
0.180
\end{array}
$$

0.17952 rounds to 0.180. Write $0.17952 \approx 0.180$

4.3 Adding and Subtracting Decimal Numbers

Use front end rounding to round each number and estimate the answer.

To find the exact answer, line up the decimal points. If needed, write in zeros as placeholders. Add or subtract as if they were whole numbers. Line up the decimal point in the answer directly below the decimal points in the problem.

Add $5.68 + 785.3 + 12 + 2.007$.

Estimate:	*Exact:*	
6	5.**680**	Use zeros as place-holders so that all numbers have three decimal places.
800	785.**300**	
10	12.**000**	
+ 2	+ 2.007	
818	804.987	Line up decimal points.

The estimate and exact answer are both in the hundreds, so the decimal point is probably in the correct place.

4.4 Multiplying Decimal Numbers

Step 1 Multiply as you would for whole numbers.

Step 2 Count the total number of decimal places in both factors.

Step 3 Write the decimal point in the product so it has the same number of decimal places as the total from Step 2. You may need to write extra zeros on the left side of the product to get enough decimal places.

Multiply 0.169×0.21.

$$
\begin{array}{r}
0.169 \quad \longleftarrow \text{3 decimal places} \\
\times \quad 0.21 \quad \longleftarrow \text{2 decimal places} \\
\hline
169 \quad \text{5 total decimal places} \\
338 \quad\quad \\
\hline
.0\,3\,5\,4\,9 \quad \longleftarrow \text{5 decimal places in product}
\end{array}
$$

Write a 0 in the product so you can count over 5 decimal places. The final answer is 0.03549.

Concepts	Examples

4.5 Dividing by Decimal Numbers

Step 1 Change the divisor to a whole number by moving the decimal point to the right.

Step 2 Move the decimal point in the dividend the same number of places to the right.

Step 3 Write the decimal point in the quotient directly above the decimal point in the dividend.

Step 4 Divide as with whole numbers.

Divide 52.8 by 0.75.

$$
\begin{array}{r}
7\,0.4 \\
0.75\overline{)5\,2.8\,0\,0} \\
\underline{5\,2\,5} \\
3\,0\,0 \\
\underline{3\,0\,0} \\
0
\end{array}
$$

Move the decimal point two places to the right in the divisor and dividend. Write zeros in the dividend so you can move the decimal point and continue dividing until the remainder is 0.

To check your answer, multiply 70.4 times 0.75. If the result matches the dividend (52.8), you solved the problem correctly.

4.6 Writing Fractions as Decimal Numbers

Divide the numerator by the denominator. If necessary, round to the place indicated.

Write $\frac{1}{8}$ as a decimal.

$\frac{1}{8}$ means $1 \div 8$. Write it as $8\overline{)1}$.

The decimal point is on the right side of 1.

$$
\begin{array}{r}
0.125 \\
8\overline{)1.000} \\
\underline{8} \\
20 \\
\underline{16} \\
40 \\
\underline{40} \\
0
\end{array}
$$

Write a decimal point and three zeros so you can continue dividing.

Therefore, $\frac{1}{8}$ is equivalent to 0.125

4.6 Comparing the Size of Fractions and Decimal Numbers

Step 1 Write any fractions as decimals.

Step 2 Write zeros so that all the numbers being compared have the same number of decimal places.

Step 3 Use < to mean "is less than," > to mean "is greater than," or list the numbers from least to greatest.

Arrange in order from least to greatest.

$$0.505 \qquad \frac{1}{2} \qquad 0.55$$

$0.505 = 505$ thousandths ⟵ 505 is in the middle.

$\frac{1}{2} = 0.5 = 0.500 = 500$ thousandths ⟵ 500 is least.

$0.55 = 0.550 = 550$ thousandths ⟵ 550 is greatest.

(least) $\frac{1}{2}$ 　 0.505 　 0.55 (greatest)

Chapter 4 *Review Exercises*

4.1 *Name the digit that has the given place value.*

1. 243.059
tenths
hundredths

2. 0.6817
ones
tenths

3. $5824.39
hundreds
hundredths

4. 896.503
tenths
tens

5. 20.73861
tenths
ten-thousandths

Write each decimal as a fraction or mixed number in lowest terms.

6. 0.5

7. 0.75

8. 4.05

9. 0.875

10. 0.027

11. 27.8

Write each decimal in words.

12. 0.8

13. 400.29

14. 12.007

15. 0.0306

Write each decimal in numbers.

16. eight and three tenths

17. two hundred five thousandths

18. seventy and sixty-six ten-thousandths

19. thirty hundredths

4.2 *Round each decimal to the place indicated.*

20. 275.635 to the nearest tenth

21. 72.789 to the nearest hundredth

22. 0.1604 to the nearest thousandth

23. 0.0905 to the nearest thousandth

24. 0.98 to the nearest tenth

Round each money amount to the nearest cent.

25. $15.8333

26. $0.698

27. $17,625.7906

Round each income or expense item to the nearest dollar.

28. Income from pancake breakfast was $350.48.

29. Members paid $129.50 in dues.

30. Refreshments cost $99.61.

31. Bank charges were $29.37.

4.3   *First use front end rounding to round each number and estimate the answer. Then find the exact answer.*

32. *Estimate:*   *Exact:*

$$
\begin{array}{r}
5.81 \\
423.96 \\
+\quad 15.09 \\
\hline
\end{array}
$$

+ ___

33. *Estimate:*   *Exact:*

$$
\begin{array}{r}
75.6 \\
1.29 \\
122.045 \\
0.88 \\
+\quad 33.7 \\
\hline
\end{array}
$$

+ ___

34. *Estimate:*   *Exact:*

$$
\begin{array}{r}
308.5 \\
-\quad 17.8 \\
\hline
\end{array}
$$

− ___

35. *Estimate:*   *Exact:*

$$
\begin{array}{r}
9.2 \\
-\quad 7.9316 \\
\hline
\end{array}
$$

− ___

36. Americans' favorite household pet is a cat. There are about 93.6 million pet cats, 77.5 million pet dogs, and 16.6 million pet birds. How many more pet cats are there than pet birds? (*Source:* American Pet Products Association.)

Estimate:

Exact:

37. Jasmin started with $406 in her bank account. She paid $315.53 to the day care center and $74.67 at the grocery store. What is the new amount in her account?

Estimate:

Exact:

38. Joey spent $1.59 for toothpaste, $5.33 for vitamins, and $18.94 for a toaster. He gave the clerk three $10 bills. How much change did he get?

Estimate:

Exact:

39. Roseanne is training for a wheelchair race. She raced 2.3 kilometers on Monday, 4 kilometers on Wednesday, and 5.25 kilometers on Friday. How far did she race altogether?

Estimate:

Exact:

4.4   *First use front end rounding to round each number and estimate the answer. Then find the exact answer.*

40. *Estimate:*   *Exact:*

$$
\begin{array}{r}
6.138 \\
\times\quad 3.7 \\
\hline
\end{array}
$$

× ___

41. *Estimate:*   *Exact:*

$$
\begin{array}{r}
42.9 \\
\times\quad 3.3 \\
\hline
\end{array}
$$

× ___

Find each product.

42. $(5.6)(0.002)$

43. $0.071(0.005)$

4.5 *Decide whether each answer is reasonable by rounding the numbers and estimating the answer. If the exact answer is not reasonable, find and correct the error.*

44. $706.2 \div 12 = 58.85$

    Estimate:

45. $26.6 \div 2.8 = 0.95$

    Estimate:

Divide. Round each quotient to the nearest thousandth if necessary.

46. $3\overline{)43.4}$

47. $\dfrac{72}{0.06}$

48. $0.00048 \div 0.0012$

4.4–4.5 *Solve each application problem.*

49. Adrienne worked 46.5 hours this week. Her hourly wage is $14.24 for the first 40 hours and 1.5 times that rate over 40 hours. Find her total earnings to the nearest dollar.

50. A book of 12 tickets costs $35.89 at the State Fair midway. What is the cost per ticket, to the nearest cent?

51. Stock in MathTronic sells for $3.75 per share. Kenneth is thinking of investing $500. How many whole shares could he buy?

52. Grapes are on sale at $0.99 per pound. How much will Ms. Lee pay for 3.5 pounds of grapes, to the nearest cent?

Simplify each expression.

53. $3.5^2 + 8.7(1.95)$

54. $11 - 3.06 \div (3.95 - 0.35)$

4.6 *Write each fraction or mixed number as a decimal. Round to the nearest thousandth when necessary.*

55. $3\dfrac{4}{5}$

56. $\dfrac{16}{25}$

57. $1\dfrac{7}{8}$

58. $\dfrac{1}{9}$

Arrange each group of numbers in order from least to greatest.

59. 3.68, 3.806, 3.6008

60. 0.215, 0.22, 0.209, 0.2102

61. $0.17, \dfrac{3}{20}, \dfrac{1}{8}, 0.159$

Mixed Review Exercises

Add, subtract, multiply, or divide as indicated.

62. $89.19 + 0.075 + 310.6 + 5$

63. 72.8×3.5

64. $1648.3 \div 0.46$ Round to the nearest thousandth.

65. $30 - 0.9102$

66. $4.38(0.007)$

67. $0.005\overline{)0.047}$

68. $72.105 + 8.2 + 95.37$

69. $81.36 \div 9$

70. $(5.6 - 1.22) + 4.8(3.15)$

71. 0.455×18

72. $(1.6)(0.58)$

73. $0.218)\overline{7.63}$

74. $21.059 - 20.8$

75. $18.3 - 3^2 \div 0.5$

Use the information in the ad to answer Exercises 76–80. Round money answers to the nearest cent. (Disregard any sales tax.)

Grand Opening Sale!
Save on Clothing for the Entire Family

Jeans for Teens
only $19.95 each
women's sizes $24.99

Athletic Shoes
regularly priced
$89.99 to $149.50
NOW just $71 to $119.60

Men's socks *NOW 3 pairs for $8.99*
Children's socks *6 pairs for $5*

Hurry in — **TWO DAYS ONLY**

76. How much would one pair of men's socks cost?

77. How much more would one pair of men's socks cost than one pair of children's socks?

78. How much would Fernando pay for a dozen pair of men's socks?

79. How much would Akiko pay for five pairs of teen jeans and four pairs of women's jeans?

80. What is the difference between the cheapest sale price for athletic shoes and the highest regular price?

To decrease your risk of clogged arteries, it is recommended that you get at least 2 milligrams of vitamin B-6 each day. Use the information in the table to answer Exercises 81–82.

SOURCES OF VITAMIN B-6 (AMOUNTS IN MILLIGRAMS)	
$\frac{1}{2}$ cup green peas	0.11
1 banana	0.68
1 baked potato with skin	0.7
$\frac{1}{2}$ cup strawberries	0.45
3 ounces skinless chicken	0.5
3 ounces water-packed tuna	0.2
$\frac{1}{2}$ cup chickpeas	0.57

Source: National Institutes of Health.

81. (a) Which food item has the highest amount of vitamin B-6?

(b) Which food item has the lowest amount?

(c) What is the difference in the amount of vitamin B-6 between the food items with the highest and lowest amounts?

82. (a) Suppose you ate a banana, 3 ounces of skinless chicken, and 1 cup of strawberries. How many milligrams of vitamin B-6 would you get?

(b) Did you get more or less than the recommended daily amount? By how much?

Write each decimal as a fraction or mixed number in lowest terms.

1. 18.4

2. 0.075

Write each decimal in words.

3. 60.007

4. 0.0208

Round each decimal to the place indicated.

5. 725.6089 to the nearest tenth

6. 0.62951 to the nearest thousandth

7. $1.4945 to the nearest cent

8. $7859.51 to the nearest dollar

First use front end rounding to round each number and estimate the answer. Then find the exact answer.

9. 7.6 + 82.0128 + 39.59

    Estimate:

    Exact:

10. 79.1 − 3.602

    Estimate:

    Exact:

11. 5.79(1.2)

    Estimate:

    Exact:

12. 20.04 ÷ 4.8

    Estimate:

    Exact:

Find the exact answer.

13. 53.1 + 4.631 + 782 + 0.031

14. 670 − 0.996

15. (0.0069)(0.007)

16. $0.15\overline{)72}$

17. Write $2\frac{5}{8}$ as a decimal. Round to the nearest thousandth, if necessary.

18. Arrange in order from least to greatest.

$$0.44, 0.451, \frac{9}{20}, 0.4506$$

19. Simplify this expression.

$$6.3^2 - 5.9 + 3.4(0.5)$$

Solve each application problem.

20. Jennifer bought a 51-inch HDTV with 3D technology. The TV had an original price of $1299.99, and she bought two pairs of 3D glasses at $49.99 each. The store gave her a $200 discount, but she had to pay $94.50 in sales tax. How much did Jennifer pay?

21. Three types of ducks that are hunted in the United States are gadwalls, wigeons, and pintails. The estimated populations of these ducks are 2.5 million gadwalls, 2.551 million wigeons, and 2.56 million pintails. List the ducks in order from the greatest number to the least. (*Source:* U.S. Fish and Wildlife Service.)

22. Mr. Yamamoto bought 1.85 pounds of cheese at $2.89 per pound. What was the total amount he paid for the cheese, to the nearest cent?

23. Loren's baby had a temperature of 102.7 degrees. Later in the day it was 99.9 degrees. How much had the baby's temperature dropped?

24. Pat bought 6.5 ft of decorative gold chain to hang a light over her dining table. She paid $24.64. What was the cost per foot, to the nearest cent?

25. Write your own application problem using decimals. Make it different from **Problems 20–24.** Then show how to solve your problem.

Chapters 1-4 *Cumulative Review Exercises*

Round each number as indicated.

1. 499,501 to the nearest thousand

2. 602.4937 to the nearest hundredth

3. $709.60 to the nearest dollar

4. $0.0528 to the nearest cent

Simplify each expression in Exercises 5–21.

5. $10 - 0.329$

6. $2\frac{3}{5} \cdot \frac{5}{9}$

7. $11\frac{1}{5} \div 8$

8. $5006 - 92$

9. $0.7 + 85 + 7.903$

10. Write your answer using R for the remainder.

$7\overline{)2831}$

11. $\frac{5}{6} + \frac{7}{8}$

12. 332×704

13. $(0.006)(5.44)$

14. $3.2(2.5)$

15. $25.2 \div 0.56$

16. $\frac{2}{3} \div 5\frac{1}{6}$

17. $5\frac{1}{4} - 4\frac{7}{12}$

18. $4.7 \div 9.3$
Round to the nearest hundredth.

19. $10 - 4 \div 2 \cdot 3$

20. $\sqrt{36} + 3(8) - 4^2$

21. Write 40.035 in words.

22. Write three hundred six ten-thousandths in numbers.

Arrange each group of numbers in order, from least to greatest.

23. 7.005, 7.5005, 7.5, 7.505

24. $\frac{7}{8}$, 0.8, $\frac{21}{25}$, 0.8015

First round the numbers and estimate the answer to each application problem. Then find the exact answer.

25. Lameck had three $20 bills. He spent $47.96 on gasoline and $0.87 for a candy bar at the convenience store. How much money does he have left?

Estimate:

Exact:

26. Manuela's daughter is 50 inches tall. Last year she was $46\frac{5}{8}$ inches tall. How much has she grown? (When estimating, round to the nearest whole number.)

Estimate:

Exact:

27. Sharon records textbooks on CDs for students who are blind. Her hourly wage is $11.63. How much did she earn working 16.5 hours last week, to the nearest cent?

Estimate:

Exact:

28. The Farnsworth Elementary School has eight classrooms with 22 students in each one and 12 classrooms with 26 students in each one. How many students attend the school?

Estimate:

Exact:

29. Toshihiro bought $2\frac{1}{3}$ yd of cotton fabric and $3\frac{7}{8}$ yd of wool fabric. How many yards did he buy in all?

Estimate:

Exact:

30. Kimberly had $29.44 in her bank account. She paid $40 at the gas station and deposited a $220.06 paycheck into her account, but not in time to prevent a $35 charge for insufficient funds. What is the new balance in her account?

Estimate:

Exact:

31. Paulette bought 2.7 pounds of apples for $6.18. What was the cost per pound, to the nearest cent?

Estimate:

Exact:

32. Carter Community College received a $78,000 grant from a local hospital to help students pay tuition for nursing classes. How much money could be given to each of 107 students? Round to the nearest dollar.

Estimate:

Exact:

Use the information in the table to answer Exercises 33–36. All measurements are in inches.

Children's Hats	Head Size	Order Size:
Measure around head above eyebrow ridges.	16½"–18"	XXS
	18¼"–19"	XS
	19¼"–20"	S
	20¼"–21⅛"	M
	21½"–22¼"	L

Source: Lands' End.

33. If the distance around your child's head is $21\frac{1}{16}$ in., which hat size should you order?

34. Find the difference between the smaller and larger measurements for each of the hat sizes.

35. Change the measurements for the medium (M) size hat to decimals. Then find the difference in the measurements. Show why this answer is equivalent to your answer for the medium (M) size in **Exercise 34.**

36. Did you prefer doing the subtraction using fractions (as in **Exercise 34**) or using decimals (as in **Exercise 35**)? Explain your reasoning.

Math in the Media

LAWN FERTILIZER

Gotta Be Green

A lot's being written about personal responsibility these days, and the idea seems to be ending up on the front lawn—literally! Each spring, homeowners across the country gear up to green up their lawns, and the increased use of fertilizer has a lot of environmentalists concerned about the potential effects of chemical runoff into nearby rivers and streams.

Every year, according to a study conducted by the University of Minnesota's Department of Agriculture, each household in the Minneapolis/St. Paul metro area uses an average of 36 pounds of lawn fertilizer. That adds up to 25,529,295 pounds, or 12,765 tons. Add to that another 193,000 pounds of weed killer and you're looking at the total picture for keeping it green in the Twin Cities.

Source: Minneapolis Star Tribune.

1. Refer to the article.
   (a) How many pounds of lawn fertilizer are used each year in the *entire metro area?*

   (b) Do a division on your calculator to find the number of *households* in the metro area.

   (c) Why does it make sense to round your answer to part (b)? How would you round it?

2. There are 2000 pounds in one ton.
   (a) Find the number of tons equivalent to 25,529,295 pounds of fertilizer.

   (b) Does your answer match the figure given in the article? If not, what did the author of the article do to get 12,765 tons?

   (c) Is the author's figure accurate? Why or why not?

3. **(a)** When the average amount of lawn fertilizer per household was calculated, the answer was probably not exactly 36 pounds. List six different values that are less than 36 that would round to 36. List two values with one decimal place; two values with two decimal places; and two values with three decimal places.

   (b) List six different values that are greater than 36 that would round to 36. List two values each with one, two, and three decimal places.

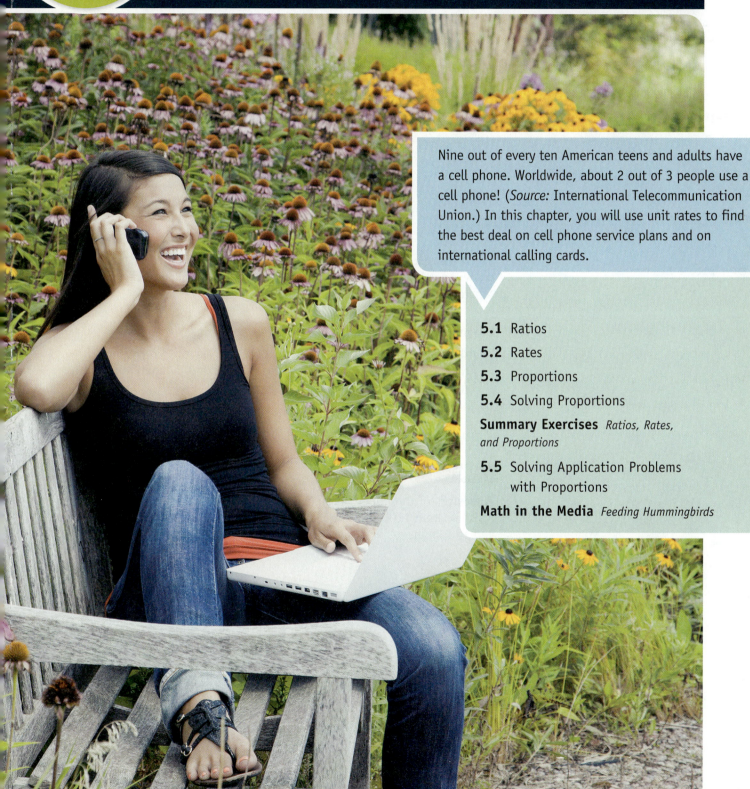

5 Ratio and Proportion

Nine out of every ten American teens and adults have a cell phone. Worldwide, about 2 out of 3 people use a cell phone! (*Source:* International Telecommunication Union.) In this chapter, you will use unit rates to find the best deal on cell phone service plans and on international calling cards.

5.1 Ratios

5.2 Rates

5.3 Proportions

5.4 Solving Proportions

Summary Exercises *Ratios, Rates, and Proportions*

5.5 Solving Application Problems with Proportions

Math in the Media *Feeding Hummingbirds*

5.1 Ratios

1 Write ratios as fractions.

2 Solve ratio problems involving decimals or mixed numbers.

3 Solve ratio problems after converting units.

A **ratio** compares two quantities. You can compare two numbers, such as 8 and 4, or two measurements that have the *same* type of units, such as 3 *days* and 12 *days*. (*Rates* compare measurements with different types of units and are covered in the next section.)

Ratios can help you see important relationships. For example, if the ratio of your monthly expenses to your monthly income is 10 to 9, then you are spending $10 for every $9 you earn and going deeper into debt.

OBJECTIVE **1** **Write ratios as fractions.** A ratio can be written in three ways.

Writing a Ratio

The ratio of $7 **to** $3 can be written:

$$7 \text{ to } 3 \quad \text{or} \quad 7\mathbf{:}3 \quad \text{or} \quad \frac{7}{3} \;\leftarrow \text{Fraction bar indicates ``to.''}$$

"**:**" indicates "**to**"

Writing a ratio as a fraction is the most common method, and the one we will use here. All three ways are read, "the ratio of 7 **to** 3." The word **to** separates the quantities being compared.

Writing a Ratio as a Fraction

Order is important when writing a ratio. The quantity mentioned **first** is the **numerator**. The quantity mentioned **second** is the **denominator**. For example:

$$\text{The ratio of } \mathbf{5} \text{ to } \mathbf{12} \text{ is written } \frac{5}{12}$$

EXAMPLE 1 **Writing Ratios**

Ancestors of the Pueblo Indians built multistory apartment towns in New Mexico about 1100 years ago. A room might measure 14 ft long, 11 ft wide, and 15 ft high.

15 ft

11 ft

14 ft

Continued on Next Page

Write each ratio as a fraction, using the room measurements.

(a) Ratio of length to width

The ratio of **length** **to width** is $\dfrac{14 \text{ ft}}{11 \text{ ft}} = \dfrac{14}{11}$

> Do *not* rewrite the ratio as $1\frac{3}{11}$.

Numerator (mentioned first)   Denominator (mentioned second)

You can divide out common *units* just like you divided out common *factors* when writing fractions in lowest terms. (See **Chapter 2.**) However, do *not* rewrite the fraction as a mixed number. Keep it as the ratio of 14 to 11.

(b) Ratio of width to height

> Divide out the common units (ft).

The ratio of width **to** height is $\dfrac{11 \text{ ft}}{15 \text{ ft}} = \dfrac{11}{15}$

CAUTION

Remember, the order of the numbers is important in a ratio. Look for the words "ratio of *a* to *b*." Write the ratio as $\frac{a}{b}$, **not** $\frac{b}{a}$. The quantity mentioned first is the numerator.

···· **Work Problem ❶ at the Side. ▶**

Any ratio can be written as a fraction. Therefore, you can write a ratio in *lowest terms,* just as you do with any fraction.

EXAMPLE 2   Writing Ratios in Lowest Terms

Write each ratio in lowest terms.

(a) 60 days of sun to 20 days of rain

The ratio is $\frac{60 \text{ days}}{20 \text{ days}}$. Divide out the common units. Then write this ratio in lowest terms by dividing the numerator and denominator by 20.

$$\frac{60 \text{ days}}{20 \text{ days}} = \frac{60}{20} = \frac{60 \div 20}{20 \div 20} = \frac{3}{1} \quad \left\{ \begin{array}{l} \text{Ratio in} \\ \text{lowest terms} \end{array} \right.$$

So, the ratio of 60 days to 20 days is 3 to 1 or, written as a fraction, $\frac{3}{1}$. In other words, for every 3 days of sun, there is 1 day of rain.

CAUTION

In the fractions chapters you would have rewritten $\frac{3}{1}$ as 3. But a *ratio* compares *two* quantities, so you need to keep both parts of the ratio and write it as $\frac{3}{1}$.

(b) 50 ounces of medicine to 120 ounces of water

The ratio is $\frac{50 \text{ ounces}}{120 \text{ ounces}}$. Divide out the common units. Then divide the numerator and denominator by 10.

$$\frac{50 \text{ ounces}}{120 \text{ ounces}} = \frac{50}{120} = \frac{50 \div 10}{120 \div 10} = \frac{5}{12} \quad \left\{ \begin{array}{l} \text{Ratio in} \\ \text{lowest terms} \end{array} \right.$$

So, the ratio of 50 ounces to 120 ounces is $\frac{5}{12}$. In other words, for every 5 ounces of medicine, there are 12 ounces of water.

···· **Continued on Next Page**

Continued on Next Page

VOCABULARY TIP

Ratio A **ratio** compares two quantities that have the **same** units. For example, if you are comparing two lengths, both measurements must be in feet, or both in yards, and so on. In a ratio, the common units divide out.

❶ Shane spent $14 on meat, $5 on milk, and $7 on fresh fruit. Write each ratio as a fraction.

GS **(a)** The ratio of amount spent on fruit to amount spent on milk.

Spent on fruit → $\dfrac{}{5}$
Spent on milk →

(b) The ratio of amount spent on milk to amount spent on meat.

(c) The ratio of amount spent on meat to amount spent on milk.

Answers

1. (a) $7; \dfrac{7}{5}$ **(b)** $\dfrac{5}{14}$ **(c)** $\dfrac{14}{5}$

2 Write each ratio as a fraction in lowest terms.

GS **(a)** 9 hours to 12 hours

$$\frac{9 \text{ hours}}{12 \text{ hours}} = \frac{9}{12} = \frac{9 \div 3}{12 \div 3} = \frac{\rule{1cm}{0.4pt}}{\rule{1cm}{0.4pt}}$$

(b) 100 meters to 50 meters

(c) Write the ratio of width to length for this rectangle.

Length
48 ft

Width
24 ft

3 Write each ratio as a ratio of whole numbers in lowest terms.

(a) The price of Tamar's favorite brand of lipstick increased from $5.50 to $7.00. Find the ratio of the increase in price to the original price.

(b) Last week, Lance worked 4.5 hours each day. This week he cut back to 3 hours each day. Find the ratio of the decrease in hours to the original number of hours.

Answers

2. **(a)** $\frac{3}{4}$ **(b)** $\frac{2}{1}$ **(c)** $\frac{1}{2}$

3. **(a)** $\frac{1.50 \times 100}{5.50 \times 100} = \frac{150 \div 50}{550 \div 50} = \frac{3}{11}$

(b) $\frac{1.5 \times 10}{4.5 \times 10} = \frac{15 \div 15}{45 \div 15} = \frac{1}{3}$

(c) 15 people in a large van to 6 people in a small van

The ratio is $\dfrac{15 \text{ people}}{6 \text{ people}} = \dfrac{15}{6} = \dfrac{15 \div 3}{6 \div 3} = \dfrac{5}{2}$ ← $\left\{ \begin{array}{l} \text{Ratio in} \\ \text{lowest terms} \end{array} \right.$

> **Note**
>
> Although $\frac{5}{2} = 2\frac{1}{2}$, ratios are *not* written as mixed numbers. Nevertheless, in **Example 2(c)** above, the ratio $\frac{5}{2}$ does mean the large van holds $2\frac{1}{2}$ times as many people as the small van.

◀ **Work Problem 2 at the Side.**

OBJECTIVE **2** **Solve ratio problems involving decimals or mixed numbers.** Sometimes a ratio compares two decimal numbers or two fractions. It is easier to understand if we rewrite the ratio as a ratio of two whole numbers.

EXAMPLE 3 **Using Decimal Numbers in a Ratio**

The price of a Sunday newspaper increased from $1.50 to $1.75. Find the ratio of the increase in price to the original price.

The words increase in price are mentioned first, so the increase will be the numerator. How much did the price go up? Use subtraction.

new price − original price = increase

$1.75 − $1.50 = $0.25 ← $\left\{ \begin{array}{l} \text{Subtract first to} \\ \text{find how much the} \\ \text{price went up.} \end{array} \right.$

The words the original price are mentioned second, so the original price of $1.50 is the denominator.

The ratio of increase in price to original price is shown below.

$\dfrac{0.25}{1.50}$ ← increase in price
← original price

Now rewrite the ratio as a ratio of whole numbers. Recall that if you multiply both the numerator and denominator of a fraction by the same number, you get an equivalent fraction. The decimals in this example are hundredths, so multiply by 100 to get whole numbers. (If the decimals are tenths, multiply by 10. If thousandths, multiply by 1000.) Then write the ratio in lowest terms.

$$\frac{0.25}{1.50} = \frac{0.25 \times 100}{1.50 \times 100} = \frac{25}{150} = \frac{25 \div 25}{150 \div 25} = \frac{1}{6}$$ ← $\left\{ \begin{array}{l} \text{Ratio in} \\ \text{lowest terms} \end{array} \right.$

Ratio as two
whole numbers

◀ **Work Problem 3 at the Side.**

EXAMPLE 4 **Using Mixed Numbers in Ratios**

Write each ratio as a comparison of whole numbers in lowest terms.

(a) 2 days to $2\frac{1}{4}$ days

Write the ratio as follows. Divide out the common units.

$$\frac{2 \text{ days}}{2\frac{1}{4} \text{ days}} = \frac{2}{2\frac{1}{4}}$$

·········· **Continued on Next Page**

Next, write 2 as $\frac{2}{1}$ and $2\frac{1}{4}$ as the improper fraction $\frac{9}{4}$.

Think: $2 \cdot 4 = 8$ and $8 + 1 = 9$ so $2\frac{1}{4} = \frac{9}{4}$.

$$\frac{2}{2\frac{1}{4}} = \frac{\frac{2}{1}}{\frac{9}{4}}$$

Rewrite the problem in horizontal format, using the "÷" symbol for division. Finally, multiply by the reciprocal of the divisor, as you did in **Chapter 2.**

$$\frac{\frac{2}{1}}{\frac{9}{4}} = \frac{2}{1} \div \frac{9}{4} = \frac{2}{1} \cdot \frac{4}{9} = \frac{8}{9}$$

Reciprocals

The ratio, in lowest terms, is $\frac{8}{9}$.

(b) $3\frac{1}{4}$ to $1\frac{1}{2}$

Write the ratio as $\dfrac{3\frac{1}{4}}{1\frac{1}{2}}$. Then write $3\frac{1}{4}$ and $1\frac{1}{2}$ as improper fractions.

Think: $3 \cdot 4 = 12$ and $12 + 1 = 13$ so $3\frac{1}{4} = \frac{13}{4}$.

$$3\frac{1}{4} = \frac{13}{4} \quad \text{and} \quad 1\frac{1}{2} = \frac{3}{2}$$

Think: $1 \cdot 2 = 2$ and $2 + 1 = 3$ so $1\frac{1}{2} = \frac{3}{2}$.

The ratio is shown here.

$$\frac{3\frac{1}{4}}{1\frac{1}{2}} = \frac{\frac{13}{4}}{\frac{3}{2}}$$

Rewrite as a division problem in horizontal format, using the "÷" symbol. Then multiply by the reciprocal of the divisor.

$$\frac{13}{4} \div \frac{3}{2} = \frac{13}{\underset{2}{4}} \cdot \frac{\overset{1}{2}}{3} = \frac{13}{6} \quad \leftarrow \begin{cases} \text{Ratio in} \\ \text{lowest terms} \end{cases}$$

Work Problem ❹ at the Side. ▶

OBJECTIVE ❸ Solve ratio problems after converting units. When a ratio compares measurements, both measurements must be in the *same* units. For example, *feet* must be compared to *feet, hours* to *hours, pints* to *pints*, and *inches* to *inches.*

EXAMPLE 5   Ratio Applications Using Measurement

(a) Write the ratio of the length of the shorter board on the left to the length of the longer board on the right. Compare in inches.

First, express 2 ft in inches. Because 1 ft has 12 in., 2 ft is

$$2 \cdot \mathbf{12 \ in.} = 24 \ in.$$

Continued on Next Page

---

❹ Write each ratio as a ratio of whole numbers in lowest terms.

GS (a) $3\frac{1}{2}$ to 4

$$\frac{3\frac{1}{2}}{4} = \frac{\frac{7}{2}}{\frac{4}{1}}$$

$$= \frac{7}{2} \div \frac{4}{1} = \frac{7}{2} \cdot \frac{1}{4}$$

Reciprocals

$$= \frac{\rule{1.2em}{0.4pt}}{\rule{1.2em}{0.4pt}}$$

(b) $5\frac{5}{8}$ pounds to $3\frac{3}{4}$ pounds

(c) $3\frac{1}{2}$ in. to $\frac{7}{8}$ in.

Answers
4. (a) $\frac{7}{8}$ (b) $\frac{3}{2}$ (c) $\frac{4}{1}$

5 Write each ratio as a fraction in lowest terms. (*Hint:* Recall that it is usually easier to write the ratio using the smaller measurement unit.)

GS **(a)** 9 in. to 6 ft

Change 6 ft to inches.

6 • 12 in. = 72 in.

$$\frac{9 \text{ in.}}{6 \text{ ft}} = \frac{9 \text{ in.}}{\underline{\quad} \text{ in.}}$$

$$= \frac{9 \div \underline{\quad}}{\underline{\quad} \div \underline{\quad}} =$$

(b) 2 days to 8 hours

(c) 7 yd to 14 ft

(d) 3 quarts to 3 gallons

(e) 25 minutes to 2 hours

(f) 4 pounds to 12 ounces

Answers

5. (a) 72 in.; $\frac{9 \div 9}{72 \div 9} = \frac{1}{8}$ **(b)** $\frac{6}{1}$ **(c)** $\frac{3}{2}$

(d) $\frac{1}{4}$ **(e)** $\frac{5}{24}$ **(f)** $\frac{16}{3}$

On the previous page, the length of the board on the left is 24 in., so the ratio of the lengths is

$$\frac{2 \text{ ft}}{30 \text{ in.}} = \frac{24 \text{ in.}}{30 \text{ in.}} = \frac{24}{30}$$

> Once the units match, you can divide them out.

Write the ratio in lowest terms.

$$\frac{24}{30} = \frac{24 \div 6}{30 \div 6} = \frac{4}{5} \quad \leftarrow \left\{ \begin{array}{l} \text{Ratio in} \\ \text{lowest terms} \end{array} \right.$$

The shorter board is $\frac{4}{5}$ the length of the longer board.

Note

Notice in the example above that we wrote the ratio using the smaller unit (inches are smaller than feet). Using the smaller unit will help you avoid working with fractions. If we wrote the ratio using feet, then

$$30 \text{ in.} = 2\frac{1}{2} \text{ ft}$$

So the ratio in feet is shown below.

$$\frac{2 \text{ ft}}{2\frac{1}{2} \text{ ft}} = \frac{2}{1} \div \frac{5}{2} = \frac{2}{1} \cdot \frac{2}{5} = \frac{4}{5} \quad \leftarrow \text{Same result}$$

The ratio is the same, but it takes more steps to get the answer. Using the smaller unit is usually easier.

(b) Write the ratio of 28 days to 3 weeks.

Since it is easier to write the ratio using the smaller measurement unit, compare in *days* because days are shorter than weeks.

First express 3 weeks in days. Because 1 week has 7 days, 3 weeks is

$$3 \cdot 7 \text{ days} = 21 \text{ days}.$$

So the ratio in days is shown below.

$$\frac{28 \text{ days}}{3 \text{ weeks}} = \frac{28 \text{ days}}{21 \text{ days}} = \frac{28}{21} = \frac{28 \div 7}{21 \div 7} = \frac{4}{3} \quad \leftarrow \left\{ \begin{array}{l} \text{Ratio in} \\ \text{lowest terms} \end{array} \right.$$

The following table will help you set up ratios that compare measurements. You will work with these measurements again in **Chapter 7.**

Measurement Comparisons

Length	**Capacity (Volume)**
12 inches = 1 foot	2 cups = 1 pint
3 feet = 1 yard	2 pints = 1 quart
5280 feet = 1 mile	4 quarts = 1 gallon

Weight	**Time**
16 ounces = 1 pound	60 seconds = 1 minute
2000 pounds = 1 ton	60 minutes = 1 hour
	24 hours = 1 day
	7 days = 1 week

◄ **Work Problem** **5** **at the Side.**

5.1 Exercises

 FOR EXTRA HELP   Download the MyDashBoard App   MyMathLab®

1. **CONCEPT CHECK** What is a ratio? Write an explanation. Then make up an example of a ratio that is different from the ones in the textbook.

2. **CONCEPT CHECK** Both quantities in a ratio have the same units. That means you can do what to the units?

3. **CONCEPT CHECK** To rewrite the ratio $\frac{80}{20}$ in lowest terms, divide both the numerator and the denominator by _____. The ratio in lowest terms is _____.

4. **CONCEPT CHECK** To rewrite the ratio $\frac{20}{75}$ in lowest terms, divide both the numerator and the denominator by _____. The ratio in lowest terms is _____.

Write each ratio as a fraction in lowest terms. ***See Examples 1 and 2.***

5. 8 days to 9 days

6. $11 to $15

7. $100 to $50

8. 35¢ to 7¢

9. 30 minutes to 90 minutes

10. 9 pounds to 36 pounds

11. 80 miles to 50 miles

12. 300 people to 450 people

13. 6 hours to 16 hours

14. 45 books to 35 books

Write each ratio as a ratio of whole numbers in lowest terms. ***See Examples 3 and 4.***

15. $4.50 to $3.50

16. $0.08 to $0.06

17. 15 to $2\frac{1}{2} = \dfrac{15}{2\frac{1}{2}} = \dfrac{\frac{15}{1}}{\frac{5}{2}}$

$= \dfrac{15}{1} \div \dfrac{5}{2} = \dfrac{\overset{3}{15}}{1} \cdot \dfrac{2}{\underset{1}{5}} = \underline{\quad\quad}$

18. 5 to $1\frac{1}{4} = \dfrac{5}{1\frac{1}{4}} = \dfrac{\frac{5}{1}}{\frac{5}{4}}$

$= \dfrac{5}{1} \div \dfrac{5}{4} = \dfrac{\overset{1}{5}}{1} \cdot \dfrac{4}{\underset{1}{5}} = \underline{\quad\quad}$

19. $1\frac{1}{4}$ to $1\frac{1}{2}$

20. $2\frac{1}{3}$ to $2\frac{2}{3}$

21. **CONCEPT CHECK** When writing the ratio of 20 ounces to 3 pounds as a fraction in lowest terms, would it be easier to use ounces or pounds? _____ Explain why.

22. **CONCEPT CHECK** When writing the ratio of 4 days to 36 hours as a fraction in lowest terms, would it be easier to use days or hours? _____ Explain why.

Write each ratio as a fraction in lowest terms. For help, use the table of measurement relationships in this section. ***See Example 5.***

23. 4 ft to 30 in.
▶ *Hint:* First, convert 4 ft to inches.

24. 8 ft to 4 yd
Hint: First, convert 4 yd to feet.

25. 5 minutes to 1 hour

26. 8 quarts to 5 pints

27. 15 hours to 2 days
▶

28. 3 pounds to 6 ounces

29. 5 gallons to 5 quarts

30. 3 cups to 3 pints

The table shows the number of greeting cards sold in the United States for various occasions. Use the information to answer Exercises 31–36. Write each ratio as a fraction in lowest terms.

Holiday/Event	Cards Sold
Valentine's Day	150 million
Mother's Day	130 million
Father's Day	95 million
Graduation	60 million
Halloween	20 million
Thanksgiving	10 million

Source: Hallmark Cards.

31. Find the ratio of Thanksgiving cards to graduation cards.

32. Find the ratio of Halloween cards to Mother's Day cards.

33. Find the ratio of Valentine's Day cards to Thanksgiving cards.

34. Find the ratio of Mother's Day cards to Father's Day cards.

35. Explain how you might use the information in the table if you owned a shop selling gifts and greeting cards.

36. Why is the ratio of Valentine's Day cards to graduation cards $\frac{5}{2}$? Give two possible reasons.

The bar graph shows worldwide sales of the most popular songs of all time. Use the graph to complete Exercises 37–40. Write each ratio as a fraction in lowest terms.

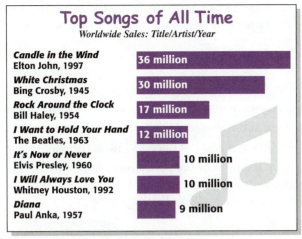

Top Songs of All Time
Worldwide Sales: Title/Artist/Year

Candle in the Wind
Elton John, 1997 — 36 million

White Christmas
Bing Crosby, 1945 — 30 million

Rock Around the Clock
Bill Haley, 1954 — 17 million

I Want to Hold Your Hand
The Beatles, 1963 — 12 million

It's Now or Never
Elvis Presley, 1960 — 10 million

I Will Always Love You
Whitney Houston, 1992 — 10 million

Diana
Paul Anka, 1957 — 9 million

Source: The Music Information Database.

37. Write a ratio that compares the top-selling song to the second best seller, and a ratio that compares the top-selling song to the third best seller.

38. Write two ratios that compare sales of Elvis Presley's song with sales of the songs just ahead and just behind it in the graph.

39. Sales of which two songs give a ratio of $\frac{3}{1}$? There may be more than one correct answer.

40. Sales of which two songs give a ratio of $\frac{5}{6}$? There may be more than one correct answer.

For each figure, find the ratio of the length of the longest side to the length of the shortest side. Write each ratio as a fraction in lowest terms.

41.

7 ft, 6 ft, 5 ft

42.

25 kilometers, 5 kilometers, 24.5 kilometers

43.

1.8 meters, 0.3 meter, 0.3 meter, 1.8 meters

44.

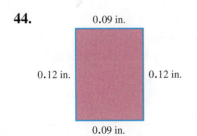

0.09 in., 0.12 in., 0.12 in., 0.09 in.

45.

5 in., $7\frac{3}{4}$ in., $4\frac{1}{4}$ in., $9\frac{1}{2}$ in., 5 in.

46.

$3\frac{1}{4}$ ft, $5\frac{3}{4}$ ft, $6\frac{3}{4}$ ft, $1\frac{3}{4}$ ft, $1\frac{1}{2}$ ft

Write each ratio as a fraction in lowest terms.

47. The price of automobile engine oil has gone from $10 to $12.50 for the 5 quarts needed for an oil change. Find the ratio of the increase in price to the original price.

48. The price that a pharmacy pays for an antibiotic decreased from $8.80 to $5.60 for 10 tablets. Find the ratio of the decrease in price to the original price.

49. The first time a movie was made in Minnesota, the cast and crew spent $59\frac{1}{2}$ days filming winter scenes. The next year, another movie was filmed in $8\frac{3}{4}$ weeks. Find the ratio of the first movie's filming time to the second movie's time. Compare in weeks.

50. The percheron, a large draft horse, measures about $5\frac{3}{4}$ ft at the shoulder. The prehistoric ancestor of the horse measured only $15\frac{3}{4}$ in. at the shoulder. Find the ratio of the percheron's height to its prehistoric ancestor's height. Compare in inches. (*Source: Eyewitness Books: Horse.*)

Relating Concepts (Exercises 51–54) For Individual or Group Work

Use your knowledge of ratios to **work Exercises 51–54 in order.**

51. In this painting, what is the ratio of the length of the longest side to the length of the shortest side? What other measurements could the painting have and still maintain the same ratio?

52. The ratio of my son's age to my daughter's age is 4 to 5. One possibility is that my son is 4 years old and my daughter is 5 years old. Find six other possibilities that fit the 4 to 5 ratio.

53. Amelia said that the ratio of her age to her mother's age is 5 to 3. Is this possible? Explain your answer.

54. Would you prefer that the ratio of your income to your friend's income be 1 to 3 or 3 to 1? Explain your answers.

5.2 Rates

A *ratio* compares two measurements with the **same** units, such as 9 feet **to** 12 feet (both length measurements). But many of the comparisons we make use measurements with **different** units.

160 dollars **for** 8 hours   (money to time)

450 miles **on** 15 gallons   (distance to capacity)

This type of comparison is called a **rate.**

OBJECTIVES

1 Write rates as fractions.

2 Find unit rates.

3 Find the best buy based on cost per unit.

OBJECTIVE **1** **Write rates as fractions.** Suppose you hiked 14 miles **in** 4 hours. The *rate* at which you hiked can be written as a fraction in lowest terms.

$$\frac{14 \text{ miles}}{4 \text{ hours}} = \frac{14 \text{ miles} \div 2}{4 \text{ hours} \div 2} = \frac{7 \text{ miles}}{2 \text{ hours}} \quad \left\{\begin{array}{l}\text{Rate in}\\\text{lowest terms}\end{array}\right.$$

In a rate, you often find these words separating the quantities you are comparing.

in   **for**   **on**   **per**   **from**

> **CAUTION**
>
> When writing a rate, always include the units, such as miles, hours, dollars, and so on. Because the units in a rate are different, the units do *not* divide out.

VOCABULARY TIP

Ratio vs. Rate **Ratios** compare two quantities that have the **same** units, so the units divide out. However, **rates** compare two quantities with **different** units, so you must write the units in a rate.

1 Write each rate as a fraction in lowest terms.

GS **(a)** $6 for 30 packets

$$\frac{\$6}{30 \text{ packets}} = \frac{\$ \underline{\quad}}{\underline{\quad} \text{ packets}}$$

To write the rate in lowest terms, divide both numerator and denominator by ____.

(b) 500 miles in 10 hours

(c) 4 teachers for 90 students

EXAMPLE 1 **Writing Rates in Lowest Terms**

Write each rate as a fraction in lowest terms.

(a) 5 gallons of chemical **for** $60.

$$\frac{5 \text{ gallons} \div 5}{60 \text{ dollars} \div 5} = \frac{1 \text{ gallon}}{12 \text{ dollars}} \quad \begin{array}{l}\text{Write the units:}\\\text{gallons and dollars.}\end{array}$$

(b) $1500 wages **in** 10 weeks

$$\frac{1500 \text{ dollars} \div 10}{10 \text{ weeks} \div 10} = \frac{150 \text{ dollars}}{1 \text{ week}}$$

> Be sure to write the units in a *rate*: dollars, miles, gallons, and so on.

(c) 2225 miles **on** 75 gallons of gas

$$\frac{2225 \text{ miles} \div 25}{75 \text{ gallons} \div 25} = \frac{89 \text{ miles}}{3 \text{ gallons}}$$

Work Problem **1** **at the Side.** ▶

OBJECTIVE **2** **Find unit rates.** When the *denominator* of a rate is 1, it is called a **unit rate.** We use unit rates frequently. For example, you earn $12.75 for *1 hour* of work. This unit rate is written:

$12.75 **per** hour   or $12.75**/hour.**

Use **per** or a slash mark (/) when writing unit rates.

Answers

1. (a) $\frac{\$1}{5 \text{ packets}}$; divide by 6.

(b) $\frac{50 \text{ miles}}{1 \text{ hour}}$ **(c)** $\frac{2 \text{ teachers}}{45 \text{ students}}$

VOCABULARY TIP

Per means "for each." For example, if your car gets 30 miles **per** gallon, it gets 30 miles *for each gallon* of gas. Use **per** or a slash mark, like this: 30 miles/gallon.

2 Find each unit rate.

(a) $4.35 for 3 pounds of apples

$$\frac{\$4.35}{3 \text{ pounds}} \left\{ \begin{array}{l} \text{Fraction bar} \\ \text{indicates} \\ \text{division.} \end{array} \right.$$

$$\begin{array}{r} 1. \\ 3\overline{)4.35} \end{array} \quad \begin{array}{l} \text{Finish the} \\ \text{division.} \end{array}$$

(b) 304 miles on 9.5 gallons of gas

(c) $850 in 5 days

(d) 24-pound turkey for 15 people

Answers

2. (a) $1.45/pound **(b)** 32 miles/gallon
   (c) $170/day **(d)** 1.6 pounds/person

EXAMPLE 2   **Finding Unit Rates**

Find each unit rate.

(a) 337.5 miles on 13.5 gallons of gas
Write the rate as a fraction.

$$\frac{337.5 \text{ miles}}{13.5 \text{ gallons}} \leftarrow \text{The fraction bar indicates division.}$$

Divide 337.5 by 13.5 to find the unit rate.

$$\begin{array}{r} 2\,5. \\ 13.5\overline{)337.5} \end{array}$$

$$\frac{337.5 \text{ miles} \div 13.5}{13.5 \text{ gallons} \div 13.5} = \frac{25 \text{ miles}}{1 \text{ gallon}}$$

> Be sure to write **miles** and **gallon** in your answer.

The unit rate is 25 miles **per** gallon, or 25 miles/gallon.

(b) 549 miles in 18 hours

$$\frac{549 \text{ miles}}{18 \text{ hours}} \qquad \text{Divide: } \begin{array}{r} 30.5 \\ 18\overline{)549.0} \end{array}$$

The unit rate is 30.5 miles/hour.

(c) $810 in 6 days

$$\frac{810 \text{ dollars}}{6 \text{ days}} \qquad \text{Divide: } \begin{array}{r} 135 \\ 6\overline{)810} \end{array}$$

> Use *per* or a slash mark to write unit rates.

The unit rate is $135/day.

◀ **Work Problem 2 at the Side.**

OBJECTIVE 3 **Find the best buy based on cost per unit.** When shopping for groceries, household supplies, and health and beauty items, you will find many different brands and package sizes. You can save money by finding the lowest *cost per unit.*

Cost per Unit

Cost per unit is a rate that tells how much you pay for *one* item or *one* unit. Examples are $3.75 per gallon, $47 per shirt, and $2.98 per pound.

EXAMPLE 3   **Determining the Best Buy**

The local store charges the following prices for pancake syrup. Find the best buy.

Continued on Next Page

The best buy is the container with the *lowest* cost per unit. All the containers are measured in *ounces* (oz), so you first need to find the *cost per ounce* for each one. Divide the price of the container by the number of ounces in it. Round to the nearest thousandth, if necessary.

Let the *order* of the *words* help you set up the rate.

cost ⟶ **$1.28**

per (means divide) ⟶ ───────

ounce ⟶ **12 ounces**

Size	Cost Per Unit (Rounded)
12 ounces	$\dfrac{\$1.28}{12 \text{ ounces}} \approx \$0.107$ per ounce (highest)
24 ounces	$\dfrac{\$1.81}{24 \text{ ounces}} \approx \$0.075$ per ounce (lowest)
36 ounces	$\dfrac{\$2.73}{36 \text{ ounces}} \approx \$0.076$ per ounce

The lowest cost per ounce is $0.075, so the 24-ounce container is the best buy.

Note

Earlier we rounded money amounts to the nearest hundredth (nearest cent). But when comparing unit costs, rounding to the nearest thousandth will help you see the difference between very similar unit costs. Notice that the 24-ounce and 36-ounce syrup containers above would both have rounded to $0.08 per ounce if we had rounded to hundredths.

⋯⋯⋯⋯⋯⋯⋯⋯⋯⋯⋯ **Work Problem ❸ at the Side. ▶**

⌨ **Calculator Tip**

When using a calculator to find unit prices, remember that division is *not* commutative. In **Example 3** you wanted to find cost per ounce. Let the *order* of the *words* help you enter the numbers in the correct order.

cost   **per**   **ounce**

↓   ↓   ↓

Enter the cost.   *Per* means divide.   Enter number of ounces.

↓   ↓   ↓

$2.73   ÷   **36**   =   **0.076 (rounded)**

If you entered 36 ÷ 2.73 =, you'd get the number of *ounces* per *dollar*. How could you use that information to find the best buy? (*Answer:* The best buy would be to get the greatest number of ounces per dollar.)

Finding the best buy is sometimes a complicated process. Things that affect the cost per unit can include "cents off" coupons and differences in how much use you'll get out of each unit.

❸ Find the best buy (lowest cost per unit) for each purchase.

GS **(a)** 2 quarts for $3.25

3 quarts for $4.95

4 quarts for $6.48

Divide to find each unit cost.

cost ⟶ $3.25

per ⟶ ─────

quart ⟶ 2 quarts

Unit cost is $____ per quart.

cost ⟶ $

per ⟶ ─────

quart ⟶ 3 quarts

Unit cost is $____ per ____.

cost ⟶

per ⟶ ───

quart ⟶

Unit cost is $____ per ____.

Now compare the three unit costs. The *lowest* one is the best buy. Circle it.

(b) 6 cans of cola for $1.99

12 cans of cola for $3.49

24 cans of cola for $7

4 Solve each problem.

(a) Some batteries claim to last longer than others. If you believe these claims, which brand is the best buy?

Four-pack of AA-size batteries for $2.79

One AA-size battery for $1.19; lasts twice as long

(b) Which tube of toothpaste is the best buy? You have a coupon for 85¢ off Brand C and a coupon for 50¢ off Brand D.

Brand C is $3.89 for 6 ounces.

Brand D is $1.89 for 2.5 ounces.

> **EXAMPLE 4   Solving Best Buy Applications**

Solve each application problem.

(a) There are many brands of liquid laundry detergent. If you feel they all do a good job of cleaning your clothes, you can base your purchase on cost per unit. But some brands are "concentrated" so you can use less detergent for each load of clothes. The labels from two bottles are shown below. Which of the choices is the best buy?

To find Sudzy's unit cost, divide $3.99 by 64 ounces, not 50 ounces. Read the label. You're getting as many clothes washed as if you bought 64 ounces. Similarly, to find White-O's unit cost, divide $9.89 by 256 ounces (twice 128 ounces, or 2 • 128 ounces = 256 ounces).

$$\text{Sudzy} \quad \frac{\$3.99}{64 \text{ ounces}} \approx \$0.062 \text{ per ounce}$$

> The best buy is the *lower* cost per ounce.

$$\text{White-O} \quad \frac{\$9.89}{256 \text{ ounces}} \approx \$0.039 \text{ per ounce}$$

White-O has the lower cost per ounce and is the better buy. (However, if you try it and it really doesn't get out all the stains, Sudzy may be worth the extra cost.)

(b) "Cents-off" coupons also affect the best buy. Suppose you are looking at these choices for "extra-strength" pain reliever. Both brands have the same amount of pain reliever in each tablet.

Brand X is $2.29 for 50 tablets.

Brand Y is $10.75 for 200 tablets.

You have a 40¢ coupon for Brand X and a 75¢ coupon for Brand Y. Which choice is the best buy?

To find the best buy, first subtract the coupon amounts, then divide to find the lower cost per ounce.

Brand X costs $2.29 − $0.40 = $1.89

$$\frac{\$1.89}{50 \text{ tablets}} \approx \$0.038 \text{ per tablet}$$

Brand Y costs $10.75 − $0.75 = $10.00

> Look for the *lower* cost per tablet.

$$\frac{\$10.00}{200 \text{ tablets}} = \$0.05 \text{ per tablet}$$

Brand X has the lower cost per tablet and is the better buy.

◀ **Work Problem** **4** **at the Side.**

5.2 Exercises

FOR EXTRA HELP

Download the MyDashBoard App

MyMathLab®

Write each rate as a fraction in lowest terms. **See Example 1.**

1. 10 cups for 6 people

2. $12 for 30 pens

3. 15 feet in 35 seconds

4. 100 miles in 30 hours

5. 72 miles on 4 gallons

6. 132 miles on 8 gallons

Find each unit rate. **See Example 2.**

7. CONCEPT CHECK   To find a unit rate, do you add, subtract, multiply, or divide? _____ Show the correct set-up to find this unit rate: $5.85 for 3 boxes.

8. CONCEPT CHECK   Circle the correct answer. When finding this unit rate, 15 hospital rooms for 3 nurses, the final answer will be:

5 to 1   5 rooms for 1 nurse   5 nurses for 1 room

9. $60 in 5 hours

10. $2500 in 20 days

11. 7.5 pounds for 6 people

12. 44 bushels from 8 trees

Earl kept the following record of the gas he bought for his car. For each entry, find the number of miles he traveled and the unit rate. Round your answers to the nearest tenth.

	Date	Odometer at Start	Odometer at End	Miles Traveled	Gallons Purchased	Miles per Gallon
13.	2/4	27,432.3	27,758.2		15.5	
14.	2/9	27,758.2	28,058.1		13.4	
15.	2/16	28,058.1	28,396.7		16.2	
16.	2/20	28,396.7	28,704.5		13.3	

Source: Author's car records.

Find the best buy (based on the cost per unit) for each item. Round to the nearest thousandth, if necessary. **See Example 3.** (*Source:* Cub Foods, Target, Rainbow Foods.)

17. Black pepper
GS

Complete the divisions:

$$\frac{\text{cost} \to \$2.25}{\text{ounce} \to 2 \text{ oz}} =$$

$$\frac{\text{cost} \to \$}{\text{ounce} \to 4 \text{ oz}} =$$

Lower unit cost (better buy)

is _____, so the best buy is

_____.

18. Shampoo
GS

Complete the divisions:

$$\frac{\text{cost} \to \$3.59}{\text{ounce} \to 8 \text{ oz}} =$$

$$\frac{\text{cost} \to \$}{\text{ounce} \to 12 \text{ oz}} =$$

Lower unit cost (better buy)

is _____, so the best buy is

_____.

19. Cereal

12 ounces for $2.49

14 ounces for $2.89

18 ounces for $3.96

20. Soup (same size cans)

2 cans for $2.18

3 cans for $3.57

5 cans for $5.29

21. Chunky peanut butter

12 ounces for $1.29

18 ounces for $1.79

28 ounces for $3.39

40 ounces for $4.39

22. Baked beans

8 ounces for $0.59

16 ounces for $0.99

21 ounces for $1.29

28 ounces for $1.89

23. Suppose you are choosing between two brands of chicken noodle soup. Brand A is $0.88 per can and Brand B is $0.98 per can. The cans are the same size but Brand B has more chunks of chicken in it. Which soup is the better buy? Explain your choice.

24. A small bag of potatoes costs $0.19 per pound. A large bag costs $0.15 per pound. But there are only two people in your family, so half the large bag would probably rot before you used it up. Which bag is the better buy? Explain.

Solve each application problem. **See Examples 2–4.**

25. Makesha lost 10.5 pounds in six weeks. What was her rate of loss in pounds per week?

26. Enrique's taco recipe uses three pounds of meat to feed 10 people. Give the rate in pounds per person.

27. Russ works 7 hours to earn $85.82. What is his pay rate per hour?

28. Find the cost of 1 gallon of Hawaiian Punch beverage if 18 gallons for a graduation party cost $55.62.

The table lists information about three long-distance calling cards. The connection fee is charged each time you make a call, no matter how long the call lasts. Use the table to answer Exercises 29–32. Round answers to the nearest thousandth when necessary.

LONG-DISTANCE CALLING CARDS (U.S.)

Card Name	Cost per Minute	Connection Fee
Penny Saver	$0.005	$0.39
Most Minutes	0.0025	0.49
USA Card	0.01	0.25

Source: www.noblecom.com

29. (a) Find the *actual* total cost, including the connection charge, for a five-minute call using each card.

    (b) Find the cost per minute for this call using each card and select the best buy.

30. (a) Find the *actual* total cost, including the connection charge, for a 90-minute call using each card.

    (b) Find the cost per minute for this call using each card and select the best buy.

31. Find the *actual* total cost and the per minute cost for a 30-minute call using each card. What do you see when you compare the unit rates for this call?

32. All the cards round calls up to the next full minute.
    (a) Suppose you call the wrong number. How much would you pay for this 40-*second* call on each card?

    (b) How much would you save on this call by using the USA Card instead of Most Minutes?

33. If you believe the claims that some batteries last longer, which is the better buy?

34. Which is the better buy, assuming these laundry detergents both clean equally well?

35. Three brands of cornflakes are available. Brand G is priced at $2.39 for 10 ounces. Brand K is $3.99 for 20.3 ounces and Brand P is $3.39 for 16.5 ounces. You have a coupon for 50¢ off Brand P and a coupon for 60¢ off Brand G. Which cereal is the best buy based on cost per unit?

36. Two brands of facial tissue are available. Brand K is priced at $5 for three boxes of 175 tissues each. Brand S is priced at $1.29 per box of 125 tissues. You have a coupon for 20¢ off one box of Brand S and a coupon for 45¢ off one box of Brand K. How can you get the best buy on one box of tissue?

Relating Concepts (Exercises 37–41) For Individual or Group Work

On the first page of this chapter, we said that unit rates can help you get the best deal on cell phone service. Use the information in the table to **work Exercises 37–41 in order.** Round all money answers to the nearest cent.

CELL PHONE PLANS

Plan	Anytime Minutes	Monthly Charge*	Cost per Minute (to nearest cent)	Overage†
A	450	$39.95		45¢
B	900	$59.95		40¢

*Does not include taxes or special fees.
†Cost per extra minute if you use more minutes than your plan covers.
Source: www.Verizon.com

37. Find the cost per minute for each plan. Round your answers to the nearest cent and write them in the table above. Which plan is the better buy based on cost per minute?

38. In a 30-day month, what is the average number of minutes you can use your phone each day on:
    (a) Plan A
    (b) Plan B

39. Suppose you are on Plan A. Last month you used 2 more minutes each day than the average number of minutes you calculated in **Exercise 38.** Find the new cost per minute for last month. (Assume 30 days in last month.)

40. Suppose you are on Plan B. Last month you used 3 more minutes each day than the average number of minutes you calculated in **Exercise 38.** Find the new cost per minute for last month. (Assume 30 days in last month.)

41. The company also offers Plan C, with unlimited minutes for $69.95 per month. If you had Plan C, find the cost per minute for the situations described in **Exercises 39 and 40.**

5.3 Proportions

OBJECTIVES

1. Write proportions.
2. Determine whether proportions are true or false.
3. Find cross products.

OBJECTIVE **1** **Write proportions.** A **proportion** states that two ratios (or rates) are equal. For example,

$$\frac{\$20}{4 \text{ hours}} = \frac{\$40}{8 \text{ hours}}$$

is a proportion that says the rate $\dfrac{\$20}{4 \text{ hours}}$ is equal to the rate $\dfrac{\$40}{8 \text{ hours}}$. As the amount of money doubles, the number of hours also doubles. This proportion is read:

20 dollars **is to** 4 hours **as** 40 dollars **is to** 8 hours.

EXAMPLE 1   **Writing Proportions**

Write each proportion.

(a) 6 ft is to 11 ft **as** 18 ft is to 33 ft.

$$\frac{6 \text{ ft}}{11 \text{ ft}} = \frac{18 \text{ ft}}{33 \text{ ft}} \quad \text{so} \quad \frac{6}{11} = \frac{18}{33}$$

The common units (ft) divide out and are not written.

(b) \$9 is to 6 liters **as** \$3 is to 2 liters.

$$\frac{\$9}{6 \text{ liters}} = \frac{\$3}{2 \text{ liters}}$$

The units do *not* match so you must write them in the proportion.

························· **Work Problem** **1** **at the Side.** ▶

OBJECTIVE **2** **Determine whether proportions are true or false.** There are two ways to see whether a proportion is true. One way is to *write both of the ratios in lowest terms.*

EXAMPLE 2   **Writing Both Ratios in Lowest Terms**

Determine whether each proportion is true or false by writing both ratios in lowest terms.

(a) $\dfrac{5}{9} = \dfrac{18}{27}$

Write each ratio in lowest terms.

$$\frac{5}{9} \leftarrow \text{Already in lowest terms} \qquad \frac{18 \div 9}{27 \div 9} = \frac{2}{3} \leftarrow \text{Lowest terms}$$

Because $\frac{5}{9}$ is *not* equivalent to $\frac{2}{3}$, the proportion is *false.*

(b) $\dfrac{16}{12} = \dfrac{28}{21}$

Write each ratio in lowest terms.

Both ratios in lowest terms.

$$\frac{16 \div 4}{12 \div 4} = \frac{4}{3} \quad \text{and} \quad \frac{28 \div 7}{21 \div 7} = \frac{4}{3}$$

Both ratios are equivalent to $\frac{4}{3}$, so the proportion is *true.*

························· **Work Problem** **2** **at the Side.** ▶

1 Write each proportion.

(a) \$7 is to 3 cans as \$28 is to 12 cans

$$\frac{\$7}{3 \text{ cans}} = \frac{}{}$$

Complete the proportion.

(b) 9 meters is to 16 meters as 18 meters is to 32 meters

(c) 5 is to 7 as 35 is to 49

(d) 10 is to 30 as 60 is to 180

2 Determine whether each proportion is true or false by writing both ratios in lowest terms.

(a) $\dfrac{6}{12} = \dfrac{15}{30}$

(b) $\dfrac{20}{24} = \dfrac{3}{4}$

(c) $\dfrac{25}{40} = \dfrac{30}{48}$

(d) $\dfrac{35}{45} = \dfrac{12}{18}$

Answers

1. (a) $\dfrac{\$7}{3 \text{ cans}} = \dfrac{\$28}{12 \text{ cans}}$ **(b)** $\dfrac{9}{16} = \dfrac{18}{32}$

  (c) $\dfrac{5}{7} = \dfrac{35}{49}$ **(d)** $\dfrac{10}{30} = \dfrac{60}{180}$

2. (a) $\dfrac{1}{2} = \dfrac{1}{2}$; true **(b)** $\dfrac{5}{6} \ne \dfrac{3}{4}$; false

  (c) $\dfrac{5}{8} = \dfrac{5}{8}$; true **(d)** $\dfrac{7}{9} \ne \dfrac{2}{3}$; false

OBJECTIVE **3** **Find cross products.** Another way to test whether the ratios in a proportion are equivalent is to compare *cross products*.

Using Cross Products to Determine Whether a Proportion Is True

To see whether a proportion is true, first multiply along one diagonal, then multiply along the other diagonal, as shown here.

$$5 \cdot 4 = 20$$

$$\frac{2}{5} = \frac{4}{10}$$

Cross products are equal.

$$2 \cdot 10 = 20$$

In this case the **cross products** are both 20. When cross products are *equal,* the proportion is *true.* If the cross products are *unequal,* the proportion is *false.*

Note

The cross products test is based on rewriting both fractions with a common denominator of $5 \cdot 10$, or 50.

$$\frac{2 \cdot 10}{5 \cdot 10} = \frac{20}{50} \quad \text{and} \quad \frac{4 \cdot 5}{10 \cdot 5} = \frac{20}{50}$$

We see that $\frac{2}{5}$ and $\frac{4}{10}$ are equivalent because both can be rewritten as $\frac{20}{50}$. The cross products test takes a shortcut by comparing only the two numerators ($20 = 20$).

EXAMPLE 3   Using Cross Products

Use cross products to see whether each proportion is true or false.

(a) $\dfrac{3}{5} = \dfrac{12}{20}$

Multiply along one diagonal, then multiply along the other diagonal.

$$5 \cdot 12 = 60$$

$$\frac{3}{5} = \frac{12}{20}$$

Equal cross products; proportion is *true.*

$$3 \cdot 20 = 60$$

The cross products are *equal,* so the proportion is *true.*

CAUTION

Use cross products *only* when working with *proportions.* Do **not** use cross products when multiplying fractions, adding fractions, or writing fractions in lowest terms.

Continued on Next Page

(b) $\dfrac{2\dfrac{1}{3}}{3\dfrac{1}{3}} = \dfrac{9}{16}$

Find the cross products.

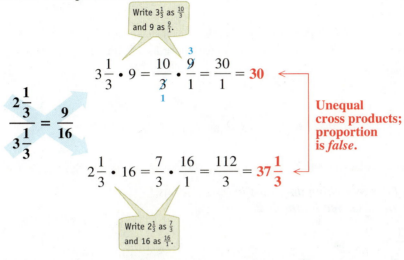

The cross products are *unequal*, so the proportion is *false*.

Note

The numbers in a proportion do *not* have to be whole numbers. They can be fractions, mixed numbers, decimal numbers, and so on.

······························ **Work Problem ❸ at the Side.** ▶

---

❸ Find the cross products to see whether each proportion is true or false.

GS **(a)** $\dfrac{5}{9} = \dfrac{10}{18}$

$(9)(10) =$ _____

$(5)(18) =$ _____

(b) $\dfrac{32}{15} = \dfrac{16}{8}$

(c) $\dfrac{10}{17} = \dfrac{20}{34}$

GS **(d)** $\dfrac{2.4}{6} = \dfrac{5}{12}$

$(6)(5) =$ _____

$(2.4)(12) =$ _____

(e) $\dfrac{3}{4.25} = \dfrac{24}{34}$

(f) $\dfrac{1\dfrac{1}{6}}{2\dfrac{1}{3}} = \dfrac{4}{8}$

Answers

3. **(a)** $(9)(10) = 90$; $(5)(18) = 90$; true
   (b) $240 \neq 256$; false
   (c) $340 = 340$; true
   (d) $(6)(5) = 30$; $(2.4)(12) = 28.8$; false
   (e) $102 = 102$; true
   (f) $9\dfrac{1}{3} = 9\dfrac{1}{3}$; true

5.3 Exercises

Write each proportion. See Example 1.

1. $9 is to 12 cans as $18 is to 24 cans.

2. 28 people is to 7 cars as 16 people is to 4 cars.

3. 200 adults is to 450 children as 4 adults is to 9 children.

4. 150 trees is to 1 acre as 1500 trees is to 10 acres.

5. 120 ft is to 150 ft as 8 ft is to 10 ft.

6. $6 is to $9 as $10 is to $15.

Determine whether each proportion is true or false by writing the ratios in lowest terms. Show the simplified ratios and then write true *or* false. *See Example 2.*

7. $\dfrac{6}{10} = \dfrac{3}{5}$

8. $\dfrac{1}{4} = \dfrac{9}{36}$

9. $\dfrac{5}{8} = \dfrac{25}{40}$

10. $\dfrac{2}{3} = \dfrac{20}{27}$

11. $\dfrac{150}{200} = \dfrac{200}{300}$

12. $\dfrac{100}{120} = \dfrac{75}{100}$

13. $\dfrac{42}{15} = \dfrac{28}{10}$

14. $\dfrac{18}{16} = \dfrac{36}{32}$

15. $\dfrac{32}{18} = \dfrac{48}{27}$

16. $\dfrac{15}{48} = \dfrac{10}{24}$

17. $\dfrac{7}{6} = \dfrac{54}{48}$

18. $\dfrac{28}{21} = \dfrac{44}{33}$

19. CONCEPT CHECK What is a proportion? Explain. Then draw arrows on the proportion below and show what numbers to multiply to find the cross products.

$$\dfrac{30}{6} = \dfrac{45}{9}$$

20. CONCEPT CHECK Explain how to tell if a proportion is true or false after you have found the cross products.

Use cross products to determine whether each proportion is true or false. Show the cross products and then circle true *or* false. *See Example 3.*

21. $\dfrac{2}{9} = \dfrac{6}{27}$

   True   False

22. $\dfrac{20}{25} = \dfrac{4}{5}$

   True   False

23. $\dfrac{20}{28} = \dfrac{12}{16}$

   True   False

24. $\dfrac{16}{40} = \dfrac{22}{55}$

  True   False

25. $\dfrac{110}{18} = \dfrac{160}{27}$

  True   False

26. $\dfrac{600}{420} = \dfrac{20}{14}$

  True   False

27. $\dfrac{3.5}{4} = \dfrac{7}{8}$

  True   False

28. $\dfrac{36}{23} = \dfrac{9}{5.75}$

  True   False

29. $\dfrac{18}{16} = \dfrac{2.8}{2.5}$

  True   False

30. $\dfrac{0.26}{0.39} = \dfrac{1.3}{1.9}$

  True   False

31. $\dfrac{6}{3\frac{2}{3}} = \dfrac{18}{11}$

  True   False

32. $\dfrac{16}{13} = \dfrac{2}{1\frac{5}{8}}$

  True   False

33. $\dfrac{2\frac{5}{8}}{3\frac{1}{4}} = \dfrac{21}{26}$

  True   False

34. $\dfrac{28}{17} = \dfrac{9\frac{1}{3}}{5\frac{2}{3}}$

  True   False

35. $\dfrac{\frac{2}{3}}{2} = \dfrac{2.7}{8}$ *Hint:* 2.7 is equivalent to $2\frac{7}{10}$.

  True   False

36. $\dfrac{3.75}{1\frac{1}{4}} = \dfrac{7.5}{2\frac{1}{2}}$

Hint: $1\frac{1}{4}$ is equivalent to 1.25 and $2\frac{1}{2}$ is equivalent to 2.5

  True   False

37. $\dfrac{2\frac{3}{10}}{8.05} = \dfrac{\frac{1}{4}}{0.9}$

  True   False

38. $\dfrac{3}{\frac{5}{6}} = \dfrac{1.5}{\frac{7}{12}}$

  True   False

39. CONCEPT CHECK Suppose Joe Mauer of the Minnesota Twins had 68 hits in 200 times at bat and Joey Votto of the Cincinnati Reds was at bat 450 times and got 153 hits. Paula is trying to convince Jenny that the two men hit equally well. Show how you could use a proportion and cross products to see whether Paula is correct.

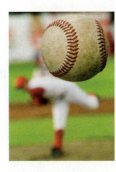

40. CONCEPT CHECK Jay worked 3.5 hours and packed 91 cartons. Craig packed 126 cartons in 5.25 hours. To see whether the men worked equally fast, Barry set up this proportion:

$$\dfrac{3.5}{91} = \dfrac{126}{5.25}$$

Explain what is wrong with Barry's proportion and write a correct one. Is the correct proportion true or false?

5.4 Solving Proportions

OBJECTIVE **1** **Find the unknown number in a proportion.** Four numbers are used in a proportion. If any three of these numbers are known, the fourth can be found. For example, find the unknown number that will make this proportion true.

$$\frac{3}{5} = \frac{x}{40}$$

The x represents the unknown number. Start by finding the cross products.

$$\frac{3}{5} \diagdown \frac{x}{40}$$

$$5 \cdot x \atop 3 \cdot 40 \Big\} \text{Cross products}$$

To make the proportion true, the cross products must be equal.

$$5 \cdot x = \underbrace{3 \cdot 40}$$

$$5 \cdot x = 120$$

The equal sign says that $5 \cdot x$ and 120 are equal. If $5 \cdot x$ and 120 are *both* divided by 5, the results will still be equal.

$$\frac{5 \cdot x}{5} = \frac{120}{5} \quad \leftarrow \text{Divide both sides by 5.}$$

On the left side, divide out the common factor of 5; slashes indicate the divisions.

$$\frac{\overset{1}{\cancel{5}} \cdot x}{\underset{1}{\cancel{5}}} = 24$$

On the right side, divide 120 by 5 to get 24.

Multiplying by 1 does *not* change a number, so in the numerator on the left side, $1 \cdot x$ is the same as x.

$$\frac{x}{1} = 24$$

Dividing by 1 does *not* change a number, so on the left side, $\frac{x}{1}$ is the same as x.

$$x = 24 \quad \text{The solution is 24.}$$

The unknown number in the proportion is 24. The complete proportion is shown below.

$$\frac{3}{5} = \frac{24}{40} \quad \text{x is 24.}$$

Check by finding the cross products. If they are equal, you solved the problem correctly. If they are unequal, rework the problem.

$$\frac{3}{5} \diagdown \frac{24}{40}$$

$$5 \cdot 24 = \mathbf{120} \atop 3 \cdot 40 = \mathbf{120} \Big\} \text{Equal; proportion is true.}$$

The cross products are equal, so the solution, $x = 24$, is correct.

> **CAUTION**
>
> **The solution is 24,** which is the unknown number in the proportion. 120 is **not** the solution; it is the cross product you get when *checking* the solution.

Solve a proportion for an unknown number by using the following steps.

Finding an Unknown Number in a Proportion

Step 1 Find the cross products.

Step 2 Show that the cross products are equal.

Step 3 Divide both products by the number multiplied by x (the number next to x).

Step 4 Check by writing the solution in the *original* proportion and finding the cross products.

EXAMPLE 1 **Solving Proportions for Unknown Numbers**

Find the unknown number in each proportion. Round answers to the nearest hundredth when necessary.

(a) $\dfrac{16}{x} = \dfrac{32}{20}$

Recall that ratios can be rewritten in lowest terms. If desired, you can do that *before* finding the cross products. In this example, write $\frac{32}{20}$ in lowest terms as $\frac{8}{5}$, which gives the proportion $\dfrac{16}{x} = \dfrac{8}{5}$.

Step 1
$$\dfrac{16}{x} = \dfrac{8}{5}$$
$x \cdot 8$
$16 \cdot 5$
Find the cross products.

Step 2 $x \cdot 8 = 16 \cdot 5$ ← Show that cross products are equal.

$x \cdot 8 = 80$

Step 3 $\dfrac{x \cdot \overset{1}{8}}{8} = \dfrac{80}{8}$ ← Divide both sides by 8.

The solution is 10 → $x = 10$ ← Find x. (No rounding necessary)

The unknown number in the proportion is 10.

Step 4 Write the solution in the *original* proportion and check by finding cross products.

$10 \cdot 32 = \mathbf{320}$

x is 10. → $\dfrac{16}{10} = \dfrac{32}{20}$ Equal; proportion is true.

$16 \cdot 20 = \mathbf{320}$

The solution is 10, not 320.

The cross products are equal, so **10 is the correct solution.**

Note

It is not necessary to write the ratios in lowest terms before solving. However, if you do, you will work with smaller numbers.

Continued on Next Page

1 Find the unknown numbers. Round to hundredths when necessary. Check your solutions by finding the cross products.

GS **(a)**

$$\frac{1}{2} = \frac{x}{12} \quad \begin{array}{l} 2 \cdot x \\ 1 \cdot 12 = 12 \end{array} \left. \begin{array}{l} \text{Cross} \\ \text{products} \end{array} \right.$$

$2 \cdot x = 12$ Now finish the solution. Divide both sides by 2.

(b) $\dfrac{6}{10} = \dfrac{15}{x}$

(c) $\dfrac{28}{x} = \dfrac{21}{9}$

(d) $\dfrac{x}{8} = \dfrac{3}{5}$

(e) $\dfrac{14}{11} = \dfrac{x}{3}$

Answers

1. **(a)** $\dfrac{\overset{1}{\cancel{2}} \cdot x}{\underset{1}{\cancel{2}}} = \dfrac{12}{2}; x = 6$ **(b)** $x = 25$

   (c) $x = 12$ **(d)** $x = 4.8$

   (e) $x \approx 3.82$ (rounded to nearest hundredth)

(b) $\dfrac{7}{12} = \dfrac{15}{x}$

Step 1 $\dfrac{7}{12} \bowtie \dfrac{15}{x}$ $12 \cdot 15 = 180$ ← Find the cross products. $7 \cdot x$ ←

Step 2 $7 \cdot x = 180$ ← Show that cross products are equal.

Step 3 $\dfrac{\overset{1}{\cancel{7}} \cdot x}{\underset{1}{\cancel{7}}} = \dfrac{180}{7}$ ← Divide both sides by 7.

The rounded solution is 25.71 $x \approx 25.71$ ← Rounded to nearest hundredth

When the division does not come out even, check for directions on how to round your answer. Divide out one more place, then round.

$$\begin{array}{r} 25.714 \\ 7\overline{)180.000} \end{array} \quad \begin{array}{l} \leftarrow \text{ Divide out to thousandths so} \\ \quad \text{you can round to hundredths.} \end{array}$$

The unknown number in the proportion is 25.71 (rounded).

Step 4   Write the solution in the original proportion and check by finding the cross products.

$$\dfrac{7}{12} \bowtie \dfrac{15}{25.71}$$

$12 \cdot 15 = \mathbf{180}$ ←

Very close, but *not* equal due to rounding the solution

$7 \cdot 25.71 = \mathbf{179.97}$ ←

The cross products are slightly different because you rounded the value of x. However, they are close enough to see that the problem was done correctly and that **25.71 is the approximate solution.**

◄ **Work Problem 1 at the Side.**

OBJECTIVE ▶ **2** **Find the unknown number in a proportion with mixed numbers or decimals.** The next example shows how to work with mixed numbers or decimals in a proportion.

EXAMPLE 2   Solving Proportions with Mixed Numbers and Decimals

Find the unknown number in each proportion.

(a) $\dfrac{2\frac{1}{5}}{6} = \dfrac{x}{10}$ $\dfrac{2\frac{1}{5}}{6} \bowtie \dfrac{x}{10}$

$6 \cdot x$ ← Find the cross products. $2\frac{1}{5} \cdot 10$ ←

Find $2\frac{1}{5} \cdot 10$.

$$2\frac{1}{5} \cdot 10 = \frac{11}{5} \cdot \frac{10}{1} = \frac{11}{\cancel{5}} \cdot \frac{\overset{2}{\cancel{10}}}{1} = \frac{22}{1} = 22$$

Changed to improper fraction

••••• **Continued on Next Page**

Show that the cross products are equal.

$$6 \cdot x = 22$$

Divide both sides by 6. $\dfrac{\overset{1}{\cancel{6}} \cdot x}{\cancel{6}_{1}} = \dfrac{22}{6}$

Write the solution as a mixed number in lowest terms.

$$x = \frac{22 \div 2}{6 \div 2} = \frac{11}{3} = 3\frac{2}{3}$$ ◄ [The solution is $3\frac{2}{3}$.]

The unknown number is $3\frac{2}{3}$.

Write the solution in the proportion and check by finding the cross products.

$$6 \cdot 3\frac{2}{3} = \frac{\overset{2}{\cancel{6}}}{1} \cdot \frac{11}{\cancel{3}_{1}} = \frac{22}{1} = 22 \leftarrow$$

$$\frac{2\frac{1}{5}}{6} \bowtie \frac{3\frac{2}{3}}{10} \qquad \text{Equal}$$

$$2\frac{1}{5} \cdot 10 = \frac{11}{\cancel{5}} \cdot \frac{\overset{2}{\cancel{10}}}{1} = \frac{22}{1} = 22 \leftarrow$$

[The solution is $3\frac{2}{3}$, not 22.]

The cross products are equal, so $3\frac{2}{3}$ is the correct solution.

(b) $\dfrac{1.5}{0.6} = \dfrac{2}{x}$

Show that cross products are equal.

$$(1.5)(x) = (0.6)(2)$$
$$(1.5)(x) = 1.2$$

Divide both sides by 1.5 $\dfrac{(\cancel{1.5})(x)}{\cancel{1.5}_{1}} = \dfrac{1.2}{1.5}$

$$x = \frac{1.2}{1.5}$$

Complete the division.

[The solution is 0.8] $x = 0.8 \qquad 1.5\overline{)1.20}^{.8}$

So the unknown number is 0.8. Write the solution in the original proportion and check it by finding the cross products.

$$(0.6)(2) = 1.2 \leftarrow$$

$$\frac{1.5}{0.6} \bowtie \frac{2}{0.8} \qquad \text{Equal}$$

$$(1.5)(0.8) = 1.2 \leftarrow$$

[The solution is 0.8, not 1.2]

The cross products are equal, so **0.8** is the correct solution.

·········· **Work Problem ❷ at the Side. ▶**

---

❷ Find the unknown numbers. Round to hundredths on the decimal problems, if necessary. Check your solutions by finding the cross products.

GS (a) $\dfrac{3\frac{1}{4}}{2} \bowtie \dfrac{x}{8}$ $2 \cdot x$

$$3\frac{1}{4} \cdot 8 = \frac{13}{4} \cdot \frac{\overset{2}{\cancel{8}}}{1} = \frac{26}{1}$$

$2 \cdot x = 26$ Finish the solution.

(b) $\dfrac{x}{3} = \dfrac{1\frac{2}{3}}{5}$

(c) $\dfrac{0.06}{x} = \dfrac{0.3}{0.4}$

(d) $\dfrac{2.2}{5} = \dfrac{13}{x}$

(e) $\dfrac{x}{6} = \dfrac{0.5}{1.2}$

(f) $\dfrac{0}{2} = \dfrac{x}{7.092}$

Answers

2. (a) $x = 13$ (b) $x = 1$ (c) $x = 0.08$
   (d) $x \approx 29.55$ (rounded to nearest hundredth)
   (e) $x = 2.5$ (f) $x = 0$

5.4 Exercises

FOR EXTRA HELP

   MyMathLab®

1. **CONCEPT CHECK** In $\frac{7}{10} = \frac{5}{x}$, what is the first step when finding the unknown number? Show how to do this step.

2. **CONCEPT CHECK** In $\frac{7}{10} = \frac{5}{x}$, the cross products are $7 \cdot x$ and 50. What is the next step when finding the unknown number? Show how to do this step.

*Find the unknown number in each proportion. Round your answers to hundredths, if necessary. Check your answers by finding the cross products. **See Examples 1 and 2.***

3.
$$\frac{1}{3} = \frac{x}{12}$$
$\qquad 3 \cdot x$
$\qquad 1 \cdot 12$

$3 \cdot x = \underbrace{1 \cdot 12}$

$3 \cdot x = \underline{\quad}$

$\dfrac{\overset{1}{\cancel{3}} \cdot x}{\underset{1}{\cancel{3}}} = \dfrac{\underline{\quad}}{3}$

$x = \underline{\quad}$

4.
$$\frac{x}{6} = \frac{15}{18}$$
$\qquad 6 \cdot 15$
$\qquad x \cdot 18$

$x \cdot 18 = \underbrace{6 \cdot 15}$

$x \cdot 18 = \underline{\quad}$

$\dfrac{x \cdot \overset{1}{\cancel{18}}}{\underset{1}{\cancel{18}}} = \dfrac{\underline{\quad}}{18}$

$x = \underline{\quad}$

5. $\dfrac{15}{10} = \dfrac{3}{x}$

6. $\dfrac{5}{x} = \dfrac{20}{8}$

7. $\dfrac{x}{11} = \dfrac{32}{4}$

8. $\dfrac{12}{9} = \dfrac{8}{x}$

9. $\dfrac{42}{x} = \dfrac{18}{39}$

10. $\dfrac{49}{x} = \dfrac{14}{18}$

11. $\dfrac{x}{25} = \dfrac{4}{20}$

12. $\dfrac{6}{x} = \dfrac{4}{8}$

13. $\dfrac{8}{x} = \dfrac{24}{30}$

14. $\dfrac{32}{5} = \dfrac{x}{10}$

15. $\dfrac{99}{55} = \dfrac{44}{x}$

16. $\dfrac{x}{12} = \dfrac{101}{147}$

17. $\dfrac{0.7}{9.8} = \dfrac{3.6}{x}$

18. $\dfrac{x}{3.6} = \dfrac{4.5}{6}$

19. $\dfrac{250}{24.8} = \dfrac{x}{1.75}$

20. $\dfrac{4.75}{17} = \dfrac{43}{x}$

Find the unknown number in each proportion. Write your answers as whole or mixed numbers when possible. **See Example 2.**

21. $\dfrac{15}{1\frac{2}{3}} = \dfrac{9}{x}$

22. $\dfrac{x}{\frac{3}{10}} = \dfrac{2\frac{2}{9}}{1}$

23. $\dfrac{2\frac{1}{3}}{1\frac{1}{2}} = \dfrac{x}{2\frac{1}{4}}$

24. $\dfrac{1\frac{5}{6}}{x} = \dfrac{\frac{3}{14}}{\frac{6}{7}}$

Solve each proportion two different ways. First change all the numbers to decimal form and solve. Then change all the numbers to fraction form and solve; write your answers in lowest terms.

25. $\dfrac{\frac{1}{2}}{x} = \dfrac{2}{0.8}$

26. $\dfrac{\frac{3}{20}}{0.1} = \dfrac{0.03}{x}$

27. $\dfrac{x}{\frac{3}{50}} = \dfrac{0.15}{1\frac{4}{5}}$

28. $\dfrac{8\frac{4}{5}}{1\frac{1}{10}} = \dfrac{x}{0.4}$

Relating Concepts (Exercises 29–30) For Individual or Group Work

Work Exercises 29–30 in order. *First prove that the proportions are* **not** *true. Then create four true proportions for each exercise by changing one number at a time.*

29. $\dfrac{10}{4} = \dfrac{5}{3}$

30. $\dfrac{6}{8} = \dfrac{24}{30}$

Summary Exercises *Ratios, Rates, and Proportions*

Use the bar graph of the countries with the most websites to complete Exercises 1–4.
Write each ratio as a fraction in lowest terms.

1. Write the ratio of the number of websites in Japan to the number of websites in Germany.

2. What is the ratio of the number of websites in Brazil to the number of websites in Italy?

3. Find the ratio of the combined number of websites in Italy and Germany compared to the number of websites in Japan.

4. Write the ratio of the number of websites in the USA compared to the total number of websites in all the other countries shown in the graph.

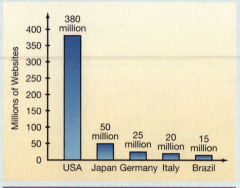

COUNTRIES WITH THE MOST WEBSITES
Number of websites, rounded to the nearest 5 million.

Source: Scholastic Book of World Records.

The bar graph shows the number of Americans who play various instruments. Use the graph to complete Exercises 5 and 6. Write each ratio as a fraction in lowest terms.

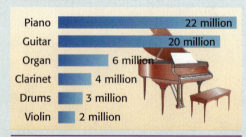

AMERICANS MAKE MUSIC BY THE MILLIONS
How many people play each type of instrument?

Piano	22 million
Guitar	20 million
Organ	6 million
Clarinet	4 million
Drums	3 million
Violin	2 million

Source: America by the Numbers.

5. Write five ratios that compare the least popular instrument to each of the other instruments.

6. Which two instruments give each of these ratios?

(a) $\frac{5}{1}$

(b) $\frac{2}{1}$

There may be more than one correct answer.

CONCEPT CHECK   *The table lists data on the top three individual scoring NBA basketball games of all time. Use the data to answer Exercises 7 and 8. Round answers to the nearest tenth.*

7. What was Wilt Chamberlain's scoring rate in points per minute and in minutes per point?

8. Find Kobe Bryant's scoring rate in points per minute and minutes per point.

Player	Date	Points	Min.
Wilt Chamberlain	3/2/62	100	48
Kobe Bryant	1/22/06	81	48
David Thompson	9/4/78	73	43

Source: www.sportcity.com

Wilt Chamberlain (#13)

9. Lucinda's paycheck showed gross pay of $652.80 for 40 hours of work and $195.84 for 8 hours of overtime work. Find her regular hourly pay rate and her overtime rate.

10. Satellite TV service is being offered to new subscribers at $29.99 per month for 150 channels, $34.99 per month for 210 channels, or $39.99 per month for 225 channels. What is the monthly cost per channel under each plan, to the nearest cent? (*Source:* DIRECTV advertisement.)

11. Find the best buy on gourmet coffee beans. Round to the nearest cent.

   11 ounces for $6.79

   12 ounces for $7.24

   16 ounces for $10.99

   (*Source:* Cub Foods.)

12. Which brand of cat food is the best buy? You have a coupon for $2 off Brand P, and another for $1 off Brand N.

   Brand N is $3.75 for 3.5 pounds.

   Brand P is $5.99 for 7 pounds.

   Brand R is $10.79 for 18 pounds.

Use either the method of writing in lowest terms or the method of finding cross products to decide whether each proportion is true or false. Show your work and then write true *or* false.

13. $\dfrac{28}{21} = \dfrac{44}{33}$

14. $\dfrac{2.3}{8.05} = \dfrac{0.25}{0.9}$

15. $\dfrac{2\frac{5}{8}}{3\frac{1}{4}} = \dfrac{21}{26}$

Solve each proportion to find the unknown number. Round your answers to hundredths when necessary.

16. $\dfrac{7}{x} = \dfrac{25}{100}$

17. $\dfrac{15}{8} = \dfrac{6}{x}$

18. $\dfrac{x}{84} = \dfrac{78}{36}$

19. $\dfrac{10}{11} = \dfrac{x}{4}$

20. $\dfrac{x}{17} = \dfrac{3}{55}$

21. $\dfrac{2.6}{x} = \dfrac{13}{7.8}$

22. $\dfrac{0.14}{1.8} = \dfrac{x}{0.63}$

23. $\dfrac{\frac{1}{3}}{8} = \dfrac{x}{24}$

24. $\dfrac{6\frac{2}{3}}{4\frac{1}{6}} = \dfrac{\frac{6}{5}}{x}$

5.5 Solving Application Problems with Proportions

OBJECTIVE **1** **Use proportions to solve application problems.**
Proportions can be used to solve a wide variety of problems. Watch for problems in which you are given a ratio or rate and then are asked to find part of a corresponding ratio or rate. Remember that a ratio or rate compares two quantities and often includes one of the following indicator words.

in   **for**   **on**   **per**   **from**   **to**

Use the six problem-solving steps you learned in **Chapter 1.**

Step 1   **Read** the problem.   *Step 4*   **Solve** the problem.

Step 2   **Work out a plan.**   *Step 5*   **State the answer.**

Step 3   **Estimate** a reasonable answer.   *Step 6*   **Check** your work.

EXAMPLE 1   **Solving a Proportion Application**

Mike's car can travel 163 **miles on** 6.4 **gallons** of gas. How far can it travel on a full tank of 14 **gallons** of gas? Round to the nearest whole mile.

Step 1   **Read** the problem. The problem asks for the number of miles the car can travel on 14 gallons of gas.

Step 2   **Work out a plan.** Decide what is being compared. This example compares **miles** to **gallons**. Write a proportion using the two rates. Be sure that *both* rates compare miles to gallons in the same order. In other words, miles is in both numerators and gallons is in both denominators. Use a letter to represent the unknown number.

Matching units

This rate compares **miles** to **gallons**. $\dfrac{163 \text{ miles}}{6.4 \text{ gallons}} = \dfrac{x \text{ miles}}{14 \text{ gallons}}$ This rate compares **miles** to **gallons**.

Matching units

Step 3   **Estimate** a reasonable answer. To estimate the answer, notice that 14 gallons is a little more than *twice as much* as 6.4 gallons, so the car should travel a little more than *twice as far.* So use 2 • 163 miles = 326 miles as the estimate.

Step 4   **Solve** the problem. Ignore the units while solving for *x*.

$$\frac{163 \text{ miles}}{6.4 \text{ gallons}} = \frac{x \text{ miles}}{14 \text{ gallons}}$$

$$(6.4)(x) = (163)(14) \qquad \text{Show that cross products are equal.}$$

$$(6.4)(x) = 2282$$

$$\frac{\overset{1}{(6.4)}(x)}{\underset{1}{6.4}} = \frac{2282}{6.4} \qquad \text{Divide both sides by 6.4}$$

$$x = 356.5625 \qquad \text{Round to 357.}$$

> Check the problem for rounding directions; this one asks for nearest whole number.

· **Continued on Next Page**

Step 5 **State the answer.** Rounded to the nearest mile, the car can travel about **357 miles** on a full tank of gas.

> Be sure to write **miles** in your answer.

Step 6 **Check** your work. The answer, 357 miles, is a little more than the estimate of 326 miles, so it is reasonable.

> ### CAUTION
>
> When setting up a proportion, do *not* mix up the units in the rates.
>
> compares **miles** } $\dfrac{163 \text{ miles}}{6.4 \text{ gallons}} \neq \dfrac{14 \text{ gallons}}{x \text{ miles}}$ { compares **gallons**
> to **gallons**   to **miles**
>
> These rates do *not* compare things in the same order and *cannot* be set up as a proportion.

·········· **Work Problem ❶ at the Side.** ▶

EXAMPLE 2 Solving a Proportion Application

A newspaper report says that 7 out of 10 people surveyed watch the news on TV. At that rate, how many of the 3200 people in town would you expect to watch the news?

Step 1 **Read** the problem. The problem asks how many of the 3200 people in town would be expected to watch TV news.

Step 2 **Work out a plan.** You are comparing people who watch the news to people surveyed. Set up a proportion using the two rates described in the example. Be sure that both rates make the same comparison. "People who watch the news" is mentioned first, so it should be in the numerator of *both* rates.

People who watch news → $\dfrac{7}{10} = \dfrac{x}{3200}$ ← People who watch news
Total group → ... ... ← Total group
(people surveyed) ... ... (people in town)

Step 3 **Estimate** a reasonable answer. To estimate the answer, notice that 7 out of 10 people is more than half the people, but less than all the people. Half of 3200 people is 3200 ÷ 2 = 1600, so our estimate is between 1600 and 3200 people.

Step 4 **Solve** the problem. Solve for the unknown number in the proportion.

$$\frac{7}{10} = \frac{x}{3200}$$

$10 \cdot x = 7 \cdot 3200$ Show that cross products are equal.

$10 \cdot x = 22{,}400$

$\dfrac{\overset{1}{\cancel{10}} \cdot x}{\underset{1}{\cancel{10}}} = \dfrac{22{,}400}{10}$ Divide both sides by 10.

$x = 2240$ ← No rounding is needed here.

·········· **Continued on Next Page**

---

❶ Set up and solve a proportion for each problem.

GS (a) If 2 pounds of fertilizer will cover 50 square feet of garden, how many pounds are needed for 225 square feet?

$$\frac{2 \text{ pounds}}{50 \text{ square feet}} = \frac{\underline{\quad} \text{ pounds}}{\underline{\quad} \text{ square feet}}$$

(b) A U.S. map has a scale of 1 inch to 75 miles. Lake Superior is 4.75 inches long on the map. What is the lake's actual length to the nearest whole mile ?

(c) A cough syrup is given at the rate of 30 milliliters for each 100 pounds of body weight. How much should be given to a 34-pound child? Round to the nearest whole milliliter.

Answers

1. (a) $\dfrac{2 \text{ pounds}}{50 \text{ square feet}} = \dfrac{x \text{ pounds}}{225 \text{ square feet}}$
$x = 9$ pounds

(b) $\dfrac{1 \text{ inch}}{75 \text{ miles}} = \dfrac{4.75 \text{ inches}}{x \text{ miles}}$
$x = 356.25$ miles, rounds to 356 miles

(c) $\dfrac{30 \text{ milliliters}}{100 \text{ pounds}} = \dfrac{x \text{ milliliters}}{34 \text{ pounds}}$
$x \approx 10$ milliliters (rounded)

2 Solve each problem to find a reasonable answer. Then flip one side of your proportion to see what answer you get with an **incorrect** set-up. Explain why the second answer is **unreasonable.**

(GS) **(a)** A survey showed that 2 out of 3 people would like to lose weight. At this rate, how many people in a group of 150 want to lose weight?

Lose weight → $\dfrac{2}{3} = \dfrac{x}{}$ ← Lose weight

People → surveyed ← People in group

(b) In one state, 3 out of 5 college students receive financial aid. At this rate, how many of the 4500 students at Central Community College receive financial aid?

Step 5 **State the answer.** You would expect **2240 people** in town to watch the news on TV.

> Be sure to write **people** in your answer.

Step 6 **Check** your work. The answer, 2240 people, is between 1600 and 3200, as called for in the estimate.

CAUTION

Always check that your answer is reasonable. If it is not, look at the way your proportion is set up. Be sure you have matching units in the numerators and matching units in the denominators.

For example, suppose you had set up the last proportion ***incorrectly,*** as shown here.

$$\frac{7}{10} = \frac{3200}{x} \quad \leftarrow \text{Incorrect set-up}$$

$$7 \cdot x = 10 \cdot 3200$$

$$\frac{\overset{1}{\cancel{7}} \cdot x}{\underset{1}{\cancel{7}}} = \frac{32{,}000}{7}$$

$$x \approx 4571 \text{ people} \leftarrow \text{Unreasonable answer}$$

This answer is ***unreasonable*** because there are only 3200 people in the town; it is ***not*** possible for 4571 people to watch the news.

◀ **Work Problem** **2** **at the Side.**

5.5 Exercises

FOR EXTRA HELP

  Download the MyDashBoard App

▶ MyMathLab®

CONCEPT CHECK *Complete the proportion for each situation in Exercises 1 and 2.*
Do not solve the proportion.

1. When Linda makes potato salad, she uses 6 potatoes and 4 hard-boiled eggs. How many eggs will she need if she uses 12 potatoes? Write the rest of the units in the proportion below.

$$\frac{6 \text{ potatoes}}{4 \underline{\hspace{1cm}}} = \frac{12 \underline{\hspace{1cm}}}{x \underline{\hspace{1cm}}}$$

2. A potato chip bag says that 1 serving is 15 chips. Joe ate 70 chips. How many servings did he eat? Write the rest of the units in the proportion below.

$$\frac{1 \underline{\hspace{1cm}}}{15 \text{ chips}} = \frac{x \underline{\hspace{1cm}}}{70 \underline{\hspace{1cm}}}$$

Set up and solve a proportion for each application problem. **See Example 1.**

3. Caroline can sketch four cartoon strips in five hours. How long will it take her to sketch 18 strips?

▶ Compares
cartoon strips } $\dfrac{4 \text{ strips}}{5 \text{ hours}} = \dfrac{\underline{\hspace{0.5cm}} \text{ strips}}{\underline{\hspace{0.5cm}} \text{ hours}}$
to hours

4. The Cosmic Toads recorded eight songs on their first album in 26 hours. At this same rate, how long will it take them to record 14 songs for their second album?

Compares
songs } $\dfrac{8 \text{ songs}}{26 \text{ hours}} = \dfrac{\underline{\hspace{0.5cm}} \text{ songs}}{\underline{\hspace{0.5cm}} \text{ hours}}$
to hours

5. Sixty newspapers cost $27. Find the cost of 16 newspapers.

6. Twenty-two guitar lessons cost $528. Find the cost of 12 lessons.

7. If three pounds of fescue grass seed cover about 350 square feet of ground, how many pounds are needed for 4900 square feet?

8. Anna earns $1242.08 in 14 days. How much does she earn in 260 days?

9. Tom makes $672.80 in 5 days. How much does he make in 3 days?

10. If 5 ounces of a medicine must be mixed with 8 ounces of water, how many ounces of medicine would be mixed with 20 ounces of water?

11. The bag of rice noodles below makes 7 servings. At that rate, how many ounces of noodles do you need for 12 servings, to the nearest ounce?

12. This can of sweet potatoes is enough for 4 servings. How many ounces are needed for 9 servings, to the nearest ounce?

13. Three quarts of a latex enamel paint will cover about 270 square feet of wall surface. How many quarts will you need to cover 350 square feet of wall surface in your kitchen and 100 square feet of wall surface in your bathroom?

14. One gallon of clear gloss wood finish covers about 550 square feet of surface. If you need to apply three coats of finish to 400 square feet of surface, how many gallons do you need, to the nearest tenth?

Use the floor plan shown to complete Exercises 15–18. On the plan, one inch represents four feet.

15. What is the actual length and width of the kitchen?

16. What is the actual length and width of the family room?

17. What is the actual length and width of the dining area?

18. What is the actual length and width of the entire floor plan?

The table below lists recommended amounts of food to order for 25 party guests. Use the table to answer Exercises 19 and 20. (Source: Cub Foods.)

FOOD FOR 25 GUESTS

Item	Amount
Fried chicken	40 pieces
Lasagna	14 pounds
Deli meats	4.5 pounds
Sliced cheese	$2\frac{1}{3}$ pounds
Bakery buns	3 dozen
Potato salad	6 pounds

19. How much of each food item should Nathan and Amanda order for a graduation party with 60 guests?

20. Taisha is having 20 neighbors over for a Fourth of July picnic. How much food should she buy?

21. CONCEPT CHECK Eight out of 10 students voted to take a fifteen-minute break in the middle of a two-hour class. There are 35 students in the class. How many students voted for the break? Carl set up the proportion below. Explain how to tell if Carl's answer is reasonable or not.

$$\frac{8}{10} = \frac{35}{x} \qquad 8 \cdot x = 350$$

$$\frac{\overset{1}{\cancel{8}} \cdot x}{\underset{1}{\cancel{8}}} = \frac{350}{8}$$

$$x \approx 44 \text{ students (rounded)}$$

22. CONCEPT CHECK Many experts recommend 30 minutes of exercise each day. If Vera follows that advice for 60 days, how many minutes of exercise will she get? Vera set up the proportion below. Explain how to tell if Vera's answer is reasonable or not.

$$\frac{1}{30} = \frac{x}{60} \qquad 30 \cdot x = 60$$

$$\frac{\overset{1}{\cancel{30}} \cdot x}{\underset{1}{\cancel{30}}} = \frac{60}{30}$$

$$x = 2 \text{ minutes}$$

Set up a proportion to solve each problem. Check to see whether your answer is reasonable. Then flip one side of your proportion to see what answer you get with an incorrect set-up. Explain why the second answer is unreasonable. See Example 2.

23. About 7 out of 10 people entering a community college need to take a refresher math course. If there are 2950 entering students, how many will probably need refresher math? (*Source:* Minneapolis Community and Technical College.)

24. In a survey, only 3 out of 100 people like their eggs poached. At that rate, how many of the 60 customers who ordered eggs at Soon-Won's restaurant this morning asked to have them poached? Round to the nearest whole person.

25. About 1 out of 3 people choose vanilla as their favorite ice cream flavor. If 250 people attend an ice cream social, how many would you expect to choose vanilla? Round to the nearest whole person.

26. In a test of 200 sewing machines, only one had a defect. At that rate, how many of the 5600 machines shipped from the factory have defects?

Set up and solve a proportion for each problem.

27. The stock market report says that 5 stocks went up for every 6 stocks that went down. If 750 stocks went down yesterday, how many went up?

28. The human body contains 90 pounds of water for every 100 pounds of body weight. How many pounds of water are in a child who weighs 80 pounds?

29. The ratio of the length of an airplane wing to its width is 8 to 1. If the length of a wing is 32.5 meters, how wide must it be? Round to the nearest hundredth.

30. The Rosebud School District wants a student-to-teacher ratio of 19 to 1. How many teachers are needed for 1850 students? Round to the nearest whole number.

31. The number of calories you burn is proportional to your weight. A 150-pound person burns 222 calories during 30 minutes of tennis. How many calories would a 210-pound person burn, to the nearest whole number? (*Source: Wellness Encyclopedia.*)

32. Refer to **Exercise 31.** A 150-pound person burns 189 calories during 45 minutes of grocery shopping. How many calories would a 115-pound person burn, to the nearest whole number? (*Source: Wellness Encyclopedia.*)

33. At 3 P.M., Coretta's shadow is 1.05 meters long. Her height is 1.68 meters. At the same time, a tree's shadow is 6.58 meters long. How tall is the tree? Round to the nearest hundredth.

34. Refer to **Exercise 33.** Later in the day, Coretta's shadow was 2.95 meters long. How long a shadow did the tree have at that time? Round to the nearest hundredth.

35. Can you set up a proportion to solve this problem? Explain why or why not. Jim is 25 years old and weighs 180 pounds. How much will he weigh when he is 50 years old?

36. Write your own application problem that can be solved by setting up a proportion. Also show the proportion and the steps needed to solve your problem.

37. A survey of college students shows that 4 out of 5 drink coffee. Of the students who drink coffee, 1 out of 8 adds cream to it. How many of the 56,100 students at Ohio State University would be expected to use cream in their coffee?

38. About 9 out of 10 adults think it is a good idea to exercise regularly. But of the ones who think it is a good idea, only 1 in 6 actually exercises at least three times a week. At this rate, how many of the 300 employees in our company exercise regularly?

39. The nutrition information on a bran cereal box says that a $\frac{1}{3}$-cup serving provides 80 calories and 8 grams of dietary fiber. At that rate, how many calories and grams of fiber are in a $\frac{1}{2}$-cup serving? (*Source:* Kraft Foods, Inc.)

40. A $\frac{3}{4}$-cup serving of whole grain penne pasta has 210 calories and 6 grams of dietary fiber. How many calories and grams of fiber would be in a 1-cup serving? (*Source:* Barilla Pasta.)

Relating Concepts (Exercises 41–42) For Individual or Group Work

*A box of instant mashed potatoes has the list of ingredients shown in the table. Use this information to **work Exercises 41–42 in order.***

Ingredient	For 12 Servings
Water	$3\frac{1}{2}$ cups
Margarine	6 Tbsp
Milk	$1\frac{1}{2}$ cups
Potato flakes	4 cups

Source: General Mills.

41. Find the amount of each ingredient needed for 6 servings. Show *two* different methods for finding the amounts. One method should use proportions.

42. Find the amount of each ingredient needed for 18 servings. Show *two* different methods for finding the amounts, one using proportions and one using your answers from **Exercise 41.**

Chapter 5 Summary

Key Terms

5.1

ratio A ratio compares two quantities having the same units. For example, the ratio of 6 apples to 11 apples is written in fraction form as $\frac{6}{11}$. The common units (apples) divide out.

5.2

rate A rate compares two measurements with different units. Examples are 96 dollars for 8 hours, or 450 miles on 18 gallons.

unit rate A unit rate has 1 in the denominator.

cost per unit Cost per unit is a rate that tells how much you pay for one item or one unit. The lowest cost per unit is the best buy.

5.3

proportion A proportion states that two ratios or rates are equal.

cross products Multiply along one diagonal and then multiply along the other diagonal to find the cross products of a proportion. If the cross products are equal, the proportion is true.

Test Your Word Power

See how well you have learned the vocabulary in this chapter.

1. A **ratio**
   A. can be written only as a fraction
   B. compares two quantities that have the same units
   C. compares two quantities that have different units.

2. A **rate**
   A. can be written only as a decimal
   B. compares two quantities that have the same units
   C. compares two quantities that have different units.

3. A **unit rate**
   A. has a numerator of 1
   B. has a denominator of 1
   C. is found by cross multiplying.

4. **Cost per unit** is
   A. the best buy
   B. a ratio written in lowest terms
   C. the price of one item or one unit.

5. A **proportion**
   A. shows that two ratios or rates are equal
   B. contains only whole numbers or decimals
   C. always has one unknown number.

6. **Cross products** are
   A. equal when a proportion is false
   B. used to find the best buy
   C. equal when a proportion is true.

Answers to Test Your Word Power

1. B; *Example:* The ratio of 3 miles to 4 miles is $\frac{3}{4}$; the common units (miles) divide out.

2. C; *Example:* $4.50 for 3 pounds is a rate comparing dollars to pounds.

3. B; *Example:* $\frac{\$1.79}{1\ \text{pound}}$ is a unit rate. We write it as $1.79 per pound or $1.79/pound.

4. C; *Example:* $3.95 per gallon tells the price of one gallon (one unit).

5. A; *Example:* The proportion $\frac{5}{6} = \frac{25}{30}$ says that $\frac{5}{6}$ is equal to $\frac{25}{30}$.

6. C; *Example:* The cross products for $\frac{5}{6} = \frac{25}{30}$ are $6 \cdot 25 = 150$ and $5 \cdot 30 = 150$.

Quick Review

Concepts	Examples

5.1 Writing a Ratio

A ratio compares two quantities that have the same units. A ratio is usually written as a fraction with the number that is mentioned first in the numerator. The common units divide out and are not written in the answer. Check that the fraction is in lowest terms.

Write this ratio as a fraction in lowest terms.

60 ounces of medicine **to** 160 ounces of medicine

$$\frac{60 \text{ ounces}}{160 \text{ ounces}} = \frac{60 \div 20}{160 \div 20} = \frac{3}{8} \quad \left\{ \begin{array}{l} \text{Ratio in lowest} \\ \text{terms} \end{array} \right.$$

Divide out common units.

5.1 Using Mixed Numbers in a Ratio

If a ratio has mixed numbers, change the mixed numbers to improper fractions. Rewrite the problem in horizontal format using the "÷" symbol for division. Finally, multiply by the reciprocal of the divisor.

Write as a ratio of whole numbers in lowest terms.

$$2\frac{1}{2} \quad \text{to} \quad 3\frac{3}{4}$$

$$\frac{2\frac{1}{2}}{3\frac{3}{4}} = \frac{\frac{5}{2}}{\frac{15}{4}}$$

Reciprocal

$$= \frac{5}{2} \div \frac{15}{4} = \frac{5}{2} \cdot \frac{4}{15}$$

$$= \frac{\overset{1}{5}}{2} \cdot \frac{\overset{2}{4}}{\underset{3}{15}} = \frac{2}{3} \quad \leftarrow \text{Ratio in lowest terms}$$

5.1 Using Measurements in Ratios

When a ratio compares measurements, both measurements must be in the *same* units. It is usually easier to compare the measurements using the smaller unit, for example, inches instead of feet.

Write as a ratio in lowest terms.

8 in. to 6 ft

Compare using the smaller unit, inches. Because 1 ft has 12 in., 6 ft is

$$6 \cdot 12 \text{ in.} = 72 \text{ in.}$$

The ratio is shown below.

$$\frac{8 \text{ in.}}{72 \text{ in.}} = \frac{8 \div 8}{72 \div 8} = \frac{1}{9}$$

Divide out common units.

5.2 Writing Rates

A rate compares two measurements with different units. The units do *not* divide out, so you must write them as part of the rate.

Write the rate as a fraction in lowest terms.

475 miles in 10 hours

$$\frac{475 \text{ miles} \div 5}{10 \text{ hours} \div 5} = \frac{95 \text{ miles}}{2 \text{ hours}} \quad \begin{array}{l} \leftarrow \text{Must write units:} \\ \leftarrow \text{miles and hours} \end{array}$$

5.2 Finding a Unit Rate

A unit rate has 1 in the denominator. To find the unit rate, divide the numerator by the denominator. Write unit rates using the word **per** or a **/** mark.

Write as a unit rate: $1278 in 9 days.

$$\frac{\$1278}{9 \text{ days}} \quad \leftarrow \text{The fraction bar indicates division.}$$

$$9\overline{)1278}^{\,142} \quad \text{so} \quad \frac{\$1278 \div 9}{9 \text{ days} \div 9} = \frac{\$142}{1 \text{ day}}$$

Write the answer as $142 **per** day or $142/day.

Concepts	**Examples**

5.2 Finding the Best Buy

The best buy is the item with the lowest cost per unit. Divide the price by the number of units. Round to thousandths, if necessary. Then compare to find the lowest cost per unit.

Find the best buy on grapes. You have a coupon for 50¢ off on 2 pounds or 75¢ off on 3 pounds.

$$2 \text{ pounds for } \$2.75$$

$$3 \text{ pounds for } \$4.15$$

Find the cost per unit (cost per pound) after subtracting the coupon.

$$2 \text{ pounds cost } \$2.75 - \$0.50 = \$2.25$$

$$\frac{\$2.25}{2} = \$1.125 \text{ per pound}$$

$$3 \text{ pounds cost } \$4.15 - \$0.75 = \$3.40$$

$$\frac{\$3.40}{3} \approx \$1.133 \text{ per pound}$$

The lower cost per pound is $1.125, so 2 pounds of grapes is the best buy.

5.3 Writing Proportions

A proportion states that two ratios or rates are equal. The proportion "5 is to 6 as 25 is to 30" is written as shown below.

$$\frac{5}{6} = \frac{25}{30}$$

To see whether a proportion is true or false, multiply along one diagonal, then multiply along the other diagonal. If the two cross products are equal, the proportion is true. If the two cross products are unequal, the proportion is false.

Write as a proportion: 8 is to 40 as 32 is to 160

$$\frac{8}{40} = \frac{32}{160}$$

Is this proportion true or false?

$$\frac{6}{8\frac{1}{2}} = \frac{24}{34}$$

Find the cross products.

$$\frac{6}{8\frac{1}{2}} = \frac{24}{34}$$

$$8\frac{1}{2} \cdot 24 = \frac{17}{2} \cdot \frac{\overset{12}{24}}{\underset{1}{1}} = \mathbf{204}$$

$$6 \cdot 34 = \mathbf{204} \longleftarrow \text{Equal}$$

The cross products are equal, so the proportion is true.

5.4 Solving Proportions

Solve for an unknown number in a proportion by using the steps shown on the next page.

Find the unknown number.

$$\frac{12}{x} = \frac{6}{8}$$

Write $\frac{6}{8}$ in lowest terms as $\frac{3}{4}$.

$$\frac{12}{x} = \frac{3}{4}$$

(Continued)

Concepts	Examples

5.4 Solving Proportions (*Continued*)

Step 1 Find the cross products. (If desired, you can rewrite the ratios in lowest terms before finding the cross products.)

Step 2 Show that the cross products are equal.

Step 3 Divide both products by the number multiplied by x (the number next to x).

Step 4 Check by writing the solution in the *original* proportion and finding the cross products.

Step 1
$$\frac{12}{x} = \frac{3}{4}$$
$x \cdot 3$ — Find cross products.
$12 \cdot 4$

Step 2
$x \cdot 3 = 12 \cdot 4$ — Show that cross products are equal.
$x \cdot 3 = 48$

Step 3
$$\frac{x \cdot \overset{1}{3}}{\underset{1}{3}} = \frac{48}{3}$$ — Divide both sides by 3.
$x = 16$ The solution is 16.

Step 4

x is 16. → $\dfrac{12}{16} = \dfrac{6}{8}$

$16 \cdot 6 = 96$
$12 \cdot 8 = 96$ } Equal

The cross products are equal, so **16 is the correct solution** (**not** 96).

5.5 Solving Application Problems with Proportions

Decide what is being compared. Set up and solve a proportion using the two rates described in the problem. Be sure that *both* rates compare things in the *same order*. Use a letter, like x, to represent the unknown number.

Use the six problem-solving steps.

Step 1 **Read** the problem carefully.

Step 2 **Work out a plan.**

Step 3 **Estimate** a reasonable answer.

Step 4 **Solve** the problem.

If 3 pounds of grass seed cover 450 square feet of lawn, how much seed is needed for 1500 square feet of lawn?

Step 1 The problem asks for the pounds of grass seed needed for 1500 square feet of lawn.

Step 2 Pounds of seed is compared to square feet of lawn. Set up and solve a proportion using the two given rates. Be sure that pounds of seed is in both numerators and square feet of lawn is in both denominators.

Step 3 Because 1500 square feet is about three times as much lawn as 450 square feet, about three times as much seed is needed. So, $3 \cdot 3$ pounds $= 9$ pounds as our estimate.

Step 4 With the proportion set up correctly, solve for the unknown number.

Matching units

$$\frac{3 \textbf{ pounds}}{450 \textbf{ square feet}} = \frac{x \textbf{ pounds}}{1500 \textbf{ square feet}}$$

Matching units

Concepts	Examples
5.5 **Solving Application Problems with Proportions** (*Continued*)	Both sides compare pounds to square feet. Ignore the units while finding the cross products and solving for x.

Both sides compare pounds to square feet. Ignore the units while finding the cross products and solving for x.

$$450 \cdot x = \underbrace{3 \cdot 1500}$$ Show that cross products are equal.

$$450 \cdot x = 4500$$

$$\frac{\overset{1}{\cancel{450}} \cdot x}{\underset{1}{\cancel{450}}} = \frac{4500}{450}$$ Divide both sides by 450.

$$x = 10$$

Step 5 State the answer.

Step 5 **10 pounds** of grass seed are needed.

Step 6 **Check** your work.

Step 6 The exact answer, 10 pounds of seed, is close to our estimate of 9 pounds, so it is reasonable.

Chapter 5   Review Exercises

5.1   *Write each ratio as a fraction in lowest terms. Change to the same units when necessary, using the table of measurement comparisons in the first section of this chapter. Use the information in the graph to answer Exercises 1–3.*

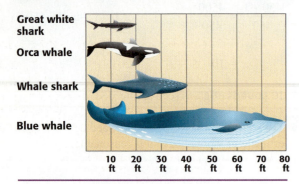

AVERAGE LENGTH OF SHARKS AND WHALES

Great white shark

Orca whale

Whale shark

Blue whale

10 ft   20 ft   30 ft   40 ft   50 ft   60 ft   70 ft   80 ft

Source: Grolier Multimedia Encyclopedia.

1. Ratio of orca whale's length to whale shark's length

2. Ratio of blue whale's length to great white shark's length

3. Which two animals' lengths give a ratio of $\frac{1}{2}$? There are several answers.

4. $2.50 to $1.25

5. $0.30 to $0.45

6. $1\frac{2}{3}$ cups to $\frac{2}{3}$ cup

7. $2\frac{3}{4}$ miles to $16\frac{1}{2}$ miles

8. 5 hours to 100 minutes

9. 9 in. to 2 ft

10. 1 ton to 1500 pounds

11. 8 hours to 3 days

12. Jake sold $350 worth of his kachina figures. Ramona sold $500 worth of her pottery. What is the ratio of Ramona's sales to Jake's sales?

13. Ms. Wei's new car gets 35 miles per gallon. Her old car got 25 miles per gallon. Find the ratio of the new car's mileage to the old car's mileage.

14. This fall, 6000 students are taking math courses and 7200 students are taking English courses. Find the ratio of math students to English students.

5.2 *Write each rate as a fraction in lowest terms.*

15. $88 for 8 dozen

16. 96 children in 40 families

17. When entering data into his computer, Patrick can type four pages in 20 minutes. Give his rate in pages per minute and minutes per page.

18. Elena made $60 in three hours. Give her earnings in dollars per hour and hours per dollar.

Find the best buy.

19. Minced onion

8 ounces for $4.98

3 ounces for $2.49

2 ounces for $1.89

20. Dog food; you have a coupon for $1 off on 8 pounds or more.

35.2 pounds for $36.96

17.6 pounds for $18.69

3.5 pounds for $4.25

5.3 *Use either the method of writing in lowest terms or the method of finding cross products to decide whether each proportion is true or false. Show your work and then write* true *or* false.

21. $\dfrac{6}{10} = \dfrac{9}{15}$

22. $\dfrac{6}{48} = \dfrac{9}{36}$

23. $\dfrac{47}{10} = \dfrac{98}{20}$

24. $\dfrac{64}{36} = \dfrac{96}{54}$

25. $\dfrac{1.5}{2.4} = \dfrac{2}{3.2}$

26. $\dfrac{3\frac{1}{2}}{2\frac{1}{3}} = \dfrac{6}{4}$

5.4 *Find the unknown number in each proportion. Round answers to the nearest hundredth, if necessary.*

27. $\dfrac{4}{42} = \dfrac{150}{x}$

28. $\dfrac{16}{x} = \dfrac{12}{15}$

29. $\dfrac{100}{14} = \dfrac{x}{56}$

30. $\dfrac{5}{8} = \dfrac{x}{20}$

31. $\dfrac{x}{24} = \dfrac{11}{18}$

32. $\dfrac{7}{x} = \dfrac{18}{21}$

33. $\dfrac{x}{3.6} = \dfrac{9.8}{0.7}$

34. $\dfrac{13.5}{1.7} = \dfrac{4.5}{x}$

35. $\dfrac{0.82}{1.89} = \dfrac{x}{5.7}$

5.5 *Set up and solve a proportion for each application problem.*

36. The ratio of cats to dogs at the animal shelter is 3 to 5. If there are 45 dogs, how many cats are there?

37. Danielle had 8 hits in 28 times at bat during last week's games. If she continues to hit at the same rate, how many hits will she gets in 161 times at bat?

38. If 3.5 pounds of ground beef cost $9.77, what will 5.6 pounds cost? Round to the nearest cent.

39. About 4 out of 10 students are expected to vote in campus elections. There are 8247 students. How many are expected to vote? Round to the nearest whole number.

40. The scale on Brian's model railroad is 1 in. to 16 ft. One of the scale model boxcars is 4.25 in. long. What is the length of a real boxcar in feet?

41. Marvette makes necklaces to sell at a local gift shop. She made 2 dozen necklaces in $16\frac{1}{2}$ hours. How long will it take her to make 40 necklaces?

42. A 180-pound person burns 284 calories playing basketball for 25 minutes. At this rate, how many calories would the person burn in 45 minutes, to the nearest whole number? (*Source: Wellness Encyclopedia.*)

43. In the hospital pharmacy, Michiko sees that a medicine is to be given at the rate of 3.5 milligrams for every 50 pounds of body weight. How much medicine should be given to a patient who weighs 210 pounds?

Mixed Review Exercises

Find the unknown number in each proportion. Round answers to the nearest hundredth, if necessary.

44. $\dfrac{x}{45} = \dfrac{70}{30}$

45. $\dfrac{x}{52} = \dfrac{0}{20}$

46. $\dfrac{64}{10} = \dfrac{x}{20}$

47. $\dfrac{15}{x} = \dfrac{65}{100}$

48. $\dfrac{7.8}{3.9} = \dfrac{13}{x}$

49. $\dfrac{34.1}{x} = \dfrac{0.77}{2.65}$

Find cross products to decide whether each proportion is true or false. Show the cross products and then circle true *or* false.

50. $\dfrac{55}{18} = \dfrac{80}{27}$

　　　True　　　False

51. $\dfrac{5.6}{0.6} = \dfrac{18}{1.94}$

　　　True　　　False

52. $\dfrac{\frac{1}{5}}{2} = \dfrac{1\frac{1}{6}}{11\frac{2}{3}}$

　　　True　　　False

Write each ratio as a fraction in lowest terms. Change to the same units when necessary.

53. 4 dollars to 10 quarters

54. $4\frac{1}{8}$ in. to 10 in.

55. 10 yd to 8 ft

56. $3.60 to $0.90

57. 12 eggs to 15 eggs

58. 37 meters to 7 meters

59. 3 pints to 4 quarts

60. 15 minutes to 3 hours

61. $4\frac{1}{2}$ miles to $1\frac{3}{10}$ miles

62. Nearly 7 out of 8 fans buy something to drink at rock concerts. How many of the 28,500 fans at today's concert would be expected to buy a beverage? Round to the nearest hundred fans.

63. Emily spent $150 on car repairs and $400 on car insurance. What is the ratio of the amount spent on insurance to the amount spent on repairs?

64. Antonio is choosing among three packages of plastic wrap. Is the best buy 25 ft for $0.78; 75 ft for $1.99; or 100 ft for $2.59? He has a coupon for 50¢ off either of the larger two packages.

65. On this scale drawing of a backyard patio, 0.5 in. represents 6 ft. If the patio measures 1.75 in. long and 1.25 in. wide on the drawing, what will be the actual length and width of the patio when it is built?

0.5 in. = 6 ft

66. A lawn mower uses 0.8 gallon of gas every 3 hours. How long can the mower run on 2 gallons of gas?

67. An antibiotic is given at the rate of $1\frac{1}{2}$ teaspoons for every 24 pounds of body weight. How much should be given to an infant who weighs 8 pounds?

68. Charles made 251 points during 169 minutes of playing time last year. At that same rate, how many points would you expect him to make if he plays 14 minutes in tonight's game? Round to the nearest whole number.

69. Refer to **Exercise 67.** Explain each step you took in solving the problem. Be sure to tell how you decided which way to set up the proportion and how you checked your answer.

70. A vitamin supplement for cats is given at the rate of 1000 milligrams for a 5-pound cat. (*Source:* St. Jon Pet Care Products.)

(a) How much should be given to a 7-pound cat?

(b) How much should be given to an 8-ounce kitten?

Chapter 5 *Test* CHAPTER **Test Prep** VIDEO | The Chapter Test Prep Videos with test solutions are available on DVD, in MyMathLab, and on YouTube —search "LialBasicCollegeMath" and click on "Channels."

Write each rate or ratio as a fraction in lowest terms. Change to the same units when necessary.

1. 16 fish to 20 fish

2. 300 miles on 15 gallons

3. $15 for 75 minutes

4. 3 hours to 40 minutes

5. The little theater at our college has 320 seats. The auditorium has 1200 seats. Find the ratio of auditorium seats to theater seats.

6. Use the information in the table about Quizno's honey mustard chicken sub sandwich to find the best buy.

Size	Length of Sub	Price
Small	5 inches	$ 5.99
Regular	8 inches	$ 6.89
Large	11 inches	$10

Source: Quizno's

7. Find the best buy on spaghetti sauce. You have a coupon for 75¢ off Brand X and a coupon for 50¢ off Brand Y.

26 ounces of Brand X for $3.89

16 ounces of Brand Y for $1.89

14 ounces of Brand Z for $1.29

8. Suppose the ratio of your income last year to your income this year is 3 to 2. Explain what this means. Give an example of the dollars earned last year and this year that fits the 3 to 2 ratio.

Determine whether each proportion is true or false. Show your work and then write true *or* false.

9. $\dfrac{6}{14} = \dfrac{18}{45}$

10. $\dfrac{8.4}{2.8} = \dfrac{2.1}{0.7}$

Find the unknown number in each proportion. In Problems 11–13, round the answers to the nearest hundredth, if necessary.

11. $\dfrac{5}{9} = \dfrac{x}{45}$

12. $\dfrac{3}{1} = \dfrac{8}{x}$

13. $\dfrac{x}{20} = \dfrac{6.5}{0.4}$

14. $\dfrac{2\frac{1}{3}}{x} = \dfrac{\frac{8}{9}}{4}$

Set up and solve a proportion for each application problem.

15. Pedro entered 18 orders into his computer in thirty minutes at his job. At that rate, how many orders could he enter in forty minutes?

16. Just 0.8 ounce of wildflower seeds is enough for 50 square feet of ground. What weight of seeds is needed for a garden with 225 square feet? (*Source: White Swan Ltd.*)

17. About 2 out of every 15 people are left-handed. How many of the 650 students in our school would you expect to be left-handed? Round to the nearest whole number.

18. A student set up the proportion for **Problem 17** this way and arrived at an answer of 4875.

$$\frac{2}{15} = \frac{650}{x} \qquad \text{CHECK} \quad \frac{2}{15} \times \frac{650}{4875} \qquad \begin{array}{l} 15 \cdot 650 \\ = 9750 \\[6pt] 2 \cdot 4875 \\ = 9750 \end{array}$$

Because the cross products are equal, the student said the answer is correct. Is the student right? Explain why or why not.

19. A medication is given at the rate of 8.2 grams for every 50 pounds of body weight. How much should be given to a 145-pound person? Round to the nearest tenth of a gram.

20. On a scale model, 1 in. represents 8 ft. If a building in the model is 7.5 in. tall, what is the actual height of the building in feet?

Chapters 1–5 Cumulative Review Exercises

Round each number as indicated.

1. 9903 to the nearest hundred

2. 617.0519 to the nearest tenth

3. $99.81 to the nearest dollar

4. $3.0555 to the nearest cent

Simplify.

5. $30 - 0.66$

6. Write the answer using R for the remainder. $33\overline{)20{,}157}$

7. $(1.9)(0.004)$

8. $3020 - 708$

9. $0.401 + 62.98 + 5$

10. $1.39 \div 0.025$

11. $36 + 18 \div 6$

12. $8 \div 4 + (10 - 3^2) \cdot 4^2$

13. $88 \div \sqrt{121} \cdot 2^3$

14. $(16.2 - 5.85) - 2.35(4)$

15. Write 0.0105 in words.

16. Write sixty and seventy-one thousandths in numbers.

In a survey of 1000 adults, 550 drank coffee daily, 250 drank coffee occasionally, and 200 never drank coffee. Use this information for Exercises 17–18. Write each ratio as a fraction in lowest terms.

17. What is the ratio of those who do not drink coffee to the total number of people in the survey?

18. Find the ratio of all the adults who drink coffee to those who never drink it.

Nest boxes for birds are made with different sizes of entry holes. Then, only certain types of birds can use the nest box and it helps prevent entry by predators. Use the information in the table to answer Exercises 19–20. The diameter is the distance across the opening at its widest point.

Eastern bluebird nest box with 1.5-inch opening

Bird	Diameter of Opening
Eastern bluebird	1.5 inches
Western bluebird	$1\frac{9}{16}$ inches
Chickadee	1.25 inches
Swallow	$1\frac{3}{8}$ inches
Wren	$1\frac{1}{8}$ inches

Source: Duncraft.

19. List the nest box openings in order from smallest to largest.

20. What is the difference in diameter between the openings for an eastern bluebird and a western bluebird? Write your answer as a fraction and as a decimal.

Find the unknown number in each proportion. Round your answers to the nearest hundredth, when necessary.

21. $\dfrac{9}{12} = \dfrac{x}{28}$

22. $\dfrac{7}{12} = \dfrac{10}{x}$

23. $\dfrac{6.7}{x} = \dfrac{62.8}{9.15}$

Solve each application problem.

24. The college honor society has a goal of collecting 1500 pounds of food to fill Thanksgiving baskets. So far it has collected $\frac{5}{6}$ of the goal. How many more pounds are needed?

25. Tara has a photo that is 10 centimeters wide by 15 centimeters long. If the photo is enlarged to a length of 40 centimeters, find the new width, to the nearest tenth.

26. The distance around Dunning Pond is $1\frac{1}{10}$ miles. Norma ran around the pond four times in the morning and $2\frac{1}{2}$ times in the afternoon. How far did she run in all?

27. Find the best buy on instant mashed potatoes. You have a coupon for 50¢ off either the 34-serving or 65-serving box.

A box that makes 20 servings for $2.15

A box that makes 34 servings for $3.35

A box that makes 65 servings for $10.41

28. In a survey, 5 out of 8 apartment residents said they are sometimes bothered by noise from their neighbors. How many of the 224 residents at Harris Towers would you expect to be bothered by noise?

29. The directions on a bottle of plant food call for $\frac{1}{2}$ teaspoon in two quarts of water. How much plant food is needed for five quarts? (*Source:* Schultz Company.)

Use the information in the table on international long-distance calling card rates to answer Exercises 30–34. Round answers to the nearest whole minute when necessary.

$20 International Phone Cards No Connection Fee!	
Place a call to	**Cost per minute**
Mexico	$0.02
Philippines	$0.084
Japan (to a cell phone)	$0.099
Hong Kong	$0.018
Canada	$0.10
India (to a cell phone)	$0.106
Afghanistan	$0.134

Source: noblecom.com.

30. List the per-minute rates from least to greatest.

31. If you buy a $20 calling card, how long a call could you make to a cell phone in Japan?

32. How many $20 cards would you have to buy to make six hours' worth of calls to a cell phone in India?

33. What is the ratio of minutes you can call Mexico for $20 to minutes you can call Canada for $20?

34. You have already made a 95-minute call to the Philippines. With the amount left on one $20 phone card, how long a call could you make to Afghanistan?

Math in the Media

FEEDING HUMMINGBIRDS

After getting a hummingbird feeder, the next step is to fill it! You have two choices at this point: you can either buy one of the commercial mixtures or you can make your own solution. See the recipe at the right.

The concentration of the sugar is important. The 1 to 4 ratio of sugar to water is recommended because it approximates the ratio of sugar to water found in the nectar of many hummingbird flowers.

Boiling the solution helps slow down fermentation. Sugar-and-water solutions are subject to rapid spoiling, especially in hot weather.

Source: The Hummingbird Book.

> **Recipe for Homemade Mixture:**
> 1 part sugar (not honey)
> 4 parts water
> **Boil for 1 to 2 minutes. Cool.**
> **Store extra in refrigerator.**

A recipe can be used to make as much of a mixture as you need as long as the ingredients are kept proportional. Use the recipe for a homemade mixture of sugar water for hummingbird feeders to answer these problems.

1. What is the ratio of sugar to water in the recipe? What is the ratio of water to sugar?

2. Complete each table.

Sugar	Water
1 cup	4 cups
	5 cups
	6 cups
	7 cups
2 cups	8 cups

Sugar	Water
1 cup	4 cups
	3 cups
	2 cups
	1 cup

3. How much water would you need if you used

   (a) 3 cups of sugar?

   (b) 4 cups of sugar?

   (c) $\frac{1}{3}$ cup of sugar?

4. As you change the amounts of water and sugar, should you change the length of time that you boil the mixture? Explain your answer.

6 Percent

In this chapter, you will look at the many ways percent is used in your daily life. For example, sales tax, commission rates, interest rates on savings and investments, automobile loans, home loans, and other installment loans are almost always given as percents.

6.1 Basics of Percent

OBJECTIVES

1. Learn the meaning of percent.
2. Write percents as decimals.
3. Write decimals as percents.
4. Understand 100%, 200%, and 300%.
5. Use 50%, 10%, and 1%.

Notice that the figure below has one hundred squares of equal size. Eleven of the squares are shaded. The shaded portion is $\frac{11}{100}$, or 0.11, of the total figure.

Shaded portion is 11 out of 100 parts, or $\frac{11}{100}$, or 0.11 or 11%.

The shaded portion is also 11% of the total, or "eleven parts out of 100 parts." Read **11%** as "eleven percent."

OBJECTIVE 1 Learn the meaning of percent. As we just saw, a percent is a ratio with a denominator of 100.

The Meaning of Percent

Percent means *per one hundred*. The "%" symbol is used to show the number of parts out of one hundred parts.

EXAMPLE 1 Understanding Percent

(a) If *43 out of 100* students are men, then *43 per 100* or $\frac{43}{100}$ or **43%** of the students are men.

(b) If a person pays a tax of $7 on every $100 of purchases, then the tax rate is $7 per $100. The ratio is $\frac{7}{100}$ and the percent of tax is **7%**.

◀ Work Problem **1** at the Side.

OBJECTIVE 2 Write percents as decimals. If 8% means 8 parts out of 100 parts or $\frac{8}{100}$, then *p*% means *p* parts out of 100 parts or $\frac{p}{100}$. Because $\frac{p}{100}$ is another way to write the division $p \div 100$, we have

$$p\% = \frac{p}{100} = p \div 100.$$

Writing a Percent as a Decimal

$$p\% = \frac{p}{100} \qquad \text{or} \qquad p\% = p \div 100$$

As a fraction   As a decimal

EXAMPLE 2 Writing Percents as Decimals

Write each percent as a decimal.

(a) 47%

$$p\% = p \div 100$$

0.47 is $\frac{47}{100}$, which is equivalent to 47%.

$$47\% = 47 \div 100 = 0.47 \leftarrow \text{Decimal form}$$

Continued on Next Page

VOCABULARY TIP

Percent means "per 100" or "out of 100," so any **percent** can be converted to a decimal by dividing the percent by 100. For example, 30% = $\frac{30}{100}$ = 0.30.

1 Write as percents.

(a) In a group of 100 adults, 74 keep fit by walking. What percent are walking?

(b) The sales tax is $6 per $100. What percent is this?

(c) Out of 100 Americans, 32 picked football as their favorite sport. What percent picked football?

Answers

1. (a) 74% (b) 6% (c) 32%

(b) 76%   76% = 76 ÷ 100 = 0.76 ← Decimal form

(c) 28.2%   28.2% = 28.2 ÷ 100 = 0.282 ← Decimal form

(d) 100%   100% = 100 ÷ 100 = 1.00 ← Decimal form

> **CAUTION**
>
> In **Example 2(d)** above, notice that 100% is **1.00**, or **1**, which is a whole number. Whenever you have a percent that is *100% or greater*, the equivalent decimal number will be *1 or greater than 1*.

···· **Work Problem ❷ at the Side.** ▶

The answers in **Example 2** above suggest these steps for writing a percent as a decimal.

> **Writing a Percent as a Decimal**
>
> **Step 1** Drop the percent symbol.
>
> **Step 2** Divide by 100.

> **Note**
>
> Recall from **Chapter 4** that a quick way to divide a number by 100 is to move the decimal point **two places to the left.**

EXAMPLE 3   **Writing Percents as Decimals by Moving the Decimal Point**

Write each percent as a decimal by moving the decimal point two places to the left.

(a) 17%

17% = 17.%   Decimal point starts at far right side.

0.17 ← Percent symbol is dropped. *(Step 1)*

Decimal point is moved two places to the left. *(Step 2)*

17% = 0.17

(b) 160%

> 1.60 is equivalent to 1.6 because $1\frac{60}{100}$ simplifies to $1\frac{6}{10}$.

160% = 160.% = 1.60 or 1.6   Decimal point starts at far right side.

(c) 4.9%

> 0 is attached so the decimal point can be moved two places to the left.

.049

4.9% = 0.049

···· **Continued on Next Page**

3 Write each percent as a decimal.

GS (a) 96%

$96\% = 96.\%$ Decimal point starts here.

$= .96\%$ Move decimal point two places to the left.

$= \underline{\hspace{1cm}}$

(b) 6%

(c) 24.8%

GS (d) 0.9%

$0.9\% = 0.\underline{\hspace{0.3cm}}\underline{\hspace{0.3cm}}9$

(d) 0.6%

$0.6\% = 0.006$

> Two zeros are attached so the decimal point can be moved two places to the left.

CAUTION

Look at **Example 3(d)** above, where 0.6% is less than 1%. Because 0.6% is $\frac{6}{10}$ of 1%, it is *less than 1%*. Any fraction of a percent is *less than 1%*.

◀ **Work Problem 3 at the Side.**

OBJECTIVE ▶ 3 Write decimals as percents. You can write any decimal as a percent. For example, the decimal 0.78 is the same as the fraction

$$\frac{78}{100}.$$

This fraction means 78 out of 100 parts, or 78%. The following steps give the same result.

Writing a Decimal as a Percent

Step 1   Multiply by 100.

Step 2   Attach a percent symbol.

Note

A quick way to divide or multiply a number by 100 is to move the decimal point two places to the left or two places to the right, respectively.

Multiply decimal by 100; move decimal point two places to the *right*.

Decimal   **Percent**

Divide percent by 100; move decimal point two places to the *left*.

EXAMPLE 4   **Writing Decimals as Percents by Moving the Decimal Point**

Write each decimal as a percent by moving the decimal point two places to the right.

(a) 0.21

0.21

↑ — Decimal point is moved two places to the right. *(Step 1)*

$0.21 = 21\%$ ← Percent symbol is attached. *(Step 2)*

> Remember to attach the percent (%) symbol.

↑ — Decimal point is not written with whole number percents.

Answers

3. (a) 0.96   **(b)** 0.06   **(c)** 0.248
    (d) 0; 0; 0.009

Continued on Next Page

(b) $0.529 = 52.9\%$

> Move the decimal point two places to the right. Then attach a % symbol.

(c) $1.92 = 192\%$

(d) 2.5

$2.5\underset{\smile}{0}$ 0 is attached so the decimal point can be moved two places to the right.

$2.5 = 250\%$ ← Attach % symbol.

> When necessary, attach zeros so you can move the decimal point.

(e) 3

$3. = 3.\underset{\smile}{00}$ Two zeros are attached so the decimal point can be moved two places to the right.

so $3 = 300\%$ ← Attach % symbol.

CAUTION

Look at **Examples 4(c), 4(d), and 4(e)** above, where 1.92, 2.5, and 3 are greater than 1. Because the number 1 is equivalent to 100%, all numbers greater than 1 will be *greater than 100%*.

·········· **Work Problem 4 at the Side.** ▶

OBJECTIVE ▶ 4 Understand 100%, 200%, and 300%. When working with percents, it is helpful to have several reference points. 100%, 200%, and 300% are three such helpful reference points.

　　100% means 100 parts out of 100 parts. That's *all* of the parts. If 100% of the 18 people attending last week's meeting attended this week's meeting, then 18 people (*all* of them) attended this week.

　　If attendance at the meeting this week is 200% of last week's attendance of 18 people, then this week's attendance is 36 people, or *two* times as many people (2 • 18 = 36). Likewise, if attendance is 300% of last week's attendance, then *three* times as many people, or 54 people, attended (3 • 18 = 54).

EXAMPLE 5 Finding 100%, 200%, and 300% of a Number

Answer the following.

> Notice that 100% of something is all of it (the whole thing).

(a) What is 100% of 82 people?

　　100% is *all* of the people. So, 100% of 82 people is **82 people**.

(b) What is 200% of $63?

　　200% is twice (2 times) as much money.

　　So, 200% of $63 is 2 • $63 = **$126**.

(c) What is 300% of 32 employees?

　　300% is 3 times as many employees.

　　So, 300% of 32 employees is 3 • 32 = **96 employees**.

·········· **Work Problem 5 at the Side.** ▶

---

4 Write each decimal as a percent.

(a) 0.74　　**(b)** 0.15

(c) 0.09

GS **(d)** 0.617

$0.\underset{\smile}{\_\,\_\,\_}$

So, $0.617 = \_\_\_\_ \%$.

(e) 0.834　**(f)** 5.34

(g) 2.8　　**(h)** 4

5 Answer the following.

(a) What is 100% of $7.80?

(b) What is 100% of 1850 workers?

GS **(c)** What is 200% of 24 photographs?

200% is twice (____ times).

So, 2 • 24 = _____.

(d) What is 300% of 8 miles?

Answers

4. **(a)** 74%　**(b)** 15%　**(c)** 9%
　(d) 6; 1; 7; 61.7%　**(e)** 83.4%　**(f)** 534%
　(g) 280%　**(h)** 400%
5. **(a)** $7.80　**(b)** 1850 workers
　(c) 2; 48 photographs　**(d)** 24 miles

6 Answer the following.

GS **(a)** What is 50% of 200 patients?

50% is half of the patients.

$\frac{1}{2}$ of 200 patients is _____ .

(b) What is 50% of 64 tweets?

GS **(c)** What is 10% of 3850 elm trees?

10% is $\frac{1}{10}$ of the trees. Move the decimal point ____ place to the ____ .

10% of 3850. is _____ .

(d) What is 10% of 7 pounds?

GS **(e)** What is 1% of 240 ft?

1% is $\frac{1}{100}$ of the length.

Move the decimal point ____ places to the ____ .

1% of 2 40. is _____ .

(f) What is 1% of $3000?

OBJECTIVE **5** **Use 50%, 10%, and 1%.** 50% means 50 parts out of 100 parts, which is *half* of the parts ($\frac{50}{100} = \frac{1}{2}$). So, 50% of $18 is $9 (*half* of the money).

When using 10%, we have 10 parts out of 100 parts, which is $\frac{1}{10}$ of the parts ($\frac{10}{100} = \frac{1}{10}$). To find 10% or $\frac{1}{10}$ of a number, we move the decimal point **one** place to the left. 10% of $285 is $28.50 (because $28.5. = $28.50).

To find 1% of a number ($\frac{1}{100}$), we move the decimal point **two** places to the left. 1% of $198 is $1.98 (because $1.98. = $1.98).

EXAMPLE 6 **Finding 50%, 10%, and 1% of a Number**

Answer the following.

(a) What is 50% of 24 hours? | Think: 50% of something is $\frac{50}{100}$ or $\frac{1}{2}$ of it.

50% is half of the hours.

So, 50% of 24 hours is **12 hours**.

(b) What is 10% of 280 pages?

10% is $\frac{1}{10}$ of the pages. Move the decimal point *one* place to the left.

So, 10% of 2 8 0. pages is **28 pages**.

(c) What is 1% of $540?

1% is $\frac{1}{100}$ of the money. Move the decimal point *two* places to the left.

So, 1% of $5 4 0. is **$5.40**.

◀ **Work Problem** **6** **at the Side.**

Note

Two other frequently used percents are 25% and 75%.

25% means 25 parts out of 100 parts, which is $\frac{1}{4}$ of the parts ($\frac{25}{100} = \frac{1}{4}$). For example, 25% of $32 is $\frac{1}{4}$ of the money, or $8.

75% means 75 parts out of 100 parts, which is $\frac{3}{4}$ of the parts ($\frac{75}{100} = \frac{3}{4}$). For example, 75% of $32 is $\frac{3}{4}$ of the money, or $24.

Answers

6. **(a)** 100 patients **(b)** 32 tweets
(c) one; left; 385 elm trees **(d)** 0.7 pound
(e) two; left; 2.4 ft **(f)** $30

CONCEPT CHECK *Fill in each blank with the correct response.*

1. To write a percent as a decimal, drop the
_____ symbol and then divide by
_____.

2. A quick way to divide a number by 100 is to move the
_____ two places to the _____.

Write each percent as a decimal. **See Examples 2 and 3.**

3. 12%
12.% = __.12

4. 57%
57.% = 0.__ __

5. 70%

6. 40%

7. 25%

8. 35%

9. 140%

10. 250%

11. 5.5%

12. 6.7%

13. 100%

14. 600%

15. 0.5%

16. 0.25%

17. 0.35%

18. 0.75%

CONCEPT CHECK *Fill in each blank with the correct response.*

19. To write a decimal as a percent, multiply by
_____ and then attach a _____ symbol.

20. A quick way to multiply a number by 100 is to move
the _____ two places to the _____.

Write each decimal as a percent. **See Example 4.**

21. 0.6
0.6 = 0.60
So, 0.6 = 60%

22. 0.9
0.9 = 0.__ __
So, 0.9 = __ __%

23. 0.01

24. 0.07

25. 0.375

26. 0.625

27. 2

28. 5

29. 3.7

30. 2.2

31. 0.0312

32. 0.0625

33. 4.162

34. 8.715

35. 0.0028

36. 0.0064

37. Fractions, decimals, and percents are all used to describe a part of something. The use of percents is much more common than that of fractions and decimals. Why do you suppose this is true?

38. List five uses of percent that are or will be part of your life. Consider the activities of working, shopping, saving, and planning for the future.

Over the next 10 years, the projected growth in the workforce for several occupations is shown. Write each percent as a decimal and each decimal as a percent. See Examples 2–4. (Source: Bloomberg BusinessWeek.)

39. Truck drivers 13%

40. Registered nurses 22.2%

41. Physicians and surgeons 0.218

42. Waiters and waitresses 0.064

43. Postsecondary teachers 15.1%

44. Security guards 14.2%

45. Customer service representatives 0.177

46. Carpenters 0.129

47. Cooks 14.6%

48. Management analysts 23.9%

Write each percent as a decimal and each decimal as a percent. See Examples 2–4.

49. Only 0.08 of the total population has the professional training.

50. The church building fund has 0.8 of the money needed.

51. The patient's blood pressure was 30% above normal.

52. Success with the diet was 170% greater than anticipated.

CONCEPT CHECK   *Fill in the blanks. Remember that 100% is all of something, 200% is two times as many, and 300% is three times as many. See Example 5.*

53. The Saturday morning tae kwon do class has 12 children enrolled. If 100% of the children are present, how many children are there? _____

54. When 500 adults were asked, "Do you think your taxes are too high," 100% said yes. How many said yes? _____

55. Last year we had 210 employees. This year we have 200% of that number. How many employees do we have this year? _____

56. Last season Active Sports sold 380 fishing licenses. This season they sold 200% of that number. How many fishing licenses did they sell this season?

_____

57. Last week 90 chairs were used for the meeting. This week we need 300% of that number of chairs. We'll need _____.

57. One month ago she had 28 friends on Facebook. Now she has 300% of that number. The new number of friends is _____.

Fill in the blanks. Remember that 50% is half of something, 10% is found by moving the decimal point one place to the left, and 1% is found by moving the decimal point two places to the left. **See Example 6.**

59. Jacob owes $755 for tuition. Financial aid will pay 50% of the cost. Financial aid will pay

_____.

60. Linda Redding needs $2320 for school and living expenses this semester. She applied for a student loan to cover 50% of this amount. The amount of the loan is _____.

61. Only 10% of 8200 commuters are carpooling to work. How many commuters carpool? _____

Move the decimal point one place to the left. 8200.

62. Sarah Bryn expects that 10% of the 240 dozen plants in her greenhouse will not be sold. The expected number of unsold plants is _____.

63. The naturalist said that 1% of the 2600 plants in the park are poisonous. How many plants are poisonous?

_____

64. Of the 4800 accidents, only 1% were caused by mechanical failure. How many accidents were caused by mechanical failure? _____

Move the decimal point two places to the left. 4800.

65. (a) Describe a shortcut method of finding 100% of a number.

(b) Show an example using your shortcut.

66. (a) Describe a shortcut method of finding 50% of a number.

(b) Show an example using your shortcut.

67. (a) Describe a shortcut method of finding 200% of a number.

(b) Show an example using your shortcut.

68. (a) Describe a shortcut method of finding 300% of a number.

(b) Show an example using your shortcut.

69. (a) Describe a shortcut method of finding 10% of a number.

(b) Show an example using your shortcut.

70. (a) Describe a shortcut method of finding 1% of a number.

(b) Show an example using your shortcut.

More than 7.4 million households dress up their pets (dogs and cats) in Halloween costumes. The bar graph shows the ranking of the top pet costumes and the percent of pet owners selecting each costume. Use this graph to answer Exercises 71–74. Write each answer as a percent and as a decimal. (Source: BIGresearch survey of 8877 adult pet owners.)

71. What portion of the pet owners selected the devil costume for their pet?

72. What portion of the pet owners selected the pirate costume for their pet?

73. (a) What was the third-most-popular costume?

(b) Write the portion of the pet owners who selected this costume.

74. (a) What was the second-most-popular costume?

(b) Write the portion of the pet owners who selected this costume.

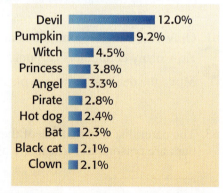

PETS GET DRESSED UP

This year, 7.4 million households plan to put their furry friends into a Halloween costume. Top outfits:

Costume	Percent
Devil	12.0%
Pumpkin	9.2%
Witch	4.5%
Princess	3.8%
Angel	3.3%
Pirate	2.8%
Hot dog	2.4%
Bat	2.3%
Black cat	2.1%
Clown	2.1%

Source: BIGresearch survey of 8877 adult pet owners.

Motorists were asked which traffic violation they felt should be given the biggest fine. The circle graph shows the percent of motorists who chose each traffic violation. Use this graph to answer Exercises 75–78. Write each answer as a percent and a decimal.

75. What portion of the motorists chose "Tailgating" for the biggest fine?

76. What portion of the motorists chose "Texting while driving" for the biggest fine?

77. (a) Which violation was chosen least often for the biggest fine?

   (b) What portion was this?

78. (a) Which violation was chosen most often for the biggest fine?

   (b) What portion was this?

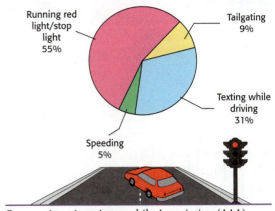

WHICH BAD DRIVING DEED DESERVES THE BIGGEST FINE?

Running red light/stop light 55%
Tailgating 9%
Texting while driving 31%
Speeding 5%

Source: American Automobile Association (AAA).

Food researchers suspect that childhood obesity starts early. The pictograph shows the percent of children 19 to 24 months old who eat each type of food at least once a day. Use this graph to answer Exercises 79–82. Write each answer as a percent and as a decimal.

79. What portion of the children eat french fries?

80. What portion of the children eat pizza?

81. (a) Which food is eaten by the lowest portion of children?

   (b) What portion is this?

82. (a) Which food is eaten by the highest portion of children?

   (b) What portion is this?

WHAT TODDLERS EAT ONCE EACH DAY

Percent of Children 19–24 Months Old

Hot Dogs 25%
Sweet Drinks 23%
French Fries 21%
Pizza 11%
Candy 10%

Type of Food

Source: Mathematica Policy Research for Gerber Products Company.

Write a percent for both the shaded and unshaded portions of each figure.

83.

95 out of 100 parts are shaded = _____ %

5 out of 100 parts are not shaded = _____ %

84.

20 out of 100 parts are shaded = _____ %

80 out of 100 parts are not shaded = _____ %

85.

86.

87.

88.

89.

90.

6.2 Percents and Fractions

OBJECTIVE ▸ 1 Write percents as fractions. Percents can be written as fractions by using what we learned in the previous section.

´OBJECTIVES

1 Write percents as fractions.

2 Write fractions as percents.

3 Use the table of percent equivalents.

> **Writing a Percent as a Fraction**
>
> $$p\% = \frac{p}{100}, \quad \text{as a fraction}$$

EXAMPLE 1   Writing Percents as Fractions

Write each percent as a fraction or mixed number in lowest terms.

(a) 25%

As we saw in the last section, 25% can be written as a decimal.

$$25\% = 25 \div 100 = 0.25 \quad \text{Percent symbol dropped}$$

Because 0.25 means 25 hundredths,

$$0.25 = \frac{25}{100} = \frac{25 \div 25}{100 \div 25} = \frac{1}{4} \leftarrow \text{Lowest terms}$$

It is not necessary, however, to write 25% as a decimal first.

$$25\% = \frac{25}{100} \quad \text{25 per 100}$$

$$= \frac{1}{4} \leftarrow \text{Lowest terms}$$

(b) 76%

The percent becomes the numerator.

Write 76% as $\frac{76}{100}$.

The *denominator* is always 100 because percent means *parts per 100*.

Write $\frac{76}{100}$ in lowest terms.

> To write a fraction in lowest terms, divide numerator and denominator by the *same number*.

$$\frac{76 \div 4}{100 \div 4} = \frac{19}{25} \leftarrow \text{Lowest terms}$$

(c) 150%

Lowest terms

$$150\% = \frac{150}{100} = \frac{150 \div 50}{100 \div 50} = \frac{3}{2} = 1\frac{1}{2} \leftarrow \text{Mixed number}$$

> **Note**
>
> Remember that percent means *per 100*.

Work Problem **1** at the Side. ▶

1 Write each percent as a fraction or mixed number in lowest terms.

GS (a) 50%

$$50\% = \frac{50}{100} = \frac{50 \div 50}{\underline{} \div 50}$$

$$= \underline{}$$

(b) 75%

(c) 48%

(d) 23%

GS (e) 125%

$$125\% = \frac{\overline{}}{100}$$

$$= \frac{\underline{} \div 25}{100 \div 25}$$

$$= \frac{\overline{}}{4} = \underline{}$$

(f) 250%

Answers

1. **(a)** $100; \frac{1}{2}$ **(b)** $\frac{3}{4}$ **(c)** $\frac{12}{25}$ **(d)** $\frac{23}{100}$

   (e) $125; 125; 5; 1\frac{1}{4}$ **(f)** $\frac{5}{2} = 2\frac{1}{2}$

2 Write each percent as a fraction in lowest terms.

(a) 37.5%

$$37.5\% = \frac{37.5}{100}$$

$$= \frac{37.5\,(10)}{100\,(\underline{})}$$

$$= \frac{375 \div \underline{}}{1000 \div 125}$$

$$= \underline{}$$

(b) 62.5%

(c) 4.5%

(d) $66\frac{2}{3}\%$

(e) $10\frac{1}{3}\%$

(f) $87\frac{1}{2}\%$

Answers

2. (a) 10; 125; $\frac{3}{8}$ **(b)** $\frac{5}{8}$ **(c)** $\frac{9}{200}$

(d) $\frac{2}{3}$ **(e)** $\frac{31}{300}$ **(f)** $\frac{7}{8}$

The next example shows how to write decimal and fraction percents as fractions.

EXAMPLE 2 **Writing Decimal or Fraction Percents as Fractions**

Write each percent as a fraction in lowest terms.

(a) 15.5%

Write 15.5 over 100.

$$15.5\% = \frac{15.5}{100}$$

> The denominator is 100 because percent means **parts per 100.**

To get a whole number in the numerator, multiply the numerator and denominator by 10. (Recall that multiplying by $\frac{10}{10}$ is the same as multiplying by 1.)

$$\frac{15.5}{100} = \frac{15.5\,(10)}{100\,(10)} = \frac{155}{1000}$$

> Move the decimal point 1 place to the *right* to multiply by 10.

Now write the fraction in lowest terms.

$$\frac{155 \div 5}{1000 \div 5} = \frac{31}{200}$$

(b) $33\frac{1}{3}\%$

Write $33\frac{1}{3}$ over 100.

$$33\frac{1}{3}\% = \frac{33\frac{1}{3}}{100}$$

When there is a mixed number in the numerator, rewrite the mixed number as an improper fraction.

$$\frac{33\frac{1}{3}}{100} = \frac{\frac{100}{3}}{100}$$

> Write $33\frac{1}{3}$ as $\frac{100}{3}$.

Next, rewrite the division problem in a horizontal form. Finally, multiply by the reciprocal of the divisor.

Reciprocals

$$\frac{\frac{100}{3}}{100} = \frac{100}{3} \div 100 = \frac{100}{3} \div \frac{100}{1} = \frac{\overset{1}{\cancel{100}}}{3} \cdot \frac{1}{\underset{1}{\cancel{100}}} = \frac{1}{3} \leftarrow \text{Lowest terms}$$

Note

In **Example 2(a)** at the top of the page, we could have changed 15.5% to $15\frac{1}{2}\%$ and then written it as the improper fraction $\frac{31}{2}$ over 100. But it is usually easier *not* to change decimals to fractions and to work with decimal percents as they are.

◄ **Work Problem** **2** at the Side.

OBJECTIVE ▶ **2** **Write fractions as percents.** We will use the formula from the beginning of this section to write fractions as percents.

$$p\% = \frac{p}{100}$$

EXAMPLE 3 **Writing Fractions as Percents**

Write each fraction as a percent. Round to the nearest tenth if necessary.

(a) $\frac{3}{5}$

Write $\frac{3}{5}$ as a percent by solving for p in the proportion below.

$$\frac{3}{5} = \frac{p}{100}$$

Find cross products and show that they are equal.

$$5 \cdot p = 3 \cdot 100$$

$$5 \cdot p = 300$$

$$\frac{\overset{1}{\cancel{5}} \cdot p}{\underset{1}{\cancel{5}}} = \frac{300}{5}$$

Divide both sides by 5.

$$p = 60$$

This result means that $\frac{3}{5} = \frac{60}{100}$ or 60%.

Note

Solving proportions can be reviewed in **Chapter 5.**

(b) $\frac{7}{8}$

Write a proportion.

$$\frac{7}{8} = \frac{p}{100}$$

$$8 \cdot p = 7 \cdot 100 \qquad \text{Show that cross products are equal.}$$

$$8 \cdot p = 700$$

$$\frac{\overset{1}{\cancel{8}} \cdot p}{\underset{1}{\cancel{8}}} = \frac{700}{8} \qquad \text{Divide both sides by 8.}$$

$$p = 87.5$$

So, $\frac{7}{8} = 87.5\%$.

Note

If you think of $\frac{700}{8}$ as an improper fraction, changing it to a mixed number gives an answer of $87\frac{1}{2}$. So $\frac{7}{8} = 87.5\%$ or $87\frac{1}{2}\%$.

Continued on Next Page

3 Write as percents. Round to the nearest tenth if necessary.

(GS) (a) $\dfrac{1}{4}$

$$\dfrac{1}{4} = \dfrac{p}{100}$$

$$4 \cdot p = 1 \cdot \underline{\quad}$$

$$\dfrac{\overset{1}{\cancel{4}} \cdot p}{\underset{1}{\cancel{4}}} = \dfrac{100}{4}$$

$$p = \underline{\quad}$$

(b) $\dfrac{3}{10}$

(c) $\dfrac{6}{25}$

(d) $\dfrac{5}{8}$

(e) $\dfrac{1}{6}$

(f) $\dfrac{2}{9}$

Answers

3. **(a)** 100; 25% **(b)** 30% **(c)** 24%
   (d) 62.5%
   (e) 16.7% (rounded); $16\frac{2}{3}$% (exact)
   (f) 22.2% (rounded); $22\frac{2}{9}$% (exact)

(c) $\dfrac{5}{6}$

Start with a proportion.

$$\dfrac{5}{6} = \dfrac{p}{100}$$

$6 \cdot p = 5 \cdot 100$ Show that cross products are equal.

$6 \cdot p = 500$

$$\dfrac{\overset{1}{\cancel{6}} \cdot p}{\underset{1}{\cancel{6}}} = \dfrac{500}{6}$$ Divide both sides by 6.

$p = 83.\overline{3}$ — A bar over the 3 indicates that the decimal keeps repeating 83.3333 forever.

$p \approx 83.3$ Round to the nearest tenth.

So, $\dfrac{5}{6} = 83.\overline{3}\% \approx 83.3\%$ (rounded) — The "$\approx$" symbol shows that 83.3% is rounded.

Note

You can change $\dfrac{500}{6}$ to a mixed number to get an exact answer of $83\frac{1}{3}\%$.

◀ **Work Problem 3 at the Side.**

OBJECTIVE 3 Use the table of percent equivalents. Knowing how to find the equivalents of fractions, decimals, and percents is important. However, the table on the next page shows common fractions and mixed numbers and their decimal and percent equivalents.

EXAMPLE 4   Using the Table of Percent Equivalents

Find the following in the table.

(a) $\frac{1}{6}$ as a percent

Find $\frac{1}{6}$ in the "fraction" column. The equivalent percent is 16.7% (rounded) or $16\frac{2}{3}\%$ (exact).

(b) 0.375 as a fraction

Look in the "decimal" column for 0.375. The equivalent fraction is $\frac{3}{8}$.

(c) $\frac{7}{8}$ as a percent

Find $\frac{7}{8}$ in the "fraction" column. The equivalent percent is 87.5% or $87\frac{1}{2}\%$.

Calculator Tip

Example 4 **(c)** above can be solved on a calculator as shown below.

$$7 \; \div \; 8 \; = \; 0.875 \; \times \; 100 \; = \; 87.5$$

Multiplying by 100 changes the decimal to a percent.
Or, if your calculator has a percent key, follow these steps.

$$7 \; \div \; 8 \; \% \; 87.5$$

Press % key.

On scientific calculators, you may need to press the **2nd** key to access the % function, and some models require pressing **=** to get the answer.

Note

When a fraction like $\frac{1}{3}$ is changed to a decimal, it is a *repeating decimal* that goes on forever, 0.333333.... In the table below these decimals are rounded to the nearest thousandth. When the decimal is then changed to a percent, it will be to the nearest tenth of a percent. Decimals that do not repeat are usually not rounded.

Work Problem ❹ at the Side. ▶

PERCENT, DECIMAL, AND FRACTION EQUIVALENTS

Percent (rounded to tenths when necessary)	Decimal	Fraction
1%	0.01	$\frac{1}{100}$
5%	0.05	$\frac{1}{20}$
10%	0.1	$\frac{1}{10}$
12.5% or $12\frac{1}{2}\%$	0.125	$\frac{1}{8}$
16.7% (rounded) or $16\frac{2}{3}\%$ (exact)	$0.1\overline{6}$ rounds to 0.167	$\frac{1}{6}$
20%	0.2	$\frac{1}{5}$
25%	0.25	$\frac{1}{4}$
33.3% (rounded) or $33\frac{1}{3}\%$ (exact)	$0.\overline{3}$ rounds to 0.333	$\frac{1}{3}$
37.5% or $37\frac{1}{2}\%$	0.375	$\frac{3}{8}$
40%	0.4	$\frac{2}{5}$
50%	0.5	$\frac{1}{2}$
60%	0.6	$\frac{3}{5}$
62.5% or $62\frac{1}{2}\%$	0.625	$\frac{5}{8}$
66.7% (rounded) or $66\frac{2}{3}\%$ (exact)	$0.\overline{6}$ rounds to 0.667	$\frac{2}{3}$
75%	0.75	$\frac{3}{4}$
80%	0.8	$\frac{4}{5}$
87.5% or $87\frac{1}{2}\%$	0.875	$\frac{7}{8}$
100%	1.0	1
150%	1.5	$1\frac{1}{2}$
200%	2.0	2

❹ Find the following fractions, mixed numbers, decimals, and percents in the table on this page. If you already know the answer or can solve for the answer quickly, don't use the table.

(a) $\frac{3}{4}$ as a percent

(b) 10% as a fraction

(c) $0.\overline{6}$ as a fraction

(d) $37\frac{1}{2}\%$ as a fraction

(e) $\frac{7}{8}$ as a percent

(f) $\frac{1}{2}$ as a percent

(g) $33\frac{1}{3}\%$ as a fraction

(h) $1\frac{1}{2}$ as a percent

Answers

4. **(a)** 75% **(b)** $\frac{1}{10}$ **(c)** $\frac{2}{3}$ **(d)** $\frac{3}{8}$ **(e)** 87.5%

(f) 50% **(g)** $\frac{1}{3}$ **(h)** 150%

6.2 Exercises

MyMathLab®

CONCEPT CHECK   *Is each percent equivalent to the given fraction? Write* true *or* false.

1. $25\% = \dfrac{1}{2}$

2. $30\% = \dfrac{3}{10}$

3. $75\% = \dfrac{3}{4}$

4. $80\% = \dfrac{5}{6}$

Write each percent as a fraction in lowest terms and as a mixed number when possible.
See Examples 1 and 2.

5. 85%
GS ▶ $0.85 = \dfrac{85}{100} = \dfrac{17}{\underline{\quad}}$

6. 45%
GS $0.45 = \dfrac{45}{100} = $

7. 62.5%

8. 87.5%

9. 6.25%
▶

10. 43.75%

11. $16\dfrac{2}{3}\%$
▶

12. $66\dfrac{2}{3}\%$

13. $6\dfrac{2}{3}\%$

14. $46\dfrac{2}{3}\%$

15. 0.5%

16. 0.8%

17. 180%
▶

18. 140%

19. 375%

20. 225%

CONCEPT CHECK   *Is each fraction equivalent to the given percent? Write* true *or* false.

21. $\dfrac{1}{2} = 50\%$

22. $\dfrac{1}{100} = 10\%$

23. $\dfrac{4}{5} = 75\%$

24. $\dfrac{3}{10} = 30\%$

Write each fraction as a percent. Round percents to the nearest tenth if necessary.
See Example 3.

25. $\dfrac{7}{10} = \dfrac{p}{100}$
GS

Solve for p.
$p = \underline{\quad}\%$

26. $\dfrac{3}{4} = \dfrac{p}{100}$
GS

Solve for p.
$p = \underline{\quad}\%$

27. $\dfrac{37}{100}$

28. $\dfrac{63}{100}$

29. $\dfrac{5}{8}$

30. $\dfrac{1}{8}$

31. $\dfrac{7}{8}$
▶

32. $\dfrac{3}{8}$

33. $\dfrac{12}{25}$
▶

34. $\dfrac{15}{25}$

35. $\dfrac{23}{50}$

36. $\dfrac{18}{50}$

37. $\dfrac{7}{20}$

38. $\dfrac{9}{20}$

39. $\dfrac{5}{6}$

40. $\dfrac{1}{6}$

41. $\dfrac{5}{9}$

42. $\dfrac{7}{9}$

43. $\dfrac{1}{7}$

44. $\dfrac{5}{7}$

45. CONCEPT CHECK   Which of the following is equal to $\frac{3}{4}$?

(a) 65%   (b) 0.75   (c) 0.5   (d) 25%

46. CONCEPT CHECK   Which of the following is equal to 80%?

(a) 0.85   (b) $\dfrac{3}{5}$ (c) $\dfrac{7}{10}$ (d) $\dfrac{4}{5}$

Complete the chart. Round decimals to the nearest thousandth and percents to the nearest tenth if necessary. **See Examples 3 and 4.**

	Fraction	Decimal	Percent
47.	_____	0.5	_____
48.	_____	0.8	_____
49.	_____	_____	87.5%
50.	_____	_____	60%
51.	$\dfrac{1}{6}$	_____	_____
52.	$\dfrac{1}{3}$	_____	_____
53.	_____	0.7	_____

Fraction	Decimal	Percent
54. _____	_____	37.5%
55. _____	_____	12.5%
56. _____	0.625	_____
57. $\dfrac{2}{3}$	_____	_____
58. $\dfrac{5}{6}$	_____	_____
59. $\dfrac{3}{50}$	_____	_____
60. $\dfrac{3}{10}$	_____	_____
61. $\dfrac{8}{100}$	_____	_____
62. _____	_____	100%
63. $\dfrac{1}{200}$	_____	_____
64. $\dfrac{1}{400}$	_____	_____

	Fraction	Decimal	Percent
65.	_____	2.5	_____
66.	_____	1.7	_____
67.	$3\frac{1}{4}$	_____	_____
68.	$2\frac{4}{5}$	_____	_____

69. Select a decimal percent and write it as a fraction. Select a different fraction and write it as a percent. Write an explanation of each step of your work.

70. Prepare a table showing fraction, decimal, and percent equivalents for five fractions and mixed numbers of your choice.

In the following application problems, write the answer as a fraction in lowest terms, as a decimal, and as a percent.

71. Many pet owners say they have used the Internet to find pet information. Of 500 people who used the Internet for this purpose, 90 said they used it when buying a pet. What portion used the Internet when buying a pet? (*Source:* American Animal Hospital Association.)

72. About $\frac{1}{3}$ of all books purchased last year were for children. Of these children's books, 27 of every 100 purchased included a coloring activity. What portion of the children's books included a coloring activity? (*Source:* Consumer Research Study on Book Publishing, the American Booksellers, and the Book Industry Study Group.)

73. Only 13 out of every 100 adults ages 19–50 consume the recommended 1000 milligrams of calcium daily. What portion consumes the recommended daily amount? (*Source:* Market Facts for Milk Mustache.)

74. In a survey on how people learn to parent, 360 parents out of 800 said they were most influenced by relatives, friends, and spouses. What portion learned to parent this way? (*Source:* Bama Research.)

75. In a recent survey, 750 workers were asked if they would hire their own boss if they were in charge. A total of 150 workers said no, they would not. What portion of the workers said no? (*Source:* Marlin's 13th annual Attitudes in the Workplace Survey.)

76. When 1500 adults were asked what was the most important factor to consider when relocating after retirement, 675 said that climate was most important. What portion consider climate to be most important? (*Source:* Longevity Alliance Retirement and Relocation Survey.)

77. An insurance office has 80 employees. If 64 of the employees have iPhones, find

(a) the portion of the employees who have iPhones.

(b) the portion of the employees who do not have iPhones.

78. For every 640 tons of food purchased, 160 tons are discarded by Americans. (*Source:* U.S. Department of Agriculture.)

(a) What portion of the food is discarded?

(b) What portion of the food is not discarded?

79. An antibiotic is used to treat 380 people. If 342 people do not have side effects from the antibiotic, find the portion that do have side effects.

80. A survey of 340 doctors found that 119 of the doctors used a basic system of electronic record keeping.

(a) What portion of the doctors used electronic record keeping?

(b) What portion did not use it?

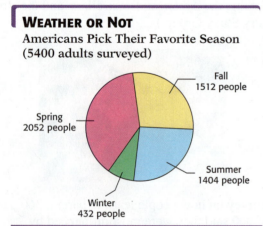

The circle graph shows the number of people in a survey of 5400 adults who picked each season as their favorite. Use this graph to answer Exercises 81–84, giving each answer as a fraction, as a decimal, and as a percent.

WEATHER OR NOT

Americans Pick Their Favorite Season (5400 adults surveyed)

Fall
1512 people

Spring
2052 people

Summer
1404 people

Winter
432 people

Source: Strategy One Survey for Back to Nature Foods.

81. What portion of the adults picked Winter as their favorite season?

82. What portion of the adults picked Spring as their favorite season?

83. Find the portion of the adults who say Summer is their favorite season.

84. Find the portion of the adults who say Fall is their favorite season.

Relating Concepts (Exercises 85–94) For Individual or Group Work

To review the basics of percent, **work Exercises 85–94 in order.**

85. 100% of a number means all of the parts or _____ parts out of _____ parts. 200% means two times as many parts and 300% means three times as many parts.

86. Fill in the blanks.

(a) 100% of 765 workers is _____.

(b) 200% of 48 letters is _____.

(c) 300% of 7 DVDs is _____.

87. 50% of a number is _____ parts out of _____ parts, which is _____ of the parts.

88. 10% of a number is _____ parts out of _____ parts and can be found quickly by moving the decimal point _____ place to the _____ .

89. 1% of a number is _____ part out of _____ parts and can be found quickly by moving the decimal point _____ places to the _____ .

90. Fill in the blanks.

(a) 50% of 1050 homes is _____.

(b) 10% of 370 printers is _____.

(c) 1% of $8 is _____.

In Exercises 91–94, use the shortcut methods for finding 1%, 10%, 50%, 100%, 200%, and 300%.

91. Describe a shortcut method for finding 15% of a number. Use your method to find 15% of $160.

92. Describe a shortcut method for finding 150% of a number. Use your method to find 150% of $160.

93. Explain a shortcut method for finding 90% of a number. Show how to find 90% of $450 using your method.

94. Explain a shortcut method for finding 210% of a number. Show how to find 210% of $800 using your method.

6.3 Using the Percent Proportion and Identifying the Components in a Percent Problem

There are two ways to solve percent problems. One method uses proportions and is discussed in this and the next section. The other method uses the percent equation and is explained later in this chapter.

OBJECTIVE ▶ 1 Learn the percent proportion. We have seen that a statement of two equivalent ratios is called a proportion.

$\frac{3}{5}$ or 3 out of 5 parts

60%

100%

For example, the fraction $\frac{3}{5}$ is the same as the ratio 3 to 5, and 60% is the same as the ratio 60 to 100. As the figure above shows, these two ratios are equivalent and make a proportion.

◀ Work Problem 1 at the Side.

The percent proportion can be used to solve percent problems.

Percent Proportion

Part is to *whole* as *percent* is to *100*.

$$\frac{\text{part}}{\text{whole}} = \frac{\text{percent}}{100} \quad \leftarrow \text{Always 100 because percent means } per\ 100$$

In the figure at the top of the page, the **whole** is 5 (the entire quantity), the **part** is 3 (the part of the whole), and the **percent** is 60. Write the percent proportion as follows.

$$\frac{\text{part}}{\text{whole}} \rightarrow \frac{3}{5} = \frac{60}{100} \leftarrow \frac{\text{percent}}{100}$$

Remember: Percent means per 100.

OBJECTIVE ▶ 2 Solve for an unknown value in a percent proportion. As shown in **Chapter 5**, if any three of the four values in a proportion are known, the fourth can be found by solving the proportion.

EXAMPLE 1 Using the Percent Proportion

Use the percent proportion to solve for *x*, representing the unknown value.

(a) part = 12, percent = 25; find the whole.

$$\frac{\text{part}}{\text{whole}} = \frac{\text{percent}}{100} \quad \text{Percent proportion}$$

Percent

$$\text{Part} \rightarrow \frac{12}{x} = \frac{25}{100} \quad \text{or} \quad \frac{12}{x} = \frac{1}{4} \qquad \frac{25}{100} \text{ in lowest terms is } \frac{1}{4}.$$
Whole (unknown) →

Continued on Next Page

① As a review of proportions, use the method of comparing cross products to decide whether each proportion is *true* or *false*. Show the cross products.

GS **(a)** $\frac{1}{2} = \frac{25}{50}$

$$\frac{1}{2} = \frac{25}{50} \qquad \begin{array}{l} 2 \cdot 25 = \underline{\quad} \\ \\ 1 \cdot 50 = \underline{\quad} \end{array}$$

The proportion is (true/false).

(b) $\frac{3}{4} = \frac{150}{200}$

(c) $\frac{7}{8} = \frac{180}{200}$

(d) $\frac{112}{41} = \frac{332}{123}$

Answers

1. **(a)** 50; 50; true **(b)** 600 = 600; true
   (c) 1440 ≠ 1400; false
   (d) 13,612 ≠ 13,776; false

First find the cross products.

$$\frac{12}{x} = \frac{1}{4}$$

The unknown in this proportion is the *whole*.

$$x \cdot 1$$
$$12 \cdot 4$$

Show that the cross products are equal.

$$x \cdot 1 = 12 \cdot 4$$
$$x = 48$$

The whole is 48.

CAUTION

In a **proportion**, you **cannot** divide out a common factor from the numerator of one ratio and the denominator of the other ratio. Dividing out common factors is done *only* when you are multiplying fractions.

(b) part = 30, whole = 50; find the percent.
Use the percent proportion.

Percent (unknown)

The unknown in this proportion is the percent.

Part → $\dfrac{30}{50} = \dfrac{x}{100}$ ← Whole   Percent proportion

$$\frac{3}{5} = \frac{x}{100} \qquad \text{Write } \tfrac{30}{50} \text{ as } \tfrac{3}{5} \text{ in lowest terms.}$$

$$5 \cdot x = 3 \cdot 100 \qquad \text{Find the cross products.}$$

$$5 \cdot x = 300$$

$$\frac{\overset{1}{\cancel{5}} \cdot x}{\underset{1}{\cancel{5}}} = \frac{300}{5} \qquad \text{Divide both sides by 5.}$$

$$x = 60$$

Remember to write the % symbol when solving for an unknown percent.

The percent is 60, written as 60%.

(c) whole = 150, percent = 18; find the part.

Percent

Part (unknown) → $\dfrac{x}{150} = \dfrac{18}{100}$ or $\dfrac{x}{150} = \dfrac{9}{50}$ ← Whole

Write $\tfrac{18}{100}$ as $\tfrac{9}{50}$ in lowest terms.

$$x \cdot 50 = 150 \cdot 9 \qquad \text{Find the cross products.}$$

$$x \cdot 50 = 1350$$

$$\frac{x \cdot \overset{1}{\cancel{50}}}{\underset{1}{\cancel{50}}} = \frac{1350}{50} \qquad \text{Divide both sides by 50.}$$

$$x = 27$$

The part is 27.

············ **Work Problem** ❷ **at the Side.** ▶

---

❷ Use the percent proportion $\left(\dfrac{\text{part}}{\text{whole}} = \dfrac{\text{percent}}{100}\right)$ and solve for the unknown value.

GS **(a)** part = 12, percent = 16

$$\frac{12}{x} = \frac{16}{100} \quad \text{or} \quad \frac{12}{x} = \frac{4}{25}$$

$$x \cdot 4 = 12 \cdot \underline{\quad}$$

$$x \cdot 4 = \underline{\quad}$$

$$\frac{x \cdot \overset{1}{\cancel{4}}}{\underset{1}{\cancel{4}}} = \frac{300}{4}$$

$$x = \underline{\quad}$$

(b) part = 30, whole = 120

(c) whole = 210,
percent = 20

(d) whole = 4000,
percent = 32

(e) part = 74, whole = 185

Answers

2. **(a)** 25; 300; 75; whole = 75
   (b) percent = 25 (so, the percent is 25%)
   (c) part = 42   **(d)** part = 1280
   (e) percent = 40 (so, the percent is 40%)

3 Identify the percent.

GS **(a)** Of the 900 blood glucose tests, 25% will be completed by Adrian.

The number preceding the _____ symbol is the percent.

So, _____ is the percent.

(b) Of the 620 preschool students, 65% will be served breakfast and lunch.

(c) Find the amount of sales tax by multiplying $590 and $6\frac{1}{2}$ percent.

(d) 8500 tons of recyclables is 42% of what number of tons?

(e) What percent of the 1450 parents use child care?

As a help in solving percent problems, keep in mind this basic idea.

> **Percent Problems**
>
> All percent problems involve a comparison between a part of something and the whole.

Solving these problems requires identifying the three components of a percent proportion: part, whole, and percent.

OBJECTIVE ▶ **3** **Identify the percent.** Look for the percent first. It is the easiest to identify.

> **Percent**
>
> The **percent** is the ratio of a part to a whole, with 100 as the denominator. In a problem, the percent appears with the word **percent** or with the symbol **"%"** after it.

EXAMPLE 2 **Finding the Percent in Percent Problems**

Find the percent in the following.

(a) 32% of the 900 men were too large for the imported car.

Percent ◀ 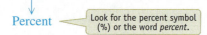 Look for the percent symbol (%) or the word *percent*.

The percent is 32. The number 32 appears with the percent symbol (%).

(b) $150 is 25 percent of what number?

Percent

The percent is 25 because 25 appears with the word *percent*.

(c) What percent of 7000 pounds is 3500 pounds?

Percent (unknown)

The word *percent* has no number with it, so the percent is the unknown part of the problem.

•• ◀ **Work Problem 3** at the Side.

OBJECTIVE ▶ **4** **Identify the whole.** Next, look for the whole.

> **Whole**
>
> The **whole** is the entire quantity. In a percent problem, the whole often appears after the word **of**.

Answers

3. **(a)** % or percent; 25 **(b)** 65 **(c)** $6\frac{1}{2}$
   (d) 42 **(e)** The percent is unknown.

EXAMPLE 3 **Finding the Whole in Percent Problems**

Identify the whole in the following.

(a) 32% **of** the 900 men were too large for the imported car.

Whole

The whole is 900. The number 900 appears after the word *of*.

(b) $150 is 25 percent **of** what number?

Whole — The whole is the unknown part of the problem.

(c) What percent **of** 7000 pounds is 3500 pounds?

Whole

··············· **Work Problem 4 at the Side.** ▶

OBJECTIVE ▶ **5 Identify the part.** Finally, look for the part.

Part

The **part** is the portion being compared with the whole.

Note

If you have trouble identifying the part, find the percent and whole first. The remaining number is the part.

EXAMPLE 4 **Finding the Part in Percent Problems**

Identify the part. Then set up the percent proportion. (Do **not** solve the proportions.)

(a) 54% **of** 700 students is 378 students.
First find the percent and the whole.

54% **of** 700 students is 378 students.

Percent; with % symbol   Whole; follows "of"   Find the percent and then the whole. The remaining number, 378, is the part.

The remaining number, 378, is the part.

54% **of** 700 students is 378 students.

Percent   Whole   Part

$$\text{Part} \rightarrow \frac{378}{700} = \frac{54}{100} \begin{array}{l} \leftarrow \text{Percent} \\ \leftarrow \text{Always 100} \end{array}$$

Whole →

(b) $150 is 25% **of** what number?

Percent   Whole (unknown)

$$\text{Part} \rightarrow \frac{150}{\text{unknown}} = \frac{25}{100} \begin{array}{l} \leftarrow \text{Percent} \\ \leftarrow \text{Always 100} \end{array}$$

Whole →

$150 is the remaining number, so the part is $150.

(c) 85% **of** 7000 is what number?

Percent   Whole   Part (unknown)

$$\text{Part} \rightarrow \frac{\text{unknown}}{7000} = \frac{85}{100} \begin{array}{l} \leftarrow \text{Percent} \\ \leftarrow \text{Always 100} \end{array}$$

Whole →

·········· **Work Problem 5 at the Side.** ▶

4 Identify the whole.

(a) **Of** the 900 blood glucose tests, 25% will be completed by Adrian.

(b) Of the 620 preschool students, 65% will be served breakfast and lunch.

(c) 8500 tons of recyclables is 42% of what number of tons?

5 Identify the part, then set up the percent proportion.

(a) **Of** the 900 blood glucose tests, 25% or 225 will be completed by Adrian.

(b) Of the 620 preschool students, 65% or 403 will be served breakfast and lunch.

(c) 8500 tons of recyclables is 42% of what number of tons?

Answers

4. (a) 900   **(b)** 620
(c) what number (an unknown)

5. (a) 225; $\dfrac{\text{Part} \rightarrow 225}{\text{Whole} \rightarrow 900} = \dfrac{25}{100} \begin{array}{l} \leftarrow \text{Percent} \\ \leftarrow \text{Always 100} \end{array}$

(b) 403; $\dfrac{\text{Part} \rightarrow 403}{\text{Whole} \rightarrow 620} = \dfrac{65}{100} \begin{array}{l} \leftarrow \text{Percent} \\ \leftarrow \text{Always 100} \end{array}$

(c) 8500;
$\dfrac{\text{Part} \rightarrow 8500}{\text{Whole} \rightarrow \text{unknown}} = \dfrac{42}{100} \begin{array}{l} \leftarrow \text{Percent} \\ \leftarrow \text{Always 100} \end{array}$

6.3 Exercises

CONCEPT CHECK *Identify the component that is unknown. Circle the correct answer. Then, set up the percent proportion. Do* **not** *solve the proportion.*

1. part = 5, percent = 10

    part   whole   percent

2. part 1.5, whole = 4.5

    part   whole   percent

3. part = 36, whole = 24

    part   whole   percent

4. whole = 72, percent = 30

    part   whole   percent

5. whole = 160, percent = 35

    part   whole   percent

6. part = 20, percent = 25

    part   whole   percent

Find the unknown value in the percent proportion $\dfrac{part}{whole} = \dfrac{percent}{100}$. *Round to the nearest tenth if necessary. If the answer is a percent, be sure to include a percent symbol (%).* **See Example 1.**

7. part = 30, percent = 20

8. part = 25, percent = 25

9. part = 28, percent = 40

10. part = 11, percent = 5

11. part = 15, whole = 60

12. part = 105, whole = 35

13. part = 9.25, whole = 27.75

14. part = 12.8, whole = 9.6

15. whole = 52, percent = 50

16. whole = 115, percent = 38

17. whole = 94.4, part = 25

18. whole = 89.6, part = 50

Solve each problem. If the answer is a percent, be sure to include a percent sign (%). **See Examples 2–4.**

19. Find the whole if the part is 46 and the percent is 40.

$$\frac{part}{whole} = \frac{percent}{100}$$

$$\frac{46}{x} = \frac{40}{100}$$

$$40 \cdot x$$
$$40 \cdot x = 4600$$
$$4600$$
$$x = \underline{\hspace{1cm}}$$

20. The percent is 45 and the whole is 160. Find the part.

$$\frac{part}{whole} = \frac{percent}{100}$$

$$\frac{x}{160} = \frac{45}{100}$$

$$7200$$
$$100 \cdot x = 7200$$
$$100 \cdot x$$
$$x = \underline{\hspace{1cm}}$$

21. The whole is 5000 and the part is 20. Find the percent.

22. Suppose the part is 15 and the whole is 2500. Find the percent.

23. Find the percent if the whole is 4300 and the part is $107\frac{1}{2}$.

24. What is the part, if the percent is $12\frac{3}{4}$ and the whole is 5600?

25. The whole is 6480 and the part is 19.44. Find the percent.

26. Suppose the part is 281.25 and the percent is $1\frac{1}{4}$. Find the whole.

CONCEPT CHECK   *Fill in each blank with the correct response.*

27. In a percent problem, the percent can be identified because it appears with the word _____ or with the _____ symbol after it.

28. In a percent problem, the whole is the _____ quantity and often appears after the word _____.

29. In a percent problem, the part is the portion being compared with the _____ .

30. To solve a percent proportion, you must cross _____ .

*In Exercises 31–42, set up the percent proportion, and write "unknown" for any value that is not given. Recall that the percent proportion is $\dfrac{part}{whole} = \dfrac{percent}{100}$. Do **not** try to solve for the unknowns. See Examples 2–4.*

31. 10% of how many bicycles is 60 bicycles?

$$\text{Part} \rightarrow \frac{60}{\text{unknown}} = \frac{}{} \begin{array}{l} \leftarrow \text{Percent} \\ \leftarrow \text{Always 100} \end{array}$$
$$\text{Whole} \rightarrow$$

32. 58% of how many preschoolers is 203 preschoolers?

$$\text{Part} \rightarrow \frac{}{} = \frac{58}{100} \begin{array}{l} \leftarrow \text{Percent} \\ \leftarrow \text{Always 100} \end{array}$$
$$\text{Whole} \rightarrow$$

33. What % of $800 is $600?

34. What % of $1500 is $1395?

35. What is 25% of $970?

36. What is 61% of 830 homes?

37. 12 injections is 20% of what number of injections?

38. 410 pallets is $33\frac{1}{3}$% of how many pallets.

39. 54.34 is 3.25% of what number?

40. 16.74 is 11.9% of what number?

41. 0.68% of $487 is what amount?

42. What amount is 6.21% of $704.35?

43. Identify the three components in a percent problem. In your own words, write a sentence telling how you will identify each of these three components.

44. Write one short sentence using numbers and words. The sentence should include a percent, a whole, and a part. Identify each of these three components.

*Set up the percent proportion for each application problem. Do **not** try to solve for any unknowns.*

45. Fry's Electronics sold 1262 computers in a recent promotion. If 730 of these computers were laptop computers, what percent of the computers were laptops?

Part → $\dfrac{730}{} = \dfrac{}{100}$ ← Percent
Whole → ← Always 100

46. Ivory Soap is $99\frac{44}{100}$ % pure. If a bar of Ivory Soap weighs 4 ounces, how many ounces are pure? (*Source:* Procter & Gamble.)

Part → $\dfrac{\text{unknown}}{} = \dfrac{}{100}$ ← Percent
Whole → ← Always 100

47. Of the 142 people attending a movie theater, 86 bought buttered popcorn. What percent bought buttered popcorn?

48. On her first check from the Pizza Hut Restaurant, 15% was withheld from Maria's total earnings of $225. What amount was withheld?

49. Of the customers buying a salad at McDonald's, 23% prefer Newman's Own Light Salad Dressing. If the total number of salad customers is 610, find the number who prefer Newman's Own Light Salad Dressing.

50. At a community college campus, it was found that 2322 of the students work full-time. If this was 27% of the students, find the total number of students on campus. (*Source*: American Association of Community Colleges.)

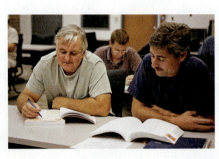

51. A survey of 8600 full-time community college students found that 4300 of them work part-time. What percent of the full-time students work part-time? (*Source*: American Association of Community Colleges.)

52. There have been 36 cups of coffee served from a banquet-sized coffee pot. If this is 30% of the capacity of the pot, find the capacity of the pot.

53. In a recent survey of 480 adults, 55% said that they would prefer to have their wedding at a religious site. How many said they would prefer the religious site? (*Source:* National Family Opinion Research.)

54. Sue Ann needs 64 credits to graduate. If she has completed 48 of the credits needed, what percent of the credits has she already completed?

55. In a poll of 822 people, 49.5% said that they get their news from television. Find the number of people who said they get their news from television. (*Source: Brill's Content.*)

56. The sales tax on a new car is $1575. If the sales tax rate is 7%, find the price of the car before the sales tax is added.

57. When asked "What co-worker behaviors annoy you the most," 12% of those surveyed said "Being perpetually late." Find the number of people in the survey if 168 people gave this response. (*Source*: Accountemps.)

58. The state troopers tested 924 cars for safety. There were 231 cars that failed the safety test for one or more reasons. Find the percent of cars that failed the test.

59. Jerry Azzaro has listed 680 antique toys on eBay. If 45% of these were antique toy trains, find the number that were toy trains.

60. In a survey of 27,400 American sports fans, ages 12 and up, 3562 picked Major League baseball as their favorite spectator sport. What percent picked Major League baseball as their favorite sport? (*Source:* ESPN Sports Poll.)

6.4 Using Proportions to Solve Percent Problems

OBJECTIVES

1. Use the percent proportion to find the part.

2. Find the whole using the percent proportion.

3. Find the percent using the percent proportion.

This is the percent proportion that you learned about in the previous section.

$$\frac{\text{part}}{\text{whole}} = \frac{\text{percent}}{100}$$

Recall that if one of the values is unknown, you can find it by solving the percent proportion.

OBJECTIVE ▶ 1 Use the percent proportion to find the part. The first example shows how to use the percent proportion to find the part.

---

1 Use the percent proportion to find the part.

(a) 8% of 400 patients

$$\frac{x}{400} = \frac{8}{100} \quad \text{or} \quad \frac{x}{400} = \frac{2}{25}$$

$$x \cdot 25 = 400 \cdot \underline{\quad}$$

$$x \cdot 25 = \underline{\quad}$$

$$\frac{x \cdot \overset{1}{\cancel{25}}}{\underset{1}{\cancel{25}}} = \frac{800}{25}$$

$$x = \underline{\quad}$$

(b) 15% of $3220

(c) 7% of 2700 miles

(d) 48% of 1580 kilowatts

EXAMPLE 1   Finding the Part with the Percent Proportion

Find 15% of $160.

Here the percent is 15 and the whole is 160. (Recall that the whole often comes after the word *of.*) Now find the part. Let *x* represent the unknown part.

$$\frac{\text{part}}{\text{whole}} = \frac{\text{percent}}{100} \quad \text{so} \quad \frac{x}{160} = \frac{15}{100} \quad \text{or} \quad \frac{x}{160} = \frac{3}{20}$$

> Write $\frac{15}{100}$ as $\frac{3}{20}$ in lowest terms.

Find the cross products in the proportion and show that they are equal.

$$x \cdot 20 = 160 \cdot 3 \quad \text{Cross products}$$

$$x \cdot 20 = 480$$

$$\frac{x \cdot \overset{1}{\cancel{20}}}{\underset{1}{\cancel{20}}} = \frac{480}{20} \quad \text{Divide both sides by 20.}$$

$$x = 24 \quad \text{The unknown part is 24.}$$

15% of $160 is **$24**.

◀ **Work Problem 1 at the Side.**

Just as with some of the fraction application problems in **Chapter 2,** the word *of* may be an indicator word meaning *multiply.* Here is an example.

$$\textbf{15\% of } 160$$
$$\downarrow$$
$$15\% \cdot 160$$

In this type of example, there is another way to find the part.

Finding the Part Using Multiplication

To find the part:

Step 1 Identify the percent. Write the percent as a decimal.

Step 2 Multiply this decimal by the whole.

Answers

1. (a) 2; 800; 32 patients (b) $483
   (c) 189 miles (d) 758.4 kilowatts

EXAMPLE 2 Finding the Part Using Multiplication

Use multiplication to find the part.

(a) Find **42% of** 830 yards.

Step 1 Here, the percent is 42. Write 42% as the decimal 0.42.

Step 2 Multiply 0.42 and the whole, which is 830.

$$\text{part} = (0.42)(830)$$

> When the percent and whole are given, the part must be found.

$$= 348.6 \text{ yd}$$

It is a good idea to estimate the answer, to make sure no mistakes were made with decimal points. Round 42% to 40% or 0.4, and round 830 to 800. Next, 40% of 800 is

$$(0.4)(800) = 320 \leftarrow \text{Estimate}$$

so the exact answer of 348.6 is reasonable.

(b) Find **25% of** 1680 cars.

Identify the percent as 25. Write 25% in decimal form as 0.25. Now, multiply 0.25 and 1680.

$$\text{part} = (0.25)(1680) = 420 \text{ cars} \quad \textit{Multiply.}$$

You can also use a shortcut to find the answer. Since 25% means 25 parts out of 100 parts, this is the same as $\frac{1}{4}$ of the whole $\left(\frac{25}{100} = \frac{1}{4}\right)$. Do you see a shortcut here? You can find $\frac{1}{4}$ of a number by dividing the number by 4. So, this shortcut gives us the exact answer, $1680 \div 4 = 420$.

(c) Find **140% of** 60 miles.

In this problem, the percent is 140. Write 140% as the decimal 1.40. Next, multiply 1.40 and 60.

$$\text{part} = (1.40)(60) = 84 \text{ miles} \quad \textit{Multiply.}$$

You can estimate the answer by realizing that 140% is close to 150% (which is $1\frac{1}{2}$) and $1\frac{1}{2}$ times 60 is 90. So, 84 miles is a reasonable answer.

(d) Find **0.4% of** 50 kilometers.

$$\text{part} = (0.004)(50) = 0.2 \text{ kilometer} \quad \textit{Multiply.}$$
↑
Write 0.4% as a decimal.

Estimate the answer by realizing that 0.4% is less than 1%.

$$1\% \text{ of } 50 \text{ kilometers} = \underset{\sim}{50.} = 0.5 \text{ kilometer}$$

So our exact answer should be *less than* 0.5 kilometer, and 0.2 kilometer fits this requirement.

············ Work Problem **2** at the Side. ▶

EXAMPLE 3 Solving for the Part in an Application Problem

Raley's Markets has 850 employees. Of these employees, 28% are students. How many of the employees are students? Use the six problem-solving steps.

···················· Continued on Next Page

2 Use multiplication to find the part.

GS **(a)** 55% of 10,000 x-rays

Write 55% in decimal form as 0.55.

$$\text{part} = (0.55)(\underline{\hspace{1cm}})$$
$$= \underline{\hspace{1cm}}$$

(b) 16% of 120 miles

(c) 135% of 60 dosages

(d) 0.5% of $238

Answers

2. (a) 10,000; 5500 x-rays **(b)** 19.2 miles
(c) 81 dosages **(d)** $1.19

3 Use the six problem-solving steps to solve each problem.

GS **(a)** One day on Jacob's mail route there were 2920 pieces of mail. If 45% **of** those were advertising pieces, find the number of advertising pieces.

Notice the blue word **of** as an indicator for multiplication. Write 45% in decimal form as ____.

part = (____) • (2920)

= ____

(b) There are 9750 students at the college. If 12% of them wear glasses or contact lenses, how many students wear glasses or contact lenses?

Step 1 **Read** the problem. The problem asks us to find the number of employees who are students.

Step 2 **Work out a plan.** Look for the word *of* as an indicator word for multiplication.

28% of the employees are students.

└───── Indicator word

The total number of employees is 850, so the whole is 850. The percent is 28. To find the number of students, find the part.

Step 3 **Estimate** a reasonable answer. You can estimate the answer by rounding 28% to 25% and 850 to 900. Remember that 25% is 25 parts out of 100, which is equivalent to $\frac{1}{4}$. So divide 900 by 4.

$$900 \div 4 = 225 \text{ students} \leftarrow \text{Estimate}$$

Step 4 **Solve** the problem.

$$\text{part} = (0.28)(850) = 238 \quad \text{Multiply.}$$

└───── Write 28% as a decimal.

> Notice that the decimal point was moved two places to the left.

Step 5 **State the answer.** Raley's Markets has 238 student employees.

Step 6 **Check.** The exact answer, 238 students, is close to our estimate of 225 students.

◀ **Work Problem 3** at the Side.

Calculator Tip

If you are using a calculator, you could solve **Example 3** above like this.

$$0.28 \ (\times) \ 850 \ (=) \ 238$$

Or, if your calculator has a % key, check the instructions for an alternate method using the % key.

OBJECTIVE 2 **Find the whole using the percent proportion.** The next example shows how to use the percent proportion to find the whole.

Note

Remember, the *whole* is the entire quantity.

EXAMPLE 4 **Finding the Whole with the Percent Proportion**

(a) 8 iPods is 4% of what number of iPods?

Here the percent is 4, the whole is unknown, and the part is 8. Use the percent proportion to find the whole. Let *x* represent the unknown whole.

$$\frac{8}{x} = \frac{4}{100} \quad \text{or} \quad \frac{8}{x} = \frac{1}{25}$$

> Write $\frac{4}{100}$ as $\frac{1}{25}$ in lowest terms.

$$x \cdot 1 = 8 \cdot 25 \quad \text{Cross products}$$

$$x = 200$$

8 iPods is 4% of **200 iPods**.

Continued on Next Page

(b) 135 tourists is 15% of what number of tourists?
The percent is 15 and the part is 135.

Part → $\dfrac{135}{x}$ = $\dfrac{15}{100}$ ← Percent

Whole (unknown) → ← Always 100

> If the part and percent are given, the *whole* must be found.

$\dfrac{135}{x} = \dfrac{3}{20}$ ← Write $\frac{15}{100}$ as $\frac{3}{20}$ in lowest terms.

$x \cdot 3 = 135 \cdot 20$ Cross products

$x \cdot 3 = 2700$

$\dfrac{x \cdot \overset{1}{\cancel{3}}}{\underset{1}{\cancel{3}}} = \dfrac{2700}{3}$ Divide both sides by 3.

$x = 900$

135 tourists is 15% of **900 tourists**.

Work Problem **4** at the Side. ▶

EXAMPLE 5 Applying the Percent Proportion

At Newark Salt Works, 78 employees are absent because of illness. If this is 5% of the total number of employees, how many employees does the company have? Use the six problem-solving steps.

Step 1 **Read** the problem. The problem asks for the total number of employees.

Step 2 **Work out a plan.** From the information in the problem, the percent is 5 and the part of the total number of employees is 78. The total number of employees or entire quantity, which is the whole, is the unknown.

Step 3 **Estimate** a reasonable answer. Round the number of employees from 78 to 80. Then, 5% is equivalent to the fraction $\frac{1}{20}$, and 80 is $\frac{1}{20}$ of the total number of employees.

$80 \cdot 20 = 1600$ employees ← Estimate

Step 4 **Solve** the problem. Use the percent proportion to find the whole (the total number of employees).

Part → $\dfrac{78}{x}$ = $\dfrac{5}{100}$ ← Percent

Whole (unknown) → ← Always 100

$\dfrac{78}{x} = \dfrac{1}{20}$ Write $\frac{5}{100}$ as $\frac{1}{20}$ in lowest terms.

$x \cdot 1 = 78 \cdot 20$ Cross products

$x = 1560$

Step 5 **State the answer.** The company has **1560 employees**.

Step 6 **Check.** The exact answer, 1560 employees, is close to our estimate of 1600 employees.

Continued on Next Page

---

4 Use the percent proportion to find the unknown whole.

(a) 75 *American Idol* contestants are only 5% of what number who auditioned?

$\dfrac{75}{x} = \dfrac{5}{100}$ Write $\frac{5}{100}$ in lowest terms.

$\dfrac{75}{x} = \dfrac{1}{\rule{1cm}{0.4pt}}$

$x \cdot 1 = 75 \cdot 20$

$x = \rule{2cm}{0.4pt}$

(b) 28 antiques is 35% of what number of antiques?

(c) 387 customers is 36% of what number of customers?

(d) 292.5 miles is 37.5% of what number of miles?

Answers

4. **(a)** 20; 1500 auditioned   **(b)** 80 antiques
   (c) 1075 customers   **(d)** 780 miles

5 Use the six problem-solving steps and the percent proportion to solve each problem.

(GS) **(a)** A freeze resulted in a loss of 52% of an avocado crop. If the loss was 182 tons, find the total number of tons in the crop.

$$\frac{182}{x} = \frac{52}{100}$$ — Write $\frac{52}{100}$ in lowest terms.

$$\frac{182}{x} = \frac{13}{\underline{\quad}}$$

$$\frac{x \cdot \overset{1}{13}}{\underset{1}{13}} = \frac{4550}{13}$$

$$x = \underline{\quad}$$

(b) A factory batch of cake mix contains 900 pounds of sugar, which is 18%, by weight, of the entire batch. What is the total weight of the batch?

Note

To estimate the answer to **Example 5** on the previous page, 5% was changed to its fraction equivalent, $\frac{1}{20}$. Because 80 (rounded) is $\frac{1}{20}$ of the total employees, 80 was multiplied by 20 to get 1600, the estimated answer.

◀ **Work Problem 5 at the Side.**

OBJECTIVE 3 **Find the percent using the percent proportion.** If the part and the whole are known, the percent proportion can be used to find the percent.

EXAMPLE 6 **Using the Percent Proportion to Find the Percent**

(a) 13 coupons is what percent of 52 coupons?
The whole is 52 (follows *of*) and the part is 13. Next, find the percent.

The whole often follows the word "of". $\quad \dfrac{\text{part}}{\text{whole}} = \dfrac{\text{percent}}{100}$

Part → $\dfrac{13}{52}$ = $\dfrac{x}{100}$ ← Percent (unknown)
Whole → $\qquad\qquad$ ← Always 100

Write $\frac{13}{52}$ as $\frac{1}{4}$ in lowest terms. $\quad \dfrac{1}{4} = \dfrac{x}{100}$

Find the cross products.

$$4 \cdot x = 1 \cdot 100 \qquad \text{Cross products}$$

$$\frac{\overset{1}{4} \cdot x}{\underset{1}{4}} = \frac{100}{4} \qquad \text{Divide both sides by 4.}$$

$$x = 25$$

13 coupons is **25%** of 52 coupons.

(b) What percent of $500 is $100?
The whole is 500 (follows *of*) and the part is 100.

$$\frac{100}{500} = \frac{x}{100} \quad \leftarrow \text{Percent (unknown)}$$

Write $\frac{100}{500}$ as $\frac{1}{5}$ in lowest terms. $\quad \dfrac{1}{5} = \dfrac{x}{100}$

$$5 \cdot x = 1 \cdot 100 \qquad \text{Cross products}$$

$$5 \cdot x = 100$$

Remember to write the % symbol in the answer. $\quad \dfrac{\overset{1}{5} \cdot x}{\underset{1}{5}} = \dfrac{100}{5} \qquad \text{Divide both sides by 5.}$

$$x = 20$$

20% of $500 is $100.

Continued on Next Page

Answers
5. (a) 25; 350 tons **(b)** 5000 pounds

> **CAUTION**
>
> When finding the percent, be sure to label your answer with the percent symbol (%).

·········· Work Problem **6** at the Side. ▶

EXAMPLE 7 Applying the Percent Proportion

A roof is expected to last 20 years before needing replacement. If the roof is now 15 years old, what percent of the roof's life has been used?

Step 1 **Read** the problem. The problem asks for the percent of the roof's life that is already used.

Step 2 **Work out a plan.** The expected life of the roof is the entire quantity or *whole*, which is 20. The *part* of the roof's life that is already used is 15. Use the percent proportion to find the percent of the roof's life used.

Step 3 **Estimate** a reasonable answer. Since the roof is 15 years old, it is $\frac{15}{20}$ or $\frac{3}{4}$ used. Remember that $\frac{3}{4}$ is equivalent to 75%, so our estimate is 75%.

Step 4 **Solve** the problem. Let x represent the unknown percent.

$$\text{Part} \rightarrow \frac{15}{20} = \frac{x}{100} \quad \text{or} \quad \frac{3}{4} = \frac{x}{100} \quad \text{Write } \tfrac{15}{20} \text{ as } \tfrac{3}{4} \text{ in lowest terms.}$$
$$\text{Whole} \rightarrow$$

$$4 \cdot x = 3 \cdot 100 \quad \text{Cross products}$$
$$4 \cdot x = 300$$
$$\frac{\overset{1}{4} \cdot x}{\underset{1}{4}} = \frac{300}{4} \quad \text{Divide both sides by 4.}$$
$$x = 75$$

Step 5 **State the answer.** **75%** of the roof's life has been used.

Step 6 **Check.** The exact answer, 75%, matches our estimate of 75%.

·········· Work Problem **7** at the Side. ▶

EXAMPLE 8 Applying the Percent Proportion

Rainfall this year was 33 inches, while normal rainfall is only 30 inches. What percent of normal rainfall is this year's rainfall?

Step 1 **Read** the problem. The problem asks us to find what percent this year's rainfall is of normal rainfall.

Step 2 **Work out a plan.** The normal rainfall is the *whole*, which is 30. This year's rainfall is *all of normal rainfall and more*, or 33 (part = 33). You need to find the percent that this year's rainfall is of normal rainfall.

·········· **Continued on Next Page**

---

6 Use the percent proportion to solve each problem.

(GS) **(a)** $21 is what percent of $105?

Write $\frac{21}{105}$ in lowest terms.

$$\frac{21}{105} = \frac{x}{100}$$
$$\frac{1}{\underline{}} = \frac{x}{100}$$
$$5 \cdot x = 100$$
$$\frac{\overset{1}{5} \cdot x}{\underset{1}{5}} = \frac{100}{5}$$
$$x = \underline{}\%$$

(b) What percent of 320 Internet companies is 48 Internet companies?

(c) What percent of 2280 court trials is 1026 trials?

7 Solve each problem.

(a) The bid price on an auction item is $289 while the minimum acceptable price is $425. The bid price is what percent of the minimum?

(b) A laboratory technician completes 80 tests in one day. If 52 of these tests were completed in the morning, what percent of the tests were completed in the morning?

Answers

6. (a) 5; 20% (b) 15% (c) 45%

7. (a) 68% (b) 65%

8 Solve each problem.

GS (a) A new Toyota Prius Hybrid gets 32 miles per gallon on the highway and 48 miles per gallon around town. What percent of the highway mileage does the car get around town?

Write $\frac{48}{32}$ in lowest terms.

$$\frac{48}{32} = \frac{x}{100}$$

$$\frac{3}{2} = \frac{x}{100}$$

$$2 \cdot x = 3 \cdot 100$$

$$\frac{\overset{1}{2} \cdot x}{\underset{1}{2}} = \frac{\rule{1.5em}{0.4pt}}{2}$$

$$x = \underline{\quad}\%$$

(b) The service department set a goal of 360 service calls this week. If they made 432 service calls, find the percent of their goal that they achieved.

Step 3 **Estimate** a reasonable answer. The increase in rainfall is 3 inches and the whole is 30 inches. The increase is $\frac{3}{30}$ or $\frac{1}{10}$ which is 10%. The whole is 100%, so 100% + 10% = 110%, our estimate.

Step 4 **Solve** the problem. Let x represent the unknown percent.

$$\frac{33}{30} = \frac{x}{100} \quad \text{or} \quad \frac{11}{10} = \frac{x}{100} \qquad \text{Write } \tfrac{33}{30} \text{ as } \tfrac{11}{10} \text{ in lowest terms.}$$

$$10 \cdot x = 11 \cdot 100 \qquad \text{Cross products}$$

$$10 \cdot x = 1100$$

$$\frac{\overset{1}{10} \cdot x}{\underset{1}{10}} = \frac{1100}{10} \qquad \text{Divide both sides by 10.}$$

$$x = 110$$

> If the part is *greater than* the whole, the percent is greater than 100.

Step 5 **State the answer.** This year's rainfall is **110%** of normal rainfall.

Step 6 **Check.** The exact answer, 110%, matches our estimate of 110%.

◀ **Work Problem 8 at the Side.**

Note

In **Example 8**, the part (33 inches) is *greater* than the whole (30 inches). This can occur when there is an increase or a gain and we are comparing this *greater* amount (part) to the original amount (whole).

Answers

8. (a) 300; 150%   **(b)** 120%

6.4 Exercises

    MyMathLab®

1. CONCEPT CHECK   To find the part using the multiplication shortcut, use the formula

$$\text{part} = \underline{\hspace{1.5cm}} \cdot \underline{\hspace{1.5cm}}.$$

2. CONCEPT CHECK   Solve for the part when the percent is 38 and the whole is 200. Circle the correct answer.

   7.6   76   760   7600

*Find the part using the multiplication shortcut. **See Example 2.***

3. 35% of 120 test tubes
   part = (0.35)(120)

   part = _____

4. 20% of 1800 rentals
   part = (0.20)(1800)

   part = _____

5. 45% of 4080 military personnel

6. 12% of 3650 websites

7. 4% of 120 ft

8. 9% of $150

9. 150% of 210 files

10. 130% of 60 trees

11. 52.5% of 1560 trucks

12. 38.2% of 4250 loads

13. 2% of $164

14. 6% of $434

15. 225% of 680 tables

16. 110% of 150 apartments

17. 17.5% of 1040 cell phones

18. 46.1% of 843 kilograms

19. 0.9% of $2400

20. 0.3% of $1400

CONCEPT CHECK *Set up a percent proportion for each application problem. Do **not** try to solve for any unknowns.*

21. 80 e-mails is 25% of what number of e-mails?

22. 32 medical exams is 5% of what number of medical exams?

*Find the whole using the percent proportion. **See Example 4.***

23. 30% of what number of hay bales is 48 hay bales?
   part is 48; percent is 30; whole is unknown

$$\frac{48}{x} = \frac{30}{100} \quad \text{or} \quad \frac{48}{x} = \frac{3}{10}$$

   $3 \cdot x = 480$ Find cross products.

$$\frac{\overset{1}{\cancel{3}} \cdot x}{\underset{1}{\cancel{3}}} = \frac{480}{3}$$ Divide both sides by 3.

   $x = \underline{\hspace{1cm}}$

24. 55% of what number of experiments is 209 experiments?

$$\frac{209}{x} = \frac{55}{100} \quad \text{or} \quad \frac{209}{x} = \frac{11}{20}$$

   $11 \cdot x = 4180$ Find cross products.

$$\frac{\overset{1}{\cancel{11}} \cdot x}{\underset{1}{\cancel{11}}} = \frac{4180}{11}$$ Divide both sides by 11.

   $x = \underline{\hspace{1cm}}$

25. 495 successful students is 90% of what number of students?

26. 168 text messages is 28% of what number of text messages?

27. 462 mountain bikes is 140% of what number of mountain bikes?

28. 1496 graduates is 110% of what number of graduates?

29. $12\frac{1}{2}\%$ of what number is 350?

(*Hint:* Write $12\frac{1}{2}\%$ as 12.5%.)

30. $5\frac{1}{2}\%$ of what number is 176?

(*Hint:* Write $5\frac{1}{2}\%$ as 5.5%.)

CONCEPT CHECK *Set up a percent proportion for each application problem. Do **not** try to solve for any unknowns.*

31. 18 bean burritos is what percent of 36 bean burritos?

32. 62 hospital rooms is what percent of 248 hospital rooms?

*Find the percent using the percent proportion. Round answers to the nearest tenth if necessary. **See Example 6.***

33. 780 hybrid cars is what percent of 1500 hybrid cars?

34. 14 tweets is what percent of 700 tweets?

35. 27 downloaded songs is what percent of 1800 downloaded songs?

part = 27; whole = 1800

$$\frac{27}{1800} = \frac{x}{100} \quad \text{or} \quad \frac{3}{200} = \frac{x}{100}$$

$$200 \cdot x = 300$$

$$\frac{\overset{1}{\cancel{200}} \cdot x}{\underset{1}{\cancel{200}}} = \frac{300}{200}$$

$$x = \underline{\quad}$$

The percent is _____.

36. 60 cartons is what percent of 2400 cartons?

$$\frac{60}{2400} = \frac{x}{100} \quad \text{or} \quad \frac{1}{40} = \frac{x}{100}$$

$$40 \cdot x = 100$$

$$\frac{\overset{1}{\cancel{40}} \cdot x}{\underset{1}{\cancel{40}}} = \frac{100}{40}$$

$$x = \underline{\quad}$$

The percent is _____.

37. What percent of $344 is $64?

38. What percent of $398 is $14?

39. What percent of 250 tires is 23 tires?

40. What percent of 105 employees is 54 employees?

41. A student turned in the following answers on a test. You can tell that two of the answers are incorrect without even working the problems. Find the incorrect answers and explain how you identified them (without actually solving the problems).

50% of $84 is __$42__.

150% of $30 is __$20__.

25% of $16 is __$32__.

100% of $217 is __$217__.

42. Write a percent problem on any topic you choose. Be sure to include only two of the three components so that you can solve for the third component. Identify each component of the problem and then solve it.

Solve each application problem. Round percent answers to the nearest tenth if necessary. **See Examples 3, 5, 7, and 8.**

43. Bonnie Boehme, who works part-time, earns $240 per week and has 22% of this amount withheld for taxes, Social Security, and Medicare. Find

(a) the amount withheld.

(b) the amount remaining after the withholdings.

44. An estimated 29.5% of automobile crashes are caused by driver distractions such as using a cell phone or texting. (*Source:* National Conference of State Legislatures.) If there are 16,450 automobile crashes in a study, find

(a) the number of crashes caused by driver distractions.

(b) the number caused by other factors. Round to the nearest whole number.

The graph below shows when people check their e-mail. This data was gathered from 3020 adults and those surveyed could give more than one response. Use this graph to answer Exercises 45–48. Round to the nearest number.

45. Find the number of people who check their e-mail while working.

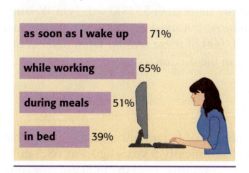

YOU'VE GOT MAIL!

I check my e-mail...

as soon as I wake up   71%

while working   65%

during meals   51%

in bed   39%

46. How many people check their e-mail while in bed?

47. Find the number of people who check their e-mail as soon as they wake up.

48. How many people check their e-mail during meals?

49. Each day in Los Angeles there are 48 million commuter trips. (*Source:* CNN.) If only 2% of these trips use public transportation, find

(a) the number of trips using public transportation.

(b) the number of trips not using public transportation.

50. At work, 68% of those surveyed said that they recycle bottles and cans while 56% recycle paper products and newspapers. (*Source*: Microsystems by Harris Interactive.) If 1750 workers were surveyed,

(a) how many recycle bottles and cans?

(b) how many recycle paper products and newspapers?

The bar graph below shows the percent of children 6–11 years of age who are neglecting dental hygiene. Use the information to answer Exercises 51–54.

DOWN IN THE MOUTH

The percent* of children ages 6–11 who are neglecting dental hygiene by:

67% — Going to bed without brushing
61% — Brushing less than one minute
49% — Not flossing
41% — Brushing only once a day

*Respondents allowed to choose multiple answers.
Source: Services for Crest.

51. What percent of the children brush their teeth before going to bed?

52. What percent of the children floss their teeth?

53. If 3400 children answered the questions for this survey, how many of the children brush less than one minute?

54. How many of the 3400 children in the survey brush only once a day?

55. A recent study examined 48,000 military jobs, such as Army attack helicopter pilot or Navy gunner's mate. It was found that only 960 of these jobs are filled by women. What percent of these jobs are filled by women? (*Source:* Rand's National Defense Research Institute.)

56. There are more than 55,000 words in *Webster's Dictionary,* but most educated people can identify only 20,000 of these words. What percent of the words in the dictionary can these people identify?

57. Ebony Durrant has 7.5% of her earnings deposited into her retirement plan. If $240 per month is deposited in the plan, find her monthly and yearly earnings.

58. The number of federal income tax returns that were filed electronically 5 years ago was 68.3 million, or 51% of all returns. Find the total number of income tax returns that were filed 5 years ago. Round to the nearest tenth of a million. (*Source:* Internal Revenue Service.)

The circle graph shows the percent of various ice cream brands purchased by the 1582 Americans in a recent survey. Use this information to answer Exercises 59–62.

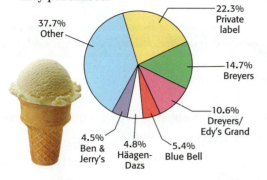

GET THE SCOOP

Last year, Americans spent a total of $4.5 billion on ice cream. These are the brands they purchased.

- 37.7% Other
- 22.3% Private label
- 14.7% Breyers
- 10.6% Dreyers/ Edy's Grand
- 4.5% Ben & Jerry's
- 4.8% Häagen-Dazs
- 5.4% Blue Bell

Source: Information Resources Inc.; NPD Group.

59. Of the specific brands purchased (not "Other" or "Private Label"), which brand was purchased most often?

60. What percent of ice cream purchases were "Private Label" or "Other" brands?

61. Find the number of people in the survey who said they purchase Häagen-Dazs. Round to the nearest whole number.

62. How many more people said they purchase Blue Bell brand than Ben & Jerry's brand? Round to the nearest whole number.

The bar graph below shows the percent of drivers in each age group who were stopped by police last year. Use the information to answer Exercises 63–66.

OH NO! FLASHING LIGHTS

Percent of drivers in each age group stopped by police:

- 16–19 POLICE 15.3%
- 20–29 POLICE 13.4%
- 30–39 POLICE 11.7%
- 40–49 POLICE 7.6%
- 50–59 POLICE 6.9%
- 60+ POLICE 4.2%

Source: Bureau of Justice Statistics.

63. What percent of drivers in the 16–19 age group were not pulled over by police?

64. What percent of drivers in the 60+ age group were not pulled over by police?

65. If 8000 drivers in the study were in the 20–29 age group, how many were pulled over by police?

66. How many of the 6000 drivers in the 30–39 age group were pulled over by police?

67. Marketing Intelligence Service says that there were 15,401 new products introduced last year. If 86% of the products introduced last year failed to reach their business objectives, find the number of products that were successful. (Round to the nearest whole number.)

68. The income earned on an investment is 8.5% of the amount invested. If the income is $12,750, find the amount of the investment.

69. Kathy West owns Banjo Shirt Company and sells handmade clothing online. If 85% of her 1540 customers paid for their orders using Pay Pal, how many of her customers used some other method of payment?

70. A family of four with a monthly income of $2900 spends 90% of its earnings and saves the rest. Find
(a) the monthly savings.
(b) the annual savings of this family.

Relating Concepts (Exercises 71–78) For Individual or Group Work

Knowing and using the percent proportion is useful when solving percent problems.
Work Exercises 71–78 in order.

71. In the percent proportion, part is to _____ as percent is to _____.

72. All percent problems involve a comparison between a part of something and the _____.

Use this Ramen Noodles label of nutrition facts to answer Exercises 73–78.
Read the label very carefully. Round to the nearest tenth of a percent.

Ramen Noodles

Nutrition Facts

Serving Size 1/2 Pkg.
(15 oz/42.5 g)
Servings Per Package 2
Calories 190
Calories from Fat 70

*Percent Daily Values (DV) are based on a 2,000 calorie diet.

Amount/serving	%DV*	Amount/serving	%DV*
Total Fat 8g	12%	Total Carbohydrates 27g	9%
Saturated Fat 4g	20%	Dietary Fiber Less Than 1g	3%
Cholesterol 0mg	0%	Sugars Less Than 1g	
Sodium 670mg	28%	Protein 4g	

Vitamin A 0% • Vitamin C 0% • Calcium 2% • Iron 4%

Calories Per Gram
Fat 9 • Carbohydrates 4 • Protein 4

Source: Ramen Noodles package.

73. How many calories per serving are from total carbohydrates?

74. The package label shows that the 27 grams (g) of carbohydrates in one serving are 9% of the recommended Daily Value (DV). What is the recommended daily value of carbohydrates?

75. Find the number of grams of total fat that are needed to meet the recommended Daily Value (DV) of fat. Round to the nearest gram.

76. The package label shows that the 4 grams (g) of saturated fat in one serving is 20% of the recommended Daily Value (DV). What is the recommended daily value of saturated fat?

77. Will a person eating two packages of Ramen Noodles in one day exceed their recommended daily value of sodium? Explain your answer.

78. How many packages of Ramen Noodles must be eaten in a day to meet the recommended daily value of fiber? Would this be possible? Would this result in good nutrition? (Round to the nearest whole package.)

6.5 Using the Percent Equation

In the last section you used a proportion to solve percent problems. In this section we show another way to solve these problems by using the **percent equation.** The percent equation is a rearrangement of the percent proportion.

OBJECTIVES

1 Use the percent equation to find the part.

2 Find the whole using the percent equation.

3 Find the percent using the percent equation.

> **Percent Equation**
>
> **part = percent • whole**
> *Be sure to write the percent as a decimal before using the equation.*

When using the percent proportion, we did *not* have to write the percent as a decimal because 100 was used in the denominator of the proportion. However, because there is no 100 in the percent *equation, we must* first write the percent as a decimal by dividing by 100.

Some of the examples solved earlier will be reworked using the percent equation. For comparison, you can look back in the previous section to see how these same problems were solved using proportions.

OBJECTIVE ▶ 1 Use the percent equation to find the part. The first example shows how to find the part.

EXAMPLE 1 Finding the Part

(a) Find 15% of $160.
Write 15% as the decimal 0.15. The whole, which comes after the word *of,* is 160. Next, use the percent equation. Let x represent the unknown part.

$$\textbf{part} = \textbf{percent} \cdot \textbf{whole}$$
$$x = (0.15)(160)$$

Multiply 0.15 and 160.   *15% must be written as the decimal 0.15*

$$x = 24$$

15% of $160 is **$24**.

(b) Find 110% of 80 cases.
Write 110% as the decimal 1.10. The whole is 80. Let x represent the unknown part.

$$\textbf{part} = \textbf{percent} \cdot \textbf{whole}$$
$$x = (1.10)(80)$$
$$x = 88$$ *Write 110% as a decimal. 110% = 1.10*

110% of 80 cases is **88 cases**.

(c) Find 0.4% of 250 patients.
Write 0.4% as the decimal 0.004. The whole is 250. Let x represent the unknown part.

$$\textbf{part} = \textbf{percent} \cdot \textbf{whole}$$
$$x = (0.004)(250)$$
$$x = 1$$ *Write 0.4% as a decimal. 0.4% = 0.004*

0.4% of 250 patients is **1 patient**.

· **Continued on Next Page**

1 Use the percent equation to find the part.

GS **(a)** 15% of 880 policyholders

part = percent • whole

part = (____) (____)
         ↑
Write 15% in decimal form.

part = ____

(b) 23% of 840 gallons

(c) 120% of $220

GS **(d)** 135% of $1080

part = percent • whole

part = (____) (____)

(e) 0.5% of 1200 fruit cups

(f) 0.25% of 1600 lab tests

To estimate the answer, think of 0.4% as approximately 0.5% or $\frac{1}{2}$ of 1%. Because 1% is $\frac{1}{100}$, you can use the shortcut from earlier in this chapter: Move the decimal point two places to the left.

$$1\% \text{ of } 250. = 2.5$$

Since 1% of 250 is 2.5, then 0.5% of 250 is 1.25 (because $2.5 \div 2 = 1.25$). So, the exact answer of 1 patient is reasonable.

◄ **Work Problem 1 at the Side.**

CAUTION

When using the percent equation, the percent must always be *changed to a decimal* before multiplying.

OBJECTIVE ▶ 2 **Find the whole using the percent equation.** The next example shows how to use the percent equation to find the whole.

Note

When the word *of* follows a percent, it is an indicator word for *multiply*.

EXAMPLE 2   Solving for the Whole

(a) 8 tables is 4% of what number of tables?

The part is 8 and the percent is 4% or the decimal 0.04. The whole is unknown.

$$8 \quad \text{is} \quad 4\% \text{ of what number?}$$

The word "of" can be used to identify the whole.

Now, use the percent equation.

$$\textbf{part = percent} \cdot \textbf{whole}$$

$$8 = (0.04)(x) \qquad \text{Let } x \text{ represent the unknown whole.}$$
$$\text{Write 4\% in decimal form as 0.04.}$$

$$\frac{8}{0.04} = \frac{\overset{1}{(0.\cancel{04})}(x)}{\underset{1}{0.\cancel{04}}} \qquad \text{Divide both sides by 0.04.}$$

$$200 = x \longleftarrow \text{Whole}$$

8 tables is 4% of **200 tables**.

(b) 135 tourists is 15% of what number of tourists?

Write 15% as 0.15. The part is 135. Use the percent equation to find the whole.

$$\textbf{part = percent} \cdot \textbf{whole}$$

$$135 = (0.15)(x) \qquad \text{Let } x \text{ represent the unknown whole.}$$
$$\text{Write 15\% in decimal form as 0.15.}$$

$$\frac{135}{0.15} = \frac{\overset{1}{(0.\cancel{15})}(x)}{\underset{1}{0.\cancel{15}}} \qquad \text{Divide both sides by 0.15.}$$

$$900 = x \longleftarrow \text{Whole}$$

135 tourists is 15% of **900 tourists**.

Continued on Next Page

(c) $8\frac{1}{2}\%$ of what number is 102?

Write $8\frac{1}{2}\%$ as 8.5%, or the decimal 0.085. The part is 102. Use the percent equation.

> Write 8.5% as a decimal.
> 8.5% = 0.085%

$$\text{part} = \text{percent} \cdot \text{whole}$$

$$102 = (0.085)(x) \quad \text{Let } x \text{ represent the unknown whole.}$$

$$\frac{102}{0.085} = \frac{\overset{1}{(0.085)}(x)}{\underset{1}{0.085}} \quad \text{Divide both sides by 0.085.}$$

$$1200 = x \longleftarrow \quad \text{Whole}$$

102 is $8\frac{1}{2}\%$ of **1200**.

Estimate the answer. Notice that $8\frac{1}{2}\%$ is close to 10%. If 102 is 10% of a number, then the number is 10 times 102, or 1020. So the exact answer, 1200, is reasonable.

CAUTION

In **Example 2(c)** above, $8\frac{1}{2}\%$ was first written as 8.5%, which is the decimal form of $8\frac{1}{2}\%$ ($8\frac{1}{2} = 8.5$). **The percent sign still remained** in 8.5%. Then 8.5% was changed to the decimal 0.085 so the equation could be solved.

················ **Work Problem ❷ at the Side.** ▶

OBJECTIVE ❸ **Find the percent using the percent equation.** The final example shows how to use the percent equation to find the percent.

EXAMPLE 3 **Finding the Percent**

(a) 13 auto mechanics is what percent of 52 auto mechanics?

Because 52 follows *of*, the whole is 52. The part is 13, and the percent is unknown. Use the percent equation.

$$\text{part} = \text{percent} \cdot \text{whole}$$

$$13 = x \cdot 52 \quad \text{Let } x \text{ represent the unknown percent.}$$

$$\frac{13}{52} = \frac{x \cdot \overset{1}{52}}{\underset{1}{52}} \quad \text{Divide both sides by 52.}$$

$$0.25 = x$$

$$0.25 \text{ is } 25\%$$

> You *must* write the decimal answer as a percent. 0.25 = 25%

13 auto mechanics is **25%** of 52 auto mechanics.

The equation can also be set up using *of* as an indicator word for multiplication and *is* as an indicator word for "is equal to."

13 is what percent of 52?

$$\downarrow\downarrow \qquad \downarrow \qquad \downarrow\downarrow$$

$$13 = \qquad x \qquad \cdot 52$$

$$13 = x \cdot 52 \quad \text{Same equation as above}$$

················ **Continued on Next Page**

---

❷ Find the whole using the percent equation.

GS **(a)** 18 dancers is 45% of what number of dancers?

$$\text{part} = \text{percent} \cdot \text{whole}$$

$$18 = (\underline{})(x)$$

$$\frac{18}{0.45} = \frac{\overset{1}{0.45}(x)}{\underset{1}{0.45}}$$

$$\underline{} = x$$

(b) 67.5 containers is 27% of what number of containers?

(c) 666 inoculations is 45% of what number of inoculations?

(d) $5\frac{1}{2}\%$ of what number of policies is 66 policies?

Answers

2. **(a)** 0.45; 40; 40 dancers **(b)** 250 containers
   (c) 1480 inoculations **(d)** 1200 policies

3 Find the percent using the percent equation.

(a) What percent of 35 Facebook friends is 7 Facebook friends?

part = percent • whole

$7 = x \cdot \underline{\quad}$

$$\frac{7}{35} = \frac{x \cdot \overset{1}{\cancel{35}}}{\underset{1}{\cancel{35}}}$$

$0.20 = x$

0.20 is $\underline{\quad}\%$

(b) 34 post office boxes is what percent of 85 post office boxes?

(c) What percent of 920 invitations is 1288 invitations?

(d) 9 world-class runners is what percent of 1125 runners?

(b) What percent of $500 is $100?
The whole is 500 and the part is 100. Let x represent the unknown percent.

part = percent • whole

$100 = x \cdot 500$ Let x represent the unknown percent.

$$\frac{100}{500} = \frac{x \cdot \overset{1}{\cancel{500}}}{\underset{1}{\cancel{500}}}$$ Divide both sides by 500.

$0.20 = x$

0.20 is 20%

> Write the decimal as a percent.
> $0.20 = 20\%$

20% of $500 is $100.

(c) What percent of $300 is $390?
The whole is 300 and the part is 390. Let x represent the unknown percent.

part = percent • whole

$390 = x \cdot 300$ Let x represent the unknown percent.

$$\frac{390}{300} = \frac{x \cdot \overset{1}{\cancel{300}}}{\underset{1}{\cancel{300}}}$$ Divide both sides by 300.

$1.3 = x$

1.3 is 130%

> Write the decimal as a percent.
> $1.3 = 1.30 = 130\%$

130% of $300 is $390.

(d) 6 ladders is what percent of 1200 ladders?
Since 1200 follows *of*, the whole is 1200. The part is 6.

part = percent • whole

$6 = x \cdot 1200$ Let x represent the unknown percent.

$$\frac{6}{1200} = \frac{x \cdot \overset{1}{\cancel{1200}}}{\underset{1}{\cancel{1200}}}$$ Divide both sides by 1200.

$0.005 = x$

0.005 is 0.5%

> Write the decimal as a percent.
> $0.005 = 0.5\%$

6 ladders is **0.5%** of 1200 ladders.

You can estimate the answer because 1% of 1200 ladders is found by moving the decimal point two places to the left in 1200, resulting in 12. Since 6 ladders is half of 12 ladders, our answer should be $\frac{1}{2}$ of 1% or 0.5%. Our exact answer matches the estimate.

> **CAUTION**
>
> When you use the percent equation to solve for an unknown percent, the answer will always be in decimal form. Notice that in **Example 3 (a)**, **(b), (c), and (d)** above, **the decimal answer had to be changed to a percent** by multiplying by 100 and attaching the percent symbol. The answers became: (a) 0.25 = 25%; (b) 0.20 = 20%; (c) 1.3 = 130%; and (d) 0.005 = 0.5%.

◀ **Work Problem 3** at the Side.

6.5 Exercises

 FOR EXTRA HELP

 Download the MyDashBoard App

MyMathLab®

CONCEPT CHECK *When solving for the part in a percent problem, the percent and the whole must be identified. Write either* percent *or* whole *next to the values taken from each problem.*

1. 46% of 780 text messages

    46 _____   780 _____

2. 14% of 1800 forums

    1800 _____   14 _____

Find the part using the percent equation. **See Example 1.**

3. 25% of 1080 blood donors

GS part = percent • whole

    $x = (0.25)(1080)$

    $x =$ _____

4. 19% of 700 MP3 players

GS part = percent • whole

    $x = (0.19)(700)$

    $x =$ _____

5. 45% of 3000 bath towels

6. 75% of 360 dosages

7. 32% of 260 quarts

8. 44% of 430 liters

9. 140% of 2500 air bags

10. 145% of 580 hamburgers

11. 12.4% of 8300 meters

12. 26.4% of 4700 miles

13. 0.8% of $520

14. 0.3% of $480

CONCEPT CHECK *When solving for the whole in a percent problem, the part and the percent must be identified. Write either* part *or* percent *next to the values taken from each problem.*

15. 70% of what number of backpackers is 476 backpackers?

    70 _____   476 _____

16. 270 lab tests is 45% of what number of lab tests?

    270 _____   45 _____

Find the whole using the percent equation. **See Example 2.**

17. 24 patients is 15% of what number of patients?

GS part = percent • whole

    $24 = (0.15)(x)$

    $\dfrac{24}{0.15} = \dfrac{(0.\overset{1}{\cancel{15}})(x)}{(0.\underset{1}{\cancel{15}})}$

    $160 = x$

18. 32 classrooms is 20% of what number of classrooms?

GS part = percent • whole

    $32 = (0.2)(x)$

    $\dfrac{32}{0.2} = \dfrac{(0.\overset{1}{\cancel{2}})(x)}{(0.\underset{1}{\cancel{2}})}$

    $160 = x$

19. 40% of what number of salads is 130 salads?

20. 75% of what number of wrenches is 675 wrenches?

21. $12\frac{1}{2}\%$ of what number of people is 135 people?

22. $18\frac{1}{2}\%$ of what number of batteries is 370 batteries?

23. $1\frac{1}{4}\%$ of what number of gallons is 3.75 gallons?

24. $2\frac{1}{4}\%$ of what number of files is 9 files?

CONCEPT CHECK *Fill in each blank with the correct response.*

25. When using the percent equation to solve for percent, you must always change the decimal answer to _____ by moving the decimal point _____ places to the _____ and attaching a _____ symbol.

26. When using the percent equation to solve for part or whole, you must always change the percent to a decimal by moving the _____ point _____ places to the _____ and dropping the _____ symbol.

Find the percent using the percent equation. ***See Example 3.***

27. 114 tuxedos is what percent of 150 tuxedos?

    part = percent • whole

$$114 = x \cdot 150$$

$$\frac{114}{150} = \frac{x \cdot \overset{1}{\cancel{150}}}{\underset{1}{\cancel{150}}}$$

$$0.76 = x$$

    change 0.76 to a percent = _____

28. 75 iPods is what percent of 125 iPods?

    part = percent • whole

$$75 = x \cdot 125$$

$$\frac{75}{125} = \frac{x \cdot \overset{1}{\cancel{125}}}{\underset{1}{\cancel{125}}}$$

$$0.6 = x$$

    0.60 is _____

29. What percent of 160 liters is 2.4 liters?

30. What percent of 600 meters is 7.5 meters?

31. 170 cartons is what percent of 68 cartons?

32. 612 orders is what percent of 425 orders?

33. When using the percent equation, the percent must always be changed to a decimal before doing any calculations. Show and explain how to change a fraction percent to a decimal. Use $2\frac{1}{2}\%$ in your explanation.

34. Suppose a problem on your homework assignment was, "Find $\frac{1}{2}\%$ of $1300." Your classmates got answers of $0.65, $6.50, $65, and $650. Which answer is correct? How and why are they getting all of these answers? Explain.

Solve each application problem.

35. A study of office workers found that 27% would like more storage space. If there are 14 million office workers, how many want more storage space? (*Source:* Steelcase Workplace Index.)

36. Most shampoos contain 75% to 90% water. If a 16-ounce bottle of shampoo contains 78% water, find the number of ounces of water in the bottle. Round to the nearest tenth of an ounce.

37. The results of a recent poll are shown below. If 12,500 people were included in the poll, find

   (a) the number who would rather admit their age.

   (b) the number who would rather admit their weight.

   (c) the number who would rather admit their salary.

WHICH WOULD YOU RATHER ADMIT: AGE, WEIGHT, OR SALARY?

Age 77%　　Weight 15%　　Salary 8%

Source: Parade magazine.

38. Three Spam Mobiles travel throughout the country promoting Spam and Spam Lite. By using these kitchens on wheels, the annual goal is to give 1.5 million taste samples to the public. If 58.6% of the goal has been met, find the number of samples that have been given. (*Source:* Hormel Foods Corporation.)

39. In the United States, 98% of all households have a refrigerator. (*Source:* American Housing Authority.) Out of 18,000 households,

   (a) how many are expected to have a refrigerator?

   (b) how many are expected to not have a refrigerator?

40. In the United States, 56% of the households have a dishwasher. (*Source:* American Housing Authority.) Out of 214,500 households,

   (a) how many are expected to have a dishwasher?

   (b) how many are expected to not have a dishwasher?

41. In the United States, 2 of the 50 states (Alaska and Louisiana) do not have any drive-in movies. The remaining states do have drive-in movies. (*Source: USA Today.*)

   (a) What percent of the states do not have drive-in movies?

   (b) What percent have drive-in movies?

42. In a recent survey of 3450 attorneys, 897 said that they take work home with them during the week and 1311 said that they take work home with them on the weekend. Find

   (a) the percent of the attorneys who take work home during the week.

   (b) the percent of attorneys who take work home on the weekend.

43. In a survey of 1250 Americans, 461 rated their health as excellent. What percent of these Americans rate their health as excellent? Round to the nearest tenth of a percent. (*Source:* National Health Interview Survey.)

44. General Nutrition Center now has 3200 stores and plans to add 450 more stores. Find the percent of additional stores that they have planned. Round to the nearest tenth of a percent.

The graph shows the type of day care used by families for their preschool children.
Assume that 7800 families were surveyed to gather the data. Use this graph to answer
Exercises 45–48.

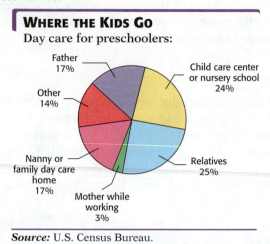

WHERE THE KIDS GO
Day care for preschoolers:

Father 17%
Other 14%
Child care center or nursery school 24%
Nanny or family day care home 17%
Relatives 25%
Mother while working 3%

Source: U.S. Census Bureau.

45. **(a)** Which type of day care was used most often?

   (b) Find the number of families who used this type of day care.

46. **(a)** Which type of day care was used least often?

   (b) What number of families used this type of day care?

47. How many families used a nanny or family day care home?

48. Find the number of families who left their preschoolers with a child care center or nursery school.

49. Last year, college seniors who graduated carried an average of $22,650 in student loan debt. This year the average amount of loan debt has increased 6%. Find the average amount of student loan debt this year. (*Source:* U.S. Department of Education.)

50. Cyndy Mason has 8.5% of her monthly earnings deposited into the credit union. If this amounts to $131.75 per month, find her annual earnings.

51. The Chevy Camaro was introduced in 1967. Sales that year were 220,917 Camaros, which was 46.2% of the number of Ford Mustangs sold in the same year. Find the number of Mustangs sold in 1967. Round to the nearest whole number.

52. Chris Goodwin is a waiter and has sales of $822.25 on Saturday. If this is 28.6% of his sales for the week, find his weekly sales.

53. J & K Mustang has increased the sale of auto parts by $32\frac{1}{2}$% over last year. If the sale of parts last year amounted to $385,200, find the amount of sales this year.

54. An ad for steel-belted radial tires promises 15% better mileage. If mileage had been 25.6 miles per gallon in the past, what mileage could be expected after these tires are installed? Round to the nearest tenth of a mile.

55. A Polaris Vac-Sweep is priced at $524 with an allowed trade-in of $125 on an old unit. If sales tax of $7\frac{3}{4}$% is charged on the price of the new Polaris unit before the trade-in, find the total cost to the customer after receiving the trade-in. (*Hint:* Trade-in is subtracted last.)

56. General Motors car sales in China were 36.7% greater than last year's sales of 629,778 cars. Find this year's sales. Round to the nearest whole number. (*Source:* General Motors Corporation.)

Summary Exercises *Using the Percent Proportion and Percent Equation*

CONCEPT CHECK   *Write each percent as a decimal and each decimal as a percent.*

1. 6.25%   **2.** 380%   **3.** 0.375   **4.** 0.006

CONCEPT CHECK   *Write each percent as a fraction or mixed number in lowest terms and each fraction as a percent.*

5. 87.5%   **6.** 160%   **7.** $\dfrac{5}{8}$ **8.** $\dfrac{1}{125}$

The circle graph shows what 1100 adults dread the most about the Thanksgiving holiday. Use this graph to answer Exercises 9–12. Write each answer as a fraction in lowest terms, as a decimal, and as a percent.

9. What portion of the adults have a fear of putting on weight?

THANKSGIVING TURNOFFS
What adults dread most about the holiday:

None of these/don't know — 253
Fear of putting on weight — 286
Too much family time — 66
The expense — 99
Traveling — 187
Hassle of cooking — 209

Source: Opinion Research Corporation.

10. What portion of the adults dread traveling?

11. Find the portion who dread too much family time.

12. Find the portion who dread the expense.

Find the part, whole, or percent as indicated. Round percent answers to the nearest tenth if necessary.

13. 6% of $780 is what amount?

14. 343 cable customers is 35% of how many cable customers?

15. What percent of 320 policies is 176 policies?

16. 0.8% of 3500 screening exams is how many exams?

17. 1016.4 acres is 280% of what number of acres?

18. What percent of 658 circuits is 18 circuits?

Solve each application problem. Round to the nearest cent or nearest tenth of a percent if necessary.

19. In a poll of 1582 people, 10.6% said that they prefer to purchase Dreyer's/Edy's brand ice cream. Find the number of people who said they prefer Dreyer's/Edy's brand ice cream. Round to the nearest whole number. (*Source:* Information Resources Inc; NPD Group.)

20. A real estate broker wants to purchase a smart phone priced at $399. If the sales tax rate is 7.75%, find the total price including the sales tax. (*Source:* Real Estate Technology.)

21. The number of outstanding student loans last year was 3,370,000. If 7.2% of these loans were in default, find the number of student loans that were in default. Round to the nearest thousand. (*Source:* U.S. Department of Education.)

22. In a recent survey of dog owners, it was found that 901, or 34%, of the owners take their dogs on vacation with them. Find the number of dog owners in the survey who do not take their dogs on vacation. (*Source:* American Animal Hospital Association.)

23. The distance around Lake Tahoe is 65 miles. If Lino Delgadillo has completed 39 miles of the Lake Tahoe run, what percent of the run remains?

24. Beutler Heating and Air Conditioning plans to lay off 45 of its 1215 workers. What percent of the workers will be laid off? (*Source:* Beutler Heating and Air Conditioning.)

25. Last year there were 98.3 million tax returns filed electronically. If this was 69% of all returns filed, how many total returns were filed? Round to the nearest tenth of a million. (*Source*: Internal Revenue Service.)

26. The number of ski-lift tickets sold last week was 3820. This week's sales are down 10.5%. Find the number of ski-lift tickets sold this week. Round to the nearest whole number.

6.6 Solving Application Problems with Percent

Percent has many applications in our daily lives. This section discusses percent as it applies to sales tax, commissions, discounts, and the percent of change (increase and decrease).

OBJECTIVES

1. Find sales tax.
2. Find commissions.
3. Find the discount and sale price.
4. Find the percent of change.

OBJECTIVE **1** **Find sales tax.** States, countries, and cities often collect taxes on sales to customers. The **sales tax** is a percent of the cost of an item. The following formula for finding sales tax is based on the percent equation.

Sales Tax Formula

$$\underbrace{\text{part}}_{\text{amount of sales tax}} = \underbrace{\text{percent}}_{\text{rate of tax}} \cdot \underbrace{\text{whole}}_{\text{cost of item}}$$

VOCABULARY TIP

Sales tax The sales tax is always part of the cost of an item. The cost of the item is the **whole,** the sales tax rate is the **percent,** and the amount of the sales tax is the **part.**

EXAMPLE 1 Solving for Sales Tax

Office Max sells a laptop computer for $499. If the sales tax rate is 5%, how much tax is paid? What is the total cost of the laptop computer? Use the six problem-solving steps.

Step 1 **Read** the problem. The problem asks for the total cost of the laptop computer, including the sales tax.

Step 2 **Work out a plan.** Use the sales tax formula to find the amount of sales tax. Write the tax rate (5%) as a decimal (0.05). The cost of the item is $499. Use the letter *a* to represent the unknown *amount* of tax. Add the sales tax to the cost of the item.

Step 3 **Estimate** a reasonable answer. Round $499 to $500. Recall that 5% is equivalent to $\frac{1}{20}$, so divide $500 by 20 to estimate the tax.

$$\$500 \div 20 = \$25 \text{ tax}$$

The total estimated cost is $500 + $25 = $525. ← Estimate

Step 4 **Solve** the problem.

$$\underbrace{\text{part}}_{\text{amount of sales tax}} = \underbrace{\text{percent}}_{\text{rate of tax}} \cdot \underbrace{\text{whole}}_{\text{cost of item}}$$

$$a = (5\%)(\$499)$$

Write 5% as a decimal 0.05. $a = (0.05)(\$499)$

$$a = \$24.95 \quad \text{Sales tax}$$

The tax paid on the laptop computer is $24.95. The customer would pay a total cost of $499 + $24.95 = $523.95.

Step 5 **State the answer.** The total cost of the laptop computer is $523.95.

Step 6 **Check.** The exact answer, $523.95, is close to our estimate of $525.

·········· **Work Problem** **1** **at the Side.** ▶

1 Suppose the sales tax rate in your state is 6%. Find the amount of the tax and the total you would pay for each item.

GS (a) $29 Little League bat

$$\text{Sales tax} = (0.06)(\$\underline{\quad})$$
$$= \$\underline{\quad}$$
$$\text{Total cost} = \$29 + \$\underline{\quad}$$
$$= \$\underline{\quad}$$

(b) $349 home theater system

(c) $1287 leather chair and ottoman

(d) $24,500 pickup truck

Answers

1. (a) $29; $1.74; $1.74; $30.74
(b) $20.94; $369.94 (c) $77.22; $1364.22
(d) $1470; $25,970

2 Find the rate of sales tax.

GS **(a)** The tax on a $320 patio set is $25.60.

$$\$25.60 = r \cdot \underline{\quad\quad}$$

$$\frac{25.6}{320} = \frac{r \cdot \overset{1}{\cancel{320}}}{\underset{1}{\cancel{320}}}$$

$$r = 0.08 = \underline{\quad\quad}$$

(b) The tax on a $96 park bench is $6.24.

(c) The tax on a $22,995 Dodge Charger is $919.80.

EXAMPLE 2 **Finding the Sales Tax Rate**

⊞ The sales tax on a $24,200 Ford Edge is $1573. Find the rate of the sales tax.

Step 1 **Read** the problem. This problem asks us to find the sales tax rate.

Step 2 **Work out a plan.** Use the sales tax formula.

sales tax = rate of tax • cost of item

Solve for the rate of tax, which is the percent. The cost of the Ford Edge (the whole) is $24,200, and the amount of sales tax (the part) is $1573. Use the letter r to represent the unknown *rate* of tax (the percent).

Step 3 **Estimate** a reasonable answer. Round $24,200 to $24,000 and round $1573 to $1600. The sales tax is $\frac{1600}{24,000}$ or $\frac{1}{15}$ of the cost of the car. So divide 1 by 15 to estimate the percent (rate) of sales tax.

$$\frac{1}{15} = 0.06\overline{6} \approx 7\% \leftarrow \text{Rounded estimate}$$

Step 4 **Solve** the problem.

sales tax = rate of tax • cost of item

$$\$1573 = (r) \quad (\$24,200)$$

$$\frac{1573}{24,200} = \frac{(r)(\overset{1}{\cancel{24,200}})}{\underset{1}{\cancel{24,200}}} \quad \text{Divide both sides by 24,200.}$$

$$0.065 = r \qquad \boxed{\begin{array}{l}\text{Write the decimal} \\ \text{as a percent.} \\ 0.065 = 6.5\%.\end{array}}$$

$$0.065 \text{ is } 6.5\%$$

Step 5 **State the answer.** The sales tax rate is 6.5% or $6\frac{1}{2}\%$.

Step 6 **Check.** The exact answer, $6\frac{1}{2}\%$, is close to our estimate of 7%.

◀ **Work Problem** **2** **at the Side.**

Note

You can use the sales tax formula to find the amount of sales tax, the cost of an item, or the rate of sales tax (the percent).

OBJECTIVE **2** **Find commissions.** Many salespeople are paid by *commission* rather than an hourly wage. If you are paid by **commission,** you are paid a certain percent of your total sales dollars. The formula below for finding the commission is based on the percent equation.

Commission Formula

part	**=**	**percent**	**•**	**whole**
↓		↓		↓

amount of commission = rate of commission • amount of sales

Answers

2. (a) $320; 8%   **(b)** 6.5% or $6\frac{1}{2}\%$ **(c)** 4%

EXAMPLE 3	Determining the Amount of Commission

Lynn Cochran had automotive tool sales of $42,500 last month. If Cochran's commission rate is 9%, find the amount of his commission.

Step 1 **Read** the problem. The problem asks for the amount of commission that Cochran earned.

Step 2 **Work out a plan.** Use the commission formula. Write the rate of commission (9%) as a decimal (0.09). The amount of Cochran's sales ($42,500) is the whole. Use the letter c to represent the unknown *amount* of commission.

Step 3 **Estimate** a reasonable answer. Round the commission rate of 9% to 10%. Round the amount of sales from $42,500 to $40,000. Since 10% is equivalent to $\frac{1}{10}$, divide $40,000 by 10 to estimate the amount of commission: $40,000 ÷ 10 gives $4000 as our estimate.

Step 4 **Solve** the problem.

amount of commission = rate of commission • amount of sales

$$c = (9\%)(\$42{,}500)$$
$$c = (0.09)(\$42{,}500)$$
$$c = \$3825 \quad \text{Amount of commission}$$

Step 5 **State the answer.** Cochran earned a commission of $3825.

Step 6 **Check.** The exact answer, $3825, is close to our estimate of $4000.

··· Work Problem ③ at the Side. ▶

EXAMPLE 4	Finding the Rate of Commission

Carol Merrigan earned a commission of $510 for selling $17,000 worth of shipping supplies. Find the rate of commission.

Step 1 **Read** the problem. We must find the rate (percent) of commission.

Step 2 **Work out a plan.** You could use the commission formula. Another approach is to use the percent proportion. The *whole* is $17,000, the *part* is $510, and the *percent* (rate of commission) is the unknown.

Step 3 **Estimate** a reasonable answer. Round the commission, $510, to $500, and round $17,000 to $20,000. The commission in fraction form is $\frac{\$500}{\$20{,}000}$, which simplifies to $\frac{1}{40}$. Changing $\frac{1}{40}$ to a percent gives $2\frac{1}{2}\%$ (rounded), as our estimate.

Step 4 **Solve** the problem.

$$\frac{\text{part}}{\text{whole}} = \frac{x}{100} \leftarrow \text{Percent (unknown)}$$

> Think: The rate of the commission is the percent.

$$\frac{510}{17{,}000} = \frac{x}{100}$$

$$17{,}000 \cdot x = 510 \cdot 100 \quad \text{Cross products}$$

$$\frac{\overset{1}{\cancel{17{,}000}} \cdot x}{\underset{1}{\cancel{17{,}000}}} = \frac{51{,}000}{17{,}000} \quad \text{Divide both sides by 17,000.}$$

$$x = 3$$

·············· **Continued on Next Page**

---

③ Find the amount of commission.

GS **(a)** Jill Owens sells dental equipment at a commission rate of 11% and has sales for the month of $38,700.

Use the letter c to represent the amount of commission.

$$c = (11\%)(\$\underline{\quad\quad})$$
$$c = (0.11)(38{,}700)$$
$$c = \$\underline{\quad\quad}$$

(b) Last month Alyssa Paige sold a home for $189,500 and earned a commission of 6%.

Answers

3. **(a)** $38,700; $4257 **(b)** $11,370

4 Find the rate of commission.

GS (a) A commission of $450 is earned on the sale of computer products worth $22,500.

$$\frac{450}{22,500} = \frac{x}{100}$$

$$22,500 \cdot x = 450 \cdot 100$$

$$\frac{\overset{1}{\cancel{22,500}} \cdot x}{\underset{1}{\cancel{22,500}}} = \frac{45,000}{22,500}$$

$$x = \underline{\qquad}$$

(b) Jamal Story earned $2898 for selling office furniture worth $32,200.

5 Find the amount of the discount and the sale price.

GS (a) An Easy-Boy leather recliner originally priced at $950 is offered at a 42% discount.

Discount = (0.42) ($950)

= $399

Sale price = $950 − $____

= $____

(b) Walmart has women's sweater sets on sale at 35% off. One sweater set was originally priced at $30.

Step 5 **State the answer.** The rate of commission is 3%.

Step 6 **Check.** The exact answer of 3% is close to our estimate of $2\frac{1}{2}\%$.

⊲ **Work Problem 4 at the Side.**

OBJECTIVE 3 **Find the discount and sale price.** Most of us prefer buying things when they are on sale. A store will reduce prices, or **discount,** to attract additional customers. Use the following formula to find the discount and the sale price.

Discount Formula and Sale Price Formula

amount of discount = rate (or percent) of discount • original price

sale price = original price − amount of discount

EXAMPLE 5 **Finding a Sale Price**

Whitings Oak Furniture Store has a home theater cabinet with an original price of $840 on sale at 15% off. Find the sale price of the cabinet.

Step 1 **Read** the problem. This problem asks for the price of a home theater cabinet after a discount of 15%.

Step 2 **Work out a plan.** The problem is solved in two steps. First, find the amount of the discount, that is, the amount that will be "taken off" (subtracted), by multiplying the original price ($840) by the rate of the discount (15%). The second step is to subtract the amount of discount from the original price. This gives you the sale price, which is what you will actually pay for the home theater cabinet.

Step 3 **Estimate** a reasonable answer. Round the original price from $840 to $800, and the rate of discount from 15% to 20%. Since 20% is equivalent to $\frac{1}{5}$, the estimated discount is $800 ÷ 5 = $160, so the estimated sale price is $800 − $160 = $640.

Step 4 **Solve** the problem. First find the exact amount of the discount.

amount of discount = rate of discount • original price

$a = (0.15)($840)$ Write 15% as a decimal.

$a = 126 Amount of discount

Now find the sale price of the home theater cabinet by subtracting the amount of the discount ($126) from the original price.

sale price = original price − amount of discount

= $840 − $126   Remember to subtract the discount from the original price to get the sale price.

= $714   Sale price

Step 5 **State the answer.** The sale price of the home theater cabinet is $714.

Step 6 **Check.** The exact answer, $714, is close to our estimate of $640.

⊲ **Work Problem 5 at the Side.**

Answers

4. (a) 2% (b) 9%
5. (a) $399; $551 (b) $10.50; $19.50

Calculator Tip

In **Example 5** on the previous page, you can use a scientific calculator to find the amount of discount and subtract the discount from the original price, all in one step.

$$840 \;\ominus\; .15 \;\otimes\; 840 \;\ominus\; 714$$

↑ Original price | Amount of discount | ↑ Sale price

A scientific calculator observes the order of operations, so it will automatically do the multiplication before the subtraction.

OBJECTIVE ▶ **4** **Find the percent of change.** We are often interested in looking at increases or decreases in sales, production, population, and many other items. This type of problem involves finding the *percent of change*. Use the following steps to find the **percent of increase.**

> **Finding the Percent of Increase**
>
> **Step 1** Use subtraction to find the amount of increase.
>
> **Step 2** Use the percent proportion to find the percent of increase.
>
> $$\frac{\text{amount of increase (part)}}{\text{original value (whole)}} = \frac{\text{percent}}{100}$$

EXAMPLE 6 Finding the Percent of Increase

Attendance at county parks climbed from 18,300 last month to 56,730 this month. Find the percent of increase.

Step 1 **Read** the problem. The problem asks for the percent of increase.

Step 2 **Work out a plan.** Subtract the attendance last month (18,300) from the attendance this month (56,730) to find the amount of increase in attendance. Next, use the percent proportion. The whole is 18,300 (last month's original attendance), the part is 38,430 (amount of increase in attendance), and the percent is unknown.

Step 3 **Estimate** a reasonable answer. Round 18,300 to 20,000 and 56,730 to 60,000. The amount of increase is $60,000 - 20,000 = 40,000$. Since 40,000 (the increase) is *twice* as large as the original amount, the estimated percent of increase is 200%.

Step 4 **Solve** the problem.

$$56,730 - 18,300 = 38,430 \quad \text{Subtract to find the } \textbf{amount of increase} \text{ in attendance.}$$

Amount of increase → $\dfrac{38,430}{18,300} = \dfrac{x}{100}$ Percent proportion

Use the *original* value of 18,300 (**not** 56,730).

Solve this proportion to find that $x = 210$.

Step 5 **State the answer.** The percent of increase is 210%.

Step 6 **Check.** The exact answer, 210%, is close to our estimate of 200%.

⋯⋯⋯⋯⋯⋯⋯⋯⋯⋯⋯⋯⋯⋯⋯⋯⋯ **Work Problem** **6** **at the Side.** ▶

VOCABULARY TIP

Percent of increase or decrease The change from the original value is the amount of increase or decrease. The amount of increase or decrease is the **part** and the original value is **always the whole.**

6 Find the percent of increase.

(a) A manufacturer of snowboards increased production from 14,100 units last year to 19,035 this year.

$$19,035 - 14,100 = \underline{\hspace{1.5cm}}$$
increase

$$\left.\begin{array}{l}\text{Increase} \rightarrow \\ \text{Original} \\ \text{value}\end{array}\right\} \;\; \frac{}{14,100} = \frac{x}{100}$$

$$\frac{\overset{1}{\cancel{14,100}} \cdot x}{\underset{1}{\cancel{14,100}}} = \frac{493,500}{14,100}$$

$$x = 35$$

Percent of increase is ____.

(b) The number of flu cases rose from 496 cases last week to 620 this week.

Answers

6. **(a)** 4935; 4935; 35% **(b)** 25%

7 Find the percent of decrease.

GS (a) The number of service calls fell from 380 last month to 285 this month.

$$380 - 285 = \underline{\hspace{1cm}} \text{ decrease}$$

$$\begin{array}{l}\text{Decrease} \rightarrow \\ \text{Original} \\ \text{value} \end{array}\Bigg\} \frac{\underline{\hspace{1cm}}}{380} = \frac{x}{100}$$

$$\frac{\overset{1}{\cancel{380}} \cdot x}{\underset{1}{\cancel{380}}} = \frac{9500}{380}$$

$$x = 25$$

Percent of decrease is ____.

(b) The number of workers applying for unemployment fell from 4850 last month to 3977 this month.

Use the following steps to find the **percent of decrease.**

Finding the Percent of Decrease

Step 1   Use subtraction to find the amount of decrease.

Step 2   Use the percent proportion to find the percent of decrease.

$$\frac{\text{amount of decrease (part)}}{\text{original value (whole)}} = \frac{\text{percent}}{100}$$

EXAMPLE 7   **Finding the Percent of Decrease**

The number of production employees this week fell to 1406 people from 1480 people last week. Find the percent of decrease.

Step 1   **Read** the problem. The problem asks for the percent of decrease.

Step 2   **Work out a plan.** Subtract the number of employees this week (1406) from the number of employees last week (1480) to find the amount of decrease. Then, use the percent proportion. The whole is 1480 (last week's *original* number of employees), the part is 74 (amount of decrease in employees), and the percent is unknown.

Step 3   **Estimate** a reasonable answer. Estimate the answer by rounding 1406 to 1400 and 1480 to 1500. The decrease is $1500 - 1400 = 100$. Since 100 is $\frac{1}{15}$ of 1500, our estimate is $1 \div 15 \approx 0.07$ or 7%.

Step 4   **Solve** the problem.

$$1480 - 1406 = 74 \quad \begin{array}{l}\text{Subtract to find the} \\ \textbf{amount of decrease} \text{ in} \\ \text{number of employees.}\end{array}$$

Amount of decrease $\rightarrow \dfrac{74}{1480} = \dfrac{x}{100}$ Percent proportion

Use the original value of 1480 (**not** 1406).

Solve this proportion to find that $x = 5$.

Step 5   **State the answer.** The percent of decrease is 5%.

Step 6   **Check.** The exact answer, 5%, is close to our estimate of 7%.

CAUTION

When solving for percent of increase or decrease, the **whole is always the original value** or **value before the change occurred.** The part is the change in values, that is, how much something went up or went down.

◀ **Work Problem 7** at the Side.

6.6 Exercises

FOR EXTRA HELP

   Download the MyDashBoard App · MyMathLab®

CONCEPT CHECK *Underline the correct answer.*

1. When solving a sales tax problem, the cost of an item is the (*tax rate/whole*), the sales tax rate is the (*percent/whole*), and the amount of sales tax is the (*part/whole*).

CONCEPT CHECK *Fill in each blank with the correct response.*

2. To find the amount of sales tax, multiply the sales tax _____ by the _____. The total cost is the cost of the item plus the _____ tax.

Find the amount of sales tax or the tax rate and the total cost (amount of sale + amount of tax = total cost). Round money answers to the nearest cent if necessary. **See Examples 1 and 2.**

	Cost of Item	Tax Rate	Amount of Tax	Total Cost
3.	$6	4%	_____	_____
4.	$45	5%	_____	_____
5.	$425	_____	$12.75	_____
6.	$84	_____	$5.88	_____
7.	$12,229	$5\frac{1}{2}\%$	_____	_____
8.	$11,789	$7\frac{1}{2}\%$	_____	_____

CONCEPT CHECK *Underline the correct answer.*

9. When solving commission problems, the whole is the (*commission/sales amount*), the percent is the (*commission rate/sales amount*), and the part is the (*commission/sales amount*).

CONCEPT CHECK *Fill in each blank with the correct response.*

10. To find the commission, multiply the rate of _____ by the _____.

Find the commission earned or the rate of commission. Round money answers to the nearest cent if necessary. **See Examples 3 and 4.**

	Sales	Rate of Commission	Commission
11.	$280	8%	_____
12.	$660	10%	_____

GS Commission = rate of commission • sales

$$= (10\%)(\$_____) = (0.\_\_\_\_)(\$660)$$

13.	$3000	_____	$600

GS

$$\frac{600}{3000} = \frac{x}{100} \quad \text{or} \quad \frac{1}{5} = \frac{x}{100}; \quad \frac{5x}{5} = \frac{100}{5}; \quad x = 20$$

14.	$7800	_____	$1170
15.	$6183.50	3%	_____
16.	$4416.70	7%	_____

CONCEPT CHECK *Underline the correct answer.*

17. When solving retail sales discount problems, the *original* price is always the (*part/whole*), the rate of discount is the (*percent/whole*), and the amount of discount is the (*percent/part*).

CONCEPT CHECK *Fill in each blank with the correct response.*

18. To find the sale price in a retail discount problem, subtract the amount of _____ from the _____ price.

*Find the amount or rate of discount and the sale price after the discount. Round money answers to the nearest cent if necessary. **See Example 5.***

	Original Price	Rate of Discount	Amount of Discount	Sale Price
19.	$199.99	10%	_____	_____
20.	$29.95	15%	_____	_____
21.	$180	_____	$54	_____
22.	$38	_____	$9.50	_____
23.	$58.40	15%	_____	_____
24.	$99.80	30%	_____	_____

25. You are trying to decide between Company A paying a 10% commission and Company B paying an 8% commission. For which company would you prefer to work? What considerations other than commission rate would be important to you?

26. Give four examples of where you might use the percent of increase or the percent of decrease in your own personal activities. Think in terms of work, school, home, hobbies, and sports.

*Solve each application problem. Round money answers to the nearest cent and rates to the nearest tenth of a percent if necessary. **See Examples 1–7.***

Country Store has a unique selection of merchandise that it sells by mail and over the Internet. Use the shipping and insurance delivery chart at the right and a sales tax rate of 5% to solve Exercises 27–30. There is no sales tax on shipping and insurance. (*Source:* Country Store Catalog.)

SHIPPING AND INSURANCE DELIVERY CHART

Up to $15.00	add $4.99
$15.01 to $25.00	add $6.99
$25.01 to $35.00	add $7.99
$35.01 to $50.00	add $8.99
$50.01 to $70.00	add $10.99
$70.01 to $99.99	add $12.99
$100.00 or more	add $14.99

27. Find the total cost of six Small Fry Handi-Pan electric skillets priced at $29.99 each.

28. A customer ordered five sets of flour-sack towels priced at $12.99 per set. What is the total cost?

29. Find the total cost of three pop-up hampers at $9.99 each and four nonstick mini doughnut pans at $10.99 each.

30. What is the total cost of five coach lamp bird feeders at $19.99 each and six garden weather centers at $14.99 each?

31. An Anderson wood-frame French door is priced at $1980 with a sales tax of $99. Find the rate of sales tax.

32. Six gallons of Dutch Boy Dirt Fighter exterior paint cost $155.94 plus sales tax of $7.80. Find the sales tax rate. (*Source:* Orchard Supply Hardware.)

33. Today there are 635,000 women motorcyclists in the United States, up from 467,400 just eight years ago. Find the percent of increase in the number of women motorcyclists. (*Source:* Motorcycle Industry Council.)

Increase $= 635{,}000 - 467{,}400 = 167{,}600$

increase → $\dfrac{167{,}600}{467{,}400} = \dfrac{x}{100}$ ← original

34. This year Jo O'Neill has earned $46,750, up from $32,500 last year. Find the percent of increase in her annual earnings.

Increase $= \$46{,}750 - \$32{,}500 = \$14{,}250$

increase → $\dfrac{14{,}250}{32{,}500} = \dfrac{x}{100}$ ← original

35. Misty Downes sold her Toyota Prius, which got 48 miles per gallon on the highway, and purchased a Chevrolet Cruze, which gets 36 miles per gallon. What is the percent of decrease in mileage? (*Source:* edmonds.com)

36. Rich Williams is considering selling his Corvette, which gets 26 miles per gallon on the highway, and purchasing a new Corvette ZR1, which gets 20 miles per gallon. If he buys the new Corvette, find the percent of decrease in mileage. (*Source:* edmonds.com)

37. A preholiday sale at Macy's offers 60% off on all women's wear. What is the sale price of a wool coat normally priced at $335?

38. What is the sale price of a $1098 Kenmore washer/dryer set with a discount of 25%?

A weekly sales report for the top four sales people at Active Sports is shown below. Use this information to answer Exercises 39–42.

Employee	Sales	Rate of Commission	Commission
Strong, A.	$18,960	3%	_____
Ferns, K.	$21,460	3%	_____
Keyes, B.	$17,680	_____	$707.20
Vargas, K.	$23,104	_____	$1152.20

39. Find the commission for Strong.

40. Find the commission for Ferns.

41. What is the rate of commission for Keyes?

42. What is the rate of commission for Vargas?

43. A Sony Micro Hi Fi Component System was priced at $390 and is on sale at 22% off. Find the discount and the sale price.

44. A Honda Pilot is offered at 12% off the manufacturer's suggested retail price. Find the discount and the sale price of this SUV, originally priced at $32,500.

45. The price of a video game was marked down from $45.50 to $35.49. Find the percent of discount.

46. In the past five years, the cost of generating electricity from the sun has been brought down from 24 cents per kilowatt hour to 8 cents (less than the newest nuclear power plants). Find the percent of decrease.

47. College students are offered a 6% discount on a dictionary that sells for $18.50. If the sales tax is 6%, find the cost of the dictionary, including the sales tax.

48. A fax machine priced at $398 is marked down 7% to promote the new model. If the sales tax is also 7%, find the cost of the fax machine, including sales tax.

49. A plumbing supply sales representative has sales of $380,450 and is paid a commission of 2% of sales. If 20% of his commission is deducted for office space and other expenses, how much does the sales representative get?

50. The local real estate agents' association collects a fee of 2% on all money received by its members. The members charge 6% of the selling price of a property as their fee. How much does the association get, if its members sell property worth a total of $8,680,000?

51. What is the total price of a ski boat with an original price of $15,321, if it is sold at a 15% discount? The sales tax rate is $7\frac{3}{4}$%.

52. A commercial security alarm system originally priced at $10,800 is discounted 22%. Find the total price of the system if the sales tax rate is $7\frac{1}{4}$%.

Relating Concepts (Exercises 53–58) For Individual or Group Work

Knowing how to use the percent equation is important when solving application problems involving sales tax. **Work Exercises 53–58 in order.**

53. The percent equation is

part = _____ • _____ .

54. The formula used to find sales tax is an application of the percent equation. The sales tax formula is

sales tax = _____ • _____ .

In the United States there are certain items on which an excise tax is charged in addition to a sales tax. A table of federal excise taxes is shown here. Use this table to answer Exercises 55–58. (Excise tax is calculated on the amount of the sale **before** sales tax is added.) Round answers to the nearest cent.

FEDERAL EXCISE TAXES*

Product or Service	Rate	Product or Service	Rate
Local telephone service	3%	Bows and arrows	11%
Cigarettes	36¢ per pack	Gasoline	18.4¢/gal
Tires	$0.0945 for each 10 pounds of the maximum rated load capacity over 3500 pounds	Diesel fuel and kerosene	24.4¢/gal
		Aviation fuel	19.4¢/gal
Air transportation	7.5%	Truck and trailer, chassis and bodies	12%
International air travel	$15.40 per person	Inland waterways fuel	20¢/gal
Air freight	6.25%	Ship passenger tax	$3/passenger
Fishing rods	10%	Vaccines	78¢/dose.

*In addition to the federal excise taxes shown here, there are a number of additional excise taxes that apply to alcoholic beverages, tobacco products, and firearms.
Source: Publication 510, I.R.S., Excise Taxes.

55. Julia Lauren purchased a fly fishing rod for $59 and a spinning rod for $36. Use the federal excise tax table and a sales tax rate of $6\frac{1}{2}$% to find the cost of the equipment, including both taxes. Sales tax is not charged on the $9.50 federal excise tax. (Round to the nearest cent.)

56. Refer to **Exercise 55.** Calculate the two taxes separately and then add them together. Now, add the two tax rates together and then find the tax. Are your answers the same? Why or why not? (*Hint:* Recall the commutative and associative properties of multiplication.)

57. The price of an international airline ticket is $1248. Use the federal excise tax table and a sales tax rate of $7\frac{3}{4}$% to find the total cost of one ticket. Sales tax is not charged on the $15.40 federal excise tax. (Round to the nearest cent.)

58. Refer to **Exercise 57.** Can the federal excise tax be added to the sales tax rate to find the total tax? Why or why not?

6.7 Simple Interest

When we open a savings account, we are actually lending money to the bank or credit union. The bank or credit union will in turn lend this money to individuals and businesses. These people then become borrowers. The bank or credit union pays a fee to the savings account holders and charges a higher fee to its borrowers. These fees are called *interest*.

Interest is a fee paid or a charge made for lending or borrowing money. The amount of money borrowed is called the **principal.** The charge for interest is often given as a percent, called the interest rate or **rate of interest.** The rate of interest is assumed to be *per year*, unless stated otherwise. Time is always expressed in years or fractions of a year.

> **OBJECTIVE ▶ 1 Find the simple interest on a loan.** In most cases, interest on a loan is computed on the *original principal* and is called **simple interest.** We use the following **interest formula** to find simple interest.

> **Formula for Simple Interest**
>
> Interest = principal • rate • time
>
> The formula is usually written using letters.
>
> $$I = p \cdot r \cdot t$$

> **Note**
>
> Simple interest is used for most short-term business loans, most real estate loans, and many automobile and consumer loans.

> **EXAMPLE 1** Finding Simple Interest for a Year

Find the interest on $5000 at 3% for 1 year.

The amount borrowed, or principal (p), is $5000. The interest rate (r) is 3%, which is 0.03 as a decimal, and the time of the loan (t) is 1 year. Use the formula.

$$I = p \cdot r \cdot t$$
$$I = (5000)(0.03)(1)$$

Notice that "1" is used as the time for 1 year.

$$I = \$150$$

The interest rate is always written in decimal form.

The interest is $150.

········· Work Problem **1** at the Side. ▶

> **EXAMPLE 2** Finding Simple Interest for More Than a Year

Find the interest on $4200 at 4% for three and a half years.

The principal (p) is $4200. The rate ($r$) is 4%, or 0.04 as a decimal, and the time (t) is $3\frac{1}{2}$ or 3.5 years. Use the formula.

$$I = p \cdot r \cdot t$$
$$I = (4200)(0.04)(3.5)$$

3.5 years is equivalent to $3\frac{1}{2}$ years because 3.5 is $3\frac{5}{10}$ which simplifies to $3\frac{1}{2}$.

$$I = 588$$

4% is changed to the decimal 0.04

The interest is $588.

········· Work Problem **2** at the Side. ▶

OBJECTIVES

1. Find the simple interest on a loan.
2. Find the total amount due on a loan.

VOCABULARY TIP

Interest is what the borrower pays for the use of the lender's money.

1 Find the interest.

(a) $1000 at 3% for 1 year

$$= p \cdot r \cdot t$$
$$= (1000)(\underline{\quad})(1)$$
$$= \underline{\quad}$$

(b) $3650 at 2% for 1 year

VOCABULARY TIP

Rate of interest The number with the % symbol or **"percent"** after it is the rate of interest.

2 Find the interest.

(a) $820 at 1% for $3\frac{1}{2}$ years

$$I = p \cdot r \cdot t$$
$$= (\underline{\quad})(\underline{\quad})(3.5)$$
$$= \underline{\quad}$$

(b) $4850 at 4% for $2\frac{1}{2}$ years

(c) $16,800 at 3% for $2\frac{3}{4}$ years

Answers

1. **(a)** 0.03; $30 **(b)** $73
2. **(a)** $820; 0.01; $28.70 **(b)** $485 **(c)** $1386

3 Find the interest.

(GS) **(a)** $1800 at 3% for 4 months

$$I = p \cdot r \cdot t$$
$$= (1800)(0.03)(\underline{\quad})$$
$$= \underline{\quad}$$

(b) $28,000 at $7\frac{1}{2}$% for 3 months

4 Find the total amount due on each loan.

(GS) **(a)** $3800 at $6\frac{1}{2}$% for 6 months

$$I = p \cdot r \cdot t$$
$$= (3800)(0.065)(\underline{\quad})$$
$$= \underline{\quad}$$

Total
amount due = principle + interest
$$= \$3800 \quad + \$123.50$$
$$= \underline{\quad}$$

(b) $12,400 at 5% for 5 years

(c) $2400 at $4\frac{1}{2}$% for $2\frac{3}{4}$ years

CAUTION

It is best to rewrite fractions of percents or fractions of years in their decimal form. In **Example 2**, $3\frac{1}{2}$ years is rewritten as 3.5 years.

Interest rates are given **per year.** For loan periods of less than one year, be careful to express time as a fraction of a year.

If time is given in months, for example, use a denominator of 12, because there are 12 months in a year. A loan for 9 months would be for $\frac{9}{12}$ of a year.

EXAMPLE 3 **Finding Simple Interest for Less Than 1 Year**

Find the interest on $840 at $4\frac{1}{2}$% for 9 months.

The principal is $840. The rate is $4\frac{1}{2}$% or 0.045 as a decimal, and the time is $\frac{9}{12}$ of a year. Use the formula $I = p \cdot r \cdot t$.

$$I = (840)(0.045)\left(\frac{9}{12}\right) \quad \text{9 months} = \frac{9}{12} \text{ of a year}$$

Rewrite $4\frac{1}{2}$% as 0.045

$$= (37.8)\left(\frac{3}{4}\right) \quad \text{Write } \frac{9}{12} \text{ in lowest terms as } \frac{3}{4}.$$

$$= \frac{(37.8)(3)}{4}$$

$$= \frac{113.4}{4} = 28.35$$

The interest is $28.35.

Calculator Tip

The calculator solution to **Example 3** above uses chain calculations.

840 ⊗ .045 ⊗ 9 ⊘ 12 ⊜ 28.35

◀ **Work Problem 3** at the Side.

OBJECTIVE ▶ 2 **Find the total amount due on a loan.** When a loan is repaid, the interest is added to the original principal to find the total amount due.

Formula for Total Amount Due

Total amount due = principal + interest

EXAMPLE 4 **Calculating the Total Amount Due**

A loan of $3240 was made at 6% for 3 months. Find the total amount due.

First find the interest. Then add the principal and the interest to find the total amount due.

$$I = (3240)(0.06)\left(\frac{3}{12}\right) \quad \text{3 months} = \frac{3}{12} \text{ of a year.}$$

$$I = \$48.60$$

The interest is $48.60.

Remember that the *total amount due* is the amount of the loan plus the interest.

Total amount due = principal + interest

$$= \$3240 \quad + \$48.60 = \$3288.60$$

The total amount due is $3288.60.

◀ **Work Problem 4** at the Side.

Answers

3. **(a)** $\frac{4}{12}$ or $\frac{1}{3}$; $18 **(b)** $525

4. **(a)** $\frac{6}{12}$ or $\frac{1}{2}$; $123.50; $3923.50
(b) $15,500 **(c)** $2697

6.7 Exercises

FOR EXTRA HELP

 Download the MyDashBoard App

 ▶ MyMathLab®

CONCEPT CHECK *Rewrite all interest rates and times in decimal form.*

1. $3\frac{1}{2}\%$

2. $7\frac{1}{2}\%$

3. $2\frac{1}{4}$ years

4. $5\frac{3}{4}$ years

Find the interest. See Examples 1 and 2.

Principal	Rate	Time in Years	Interest
5. $100	6%	1	_____
6. $200	3%	1	_____
7. $700	5%	3	_____
8. $900	2%	4	_____
9. $2300	$4\frac{1}{2}\%$	$2\frac{1}{2}$	_____
10. $4700	$5\frac{1}{2}\%$	$1\frac{1}{2}$	_____
11. $10,800	$7\frac{1}{2}\%$	$2\frac{3}{4}$	_____
12. $12,400	$1\frac{1}{2}\%$	$3\frac{3}{4}$	_____

5. GS $I = p \cdot r \cdot t$
    $= (100)(0.06)(1)$

6. GS $I = p \cdot r \cdot t$
    $= (200)(0.03)(1)$

9. GS $I = p \cdot r \cdot t$
    $= (2300)(0.045)(2.5)$

10. GS $I = p \cdot r \cdot t$
    $= (4700)(0.055)(1.5)$

CONCEPT CHECK *Change all times in months to fractions of a year and to decimal form.*

13. 3 months

14. 6 months

15. 9 months

16. 15 months

Find the interest. Round to the nearest cent if necessary. See Example 3.

Principal	Rate	Time in Months	Interest
17. $400	3%	6	_____

17. GS $I = p \cdot r \cdot t$

    $= (400)(0.03)\left(\frac{1}{2}\right)$ 6 months $= \frac{1}{2}$ year

Principal	Rate	Time in Months	Interest
18. $780	5%	24	_____

GS $I = p \cdot r \cdot t$

   $= (780)(0.05)(2)$ 24 months $= 2$ years

19. $940	3%	18	_____
20. $178	4%	12	_____
21. $1225	$5\frac{1}{2}\%$	3	_____
22. $2660	$7\frac{1}{2}\%$	3	_____
23. $15,300	$7\frac{1}{4}\%$	7	_____
24. $13,700	$3\frac{3}{4}\%$	11	_____

Find the total amount due on the following loans. Round to the nearest cent if necessary.
See Example 4.

Principal	Rate	Time	Total Amount Due
25. $200	5%	1 year	_____

GS $I = p \cdot r \cdot t$ Total amount due = principal + interest

   $= (200)(0.05)(1)$ $= \$200 +$ interest

26. $400	2%	6 months	_____

GS $I = p \cdot r \cdot t$ Total amount due = principal + interest

   $= (400)(0.02)\left(\dfrac{1}{2}\right)$ 6 months $= \dfrac{1}{2}$ year $= \$400 +$ interest

27. $740	6%	9 months	_____
28. $1180	3%	2 years	_____

	Principal	Rate	Time	Total Amount Due
29.	$1800	9%	18 months	_____
30.	$9000	6%	7 months	_____
31.	$3250	$3\frac{1}{2}\%$	6 months	_____
32.	$7600	$4\frac{1}{2}\%$	1 year	_____
33.	$16,850	$7\frac{1}{2}\%$	9 months	_____
34.	$19,450	$5\frac{1}{2}\%$	6 months	_____

35. The amount of interest paid on savings accounts and charged on loans can vary from one institution to another. However, when the amount of interest is calculated, three factors are used in the calculation. Name these three factors and describe them in your own words.

36. Interest rates are usually given as a rate per year (annual rate). Explain what must be done when time is given in months. Write your own problem where time is given in months and then show how to solve it.

Solve each application problem. Round to the nearest cent if necessary.

37. Paul Plescia has a savings account of $8642 at his credit union. If the account pays interest of 2%, how much interest will he earn in 2 years?

38. Esther Albert, a professional dancer, deposits $68,000 of her earnings at 4% for 5 years. How much interest will she earn?

39. To increase distribution of her office supply products, Stella Glitter borrowed $150,000 at 7% for 30 months. Find the amount of interest on this loan.

40. The Jidobu family invests $18,000 at 3% for 9 months. What amount of interest will the family earn?

41. To complete her fourth semester of college, Yohko Kitagawa borrowed $4650 at 5%. If she repaid the loan in 5 months, find the total amount due.

42. Sarah Brynski borrows $5500 from her dad for a used car. The loan will be paid back with 4% interest at the end of 9 months. Find the total amount due.

43. Nicholas Thomas deposits $14,800 in his school credit union account for 10 months. If the credit union pays $2\frac{1}{4}$% interest, find the amount of interest he will earn.

44. Sid and Shirley Kordell, owners of the Nut House, borrow $54,000 to update their store. If the loan is for 42 months at $7\frac{1}{4}$%, find the amount of interest they will owe.

45. After the sale of her home, the buyer still owes Tiina Luig a balance of $8800. If the interest rate on this amount is $7\frac{1}{4}$%, find the amount of interest she will earn every 3 months.

46. Pat Carper owes $1900 in taxes. She is charged a penalty of $9\frac{1}{4}$% annual interest and pays the taxes and penalty after 6 months. Find the amount of the penalty.

47. Business has been good at Rocky's Grill, and the owner deposits $5400 into an account paying $2\frac{3}{4}$% interest. Find

  (a) the amount of interest.

  (b) the total amount in the account at the end of $1\frac{1}{2}$ years.

48. Moises Guardado, the owner of MG Painting, borrowed $29,400 to purchase another truck and other equipment for his business. The loan is for $3\frac{1}{2}$ years at $5\frac{1}{4}$% interest. Find

  (a) the amount of interest.

  (b) the total amount he must repay.

49. Jake's Exotic Pets bought four new aviary environments (bird cages) at a cost of $980 each. The owner borrowed 70% of the cost of the cages at an interest rate of $7\frac{1}{2}$%. Find the total amount owed at the end of $2\frac{1}{2}$ years.

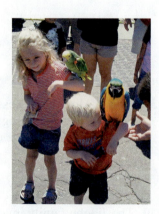

50. The owners of Baily and Daughters Excavating purchased four earth movers at a cost of $485,000 each. If they borrowed 80% of the total purchase price for $2\frac{1}{2}$ years at $10\frac{1}{2}$% interest, find the total amount due.

6.8 Compound Interest

The interest we studied in the last section was *simple interest* (interest only on the original principal). A common type of interest used with savings accounts and most investments is **compound interest** or interest paid on past interest as well as on the principal.

OBJECTIVE ▶ 1 Understand compound interest. Suppose that you make a single deposit of $1000 in a savings account that earns 5% per year. What will happen to your savings over 3 years? At the end of the first year, 1 year's interest on the original deposit is calculated. Use the simple interest formula.

> Interest = principal • rate • time

Year 1 ($1000)(0.05)(1) = **$50**

Add the interest to the $1000 to find the amount in your account at the end of the first year. $1000 + **$50** = **$1050** ◀ In year 1, interest is calculated on principal.

The interest for the second year is found on $1050, that is, the interest is **compounded.**

Year 2 (**$1050**)(0.05)(1) = **$52.50**

Add this interest to the $1050 to find the amount in your account at the end of the second year. **$1050** + **$52.50** = **$1102.50**.

The interest for the third year is found on $1102.50. ◀ In year 2 and thereafter, interest is calculated on the principal and all past interest.

Year 3 (**$1102.50**)(0.05)(1) ≈ **$55.13**

Add this interest to the **$1102.50**. So, **$1102.50** + **$55.13** = **$1157.63**.

At the end of 3 years, you will have **$1157.63** in your savings account. The $1157.63 that you have in your account is called the **compound amount.**

If you had earned only *simple* interest for 3 years, your interest would be as follows.

$$I = (\$1000)(0.05)(3)$$

$$= \$150 \leftarrow \text{Simple interest}$$

At the end of 3 years, you would have $1000 + $150 = $1150 in your account. Compounding the interest increased your earnings by $7.63 because $1157.63 − $1150 = $7.63.

With *compound* interest, the interest earned during the second year is greater than that earned during the first year, and the interest earned during the third year is greater than that earned during the second year. This happens because the interest earned each year is *added* to the principal, and the new total is used to calculate the amount of interest in the next year.

Compound Interest

Interest paid on principal plus past interest is called **compound interest.**

OBJECTIVES

1 Understand compound interest.

2 Understand compound amount.

3 Find the compound amount.

4 Use a compound interest table.

5 Find the compound amount and the amount of interest.

VOCABULARY TIP

Simple interest With simple interest, there is usually only one interest calculation for a specific period of time at a specific rate of interest.

VOCABULARY TIP

Compound interest/Compounded *Compound* means to mix, combine, or put together. When interest has been compounded, it means that it has been combined with the original principal and all previous interest on the principal.

VOCABULARY TIP

Compound amount The compound amount at the end of any compound interest period is the total of the original principal and all the interest combined.

1 Find the compound amount given the following deposits. Round to the nearest cent if necessary.

(a) $400 at 4% for 2 years

yr 1: ($400)(0.04)(1) = $16

$400 + $16 = $416

yr 2: ($416)(0.04)(1) = _____

$416 + $16.64 = _____

(b) $2000 at 2% for 3 years

2 Find the compound amount by multiplying the original deposit by 100% plus the compound interest rate. Round to the nearest cent if necessary.

(a) $1800 at 2% for 3 years

$(1800)(1.02)(1.02)(1.02) =$

_____

(b) $900 at 3% for 2 years

(c) $2500 at 5% for 4 years

Answers

1. (a) $16.64; $432.64 (b) $2122.42 (rounded)
2. (a) $1910.17 (rounded) (b) $954.81
   (c) $3038.77 (rounded)

OBJECTIVE **2** **Understand compound amount.** Find the compound amount as shown below.

EXAMPLE 1 **Finding the Compound Amount**

Mavis Chamski deposits $3400 in an account that pays 3% interest compounded annually for 4 years. Find the compound amount. Round to the nearest cent when necessary.

Year	Interest		Compound Amount
1	($3400)(0.03)(1) = $102	$3400 + $102 =	$3502
2	($3502)(0.03)(1) = $105.06	$3502 + $105.06 =	$3607.06
3	($3607.06)(0.03)(1) ≈ $108.21	$3607.06 + $108.21 =	$3715.27
4	($3715.27)(0.03)(1) ≈ $111.46	$3715.27 + $111.46 =	$3826.73

The compound amount is $3826.73.

◀ **Work Problem 1 at the Side.**

OBJECTIVE **3** **Find the compound amount.** A more efficient way of finding the compound amount is to add the interest rate to 100% and then multiply by the original deposit. Notice that in **Example 1** above, at the end of the first year, you will have $3400 (100% of the original deposit) plus 3% (of the original deposit) or 103% (because 100% + 3% = 103%).

EXAMPLE 2 **Finding the Compound Amount**

Find the compound amount in **Example 1** using multiplication.

Year 1 Year 2 Year 3 Year 4

($3400)(1.03)(1.03)(1.03)(1.03) ≈ $3826.73

Original deposit   100% + 3% = 103% = 1.03   Compound amount

This method works well with a calculator.

Our answer, $3826.73, is the same as in **Example 1** above.

◀ **Work Problem 2 at the Side.**

Note

By adding the compound interest rate to 100%, we can then multiply by the original deposit. This will give us the compound amount at the end of each compound interest period.

Calculator Tip

If you use a calculator for **Example 2** above, you can use the y^x key (exponent key). (On some calculators, you use the ⌃ key instead.)

3400 × 1.03 y^x 4 = 3826.73 (rounded)

The 4 following the y^x key represents the number of compound interest periods.

OBJECTIVE **4** **Use a compound interest table.** The calculation of compound interest can be quite tedious. For this reason, compound interest tables have been developed.

Suppose you deposit $1 in a savings account today that earns 4% compounded annually and you allow the deposit to remain for 3 years. The diagram below shows the compound amount at the end of each of the 3 years.

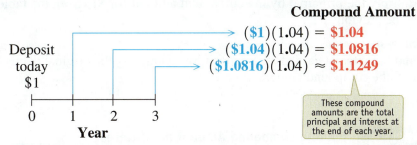

Compound Amount

($1)(1.04) = **$1.04**
($1.04)(1.04) = **$1.0816**
($1.0816)(1.04) ≈ **$1.1249**

These compound amounts are the total principal and interest at the end of each year.

Deposit today $1

0   1   2   3
Year

Using the compound amounts for $1, a table can be formed. Look at the table below and find the column headed 4%. The first three numbers for years 1, 2, and 3 are the same as those we have calculated for $1 at 4% for 3 years. This table, giving the compound amounts on a $1 deposit for given lengths of time and interest rates, can be used for finding the compound amount on any amount of deposit.

COMPOUND INTEREST TABLE

Time Periods	2.00%	2.50%	3.00%	3.50%	4.00%	4.50%	5.00%	5.50%	6.00%	Time Periods
1	1.0200	1.0250	1.0300	1.0350	1.0400	1.0450	1.0500	1.0550	1.0600	1
2	1.0404	1.0506	1.0609	1.0712	1.0816	1.0920	1.1025	1.1130	1.1236	2
3	1.0612	1.0769	1.0927	1.1087	**1.1249**	1.1412	1.1576	1.1742	1.1910	3
4	1.0824	1.1038	1.1255	1.1475	1.1699	1.1925	1.2155	1.2388	1.2625	4
5	1.1040	1.1314	1.1593	1.1877	1.2167	1.2462	1.2763	1.3070	1.3382	5
6	1.1261	1.1597	1.1941	**1.2293**	1.2653	1.3023	1.3401	1.3788	1.4185	6
7	**1.1486**	1.1887	1.2299	1.2723	1.3159	1.3609	1.4071	1.4547	1.5036	7
8	1.1717	1.2184	1.2668	1.3168	1.3686	1.4221	1.4775	1.5347	1.5938	8
9	1.1951	1.2489	1.3048	1.3629	1.4233	1.4861	1.5513	1.6191	1.6895	9
10	1.2190	1.2801	1.3439	1.4106	1.4802	1.5530	**1.6289**	1.7081	1.7908	10
11	1.2434	1.3121	1.3842	1.4600	1.5395	1.6229	1.7103	1.8021	1.8983	11
12	1.2682	1.3449	1.4258	1.5111	1.6010	1.6959	1.7959	**1.9012**	2.0122	12

EXAMPLE 3   **Using a Compound Interest Table**

Find the compound amount using the compound interest table.

A ruler or straightedge helps align the column and row.

(a) $1 is deposited at a 5% interest rate for 10 years.
Look down the column headed 5%, and across to row 10 (because 10 years = 10 time periods). At the intersection of the column and row, read the compound amount, **1.6289**.

(b) $1 is deposited at $3\frac{1}{2}$% for 6 years.
The intersection of the $3\frac{1}{2}$% (3.50%) column and row 6 shows **1.2293** as the compound amount.

········· **Work Problem 3 at the Side.** ▶

3 Find the compound amount using the compound interest table.

GS **(a)** $1 at 3% for 6 years
(1)(1.1941) = _____
↑
Look down the 3% column in the table to row 6.

(b) $1 at 2% for 8 years

(c) $1 at $4\frac{1}{2}$% for 12 years

Answers

3. **(a)** $1.19 (rounded)   **(b)** $1.17 (rounded)
    (c) $1.70 (rounded)

4 Use the compound interest table to find the compound amount and the interest.

GS **(a)** $4000 at 3% for 10 years

Table shows 1.3439.

$(4000)(1.3439) =$ _____

_____ $- 4000$

$=$ _____ interest

(b) $12,600 at $3\frac{1}{2}\%$ for 8 years

(c) $32,700 at $4\frac{1}{2}\%$ for 12 years

OBJECTIVE ▸ **5** **Find the compound amount and the amount of interest.**
Find the compound amount and interest as follows.

> **Finding the Compound Amount and the Interest**
>
> **Compound Amount**
> Multiply the principal by the compound amount for $1 (from the table on the previous page).
>
> **Interest**
> Find the interest earned on a deposit by subtracting the original deposit from the compound amount.

EXAMPLE 4   **Finding Compound Amount and Interest**

Use the compound interest table to find the compound amount and the interest.

(a) $1000 at $5\frac{1}{2}\%$ interest for 12 years

Look in the table on the previous page for $5\frac{1}{2}\%$ (5.50%) and 12 periods to find the number **1.9012** but do *not* round it. Multiply this number and the principal of $1000.

> Never round the numbers found in the table.

$$(\$1000)(1.9012) = \$1901.20$$

The account will contain $1901.20 after 12 years.

Find the interest by subtracting the original deposit from the compound amount.

$$\$1901.20 - \$1000 = \$901.20$$

(b) $6400 at 2% for 7 years

Look in the table for 2% and 7 periods to find **1.1486**. Multiply.

$$(\$6400)(1.1486) = \$7351.04 \quad \text{Compound amount}$$

Subtract the original deposit from the compound amount.

$$\$7351.04 - \$6400 = \$951.04 \quad \text{Interest}$$

> Remember: Compound amount − Principal = Interest.

A total of $951.04 in interest was earned.

◀ **Work Problem** **4** **at the Side.**

Answers

4. **(a)** $5375.60; $5375.60; $1375.60
   (b) $16,591.68; $3991.68
   (c) $55,455.93; $22,755.93

6.8 Exercises

FOR EXTRA HELP

Download the MyDashBoard App

MyMathLab®

1. CONCEPT CHECK *Write* true *or* false.

For short periods of time, five years or less, the amount of interest earned will be the same whether you use simple or compound interest.

2. CONCEPT CHECK *Underline the correct answer.*

When using compound interest, the amount of (*interest/principal*) earned at the end of a compound interest period will be added to the amount of (*interest/principal*) at the beginning of that period.

*Find the compound amount given the following deposits. Calculate the interest each year, then add it to the previous year's amount. **See Example 1.***

3. $500 at 4% for 2 years

$$I = p \cdot r \cdot t$$

Year 1 $\begin{cases} (\$500)(0.04)(1) = \$20 \\ \$500 + \$20 = \$520 \end{cases}$

Year 2 $\begin{cases} (\$520)(0.04)(1) = \$20.80 \\ \$520 + \$20.80 = \underline{\hspace{2cm}} \end{cases}$

4. $1500 at 5% for 3 years

$$I = p \cdot r \cdot t$$

Year 1 $\begin{cases} (\$1500)(0.05)(1) = \$75 \\ \$1500 + \$75 = \$1575 \end{cases}$

Year 2 $\begin{cases} (\$1575)(0.05)(1) = \$78.75 \\ \$1575 + \$78.75 = \$1653.75 \end{cases}$

Year 3 $\begin{cases} (\$1653.75)(0.05)(1) = \$82.69 \text{ (rounded)} \\ \$1653.75 + \$82.69 = \underline{\hspace{2cm}} \end{cases}$

5. $1800 at 3% for 3 years

6. $2000 at 2% for 3 years

7. $3500 at 7% for 4 years

8. $5500 at 6% for 4 years

9. CONCEPT CHECK *Fill in the blanks.*

To find the compound amount for a deposit of $2000 at 2% for 3 years, you can multiply

$(2000)\underline{\hspace{1cm}}\,\underline{\hspace{1cm}}\,\underline{\hspace{1cm}} = \$2122.42$ (rounded)

10. CONCEPT CHECK *Fill in the blanks.*

To find the compound amount for a deposit of $3000 at 5% for 2 years, you can multiply

$(\$3000)\underline{\hspace{1cm}}\,\underline{\hspace{1cm}} = \$3307.50$

 *Find each compound amount by multiplying the original deposit by 100% plus the compound rate given in the following. **See Example 2.** Round answers to the nearest cent if necessary.*

11. $1000 at 5% for 2 years

Year 1 Year 2
$(\$1000)(1.05)(1.05) = \underline{\hspace{3cm}}$

12. $500 at 4% for 3 years

Year 1 Year 2 Year 3
$(\$500)(1.04)(1.04)(1.04) = \underline{\hspace{3cm}}$

13. $1400 at 6% for 5 years

14. $2500 at 3% for 4 years

15. $1180 at 7% for 8 years

16. $12,800 at 6% for 7 years

17. $10,940 at 4% for 6 years

18. $15,710 at 2% for 8 years

CONCEPT CHECK *In the compound interest table, locate the compound amount for*
$1 for each of the following situations.

19. 3% for 6 years   **20.** 4.50% for 10 years   **21.** 2% for 8 years   **22.** 5.50% for 3 years

Use the table to find the compound amount and the interest. Interest is compounded
annually. Round answers to the nearest cent if necessary. **See Examples 3 and 4.**

23. $1000 at 4% for 5 years

Table shows 1.**2167**.

($1000)(1.2167) = $1216.70 compound amount

$1216.70 − $1000 = _____

↑ Compound amount   ↑ Original deposit   ↑ Interest

24. $10,000 at 3% for 4 years

Table shows 1.**1255**.

($10,000)(1.1255) = $11,255 compound amount

$11,255 − $10,000 = _____

↑ Compound amount   ↑ Original deposit   ↑ Interest

25. $8000 at 2% for 10 years

26. $7800 at 5% for 8 years

27. $8428.17 at $4\frac{1}{2}$% for 6 years

28. $10,472.88 at $5\frac{1}{2}$% for 12 years

29. Write a definition for compound interest. Describe in your own words what compound interest means to you.

30. What is the difference between the compound amount and the interest?

⊞ *Use the compound interest table in this section to solve each application problem.*
Round answers to the nearest cent if necessary. **See Examples 3 and 4.**

31. After winning $50,000 in the Veterans of Foreign Wars (VFW) lottery, Jerry Murray opened a money market account. He deposited all of his winnings and earned 4% interest compounded annually. Find the amount he will have (compound amount) in the account at the end of 5 years.

32. Yen Lee borrowed $32,800 from her family to open a small restaurant named Shanghai Winds. Her plan is to repay the loan at the end of 4 years at $3\frac{1}{2}\%$ interest compounded annually. Find the total amount that she must repay.

33. Al Granard lends $76,000 to the owner of Rick's Limousine Service. He will be repaid at the end of 9 years at 6% interest compounded annually. Find

(a) the total amount that he should be repaid.

(b) the amount of interest earned.

34. Recent sales of their music gave Bob and Cindy Kilpatrick $28,500 in additional income. If they invest all of it at 4% interest compounded annually for 10 years, find

(a) the total amount they will have at the end of 10 years.

(b) the amount of interest earned.

35. Jennifer Barrister deposits $30,000 at 6% interest compounded annually. Two years after she makes the first deposit, she deposits another $40,000, also at 6% compounded annually.

(a) What total amount will she have five years after her first deposit?

(b) What amount of interest will she have earned?

36. Christine Campbell invested $25,000 at 4% interest compounded annually. Three years after she made the first deposit, she deposited another $25,000, also at 4% compounded annually.

(a) What total amount will she have five years after her first deposit?

(b) What amount of interest will she earn?

Relating Concepts (Exercises 37–42) For Individual or Group Work

Knowing how to solve interest problems is important to businesspeople and consumers alike. **Work Exercises 37–40 in order.**

37. Simple interest calculation is used for most short-term business loans, most real estate loans, and many automobile and consumer loans. The formula for simple interest is

Interest = _____ • _____ • _____

or $I =$ _____ • _____ • _____ .

38. When a loan is repaid, the interest is added to the original principal. The formula for the total amount due is

Amount due = _____ + _____ .

39. Compound interest is paid on most savings accounts and many other types of investments. Compound interest is interest calculated on _____ plus past _____ .

40. The compound amount is the total amount in an account at the end of a period of time. Compound amount is the original _____ + compound _____ .

Use the compound interest table to answer Exercises 41 and 42.

41. Julie Maxey has two choices. She can invest $4350 at 6% simple interest for 6 years, or she can invest the same amount at 6% interest compounded annually for 6 years.

   (a) Find the difference in the amount of interest earned in these two accounts.

   (b) If the length of time is doubled from 6 to 12 years, will the difference in the interest earned also double?

   (c) Use your own example to determine that what you found in part (b) is true with a different interest rate and length of time.

42. One account is opened with $10,000 at 5% simple interest for 10 years. Another account is opened with $9500 at 5% interest compounded annually for 10 years.

   (a) At the end of the 10 years, which account has the higher balance and by how much?

   (b) What does this tell you about compound interest? Why?

Study Skills

Your math final exam is likely to be a **comprehensive exam.** This means that it will cover material from the **entire term.** The end of the term will be less stressful if you **make a plan** for how you will prepare for each of your exams.

First, figure out the **score you need to earn on the final exam** to get the course grade you want. Check your course syllabus for grading policies, or ask your instructor if you are not sure of them. This allows you to set a goal for yourself.

OBJECTIVES

1. Create a final exam week plan.
2. Break studying into chunks and study over several days.
3. Practice all types of problems.

> How many points do you need to earn on your mathematics final exam to get the grade you want? _____
> _____

Create a Plan

Second, create a **final exam week plan for your work and personal life.** If you need to make an adjustment in your work schedule, do it in advance, so you aren't scrambling at the last minute. If you have family members to care for, you might want to enlist some help from others so you can spend extra time studying. Try to plan in advance so you don't create additional stress for yourself. You will have to set some priorities, and studying has to be at the top of the list! Although life doesn't stop for finals, some things can be ignored for a short time. You don't want to "burn out" during final exam week; **get enough sleep and healthy food so you can perform your best.**

> What adjustments in your personal life do you need to make for final exam week? _____
> _____

Study and Review

Third, use the following suggestions to guide your studying and reviewing.

▶ **Know exactly which chapters and sections will be on the final exam.**

▶ **Divide up the chapters,** and decide how much you will review each day.

▶ Begin your reviewing **several days** before the exam.

▶ **Use returned quizzes and tests** to review earlier material (if you have them).

▶ **Practice all types of problems,** but emphasize the types that are most difficult for you. Use the **Cumulative Reviews** that are at the end of each chapter in your textbook.

▶ **Rewrite your notes or make mind maps** to create summaries.

▶ **Make study cards for all types of problems.** Be sure to use the same **direction words** (such as *simplify, solve, estimate*) that your exam will use. Carry the cards with you and review them whenever you have a few spare minutes.

Study Skills

Continued from page 459

Managing Stress

Of course, a week of final exams produces stress. **Students who develop skills for reducing and managing stress do better on their final exams and are less likely to "bomb" an exam.** You already know the damaging effect of adrenaline on your ability to think clearly. But several days (or weeks) of elevated stress is also harmful to your brain and your body. You will feel better if you make a conscious effort to reduce your stress level. Even if it takes you away from studying for a little while each day, the time will be well spent.

Reducing Physical Stress

Examples of ways to reduce **physical stress** are listed below. Can you add any of your own ideas to the list?

▶ *Laugh until your eyes water.* Watching your favorite funny movie, exchanging a joke with a friend, or viewing a comedy bit on the Internet are all ways to generate a healthy laugh. Laughing raises the level of *calming* chemicals (endorphins) in your brain.

▶ *Exercise for 20 to 30 minutes.* If you normally exercise regularly, do NOT stop during final exam week! Exercising helps relax muscles, diffuses adrenaline, and raises the level of endorphins in your body. If you don't exercise much, get some gentle exercise, such as a daily walk, to help you relax.

▶ *Practice deep breathing.* Several minutes of deep, smooth breathing will calm you. Close your eyes too.

▶ *Visualize a relaxing scene.* Choose something that you find peaceful and picture it. Imagine what it feels like and sounds like. Try to put yourself in the picture.

▶ If you feel stress in your muscles, such as in your shoulders or back, *slowly squeeze the muscles as much as you can, and then release them.* Sometimes we don't realize we are clenching our teeth or holding tension in our shoulders until we consciously work with them. Try to notice what it feels like when they are relaxed and loose. Squeezing and then releasing muscles is also something you can do during an exam if you feel yourself tightening up.

Reducing Mental Stress

Mental stress reduction is also a powerful tool both before and during an exam. In addition to these suggestions, do you have any of your own techniques?

▶ *Talk positively to yourself.* Tell yourself you will get through it.

▶ *Reward yourself.* Give yourself small breaks, a little treat—something that makes you happy—every day of final exam week.

▶ *Make a list of things to do* and feel the sense of accomplishment when you cross each item off.

▶ When you take time to relax or exercise, *make sure you are relaxing your mind too.* Use your mind for something *completely* different from the kind of thinking you do when you study. Plan your garden, play your favorite music, walk your dog, read a good book.

▶ *Visualize.* Picture yourself completing exams and projects successfully. Picture yourself taking the test calmly and confidently.

Which techniques will you try?

_____

_____

_____

Which techniques will you try?

_____

_____

_____

Chapter 6 *Summary*

Key Terms

6.1

percent Percent means "per one hundred." A percent is a ratio with a denominator of 100.

6.3

percent proportion The proportion $\dfrac{\text{part}}{\text{whole}} = \dfrac{\text{percent}}{100}$ is used to solve percent problems.

whole The whole in a percent problem is the entire quantity, the total, or the base.

part The part in a percent problem is the portion being compared with the whole.

6.5

percent equation The percent equation is: part = percent • whole. It is another way to solve percent problems.

6.6

sales tax Sales tax is a percent of the cost of an item charged as a tax.

commission Commission is a percent of the dollar value of total sales paid to a salesperson.

discount Discount is often expressed as a percent of the original price; it is then deducted from the original price, resulting in the sale price.

percent of increase or decrease Percent of increase or decrease is the amount of change (increase or decrease) expressed as a percent of the original amount.

6.7

interest Interest is a fee paid or a charge made for lending or borrowing money.

principal Principal is the amount of money on which interest is earned.

rate of interest Often referred to as "rate," it is the charge for interest and is given as a percent.

simple interest Interest that is computed only on the original principal is simple interest.

interest formula The interest formula is used to calculate interest. It is: interest = principal • rate • time, or $I = p \cdot r \cdot t$.

6.8

compound interest Compound interest is interest paid both on the past interest and on the principal.

compounding Interest that is compounded once each year is compounded annually.

compound amount Compound amount is the total amount in an account, including compound interest and the original principal.

New Symbols

%   percent (per one hundred)

New Formulas

To write percents as decimals: $p\% = p \div 100$ **To write percents as fractions:** $p\% = \dfrac{p}{100}$

Percent proportion: $\dfrac{\text{part}}{\text{whole}} = \dfrac{\text{percent}}{100}$ **Percent equation:** part = percent • whole

Amount of sales tax: amount of sales tax = rate of tax • cost of item

Amount of commission: amount of commission = rate of commission • amount of sales

Amount of discount: amount of discount = rate of discount • original price

Sale price: sale price = original price − amount of discount

Percent of increase: $\dfrac{\text{amount of increase}}{\text{original value}} = \dfrac{\text{percent}}{100}$

Percent of decrease: $\dfrac{\text{amount of decrease}}{\text{original value}} = \dfrac{\text{percent}}{100}$

Interest: Interest = principal • rate • time   or $I = p \cdot r \cdot t$

Total amount due: total amount due = principal + interest

Test Your Word Power

See how well you have learned the vocabulary in this chapter.

1 To write a **percent as a decimal,** you drop the percent symbol
  A. after finding the decimal point
  B. and move the decimal point two places to the right
  C. and move the decimal point two places to the left.

2 To write a **decimal as a percent,** you attach the percent symbol
  A. after removing the decimal point
  B. after moving the decimal point two places to the right
  C. after moving the decimal point two places to the left.

3 **Percent** means
  A. the same as interest
  B. per one thousand
  C. per one hundred.

4 When you use
$$\frac{\text{part}}{\text{whole}} = \frac{\text{percent}}{100}$$
to solve percent problems, you are using the
  A. simple interest formula
  B. percent proportion
  C. percent equation.

5 The **percent equation** is
  A. part = percent • whole
  B. $I = p \cdot r \cdot t$
  C. $p\% = \dfrac{p}{100}$

6 In a **percent of increase** problem, the increase is a percent of
  A. the largest amount
  B. the original amount
  C. the new or most recent amount.

7 In the formula $I = p \cdot r \cdot t$, the p stands for
  A. proportion
  B. principal
  C. percent.

8 The term **rate** in an interest problem represents the
  A. whole
  B. percent
  C. part.

Answers to Test Your Word Power

1. C; *Example:* 50% written as a decimal is 0.50. or 0.5.

2. B; *Example:* 0.25 written as a percent is 0.25 or 25%.

3. C; *Example:* 8% means 8 per 100.

4. B; *Example:* Part = 4, and whole = 25. To find the percent, $\frac{4}{25} = \frac{x}{100}$; $x = 16$ or 16%.

5. A; *Example:* Percent = 25, and whole = 300. To find the part, $(0.25)(300) = 75$.

6. B; *Example:* Original value = \$200, and amount of increase = \$40. To find the percent of increase, $\frac{40}{200} = \frac{x}{100}$; $x = 0.20$ or 20% increase.

7. B; *Example:* Principal (p) = \$800, rate (r) = 5%, and time (t) = 1 year. Then, $I = p \cdot r \cdot t$ so $I = (\$800)(0.05)(1) = \$40$.

8. B; *Example:* Principal (p) = \$1650, rate (r) = 4%, and time (t) = $\frac{1}{2}$ year. To find the interest, $I = (\$1650)(0.04)\left(\frac{1}{2}\right) = \$33$.

Quick Review

Concepts	Examples

6.1 Basics of Percent

Writing a Percent as a Decimal
To write a percent as a decimal, drop the percent symbol and move the decimal point two places to the left.

    50% (.50%) = 0.50 or just 0.5

    3%(.03%) = 0.03

    12.5% (12.5%) = 0.125

Writing a Decimal as a Percent
To write a decimal as a percent, move the decimal point two places to the right and attach a % symbol.

    0.75 (0.75) = 75%

    0.875(0.875) = 87.5%

    3.6 (3.60) = 360%

6.2 Writing a Fraction as a Percent

Use a proportion and solve for p to change a fraction to percent.

$$\frac{2}{5} = \frac{p}{100} \qquad \text{Proportion}$$

$$5 \cdot p = 2 \cdot 100 \qquad \text{Cross products}$$

$$5 \cdot p = 200$$

$$\frac{\overset{1}{5} \cdot p}{\underset{1}{5}} = \frac{200}{5} \qquad \text{Divide both sides by 5.}$$

$$p = 40$$

$$\frac{2}{5} = 40\% \qquad \text{Attach \% symbol.}$$

Concepts	Examples

6.3 Learning the Percent Proportion

Part is to whole as percent is to 100.

$$\frac{\text{part}}{\text{whole}} = \frac{\text{percent}}{100}$$

← Always 100 because percent means "per 100."

Use the percent proportion to solve for the unknown value.
part = 30, whole = 50; find the percent.

Percent (unknown)

Part → $\dfrac{30}{50} = \dfrac{x}{100}$ ← Always 100
Whole →

$\dfrac{3}{5} = \dfrac{x}{100}$ Write $\frac{30}{50}$ as $\frac{3}{5}$ in lowest terms.

$5 \cdot x = 3 \cdot 100$ Cross products

$5 \cdot x = 300$

$\dfrac{\overset{1}{5} \cdot x}{\underset{1}{5}} = \dfrac{300}{5}$ Divide both sides by 5.

$x = 60$

The percent is 60, which is written as 60%.

6.3 Identifying Percent, Whole, and Part in a Percent Problem

The percent appears with the word **percent** or with the symbol **%**.

The whole often appears after the word **of**. The whole is the entire quantity or total.

The part is the portion of the total. If the percent and the whole are found first, the remaining number is the part.

Find the percent, whole, and part in the following.

10% **of** the 500 pies is how many pies?
Percent   Whole   Part (unknown)

20 cats is 5% **of** what number of cats?
Part   Percent   Whole (unknown)

What percent   **of** $220   is $33?
Percent (unknown)   Whole   Part

6.4 Applying the Percent Proportion

Read the problem and identify the percent, whole, and part. Use the percent proportion to solve for the unknown quantity.

A liquid mixture in a tank contains 35% distilled water. If 28 gallons of distilled water are in the tank when it is full, find the capacity of the tank.

$$\text{percent} = 35 \quad \text{and} \quad \text{part} = 28$$

Use the percent proportion to find the whole.

Whole (unknown) → $\dfrac{\text{part}}{x} = \dfrac{\text{percent}}{100}$

$\dfrac{28}{x} = \dfrac{35}{100}$

$\dfrac{28}{x} = \dfrac{7}{20}$ Write $\frac{35}{100}$ as $\frac{7}{20}$ in lowest terms.

$x \cdot 7 = 560$ Cross products

$\dfrac{x \cdot \overset{1}{7}}{\underset{1}{7}} = \dfrac{560}{7}$ Divide both sides by 7.

$x = 80$

The capacity of the tank is 80 gallons.

Concepts	Examples

6.5 Using the Percent Equation

The percent equation is part = percent • whole. Identify the percent, whole, and part and solve for the unknown quantity. Always write the percent as a decimal before using the equation.

Solve each problem.

(a) Find 20% of 220 applicants.

$$\textbf{part (unknown)} = \text{percent} \cdot \text{whole}$$

$$x = (0.2)(220)$$

$$x = 44$$

20% of 220 applicants is 44 applicants.

(b) 8 balls is 4% of what number of balls?

$$\text{part} = \text{percent} \cdot \textbf{whole (unknown)}$$

$$8 = (0.04)(x)$$

$$\frac{8}{0.04} = \frac{\overset{1}{(0.04)}(x)}{\underset{1}{0.04}}$$

$$x = 200$$

8 balls is 4% of 200 balls.

(c) $13 is what percent of $52?

$$\text{part} = \textbf{percent (unknown)} \cdot \text{whole}$$

$$13 = x \cdot 52$$

$$\frac{13}{52} = \frac{x \cdot \overset{1}{52}}{\underset{1}{52}}$$

$$x = 0.25 = 25\%$$

$13 is 25% of $52.

6.6 Solving Application Problems with Proportions

To solve for **sales tax,** use this formula.

amount of sales tax = rate of tax • cost of item

The price of a 46-inch plasma HD television is $699, and the sales tax is 5%. Find the sales tax.

$$\textbf{amount of sales tax} = (5\%)(\$699)$$

$$= (0.05)(\$699) = \$34.95$$

To find **commissions,** use this formula.

**amount of commission =
rate of commission • amount of sales**

The sales are $92,000 with a commission rate of 3%. Find the commission.

$$\textbf{amount of commission} = (3\%)(\$92,000)$$

$$= (0.03)(\$92,000)$$

$$= \$2760$$

To find the **discount** and the **sale price,** use these formulas.

amount of discount = rate of discount • original price

sale price = original price − amount of discount

A gas oven originally priced at $480 is offered at a 25% discount. Find the amount of the discount and the sale price.

$$\textbf{discount} = (0.25)(\$480) = \textbf{\$120}$$

$$\textbf{sale price} = \$480 - \textbf{\$120} = \textbf{\$360}$$

Concepts	Examples

6.6 Solving Application Problems with Proportions (Continued)

To find the **percent of change,** subtract to find the amount of change (increase or decrease), which is the part. The whole is the *original* value or value *before* the change.

The number of parking violations rose from 1980 violations to 2277. Find the percent of increase.

$$2277 - 1980 = 297 \quad \text{Increase}$$

$$\text{Increase} \rightarrow \frac{297}{1980} = \frac{\textbf{percent}}{100} \leftarrow \text{Original value}$$

Solve the proportion to find that the percent $= 15$, so the percent of increase is 15%.

6.7 Finding Simple Interest

Use the formula $\quad I = p \cdot r \cdot t$

$$\textbf{Interest} = \textbf{principal} \cdot \textbf{rate} \cdot \textbf{time}$$

Time (t) is in years. When the time is given in months, use a fraction with 12 in the denominator because there are 12 months in a year.

$2800 is deposited at 2% for 3 months. Find the amount of interest.

$$I = p \cdot r \cdot t$$

$$= \underbrace{(2800)(0.02)}\left(\frac{3}{12}\right)$$

$$= \quad (56) \quad \left(\frac{1}{4}\right) = \frac{(56)(1)}{4} = \$14$$

6.8 Finding Compound Amount and Compound Interest

There are three methods for finding the compound amount.

1. Calculate the interest for each compound interest period, then add it back to the principal.

Find the compound amount and interest if $1500 is deposited at 5% interest for 3 years.

1.

	Interest	Compound Amount
Year 1	($1500)(0.05)(1) = **$75**	
		$1500 + **$75** = **$1575**
Year 2	($1575)(0.05)(1) = **$78.75**	
		$1575 + **$78.75** = **$1653.75**
Year 3	($1653.75)(0.05) ≈ **$82.69**	
		$1653.75 + **$82.69** = **$1736.44**

2. Multiply the original deposit by 100% plus the compound interest rate.

2. ($1500)(1.05)(1.05)(1.05) ≈ $1736.44

Original deposit   Compound amount

$$100\% + 5\% = 105\% = 1.05$$

3. Use the compound interest table to find the interest on $1. Then, multiply the table value by the principal.

The interest is found with this formula.

Interest = compound amount − original deposit

3. Locate 5% across the top of the table and 3 periods at the left. The table value is **1.1576.**

compound amount = ($1500)(**1.1576**) = **$1736.40***

interest = $1736.40 − $1500 = $236.40

* The difference in the compound amount results from rounding in the table.

Chapter 6 Review Exercises

6.1 *Write each percent as a decimal and each decimal as a percent.*

1. 35%   2. 150%   3. 99.44%   4. 0.085%

5. 3.15   6. 0.02   7. 0.875   8. 0.002

6.2 *Write each percent as a fraction or mixed number in lowest terms and each fraction as a percent.*

9. 15%   10. 37.5%   11. 175%   12. 0.25%

13. $\dfrac{3}{4}$ 14. $\dfrac{5}{8}$ 15. $3\dfrac{1}{4}$ 16. $\dfrac{1}{200}$

Complete this chart.

Fraction	Decimal	Percent
$\dfrac{1}{8}$	17. _____	18. _____
19. _____	0.25	20. _____
21. _____	22. _____	180%

6.3 *Find the unknown value in the percent proportion* $\dfrac{part}{whole} = \dfrac{percent}{100}$.

23. part = 25, percent = 10

24. whole = 480, percent = 5

Identify each component and then set up each problem using the percent proportion $\dfrac{part}{whole} = \dfrac{percent}{100}$. *Do **not** try to solve for the unknown value.*

25. 35% of 820 mailboxes is 287 mailboxes.

26. 73 DVDs is what percent of 90 DVDs?

27. Find 14% of 160 mountain bikes.

28. 418 curtains is 16% of what number of curtains?

29. A golfer lost three of his eight golf balls. What percent were lost?

30. Only 88% of the door keys cut will operate properly. If there are 1280 keys cut, find the number of keys that will operate properly.

6.4 *Find the part using the percent proportion or the multiplication shortcut.*

31. 18% of 950 programs

32. 60% of 1450 reference books

33. 0.6% of 5200 acres

34. 0.2% of 1400 kilograms

Find the whole using the percent proportion.

35. 105 crates is 14% of what number of crates?

36. 348 test tubes is 15% of what number of test tubes?

37. 677.6 miles is 140% of what number of miles?

38. 2.5% of what number of cases is 425 cases?

Find the percent using the percent proportion. Round percent answers to the nearest tenth if necessary.

39. 649 tulip bulbs is what percent of 1180 tulip bulbs?

40. What percent of 1620 dinner rolls is 85 dinner rolls?

41. What percent of 380 pairs of socks is 36 pairs?

42. What percent of 650 soup cans is 200 soup cans?

6.1–6.4 *Solve each application problem. Round percent answers to the nearest tenth if necessary.*

43. Last year there was a total of 63 shark attacks on humans in the world (6 were fatal). This year there was an increase of 25.4% in the number of these attacks. Find the number of shark attacks on humans this year. (Round to the nearest whole number.) (*Source*: MSNBC.com)

44. Each week 3200 people shop at the three Crescent City Farmers Markets in New Orleans, Louisiana. If 896 of these shoppers are seniors, what percent of the shoppers are seniors? (*Source: USA Today.*)

6.5 *Use the percent equation to answer each question.*

45. 32% of $454 is what amount?

46. 155% of 120 trucks is how many trucks?

47. 0.128 ounce is what percent of 32 ounces?

48. 304.5 meters is what percent of 174 meters?

49. 33.6 miles is 28% of what number of miles?

50. $92 is 16% of what amount?

6.6 *Find the amount of sales tax or the tax rate and the total cost. Round to the nearest cent if necessary.*

Cost of Item	Tax Rate	Amount of Tax	Total Cost
51. $630	5%	_____	_____
52. $780	_____	$58.50	_____

Find the commission earned or the rate of commission.

Sales	Rate of Commission	Commission
53. $3450	8%	_____
54. $65,300	_____	$3265

Find the amount or rate of discount and the sale price. Round to the nearest cent if necessary.

Original Price	Rate of Discount	Amount of Discount	Sale Price
55. $112.50	30%	_____	_____
56. $252	_____	$63	_____

6.7 *Find the simple interest due on each loan.*

	Principal	Rate	Time in Years	Interest
57.	$200	4%	1	_____
58.	$1080	5%	$1\frac{1}{4}$	_____

Find the simple interest paid on each investment.

	Principal	Rate	Time in Months	Interest
59.	$400	$3\frac{1}{2}\%$	3	_____
60.	$1560	$6\frac{1}{2}\%$	18	_____

Find the total amount due on each simple interest loan.

	Principal	Rate	Time	Total Amount Due
61.	$750	$5\frac{1}{2}\%$	2 years	_____
62.	$1560	3%	9 months	_____

6.8 *Find the compound amount and the interest. Interest is compounded annually. You may use the compound interest table. Round answers to the nearest cent if necessary.*

	Principal	Rate	Time in Years	Compound Amount	Interest
63.	$4000	3%	10	_____	_____
64.	$1870	4%	4	_____	_____
65.	$3600	$4\frac{1}{2}\%$	3	_____	_____
66.	$12,500	$5\frac{1}{2}\%$	5	_____	_____

Mixed Review Exercises

Find the unknown value in the percent proportion $\dfrac{part}{whole} = \dfrac{percent}{100}$.

67. whole = 80, percent = 15

68. part = 738, percent = 45

Use the percent proportion or percent equation to answer each question.

69. 12% of 194 meters is how many meters?

70. 327 cars is what percent of 218 cars?

71. 0.6% of $85 is what amount?

72. 99 employees is 5% of what number of employees?

73. 76 chickens is what percent of 190 chickens?

74. 214.484 liters is 43% of what number of liters?

Write each percent as a decimal and each decimal as a percent.

75. 55%

76. 300%

77. 5

78. 4.71

79. 8.6%

80. 0.621

81. 0.375%

82. 0.0006

Write each percent as a fraction in lowest terms and each fraction or mixed number as a percent.

83. $\dfrac{3}{4}$

84. 42%

85. 87.5%

86. $\dfrac{3}{8}$

87. $32\dfrac{1}{2}\%$

88. $\dfrac{3}{5}$

89. 0.25%

90. $3\dfrac{3}{4}$

Solve each application problem. Round percent answers to the nearest tenth and money answers to the nearest cent if necessary.

91. Eva Jacob deposits $20,500 in her credit union savings account. If she earns $6\dfrac{1}{2}\%$ simple interest for 30 months, how much interest will be earned?

92. Hap Pishke, owner of Mardi Gras Barbers, borrows $14,750 to remodel his shop. He agrees to an 8% simple interest rate and will repay the loan in 18 months. Find the total amount due.

93. A study of 1005 adults found that 965 of them have a smoke alarm in their home and 461 have a carbon monoxide detector. (*Source*: Liberty Mutual International Association of Firefighters.)

(a) Find the percent of adults who have a smoke alarm in their home. Round to the nearest tenth of a percent.

(b) Find the percent of adults who have a carbon monoxide detector in their home. Round to the nearest tenth of a percent.

94. Alan Zagorin borrows $148,000 to expand his business, located at the end of historic Route 66. The loan has an interest rate of 5% compounded annually and will be repaid in 4 years.

(a) Find the compound amount of this loan at the end of 4 years. Do not use the table.

(b) Find the amount of interest that he owes.

95. Tom Dugally, a real estate agent, sold two properties, one for $125,000 and the other for $290,000. He receives a commission of $1\frac{1}{2}$% of total sales. Find the commission that he earned.

96. Vending machines on campus must include healthy food choices such as fruits, fruit juices, and healthy snacks. Sales of healthy foods in the vending machines increased from 4320 items last month to 5107 items this month. Find the percent of increase.

97. A Sears Kenmore washer/dryer set priced at $958 is marked down 18%. If the sales tax is 8%, find the cost of the washer/dryer set, including the sales tax.

98. In a recent insurance company study of boaters who had lost items overboard, 88 boaters or 8% said that they lost their cell phones. Find the total number of boaters in the survey. (*Source:* Progressive Groups of Insurance Companies.)

99. Jack and Jill Ahearn begin to budget 25% for rent, 20% for food, 7% for education, 5% for clothing, 10% for transportation, 12% for travel and recreation, 4% for miscellaneous, and the remainder for savings. Jack takes home $2850 per month, and Jill takes home $42,300 per year. How much money will the couple save in a year?

100. The mileage on a hybrid car dropped from 42.8 miles per gallon in the city to 28.5 miles per gallon on the highway. Find the percent of decrease.

Write each percent as a decimal and each decimal as a percent.

1. 65%

2. 0.8

3. 1.75

4. 0.875

5. 300%

6. 2%

Write each percent as a fraction in lowest terms.

7. 12.5%

8. 0.25%

Write each fraction or mixed number as a percent.

9. $\dfrac{3}{5}$

10. $\dfrac{5}{8}$

11. $2\dfrac{1}{2}$

Solve each problem.

12. 32 sacks is 4% of what number of sacks?

13. $680 is what percent of $3400?

14. There are still 100,000 households in the United States that do not have electricity. If this is 0.08% of the homes, find the total number of households. (*Source: Time* magazine.)

15. The price of a diamond engagement ring is $3240 plus sales tax of $6\frac{1}{2}$%. Find the total cost of the engagement ring including sales tax.

16. An insurance company pays its salespeople on commission. If a commission of $628 is earned on insurance sales of $7850, find the rate of commission.

17. Attendance at the homecoming game decreased from 5760 fans last year to 4320 fans this year. Find the percent of decrease.

18. A problem includes last year's salary, this year's salary, and asks for the percent of increase. Explain how you would identify the part, the whole, and the percent in the problem. Show the percent proportion that you would use.

19. Write the formula used to find interest. Explain the difference in what to do if the time is expressed in months or in years. Write a problem that involves finding interest for 9 months and another problem that involves finding interest for $2\frac{1}{2}$ years. Use your own numbers for the principal and the rate. Show how to solve your problems.

Find the amount of discount and the sale price.

	Original Price	Rate of Discount	Amount of Discount	Sale Price
20.	$96	12%	_____	_____
21.	$280	32.5%	_____	_____

⊞ *Find the simple interest on each loan.*

	Principal	Rate	Time	Simple Interest
22.	$4200	6%	$1\frac{1}{2}$ years	_____
23.	$6400	9%	4 months	_____

24. Sita Lalchandani takes out a short-term loan of $19,200 to pay for her daughter's medical school expenses. The interest rate on the loan is 7%, and the loan is for 15 months. Find the total amount needed to repay the loan.

25. The River City School PTA Emergency Fund deposited $4000 at 6% interest compounded annually. Two years after the first deposit, they deposit another $5000, also at 6% interest compounded annually. Use the compound interest table.

(a) What total amount will they have 4 years after their first deposit? Round to the nearest dollar.

(b) What amount of interest will they have earned?

Chapters 1–6 Cumulative Review Exercises

First use front end rounding to round each number and estimate the answer. Then find the exact answer.

1. *Estimate:* *Exact:*

$$\begin{array}{r} 75{,}078 \\ -\ 46{,}090 \\ \hline \end{array}$$

2. *Estimate:* *Exact:*

$$\begin{array}{r} 7.8 \\ -\ 3.5029 \\ \hline \end{array}$$

3. *Estimate:* *Exact:*

$$\begin{array}{r} 6538 \\ \times\ \ \ 708 \\ \hline \end{array}$$

4. *Estimate:* *Exact:*

$$\begin{array}{r} 65.3 \\ \times\ 8.7 \\ \hline \end{array}$$

5. *Estimate:* *Exact:*

$$43\overline{)38{,}786}$$

6. *Estimate:* *Exact:*

$$0.8\overline{)6.76}$$

Use the order of operations to simplify each expression.

7. $6^2 - 3(6)$

8. $\sqrt{49} + 5 \cdot 4 - 8$

9. $9 + 6 \div 3 + 7(4)$

Round each number.

10. 7,583,281 to the nearest hundred-thousand

11. \$513.499 to the nearest dollar

12. \$362.735 to the nearest cent

Add, subtract, multiply, or divide as indicated. Write answers in lowest terms and as whole or mixed numbers when possible.

13.
$$\begin{array}{r} 5\frac{3}{4} \\ +\ 7\frac{5}{8} \\ \hline \end{array}$$

14.
$$\begin{array}{r} 8\frac{3}{8} \\ -\ 4\frac{1}{2} \\ \hline \end{array}$$

15. $36 \cdot \dfrac{4}{5}$

16. $12 \div \dfrac{3}{4}$

17. The size of a prison cell at Alcatraz Prison in the San Francisco Bay is 5 feet by 9 feet. The average size of a shark cage is 5 feet by $6\frac{1}{2}$ feet. How many more square feet are there in the floor of the prison cell than the shark cage? (*Source:* Discovery Channel, Monster Garage Factoid.)

18. To prepare for the state real estate exam, Mia Dawson studied $5\frac{1}{2}$ hours on the first day, $6\frac{1}{4}$ hours on the second day, $3\frac{3}{4}$ hours on the third day, and 7 hours on the fourth day. How many hours did she study altogether?

Write < or > to make each statement true.

19. $\dfrac{5}{8} \underline{} \dfrac{2}{3}$

20. $\dfrac{8}{15} \underline{} \dfrac{11}{20}$

21. $\dfrac{2}{3} \underline{} \dfrac{7}{12}$

Simplify each expression. Use the order of operations as needed.

22. $\dfrac{2}{3}\left(\dfrac{7}{8} - \dfrac{1}{2}\right)$

23. $\dfrac{7}{8} \div \left(\dfrac{3}{4} + \dfrac{1}{8}\right)$

24. $\left(\dfrac{5}{6} - \dfrac{5}{12}\right) - \left(\dfrac{1}{2}\right)^2 \cdot \dfrac{2}{3}$

Write each fraction as a decimal. Round to the nearest thousandth if necessary.

25. $\dfrac{3}{4}$

26. $\dfrac{3}{8}$

27. $\dfrac{7}{12}$

28. $\dfrac{11}{20}$

Find the unknown value in each proportion.

29. $\dfrac{1}{5} = \dfrac{x}{30}$ **30.** $\dfrac{224}{32} = \dfrac{28}{x}$ **31.** $\dfrac{8}{x} = \dfrac{72}{144}$ **32.** $\dfrac{x}{120} = \dfrac{7.5}{30}$

Write each percent as a decimal. Write each decimal as a percent.

33. 3%   **34.** 200%   **35.** 0.87   **36.** 3.8

Write each percent as a fraction or mixed number in lowest terms. Write each fraction or mixed number as a percent.

37. 8%   **38.** 62.5%   **39.** 175%   **40.** $\dfrac{7}{8}$ **41.** $4\dfrac{1}{5}$

Solve.

42. 65% of 3280 DVDs is how many DVDs?

43. $4\frac{1}{2}$% of what number of miles is 76.5 miles?

44. 252 hours is what percent of 180 hours?

45. Diane McKinney paid $29.90 in sales tax on a $460 purchase. What was the tax rate?

46. While enrolled in college full-time, Theresa Goldsmith is a part-time waitress. Her tips average 14%, and on a recent shift she had total sales of $837. Find the amount of her tips.

47. A box spring and mattress originally priced at $456 is discounted 45%. Find the amount of discount and the sale price.

48. A loan of $46,300 is made at 3% simple interest for 9 months. Find the total amount to be repaid.

Set up and solve a proportion for each problem.

49. Gerri Junso gives school readiness tests to 9 children in 4 hours. Find the number of children she can test in 20 hours.

50. If 12.5 ounces of Roundup weed and grass killer is needed to make 5 gallons of spray, how much Roundup is needed for 102 gallons of spray?

The number of existing single-family homes sold in four regions of the country in the same month of two separate years is shown in the figure below. Use this information to answer Exercises 51–54. Round answers to the nearest tenth of a percent if necessary.

51. Find the percent of increase in sales in the northeastern region.

52. Find the percent of increase in sales in the midwestern region.

53. What is the percent of decrease in sales in the southern region?

54. What is the percent of decrease in sales in the western region?

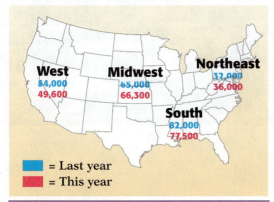

AT HOME

Existing home sales by region:

West 54,000 / 49,600
Midwest 65,000 / 66,300
Northeast 32,000 / 36,000
South 82,000 / 77,500

■ = Last year
■ = This year

Math in the Media

EDUCATIONAL TAX INCENTIVES

The government sponsors tax incentive programs to make education more affordable. To qualify for the programs, you must have an adjusted gross income below a certain level (most recently $90,000). You can find specific information at the Internal Revenue Service website (irs.gov).

- The Hope Scholarship offers 100% of the first $2000 spent for certain expenses, such as tuition and books, during the first year of college, plus 25% of the next $2000. The scholarship money is payable as a tax refund. The student cannot have completed the first two years of post-secondary education and must meet certain educational goals and workload criteria.

Suppose you are paying your own educational costs, and your adjusted gross income meets the guidelines to qualify for the Hope Scholarship. Your goals are to earn an Associate of Arts degree from a community college and then transfer to a state university to complete a Bachelor's degree. Tuition costs for resident students at American River Community College in California are used as an example of educational expenses.

Residents of the college district pay an enrollment fee of $46 per semester hour plus a parking permit fee of $30 each semester. Assume that you must study a total of 15 semester hours in developmental work in mathematics, reading, and writing, and an additional 60 semester hours to complete an Associate of Arts degree. You decide to limit your course load to 15 credit hours each semester. Assume that one course is 3 semester hours, and you will have to purchase books at an approximate cost of $95 per course.

1. How many semesters and how many courses will it take you to finish the requirements for an Associate of Arts degree?

2. What is the total cost to complete the Associate of Arts degree for (**a**) books and (**b**) tuition and fees?

3. (**a**) Calculate the total cost for enrollment fees, parking fees, and books to complete the Associate of Arts degree (5 semesters).
   (**b**) What is the maximum tax incentive payable to you under the Hope Scholarship?

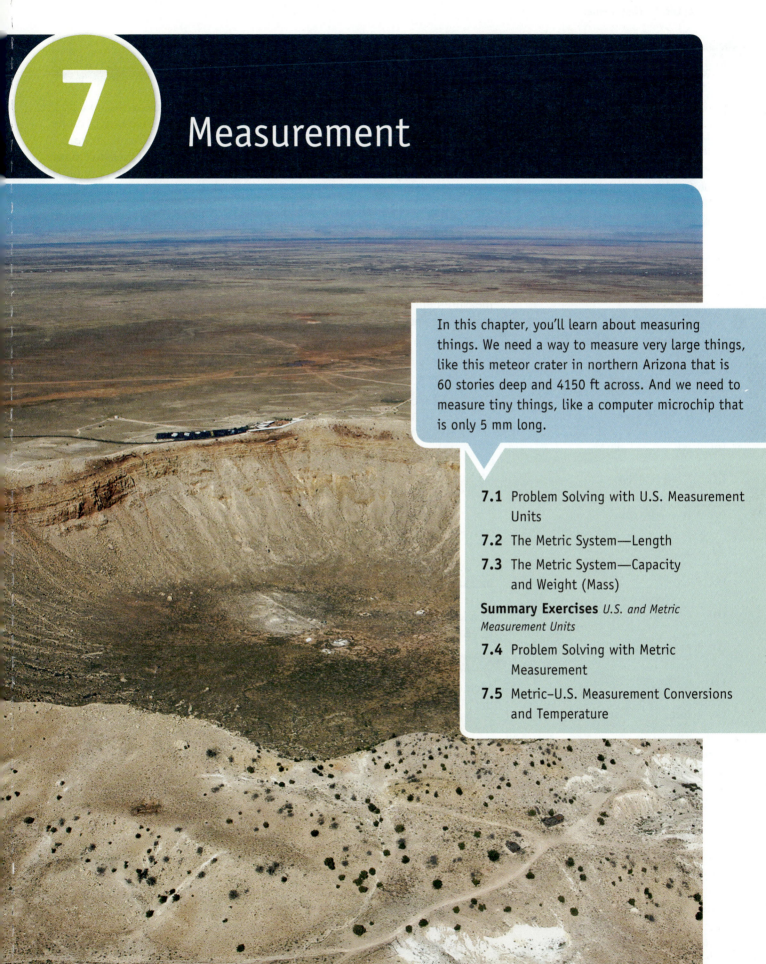

7 Measurement

In this chapter, you'll learn about measuring things. We need a way to measure very large things, like this meteor crater in northern Arizona that is 60 stories deep and 4150 ft across. And we need to measure tiny things, like a computer microchip that is only 5 mm long.

7.1 Problem Solving with U.S. Measurement Units

OBJECTIVES

1. **Learn the basic U.S. measurement units.**

2. **Convert among U.S. measurement units using multiplication or division.**

3. **Convert among U.S. measurement units using unit fractions.**

4. **Solve application problems using U.S. measurement units.**

We measure things all the time: the distance traveled on vacation, the floor area we want to cover with carpet, the amount of milk in a recipe, the weight of the bananas we buy at the store, the number of hours we work, and many more.

In the United States, we still use the **United States (U.S.) measurement system** for many everyday activities. Examples of U.S. measurement units are inches, feet, quarts, ounces, and pounds. However, science, medicine, sports, and manufacturing use the **metric system** (meters, liters, and grams). Because the rest of the world uses *only* the metric system, U.S. businesses have been changing to the metric system in order to compete internationally.

OBJECTIVE ▶ 1 Learn the basic U.S. measurement units. Until the switch to the metric system is complete, we still need to know how to use U.S. measurement units. The table below lists the relationships you should memorize. The time relationships are used in both the U.S. and metric systems.

1 After memorizing the U.S. measurement conversions, fill in the blanks.

(a) 1 c = _____ fl oz

(b) _____ qt = 1 gal

(c) 1 wk = _____ days

(d) _____ ft = 1 yd

(e) 1 ft = _____ in.

(f) _____ oz = 1 lb

(g) 1 ton = _____ lb

(h) _____ min = 1 hr

(i) 1 pt = _____ c

(j) _____ hr = 1 day

(k) 1 min = _____ sec

(l) 1 qt = _____ pt

(m) _____ ft = 1 mi

U.S. Measurement Unit Relationships	
U.S. Length Units	**U.S. Weight Units**
12 inches (in.) = 1 foot (ft)	16 ounces (oz) = 1 pound (lb)
3 feet (ft) = 1 yard (yd)	2000 pounds (lb) = 1 ton
5280 feet (ft) = 1 mile (mi)	
U.S. Capacity Units	**Time (U.S. and Metric)**
8 fluid ounces (fl oz) = 1 cup (c)	60 seconds (sec) = 1 minute (min)
2 cups (c) = 1 pint (pt)	60 minutes (min) = 1 hour (hr)
2 pints (pt) = 1 quart (qt)	24 hours (hr) = 1 day
4 quarts (qt) = 1 gallon (gal)	7 days = 1 week (wk)

As you can see, there is no simple way to convert among these various measures. The units evolved over hundreds of years and were based on a variety of "standards." For example, one yard was the distance from the tip of a king's nose to his thumb when his arm was outstretched. An inch was three dried barleycorns laid end to end.

EXAMPLE 1   Knowing U.S. Measurement Units

Memorize the U.S. measurement conversions in the table above. Then fill in the blanks.

(a) 24 hr = _____ day   Answer: 1 day

(b) 1 yd = _____ ft   Answer: 3 ft

◀ Work Problem **1** at the Side.

OBJECTIVE ▶ 2 Convert among U.S. measurement units using multiplication or division. You often need to convert from one unit of measure to another. Two methods of converting measurements are shown here. Study each way and use the method you prefer. The first method involves deciding whether to multiply or divide.

Converting among Measurement Units

1. *Multiply* when converting from a larger unit to a smaller unit.
2. *Divide* when converting from a smaller unit to a larger unit.

Answers

1. **(a)** 8 **(b)** 4 **(c)** 7 **(d)** 3 **(e)** 12
   (f) 16 **(g)** 2000 **(h)** 60 **(i)** 2 **(j)** 24
   (k) 60 **(l)** 2 **(m)** 5280

| EXAMPLE 2 | **Converting from One U.S. Unit of Measure to Another** |

Convert each measurement.

(a) 7 ft to inches

You are converting from a *larger* unit to a *smaller* unit (a *foot* is longer than an *inch*), so multiply.

Because *1 ft = 12 in.*, multiply by 12.

$$7 \text{ ft} = 7 \cdot 12 = 84 \text{ in.}$$

(b) $3\frac{1}{2}$ lb to ounces

You are converting from a *larger* unit to a *smaller* unit (a *pound* is heavier than an *ounce*), so multiply.

Because *1 lb = 16 oz*, multiply by 16.

> Divide 16 and 2 by their common factor of 2.
> 16 ÷ 2 is 8.
> 2 ÷ 2 is 1.

$$3\frac{1}{2} \text{ lb} = 3\frac{1}{2} \cdot 16 = \frac{7}{2} \cdot \frac{\overset{8}{16}}{1} = \frac{56}{1} = 56 \text{ oz}$$

(c) 20 qt to gallons

You are converting from a *smaller* unit to a *larger* unit (a *quart* is smaller than a *gallon*) so divide.

Because *4 qt = 1 gal*, divide by 4.

$$20 \text{ qt} = \frac{20}{4} = 5 \text{ gal}$$

> Remember to write **gal** in your answer.

Divide by 4.

(d) 45 min to hours

You are converting from a *smaller* unit to a *larger* unit (a *minute* is less than an *hour*), so divide.

Because *60 min = 1 hr*, divide by 60 and write the fraction in lowest terms.

$$45 \text{ min} = \frac{45}{60} = \frac{45 \div 15}{60 \div 15} = \frac{3}{4} \text{ hr} \leftarrow \text{Lowest terms}$$

Divide by 60.

·········· **Work Problem** ➋ **at the Side.** ▶

| OBJECTIVE ➌ **Convert among U.S. measurement units using unit fractions.** If you have trouble deciding whether to multiply or divide when converting measurements, use *unit fractions* to solve the problem. You'll also use this method in science courses. A **unit fraction** is equivalent to 1. Here is an example.

$$\frac{12 \text{ in.}}{12 \text{ in.}} = \frac{\overset{1}{\cancel{12 \text{ in.}}}}{\underset{1}{\cancel{12 \text{ in.}}}} = 1$$

Use the table of measurement relationships on the previous page to find that 12 in. is the same as 1 ft. So you can substitute 1 ft for 12 in. in the numerator, or you can substitute 1 ft for 12 in. in the denominator. This makes two useful unit fractions.

$$\frac{1 \text{ ft}}{12 \text{ in.}} = 1 \qquad \text{or} \qquad \frac{12 \text{ in.}}{1 \text{ ft}} = 1$$

➋ Convert each measurement using multiplication or division.

(a) $5\frac{1}{2}$ ft to inches

Because you are converting from a *larger* to a *smaller* unit, do you multiply or divide? _____

(b) 64 oz to pounds

Because you are converting from a *smaller* to a *larger* unit, you _____.

(c) 6 yd to feet

(d) 2 tons to pounds

(e) 35 pt to quarts

(f) 20 min to hours

Answers

2. **(a)** multiply; 66 in. **(b)** divide; 4 lb
(c) 18 ft **(d)** 4000 lb **(e)** $17\frac{1}{2}$ qt **(f)** $\frac{1}{3}$ hr

VOCABULARY TIP
Uni- is a prefix that means "one," so a **unit fraction** is equivalent to 1.

3 First write the unit fraction needed to make each conversion. Then complete the conversion.

GS **(a)** 36 in. to feet

unit fraction $\Big\}$ $\dfrac{1 \text{ ft}}{12 \text{ in.}}$

$$\dfrac{\overset{3}{\cancel{36} \text{ in.}}}{1} \cdot \dfrac{1 \text{ ft}}{\underset{1}{\cancel{12} \text{ in.}}} =$$

(b) 14 ft to inches

unit fraction $\Big\}$ $\dfrac{\text{in.}}{\text{ft}}$

(c) 60 in. to feet

unit fraction $\Big\}$ _____

(d) 4 yd to feet

unit fraction $\Big\}$ _____

(e) 39 ft to yards

unit fraction $\Big\}$ _____

(f) 2 mi to feet

unit fraction $\Big\}$ _____

Answers

3. **(a)** $\dfrac{3 \cdot 1 \text{ ft}}{1} = 3 \text{ ft}$ **(b)** $\dfrac{12 \text{ in.}}{1 \text{ ft}}$; 168 in.

(c) $\dfrac{1 \text{ ft}}{12 \text{ in.}}$; 5 ft **(d)** $\dfrac{3 \text{ ft}}{1 \text{ yd}}$; 12 ft

(e) $\dfrac{1 \text{ yd}}{3 \text{ ft}}$; 13 yd **(f)** $\dfrac{5280 \text{ ft}}{1 \text{ mi}}$; 10,560 ft

---

To convert from one measurement unit to another, just multiply by the appropriate unit fraction. Remember, a unit fraction is equivalent to 1. Multiplying something by 1 does *not* change its value.

Use these guidelines to choose the correct unit fraction.

Choosing a Unit Fraction

The *numerator* should use the measurement unit you want in the *answer*.
The *denominator* should use the measurement unit you want to *change*.

EXAMPLE 3 **Using Unit Fractions with U.S. Length Measurements**

(a) Convert 60 in. to feet.

Use a unit fraction with feet (the unit for your answer) in the numerator, and inches (the unit being changed) in the denominator. Because *1 ft = 12 in.*, the necessary unit fraction is

$$\dfrac{1 \text{ ft}}{12 \text{ in.}} \quad \begin{array}{l} \leftarrow \text{Unit for your } \textbf{answer} \text{ is } \textbf{feet.} \\ \leftarrow \text{Unit being } \textbf{changed} \text{ is } \textbf{inches.} \end{array}$$

Next, multiply 60 in. times this unit fraction. Write 60 in. as the fraction $\dfrac{60 \text{ in.}}{1}$ and divide out common units and factors wherever possible.

$$60 \text{ in.} \cdot \dfrac{1 \text{ ft}}{12 \text{ in.}} = \dfrac{\overset{5}{\cancel{60} \text{ in.}}}{1} \cdot \dfrac{1 \text{ ft}}{\underset{1}{\cancel{12} \text{ in.}}} = \dfrac{5 \cdot 1 \text{ ft}}{1} = 5 \text{ ft}$$

Remember to write **ft** in your answer.

These units should match.

Divide out inches.

Divide 60 and 12 by 12.

(b) Convert 9 ft to inches.

Select the correct unit fraction to change 9 ft to inches.

$$\dfrac{12 \text{ in.}}{1 \text{ ft}} \quad \begin{array}{l} \leftarrow \text{Unit for your } \textbf{answer} \text{ is } \textbf{inches.} \\ \leftarrow \text{Unit being } \textbf{changed} \text{ is } \textbf{feet.} \end{array}$$

Multiply 9 ft times the unit fraction.

$$9 \text{ ft} \cdot \dfrac{12 \text{ in.}}{1 \text{ ft}} = \dfrac{9 \cancel{\text{ ft}}}{1} \cdot \dfrac{12 \text{ in.}}{1 \cancel{\text{ ft}}} = \dfrac{9 \cdot 12 \text{ in.}}{1} = 108 \text{ in.}$$

These units should match.

Divide out feet.

CAUTION

If no units will divide out, you used the wrong unit fraction.

◀ Work Problem **3** at the Side.

EXAMPLE 4 **Using Unit Fractions with U.S. Capacity and Weight Measurements**

(a) Convert 9 pt to quarts.

First select the correct unit fraction.

$$\dfrac{1 \text{ qt}}{2 \text{ pt}} \quad \begin{array}{l} \leftarrow \text{Unit for your } \textbf{answer} \text{ is } \textbf{quarts.} \\ \leftarrow \text{Unit being } \textbf{changed} \text{ is } \textbf{pints.} \end{array}$$

Continued on Next Page

Next multiply.

Write as a mixed number.

$$9 \text{ pt} \cdot \frac{1 \text{ qt}}{2 \text{ pt}} = \frac{9 \text{ pt}}{1} \cdot \frac{1 \text{ qt}}{2 \text{ pt}} = \frac{9}{2} \text{ qt} = 4\frac{1}{2} \text{ qt}$$

These units should match. Divide out pints.

(b) Convert $7\frac{1}{2}$ gal to quarts.

Write as an improper fraction.

$$\frac{7\frac{1}{2} \text{ gal}}{1} \cdot \frac{4 \text{ qt}}{1 \text{ gal}} = \frac{15}{2} \cdot \frac{4}{1} \text{ qt}$$

Divide out gallons.

Divide 4 and 2 by their common factor of 2.
4 ÷ 2 is 2.
2 ÷ 2 is 1.

$$= \frac{15}{2} \cdot \frac{\overset{2}{4}}{1} \text{ qt}$$
$$\phantom{= \frac{15}{2}} \underset{1}{}$$

$$= 30 \text{ qt}$$

(c) Convert 36 oz to pounds.

Notice that **oz** divides out, leaving **lb**, the unit you want for the answer.

$$\frac{\overset{9}{36} \text{ oz}}{1} \cdot \frac{1 \text{ lb}}{\underset{4}{16} \text{ oz}} = \frac{9}{4} \text{ lb} = 2\frac{1}{4} \text{ lb}$$

> **Note**
>
> In **Example 4(c)** above you get $\frac{9}{4}$ lb. Recall that $\frac{9}{4}$ means $9 \div 4$. If you do $9 \div 4$ on your calculator, you get 2.25 lb. U.S. measurements usually use fractions or mixed numbers, like $2\frac{1}{4}$ lb. However, 2.25 lb is also correct and is the way grocery stores often show weights of produce, meat, and cheese.

·········· **Work Problem** ④ **at the Side.** ▶

EXAMPLE 5 Using Several Unit Fractions

Sometimes you may need to use two or three unit fractions to complete a conversion.

(a) Convert 63 in. to yards.

Use the unit fraction $\dfrac{1 \text{ ft}}{12 \text{ in.}}$ to change inches to feet and the unit fraction $\dfrac{1 \text{ yd}}{3 \text{ ft}}$ to change feet to yards. Notice how all the units divide out except yards, which is the unit you want in the answer.

$$\frac{63 \text{ in.}}{1} \cdot \frac{1 \text{ ft}}{12 \text{ in.}} \cdot \frac{1 \text{ yd}}{3 \text{ ft}} = \frac{63}{36} \text{ yd} = \frac{63 \div 9}{36 \div 9} \text{ yd} = \frac{7}{4} \text{ yd} = 1\frac{3}{4} \text{ yd}$$

··········· **Continued on Next Page**

---

④ Convert using unit fractions.

GS (a) 16 qt to gallons

unit fraction } $\dfrac{1 \text{ gal}}{4 \text{ qt}}$

$$\frac{16 \text{ qt}}{1} \cdot \frac{1 \text{ gal}}{4 \text{ qt}} = \frac{16}{4} \text{ gal} = \underline{}$$

GS (b) 3 c to pints

unit fraction } $\dfrac{\text{pt}}{\text{c}}$

(c) $3\frac{1}{2}$ tons to pounds

(d) $1\frac{3}{4}$ lb to ounces

(e) 4 oz to pounds

Answers

4. (a) 4 gal   **(b)** $\dfrac{1 \text{ pt}}{2 \text{ c}}$; $1\frac{1}{2}$ pt or 1.5 pt

(c) 7000 lb   **(d)** 28 oz   **(e)** $\dfrac{1}{4}$ lb or 0.25 lb

5 Convert using two or three unit fractions.

GS **(a)** 4 tons to ounces

$$\frac{4 \text{ tons}}{1} \cdot \frac{2000 \text{ lb}}{1 \text{ ton}} \cdot \frac{16 \text{ oz}}{1 \text{ lb}} =$$

(b) 3 mi to inches

(c) 36 pt to gallons

(d) 2 wk to minutes

You can also divide out common factors in the numbers.

$$\frac{\overset{7}{\overset{21}{\cancel{63}}}}{1} \cdot \frac{1}{\underset{4}{\cancel{12}}} \cdot \frac{1}{\underset{1}{\cancel{3}}} = \frac{7}{4} = 1\frac{3}{4} \text{ yd}$$

Instead of changing $\frac{7}{4}$ to $1\frac{3}{4}$, you can enter $7 \div 4$ on your calculator to get 1.75 yd. Both answers are correct because 1.75 is equivalent to $1\frac{3}{4}$.

(b) Convert 2 days to seconds.

Use three unit fractions. The first one changes days to hours, the next one changes hours to minutes, and the last one changes minutes to seconds. All the units divide out except seconds, which is the unit you want in your answer.

$$\frac{2 \text{ days}}{1} \cdot \frac{24 \text{ hr}}{1 \text{ day}} \cdot \frac{60 \text{ min}}{1 \text{ hr}} \cdot \frac{60 \text{ sec}}{1 \text{ min}} = 172,800 \text{ sec}$$

Divide out **days**.

Divide out **hr**.

Divide out **min**.

> Remember to write **sec** in your answer.

◀ **Work Problem** **5** **at the Side.**

OBJECTIVE **4** **Solve application problems using U.S. measurement units.** To solve application problems, we will use the steps you learned in **Chapter 1.** Those steps are summarized here.

Step 1 **Read** the problem carefully.

Step 2 **Work out a plan.**

Step 3 **Estimate** a reasonable answer.

Step 4 **Solve** the problem.

Step 5 **State the answer.**

Step 6 **Check** your work.

EXAMPLE 6 **Solving U.S. Measurement Applications**

(a) A 36 oz can of coffee is on sale at Jerry's Foods for $14.39. What is the cost per pound, to the nearest cent? (*Source:* Jerry's Foods.)

Step 1 **Read** the problem. The problem asks for the cost per *pound* of coffee.

Step 2 **Work out a plan.** The weight of the coffee is given in *ounces,* but the answer must be cost *per pound.* Convert ounces to pounds. The word *per* indicates division. You need to divide the cost by the number of pounds.

Step 3 **Estimate** a reasonable answer. To estimate, round $14.39 to $14. Then, there are 16 oz in a pound, so 36 oz are a little more than 2 pounds. Finally, $14 \div 2 = $7 per pound is our estimate.

Continued on Next Page

Answers

5. **(a)** 128,000 oz **(b)** 190,080 in.
(c) $4\frac{1}{2}$ gal or 4.5 gal **(d)** 20,160 min

Step 4 **Solve** the problem. Use a unit fraction to convert 36 oz to pounds.

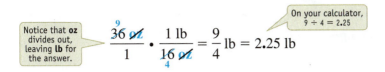

Notice that **oz** divides out, leaving **lb** for the answer.

On your calculator, $9 \div 4 = 2.25$

$$\frac{\overset{9}{\cancel{36}}\ \cancel{oz}}{1} \cdot \frac{1\ lb}{\underset{4}{\cancel{16}}\ \cancel{oz}} = \frac{9}{4}\ lb = 2.25\ lb$$

Then divide to find the *cost* per *pound*.

$$\begin{matrix} \text{Cost} \rightarrow \\ \text{per} \rightarrow \\ \text{pound} \rightarrow \end{matrix} \frac{\$14.39}{2.25\ lb} = 6.39\overline{5} \approx 6.40 \quad \text{(rounded)}$$

Step 5 **State the answer.** The coffee costs $6.40 per pound (to the nearest cent).

Step 6 **Check** your work. The exact answer of $6.40 is close to our estimate of $7.

(b) Bilal's favorite cake recipe uses $1\frac{2}{3}$ cups of milk for one cake. If he makes six cakes for a bake sale at his son's school, how many quarts of milk will he need?

Step 1 **Read** the problem. The problem asks for the number of *quarts* of milk needed for six cakes.

Step 2 **Work out a plan.** Multiply to find the number of *cups* of milk for six cakes. Then convert *cups* to *quarts* (the unit required in the answer).

Step 3 **Estimate** a reasonable answer. To estimate, round $1\frac{2}{3}$ cups to 2 cups. Then, 2 cups times $6 = 12$ cups. There are 4 cups in a quart, so 12 cups $\div 4 = 3$ quarts is our estimate.

Step 4 **Solve** the problem. First multiply. Then use unit fractions to convert.

$$1\frac{2}{3} \cdot 6 = \frac{5}{\underset{1}{\cancel{3}}} \cdot \frac{\overset{2}{\cancel{6}}}{1} = \frac{10}{1} = 10\ \text{cups} \qquad \left\{ \begin{matrix} \text{Milk needed for} \\ \text{six cakes} \end{matrix} \right.$$

$$\frac{\overset{5}{\cancel{10}}\ \cancel{cups}}{1} \cdot \frac{1\ \cancel{pt}}{\underset{1}{\cancel{2}}\ \cancel{cups}} \cdot \frac{1\ qt}{2\ \cancel{pt}} = \frac{5}{2}\ qt = 2\frac{1}{2}\ qt$$

Both **cups** and **pt** divide out, leaving **qt**, the unit you want for the answer.

Step 5 **State the answer.** Bilal needs $2\frac{1}{2}$ qt (or 2.5 qt) of milk.

Step 6 **Check** your work. The exact answer of $2\frac{1}{2}$ qt is close to our estimate of 3 qt.

Note

In Step 2 above, we *first multiplied* $1\frac{2}{3}$ cups by 6 to find the number of cups needed, then *converted* 10 cups to $2\frac{1}{2}$ quarts. It would also work to *first convert* $1\frac{2}{3}$ cups to $\frac{5}{12}$ qt, then *multiply* $\frac{5}{12}$ qt by 6 to get $2\frac{1}{2}$ qt.

Work Problem **6** at the Side. ▶

---

6 Solve each application problem using the six problem-solving steps.

GS **(a)** Kristin paid $3.29 for 12 oz of extra sharp cheddar cheese. What is the price per pound, to the nearest cent?

Step 1 Problem asks for price per _____.

Step 2 So change 12 oz to what unit? _____

(b) A moving company estimates 11,000 lb of furnishings for an average 3-bedroom house. If the company made five such moves last week, how many tons of furnishings did it move? (*Source:* North American Van Lines.)

Answers

6. **(a)** pound; pounds; $4.39 per pound (rounded)

(b) 27.5 or $27\frac{1}{2}$ tons

7.1 Exercises

MyMathLab®

CONCEPT CHECK   *Rewrite each group of measurement units in order from smallest to largest.*

1. (a) foot, mile, inch, yard

(b) pint, fluid ounce, gallon, quart, cup

2. (a) ton, ounce, pound

(b) day, hour, week, second, minute

Fill in the blanks with the measurement relationships you have memorized.
See Example 1.

3. (a) 1 yd = _____ ft;   **(b)** _____ in. = 1 ft

4. (a) 1 ft = _____ in.;   **(b)** _____ ft = 1 mi

5. (a) _____ fl oz = 1 c;   **(b)** 1 qt = _____ pt

6. (a) _____ qt = 1 gal;   **(b)** 1 pt = _____ c

7. (a) _____ lb = 1 ton;   **(b)** 1 lb = _____ oz

8. (a) _____ oz = 1 lb;   **(b)** 1 ton = _____ lb

9. (a) 1 min = _____ sec;   **(b)** _____ min = 1 hr

10. (a) 1 day = _____ hr;   **(b)** _____ sec = 1 min

Convert each measurement by multiplying or dividing. See Example 2.

11. (a) 120 sec = _____ min

(b) 4 hr = _____ min

12. (a) 180 min = _____ hr

(b) 5 min = _____ sec

13. (a) 2 qt = _____ gal

(b) $6\frac{1}{2}$ ft = _____ in.

14. (a) $4\frac{1}{2}$ gal = _____ qt

(b) 12 oz = _____ lb

15. An adult African elephant could weigh 7 to 8 tons. How many pounds could it weigh? (*Source: The Top 10 of Everything.*)

16. A reticulated python snake is the world's longest snake. It grows to a length of 18 to 33 feet. How many yards long can the snake be? (*Source: The Top 10 of Everything.*)

Convert each measurement in Exercises 17–38 using unit fractions.
See Examples 3 and 4.

17. 9 yd = _____ ft

GS

unit fraction } $\dfrac{ft}{yd}$

18. 20,000 lb = _____ tons

GS

unit fraction } $\dfrac{ton}{lb}$

19. 7 lb = _____ oz

20. 96 oz = _____ lb

21. 5 qt = _____ pt

22. 26 pt = _____ qt

23. 90 min = _____ hr

24. 45 sec = _____ min

25. 3 in. = _____ ft

26. 30 in. = _____ ft

27. 24 oz = _____ lb

28. 36 oz = _____ lb

29. 5 c = _____ pt

30. 15 qt = _____ gal

Use the information in the bar graph below to answer Exercises 31–32.

Thickness of Lake Ice Needed for Safe Walking/Driving

15 in. Pickup truck

12 in. Car

5 in. Snowmobile or ATV

4 in. Person walking

Source: Wisconsin DNR.

31. If the ice on a lake is $\frac{1}{2}$ ft thick, what will it safely support?

32. How many feet of ice are needed to safely drive a pickup truck on a lake?

33. $2\frac{1}{2}$ tons = _____ lb

34. $4\frac{1}{2}$ pt = _____ c

35. $4\frac{1}{4}$ gal = _____ qt

36. $2\frac{1}{4}$ hr = _____ min

37. After 15 years, a saguaro cactus is still only one-third to two-thirds of a foot tall, depending upon rainfall. How tall could the cactus be in inches? (*Source: Ecology of the Saguaro III.*)

38. Kevin Durant, an NBA basketball player, is $6\frac{3}{4}$ ft tall. What is his height in inches? (*Source:* NBA.com)

Use two or three unit fractions to make each conversion. **See Example 5.**

39. 6 yd = _____ in.

40. 2 tons = _____ oz

41. 112 c = _____ qt

42. 336 hr = _____ wk

43. 6 days = _____ sec

44. 5 gal = _____ c

45. $1\frac{1}{2}$ tons = _____ oz

46. $3\frac{1}{3}$ yd = _____ in.

47. CONCEPT CHECK The statement $8 = 2$ is *not* true. But with appropriate measurement units, it *is* true.

$$8 \ quarts = 2 \ gallons$$

Attach measurement units to these numbers to make the statement true.

(a) 1 _____ = 16 _____

(b) 10 _____ = 20 _____

(c) 120 _____ = 2 _____

(d) 2 _____ = 24 _____

(e) 6000 _____ = 3 _____

(f) 35 _____ = 5 _____

48. CONCEPT CHECK Explain in your own words why you can add 2 feet + 12 inches to get 3 feet, but you cannot add 2 feet + 12 pounds.

Convert each measurement. **See Example 5.**

49. $2\frac{3}{4}$ mi = _____ in.

50. $5\frac{3}{4}$ tons = _____ oz

51. $6\frac{1}{4}$ gal = _____ fl oz

52. $3\frac{1}{2}$ days = _____ sec

53. 24,000 oz = _____ ton

54. 57,024 in. = _____ mi

Solve each application problem. Show your work. **See Example 6.**

55. Geralyn bought 20 oz of strawberries for $2.29. What was the price per pound for the strawberries, to the nearest cent?

56. Zach paid $0.90 for a 0.8 oz candy bar. What was the cost per pound?

57. Dan orders supplies for the science labs. Each of the 24 stations in the chemistry lab needs 2 ft of rubber tubing. If rubber tubing sells for $8.75 per yard, how much will it cost to equip all the stations?

58. In 2009, Marquette, Michigan, had 181.0 inches of snowfall, while Reno, Nevada, had 22.5 inches. What was the difference in snowfall between the two cities, in feet? Round to the nearest tenth.
(*Source: World Almanac and Book of Facts.*)

59. Tropical cockroaches are the fastest land insects. They can run about 5 feet per second. At this rate, how long would it take the cockroach to travel one mile? (*Source: Guinness World Records.*)

Give your answer

(a) in seconds.

(b) in minutes.

60. A snail moves at an average speed of 2 feet every 3 minutes. At that rate, how long would it take the snail to travel one mile? (*Source: The Concise Nature Encyclopedia.*)

Give your answer

(a) in hours.

(b) in days.

61. At the day care center, each of the 15 toddlers drinks about $\frac{2}{3}$ cup of milk with lunch. The center is open 5 days a week.

(a) How many quarts of milk will the center need for one week of lunches?

(b) If the center buys milk in gallon containers, how many containers should be ordered for one week?

62. Farley's and Sathers Candy Company makes 995,000 pounds of gummies candy each day. (*Source:* Farley's and Sathers.)

(a) How many tons of gummies are produced during a 5-day workweek?

(b) The plant operates 24 hours per day. How many tons of gummies are produced each hour, to the nearest tenth?

Relating Concepts (Exercises 63–66) For Individual or Group Work

On the first page of this chapter, we said that the bowl-shaped crater in northern Arizona made by a meteor crash was sixty stories deep and 4150 ft across. Use this information as you **work Exercises 63–66 in order.** (*Source: The Meteor Crater Story.*)

63. (a) The distance across the crater is what part of a mile, to the nearest tenth?

(b) The crater is nearly circular. In a circle, the distance around the outside edge is about 3.14 times the distance across the circle. How far is it to walk around the edge of the crater, as measured in feet?

(c) How far is it to walk around the edge of the crater in miles, to the nearest tenth?

64. (a) The crater is 550 ft deep. The depth is how many yards, to the nearest whole number?

(b) How many inches deep is the crater?

(c) The depth of the crater is what part of a mile, to the nearest tenth?

(d) When we say that the crater is as deep as a sixty-story building, we are assuming that each story is how many feet tall, to the nearest foot?

65. On one side of the crater there are a few small juniper trees. The trees are 700 years old but only 18 inches to 30 inches tall because of the strong winds and lack of rain.

(a) How tall are the trees in feet?

(b) How many months old are the trees?

(c) At this rate of growth, how long would it take a 30-inch tree to reach a height of three feet?

66. Evidence of two huge meteor crashes has been found on the floor of the Caribbean Sea. One giant circular crater is 90 miles across and the other is 120 miles across.

(a) Using the information about circles in **Exercise 63,** what is the approximate distance around the edge of the smaller crater, to the nearest mile?

(b) Around the larger crater?

7.2 The Metric System—Length

OBJECTIVES

1 Learn the basic metric units of length.

2 Use unit fractions to convert among metric units.

3 Move the decimal point to convert among metric units.

VOCABULARY TIP

Meter means "measurement." The metric system is simply a system using certain types of measurement units. The basic unit for measuring length is the **meter**.

1 Circle the items that measure about 1 meter.

Length of a pencil

Length of a baseball bat

Height of doorknob from the floor

Height of a house

Basketball player's arm length

Length of a paper clip

Answer

1. baseball bat, height of doorknob, basketball player's arm length

---

Around 1790, a group of French scientists developed the metric system of measurement. It is an organized system based on multiples of 10, like our number system and our money. After you are familiar with metric units, you will see that they are easier to use than the hodgepodge of U.S. measurement relationships you used in the last section.

Note

The metric system information in this text is consistent with usage guidelines from the National Institute of Standards and Technology, www.nist.gov/metric.

OBJECTIVE ▶ 1 **Learn the basic metric units of length.** The basic unit of length in the metric system is the **meter** (also spelled *metre*). Use the symbol **m** for meter; do **not** put a period after it. If you put five of the pages from this textbook side by side, they would measure about 1 meter. Or, look at a yardstick—a meter is just a little longer. A yard is 36 inches long; a meter is about 39 inches long.

In the metric system, you use meters for things like measuring the length of your living room, talking about heights of buildings, or describing track and field athletic events.

◀ **Work Problem 1 at the Side.**

To make longer or shorter length units in the metric system, **prefixes** are written in front of the word *meter*. For example, the prefix *kilo* means **1000**, so a *kilo*meter is **1000** meters. The table below shows how to use the prefixes for length measurements. It is helpful to memorize the prefixes because they are also used with weight and capacity measurements. The arrows point to the units you will use most often in daily life.

Prefix	kilo-meter	hecto-meter	deka-meter	meter	deci-meter	centi-meter	milli-meter
Meaning	1000 meters	100 meters	10 meters	1 meter	$\frac{1}{10}$ of a meter	$\frac{1}{100}$ of a meter	$\frac{1}{1000}$ of a meter
Symbol	**km**	**hm**	**dam**	**m**	**dm**	**cm**	**mm**

Length units that are used most often

Here are some comparisons to help you get acquainted with the commonly used length units: km, m, cm, mm.

*Kilo*meters are used instead of miles. A kilometer is **1000** meters. It is about 0.6 mile (a little more than half a mile) or about 5 to 6 city blocks. If you participate in a 10 km run, you'll run about 6 miles.

A meter is divided into 100 smaller pieces called *centi*meters. Each centimeter is $\frac{1}{100}$ of a meter. Centimeters are used instead of inches. A centimeter is a little shorter than $\frac{1}{2}$ inch. The cover of this textbook is about 21 cm wide. A nickel is about 2 cm across. Measure the width and length of your little finger on this centimeter ruler. The tip of your little finger is probably between 1 cm and 2 cm wide.

Measure the width of your little finger.

A nickel is about 2 cm across.

centimeters (cm)

A meter is divided into 1000 smaller pieces called *milli*meters. Each millimeter is $\frac{1}{1000}$ of a meter. It takes 10 mm to equal 1 cm, so it is a very small length. The thickness of a dime is about 1 mm. Measure the width of your pen or pencil and the width of your little finger on this millimeter ruler.

Thickness of a dime is about 1 mm.

mm

10 mm
(same as 1 cm)

50 mm
(same as 5 cm)

100 mm
(same as 10 cm)

EXAMPLE 1 **Using Metric Length Units**

Write the most reasonable metric unit in each blank.
Choose from km, m, cm, and mm.

(a) The distance from home to work is 20 ____.

20 **km** because kilometers are used instead of miles. 20 km is about 12 miles.

(b) My wedding ring is 4 ____ wide.

4 **mm** because the width of a ring is very small.

(c) The newborn baby is 50 ____ long.

50 **cm**, which is half of a meter; a meter is about 39 inches, so half a meter is around 20 inches.

· Work Problem **2** at the Side. ▶

VOCABULARY TIP

Prefix **Pre-** means "before" and **fix** means "stick." So, a prefix is a word part stuck on, or attached, to the front of a word. Prefixes commonly used with **meter** include **kilo**meter (1000 m); **centi**meter ($\frac{1}{100}$ m); and **milli**meter ($\frac{1}{1000}$ m).

2 Write the most reasonable metric unit in each blank. Choose from km, m, cm, and mm.

(a) The woman's height is 168 ____.

(b) The man's waist is 90 ____ around.

(c) Louise ran the 100 ____ dash in the track meet.

(d) A postage stamp is 22 ____ wide.

(e) Michael paddled his canoe 2 ____ down the river.

(f) The pencil lead is 1 ____ thick.

(g) A stick of gum is 7 ____ long.

(h) The highway speed limit is 90 ____ per hour.

(i) The classroom was 12 ____ long.

(j) A penny is about 18 ____ across.

Answers

2. (a) cm (b) cm (c) m (d) mm (e) km
(f) mm (g) cm (h) km (i) m (j) mm

3 First write the unit fraction needed to make each conversion. Then complete the conversion.

GS **(a)** 3.67 m to cm

$$\text{unit fraction} \Big\} \frac{100 \text{ cm}}{1 \text{ m}}$$

$$\frac{3.67 \text{ m}}{1} \cdot \frac{100 \text{ cm}}{1 \text{ m}}$$

$$= \frac{3.67 \cdot 100 \text{ cm}}{1} =$$

(b) 92 cm to m

$$\text{unit fraction} \Big\} \frac{}{\text{cm}} \text{ m}$$

(c) 432.7 cm to m

$$\text{unit fraction} \Big\} \underline{}$$

(d) 65 mm to cm

$$\text{unit fraction} \Big\} \underline{}$$

(e) 0.9 m to mm

$$\text{unit fraction} \Big\} \underline{}$$

(f) 2.5 cm to mm

$$\text{unit fraction} \Big\} \underline{}$$

Answers

3. **(a)** 367 cm   **(b)** $\dfrac{1 \text{ m}}{100 \text{ cm}}$; 0.92 m

(c) $\dfrac{1 \text{ m}}{100 \text{ cm}}$; 4.327 m

(d) $\dfrac{1 \text{ cm}}{10 \text{ mm}}$; 6.5 cm

(e) $\dfrac{1000 \text{ mm}}{1 \text{ m}}$; 900 mm

(f) $\dfrac{10 \text{ mm}}{1 \text{ cm}}$; 25 mm

OBJECTIVE **2** **Use unit fractions to convert among metric units.** You can convert among metric length units using unit fractions. Keep these relationships in mind when setting up the unit fractions.

Metric Length Relationships

1 km = 1000 m so the unit fractions are:	**1 m = 1000 mm** so the unit fractions are:
$\dfrac{1 \text{ km}}{1000 \text{ m}}$ or $\dfrac{1000 \text{ m}}{1 \text{ km}}$	$\dfrac{1 \text{ m}}{1000 \text{ mm}}$ or $\dfrac{1000 \text{ mm}}{1 \text{ m}}$
1 m = 100 cm so the unit fractions are:	**1 cm = 10 mm** so the unit fractions are:
$\dfrac{1 \text{ m}}{100 \text{ cm}}$ or $\dfrac{100 \text{ cm}}{1 \text{ m}}$	$\dfrac{1 \text{ cm}}{10 \text{ mm}}$ or $\dfrac{10 \text{ mm}}{1 \text{ cm}}$

EXAMPLE 2 **Using Unit Fractions to Convert Metric Length Measurements**

Convert each measurement using unit fractions.

(a) 5 km to m

Put the unit for the answer (meters) in the numerator of the unit fraction; put the unit you want to change (km) in the denominator.

Unit fraction equivalent to 1 $\Big\{ \dfrac{1000 \text{ m}}{1 \text{ km}}$ ← Unit for your **answer** is **m**
← Unit being **changed** is **km**

Multiply. Divide out common units where possible.

$$5 \text{ km} \cdot \frac{1000 \text{ m}}{1 \text{ km}} = \frac{5 \text{ km}}{1} \cdot \frac{1000 \text{ m}}{1 \text{ km}} = \frac{5 \cdot 1000 \text{ m}}{1} = 5000 \text{ m}$$

These units should match.

Here, **km** divides out leaving **m**, the unit you want for your answer.

Do **not** write a period here.

5 km = 5000 m

The answer makes sense because a kilometer is much longer than a meter, so 5 km will contain many meters.

(b) 18.6 cm to m

Multiply by a unit fraction that allows you to divide out centimeters.

Unit fraction

$$\frac{18.6 \text{ cm}}{1} \cdot \frac{1 \text{ m}}{100 \text{ cm}} = \frac{18.6}{100} \text{ m} = 0.186 \text{ m}$$

Do **not** write a period here.

18.6 cm = 0.186 m

There are 100 cm in a meter, so 18.6 cm will be a small part of a meter. The answer makes sense.

◄ **Work Problem** **3** **at the Side.**

OBJECTIVE ▶ ③ Move the decimal point to convert among metric units.
By now you have probably noticed that conversions among metric units are made by multiplying or dividing by 10, by 100, or by 1000. A quick way to *multiply* by 10 is to move the decimal point one place to the *right*. Move it two places to the right to multiply by 100, three places to multiply by 1000. *Dividing* is done by moving the decimal point to the *left* in the same manner.

Work Problem ④ at the Side. ▶

An alternate conversion method to unit fractions is moving the decimal point using this **metric conversion line.**

Here are the steps for using the conversion line.

Using the Metric Conversion Line

Step 1 Find the unit you are given on the metric conversion line.

Step 2 Count the number of places to get from the unit you are given to the unit you want in the answer.

Step 3 Move the decimal point the **same number of places** and in the **same direction** as you did on the conversion line.

EXAMPLE 3 Using the Metric Conversion Line

Use the metric conversion line to make the following conversions.

(a) 5.702 km to m

Find **km** on the metric conversion line. To get to **m**, you move *three places* to the *right*. So move the decimal point in 5.702 *three places* to the *right*.

5.702

Move decimal point three places to the right.

Three places to the right

5.702 km = 5702 m

(b) 69.5 cm to m

Find **cm** on the conversion line. To get to **m**, move *two places* to the *left*.

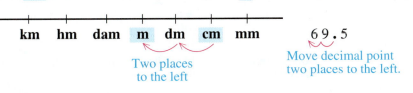

69.5

Move decimal point two places to the left.

Two places to the left

69.5 cm = 0.695 m

Continued on Next Page

---

④ Do each multiplication or division by hand or on a calculator. Compare your answer to the one you get by moving the decimal point.

GS (a) $(43.5)(10) =$ _____

43.5 gives 435

(b) $43.5 \div 10 =$ _____

43.5 gives _____

(c) $(28)(100) =$ _____

28.00 gives _____

(d) $28 \div 100 =$ _____

28. gives _____

(e) $(0.7)(1000) =$ _____

0.700 gives _____

(f) $0.7 \div 1000 =$ _____

000.7 gives _____

Answers

4. (a) 435 **(b)** 4.35; 4.35 **(c)** 2800; 2800
(d) 0.28; 0.28 **(e)** 700; 700
(f) 0.0007; 0.0007

5 Convert using the metric conversion line.

GS **(a)** 12.008 km to m

km to m is <u>three</u> places to the <u>right</u>.

12.008 km = _____ m

GS **(b)** 561.4 m to km

m to km is _____ places to the _____.

561.4 m = _____ km

(c) 20.7 cm to m

(d) 20.7 cm to mm

(e) 4.66 m to cm

6 Convert using the metric conversion line.

GS **(a)** 9 m to mm   9.000

m to mm is _____ places to the _____.

9 m = _____ mm

GS **(b)** 3 cm to m

cm to m is _____ places to the _____.

3 cm = _____ m

(c) 5 mm to cm

(d) 70 m to km

(e) 0.8 m to cm

Answers

5. **(a)** 12,008 m   **(b)** three; left; 0.5614 km
(c) 0.207 m   **(d)** 207 mm   **(e)** 466 cm
6. **(a)** three; right; 9000 mm
(b) two; left; 0.03 m   **(c)** 0.5 cm
(d) 0.07 km   **(e)** 80 cm

(c) 8.1 cm to mm

From **cm** to **mm** is *one place* to the *right*.

km   hm   dam   m   dm   **cm**   **mm**   8 . 1

One place   Move decimal point
to the right   one place to the right.

8.1 cm = 81 mm

◀ **Work Problem 5 at the Side.**

EXAMPLE 4   **Practicing Length Conversions**

Convert using the metric conversion line.

(a) 1.28 m to mm

Moving from **m** to **mm** is going *three places to the right*. In order to move the decimal point in 1.28 three places to the right, you must write a 0 as a placeholder.

1.28**0**   Zero is written in as a placeholder.

Move decimal point
three places to the right.

1.28 m = 1280 mm

> The answer is a *whole number*, so you do not need to write the decimal point.

(b) 60 cm to m

From **cm** to **m** is two places to the left. The decimal point in 60 starts at the *far right side* because 60 is a whole number. Then move it two places to the left.

60.   60.

Decimal point   Move decimal point
starts here.   two places to the left.

60 cm = 0.60 m, and 0.60 m is equivalent to 0.6 m.

> Think:
> $\frac{60}{100}$ in lowest terms is $\frac{6}{10}$

(c) 8 m to km

From **m** to **km** is three places to the left. The decimal point in 8 starts at the far right side. In order to move it three places to the left, you must write two zeros as placeholders.

                          Two zeros are written
                          in as placeholders.

8.   **00**8.

Decimal point   Move decimal point
starts here.   three places to the left.

8 m = 0.008 km

◀ **Work Problem 6 at the Side.**

7.2 Exercises

   MyMathLab®

Download the MyDashBoard App

CONCEPT CHECK *Use your knowledge of the meaning of metric prefixes to fill in the blanks.*

1. *kilo* means ____1000____ so

  1 km = _____ m

2. *deka* means ____10____ so

  1 dam = _____ m

3. *milli* means _____ so

  1 mm = _____ m

4. *deci* means _____ so

  1 dm = _____ m

5. *centi* means _____ so

  1 cm = _____ m

6. *hecto* means _____ so

  1 hm = _____ m

7. Circle the shorter length unit in each pair.

  (a) cm m   **(b)** m mm   **(c)** km cm

8. Circle the longer length unit in each pair.

  (a) m km   **(b)** mm cm   **(c)** km mm

Use this ruler to measure the width of your thumb and hand for Exercises 9–12.

9. The width of your hand in centimeters

10. The width of your hand in millimeters

11. The width of your thumb in millimeters

12. The width of your thumb in centimeters

Write the most reasonable metric length unit in each blank.
Choose from km, m, cm, and mm. **See Example 1.**

13. The child was 91 _____ tall.

14. The cardboard was 3 _____ thick.

15. Ming-Na swam in the 200 _____ backstroke race.

16. The bookcase is 75 _____ wide.

17. Adriana drove 400 _____ on her vacation.

18. The door is 2 _____ high.

19. An aspirin tablet is 10 _____ across.

20. Lamard jogs 4 _____ every morning.

21. A paper clip is about 3 _____ long.

22. My pen is 145 _____ long.

23. Dave's truck is 5 _____ long.

24. Wheelchairs need doorways that are at least 80 _____ wide.

Convert each measurement. Use unit fractions or the metric conversion line.
See Examples 2–4.

25. 7 m to cm

GS $\dfrac{7 \text{ m}}{1} \cdot \dfrac{100 \text{ cm}}{1 \text{ m}} = \dfrac{7 \cdot 100 \text{ cm}}{1} =$

26. 18 m to cm

GS $\dfrac{18 \text{ m}}{1} \cdot \dfrac{100 \text{ cm}}{1 \text{ m}} =$

27. 40 mm to m

28. 6 mm to m

29. 9.4 km to m

30. 0.7 km to m

31. 509 cm to m

32. 30 cm to m

33. 400 mm to cm

34. 25 mm to cm

35. 0.91 m to mm

36. 4 m to mm

37. CONCEPT CHECK Is 82 cm greater than or less than 1 m? What is the difference in the lengths?

38. CONCEPT CHECK Is 1022 m greater than or less than 1 km? What is the difference in the lengths?

39. Computer microchips may be only 5 mm long and 1 mm wide. Using the ruler on the previous page, draw a rectangle that measures 5 mm by 1 mm. Then convert each measurement to centimeters.

A greatly enlarged photo of a computer microchip.

40. The world's smallest butterfly has a wingspan of 15 mm. The smallest mouse is 50 mm long. Using the ruler on the previous page, draw a line that is 15 mm long and a line 50 mm long. Then convert each measurement to centimeters. (*Source: Top 10 of Everything.*)

41. The Roe River near Great Falls, Montana, is the shortest river in the world, just 61 m long. How many kilometers long is the Roe River? (*Source: Guinness World Records.*)

42. There are over 100,000 km of blood vessels in the human body. How many meters of blood vessels are in the body? (*Source: The Human Body.*)

43. The median height for U.S. females who are 20 to 29 years old is about 1.64 m. Convert this height to centimeters and to millimeters. (*Source: U.S. National Center for Health Statistics.*)

44. The median height for 20- to 29-year-old males in the United States is about 177 cm. Convert this height to meters and to millimeters. (*Source: U.S. National Center for Health Statistics.*)

45. Use two unit fractions to convert 5.6 mm to km.

46. Use two unit fractions to convert 16.5 km to mm.

7.3 The Metric System—Capacity and Weight (Mass)

We use capacity units to measure liquids, such as the amount of milk in a recipe, the gasoline in our car tank, and the water in an aquarium. (The capacity units in the U.S. system are cups, pints, quarts, and gallons.) The basic metric unit for capacity is the **liter** (also spelled *litre*). The capital letter L is the symbol for liter, to avoid confusion with the numeral 1.

OBJECTIVE 1 **Learn the basic metric units of capacity.** The liter is related to metric length in this way: A box that measures 10 cm on every side holds exactly one liter. (The volume of the box is 10 cm • 10 cm • 10 cm = 1000 cubic centimeters. Volume is discussed in **Chapter 8**.) A liter is just a little more than 1 quart.

10 cm
10 cm
10 cm
Holds exactly
1 liter (L)

About 1 liter
of milk

About 1 liter of
oil for your car

A liter is a little more than one quart
(just $\frac{1}{4}$ cup more).

In the metric system you use liters for things like buying milk and soda at the store, filling a pail with water, and describing the size of your home aquarium.

Buy a 2 L
bottle of soda.

Use a 12 L
pail to wash floors.

Watch the fish in
your 40 L aquarium.

Work Problem 1 **at the Side.** ▶

To make larger or smaller capacity units, we use the same **prefixes** as we did with length units. For example, the prefix *kilo* means 1000 so a *kilo*meter is 1000 meters. In the same way, a *kilo*liter is 1000 liters.

1 Circle the things that can be measured in liters.

Amount of water in the bathtub

Length of the bathtub

Width of your car

Amount of gasoline you buy for your car

Weight of your car

Height of a pail

Amount of water in a pail

Prefix	kilo- liter	hecto- liter	deka- liter	liter	deci- liter	centi- liter	milli- liter
Meaning	1000 liters	100 liters	10 liters	1 liter	$\frac{1}{10}$ of a liter	$\frac{1}{100}$ of a liter	$\frac{1}{1000}$ of a liter
Symbol	kL	hL	daL	L	dL	cL	mL

↑ ↑
Capacity units used most often

2 Write the most reasonable metric unit in each blank. Choose from L and mL.

(a) I bought 8 _____ of soda at the store.

(b) The nurse gave me 10 _____ of cough syrup.

(c) This is a 100 _____ garbage can.

(d) It took 10 _____ of paint to cover the bedroom walls.

(e) My car's gas tank holds 50 _____.

(f) I added 15 _____ of oil to the pancake mix.

(g) The can of orange soda holds 350 _____.

(h) My friend gave me a 30 _____ bottle of expensive perfume.

Answers

2. (a) L (b) mL (c) L (d) L (e) L
(f) mL (g) mL (h) mL

The capacity units you will use most often in daily life are liters (L) and *milli*liters (mL). A tiny box that measures 1 cm on every side holds exactly one milliliter. (In medicine, this small amount is also called 1 cubic centimeter, or 1 cc for short.) It takes 1000 mL to make 1 L. Here are some comparisons.

Holds exactly 1 milliliter (mL)	5 mL in a teaspoon	About 250 mL in one cup

EXAMPLE 1 Using Metric Capacity Units

Write the most reasonable metric unit in each blank. Choose from L and mL.

(a) The bottle of shampoo held 500 _____.
500 **mL** because 500 L would be about 500 quarts, which is too much.

(b) I bought a 2 _____ carton of orange juice.
2 **L** because 2 mL would be less than a teaspoon.

◄ **Work Problem 2** at the Side.

OBJECTIVE ► **2** **Convert among metric capacity units.** Just as with length units, you can convert between milliliters and liters using unit fractions.

Metric Capacity Relationships
1 L = 1000 mL, so the unit fractions are:
$$\frac{1\ L}{1000\ mL} \quad \text{or} \quad \frac{1000\ mL}{1\ L}$$

Or you can use a metric conversion line to decide how to move the decimal point.

The blue prefixes are the same ones you used with meters.

1000	100	10	1	$\frac{1}{10}$	$\frac{1}{100}$	$\frac{1}{1000}$
kL	hL	daL	L	dL	cL	mL

EXAMPLE 2 Converting among Metric Capacity Units

Convert using the metric conversion line or unit fractions.

(a) 2.5 L to mL

Using the metric conversion line:

From **L** to **mL** is *three places* to the *right*.

2.5**00** ← Write two zeros as placeholders.

2.5 L = 2500 mL

Using unit fractions:

Multiply by a unit fraction that allows you to divide out liters.

$$\frac{2.5\ \cancel{L}}{1} \cdot \frac{1000\ mL}{1\ \cancel{L}} = 2500\ mL$$

L divides out, leaving mL for your answer.

Do **not** write a period here.

················ Continued on Next Page

(b) 80 mL to L

Using the metric conversion line:

From **mL** to **L** is *three places* to the *left*.

80.⟍   080.

Decimal point starts here.   Move decimal point three places to the left.

80 mL = 0.080 L or 0.08 L

Using unit fractions:

Multiply by a unit fraction that allows you to divide out mL.

$$\frac{80 \text{ mL}}{1} \cdot \frac{1 \text{ L}}{1000 \text{ mL}}$$

mL divides out, leaving **L** for your answer.

$$= \frac{80}{1000} \text{ L} = 0.08 \text{ L}$$

Do **not** write a period here.

·········· **Work Problem ❸ at the Side.** ▶

OBJECTIVE ❸ **Learn the basic metric units of weight (mass).** The **gram** is the basic metric unit for *mass*. Although we often call it "weight," there is a difference.

Weight is a measure of the pull of gravity; the farther you are from the center of Earth, the less you weigh. In outer space you become weightless, but your mass, the amount of matter in your body, stays the same regardless of where you are.

In science courses, it will be important to distinguish between the weight of an object and its mass. But for everyday purposes, we will use the word *weight*.

The gram is related to metric length in this way: The weight of the water in a box measuring 1 cm on every side is 1 gram. This is a very tiny amount of water (1 mL) and a very small weight. One gram is also the weight of a dollar bill or a single raisin. A nickel weighs 5 grams. A plain, regular-sized hamburger (with bun) weighs from 175 to 200 grams.

The 1 mL of water in this tiny box weighs 1 gram.

A nickel weighs 5 grams.

A dollar bill weighs 1 gram.

A plain hamburger weighs 175 to 200 grams.

Work Problem ❹ at the Side. ▶

❸ Convert using the metric conversion line or unit fractions.

GS **(a)** 9 L to mL

On the conversion line, L to mL is three places to the right **or** use the unit fraction

$$\frac{1000 \text{ mL}}{1 \text{ L}}.$$

9 L = _____ mL

(b) 0.75 L to mL

(c) 500 mL to L

(d) 5 mL to L

(e) 2.07 L to mL

(f) 3275 mL to L

VOCABULARY TIP

gram   The weight of a nickel (**5** cents) is **5** grams.

❹ Circle the things that weigh about 1 gram.

A small paper clip

A pair of scissors

One playing card from a deck of cards

A calculator

An average-sized apple

The check you wrote to the cable company

Answers
3. **(a)** 9000 mL   **(b)** 750 mL   **(c)** 0.5 L
   (d) 0.005 L   **(e)** 2070 mL   **(f)** 3.275 L
4. paper clip, playing card, check

5 Write the most reasonable metric unit in each blank. Choose from kg, g, and mg.

(a) A thumbtack weighs 800 ____.

(b) A teenager weighs 50 ____.

(c) This large cast-iron frying pan weighs 1 ____.

(d) Jerry's basketball weighed 600 ____.

(e) Tamlyn takes a 500 ____ calcium tablet every morning.

(f) On his diet, Greg can eat 90 ____ of meat for lunch.

(g) One strand of hair weighs 2 ____.

(h) One banana might weigh 150 ____.

Answers

5. **(a)** mg **(b)** kg **(c)** kg **(d)** g **(e)** mg
(f) g **(g)** mg **(h)** g

To make larger or smaller weight units, we use the same **prefixes** as we did with length and capacity units. For example, the prefix *kilo-* means 1000 so a *kilo*meter is 1000 meters, a *kilo*liter is 1000 liters, and a *kilo*gram is 1000 grams.

Prefix	kilo-gram	hecto-gram	deka-gram	gram	deci-gram	centi-gram	milli-gram
Meaning	1000 grams	100 grams	10 grams	1 gram	$\frac{1}{10}$ of a gram	$\frac{1}{100}$ of a gram	$\frac{1}{1000}$ of a gram
Symbol	**kg**	**hg**	**dag**	g	**dg**	**cg**	**mg**

Weight (mass) units that are used most often

The units you will use most often in daily life are kilograms (kg), grams (g), and milligrams (mg). *Kilo*grams are used instead of pounds. A kilogram is 1000 grams. It is about 2.2 pounds. Two packages of butter plus one stick of butter weigh about 1 kg. An average newborn baby weighs 3 to 4 kg; a college football player might weigh 100 to 130 kg.

1 kilogram is about **2**.2 pounds.   100 to 130 kg   3 to 4 kg

Extremely small weights are measured in *milli*grams. It takes 1000 mg to make 1 g. Recall that a dollar bill weighs about 1 g. Imagine cutting it into 1000 pieces; the weight of one tiny piece would be 1 mg. Dosages of medicine and vitamins are given in milligrams. You will also use milligrams in science classes.

Cut a dollar bill into 1000 pieces. One tiny piece weighs 1 milligram.

EXAMPLE 3 **Using Metric Weight Units**

Write the most reasonable metric unit in each blank. Choose from kg, g, and mg.

(a) Ramon's suitcase weighed 20 _____.
Write 20 **kg** because kilograms are used instead of pounds; 20 kg is about 44 pounds.

(b) LeTia took a 350 _____ aspirin tablet.
Write 350 **mg** because 350 g would be more than the weight of a hamburger, which is too much.

(c) Jenny mailed a letter that weighed 30 _____.
Write 30 **g** because 30 kg would be much too heavy and 30 mg is less than the weight of a dollar bill.

◀ **Work Problem 5** at the Side.

OBJECTIVE ▶ 4 Convert among metric weight (mass) units. As with length and capacity, you can convert among metric weight units by using unit fractions. The unit fractions you need are shown here.

Metric Weight (Mass) Relationships	
1 kg = 1000 g, so the unit fractions are:	**1 g = 1000 mg,** so the unit fractions are:
$\dfrac{1 \text{ kg}}{1000 \text{ g}}$ or $\dfrac{1000 \text{ g}}{1 \text{ kg}}$	$\dfrac{1 \text{ g}}{1000 \text{ mg}}$ or $\dfrac{1000 \text{ mg}}{1 \text{ g}}$

Or you can use a metric conversion line to decide how to move the decimal point.

The blue prefixes are the same ones you used with meters and liters.

1000	100	10	1	$\frac{1}{10}$	$\frac{1}{100}$	$\frac{1}{1000}$
kg	**hg**	**dag**	**g**	**dg**	**cg**	**mg**

EXAMPLE 4   Converting among Metric Weight Units

Convert using the metric conversion line or unit fractions.

(a) 7 mg to g

Using the conversion line:
From **mg** to **g** is *three places* to the *left*.

7.   007.

Decimal point starts here.   Move decimal point three places to the left.

7 mg = 0.007 g

Do **not** write a period here.

Using unit fractions:

Multiply by a unit fraction that allows you to divide out mg.

$$\frac{7 \text{ mg}}{1} \cdot \frac{1 \text{ g}}{1000 \text{ mg}} = \frac{7}{1000} \text{ g}$$

$$= 0.007 \text{ g}$$

Three decimal places for thousandths.

(b) 13.72 kg to g

Using the conversion line:
From **kg** to **g** is *three places* to the *right*.

13.720   Decimal point moves three places to the right.

13.72 kg = 13,720 g

This is a comma (**not** a decimal point).

Using unit fractions:

Multiply by a unit fraction that allows you to divide out kg.

$$\frac{13.72 \text{ kg}}{1} \cdot \frac{1000 \text{ g}}{1 \text{ kg}} = 13,720 \text{ g}$$

This is a comma (**not** a decimal point).

·········· **Work Problem 6 at the Side.** ▶

---

6 Convert using the metric conversion line or unit fractions.

GS (a) 10 kg to g

On the conversion line, kg to g is _____ places to the _____ **or** use the unit fraction $\dfrac{1000 \text{ g}}{1 \text{ kg}}$.

10 kg = _____ g

(b) 45 mg to g

(c) 6.3 kg to g

(d) 0.077 g to mg

(e) 5630 g to kg

(f) 90 g to kg

Answers

6. **(a)** three; right; 10,000 g **(b)** 0.045 g
(c) 6300 g **(d)** 77 mg **(e)** 5.63 kg
(f) 0.09 kg

VOCABULARY TIP

Length; capacity; weight Think:
  meter/mile; liter/liquid;
  gram/gravity (weight).

7 First decide which type of units are needed: length, capacity, or weight. Then write the most appropriate unit in the blank. Choose from km, m, cm, mm, L, mL, kg, g, and mg.

(a) Gail bought a 4 ____ can of paint.

Use _____ units.

(b) The bag of chips weighed 450 ____.

Use _____ units.

(c) Give the child 5 ____ of cough syrup.

Use _____ units.

(d) The width of the window is 55 ____.

Use _____ units.

(e) Akbar drives 18 ____ to work.

Use _____ units.

(f) The laptop computer weighs 2 ____.

Use _____ units.

(g) A credit card is 55 ____ wide.

Use _____ units.

Answers

7. (a) L; capacity (b) g; weight
   (c) mL; capacity (d) cm; length
   (e) km; length (f) kg; weight
   (g) mm; length

OBJECTIVE **5** **Distinguish among basic metric units of length, capacity, and weight (mass).** As you encounter things to be measured at home, on the job, or in your classes at school, be careful to use the correct type of measurement unit.

Use *length units* (kilometers, meters, centimeters, millimeters) to measure:

how long	how high	how far away
how wide	how tall	how far around (perimeter)
how deep	distance	

Use *capacity units* (liters, milliliters) to measure liquids (things that can be poured) such as:

water	shampoo	gasoline
milk	perfume	oil
soft drinks	cough syrup	paint

Also use liters and milliliters to describe how much liquid something can hold, such as an eyedropper, measuring cup, pail, or bathtub.

Use *weight units* (kilograms, grams, milligrams) to measure:

the weight of something   how heavy something is

In **Chapter 8** you will use square units (such as square meters) to measure area, and cubic units (such as cubic centimeters) to measure volume.

EXAMPLE 5   **Using a Variety of Metric Units**

First decide which type of units are needed: length, capacity, or weight. Then write the most appropriate metric unit in the blank. Choose from km, m, cm, mm, L, mL, kg, g, and mg.

(a) The letter needs another stamp because it weighs 40 _____.

Use _____ units.

Use **weight** units because of the word "weighs."

The letter weighs 40 **g** because 40 mg is less than the weight of a dollar bill and 40 kg would be about 88 pounds.

(b) The swimming pool is 3 _____ deep at the deep end.

Use _____ units.

Use **length** units because of the word "deep."

The pool is 3 **m** deep because 3 cm is only about an inch and 3 km is more than a mile.

(c) This is a 340 _____ can of juice.

Use _____ units.

Use **capacity** units because juice is a liquid.

It is a 340-**mL** can because 340 liters would be more than 340 quarts.

◀ Work Problem **7** at the Side.

7.3 Exercises

  MyMathLab®

CONCEPT CHECK   *Fill in the blank with the best metric unit to measure each item.*
Choose from liters, milliliters, kilograms, grams, and milligrams.

1. (a) For a dose of cough syrup, use _____.

  (b) For a large carton of milk, use _____.

  (c) For a vitamin pill, use _____.

  (d) For a heavyweight wrestler, use _____.

2. (a) For a serving of vegetables, use _____.

  (b) For a large bottle of soda, use _____.

  (c) For a travel size bottle of shampoo, use
  _____.

  (d) For a pain reliever tablet, use _____.

Write the most reasonable metric unit in each blank. Choose from L, mL, kg, g, and mg.
See Examples 1 and 3.

3. The glass held
250 _____ of water.

4. Hiromi used 12 _____ of water
to wash the kitchen floor.

5. Dolores can make 10 _____ of soup in that pot.

6. Jay gave 2 _____ of vitamin drops to the baby.

7. Our yellow Labrador
dog grew up to weigh
40 _____.

8. A small safety pin
weighs 750 _____.

9. Lori caught a small sunfish weighing 150 _____.

10. One dime weighs 2 _____.

11. Andre donated 500 _____ of blood today.

12. Barbara bought the 2 _____ bottle of cola.

13. The patient received a 250 _____ tablet of medication
each hour.

14. The 8 people on the elevator weighed a
total of 500 _____.

15. The gas can for the lawn mower
holds 4 _____.

16. Kevin poured 10 _____
of vanilla into
the mixing bowl.

17. Pam's backpack weighs 5 _____
when it is full of books.

18. One grain of salt
weighs 2 _____.

CONCEPT CHECK *Today, medical prescriptions are usually given in the metric system. But sometimes a mistake is made. Indicate whether each dose is* reasonable *or* unreasonable. *If a dose is unreasonable, indicate whether it is* too much *or* too little.

19. Drink 4.1 L of Kaopectate after each meal.

20. Drop 1 mL of solution into the eye twice a day.

21. Soak your feet in 5 kg of Epsom salts per liter of water.

22. Inject 0.5 L of insulin each morning.

23. Take 15 mL of cough syrup every four hours.

24. Take 200 mg of vitamin C each day.

25. Take 350 mg of aspirin three times a day.

26. Buy a tube of ointment weighing 0.002 g.

27. CONCEPT CHECK Describe how you decide which unit fraction to use when converting 6.5 kg to grams.

28. CONCEPT CHECK Write an explanation of each step you use to convert 20 mg to grams using the metric conversion line.

Convert each measurement. Use unit fractions or the metric conversion line.
See Examples 2 and 4.

29. 15 L to mL

$$\text{unit} \atop \text{fraction} \Big\} \frac{1000 \text{ mL}}{1 \text{ L}} \quad \textbf{or}$$

15 L = _____ mL

On conversion line,
L to mL is _____
places to the _____.

30. 6 L to mL

$$\text{unit} \atop \text{fraction} \Big\} \frac{1000 \text{ mL}}{1 \text{ L}} \quad \textbf{or}$$

6 L = _____ mL

On the conversion line,
L to mL is _____
places to the _____.

31. 3000 mL to L

32. 18,000 mL to L

33. 925 mL to L

34. 200 mL to L

35. 8 mL to L

36. 25 mL to L

37. 4.15 L to mL

38. 11.7 L to mL

39. 8000 g to kg

40. 25,000 g to kg

41. 5.2 kg to g

42. 12.42 kg to g

43. 0.85 g to mg

44. 0.2 g to mg

45. 30,000 mg to g

46. 7500 mg to g

47. 598 mg to g

48. 900 mg to g

49. 60 mL to L

50. 6.007 kg to g

51. 3 g to kg

52. 12 mg to g

53. 0.99 L to mL

54. 13,700 mL to L

Write the most appropriate metric unit in each blank. Choose from km, m, cm, mm, L, mL, kg, g, and mg. **See Example 5.**

55. The masking tape is 19 _____ wide.

56. The roll has 55 _____ of tape on it.

57. ▶ Buy a 60 _____ jar of acrylic paint for art class.

58. One onion weighs 200 _____.

59. My waist measurement is 65 _____.

60. Add 2 _____ of windshield washer fluid to your car.

61. ▶ A single postage stamp weighs 90 _____.

62. The hallway is 10 _____ long.

Solve each application problem. Show your work. (Source: The Human Body.)

63. Human skin has about 3 million sweat glands, which release an average of 300 mL of sweat per day. How many liters of sweat are released each day?

64. In hot climates, the sweat glands in a person's skin may release up to 3.5 L of sweat in one day. How many milliliters is that?

65. The average weight of an adult human brain is 1.34 kg. An adult human's skin weighs about 5 kg. How many grams does each body part weigh?

66. A healthy human heart pumps between 70 mL and 230 mL of blood per beat depending on the person's activity level. How many liters of blood does the heart pump per beat?

67. On average, we breathe in and out roughly 900 mL of air every 10 seconds. How many liters of air is that?

68. In the Victorian era, people believed that heavier brains meant greater intelligence. They were impressed that Otto von Bismarck's brain weighed 1907 g, which is how many kilograms?

69. A small adult cat weighs from 3000 g to 4000 g. How many kilograms is that? (*Source: Lyndale Animal Hospital.*)

70. If the letter you are mailing weighs 29 g, you must put additional postage on it. How many kilograms does the letter weigh? (*Source: U.S. Postal Service.*)

71. CONCEPT CHECK Is 1005 mg greater than or less than 1 g? What is the difference in the weights?

72. CONCEPT CHECK Is 990 mL greater than or less than 1 L? What is the difference in the amounts?

73. One nickel weighs 5 g. How many nickels are in 1 kg of nickels?

74. The ratio of the total length of all the fish to the amount of water in an aquarium can be 3 cm of fish for every 4 L of water. What is the total length of all the fish you can put in a 40 L aquarium? (*Source: Tropical Aquarium Fish.*)

Relating Concepts (Exercises 75–78) For Individual or Group Work

*Recall that the prefix **kilo** means 1000, so a **kilo**meter is 1000 meters. You'll learn about other prefixes for numbers greater than 1000 as you **work Exercises 75–78 in order.***

75. (a) The prefix *mega* means one million. Use the symbol M (capitalized) for *mega*. So a megameter (Mm) is how many meters?

   1 Mm = _____ m

   (b) Figure out a unit fraction that you can use to convert megameters to meters. Then use it to convert 3.5 Mm to meters.

76. (a) The prefix *giga* means one billion. Use the symbol G (capitalized) for *giga*. So a gigameter (Gm) is how many meters?

   1 Gm = _____ m

   (b) Figure out a unit fraction you can use to convert meters to gigameters. Then use it to convert 2500 m to gigameters.

77. (a) The prefix *tera* means one trillion. Use the symbol T (capitalized) for *tera*. So a terameter (Tm) is how many meters?

   1 Tm = _____ m

   (b) Think carefully before you fill in the blanks:

   1 Tm = _____ Gm

   1 Tm = _____ Mm

78. A computer's memory is measured in *bytes*. A byte can represent a single letter, a digit, or a punctuation mark. The memory for a laptop computer may be measured in megabytes (abbreviated MB) or gigabytes (abbreviated GB). Using the meanings of *mega* and *giga*, you would think that

   1 MB = _____ bytes and

   1 GB = _____ bytes.

   However, because computers use a base 2 or binary system, 1 MB is actually 2^{20} and 1 GB is 2^{30}. Use your calculator to find the actual values.

   2^{20} = _____ 2^{30} = _____

Summary Exercises *U.S. and Metric Measurement Units*

CONCEPT CHECK   *The most commonly used U.S. and metric measurement units are listed in mixed-up order. Write each unit in the correct box in the table below. Then write the abbreviation or symbol next to each unit.*

pound	yard	liter	kilogram	ton	gallon	pint
millimeter	gram	quart	meter	inch	cup	centimeter
milliliter	mile	foot	milligram	kilometer	ounce	fluid ounce

1. U.S. Measurement Units and Abbreviations			2. Metric Measurement Units and Symbols		
Length	**Weight**	**Capacity**	**Length**	**Weight**	**Capacity**

Fill in the blanks with the measurement relationships that you memorized.

3. (a) 1 _____ = 12 in.

   (b) 3 ft = 1 _____

   (c) 1 mi = _____ ft

4. (a) 60 sec = 1 _____

   (b) 1 hr = _____ min

   (c) _____ hr = 1 day

5. (a) 1 cup = _____ fl oz

   (b) 4 qt = 1 _____

   (c) _____ pt = 1 qt

6. (a) 16 oz = 1 _____

   (b) _____ lb = 1 ton

   (c) 1 lb = _____ oz

Write the most reasonable metric unit in each blank. Choose from km, m, cm, mm, L, mL, kg, g, and mg.

7. My water bottle holds 450 _____.

8. Michael won the 200 _____ race today.

9. The child weighed 23 _____.

10. Jifar took a 375 _____ aspirin tablet.

11. The red pen is 14 _____ long.

12. A quarter is about 25 _____ across.

13. Merlene made 12 _____ of fruit punch for her daughter's birthday party.

14. This cereal has 4 _____ of protein in each serving.

Convert each measurement using unit fractions or the metric conversion line.
Show your work.

15. 45 cm to meters

16. $\frac{3}{4}$ min to seconds

17. 0.6 L to milliliters

18. 8 g to milligrams

19. 300 mm to centimeters

20. 45 in. to feet

21. 50 mL to liters

22. 18 qt to gallons

23. 7.28 kg to grams

24. $2\frac{1}{4}$ lb to ounces

25. 9 g to kilograms

26. 5 yd to inches

Solve Exercises 27–30 using the data in the table at the right on some of the world's
tallest people. (Source: Guinness World Records.)

27. What is the height of the tallest person in centimeters? In millimeters?

WORLD'S TALLEST PEOPLE

Rank	Name/Dates/Country	Height
1st	Robert Wadlow (1918–1940) USA	2.72 m
2nd	John Rogan (1868–1905) USA	268 cm
3rd	John Carroll (1932–1969) USA	264 cm
4th	Sultan Kösen (1982–) Turkey	2.5 m
10th	Jeng Jinlian (tallest female) (1964–1982) China	248 cm

Robert Wadlow was
2.72 m tall.

28. Find the 3rd tallest person's height in meters and in kilometers.

29. How much taller is the tallest person than the second tallest, in meters? Convert this difference to centimeters and millimeters.

30. What is the difference in height between the tallest and 10th tallest people, in centimeters? Convert the height difference to meters and millimeters.

Solve each application problem. Show your work.

31. Vernice bought a 12 oz bag of chips for $3.89 today. What was the price per pound to the nearest cent?

32. Miami, Florida gets an average of 56 in. of rain every year. At that rate, how many feet of rain does Miami get during two years, to the nearest tenth?

7.4 Problem Solving with Metric Measurement

OBJECTIVE ▶ **1** **Solve application problems involving metric measurements.** One advantage of the metric system is the ease of comparing measurements in application situations. Just be sure that you are comparing similar units: mg to mg, km to km, and so on.

Once again, we will use the six problem-solving steps that you learned in **Chapter 1.**

EXAMPLE 1 Solving a Metric Application

Ground turkey is on sale at $8.99 per kilogram. Jake bought 350 g of the turkey. How much did he pay, to the nearest cent?

Step 1 **Read** the problem. The problem asks for the cost of 350 g of turkey.

Step 2 **Work out a plan.** The price is $8.99 per *kilogram*, but the amount Jake bought is given in *grams*. Convert grams to kilograms (the unit in the price). Then multiply the weight by the cost per kilogram.

Step 3 **Estimate** a reasonable answer. Round the cost of 1 kg from $8.99 to $9. There are 1000 g in a kilogram, so 350 g is about $\frac{1}{3}$ of a kilogram. Jake is buying about $\frac{1}{3}$ of a kilogram, so $\frac{1}{3}$ of $9 = $3 is our estimate.

Step 4 **Solve** the problem. Use a unit fraction to convert 350 g to kilograms.

g divides out, leaving **kg** for your answer.

$$\frac{350 \text{ g}}{1} \cdot \frac{1 \text{ kg}}{1000 \text{ g}} = \frac{350}{1000} \text{ kg} = 0.35 \text{ kg}$$

Now multiply 0.35 kg times the cost per kilogram.

$$\frac{0.35 \text{ kg}}{1} \cdot \frac{\$8.99}{1 \text{ kg}} = \$3.1465 \approx \$3.15 \quad \text{(rounded)}$$

Nearest cent is the nearest hundredth

Step 5 **State the answer.** Jake paid $3.15, rounded to the nearest cent.

Step 6 **Check** your work. The exact answer of $3.15 is close to our estimate of $3.

·········· **Work Problem** **1** **at the Side.** ▶

EXAMPLE 2 Solving a Metric Application

Olivia has 2.6 m of lace. How many centimeters of lace can she use to trim each of six hair ornaments? Round to the nearest tenth of a centimeter.

Step 1 **Read** the problem. The problem asks for the number of centimeters of lace for each of six hair ornaments.

Step 2 **Work out a plan.** The given amount of lace is in *meters,* but the answer must be in *centimeters.* Convert meters to centimeters, then divide by 6 (the number of hair ornaments).

········· **Continued on Next Page**

---

1 **GS** Solve this problem using the six problem-solving steps.

Satin ribbon is on sale at $0.89 per meter. How much will 75 cm cost, to the nearest cent?

Step 1 The problem asks for the cost of 75 ____ of ribbon.

Step 2 The price is $0.89 per ____. So, convert cm to ____ (the unit in the price).

Step 3 There are 100 cm in a meter, so 75 cm will cost a little less than $0.89 (the price of a meter of ribbon).

Step 4 Finish the solution.

$$\frac{75 \text{ cm}}{1} \cdot \frac{1 \text{ m}}{100 \text{ cm}} = \frac{75}{100} \text{ m} = \underline{}$$

Multiply the number of meters of ribbon by the cost per meter.

$$\frac{\underline{} \text{ m}}{1} \cdot \frac{\$\underline{}}{1 \text{ m}} = \$\underline{}$$

Step 5 **State the answer.** 75 cm of ribbon costs ____

Step 6 **Check** your work.

Answer

1. *Step 1* cm;   *Step 2* m; m
Step 4 0.75 m;
$$\frac{0.75 \text{ m}}{1} \cdot \frac{\$0.89}{1 \text{ m}} = \$0.6675 \approx \$0.67 \text{ (rounded)}$$
Step 5 $0.67
Step 6 The exact answer is a little less than $0.89, which fits the estimate.

2 Lucinda's doctor wants her to take 1.2 g of medication each day in three equal doses. How many milligrams should be in each dose? Show your work.

Step 3 **Estimate** a reasonable answer. To estimate, round 2.6 m of lace to 3 m. Then, 3 m = 300 cm, and 300 cm ÷ 6 = 50 cm is our estimate.

Step 4 **Solve** the problem. On the metric conversion line, moving from **m** to **cm** is two places to the right, so move the decimal point in 2.6 m two places to the right. Then divide by 6.

$$2.60 \text{ m} = 260 \text{ cm} \qquad \frac{260 \text{ cm}}{6 \text{ ornaments}} \approx 43.3 \text{ cm per ornament}$$

Step 5 **State the answer.** Olivia can use about 43.3 cm of lace on each ornament. | Be sure to write **cm** in your answer.

Step 6 **Check** your work. The exact answer of 43.3 cm is close to our estimate of 50 cm.

·········· ◄ **Work Problem ② at the Side.**

> **Note**
>
> In **Example 1** we used a unit fraction to convert the measurement; in **Example 2** we moved the decimal point. Use whichever method you prefer.
>
> Also, there is more than one way to solve an application problem. Another way to solve **Example 2** is to divide 2.6 m by 6 to get 0.4333 m of lace for each ornament. Then convert 0.4333 m to 43.3 cm (rounded to the nearest tenth).

3 Andrea has two pieces of fabric. One measures 2 m 35 cm and the other measures 1 m 85 cm. How many meters of fabric does she have in all? Show your work.

EXAMPLE 3 Solving a Metric Application

Rubin measured a board and found that the length was 3 m plus an additional 5 cm. He cut off a piece measuring 1 m 40 cm for a shelf. Find the length in meters of the remaining piece of board.

Step 1 **Read** the problem. Part of a board is cut off. The problem asks what length of board, in meters, is left over. It may help to make a drawing of the board and label the lengths given in the problem.

Step 2 **Work out a plan.** The lengths involve two units, m and cm. Rewrite both lengths in meters (the unit called for in the answer), and then subtract.

Step 3 **Estimate** a reasonable answer. To estimate, 3 m 5 cm can be rounded to 3 m, because 5 cm is less than half of a meter (less than 50 cm). Round 1 m 40 cm down to 1 m. Then, 3 m − 1 m = 2 m is our estimate.

Step 4 **Solve** the problem. Rewrite the lengths in meters. Then subtract.

$$
\begin{array}{rl}
3 \text{ m} \rightarrow & 3.00 \text{ m} \\
\text{plus 5 cm} \rightarrow & + 0.05 \text{ m} \\
\hline
& 3.05 \text{ m}
\end{array}
\qquad
\begin{array}{rl}
1 \text{ m} \rightarrow & 1.0 \text{ m} \\
\text{plus 40 cm} \rightarrow & + 0.4 \text{ m} \\
\hline
& 1.4 \text{ m}
\end{array}
$$

$$
\begin{array}{rl}
& 3.05 \text{ m} \leftarrow \text{Board} \\
\text{Subtract to find} & - 1.40 \text{ m} \leftarrow \text{Shelf} \\
\text{leftover length.} & \overline{\quad 1.65 \text{ m}} \leftarrow \text{Leftover piece}
\end{array}
$$

Step 5 **State the answer.** The length of the remaining piece is 1.65 m.

Step 6 **Check** your work. The exact answer of 1.65 m is close to our estimate of 2 m.

·········· ◄ **Work Problem ③ at the Side.**

Answers
2. 400 mg per dose
3. 4.2 m

7.4 Exercises

FOR EXTRA HELP

Download the MyDashBoard App

MyMathLab®

1. **CONCEPT CHECK** *Step 5* in the problem-solving steps is **State the answer**. Which is an important thing to include in the answer?

(a) Your estimated answer.

(b) How many steps you followed.

(c) The units, such as cm, kg, $, etc.

2. **CONCEPT CHECK** *Step 6* in the problem-solving steps is **Check** your work. How can you tell if your answer is reasonable?

(a) Take a guess.

(b) Compare your exact answer to the estimated answer.

(c) Use your calculator.

Solve each application problem. Show your work. Round money answers to the nearest cent.
See Examples 1–3.

3. Bulk rice at the food co-op is on special at $0.98 per kilogram. Pam scooped some rice into a bag and put it on the scale. How much will she pay for 850 g of rice?

The price is $0.98 per _____, so convert 850 g to _____.

4. Lanh is buying a piece of plastic tubing measuring 315 cm for the science lab. The price is $5.50 per meter. How much will Lanh pay?

The price is $5.50 per _____, so convert 315 cm to _____.

5. A miniature Yorkshire terrier, one of the smallest dogs, may weigh only 500 g. But a St. Bernard, the heaviest dog, could easily weigh 90 kg. What is the difference in the weights of the two dogs, in kilograms? (*Source: Big Book of Knowledge.*)

6. The world's longest insect is the giant stick insect of Borneo, measuring 57 cm. The fairy fly, the smallest insect, is just 0.2 mm long. How much longer is the giant stick insect, in millimeters?
(*Source:* worldwildlife.org)

7. An adult human body contains about 5 L of blood. If each beat of the heart pumps 70 mL of blood, how many times must the heart beat to pass all the blood through the heart? Round to the nearest whole number of beats. (*Source: The Human Body.*)

8. A floor tile measures 30 cm by 30 cm and weighs 185 g. How many kilograms would a stack of 24 tiles weigh? How much would five stacks of tiles weigh? (*Source: The Tile Shop.*)

9. Each piece of lead for a mechanical pencil has a thickness of 0.5 mm and is 60 mm long. Find the total length in centimeters of the lead in a package with 30 pieces. If the price of the package is $3.29, find the cost per centimeter for the lead. (*Source: Pentel.*)

60 mm

10. An apartment building caretaker puts 750 mL of chlorine into the swimming pool every day. How many liters should he order to have a one-month (30-day) supply on hand? If chlorine is sold in containers that hold 4 L, how many containers should be ordered for one month? How much chlorine will be left over at the end of the month?

11. Rosa is building a bookcase. She has one board that is 2 m 8 cm long and another that is 2 m 95 cm long. What is the total length of the two boards in meters?

12. Janet has a piece of fabric that is 10 m 30 cm in length. She wants to make curtains for three windows that are all the same size. What length of fabric can she use for each window, to the nearest tenth of a meter?

13. In a chemistry lab, each of the 45 students needs 85 mL of acid. How many 1 L bottles of acid need to be ordered?

14. James needs two 1.3 m pieces and two 85 cm pieces of wood molding to frame a picture. The price is $5.89 per meter plus 7% sales tax. How much will James pay?

Use the bar graph below to answer Exercises 15 and 16.

CAFFEINE AMOUNTS
Average milligrams of caffeine per 8 oz cup or equivalent

Double espresso 160 mg
Drip coffee 90 mg
Cola 45 mg
25 mg Chocolate bar
5 mg Decaffeinated coffee

Source: Celestial Seasonings.

15. If Agnete usually drinks three 8 oz cups of drip coffee each day, how many grams of caffeine will she consume in one week?

16. Lorenzo's doctor suggested that he cut down on caffeine. So Lorenzo switched from drinking four 8 oz cups of cola every day to drinking two 8 oz cups of decaffeinated coffee. How many fewer grams of caffeine is he consuming each week?

17. The average winter snowfall in Boston is 106 cm. Lander, Wyoming usually gets 2.6 m of snow each winter. Compare the yearly snowfall in the two cities.

   (a) What is the difference for one year in meters?

   (b) What is the difference over two years in centimeters?
   (*Source:* National Climatic Data Center.)

18. New football stadiums have Field Turf instead of grass. Use the drawing below to find the total thickness in centimeters of the top two layers of Field Turf.

64 mm fiber grass

4.5 cm rubber and sand filler

asphalt base

Source: Sports Facilities Commission.

Relating Concepts (Exercises 19–22) For Individual or Group Work

It is difficult to weigh very light objects, such as a single sheet of paper or a single staple (unless you have an expensive scientific scale). But you can weigh a large number of the items and then divide to find the weight of one item. Before dividing, subtract the weight of the box or wrapper that the items are packaged in to find the net weight.
Work Exercises 19–22 in order, to complete the table.

Item	Total Weight in Grams	Weight of Packaging	Net Weight	Weight of One Item in Grams	Weight of One Item in Milligrams
19. Box of 50 envelopes	255 g	40 g	_____	_____	_____
20. Box of 1000 staples	350 g	20 g	_____	_____	_____
21. Package of paper (500 sheets)	_____	50 g	_____	_____	3000 mg
22. Box of 100 small paper clips	_____	5 g	_____	_____	500 mg

7.5 Metric–U.S. Measurement Conversions and Temperature

OBJECTIVE ▶ **1** **Use unit fractions to convert between metric and U.S. measurement units.** Until the United States has switched completely to the metric system, it will be necessary to make conversions between U.S. and metric units. *Approximate* conversions can be made with the help of the table below, in which the values have been rounded to the nearest hundredth or thousandth. (The only value that is exact, not rounded, is 1 inch = 2.54 cm.)

Metric to U.S. Units	U.S. to Metric Units
1 kilometer ≈ 0.62 mile	1 mile ≈ 1.61 kilometers
1 meter ≈ 1.09 yards	1 yard ≈ 0.91 meter
1 meter ≈ 3.28 feet	1 foot ≈ 0.30 meter
1 centimeter ≈ 0.39 inch	1 inch = 2.54 centimeters
1 liter ≈ 0.26 gallon	1 gallon ≈ 3.79 liters
1 liter ≈ 1.06 quarts	1 quart ≈ 0.95 liter
1 kilogram ≈ 2.20 pounds	1 pound ≈ 0.45 kilogram
1 gram ≈ 0.035 ounce	1 ounce ≈ 28.35 grams

EXAMPLE 1 Converting between Metric and U.S. Units of Length

Convert 10 m to yards using unit fractions. Round your answer to the nearest tenth if necessary.

We're changing from a *metric length* unit to a *U.S. length* unit. In the "Metric to U.S. Units" side of the table, you see that 1 meter ≈ 1.09 yards. Two unit fractions can be written using that information.

$$\frac{1 \text{ m}}{1.09 \text{ yd}} \quad \text{or} \quad \frac{1.09 \text{ yd}}{1 \text{ m}}$$

Multiply by the unit fraction that allows you to divide out meters (that is, meters is in the denominator).

$$10 \text{ m} \cdot \frac{1.09 \text{ yd}}{1 \text{ m}} = \frac{10 \text{ m}}{1} \cdot \frac{1.09 \text{ yd}}{1 \text{ m}} = \frac{(10)(1.09 \text{ yd})}{1} = 10.9 \text{ yd}$$

These units should match.

Meters (**m**) divide out leaving **yd**, the unit you want for the answer.

10 m ≈ 10.9 yd

Note

In **Example 1** above, you could also use the numbers from the "U.S. to Metric Units" side of the table that involve meters and yards:

$$\frac{10 \text{ m}}{1} \cdot \frac{1 \text{ yd}}{0.91 \text{ m}} = \frac{10}{0.91} \text{ yd} \approx 10.99 \text{ yd}$$

The answer is slightly different because the values in the table are rounded. Also, you have to divide instead of multiply, which is usually more difficult to do without a calculator. We will use the first method in this chapter.

·········· **Work Problem 1** at the Side. ▶

OBJECTIVES

1 Use unit fractions to convert between metric and U.S. measurement units.

2 Learn common temperatures on the Celsius scale.

3 Use formulas to convert between Celsius and Fahrenheit temperatures.

1 Convert using unit fractions. Round your answers to the nearest tenth.

(a) 23 m to yards (Look at the "Metric to U.S. Units" side of the table.)

(b) 40 cm to inches

(c) 5 mi to kilometers (Look at the "U.S. to Metric Units" side of the table.)

(d) 12 in. to centimeters

Answers

1. (a) 23 m ≈ 25.1 yd **(b)** 40 cm ≈ 15.6 in.
   (c) 5 mi ≈ 8.1 km **(d)** 12 in. ≈ 30.5 cm

❓ Convert. Use the values from the table on the previous page to make unit fractions. Round answers to the nearest tenth.

GS **(a)** 17 kg to pounds. From the table, 1 kg ≈ 2.20 lb

$$\frac{17 \ \cancel{kg}}{1} \cdot \frac{2.20 \ lb}{1 \ \cancel{kg}} =$$

(b) 5 L to quarts

(c) 90 g to ounces

(d) 3.5 gal to liters

(e) 145 lb to kilograms

(f) 8 oz to grams

| EXAMPLE 2 | Converting between Metric and U.S. Units of Weight and Capacity |

Convert using unit fractions. Round your answers to the nearest tenth.

(a) 3.5 kg to pounds
  Look in the "Metric to U.S. Units" side of the table on the previous page to see that 1 kilogram ≈ 2.20 pounds. Use this information to write a unit fraction that allows you to divide out kilograms.

$$\frac{3.5 \ \cancel{kg}}{1} \cdot \frac{2.20 \ lb}{1 \ \cancel{kg}} = \frac{(3.5)\,(2.20 \ lb)}{1} = 7.7 \ lb$$

3.5 kg ≈ 7.7 lb ◁ The conversion value is approximate, so use the ≈ symbol in your answer.

(b) 18 gal to liters
  Look in the "U.S. to Metric Units" side of the table to see that 1 gallon ≈ 3.79 liters. Write a unit fraction that allows you to divide out gallons.

gal divides out, leaving **L** for your answer. ▷
$$\frac{18 \ \cancel{gal}}{1} \cdot \frac{3.79 \ L}{1 \ \cancel{gal}} = \frac{(18)\,(3.79 \ L)}{1} = 68.22 \ L$$

68.22 rounded to the nearest tenth is 68.2
18 gal ≈ 68.2 L

(c) 300 g to ounces
  In the "Metric to U.S. Units" side of the table, 1 gram ≈ 0.035 ounce.

$$\frac{300 \ \cancel{g}}{1} \cdot \frac{0.035 \ oz}{1 \ \cancel{g}} = \frac{(300)\,(0.035 \ oz)}{1} = 10.5 \ oz$$
◁ **g** divides out, leaving **oz** for your answer.

300 g ≈ 10.5 oz

CAUTION

Because the metric and U.S. measurement systems were developed independently, almost all comparisons are approximate. Your answers should be written with the "≈" symbol to show they are approximate.

◁ **Work Problem ❷ at the Side.**

OBJECTIVE ❷ Learn common temperatures on the Celsius scale. In the metric system, temperature is measured on the **Celsius** scale. On the Celsius scale, water freezes at 0 °C and boils at 100 °C. The small raised circle stands for "degrees" and the capital **C** is for Celsius. Read the temperatures like this:

Water freezes at 0 degrees Celsius (0 °C).   ◁ In the metric system, °C is the symbol for "degrees Celsius."
Water boils at 100 degrees Celsius (100 °C).

The U.S. temperature system, used only in the United States, is measured on the **Fahrenheit** scale. On this scale:

Water freezes at 32 degrees Fahrenheit (32 °F).   ◁ In the U.S. system, °F stands for "degrees Fahrenheit."
Water boils at 212 degrees Fahrenheit (212 °F).

Answers

2. **(a)** $\dfrac{(17)\,(2.20 \ lb)}{1}$ = 37.4 lb; 17 kg ≈ 37.4 lb
  (b) 5 L ≈ 5.3 qt   **(c)** 90 g ≈ 3.2 oz
  (d) 3.5 gal ≈ 13.3 L   **(e)** 145 lb ≈ 65.3 kg
  (f) 8 oz ≈ 226.8 g

The thermometer below shows some typical temperatures in both Celsius and Fahrenheit. For example, comfortable room temperature is about 20 °C or 68 °F, and normal body temperature is about 37 °C or 98.6 °F.

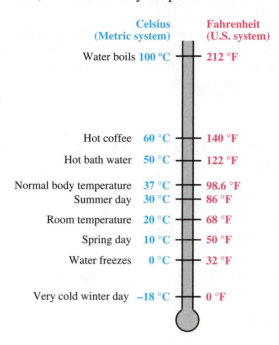

	Celsius (Metric system)	Fahrenheit (U.S. system)
Water boils	100 °C	212 °F
Hot coffee	60 °C	140 °F
Hot bath water	50 °C	122 °F
Normal body temperature	37 °C	98.6 °F
Summer day	30 °C	86 °F
Room temperature	20 °C	68 °F
Spring day	10 °C	50 °F
Water freezes	0 °C	32 °F
Very cold winter day	−18 °C	0 °F

Note

The freezing and boiling temperatures are exact. The other temperatures are approximate. Even normal body temperature varies slightly from person to person.

EXAMPLE 3   Using Celsius Temperatures

Circle the metric temperature that is most reasonable for each situation.

(a) Warm summer day   29 °C   64 °C   90 °C

29 °C is reasonable. 64 °C and 90 °C are too hot; they're both above the temperature of hot bath water (above 122 °F).

(b) Inside a freezer   −10 °C   3 °C   25 °C

−10 °C is the reasonable temperature because it is the only one below the freezing point of water (0 °C). Your frozen foods would start thawing at 3 °C or 25 °C.

Work Problem **3** at the Side. ▶

OBJECTIVE ▶ **3** **Use formulas to convert between Celsius and Fahrenheit temperatures.** You can use these formulas to convert between Celsius and Fahrenheit temperatures.

Celsius–Fahrenheit Conversion Formulas

Converting from Fahrenheit (F) to Celsius (C)

$$C = \frac{5(F - 32)}{9}$$

Converting from Celsius (C) to Fahrenheit (F)

$$F = \frac{9 \cdot C}{5} + 32$$

3 Circle the metric temperature that is *most* reasonable for each situation.

(a) Set the living room thermostat at:

11 °C   21 °C   71 °C

Hint: Look at the blue side of the thermometer in the text.

(b) The baby has a fever of:

29 °C   39 °C   49 °C

(c) Wear a sweater outside because it's:

15 °C   28 °C   50 °C

(d) My iced tea is:

−5 °C   5 °C   30 °C

(e) Let's go swimming! It's:

95 °C   65 °C   35 °C

(f) Inside a refrigerator (not the freezer) it's:

−15 °C   0 °C   3 °C

(g) There's a blizzard outside. It's:

12 °C   4 °C   −20 °C

(h) I need hot water to get these clothes clean. The water should be:

55 °C   105 °C   200 °C

Answers

3. **(a)** 21 °C   **(b)** 39 °C   **(c)** 15 °C
(d) 5 °C   **(e)** 35 °C   **(f)** 3 °C
(g) −20 °C   **(h)** 55 °C

4 Convert to Celsius.

GS **(a)** 59 °F $C = \dfrac{5\,(F - 32)}{9}$

$C = \dfrac{5\,(59 - 32)}{9}$ Finish the work.

(b) 41 °F

(c) 212 °F

(d) 98.6 °F

5 Convert to Fahrenheit.

GS **(a)** 100 °C $F = \dfrac{9 \cdot C}{5} + 32$

$F = \dfrac{9 \cdot 100}{5} + 32$ Finish the work.

(b) 25 °C

(c) 80 °C

(d) 5 °C

Answers

4. **(a)** $C = \dfrac{5(\overset{3}{\cancel{27}})}{\underset{1}{\cancel{9}}} = 15\,°C$ **(b)** 5 °C

  (c) 100 °C **(d)** 37 °C

5. **(a)** $F = \dfrac{9 \cdot \overset{20}{\cancel{100}}}{\underset{1}{\cancel{5}}} + 32 = 180 + 32 = 212\,°F$

  (b) 77 °F **(c)** 176 °F **(d)** 41 °F

As you use these formulas, be sure to follow the order of operations from **Chapter 1.**

Order of Operations

1. Do all operations inside *parentheses* or *other grouping symbols.*
2. Simplify any expressions with *exponents* and find any *square roots.*
3. *Multiply* or *divide,* proceeding from left to right.
4. *Add* or *subtract,* proceeding from left to right.

EXAMPLE 4   **Converting Fahrenheit to Celsius (U.S. to Metric)**

Convert 68 °F to Celsius.

Use the formula and follow the order of operations.

$$C = \dfrac{5\,(F - 32)}{9}$$ Fahrenheit to Celsius formula

$$= \dfrac{5\,(68 - 32)}{9}$$ Work inside parentheses first: 68 − 32 is 36.

$$= \dfrac{5\,(36)}{9}$$

$$= \dfrac{5\,(\overset{4}{\cancel{36}})}{\underset{1}{\cancel{9}}}$$ Divide 36 and 9 by their common factor: 36 ÷ 9 is 4, and 9 ÷ 9 is 1. Multiply 5(4) in the numerator.

$$= 20$$

Thus, 68 °F = 20 °C.

◀ **Work Problem 4 at the Side.**

EXAMPLE 5   **Converting Celsius to Fahrenheit (Metric to U.S.)**

Convert 15 °C to Fahrenheit.

Use the formula and follow the order of operations.

$$F = \dfrac{9 \cdot C}{5} + 32$$ Celsius to Fahrenheit formula

$$= \dfrac{9 \cdot 15}{5} + 32$$

$$= \dfrac{9 \cdot \overset{3}{\cancel{15}}}{\underset{1}{\cancel{5}}} + 32$$ Divide 15 and 5 by their common factor: 15 ÷ 5 is 3, and 5 ÷ 5 is 1. Multiply 9 • 3 in the numerator.

$$= 27 + 32$$ Add.

$$= 59$$

Thus, 15 °C = 59 °F.

◀ **Work Problem 5 at the Side.**

7.5 Exercises

FOR
EXTRA
HELP

 Download the MyDashBoard App

 MyMathLab®

CONCEPT CHECK   *Use the table on the first page of this section to answer Exercises 1 and 2.*

1. **(a)** Use only the "Metric to U.S." side of the table. Write **two** unit fractions using the information for converting between kilograms and pounds.

   (b) When converting 5 kg to pounds, use the unit fraction from part (a) that allows you to divide out which unit? _____ Would you use the unit fraction with kg in the *numerator* or the *denominator*? _____

2. **(a)** Use only the "U.S. to Metric" side of the table. Write **two** unit fractions using the information for converting between quarts and liters.

   (b) When converting 3 qt to liters, use the unit fraction from part (a) that allows you to divide out which unit? _____. Would you use the unit fraction with qt in the *numerator* or the *denominator*? _____

Use the table on the first page of this section and unit fractions to make approximate conversions from metric to U.S. units or U.S. to metric units. Round your answers to the nearest tenth. **See Examples 1 and 2.**

3. 20 m to yards

   $$\frac{20 \ \cancel{m}}{1} \cdot \frac{1.09 \ yd}{1 \ \cancel{m}} =$$

4. 8 km to miles

   $$\frac{8 \ \cancel{km}}{1} \cdot \frac{0.62 \ mi}{1 \ \cancel{km}} =$$

5. 80 m to feet

6. 85 cm to inches

7. 16 ft to meters

8. 3.2 yd to meters

9. 150 g to ounces

10. 2.5 oz to grams

11. 248 lb to kilograms

12. 7.68 kg to pounds

13. 28.6 L to quarts

14. 15.75 L to gallons

15. For the 2000 Olympics, the 3M Company used 5 g of pure gold to coat Michael Johnson's track shoes. (*Source:* 3M Company.)

(a) How many ounces of gold were used, to the nearest tenth?

(b) Was this enough extra weight to slow him down?

16. The label on a Van Ness auto feeder for cats and dogs says it holds 1.4 kg of dry food. How many pounds of food does it hold, to the nearest tenth? (*Source:* Van Ness Plastics.)

17. The heavy-duty wash cycle in a dishwater uses 8.4 gal of water. How many liters does it use, to the nearest tenth? (*Source:* Frigidaire.)

18. The rinse-and-hold cycle in a dishwasher uses only 4.5 L of water. How many gallons does it use, to the nearest tenth? (*Source:* Frigidaire.)

19. The smallest pet fish are dwarf gobies, which are half an inch long. How many centimeters long is a dwarf gobie, to the nearest tenth? (*Hint:* First write half an inch in decimal form.)

20. The fastest nerve signals in the human body travel 120 meters per second. How many feet per second do the signals travel? (*Source: The Human Body.*)

CONCEPT CHECK   *Circle the most reasonable metric temperature for each situation.*
See Example 3.

21. A snowy day

12 °C   28 °C   −8 °C

22. Brewing coffee

80 °C   180 °C   15 °C

23. A high fever

21 °C   40 °C   103 °C

24. Swimming pool water

90 °C   78 °C   25 °C

25. Oven temperature

150 °C   50 °C   30 °C

26. Light jacket weather

0 °C   10 °C   −10 °C

Use the conversion formulas from this section and the order of operations to convert Fahrenheit temperatures to Celsius, or Celsius temperatures to Fahrenheit. Round your answers to the nearest degree if necessary. **See Examples 4 and 5.**

27. 60 °F   Complete the conversion.

$$C = \frac{5\,(60 - 32)}{9} = \frac{5\,(28)}{9} =$$

28. 80 °F   Complete the conversion.

$$C = \frac{5\,(80 - 32)}{9} = \frac{5\,(48)}{9} =$$

29. 104 °F

30. 36 °F

31. 8 °C

32. 18 °C

33. 35 °C

34. 0 °C

Solve each application problem. Round your answers to the nearest degree if necessary.

35. The highest temperature ever recorded on Earth was 136 °F at El Azizia, Libya, in 1922. Convert this temperature to Celsius. (*Source: World Almanac and Book of Facts.*)

36. Hummingbirds have a normal body temperature of 107 °F. But on cold nights they go into a state of torpor where their body temperature drops to 39 °F. What are these temperatures in Celsius? (*Source: Wildbird.*)

37. (a) Here is the tag on a pair of Sorel boots. In what kind of weather would you wear these boots?

> **Comfort range**
> **24 °C to 4 °C**

Source: Sorel.

(b) For what Fahrenheit temperatures are the boots designed?

(c) What range of metric temperatures do you have in January where you live? Would you be comfortable in these boots?

38. Sleeping bags made by Eddie Bauer are sold around the world. Each type of sleeping bag is designed for outdoor camping in certain temperatures.

Junior bag	5 °C or warmer
Removable liner bag	0 °C to 15 °C
Conversion bag	−7 °C to 0 °C

Source: Eddie Bauer.

(a) At what Fahrenheit temperatures should you use the Junior bag?

(b) What Fahrenheit temperatures is the removable liner bag designed for?

Relating Concepts (Exercises 39–46) For Individual or Group Work

The information below appeared in American newspapers. However, both Newfoundland (part of Canada) and Ireland use the metric system. Their newspapers would have reported all the measurements in metric units. Complete the conversions to metric, rounding answers to the nearest tenth.

Q. A recent news brief reported on some men who flew a model airplane from Newfoundland to Ireland. Can you provide some details of the flight?

Newfoundland
Europe
Ireland
Africa

A. The model plane is 6 feet long and weighs 11 pounds. Made of balsa wood and Mylar, it crossed the Atlantic—the flight path took it 1,888.3 miles—in 38 hours, 23 minutes. It soared at a cruising altitude of 1,000 feet. The plane used a souped-up piston engine and carried less than a gallon of fuel, as mandated by rules of the Federation Aeronautique Internationale, the governing body of model airplane building. When it landed in County Galway, Ireland, it had less than 2 fluid ounces of fuel left. The plane was built by Maynard Hill of Silver Spring, Maryland.

(*Source*: nctimes.com)

39. Length of model plane

40. Weight of plane

41. Length of flight path

42. Time of flight

43. Cruising altitude

44. Fuel at the start, in milliliters

45. Fuel left after landing, in milliliters
(*Hint:* First convert 2 fl oz to quarts.)

46. What *percent* of the fuel was left at the end of the flight?

Chapter 7 *Summary*

Key Terms

7.1

U.S. measurement system The U.S. measurement system is used for many daily activities only in the United States. Commonly used units include quarts, pounds, feet, miles, and degrees Fahrenheit.

metric system The metric system of measurement is an international system used in manufacturing, science, medicine, sports, and other fields. Commonly used units in this system include meters, liters, grams, and degrees Celsius.

unit fraction A unit fraction involves measurement units and is equivalent to 1. Unit fractions are used to convert among different measurements.

7.2

meter The meter is the basic unit of length in the metric system. The symbol **m** is used for meter. One meter is a little longer than a yard.

prefixes Attaching a prefix such as *kilo-* or *milli-* to the words meter, liter, or gram gives names of larger or smaller units. For example, the prefix *kilo-* means 1000 so a *kilo*meter is 1000 meters.

metric conversion line The metric conversion line is a line showing the various metric measurement prefixes and their size relationship to each other.

7.3

liter The liter is the basic unit of capacity in the metric system. The symbol **L** is used for liter. One liter is a little more than one quart.

gram The gram is the basic unit of weight (mass) in the metric system. The symbol **g** is used for gram. One gram is the weight of 1 milliliter of water or one dollar bill.

7.5

Celsius The Celsius scale is used to measure temperature in the metric system. Water boils at 100 °C and freezes at 0 °C.

Fahrenheit The Fahrenheit scale is used to measure temperature in the U.S. system. Water boils at 212 °F and freezes at 32 °F.

New Symbols

Frequently used metric length units

   km kilometer
   m meter
   cm centimeter
  mm millimeter

Frequently used metric capacity units

   L liter
  mL milliliter

Frequently used metric weight (mass) units

  kg kilogram
   g gram
  mg milligram

°C degrees Celsius
   (metric temperature unit)
°F degrees Fahrenheit
   (U.S. temperature unit)

New Formulas

Converting from Celsius to Fahrenheit (metric to U.S.):

$$F = \frac{9 \cdot C}{5} + 32$$

Converting from Fahrenheit to Celsius (U.S. to metric):

$$C = \frac{5(F - 32)}{9}$$

Test Your Word Power

See how well you have learned the vocabulary in this chapter.

1 The **metric system**
   A. uses meters, liters, and degrees Fahrenheit
   B. is based on multiples of 10
   C. is used only in the United States.

2 **U.S. measurement units**
   A. are used throughout the world
   B. are based on multiples of 12
   C. include feet, inches, quarts, and pounds.

3 A **unit fraction**
   A. has the unit you want to change in the numerator
   B. has a denominator of 1
   C. is equivalent to 1.

4 A **gram** is
   A. the weight of 1 mL of water
   B. equivalent to 1000 kg
   C. approximately equal to 2.2 pounds.

5 A **meter** is
   A. equivalent to 1000 cm
   B. approximately equal to $\frac{1}{2}$ inch
   C. abbreviated m with no period after it.

6 The **Celsius** temperature scale
   A. shows water freezing at 32°
   B. is used in the U.S. system of measurement
   C. shows water boiling at 100°.

Answers to Test Your Word Power

1. B; *Examples:* 10 meters = 1 dekameter; 100 meters = 1 hectometer; 1000 meters = 1 kilometer.

2. C; *Examples:* Feet and inches are used to measure length, quarts to measure capacity, pounds to measure weight.

3. C; *Example:* Because 12 in. = 1 ft, the unit fraction $\frac{12 \text{ in.}}{1 \text{ ft}}$ is equivalent to $\frac{12 \text{ in.}}{12 \text{ in.}} = 1$.

4. A; *Example:* A small box measuring 1 cm on every edge holds exactly 1 mL of water, and the water weighs 1 g.

5. C; *Example:* A measurement of 16 meters is written 16 m (without a period).

6. C; *Example:* In the metric system, water freezes at 0 °C and boils at 100 °C. The U.S. system uses the Fahrenheit temperature scale in which water freezes at 32 °F and boils at 212 °F.

Quick Review

Concepts	Examples

7.1 U.S. Measurement Units

Memorize the basic measurement relationships. Then, to convert units, multiply when changing from a larger unit to a smaller unit; divide when changing from a smaller unit to a larger unit.

Convert each measurement.
(a) 5 ft to inches (larger unit to smaller unit)

$$5 \text{ ft} = 5 \cdot \mathbf{12} = 60 \text{ in.}$$

(b) 3 lb to ounces (larger unit to smaller unit)

$$3 \text{ lb} = 3 \cdot \mathbf{16} = 48 \text{ oz}$$

(c) 15 qt to gallons (smaller unit to larger unit)

$$15 \text{ qt} = \frac{15}{\mathbf{4}} = 3\frac{3}{4} \text{ gal}$$

7.1 Using Unit Fractions

Another, more useful, conversion method is multiplying by a unit fraction.
The unit you want in the answer should be in the numerator.
The unit you want to change should be in the denominator.

Convert 32 oz to pounds.

$$32 \text{ oz} \cdot \frac{1 \text{ lb}}{16 \text{ oz}} \quad \Big\} \text{ Unit fraction}$$

These units should match.

$$= \frac{\overset{2}{\cancel{32} \text{ oz}}}{1} \cdot \frac{1 \text{ lb}}{\underset{1}{\cancel{16} \text{ oz}}} \quad \begin{array}{l} \text{Divide out ounces.} \\ \text{Divide out common factors.} \end{array}$$

$$= 2 \text{ lb}$$

Concepts	Examples

7.1 Solving U.S. Measurement Application Problems

To solve application problems, use the six problem-solving steps.

Step 1 **Read** the problem carefully.

Step 2 **Work out a plan.**

Step 3 **Estimate** a reasonable answer.

Step 4 **Solve** the problem.

Step 5 **State the answer.**
Step 6 **Check** your work.

Use the six steps to solve this problem.
Mr. Green has 10 yd of rope. He is cutting it into eight pieces so his sailing class can practice knot tying. How many feet of rope will each of his eight students get?

Step 1 The problem asks how many feet of rope can be given to each of eight students.

Step 2 Convert 10 yd to feet (the unit required in the answer). Then divide by eight students.

Step 3 There are 3 ft in one yard, so there are 30 ft in 10 yd. Then 30 ft $\div$ 8 $\approx$ 4 ft is our estimate.

Step 4 Use a unit fraction to convert 10 yd to feet, then divide.

$$\frac{10 \text{ yd}}{1} \cdot \frac{3 \text{ ft}}{1 \text{ yd}} = 30 \text{ ft}$$

$$\frac{30 \text{ ft}}{8 \text{ students}} = 3\frac{3}{4} \text{ ft or } 3.75 \text{ ft per student}$$

Step 5 Each student gets $3\frac{3}{4}$ ft or 3.75 ft of rope.
Step 6 The exact answer of 3.75 ft is close to our estimate of 4 ft.

7.2 Basic Metric Length Units

Use approximate comparisons to judge which length units are appropriate:

1 mm is the thickness of a dime.
1 cm is about $\frac{1}{2}$ inch.
1 m is a little more than 1 yard.
1 km is about 0.6 mile.

Write the most reasonable metric unit in each blank. Choose from km, m, cm, and mm.

The room is 6 __m__ long.
A paper clip is 30 __mm__ long.
He drove 20 __km__ to work.

7.2–7.3 Converting within the Metric System

Using Unit Fractions
One conversion method is to multiply by a unit fraction. Use a fraction with the unit you want in the answer in the numerator and the unit you want to change in the denominator.

Convert.
(a) 9 g to kg

$$\frac{9 \text{ g}}{1} \cdot \frac{1 \text{ kg}}{1000 \text{ g}} = \frac{9}{1000} \text{ kg} = 0.009 \text{ kg}$$

9 g = 0.009 kg

(b) 3.6 m to cm

$$\frac{3.6 \text{ m}}{1} \cdot \frac{100 \text{ cm}}{1 \text{ m}} = 360 \text{ cm}$$

3.6 m = 360 cm

(Continued)

Concepts	Examples

7.2-7.3 Converting within the Metric System (*Continued*)

Using the Metric Conversion Line
Another conversion method is to find the unit you are given on the metric conversion line. Count the number of places to get from the unit you are given to the unit you want. Move the decimal point the same number of places and in the same direction.

1000	100	10	1	$\frac{1}{10}$	$\frac{1}{100}$	$\frac{1}{1000}$
kilo	**hecto**	**deka**	meter	**deci**	**centi**	**milli**
			liter			
			gram			

Convert.
(a) 68.2 kg to g
From **kg** to **g** is three places to the right.

    6 8.2 0 0   Decimal point is moved three places to the right.

68.2 kg = 68,200 g

(b) 300 mL to L
From **mL** to **L** is three places to the left.

    3 0 0.   Decimal point is moved three places to the left.

300 mL = 0.3 L

(c) 825 cm to m
From **cm** to **m** is two places to the left.

    8 2 5.   Decimal point is moved two places to the left.

825 cm = 8.25 m

7.3 Basic Metric Capacity Units

Use approximate comparisons to judge which capacity units are appropriate:

1 L is a little more than 1 quart.

1 mL is the amount of water in a cube that is 1 cm on each side.

5 mL is about one teaspoon.

250 mL is about one cup.

Write the most reasonable metric unit in each blank. Choose from L and mL.

The pail holds 12 __L__.

The milk carton from the vending machine holds 250 __mL__.

7.3 Basic Metric Weight (Mass) Units

Use approximate comparisons to judge which weight units are appropriate:

1 kg is about 2.2 pounds.

1 g is the weight of 1 mL of water or one dollar bill.

1 mg is $\frac{1}{1000}$ of a gram; very tiny!

Write the most reasonable metric unit in each blank. Choose from kg, g, and mg.

The wrestler weighed 95 __kg__.

She took a 500 __mg__ aspirin tablet.

One banana weighs 150 __g__.

Concepts	Examples

7.4 Solving Metric Measurement Application Problems

Convert units so you are comparing kg to kg, cm to cm, and so on. When a measurement involves two units, such as 6 m 20 cm, write it in terms of the unit called for in the answer (6.2 m or 620 cm).

Use the six problem-solving steps.

Step 1 **Read** the problem carefully.

Step 2 **Work out a plan.**

Step 3 **Estimate** a reasonable answer.

Step 4 **Solve** the problem.

Step 5 **State the answer.**

Step 6 **Check** your work.

Use the six steps to solve this problem.

George cut 1 m 35 cm off of a 3 m board. How long was the leftover piece, in meters?

Step 1 The problem asks for the length of the leftover piece in meters.

Step 2 Convert the cut-off measurement to meters (the unit required in the answer), and then subtract to find the "leftover."

Step 3 To estimate, round 1 m 35 cm to 1 m, because 35 cm is less than half a meter. Then, 3 m − 1 m = 2 m is our estimate.

Step 4 Convert the cut-off measurement to meters, then subtract.

$$
\begin{array}{rl}
1\text{ m} \rightarrow & 1.00\text{ m} \\
\text{plus }35\text{ cm} \rightarrow & +\ 0.35\text{ m} \\
\hline
& 1.35\text{ m}
\end{array}
\qquad
\begin{array}{rl}
3.00\text{ m} & \leftarrow \text{Board} \\
-\ 1.35\text{ m} & \leftarrow \text{Cut off} \\
\hline
1.65\text{ m} & \leftarrow \text{Left}
\end{array}
$$

Step 5 The leftover piece is 1.65 m long.

Step 6 The exact answer of 1.65 m is close to our estimate of 2 m.

7.5 Converting between Metric and U.S. Measurement Units

Write a unit fraction using the values in the table of conversion factors in the last section of this chapter. Because the values in the table are rounded, your answers will be approximate.

Convert. Round answers to the nearest tenth.

(a) 23 m to yards
From the table, 1 meter ≈ 1.09 yards.

$$\frac{23\ \cancel{m}}{1} \cdot \frac{1.09\text{ yd}}{1\ \cancel{m}} = 25.07\text{ yd}$$

25.07 rounds to 25.1, so 23 m ≈ 25.1 yd.

(b) 4 oz to grams
From the table, 1 ounce ≈ 28.35 grams.

$$\frac{4\ \cancel{oz}}{1} \cdot \frac{28.35\text{ g}}{1\ \cancel{oz}} = 113.4\text{ g}$$

4 oz ≈ 113.4 g

Concepts	Examples

7.5 Common Celsius (Metric) Temperatures

Use approximate and exact comparisons to judge which temperatures are appropriate:

Exact comparisons:

0 °C is the freezing point of water

100 °C is the boiling point of water

Approximate comparisons:

10 °C for a spring day

20 °C for room temperature

30 °C for a summer day

37 °C for normal body temperature

Circle the metric temperature that is most reasonable.

(a) Hot summer day:

(35 °C)   90 °C   110 °C

(b) The first snowy day in winter:

−20 °C   (0 °C)   15 °C

7.5 Converting between Fahrenheit and Celsius Temperatures

Use this formula to convert from Fahrenheit (F) to Celsius (C); that is, from U.S. to metric temperatures.

$$C = \frac{5(F-32)}{9}$$

Use this formula to convert from Celsius (C) to Fahrenheit (F); that is, from metric to U.S. temperatures.

$$F = \frac{9 \cdot C}{5} + 32$$

Convert 176 °F to Celsius.

$$C = \frac{5\,(\mathbf{176}-32)}{9}$$ Replace F with 176. Work inside parentheses; $176-32$ is 144.

$$= \frac{5\,(\overset{16}{\cancel{144}})}{\underset{1}{\cancel{9}}}$$ Divide out common factors. Then multiply in the numerator.

$$= 80$$

176 °F = 80 °C

Convert 80 °C to Fahrenheit.

$$F = \frac{9 \cdot \mathbf{80}}{5} + 32$$ Replace C with 80.

$$= \frac{9 \cdot \overset{16}{\cancel{80}}}{\underset{1}{\cancel{5}}} + 32$$ Divide out common factors. Then multiply $9 \cdot 16$ in the numerator.

$$= 144 + 32$$ Add.

$$= 176$$

80 °C = 176 °F

Chapter 7 Review Exercises

7.1 *Fill in the blanks with the measurement relationships you have memorized.*

1. 1 lb = _____ oz

2. _____ ft = 1 yd

3. 1 ton = _____ lb

4. _____ qt = 1 gal

5. 1 hr = _____ min

6. 1 c = _____ fl oz

7. _____ sec = 1 min

8. _____ ft = 1 mi

9. _____ in. = 1 ft

Convert using unit fractions.

10. 4 ft = _____ in.

11. 6000 lb = _____ tons

12. 64 oz = _____ lb

13. 18 hr = _____ day

14. 150 min = _____ hr

15. $1\frac{3}{4}$ lb = _____ oz

16. $6\frac{1}{2}$ ft = _____ in.

17. 7 gal = _____ c

18. 4 days = _____ sec

19. The average depth of the world's oceans is 12,460 ft. (*Source: Handy Ocean Answer Book.*)

    (a) What is the average depth in yards?

    (b) What is the average depth in miles, to the nearest tenth?

20. During the first year of a program to recycle office paper, a company recycled 123,260 pounds of paper. The company received $40 per ton for the paper. How much money did the company make? Use the six problem-solving steps. (*Source:* I. C. System.)

7.2 *Write the most reasonable metric length unit in each blank. Choose from km, m, cm, and mm.*

21. My thumb is 20 _____ wide.

22. Her waist measurement is 66 _____.

23. The two towns are 40 _____ apart.

24. A basketball court is 30 _____ long.

25. The height of the picnic bench is 45 _____.

26. The eraser on the end of my pencil is 5 _____ long.

Height

Convert using unit fractions or the metric conversion line.

27. 5 m to cm

28. 8.5 km to m

29. 85 mm to cm

30. 370 cm to m   **31.** 70 m to km   **32.** 0.93 m to mm

7.3 *Write the most reasonable metric unit in each blank. Choose from L, mL, kg, g, and mg.*

33. The eyedropper holds 1 _____.

34. I can heat 3 _____ of water in this saucepan.

35. Loretta's hammer weighed 650 _____.

36. Yongshu's large suitcase weighed 20 _____ when it was packed.

37. My fish tank holds 80 _____ of water.

38. I'll buy the 500 _____ bottle of mouthwash.

39. Mara took a 200 _____ antibiotic pill.

40. This piece of chicken weighs 100 _____.

Convert using unit fractions or the metric conversion line.

41. 5000 mL to L   **42.** 8 L to mL   **43.** 4.58 g to mg

44. 0.7 kg to g   **45.** 6 mg to g   **46.** 35 mL to L

7.4 *Solve each application problem. Show your work. Use the six problem-solving steps.*

47. Each serving of punch at the wedding reception will be 180 mL. How many liters of punch are needed for 175 servings?

48. Jason is serving a 10 kg turkey to 28 people. How many grams of meat is he allowing for each person? Round to the nearest whole gram.

49. Yerald weighed 92 kg. Then he lost 4 kg 750 g. What is his weight now, in kilograms?

50. Young-Mi bought 950 g of onions. The price was $1.49 per kilogram. How much did she pay, to the nearest cent?

7.5 *Use the table in the last section of this chapter and unit fractions to make approximate conversions. Round your answers to the nearest tenth if necessary.*

51. 6 m to yards   **52.** 30 cm to inches

53. 108 km to miles

54. 800 mi to kilometers

55. 23 qt to liters

56. 41.5 L to quarts

*Write the appropriate **metric** temperature in each blank.*

57. Water freezes at _____.

58. Water boils at _____.

59. Normal body temperature is about _____.

60. Comfortable room temperature is about _____.

Use the conversion formulas in the last section of this chapter to convert each temperature to Fahrenheit or to Celsius. Round to the nearest degree if necessary.

61. 77 °F

62. 92 °F

63. 6 °C

64. Water coming into a dishwasher should be at least 49 °C to clean the dishes properly. What Fahrenheit temperature is that? (*Source:* Frigidaire.)

Mixed Review Exercises

Write the most reasonable metric unit in each blank. Choose from km, m, cm, mm, L, mL, kg, g, and mg.

65. I added 1 _____ of oil to my car.

66. The box of books weighed 15 _____ .

67. Larry's shoe is 30 _____ long.

68. Jan used 15 _____ of shampoo on her hair.

69. My fingernail is 10 _____ wide.

70. I walked 2 _____ to school.

71. The tiny bird weighed 15 _____ .

72. The new library building is 18 _____ wide.

73. The cookie recipe uses 250 _____ of milk.

74. Renee's pet mouse weighs 30 _____ .

75. One postage stamp weighs 90 _____ .

76. I bought 30 _____ of gas for my car.

Convert the following using unit fractions, the metric conversion line, or the temperature conversion formulas.

77. 10.5 cm to millimeters

78. 45 min to hours

79. 90 in. to feet

80. 1.3 m to centimeters

81. 25 °C to Fahrenheit

82. $3\frac{1}{2}$ gal to quarts

83. 700 mg to grams

84. 0.81 L to milliliters

85. 5 lb to ounces

86. 60 kg to grams

87. 1.8 L to milliliters

88. 86 °F to Celsius

89. 0.36 m to centimeters

90. 55 mL to liters

Solve each application problem. Show your work.

91. Peggy had a board measuring 2 m 4 cm. She cut off 78 cm. How long is the board now, in meters?

2 m 4 cm

78 cm

92. During the 12-day Minnesota State Fair, one of the biggest in the United States, Sweet Martha's booth sells an average of 3000 pounds of cookies per day. How many tons of cookies are sold in all? (*Source: Minneapolis Star Tribune.*)

93. Olivia is sending a recipe to her mother in Mexico. Among other things, the recipe calls for 4 oz of rice and a baking temperature of 350 °F. Convert these measurements to metric, rounding to the nearest gram and nearest degree.

94. While on vacation in Canada, Jalo became ill and went to a health clinic. They said he weighed 80.9 kg and was 1.83 m tall. Find his weight in pounds and height in feet. Round to the nearest tenth.

The largest two-axle trucks in the world are mining trucks used to haul huge loads of rock and iron ore. Some information about these $2 million trucks is given below. Fill in the blank spaces in the table. Then use the information to answer Exercises 101–102. Round answers to the nearest tenth.

WORLD'S LARGEST TWO-AXLE TRUCK

	Mining Truck	Measurements	
		Metric Units	U.S. Units
95.	Length of truck	_____ m	44 ft
96.	Height of truck	6.7 m	_____ ft
97.	Height of tire	_____ m	4 yd
98.	Width of tire tread	102 cm	_____ in.
99.	Weight of load truck can carry	220,000 kg	_____ lb
100.	Fuel needed to travel 1 mile	_____ to _____ L	5 to 6 gal

Source: Hull Rust Mahoning Mine.

101. Using U.S. measurements:

(a) The truck can carry a load weighing how many tons?

(b) How many inches high is each tire?

102. Using metric measurements:

(a) What is the width of the tire tread in meters?

(b) How tall is the truck in centimeters?

The Chapter Test Prep Videos with test solutions are available on DVD, in MyMathLab, and on You Tube—search "LialBasicCollegeMath" and click on "Channels."

Convert each measurement.

1. 9 gal = _____ qt

2. 45 ft = _____ yd

3. 135 min = _____ hr

4. 9 in. = _____ ft

5. $3\frac{1}{2}$ lb = _____ oz

6. 5 days = _____ min

Write the most reasonable metric unit in each blank. Choose from km, m, cm, mm, L, mL, kg, g, and mg.

7. My husband weighs 75 _____ .

8. I hiked 5 _____ this morning.

9. She bought 125 _____ of cough syrup.

10. This apple weighs 180 _____ .

11. This page is about 21 _____ wide.

12. My watch band is 10 _____ wide.

13. I bought 10 _____ of soda for the picnic.

14. The bracelet is 16 _____ long.

Convert the following measurements. Show your work.

15. 250 cm to meters

16. 4.6 km to meters

17. 5 mm to centimeters

18. 325 mg to grams

19. 16 L to milliliters

20. 0.4 kg to grams

21. 10.55 m to centimeters

22. 95 mL to liters

23. The rainiest place in the world is Mount Waialeale in Hawaii, which receives 460 inches of rain each year. What is the average rainfall per month, in feet, to the nearest tenth? (*Source:* National Geographic Society.)

24. A 6-inch Subway Veggie Delite sandwich has 310 mg of sodium. A 6-inch Spicy Italian sandwich has 1.7 g of sodium. (*Source:* SUBWAY.)

    (a) How much more sodium is in the Spicy Italian sandwich, in milligrams, than in the Veggie Delite?

    (b) The recommended amount of sodium is no more than 1500 mg daily. How much more or less sodium does the Spicy Italian sandwich have than the recommended daily amount?

Pick the metric temperature that is most reasonable in each situation.

25. The water is almost boiling.

    210 °C   155 °C   95 °C

26. The tomato plants may freeze tonight.

    30 °C   20 °C   0 °C

Use the U.S./metric conversion table from the last section of this chapter and unit fractions to convert each measurement. Round your answers to the nearest tenth if necessary.

27. 6 ft to meters

28. 125 lb to kilograms

29. 50 L to gallons

30. 8.1 km to miles

In Problems 31 and 32, Use the conversion formulas to convert each temperature. Round your answers to the nearest degree if necessary.

31. 74 °F to Celsius

32. 2 °C to Fahrenheit

33. Denise is making five matching pillows. She needs 120 cm of braid to trim each pillow. If the braid costs $3.98 per yard, how much will she spend to trim the pillows, to the nearest cent? (First find the number of meters of braid Denise needs.) Show your work.

34. Describe two benefits the United States would achieve by switching entirely to the metric system.

Chapters 1–7 Cumulative Review Exercises

First use front end rounding to round each number and estimate the answer.
Then find the exact answer.

1. *Estimate:*   *Exact:*

$$\times \underline{\quad\quad} \quad\quad \begin{array}{r} 56.52 \\ \times \quad 4.7 \\ \hline \end{array}$$

2. *Estimate:*   *Exact:*

$$)\overline{\quad\quad} \quad\quad 37)\overline{19{,}610}$$

3. *Estimate:*   *Exact:*

$$1\dfrac{7}{10}$$
$$+ \underline{\quad\quad} \quad\quad + 3\dfrac{4}{5}$$
$$\underline{\quad\quad} \quad\quad \underline{\quad\quad\quad}$$

Simplify. Write answers to fraction problems in lowest terms and as whole or mixed numbers when possible.

4. $3 - 2\dfrac{5}{16}$

5. $12 \cdot 2\dfrac{2}{9}$

6. $0.86 \div 0.066$
Round your answer to the nearest tenth.

7. $\dfrac{3}{8} + \dfrac{5}{6}$

8. $8 - 0.9207$

9. $3\dfrac{3}{4} \div 6$

10. $(2.54)(0.003)$

11. $24 - 12 \div 6\,(8) + (25 - 25)$

12. $3^2 + 2^5 \cdot \sqrt{64}$

13. Arrange in order from least to greatest.

0.67   0.067   0.6   0.6007

14. Find the best buy on disposable diapers.
Package of 27 for $11.71
Package of 31 for $19.98
Package of 56 for $29.99

Solve each proportion or percent problem. Round your answers to hundredths if necessary.

15. $\dfrac{x}{16} = \dfrac{3}{4}$

16. $\dfrac{0.9}{0.75} = \dfrac{2}{x}$

17. $4 is what percent of $80?

18. 36 hours is 120% of what number of hours?

Convert the following measurements. Use the U.S./metric conversion table when necessary.

19. $2\dfrac{1}{2}$ ft to inches

20. 105 sec to minutes

21. 2.8 m to centimeters

22. 198 km to miles

Write the most reasonable metric unit in each blank. Choose from km, m, cm, mm, L, mL, kg, g, and mg.

23. Ron bought the tube of toothpaste weighing 100 _____.

24. Tia added 125 _____ of milk to her cereal.

25. The hallway is 3 _____ wide.

26. Joe's hammer weighed 1 _____.

Solve each application problem.

27. Calbert works at a Walmart store and used his employee discount to buy a $189.94 digital camera at 10% off. Find the amount of his discount, to the nearest cent, and the sale price. (*Source:* Walmart.)

28. Danielle ordered 45 prints online from her digital camera. Because it was her first order, she got 20 free prints. The rest were $0.12 each plus $6\frac{1}{2}\%$ sales tax. She also paid $2.95 for shipping. What was the total amount she paid and the cost per print, to the nearest cent?

29. Bags of slivered almonds weigh 4 oz each. They are packed in a carton that weighs 12 oz. How many pounds would a carton containing 48 bags weigh?

30. On the Illinois map, one centimeter represents 12 km. The center of Springfield is 7.8 cm from the center of Bloomington on the map.

(a) What is the actual distance in kilometers?

(b) What is the actual distance in miles, to the nearest tenth?

31. Dimitri took out a $3\frac{1}{2}$ year car loan for $18,750 at 4.3% simple interest. Find the interest and the total amount due on the loan, to the nearest cent.

32. On a 35-problem math test, Juana solved 31 problems correctly. What percent of the problems were correct? Round to the nearest tenth of a percent.

33. Mark bought 650 g of maple sugar candy on his vacation in Montreal. The candy is priced at $14.98 per kilogram. How much did Mark pay, to the nearest cent?

34. The Jackson family is making three kinds of holiday cookies that require brown sugar. The recipes call for $2\frac{1}{4}$ cups, $1\frac{1}{2}$ cups, and $\frac{3}{4}$ cup, respectively. They bought two packages of brown sugar, each holding $2\frac{1}{3}$ cups. The amount bought is how much more or less than the amount needed?

35. Akuba is knitting a scarf. Six rows of knitting result in 5 cm of scarf. At that rate, how many rows will she have to knit to make a 100 cm scarf?

36. A survey of the 5600 students on our campus found that $\frac{3}{8}$ of the students work 20 hours or more per week. How many students work 20 hours or more?

37. A spray-on bandage applies a transparent film over a wound that protects it from bacteria. The cost is $6 for about 40 applications. What is the cost per application? (*Source:* Curad.)

38. The average hospital stay is now about 5 days, compared to about 8 days in 1970. What is the percent of decrease in the length of hospital stays? (*Source:* Centers for Disease Control and Prevention.)

8 Geometry

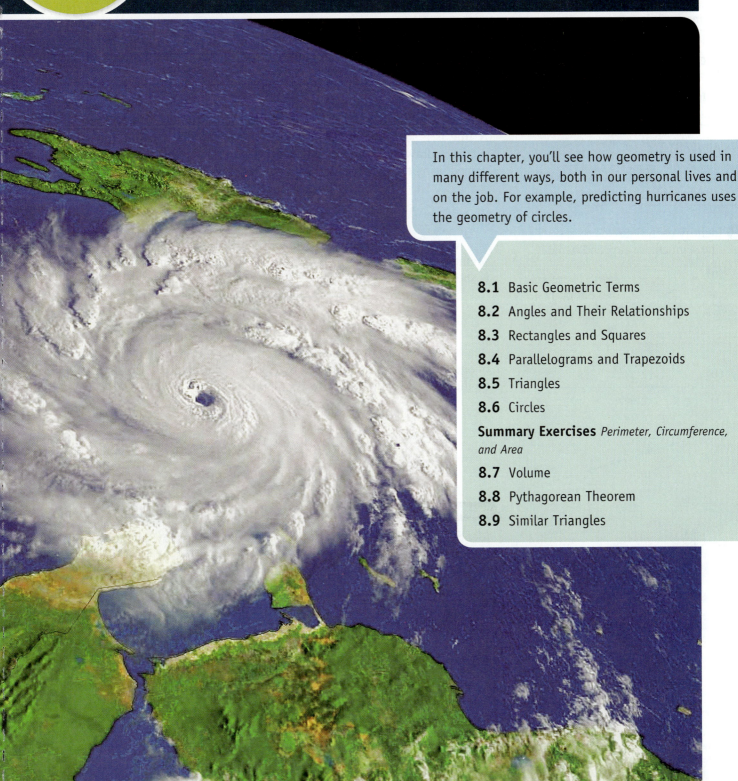

In this chapter, you'll see how geometry is used in many different ways, both in our personal lives and on the job. For example, predicting hurricanes uses the geometry of circles.

8.1 Basic Geometric Terms

OBJECTIVES

1 Identify and name lines, line segments, and rays.

2 Identify parallel and intersecting lines.

3 Identify and name angles.

4 Classify angles as right, acute, straight, or obtuse.

5 Identify perpendicular lines.

VOCABULARY TIP

Ray and *Radius* come from the same root word, which means "spoke," like the spoke of a bicycle wheel.

1 Identify each figure as a line, line segment, or ray, and name it.

(a) *Hint:* There are two endpoints.

(b)

(c)

(d)

Answers

1. (a) line segment named $\overline{EF}$ or $\overline{FE}$
   (b) ray named $\overrightarrow{SR}$
   (c) line named $\overleftrightarrow{WX}$ or $\overleftrightarrow{XW}$
   (d) line segment named $\overline{CD}$ or $\overline{DC}$

Geometry was developed centuries ago when people needed a way to measure land. The name *geometry* comes from the Greek *geo,* meaning earth, and *metron,* meaning measure. Today we still use geometry to measure land. It is also important in architecture, construction, navigation, art and design, physics, chemistry, and astronomy. You can use it at home when you buy carpet or wallpaper, hang a picture, or do home repairs.

Geometry starts with the idea of a point. A **point** is a location in space. It has no length or width. A point is represented by a dot and is named by writing a capital letter next to the dot, as shown below.

• *P*

OBJECTIVE 1 Identify and name lines, line segments, and rays.
A **line** is a straight row of points that goes on forever in both directions. A line is drawn by using arrowheads to show that it never ends. The line is named using the letters of any two points on the line.

Notice the arrowheads on *both* ends.

Line *AB*, written $\overleftrightarrow{AB}$

A piece of a line that has two endpoints is called a **line segment.** A line segment is named using its endpoints. The segment with endpoints *P* and *Q* is shown below. It can be named $\overline{PQ}$ or $\overline{QP}$.

Notice there are *no* arrowheads.

Line segment *PQ*, written $\overline{PQ}$

A **ray** is a part of a line that has only one endpoint and goes on forever in one direction. A ray is named by using the endpoint and some other point on the ray. The endpoint is always mentioned first.

Notice the arrowhead on *one* end only.

Ray *RS*, written $\overrightarrow{RS}$

EXAMPLE 1 Identifying and Naming Lines, Rays, and Line Segments

Identify each figure below as a line, line segment, or ray, and name it using the appropriate symbol.

(a)   (b)   (c)

Figure **(a)** has two endpoints, so it is a *line segment* named $\overline{RS}$ or $\overline{SR}$.
Figure **(b)** starts at point *A* and goes on forever in one direction, so it is a *ray* named $\overrightarrow{AB}$.
Figure **(c)** goes on forever in both directions, so it is a *line* named $\overleftrightarrow{EF}$ or $\overleftrightarrow{FE}$.

◀ Work Problem 1 at the Side.

OBJECTIVE ▶ **2** **Identify parallel and intersecting lines.** A *plane* is an infinitely large flat surface. A floor or a wall is a part of a plane. Lines that are in the *same plane,* but that never intersect (never cross), are called **parallel lines,** while lines that cross are called **intersecting lines.** (Think of an intersection, where two streets cross each other.)

> **EXAMPLE 2** **Identifying Parallel and Intersecting Lines**
>
> Label each pair of the lines as appearing to be parallel or as intersecting.
>
>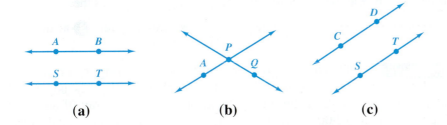
>
> (a)   (b)   (c)
>
> The lines in Figures **(a)** and **(c)** do not intersect; they appear to be *parallel lines.* The lines in Figure **(b)** cross at *P,* so they are *intersecting lines.*
>
> **CAUTION**
>
> Appearances may be deceiving! Do not assume that lines are parallel unless it is stated that they are parallel.
>
> ⋯⋯⋯⋯⋯⋯⋯⋯⋯⋯⋯⋯⋯⋯⋯⋯⋯⋯⋯ **Work Problem** **2** **at the Side.** ▶

2 Label each pair of lines as appearing to be parallel or as intersecting.

GS **(a)**

These lines cross each other so they are called
_____ lines.

GS **(b)**

These lines do **not** intersect; they appear to be _____ lines.

(c)

OBJECTIVE ▶ **3** **Identify and name angles.** An **angle** is made up of two rays that start at a common endpoint. This common endpoint is called the *vertex.*

$\overrightarrow{PQ}$ and $\overrightarrow{PR}$ are called the *sides* of the angle. The angle can be named in four different ways, as shown below.

$$\angle 1 \qquad \angle P \qquad \angle QPR \qquad \angle RPQ$$

Vertex alone   Vertex in the middle

Naming an Angle

When naming an angle, the vertex is written alone or it is written in the middle of two other points. If two or more angles have the **same vertex,** as in **Example 3** on the next page, do *not* use the vertex alone to name an angle.

Answers

2. **(a)** intersecting **(b)** parallel
   (c) parallel

3 **(a)** Name the highlighted angle in three different ways.

(b) Darken the rays that make up $\angle ZTW$.

(c) Name this angle in four different ways.

Answers

3. **(a)** $\angle 3$, $\angle CQD$, $\angle DQC$

(b)

(c) $\angle 1$, $\angle R$, $\angle MRN$, $\angle NRM$

EXAMPLE 3 **Identifying and Naming an Angle**

Name the highlighted angle in three different ways.

Do **not** name this $\angle P$ because point P is the vertex of four different angles.

The angle can be named $\angle BPA$, $\angle APB$, or $\angle 2$. It cannot be named $\angle P$, using the vertex alone, because four different angles have P as their vertex.

◀ **Work Problem** **3** **at the Side.**

OBJECTIVE **4** **Classify angles as right, acute, straight, or obtuse.**
Angles can be measured in **degrees**. The symbol for degrees is a small, raised circle °. Think of the minute hand on a clock as a ray of an angle. Suppose it is at 12:00. During one hour of time, the minute hand moves around in a complete circle. It moves 360 *degrees*, or 360°. In half an hour, at 12:30, the minute hand has moved halfway around the circle or 180°. An angle of 180° is called a **straight angle.** When two rays go in opposite directions and form a straight line, then the rays form a straight angle.

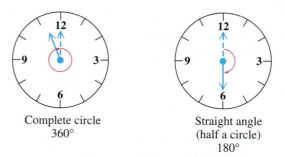

Complete circle
360°

Straight angle
(half a circle)
180°

In a quarter of an hour, at 12:15, the minute hand has moved $\frac{1}{4}$ of the way around the circle, or 90°. An angle of 90° is called a **right angle.** Sometimes you hear it called a *square angle*. The minute hands at 12:00 and 12:15 form one corner of a square. So, to show that an angle is a **right angle**, we draw a **small square** at the vertex.

A small square at the vertex identifies a right angle.

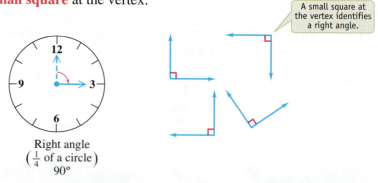

Right angle
$\left(\frac{1}{4}$ of a circle$\right)$
90°

An angle that measures 1° is shown below. You can see that an angle of 1° is very small.

1° angle

Some other terms used to describe angles are shown below.

Acute angles measure less than 90°.

Examples of acute angles

Obtuse angles measure more than 90° but less than 180°.

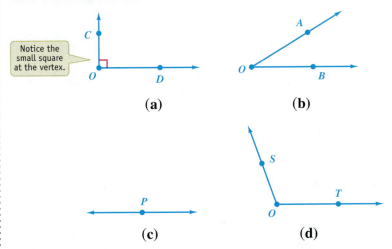

Examples of obtuse angles

In **Chapter 10,** you will use a tool called a *protractor* to measure the number of degrees in an angle.

Classifying Angles: Four Types of Angles

Acute angles measure less than 90°.

Right angles measure *exactly* 90°.

Obtuse angles measure more than 90° but less than 180°.

Straight angles measure *exactly* 180°.

Note

Angles can also be measured in radians, which you will learn about in a later math course.

EXAMPLE 4 **Classifying Angles**

Label each angle as acute, right, obtuse, or straight.

Notice the small square at the vertex.

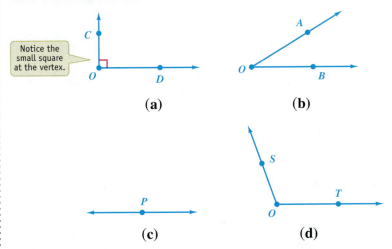

(a) **(b)**

(c) **(d)**

Figure **(a)** shows a *right angle* (exactly 90° and identified by a small square at the vertex).

Figure **(b)** shows an *acute angle* (less than 90°).

Figure **(c)** shows a *straight angle* (exactly 180°).

Figure **(d)** shows an *obtuse angle* (more than 90° but less than 180°).

▸ **Work Problem ④ at the Side.** ▶

④ Label each figure as an acute, right, obtuse, or straight angle.

GS **(a)**

Notice the small red square at the vertex. This tells you that the angle measures 90°, so it is a _____ angle.

(b)

(c)

(d)

Answers

4. **(a)** right **(b)** straight **(c)** obtuse
   (d) acute

5 Which pair of lines is perpendicular? (*Hint:* Look for a right angle.)

How can you describe the other pair of lines?

(a)

(b)

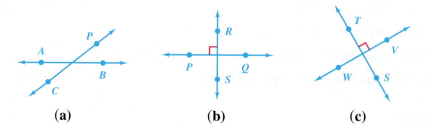

OBJECTIVE **5** **Identify perpendicular lines.** Two lines are called **perpendicular lines** if they intersect to form a right angle.

The small red square indicates a 90° (right) angle.

$\overleftrightarrow{CB}$ and $\overleftrightarrow{ST}$ are **perpendicular lines** because they intersect at right angles, as indicated by the small red square in the figure.

Perpendicular lines can be written in the following way: $\overleftrightarrow{CB} \perp \overleftrightarrow{ST}$.

EXAMPLE 5 **Identifying Perpendicular Lines**

Which pairs of lines are perpendicular?

(a)   **(b)**   **(c)**

The lines in Figures **(b)** and **(c)** are *perpendicular* to each other because they intersect at right angles.

The lines in Figure **(a)** are *intersecting lines,* but they are *not* perpendicular because they do *not* form a right angle.

◀ **Work Problem** **5** **at the Side.**

8.1 Exercises

FOR EXTRA HELP

Download the MyDashBoard App

MyMathLab®

1. CONCEPT CHECK Explain the difference between a line, a line segment, and a ray. Draw a picture of each one.

2. CONCEPT CHECK Explain the difference between acute, obtuse, straight, and right angles. Draw a picture of each type of angle.

Identify each figure as a line, line segment, or ray and name it using the appropriate symbol. ***See Example 1.***

3.

4.

5.

6.

7.

8.

Label each pair of lines as appearing to be parallel, as perpendicular, or as intersecting. ***See Examples 2 and 5.***

9.

10.

11.

12.

13.

14.

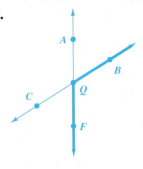

Name each highlighted angle by using the three-letter form of identification. ***See Example 3.***

15.

Which letter names the vertex? _____ The vertex letter must be in the middle when naming an angle, so this is ∠ _____ or ∠ _____.

16.

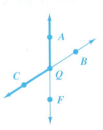

Which letter names the vertex? _____ The vertex letter must be in the middle when naming an angle, so this is ∠ _____ or ∠ _____.

17.

18.

*Label each angle as acute, right, obtuse, or straight. For right and straight angles, indicate the number of degrees in the angle. **See Example 4.***

19.

Hint: Notice the small red square at the vertex.

20.

Because the angle measure is greater than a square corner of _____ and less than a straight line of _____, it is a(n) _____ angle.

21.

22.

23.

24.

25. CONCEPT CHECK Explain what is happening in each sentence.

(a) The road was so slippery that my car did a 360.

(b) After the election, the governor's view on taxes took a 180° turn.

26. CONCEPT CHECK Find at least four examples of right angles in your home, at work, or on the street. Make a sketch of each example and label the right angle.

Relating Concepts (Exercises 27–32) For Individual or Group Work

*Use the figure below to **work Exercises 27–32 in order.** Decide whether each statement is true or false. If it is true, explain why. If it is false, rewrite it to make it a true statement.*

27. $\angle UST$ is 90°.

28. $\overleftrightarrow{SQ}$ and $\overleftrightarrow{PQ}$ are perpendicular.

29. The measure of $\angle USQ$ is less than the measure of $\angle PQR$.

30. $\overleftrightarrow{ST}$ and $\overleftrightarrow{PR}$ are intersecting.

31. $\overleftrightarrow{QU}$ and $\overleftrightarrow{TS}$ are parallel.

32. $\angle UST$ and $\angle UQR$ measure the same number of degrees.

8.2 Angles and Their Relationships

OBJECTIVES

1. Identify complementary angles and supplementary angles and find the measure of a complement or supplement of a given angle.

2. Identify congruent angles and vertical angles and use this knowledge to find the measures of angles.

OBJECTIVE ▶ 1 Identify complementary angles and supplementary angles and find the measure of a complement or supplement of a given angle. Two angles are called **complementary angles** if the sum of their measures is 90°. If two angles are complementary, each angle is the *complement* of the other.

EXAMPLE 1 Identifying Complementary Angles

Identify each pair of complementary angles.

$\angle MPN$ (40°) and $\angle NPC$ (50°) are complementary angles because

$$40° + 50° = \mathbf{90°}$$

$\angle CAB$ (30°) and $\angle FHG$ (60°) are complementary angles because

$$30° + 60° = \mathbf{90°}$$

············· **Work Problem 1 at the Side. ▶**

EXAMPLE 2 Finding the Complement of Angles

Find the complement of each angle.

> Subtract from 90° to find the complement.

(a) 30°
Find the complement of 30° by subtracting. $\mathbf{90° - 30° = 60°}$ ← Complement

(b) 75°
Find the complement of 75° by subtracting. $\mathbf{90° - 75° = 15°}$ ← Complement

············· **Work Problem 2 at the Side. ▶**

Two angles are called **supplementary angles** if the sum of their measures is 180°. If two angles are supplementary, each angle is the *supplement* of the other.

EXAMPLE 3 Identifying Supplementary Angles

Identify each pair of supplementary angles.

$\angle BOA$ and $\angle BOC$, because $65° + 115° = \mathbf{180°}$
$\angle BOA$ and $\angle ERF$, because $65° + 115° = \mathbf{180°}$
$\angle BOC$ and $\angle MPN$, because $115° + 65° = \mathbf{180°}$
$\angle MPN$ and $\angle ERF$, because $65° + 115° = \mathbf{180°}$

············· **Work Problem 3 at the Side. ▶**

VOCABULARY TIP

Complementary angles Hint: The letter "**C**" in **c**omplementary can remind you of "**c**orner," like a 90° angle. **Complementary angles** add up to 90 degrees.

---

1 Identify each pair of complementary angles.

2 Find the complement of each angle.

GS (a) 35° $90° - 35° =$ _____

(b) 80°

3 Identify each pair of supplementary angles.

Answers

1. $\angle COD$ and $\angle DOE$; $\angle RST$ and $\angle XPY$
2. **(a)** 55° **(b)** 10°
3. $\angle CRF$ and $\angle BRF$; $\angle CRE$ and $\angle ERB$; $\angle BRF$ and $\angle BRE$; $\angle CRE$ and $\angle CRF$

VOCABULARY TIP

Supplementary angles *Hint:* The letter "**S**" in **s**upplementary can remind you of "**s**traight," like a 180° angle. **Supplementary angles** add up to 180 degrees.

4 Find the supplement of each angle.

GS **(a)** 175°

$$180° - 175° = \underline{\quad}$$

(b) 30°

5 Identify the angles that are congruent.

6 Identify the vertical angles. What is special about vertical angles?

Answers

4. **(a)** 5° **(b)** 150°
5. ∠BOC ≅ ∠AOD; ∠AOB ≅ ∠DOC
6. ∠SPB and ∠MPD; ∠BPD and ∠SPM; vertical angles are congruent (they measure the same number of degrees).

EXAMPLE 4   Finding the Supplement of Angles

Find the supplement of each angle.

> Subtract from 180° to find the supplement.

(a) 70°
Find the supplement of 70° by subtracting. **180° − 70° = 110°** ← Supplement

(b) 140°
Find the supplement of 140° by subtracting. **180° − 140° = 40°** ← Supplement

◀ **Work Problem 4 at the Side.**

OBJECTIVE 3 Identify congruent angles and vertical angles and use this knowledge to find the measures of angles. Two angles are called **congruent angles** if they measure the same number of degrees. If two angles are congruent, this is written as ∠A ≅ ∠B and read as, "angle A **is congruent to** angle B." Here is an example.

> The symbol ≅ means "is congruent to."

Congruent angles: ∠A ≅ ∠B

EXAMPLE 5   Identifying Congruent Angles

Identify the angles that are congruent.

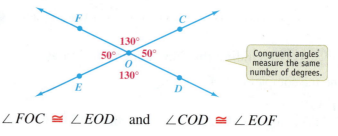

> Congruent angles measure the same number of degrees.

∠FOC ≅ ∠EOD   and   ∠COD ≅ ∠EOF

◀ **Work Problem 5 at the Side.**

Angles that share a common side and a common vertex are called *adjacent* angles, such as ∠FOC and ∠COD in **Example 5** above. Angles that do *not* share a common side are called *nonadjacent* angles. Two nonadjacent angles formed by two intersecting lines are called **vertical angles.**

EXAMPLE 6   Identifying Vertical Angles

Identify the vertical angles in this figure.

∠AOF and ∠COE are vertical angles because they do not share a common side and they are formed by two intersecting lines (⃡CF and ⃡EA).

∠COA and ∠EOF are also vertical angles.

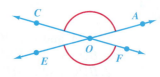

◀ **Work Problem 6 at the Side.**

Look back at **Example 5** on the previous page. Notice that the two *congruent* angles that measure 130° are also *vertical* angles. Also, the two congruent angles that measure 50° are vertical angles. This illustrates the following property.

> **Vertical Angles Are Congruent**
>
> If two angles are *vertical* angles, they are *congruent,* that is, they measure the same number of degrees.

EXAMPLE 7 Finding the Measures of Vertical Angles

In the figure below, find the measure of each unlabeled angle. Then write the measure of each angle on the figure.

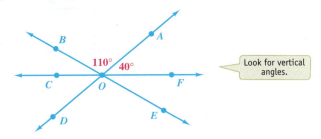

Look for vertical angles.

(a) ∠COD

∠COD and ∠AOF are vertical angles, so they are congruent. This means they measure the same number of degrees.

The measure of ∠AOF is 40° so the measure of ∠COD is 40° also.

(b) ∠DOE

∠DOE and ∠BOA are vertical angles, so they are congruent.

The measure of ∠BOA is 110° so the measure of ∠DOE is 110° also.

(c) ∠COB

Look at ∠COB, ∠BOA, and ∠AOF. Notice that $\overrightarrow{OC}$ and $\overrightarrow{OF}$ go in opposite directions. Therefore, ∠COF is a straight angle and measures 180°. To find the measure of ∠COB, subtract the sum of the other two angles from 180°.

$$180° - (110° + 40°) = 180° - (150°) = 30°$$

The measure of ∠COB is 30°.

(d) ∠EOF

∠EOF and ∠COB are vertical angles, so they are congruent. We know from part (c) above that the measure of ∠COB is 30° so the measure of ∠EOF is 30° also.

The figure with all the angle measures is shown below.

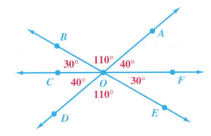

···· **Work Problem ❼ at the Side.** ▶

---

❼ In the figure below, find the number of degrees in each unlabeled angle. Write the angle measures on the figure.

GS **(a)** ∠TOS

∠TOS has a small square at the vertex, so it is a right angle and measures _____.

GS **(b)** ∠QOR

∠QOR and ∠TOS are vertical angles, so they are congruent. ∠QOR also measures _____.

(c) ∠POQ

(d) ∠VOR

(e) ∠POS

Answers

7. **(a)** 90° **(b)** 90° **(c)** 52°
(d) 38° **(e)** 38°

8.2 Exercises

   MyMathLab®

1. **CONCEPT CHECK** Can you find the supplement of a 5° angle? Explain why or why not. If it is possible to find a supplement, show how to do it.

2. **CONCEPT CHECK** Can you find the complement of a 135° angle? Explain why or why not. If it is possible to find a complement, show how to do it.

Identify each pair of complementary angles. See Example 1.

3.

4.

Identify each pair of supplementary angles. See Example 3.

5.

6.

Find the complement of each angle. See Example 2.

7. 40°
GS 90° − 40° = _____

8. 35°
GS 90° − 35° = _____

9. 86°

10. 59°

Find the supplement of each angle. See Example 4.

11. 130°
GS 180° − 130° = _____

12. 75°
GS 180° − 75° = _____

13. 90°

14. 5°

15. **CONCEPT CHECK** Can two obtuse angles be supplementary? Explain why or why not.

16. **CONCEPT CHECK** Can two acute angles be complementary? Explain why or why not.

*In each figure, identify the angles that are congruent. **See Examples 5 and 6.***

17.

18.

*Use your knowledge of vertical angles to answer Exercises 19 and 20. **See Example 7.***

19. In the figure below, ∠AOH measures 37° and ∠COE measures 63°. Find the measure of each of the other angles.

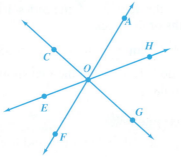

20. In the figure below, ∠POU measures 105° and ∠UOT measures 40°. Find the measure of each of the other angles.

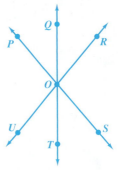

In each figure, $\overrightarrow{BA}$ is parallel to $\overrightarrow{CD}$. Identify two pairs of congruent angles and the number of degrees in each congruent angle.

21.

22.

8.3 Rectangles and Squares

OBJECTIVES

1 Find the perimeter and area of a rectangle.

2 Find the perimeter and area of a square.

3 Find the perimeter and area of a composite figure.

A **rectangle** is a figure with four sides that meet to form 90° angles. Each set of opposite sides is *parallel* and *congruent* (has the same length).

Each longer side of a rectangle is called the length (*l*) and each shorter side is called the width (*w*).

◀ **Work Problem ① at the Side.**

OBJECTIVE ▶ 1 **Find the perimeter and area of a rectangle.** The distance around the outside edges of a flat figure is the **perimeter** of the figure. Think of how much fence you would need to put around the sides of a garden plot, or how far you would walk if you walked around the outside edges of your living room. In either case you would add up the lengths of the sides. Look at the rectangle above that has the lengths of the sides labeled. To find its perimeter, you add the lengths of the sides.

Perimeter = **12 cm** + **12 cm** + **7 cm** + **7 cm** = 38 cm

Because the two long sides are both 12 cm, and the two short sides are both 7 cm, you can also use this formula.

> **Finding the Perimeter of a Rectangle**
>
> Perimeter of a rectangle = length + length + width + width
>
> $$P = (2 \cdot \text{length}) + (2 \cdot \text{width})$$
>
> $$P = 2 \cdot l + 2 \cdot w$$

> **EXAMPLE 1** Finding the Perimeter of a Rectangle
>
> Find the perimeter of each rectangle.
>
> **(a)**

The length of this rectangle is **27 m** and the width is **11 m.**

Use the formula $P = 2 \cdot l + 2 \cdot w$

$P = 2 \cdot$ *l* $+ 2 \cdot$ *w*   Replace *l* with 27 m and *w* with 11 m.

$P = 2 \cdot \textbf{27 m} + 2 \cdot \textbf{11 m}$ Do the multiplications first.

$P = 54 \text{ m} + 22 \text{ m}$ Add last.

$P = 76 \text{ m}$

The perimeter of the rectangle (the distance you would walk around the outside edges of the rectangle) is 76 m.

⋯ **Continued on Next Page**

---

1 Identify all the rectangles.

(a)

(b)

(c)

(d)

(e)

(f)

(g)

Answers

1. **(a)**, **(b)**, and **(e)** are rectangles; **(c)**, **(d)**, **(f)**, and **(g)** are not.

As a check, you can add up the lengths of the four sides.

$$P = \textbf{27 m} + \textbf{27 m} + \textbf{11 m} + \textbf{11 m}$$

$$P = 76\ \text{m} \leftarrow \text{Same result as using the formula}$$

(b) A rectangle 8.9 ft by 12.3 ft
You can use the formula, as shown below.

$$P = 2 \cdot \quad l \quad + 2 \cdot \quad w$$

$$P = 2 \cdot \textbf{12.3 ft} + 2 \cdot \textbf{8.9 ft}$$

$$P = \quad 24.6\ \text{ft} \quad + \quad 17.8\ \text{ft}$$

$$P = 42.4\ \text{ft}$$

> Be sure to write **ft** in the answer.

Or, you can add up the lengths of the four sides.

$$P = \textbf{12.3 ft} + \textbf{12.3 ft} + \textbf{8.9 ft} + \textbf{8.9 ft}$$

$$P = 42.4\ \text{ft} \leftarrow \text{Same result as using the formula}$$

Either method will give you the correct result.

········· **Work Problem 2 at the Side.** ▶

The *perimeter* of a rectangle is the distance around the *outside edges*. The **area** of a rectangle is the amount of surface *inside* the rectangle. We measure area by seeing how many squares of a certain size are needed to cover the surface inside the rectangle. Think of covering the floor of a rectangular living room with carpet. Carpet is measured in square yards, that is, square pieces that measure 1 yard along each side. Here is a drawing of a living room floor.

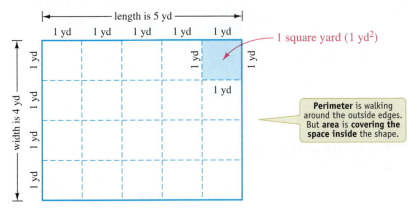

> **Perimeter** is walking around the outside edges. But **area** is **covering the space inside** the shape.

You can see from the drawing that it takes 20 squares to cover the floor. We say that the area of the floor is 20 *square yards.* A shorter way to write square yards is yd².

20 **square yards** can be written as 20 **yd²**.

To find the number of squares, you can count them, or you can multiply the number of squares in the length (5) times the number of squares in the width (4) to get 20. The formula is given below.

Finding the Area of a Rectangle

Area of a rectangle = length • width

$$A = l \cdot w$$

Remember to use **square units** when measuring area.

---

2 Find the perimeter of each rectangle by using the formula or by adding the lengths of the sides.

GS (a)

Try adding the lengths of the sides:

$$P = 17\ \text{cm} + 17\ \text{cm} + \underline{\hspace{1.5cm}}$$

$$+ \underline{\hspace{1.5cm}}$$

$$P = \underline{\hspace{1.5cm}}$$

GS (b)

Try using the formula:

$$P = 2 \cdot \underline{\hspace{1cm}} + 2 \cdot \underline{\hspace{1cm}}$$

$$P = \underline{\hspace{1.5cm}} + \underline{\hspace{1.5cm}}$$

$$P = \underline{\hspace{1.5cm}}$$

(c) 6 m wide and 11 m long

(d) 0.9 km by 2.8 km

Answers

2. **(a)** $P = 17\ \text{cm} + 17\ \text{cm} + 10\ \text{cm} + 10\ \text{cm}$;
   $P = 54\ \text{cm}$
   (b) $P = 2 \cdot 10.5\ \text{ft} + 2 \cdot 7\ \text{ft}$
   $P = 21\ \text{ft} + 14\ \text{ft}$; $P = 35\ \text{ft}$
   (c) $P = 34\ \text{m}$ **(d)** $P = 7.4\ \text{km}$

Squares of many sizes can be used to measure area. For smaller areas, you might use the ones shown below.

Area is measured in **square units.**

1 square inch (1 in.²)   **Actual-size drawings**

Other sizes of squares that are often used to measure area are listed here, but they are too large to draw on this page.

1 square meter (1 m²)	1 square foot (1 ft²)
1 square kilometer (1 km²)	1 square yard (1 yd²)
	1 square mile (1 mi²)

CAUTION

The raised 2 in 4² means that you multiply 4 • 4 to get 16. The raised 2 in cm² or yd² is a short way to write the word *square*. When you see 5 cm², say "five square centimeters." Do *not* multiply 5 • 5. The exponent applies to cm, *not* to the number.

EXAMPLE 2   **Finding the Area of a Rectangle**

Find the area of each rectangle.

(a)

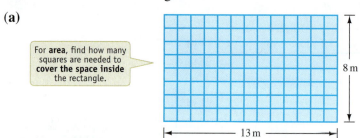

For **area**, find how many squares are needed to **cover the space inside** the rectangle.

8 m

13 m

    The length of this rectangle is 13 m and the width is 8 m. Use the formula $A = l • w$.

$$A = \quad l \quad • \quad w \qquad \text{Replace } l \text{ with 13 m and } w \text{ with 8 m.}$$

$$A = \mathbf{13\ m} • \mathbf{8\ m} \qquad \text{Multiply.}$$

$$A = 104 \text{ square meters}$$

Write **m²** in the answer.

"Square meters" can be written as m², so the area is 104 m².

(b) A rectangle measuring 7 cm by 21 cm

    First make a sketch of the rectangle. The length is 21 cm (the longer measurement) and the width is 7 cm. Then use the formula for the area of a rectangle, $A = l • w$.

7 cm

21 cm

$$A = 21 \text{ cm} • 7 \text{ cm}$$

$$A = 147 \ \mathbf{cm^2}$$

Square units for area; write **cm²** in the answer.

The area of the rectangle is 147 cm².

Continued on Next Page

CAUTION

The units for **area** will always be **square units** (cm^2, m^2, yd^2, mi^2, and so on). The units for **perimeter** will always be **linear units** (cm, m, yd, mi, and so on) *not* square units.

..... **Work Problem ❸ at the Side.** ▶

OBJECTIVE ❷ **Find the perimeter and area of a square.** A **square** is a rectangle with all sides the same length. Two squares are shown below. Notice the 90° angles.

90° angles 20 in. 20 in. 20 in. 20 in.

In a square, if one right angle is shown, the other three are also right angles.

9 m 9 m 9 m 9 m

To find the *perimeter* of (distance around) the square on the right, you could add 9 m + 9 m + 9 m + 9 m to get 36 m. A shorter way is to multiply the length of one side times 4, because all four sides are the same length.

Finding the Perimeter of a Square

Perimeter of a square = side + side + side + side

or, $P = 4 \cdot side$

$P = 4 \cdot s$

As with a rectangle, you can multiply length times width to find the *area* of (surface inside) a square. Because the length and the width are the same in a square, the formula is written as shown below.

Finding the Area of a Square

Area of a square = side • side

$A = s \cdot s$

$A = s^2$

Remember to use *square units* when measuring area.

EXAMPLE 3 **Finding the Perimeter and Area of a Square**

(a) Find the perimeter of the square shown above where each side measures 9 m.

Use the formula.   Or add up the four sides.

$P = 4 \cdot s$ $P = 9\,m + 9\,m + 9\,m + 9\,m$

$P = 4 \cdot 9\,m$ $P = 36\,m$ ← This is **perimeter**, so write **m** in the answer.

$P = 36\,m$

↑_____ Same answer _____↑

..... **Continued on Next Page** .....

Area The **area** is the amount of surface **inside** a flat shape. Write the area using **square units,** such as square inches or square feet—for example, 10 **in.**^{2} or 12 **ft**^{2}.

❸ Find the area of each rectangle.

GS **(a)**

9 ft 4 ft 4 ft 9 ft

$A = l \cdot w$

$A = \mathbf{9\ ft} \cdot \mathbf{4\ ft}$

$A = $ _____

Write **ft**^{2} as part of your answer.

GS **(b)** A rectangle is 6 m long and 0.5 m wide. (First make a sketch of the rectangle and label the lengths of the sides.)

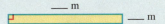

___ m   ___ m

(c) A rectangular patio measures 3.5 yd by 2.5 yd. (First make a sketch of the patio and label the lengths of the sides.)

Answers

3. **(a)** $A = 36\ ft^2$
   (b) $A = 3\ m^2$

6 m   0.5 m

   (c) $A = 8.75\ yd^2$

2.5 yd

3.5 yd

4 Find the perimeter and area of each square.

(a)

3 ft

3 ft

For perimeter of a square,

$P = 4 \cdot s$

$P = 4 \cdot$ _____

$P =$ _____

For area of a square,

$A = s^2$ or $A = s \cdot s$

$A =$ _____ $\cdot$ _____

$A =$ _____

(b) 10.5 cm on each side. (Make a sketch of the square.)

(c) 2.1 mi on a side. (Make a sketch of the square.)

Answers

4. **(a)** $P = 4 \cdot 3$ ft; $P = 12$ ft;
$A = 3$ ft $\cdot$ 3 ft; $A = 9$ ft^2
(b) $P = 42$ cm;
$A = 110.25$ cm^2

10.5 cm

10.5 cm

(c) $P = 8.4$ mi;
$A = 4.41$ mi^2

2.1 mi

2.1 mi

(b) Find the area of the same square where each side measures 9 m.

$$A = s^2$$
$$A = s \cdot s$$
$$A = \mathbf{9\ m} \cdot \mathbf{9\ m}$$
$$A = \mathbf{81\ m^2}$$

*This is **area**, so write m^2 in the answer.*

CAUTION

Be careful! s^2 means $s \cdot s$. It does **not** mean $s \cdot 2$. In **Example 3(b)** above, s is 9 m, so s^2 is 9 m $\cdot$ 9 m $= 81$ m^2. It is **not** 9m $\cdot$ 2 $= 18$ m.

◀ **Work Problem** **4** **at the Side.**

OBJECTIVE **3** **Find the perimeter and area of a composite figure.** As with any other shape, you can find the perimeter of (distance around) an irregular shape by adding up the lengths of the sides. To find the area (surface inside the shape), try to break it up into pieces that are squares or rectangles. Find the area of each piece and then add them together.

CAUTION

Perimeter is the **distance around the outside edges** of a flat shape. It is always measured in *linear units* such as cm, m, yd, and so on.

Area is the amount of **surface inside** a flat shape. It is always measured in *square units* such as cm^2, m^2, yd^2, and so on.

EXAMPLE 4 **Finding the Perimeter and Area of a Composite Figure**

The floor of a room has the shape shown below.

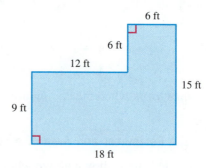

(a) Suppose you want to put a new wallpaper border along the top of all the walls. How much material do you need?

Find the **perimeter** of the room by adding up the lengths of the sides.

$$P = 9 \text{ ft} + 12 \text{ ft} + 6 \text{ ft} + 6 \text{ ft} + 15 \text{ ft} + 18 \text{ ft}$$
$$P = 66 \text{ ft}$$

You need 66 ft of wallpaper border.

············· **Continued on Next Page**

(b) The carpet you like costs $20.50 per square yard. How much will it cost to carpet the room?

First change the measurements from feet to yards, because the carpet is sold in square yards. There are 3 ft in 1 yd, so multiply by the unit fraction that allows you to divide out feet. Let's start with 9 ft.

$$\frac{\overset{3}{\cancel{9}} \, \cancel{\text{ft}}}{1} \cdot \frac{1 \text{ yd}}{\underset{1}{\cancel{3}} \, \cancel{\text{ft}}} = 3 \text{ yd}$$

— Divide out ft.
— Divide 9 and 3 by 3.

Use the same unit fraction to change the other measurements to yards.

$$\frac{\overset{4}{\cancel{12}} \, \cancel{\text{ft}}}{1} \cdot \frac{1 \text{ yd}}{\underset{1}{\cancel{3}} \, \cancel{\text{ft}}} = 4 \text{ yd} \qquad \frac{\overset{2}{\cancel{6}} \, \cancel{\text{ft}}}{1} \cdot \frac{1 \text{ yd}}{\underset{1}{\cancel{3}} \, \cancel{\text{ft}}} = 2 \text{ yd}$$

$$\frac{\overset{5}{\cancel{15}} \, \cancel{\text{ft}}}{1} \cdot \frac{1 \text{ yd}}{\underset{1}{\cancel{3}} \, \cancel{\text{ft}}} = 5 \text{ yd} \qquad \frac{\overset{6}{\cancel{18}} \, \cancel{\text{ft}}}{1} \cdot \frac{1 \text{ yd}}{\underset{1}{\cancel{3}} \, \cancel{\text{ft}}} = 6 \text{ yd}$$

Next, break up the room into two pieces. Use just the measurements for the length and width of each piece.

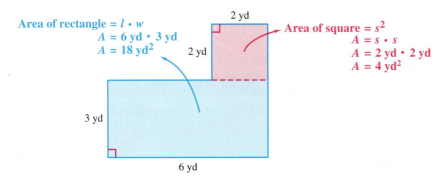

Area of rectangle $= l \cdot w$
$A = 6 \text{ yd} \cdot 3 \text{ yd}$
$A = 18 \text{ yd}^2$

Area of square $= s^2$
$A = s \cdot s$
$A = 2 \text{ yd} \cdot 2 \text{ yd}$
$A = 4 \text{ yd}^2$

Total area $= \mathbf{18 \text{ yd}^2 + 4 \text{ yd}^2} = 22 \text{ yd}^2$

This is **area**, so it is measured in **yd²**.

Multiply to find the cost of the carpet.

$$\text{Cost} = \frac{22 \, \cancel{\text{yd}^2}}{1} \cdot \frac{\$20.50}{1 \, \cancel{\text{yd}^2}} = \$451.00$$

It will cost $451.00 to carpet the room.

You could have cut the room into two rectangles as shown below. The total area is the same.

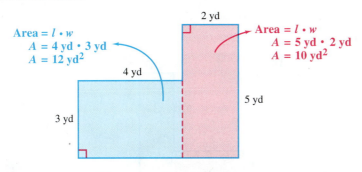

Area $= l \cdot w$
$A = 4 \text{ yd} \cdot 3 \text{ yd}$
$A = 12 \text{ yd}^2$

Area $= l \cdot w$
$A = 5 \text{ yd} \cdot 2 \text{ yd}$
$A = 10 \text{ yd}^2$

Total area $= \mathbf{12 \text{ yd}^2 + 10 \text{ yd}^2} = 22 \text{ yd}^2$ ← Same answer as above

Work Problem ❺ at the Side. ▶

---

❺ Carpet costs $19.95 per square yard. Find the cost of carpeting each room. Round your answers to the nearest cent if necessary.

GS (a)

6.5 yd

5 yd

$A = l \cdot w$
$A = 6.5 \text{ yd} \cdot 5 \text{ yd}$
$A = \underline{\quad} \text{ yd}^2$

$$\text{Cost} = \frac{\underline{\quad} \text{ yd}^2}{1} \cdot \frac{\$19.95}{1 \text{ yd}^2}$$

$$\text{Cost} = \underline{\qquad}$$

(b)

21 ft

21 ft

12 ft

12 ft

9 ft

9 ft

(c) A rectangular classroom is 24 ft long and 18 ft wide. (Make a sketch of the classroom.)

Answers

5. (a) $A = 32.5 \text{ yd}^2$; Cost $= \$648.38$ (rounded)
(b) 37 yd^2; Cost $= \$738.15$
(c)

18 ft

24 ft

48 yd^2;
Cost $= \$957.60$

8.3 Exercises

FOR EXTRA HELP

  MyMathLab®

1. **CONCEPT CHECK** Look at the figure in **Exercise 3** below. Explain the difference between finding the **perimeter** of the figure and finding the **area** of the figure. Will the units in the answer for the perimeter be **yd** or **yd²**? Will the units in the answer for the area be **yd** or **yd²**?

2. **CONCEPT CHECK** Explain what is special about a rectangle. Explain what is special about a square. Make a drawing of each figure and label all the sides with measurements that you choose.

Find the perimeter and area of each rectangle or square. See Examples 1–3.

3.

8 yd
6 yd   6 yd
8 yd

$P = 2 \cdot l + 2 \cdot w$

$P = 2 \cdot 8 \text{ yd} + 2 \cdot 6 \text{ yd}$

$P = \underline{\hspace{1cm}} + \underline{\hspace{1cm}}$

$P = \underline{\hspace{1cm}}$

$A = l \cdot w$

$A = 8 \text{ yd} \cdot 6 \text{ yd} = \underline{\hspace{1cm}}$

4.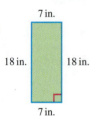
7 in.
18 in.   18 in.
7 in.

$P = 2 \cdot l + 2 \cdot w$

$P = 2 \cdot 18 \text{ in.} + 2 \cdot 7 \text{ in.}$

$P = \underline{\hspace{1cm}} + \underline{\hspace{1cm}}$

$P = \underline{\hspace{1cm}}$

$A = l \cdot w$

$A = 18 \text{ in.} \cdot 7 \text{ in.} = \underline{\hspace{1cm}}$

5.
0.9 km 0.9 km
0.9 km 0.9 km

6.
7.5 m
7.5 m

Draw a sketch of each square or rectangle and label the lengths of the sides. Then find the perimeter and the area. (Sketches may vary; show your sketches to your instructor.)

7. 10 ft by 10 ft

8. 8 cm by 17 cm

9. A storage building that is 76.1 ft by 22 ft

10. A science lab measuring 12 m by 12 m

11. A square nature preserve 3 mi wide

12. A square of cardboard 20.3 cm on a side

Find the perimeter and area of each figure. See Example 4.

13.
7 m
3 m
5 m
12 m
9 m
2 m

14.
4 ft
9 ft
12 ft
8 ft
3 ft
12 ft

15.

17 m
12 m
4 m
28 m 4 m
4 m
12 m
17 m

16.

3.5 cm
3 cm
1.5 cm
8 cm
5 cm
5 cm

First find the length of the unlabeled side in each figure. Then find the perimeter and area of each figure.

17.

6 in.
6 in. 6 in.
6 in. 6 in.
9 in.
15 in.
?
6 in.
6 in.

18.

12 ft
18 ft 20 ft
16 ft 10 ft
40 ft
32 ft
?

Solve each application problem. In Exercises 19–24, draw a sketch for each problem and label it with the appropriate measurements. (Sketches may vary; show your sketches to your instructor.)

19. Gymnastic floor exercises are performed on a square mat that is 12 meters on a side. Find the perimeter and area of a mat. (*Source:* nist.gov)

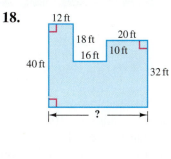

20. A regulation volleyball court is 18 meters by 9 meters. Find the perimeter and area of a regulation court. (*Source:* nist.gov)

21. SooJin's laptop computer has a rectangular screen that is 12.3 in. wide and 7.8 in. high. The screen on her notebook computer measures 10.3 in. by 7.5 in. Find the area of each screen to the nearest whole number. Then find the difference in the areas.

22. A page in this book measures 27.5 cm from top to bottom and 21 cm from side to side. Find the perimeter and the area of the page.

23. Tyra's kitchen is 4.4 m wide and 5.1 m long. She is pasting a decorative border strip that costs $4.99 per meter around the top edge of all the walls. How much will she spend?

24. Mr. and Mrs. Gomez are buying carpet for their square-shaped bedroom that is 5 yd wide. The carpet is $23 per square yard and padding and installation is another $6 per square yard. How much will they spend in all?

25. Regulation soccer fields for teens and adults can measure from 50 to 80 yards wide and 100 to 120 yards long, depending on the age and skill level of the players. Find the area of the smallest soccer field and the area of the largest soccer field. What is the difference in the playing room between the smallest and largest fields?

26. The table below shows information on two tents for camping.

Tents	Coleman Family Dome	Eddie Bauer Dome Tent
Dimensions	13 ft × 13 ft	12 ft × 12 ft
Sleeps	8 campers	6 campers
Sale price	$127	$99

Source: Target.com

(a) For the Coleman tent, find the perimeter, area, and number of square feet of floor space for each camper. Round to the nearest whole number.

(b) Find the same information for the Eddie Bauer tent.

27. A regulation football field is rectangular, 100 yd long (excluding end zones), and has an area of 5300 yd². Find the width of the field. (*Source:* National Football League.)

28. There are 15,130 ft² of ice on the surface of the rectangular playing area for a major league hockey game (excluding the area behind the goal lines). If the playing area is 85 ft wide, how long is it? (*Source:* National Hockey League.)

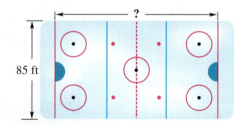

29. A rectangular lot is 124 ft by 172 ft. County rules require that nothing be built on land within 12 ft of any edge of the lot. First, add labels to the sketch of the lot, showing the land that cannot be built on. Then find the area of the land that cannot be built on.

30. Find the cost of fencing needed for this rectangular field. Fencing along the country roads costs $4.25 per foot. Fencing for the other two sides costs $2.75 per foot.

Relating Concepts (Exercises 31–36) For Individual or Group Work

Use your knowledge of perimeter and area to **work Exercises 31–36 in order.**

31. Suppose you have 12 ft of fencing to make a square or rectangular garden plot. Draw sketches of *all* the possible plots that use exactly 12 ft of fencing and label the lengths of the sides. Use only *whole number* lengths. (*Hint:* There are three possibilities.)

32. (a) Find the area of each plot in **Exercise 31.**

    (b) Which plot has the greatest area?

33. Repeat **Exercise 31** using 16 ft of fencing. Be sure to draw *all* possible plots that have whole number lengths for the sides.

34. (a) Find the area of each plot in **Exercise 33.**

    (b) Compare your results to those from **Exercise 32.** What do you notice about the plots with the greatest area?

35. (a) Draw a sketch of a rectangular plot 3 ft by 2 ft. Find the perimeter and area.

    (b) Suppose you *double* the length of the plot and *double* the width. Draw a sketch of the enlarged plot and find the perimeter and area.

    (c) The *perimeter* of the enlarged plot is how many times greater than the perimeter of the original plot? The *area* of the enlarged plot is how many times greater than the original area?

36. (a) Refer to **Exercise 35(a).** Suppose you *triple* the length and width of the original plot. Draw a sketch of the enlarged plot and find the perimeter and area.

    (b) How many times greater is the *perimeter* of the enlarged plot? How many times greater is the *area* of the enlarged plot?

    (c) Suppose you make the length and width *four times greater* in the enlarged plot. What would you predict will happen to the perimeter and area, compared to the original plot?

8.4 Parallelograms and Trapezoids

OBJECTIVES

1. Find the perimeter and area of a parallelogram.

2. Find the perimeter and area of a trapezoid.

1 Find the perimeter of each parallelogram.

(a)

$$P = 27 \text{ yd} + 27 \text{ yd} + 15 \text{ yd}$$
$$+ \underline{\quad} \text{ yd}$$

$$P = \underline{\qquad}$$

(b)

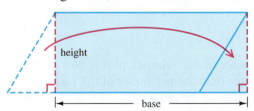

6.91 km

10.3 km

A **parallelogram** is a four-sided figure with opposite sides parallel, such as the ones shown below. Notice that opposite sides have the same length.

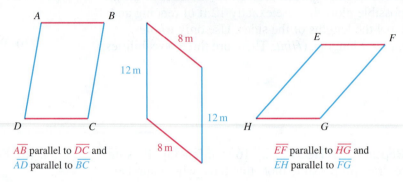

$\overline{AB}$ parallel to $\overline{DC}$ and $\overline{AD}$ parallel to $\overline{BC}$

$\overline{EF}$ parallel to $\overline{HG}$ and $\overline{EH}$ parallel to $\overline{FG}$

OBJECTIVE **1** **Find the perimeter and area of a parallelogram.** Perimeter is the distance around a flat shape, so the easiest way to find the perimeter of a parallelogram is to add the lengths of the four sides.

EXAMPLE 1   Finding the Perimeter of a Parallelogram

Find the perimeter of the middle parallelogram above.

$$P = 12 \text{ m} + 12 \text{ m} + 8 \text{ m} + 8 \text{ m} = 40 \text{ m}$$

This is **perimeter,** so write **m** in the answer, **not** m².

◄ **Work Problem** **1** **at the Side.**

To find the area of a parallelogram, first draw a dashed line inside the figure as shown here.

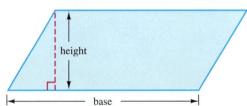

Try this yourself by tracing this parallelogram onto a piece of paper.

The length of the dashed line is the *height* of the parallelogram. It forms a *right angle* with the base. The height is the shortest distance between the base and the opposite side.

Now cut off the triangle created on the left side of the parallelogram above and move it to the right side, as shown below.

The parallelogram has been made into a rectangle. You can see that the area of the parallelogram and the rectangle are the same.

Equal areas
- Area of the   rectangle   = length • width
- Area of the parallelogram   =   base • height

Answers

1. (a) 15 yd; $P = 84$ yd
   (b) $P = 34.42$ km

Finding the Area of a Parallelogram

Area of a parallelogram = base • height

$$A = b \cdot h$$

Remember to use *square units* when measuring area.

EXAMPLE 2 Finding the Area of Parallelograms

Find the area of each parallelogram.

(a)

The base is 24 cm and the height is 19 cm. Use the formula $A = b \cdot h$.

$A = \quad b \quad \cdot \quad h$

$A = 24 \text{ cm} \cdot 19 \text{ cm}$

$A = 456 \text{ cm}^2$ ◁ This is **area**; write cm² in the answer.

(b)

$A = 47 \text{ m} \cdot 24 \text{ m}$

$A = 1128 \text{ m}^2$ ◁ Write **m²** in the answer.

Notice that you do *not* use the 30 m sides when finding the area. But you would use them when finding the *perimeter* of the parallelogram.

· **Work Problem 2 at the Side.** ▶

OBJECTIVE ▶ 2 Find the perimeter and area of a trapezoid. A **trapezoid** is a four-sided figure with exactly one pair of parallel sides, such as the figures shown below. Unlike parallelograms, opposite sides of a trapezoid might *not* have the same length.

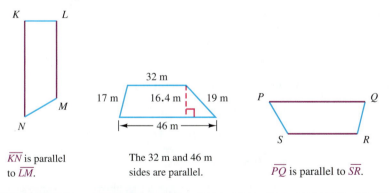

$\overline{KN}$ is parallel to $\overline{LM}$.

The 32 m and 46 m sides are parallel.

$\overline{PQ}$ is parallel to $\overline{SR}$.

EXAMPLE 3 Finding the Perimeter of a Trapezoid

Find the perimeter of the middle trapezoid above.

You can find the perimeter of any flat shape by adding the lengths of the sides.

$P = 17 \text{ m} + 32 \text{ m} + 19 \text{ m} + 46 \text{ m}$ ◁ The height of 16.4 m is **not** one of the sides in the perimeter.

$P = 114 \text{ m}$

Notice that the height (16.4 m) is *not* part of the perimeter, because the height is *not* one of the *outside edges* of the shape.

· **Work Problem 3 at the Side.** ▶

---

2 Find the area of each parallelogram.

GS **(a)**

$A = b \cdot h$

$A = 50 \text{ ft} \cdot 42 \text{ ft}$

$A = \underline{\qquad}$

(b)

---

3 Find the perimeter of each trapezoid.

(a)

(b)

Answers

2. (a) $A = 2100 \text{ ft}^2$ (b) $A = 8.74 \text{ cm}^2$
3. (a) $P = 28.6 \text{ in.}$ (b) $P = 5.83 \text{ km}$

4 Find the area of each trapezoid.

(a)

$A = \dfrac{1}{2} \cdot \quad h \quad \cdot (b \quad + \quad B)$

$A = \dfrac{1}{2} \cdot 30 \text{ ft} \cdot (40 \text{ ft} + 60 \text{ ft})$

$A = \dfrac{1}{\underset{1}{2}} \cdot \overset{15}{30} \text{ ft} \cdot (100 \text{ ft})$

$A = \underline{\hspace{2cm}}$

(b)

(c) A trapezoid with height 4.7 m, short base 9 m, and long base 10.5 m (First draw a sketch and label the bases and height.)

Answers

4. (a) $A = 1500 \text{ ft}^2$ (b) $A = 198 \text{ cm}^2$
   (c) $A = 45.825 \text{ m}^2$

Use this formula to find the *area* of a trapezoid.

Finding the Area of a Trapezoid

$$\text{Area} = \frac{1}{2} \cdot \text{height} \cdot (\text{short base} + \text{long base})$$

$$A = \frac{1}{2} \cdot h \cdot (b + B)$$

or $\quad A = 0.5 \cdot h \cdot (b + B)$

Remember to use *square units* when measuring area.

EXAMPLE 4 Finding the Area of a Trapezoid

Find the area of this trapezoid. The short base and long base are the *parallel* sides.

Look for the **parallel sides** when finding the **area** of a trapezoid.

The height (**h**) is **7 ft**, the short base (**b**) is **10 ft**, and the long base (**B**) is **16 ft**. You do *not* need the lengths of the other two sides to find the area.

$A = \dfrac{1}{2} \cdot h \cdot (b + B)$

$A = \dfrac{1}{2} \cdot 7 \text{ ft} \cdot (10 \text{ ft} + 16 \text{ ft})$ Work inside parentheses first.

$A = \dfrac{1}{\underset{1}{2}} \cdot 7 \text{ ft} \cdot (\overset{13}{26} \text{ ft})$

This is **area**, so write **ft²** in the answer.

$A = 91 \text{ ft}^2$

You can also use 0.5, the decimal equivalent for $\frac{1}{2}$, in the formula.

$A = 0.5 \cdot h \cdot (b + B)$

$A = 0.5 \cdot 7 \cdot (10 + 16)$

$A = 0.5 \cdot 7 \cdot \quad 26$

$A = 91 \text{ ft}^2$ Same answer as above

Calculator Tip

Use the parentheses keys on your scientific calculator to work **Example 4** above.

0.5 ⊗ 7 ⊗ (10 ⊕ 16) ⊜ 91

What happens if you do *not* use the parentheses keys? What order of operations will the calculator follow then? (Answer: The calculator will multiply 0.5 times 7 times 10, and then add 16, giving an *incorrect* answer of 51.)

◄ **Work Problem ④ at the Side.**

| EXAMPLE 5 | Finding the Area of a Composite Figure |

Find the area of this figure.

Break the figure into two pieces, a parallelogram (top part) and a trapezoid (bottom part). Find the area of each piece, and then add the areas.

Area of parallelogram

$A = b \cdot h$

$A = 50 \text{ cm} \cdot 20 \text{ cm}$

$A = \textbf{1000 m}^2$

Area of trapezoid

$A = \dfrac{1}{2} \cdot h \cdot (b + B)$

$A = 0.5 \cdot 15 \text{ m} \cdot (50 \text{ m} + 58 \text{ m})$

$A = \textbf{810 m}^2$

Total area = **1000 m² + 810 m² = 1810 m²**   `This is area, so write m² in the answer.`

The area of the figure is 1810 m².

·················· Work Problem ❺ at the Side. ▶

| EXAMPLE 6 | Applying Knowledge of Area |

Suppose the figure in **Example 5** above represents the floor plan of a hotel lobby. What is the cost of labor to install tile on the floor if the labor charge is $35.11 per square meter?

From **Example 5,** the floor area is 1810 m². To find the labor cost, multiply the number of square meters times the cost of labor per square meter.

$$\text{Cost} = \frac{1810 \ \cancel{m^2}}{1} \cdot \frac{\$35.11}{1 \ \cancel{m^2}}$$

$$\text{Cost} = \$63,549.10$$

The cost of the labor is $63,549.10.

·················· Work Problem ❻ at the Side. ▶

❺ Find the area of each floor.

(a)

Hint: The left side of the floor is a parallelogram; the right side is a trapezoid.

(b)

❻ Find the cost of carpeting the floors in **Margin Problem 5** above. The cost of carpet is as follows:

(a) Floor (a) is $18.50 per square meter.

(b) Floor (b) is $28 per square yard.

Answers

5. **(a)** $A = 40 \text{ m}^2 + 44 \text{ m}^2 = 84 \text{ m}^2$
   (b) $A = 25 \text{ yd}^2 + 37.5 \text{ yd}^2 = 62.5 \text{ yd}^2$
6. **(a)** $1554   **(b)** $1750

8.4 Exercises

1. **CONCEPT CHECK** Explain what is special about a parallelogram. Draw a parallelogram and label the lengths of all the sides.

2. **CONCEPT CHECK** Suppose you want to find the perimeter of a figure with five sides, like the Pentagon building in Washington, D.C. Explain how you would do it. Draw a figure with five sides, label the lengths of all the sides, and find the perimeter.

Find the perimeter of each figure. ***See Examples 1 and 3.***

3.

$P = 46 \text{ m} + \underline{\hspace{1cm}}$

$+ \underline{\hspace{1cm}} + \underline{\hspace{1cm}}$

$P = \underline{\hspace{1cm}}$

4.

$P = 1000 \text{ ft} + \underline{\hspace{1cm}}$

$+ \underline{\hspace{1cm}} + \underline{\hspace{1cm}}$

$P = \underline{\hspace{1cm}}$

5.

51.8 m, 48.3 m, 51.8 m, 51.8 m, 51.8 m

6.

12.6 in., 11.1 in., 14.7 in., 24.8 in., 11.9 in.

7.

0.8 km, 0.4 km, 3 km, 0.95 km, 1.31 km

8.

7.33 cm, 2.8 cm, 3 cm, 4.3 cm, 4.17 cm

Find the area of each figure. ***See Examples 2 and 4.***

9.
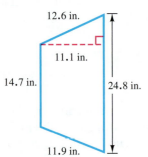

$A = b \cdot h$

10.

$A = b \cdot h$

11.

12.

13.

14.

CONCEPT CHECK *Find **two** errors in each student's solution below. Write a sentence explaining each error. Then show how to work the problem correctly.*

15.

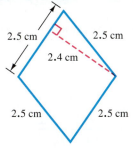

2.5 cm 2.5 cm
2.4 cm
2.5 cm 2.5 cm

$P = 2.5\text{ cm} + 2.4\text{ cm} + 2.5\text{ cm} + 2.5\text{ cm} + 2.5\text{ cm}$
$P = 12.4\text{ cm}^2$

16.

13 ft
13 ft
12 ft 11.6 ft
22 ft

$A = (0.5)(11.6\text{ ft}) \cdot (12\text{ ft} + 13\text{ ft})$
$A = 145\text{ ft}$

Solve each application problem. First label the bases and heights on the sketches.
See Example 6.

17. The backyard of a new home is shaped like a trapezoid with a height of 45 ft and bases of 80 ft and 110 ft. What is the cost of putting sod on the yard if the landscaper charges $0.33 per square foot for sod?

18. The Dockland office building in Hamburg, Germany, is seven stories tall. Each of the seven floors is in the shape of a parallelogram with a height of 22 m and a base of 88 m. Find the total area of all seven floors.

19. A piece of fabric for a quilt design is in the shape of a parallelogram. The base is 5 in. and the height is 3.5 in. What is the total area of the 25 parallelogram pieces needed for the quilt?

20. An accountant is paying $832 per month to rent an office in an old building. Her office is shaped like a trapezoid, with bases of 32 ft and 20 ft and a height of 20 ft. How much rent is she paying per square foot?

Find the area of each figure. See Example 5.

21.

0.8 m
1.8 m
1.3 m
1.1 m 1.3 m
0.8 m 1.3 m
1.8 m

22.

87.3 cm
46.2 cm
61.7 cm
32 cm
92.3 cm

23.

96 ft
96 ft 96 ft
96 ft
72 ft
96 ft
96 ft 96 ft
96 ft

8.5 Triangles

OBJECTIVES

1. Find the perimeter of a triangle.
2. Find the area of a triangle.
3. Given the measures of two angles in a triangle, find the measure of the third angle.

VOCABULARY TIP

Tri- means three. Just as a **tri**cycle has three wheels, a **tri**angle has three sides.

---

1 Find the perimeter of each triangle.

GS (a)

$P = 31 \text{ mm} + 25 \text{ mm}$

$+ \underline{\hspace{2cm}}$

$P = \underline{\hspace{2cm}}$

(b)

(c) A triangle with sides of $6\frac{1}{2}$ yd, $9\frac{3}{4}$ yd, and $11\frac{1}{4}$ yd

---

Answers

1. (a) 16 mm; $P = 72$ mm
   (b) $P = 53.8$ m   (c) $P = 27\frac{1}{2}$ yd or 27.5 yd

A **triangle** is a figure with exactly three sides. Some examples are shown below.

OBJECTIVE ▶ 1 Find the perimeter of a triangle. To find the perimeter of a triangle (the distance around the edges), add the lengths of the three sides.

> **EXAMPLE 1   Finding the Perimeter of a Triangle**
>
> Find the perimeter of the triangle above on the right.
>
> $$P = 8 \text{ ft} + 15 \text{ ft} + 17 \text{ ft}$$
>
> $$P = 40 \text{ ft}$$ ◀ This is **perimeter**, so write **ft** in the answer.

·· ◀ **Work Problem 1** at the Side.

As with parallelograms, you can find the *height* of a triangle by measuring the distance from one vertex of the triangle to the opposite side (the base). The height line must be *perpendicular* to the base; that is, it must form a right angle with the base. Sometimes you have to extend the base in order to draw the height perpendicular to it, as shown below in the figure on the right.

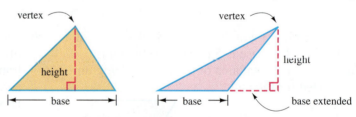

If you cut out two identical triangles and turn one upside down, you can fit them together to form a parallelogram, as shown below.

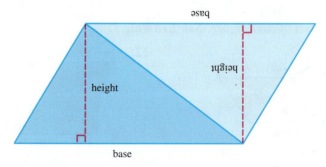

Recall from the last section that the area of the parallelogram is *base* times *height*. Because each triangle is *half* of the parallelogram, the area of one triangle is

$$\frac{1}{2} \text{ of base times height}.$$

OBJECTIVE **2** **Find the area of a triangle.** Use the following formula to find the *area* of a triangle.

Finding the Area of a Triangle

$$\text{Area of a triangle} = \frac{1}{2} \cdot \text{base} \cdot \text{height}$$

$$A = \frac{1}{2} \cdot b \cdot h$$

$$\text{or} \quad A = 0.5 \cdot b \cdot h$$

Remember to use *square units* when measuring area.

EXAMPLE 2 **Finding the Area of Triangles**

Find the area of each triangle.

(a)

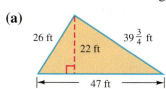

The base is 47 ft and the height is 22 ft. You do *not* need the 26 ft or $39\frac{3}{4}$ ft sides to find the area.

$$A = \frac{1}{2} \cdot b \cdot h$$

$$A = \frac{1}{2} \cdot 47 \text{ ft} \cdot \overset{11}{\cancel{22}} \text{ ft} \qquad \begin{array}{l}\text{Divide out}\\\text{common}\\\text{factor of 2.}\end{array}$$

> This is *area*, so write **ft²** in the answer.

$$A = 517 \text{ ft}^2 \qquad \begin{array}{l}\text{Square units}\\\text{for area}\end{array}$$

(b)

The base must be extended to draw the height. However, still use 45.6 cm for *b* in the formula. Because the measurements are decimal numbers, it is easier to use 0.5 (the decimal equivalent of $\frac{1}{2}$) in the formula.

$$A = 0.5 \cdot 45.6 \text{ cm} \cdot 18.5 \text{ cm}$$

$$A = 421.8 \text{ cm}^2$$

(c)

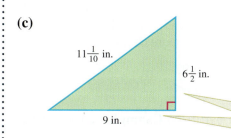

The measurements for this triangle are mixed numbers. Change them to improper fractions or use the equivalent decimal numbers.

> Two sides of the triangle are perpendicular to each other, so use those sides as the base and the height.

Using fractions
$$A = \frac{1}{2} \cdot 9 \text{ in.} \cdot 6\frac{1}{2} \text{ in.} = \frac{1}{2} \cdot \frac{9 \text{ in.}}{1} \cdot \frac{13 \text{ in.}}{2} = \frac{117}{4} \text{ in.}^2 = 29\frac{1}{4} \text{ in.}^2$$

Using decimals
$$A = 0.5 \cdot 9 \text{ in.} \cdot 6.5 \text{ in.} = 29.25 \text{ in.}^2 \leftarrow \text{Equivalent}$$

·········· **Work Problem 2 at the Side.** ▶

---

2 Find the area of each triangle.

GS **(a)**

$$A = \frac{1}{2} \cdot \quad b \quad \cdot \quad h$$

$$A = \frac{1}{\underset{1}{\cancel{2}}} \cdot \overset{13}{\cancel{26}} \text{ m} \cdot 20 \text{ m}$$

$$A = \underline{}$$

(b)

(c)

GS **(d)**

Hint: The base and height are perpendicular. So the base is $9\frac{1}{2}$ ft, or use 9.5 ft. The height is 7 ft.

Answers

2. **(a)** $A = 260 \text{ m}^2$ **(b)** $A = 1.785 \text{ cm}^2$
(c) $A = 16.5 \text{ in.}^2$
(d) $A = 33.25 \text{ ft}^2$ or $33\frac{1}{4} \text{ ft}^2$

3 Find the area of the shaded part
GS in this figure.

25 m

25 m

10 m

10 m

25 m

25 m

The entire figure is a square.
Find the area of the square.

$A = 25 \text{ m} \cdot 25 \text{ m}$

$A = \underline{\hspace{1cm}}$ for entire figure.

Now find the area of the white
triangles. Then subtract to find
the shaded area.

4 Suppose the figure in **Problem 3**
above is an auditorium floor
plan. The shaded part will be
covered with carpet costing
$27 per square meter. The rest
will be covered with vinyl floor
covering costing $18 per square
meter. What is the total cost of
covering the floor?

Answers

3. $A = 625 \text{ m}^2$ for entire figure.
   For shaded portion,
   $A = 625 \text{ m}^2 - 125 \text{ m}^2 - 125 \text{ m}^2 = 375 \text{ m}^2$.
4. $10,125 + $2250 + $2250 = $14,625

EXAMPLE 3 Using the Concept of Area

Find the area of the
shaded part in
this figure.

32
cm

40 cm

30 cm

The *entire* figure is a rectangle. Find the area of the rectangle.

$$A = l \cdot w$$

$$A = 30 \text{ cm} \cdot 40 \text{ cm}$$

$$A = 1200 \text{ cm}^2$$

The *unshaded* part is a triangle. Find the area of the triangle.

$$A = \frac{1}{\underset{1}{2}} \cdot \overset{15}{30} \text{ cm} \cdot 32 \text{ cm}$$

$$A = 480 \text{ cm}^2$$

Subtract to find the area of the shaded part.

$$\overset{\text{Entire area}}{A = \overbrace{1200 \text{ cm}^2}} - \overset{\text{Unshaded part}}{\overbrace{480 \text{ cm}^2}} = \overset{\text{Shaded part}}{\overbrace{720 \text{ cm}^2}}$$

The area of the shaded part of the figure is 720 cm^2. ◁ This is *area*, so write cm^2 in the answer.

········ ◀ **Work Problem 3 at the Side.**

EXAMPLE 4 Applying the Concept of Area

The Department of Transportation cuts triangular signs out of rectangular
pieces of metal using the measurements shown above in **Example 3**. If the
metal costs $0.02 per square centimeter, how much does the metal cost for
the sign? What is the cost of the metal that is *not* used?

From **Example 3** above, the area of the triangle (the sign) is 480 cm^2.
Multiply the area of the sign times the cost per square centimeter.

$$\text{cost of sign} = \frac{480 \text{ cm}^2}{1} \cdot \frac{\$0.02}{1 \text{ cm}^2} = \$9.60$$

The metal that is *not* used is the *shaded* part from **Example 3**. The unused
area is 720 cm^2.

$$\text{cost of unused metal} = \frac{720 \text{ cm}^2}{1} \cdot \frac{\$0.02}{1 \text{ cm}^2} = \$14.40$$

The cost of the metal for the triangular sign is $9.60. The unused metal costs
$14.40.

········ ◀ **Work Problem 4 at the Side.**

OBJECTIVE **3** **Given the measures of two angles in a triangle, find the measure of the third angle.** The *tri* in *tri*angle means *three*. So the name tells you that a triangle has three angles. The sum of the measures of the three angles in any triangle is *always* 180°. You can see it by drawing a triangle, cutting off the three angles, and rearranging them to make a straight angle (180°).

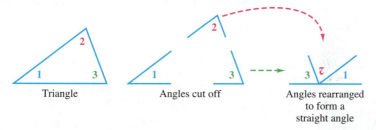

Triangle   Angles cut off   Angles rearranged
to form a
straight angle

Finding the Unknown Angle Measurement in a Triangle

Step 1 Add the number of degrees in the measures of the two given angles.

Step 2 Subtract the sum from 180°.

EXAMPLE 5   **Finding an Angle Measurement in Triangles**

Find the number of degrees in the indicated angle.

(a) Angle *R*

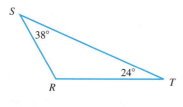

Step 1 Add the two angle measurements you are given.

$$38° + 24° = 62°$$

Step 2 Subtract the sum from 180°.

$$180° - 62° = 118°$$

∠*R* measures 118°.

(b) Angle *F*

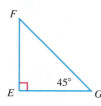

∠*E* is a right angle, so it measures 90°.

Step 1 $90° + 45° = 135°$

Step 2 $180° - 135° = 45°$

∠*F* measures 45°.

> Write a small, raised circle for "degrees."

·················· **Work Problem 5 at the Side.** ▶

---

5 Find the number of degrees in the third angle of each triangle.

GS **(a)**

Step 1 48° + 59° = ____

Step 2 180° − ____ = ____

∠*G* measures ____

(b)

(c)

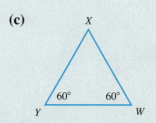

Answers

5. (a) 107°; 180° − 107° = 73°;
∠*G* measures 73°
(b) ∠*C* measures 35°
(c) ∠*X* measures 60°

8.5 Exercises

FOR EXTRA HELP

Download the MyDashBoard App

MyMathLab®

1. CONCEPT CHECK Look at **Exercise 3** below. Which measurements will you use to find the area of the triangle? Will the units in your answer be m or m²?

2. CONCEPT CHECK Look at **Exercise 3** below. Which measurements will you use to find the perimeter of the triangle? Will the units in your answer be m or m²?

Find the perimeter and area of each triangle. See Examples 1 and 2.

3.

58 m
66 m
72 m 72 m

4.

9 yd
8 yd
13 yd 12 yd

5.

25.3 cm
15.6 cm 18 cm
11 cm

6.
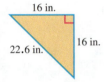
16 in.
22.6 in. 16 in.

7.

9 yd 6 yd 7 yd
$10\frac{1}{4}$ yd

8.

$6\frac{1}{4}$ ft 10 ft 18 ft
9 ft

9.

35.5 cm
21.3 cm
28.4 cm

10.
7.2 ft 7.2 ft
6.2 ft
7.2 ft

Find the shaded area in each figure. See Example 3.

11.
GS

10.8 m 10.8 m
9 m
12 m
12 m 12 m
12 m

Find the area of the triangle.

$A = 0.5 \cdot 12 \text{ m} \cdot \underline{\hspace{1cm}}$

$A = \underline{\hspace{1cm}}$

Find the area of the square.

$A = 12 \text{ m} \cdot \underline{\hspace{1cm}}$

$A = \underline{\hspace{1cm}}$

Now add to find the area of the entire figure.

12.
GS

34 m 22 m
20 m
27 m 19 m
22 m 20 m

Find the area of the triangle on the left.

$A = 0.5 \cdot 27 \text{ m} \cdot \underline{\hspace{1cm}}$

$A = \underline{\hspace{1cm}}$

Find the area of the parallelogram on the right.

$A = 20 \text{ m} \cdot \underline{\hspace{1cm}}$

$A = \underline{\hspace{1cm}}$

Now add to find the area of the entire figure.

13.

14.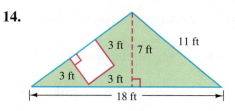

*Find the number of degrees in the third angle of each triangle. **See Example 5.***

15.

16.

17.

18.

19. CONCEPT CHECK Can a triangle have two right angles? Explain your answer.

20. CONCEPT CHECK In your own words, explain where the $\frac{1}{2}$ comes from in the formula for area of a triangle. Draw a sketch to illustrate your explanation.

*Solve each application problem. **See Example 4.***

21. A triangular tent flap measures $3\frac{1}{2}$ ft along the base and has a height of $4\frac{1}{2}$ ft. How much canvas is needed to make the flap?

22. A wooden sign in the shape of a right triangle has perpendicular sides measuring 1.5 m and 1.2 m. How much surface area does the sign have?

23. A triangular space between three streets has the measurements shown below.

   (a) How much new curbing will be needed to go around the space?

   (b) How much sod will be needed to cover the space?

24. Each gable end of a new house has a span of 36 ft and a rise of 9.5 ft. What is the total area of both gable ends of the house?

25. All sides of the house below are congruent and all roof sections are congruent.

   (a) Find the area of one side of the house.

   (b) Find the area of one roof section.

26. The sketch shows the plan for an office building. The shaded area will be a parking lot. What is the cost of building the parking lot if the contractor charges $35 per square yard for materials and labor?

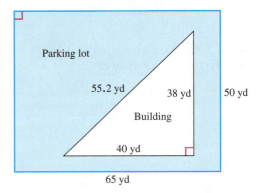

27. A city lot with an unusual shape is shown below.

   (a) How much frontage (distance along streets) does the lot have?

   (b) What is the area of the lot?

28. A car dealership wants three nylon pennants to hang in its front window. The height of the two smaller pennants is 3.5 ft, and the larger pennant has a height of 6.5 ft. How much nylon fabric is needed for all the pennants?

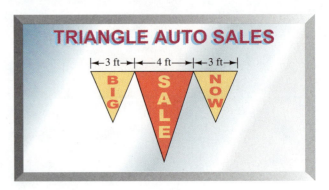

8.6 Circles

OBJECTIVE ▸ 1 Find the radius and diameter of a circle. Suppose you start with one dot on a piece of paper. Then you draw many dots that are each 2 cm away from the first dot. If you draw enough dots (points) you'll end up with a *circle*. Each point on the circle is exactly 2 cm away from the *center* of the circle. The 2 cm distance is called the *radius, r*, of the circle. The distance across the circle (passing through the center) is called the *diameter, d*, of the circle.

OBJECTIVES

1. Find the radius and diameter of a circle.

2. Find the circumference of a circle.

3. Find the area of a circle.

4. Become familiar with Latin and Greek prefixes used in math terminology.

VOCABULARY TIP

Diameter **Dia** means "through," and **meter** means "measure." So, the **diameter** of a circle is a line through the circle, passing through the center. Recall that **radius** comes from the root word meaning "spoke," like the spoke of a wheel. So the **radius** is a line from the center of a circle out to the edge of the circle.

Circle, Radius, and Diameter

A **circle** is a two-dimensional (flat) figure with all points the same distance from a fixed center point.

The **radius** (*r*) is the distance from the center of the circle to any point on the circle.

The **diameter** (*d*) is the distance across the circle passing through the center.

Using the circle above on the right as a model, you can see some relationships between the radius and diameter.

Finding the Diameter and Radius of a Circle

$$\text{diameter} = 2 \cdot \text{radius}$$
$$d = 2 \cdot r$$

and $\quad \text{radius} = \dfrac{\text{diameter}}{2} \quad$ or $\quad r = \dfrac{d}{2}$

EXAMPLE 1 Finding the Diameter and Radius of a Circle

Find the unknown length of the diameter or radius in each circle.

(a)

Because the radius is 9 cm, the diameter is twice as long.

$$d = 2 \cdot \mathbf{r}$$
$$d = 2 \cdot \mathbf{9\ cm}$$
$$d = 18\ cm$$

> Write **cm** in your answer, **not** cm².

····· **Continued on Next Page**

1 Find the unknown length of the diameter or radius in each circle.

(a)

40 ft

radius (*r*) is unknown

$$r = \frac{d}{2}$$

$$r = \frac{40 \text{ ft}}{2} = \underline{\qquad}$$

(b)

11 cm

(c)

32 yd

(d)

9.5 m

Answers

1. (a) $r = 20$ ft   **(b)** $r = 5.5$ cm
(c) $d = 64$ yd   **(d)** $d = 19$ m

(b)

r = ?

17 m

The radius is half the diameter.

$$r = \frac{d}{2}$$

$$r = \frac{17 \text{ m}}{2}$$

$$r = 8.5 \text{ m} \quad \text{or} \quad 8\frac{1}{2} \text{ m}$$

> $8.5 = 8\frac{1}{2}$ because $8\frac{5}{10}$ simplifies to $8\frac{1}{2}$.

◄ **Work Problem 1 at the Side.**

OBJECTIVE 2 **Find the circumference of a circle.** The perimeter of a circle is called its **circumference.** Circumference is the distance around the edge of a circle.

The diameter of the can in the drawing is about 10.6 cm, and the circumference of the can is about 33.3 cm. Dividing the circumference of the circle by the diameter gives an interesting result.

$$\frac{\text{circumference}}{\text{diameter}} = \frac{33.3}{10.6} \approx 3.14 \qquad \text{Rounded to the nearest hundredth}$$

Dividing the circumference of *any* circle by its diameter *always* gives an answer close to 3.14. This means that going around the edge of any circle is a little more than 3 times as far as going straight across the circle.

This ratio of circumference to diameter is called π, (the Greek letter **pi**, pronounced PIE). There is no decimal that is exactly equal to π, but here is the *approximate* value.

$$\pi \approx 3.14159265359$$

Rounding the Value of *Pi* (π)

We usually round π to 3.14. Therefore, calculations involving π will give approximate answers and should be written using the $\approx$ symbol.

Use the following formulas to find the *circumference* of a circle.

Finding the Circumference (Distance around a Circle)

$$\text{Circumference} = \pi \cdot \text{diameter}$$

$$C = \pi \cdot d$$

or, because $d = 2 \cdot r$, then $C = \pi \cdot 2 \cdot r$ usually written $C = 2 \cdot \pi \cdot r$

Remember to use linear units such as ft, yd, m, and cm when measuring circumference (**not** square units).

EXAMPLE 2   **Finding the Circumference of Circles**

Find the circumference of each circle. Use 3.14 as the approximate value for π. Round answers to the nearest tenth.

(a)

The *diameter* is 38 m, so use the formula with *d* in it.

$$C = \pi \cdot d$$

$$C \approx 3.14 \cdot 38 \text{ m}$$

Write **m** in the answer, **not** m².

$$C \approx 119.3 \text{ m} \quad \text{Rounded}$$

(b)

In this example, the *radius* is labeled, so it is easier to use the formula with *r* in it.

$$C = 2 \cdot \pi \cdot r$$

$$C \approx 2 \cdot 3.14 \cdot 11.5 \text{ cm}$$

Write **cm** in the answer, **not** cm².

$$C \approx 72.2 \text{ cm} \quad \text{Rounded}$$

🖩 Calculator Tip

Many *scientific* calculators have a $\boxed{\pi}$ key. Try pressing it. With a 10-digit display, you'll see the value of π to the nearest billionth.

$\boxed{\pi}$ $\boxed{=}$ **3.141592654**

But this is still an approximate value, although it is more precise than rounding π to 3.14 Try finding the circumference in **Example 2(a)** above using the $\boxed{\pi}$ key.

$\boxed{\pi}$ $\boxed{\times}$ 38 $\boxed{=}$ Answer is 119.3805208; round to 119.4

When you used 3.14 as the approximate value of π, the result rounded to 119.3 so the answers are slightly different. In this book, we will use 3.14 instead of the $\boxed{\pi}$ key. Our measurements of radius and diameter are given as whole numbers or with tenths, so it is acceptable to round π to hundredths. And, some students may be using standard calculators without a $\boxed{\pi}$ key or doing the calculations by hand.

·············· **Work Problem ❷ at the Side.** ▶

OBJECTIVE ▶ ❸ Find the area of a circle. To find the formula for the area of a circle, start by cutting two circles into many pie-shaped pieces.

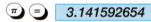

Circumference (distance around) is $2 \cdot \pi \cdot r$.

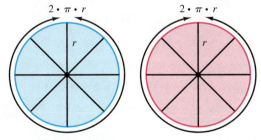

Unfold the circles, much as you might "unfold" a peeled orange, and put them together as shown here.

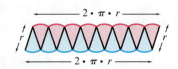

❷ Find the circumference of each circle. Use 3.14 as the approximate value for π. Round answers to the nearest tenth.

GS (a)

diameter is known; use

$$C = \pi \cdot d$$

$$C \approx 3.14 \cdot 150 \text{ ft}$$

$$C \approx \underline{\qquad}$$

(b)

radius is known, use

$$C = 2 \cdot \pi \cdot r$$

(c) diameter 0.9 km

(d) radius 4.6 m

Answers

2. **(a)** $C \approx 471$ ft   **(b)** $C \approx 44.0$ in. (rounded)
   (c) $C \approx 2.8$ km (rounded)
   (d) $C \approx 28.9$ m (rounded)

3 Find the area of each circle. Use 3.14 for π. Round your answers to the nearest tenth.

 (a)

radius is 4 ft

4 ft

$A = \pi \cdot r^2$

$A = \pi \cdot r \cdot r$

$A \approx 3.14 \cdot 4 \text{ ft} \cdot 4 \text{ ft}$

$A \approx$ _____

(b)

12 in.

Hint: The diameter is 12 in. so $r =$ ____ in.

 (c)

1.8 km

 (d)

8.4 cm

The figure is approximately a parallelogram with height r (the radius of the original circle) and base $2 \cdot \pi \cdot r$ (the circumference of the original circle). The area of the "parallelogram" is base times height.

Area = $b \quad \cdot h$

Area = $2 \cdot \pi \cdot r \cdot r$

Area = $2 \cdot \pi \cdot r^2 \leftarrow$ Recall that $r \cdot r$ is r^2.

Because the "parallelogram" was formed from *two* circles, the area of *one* circle is half as much.

$$\frac{1}{2} \cdot 2 \cdot \pi \cdot r^2 = 1 \cdot \pi \cdot r^2 \quad \text{or simply} \quad \pi \cdot r^2$$

Finding the Area of a Circle

Area of a circle = $\pi \cdot$ radius $\cdot$ radius

$$A = \pi \cdot r^2$$

Remember to use *square units* when measuring area.

EXAMPLE 3 **Finding the Area of Circles**

Find the area of each circle. Use 3.14 for π. Round your answers to the nearest tenth.

(a) A circle with a radius of 8.2 cm
Use the formula $A = \pi \cdot r^2$, which means $\pi \cdot r \cdot r$.

$A = \pi \cdot r \cdot r$

$A \approx 3.14 \cdot 8.2 \text{ cm} \cdot 8.2 \text{ cm}$

$A \approx 211.1 \text{ cm}^2$ Rounded

> This is **area**, so write cm² in the answer.

(b)

10 ft

To use the area formula, you need to know the *radius* (r). In this circle, the *diameter* is 10 ft. First find the radius.

> You *cannot* use the *diameter* measurement in the area formula, so find the *radius*.

$r = \dfrac{d}{2}$

$r = \dfrac{10 \text{ ft}}{2} = 5 \text{ ft}$

Now find the area.

$A \approx 3.14 \cdot 5 \text{ ft} \cdot 5 \text{ ft}$

$A \approx 78.5 \text{ ft}^2$

> Write square units (**ft²**) in your answer.

CAUTION

When finding *circumference*, you can start with either the radius or the diameter. When finding *area*, you must use the *radius*. If you are given the diameter, divide it by 2 to find the radius. Then find the area.

◀ **Work Problem 3** at the Side.

Calculator Tip

You can use your calculator to find the area of the circle in **Example 3(a)** on the previous page. The first method works on both scientific and standard calculators:

3.14 ⊗ 8.2 ⊗ 8.2 ⊜   Answer is 211.1336

You round the answer to 211.1 (nearest tenth)

On a *scientific* calculator you can also use the ⓧ² key, which automatically squares the number you enter (that is, multiplies the number times itself):

3.14 ⊗ 8.2 ⓧ² 67.24 ⊜   Answer is 211.1336

Appears automatically on some calculators;
8.2 × 8.2 is 67.24

Next, we find the area of a *semicircle,* which is half the area of a circle.

EXAMPLE 4   Finding the Area of a Semicircle

Find the area of the semicircle. Use 3.14 for π. Round your answer to the nearest tenth.

First, find the area of a whole circle with a radius of 12 ft.

$$A = \pi \cdot r \cdot r$$

$$A \approx 3.14 \cdot 12 \text{ ft} \cdot 12 \text{ ft}$$

$$A \approx 452.16 \text{ ft}^2 \leftarrow \text{Do not round yet.}$$

Divide the area of the whole circle by 2 to find the area of the semicircle.

$$\frac{452.16 \text{ ft}^2}{2} = 226.08 \text{ ft}^2$$

The *last* step is rounding 226.08 to the nearest tenth.

Area of semicircle $\approx 226.1 \text{ ft}^2$ Rounded   *This is area, so write ft² in the answer.*

Work Problem ④ at the Side. ▶

EXAMPLE 5   Applying the Concept of Circumference

A circular rug is 8 ft in diameter. The cost of fringe for the edge is $2.25 per foot. What will it cost to add fringe to the rug? Use 3.14 for π.

$$\text{Circumference} = \pi \cdot d$$

$$C \approx 3.14 \cdot 8 \text{ ft}$$

$$C \approx 25.12 \text{ ft}$$

$$\text{cost} = \text{cost per foot} \cdot \text{circumference}$$

$$\text{cost} = \frac{\$2.25}{1 \text{ ft}} \cdot \frac{25.12 \text{ ft}}{1}$$ *The common units divide out.*

$$\text{cost} = \$56.52$$

The cost of adding fringe to the rug is $56.52.

Work Problem ⑤ at the Side. ▶

VOCABULARY TIP

Semicircle **Semi** means "half."

④ Find the area of each semicircle. Use 3.14 for π. Round your answers to the nearest tenth.

(a)

24 m

(b)

|← 35.4 ft →|

(c)

9.8 m

⑤ Find the cost of binding around the edge of a circular rug that is 3 m in diameter. The binder charges $4.50 per meter. Use 3.14 for π.

Answers

4. All answer are rounded.
   (a) $A \approx 904.3 \text{ m}^2$ (b) $A \approx 491.9 \text{ ft}^2$
   (c) $A \approx 150.8 \text{ m}^2$
5. $42.39

6 Find the cost of covering the underside of the rug in **Margin Problem 5** with a nonslip rubber backing. The rubber backing costs $3.89 per square meter.

7 (a) Here are some prefixes you have seen in this textbook. List at least one math term and one nonmathematical word that use each prefix.

dia- (through):

fract- (break):

par- (beside):

per- (divide):

peri- (around);

rad- (ray):

rect- (right):

sub- (below):

(b) How could you use your knowledge of prefixes to remember the difference between perimeter and area?

EXAMPLE 6 Applying the Concept of Area

Find the cost of covering the rug in **Example 5** (on the previous page) with a plastic cover. The material for the cover costs $1.50 per square foot. Use 3.14 for π.

First find the radius. $\quad r = \dfrac{d}{2} = \dfrac{8 \text{ ft}}{2} = 4 \text{ ft}$

Then find the area. $\quad A = \pi \cdot r^2$

$$A \approx 3.14 \cdot 4 \text{ ft} \cdot 4 \text{ ft}$$
$$A \approx 50.24 \text{ ft}^2$$
$$\text{cost} = \frac{\$1.50}{1 \text{ ft}^2} \cdot \frac{50.24 \text{ ft}^2}{1} = \$75.36$$

The cost of the plastic cover is $75.36.

◀ **Work Problem 6** at the Side.

OBJECTIVE 4 Become familiar with Latin and Greek prefixes used in math terminology. Many English words are built from Latin or Greek root words and prefixes. Knowing the meaning of the more common ones can help you figure out the meaning of terms in many subject areas, including mathematics.

EXAMPLE 7 Using Prefixes to Understand Math Terms

(a) Listed below are some Latin and Greek root words and prefixes with their meanings in parentheses. You've already seen math terms in this textbook that use these prefixes. List at least one math term and one nonmathematical word that use each prefix or root word.

cent- (100): **cent**imeter; **cent**ury

circum- (around): **circum**ference; **circum**vent

de- (down): **de**nominator; **de**cline

dec- (10): **dec**imal; **Dec**ember (originally the 10th month in the old calendar)

There are many answers. These are some of the possibilities.

(b) Suppose you have trouble remembering which part of a fraction is the denominator. How could your knowledge of prefixes help in this situation?

The *de-* prefix in **de**nominator means "down" so the denominator is the number *down* below the fraction bar.

◀ **Work Problem 7** at the Side.

Note

Here are some additional prefixes and root words and their meanings that you will see in the rest of **Chapter 8,** in **Chapter 9,** and in other math classes. Examples of a math term and a nonmathematical word are shown for each one.

equ- (equal): **equ**ation; **equ**inox

hemi- (half): **hemi**sphere; **hemi**trope

lateral (side): **quadri**lateral; **bi**lateral

re- (back or again): **re**ciprocal; **re**duce

Answers

6. $27.48 (rounded)
7. (a) Some possibilities are:
diameter; diagonal
fraction; fracture
parallel; paramedic
percent; per capita
perimeter; periscope
radius; radiate
rectangle; rectify
subtract; submarine
(b) *Peri-* in perimeter means "around," so perimeter is the distance *around* the edges of a shape.

8.6 Exercises

  MyMathLab®

Download the MyDashBoard App

1. **CONCEPT CHECK** Fill in the blanks. Choose from height, perimeter, area, base, ft, and ft².

   Finding the circumference of a circle is like finding the _____ of a triangle. If the circle's radius and diameter are measured in feet, then the units for the circumference will be _____.

2. **CONCEPT CHECK** Fill in the blanks. Choose from radius, diameter, circumference, divide by 2, and multiply by 2.

   To use the formula for finding the area of a circle, you need to know the length of the _____. If you are given the diameter of a circle, what should you do to it to find the radius? _____

*Find the unknown length in each circle. **See Example 1.***

3.

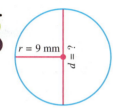

$r = 9$ mm, $d = ?$

$d = 2 \cdot r$

$d = 2 \cdot$ _____

$d =$ _____

4.

$d = ?$, $r = 15$ mi

$d = 2 \cdot r$

$d = 2 \cdot$ _____

$d =$ _____

5.

$r = ?$, $d = 0.7$ km

6.

$d = 6.1$ cm, $r = ?$

*Find the circumference and area of each circle. Use 3.14 as the approximate value for π. Round your answers to the nearest tenth. **See Examples 2 and 3.***

7.

11 ft

8.

41 cm

9.

2.6 m

10.

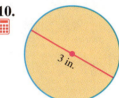

3 in.

*Find the circumference and area of circles having the following diameters. Use 3.14 as the approximate value of π. Round your answers to the nearest tenth. **See Examples 2 and 3.***

11. $d = 15$ cm

   $C = \pi \cdot d$

   $C \approx 3.14 \cdot 15$ cm

   $C \approx$ _____

   To find the **area,** first find the **radius.** Then use $A = \pi \cdot r \cdot r$.

12. $d = 39$ ft

   $C = \pi \cdot d$

   $C \approx 3.14 \cdot 39$ ft

   $C \approx$ _____

   To find the **area,** first find the **radius.** Then use $A = \pi \cdot r \cdot r$.

13. $d = 7\frac{1}{2}$ ft

14. $d = 4\frac{1}{2}$ yd

*Find each shaded area. Note that Exercises 15–18 all contain semicircles. Use 3.14 as the approximate value of π. Round your answers to the nearest tenth if necessary. **See Example 4.***

15.

7 in.

16.

15 yd

17.

18.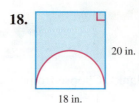

19. CONCEPT CHECK How would you explain π to a friend who is not in your math class? Write an explanation.

20. CONCEPT CHECK Make up a test question that requires the use of π. Show how to solve your problem.

Solve each application problem. Use 3.14 as the approximate value of π. Round your answers to the nearest tenth. **See Examples 5 and 6.**

21. An irrigation system moves around a center point to water a circular area for crops. If the irrigation system is 50 yd long, how large is the watered area?

22. If you swing a ball held at the end of a string 20 cm long, how far will the ball travel on each turn?

23. A Michelin Cross Terrain SUV tire has an overall diameter of 29.10 inches. How far will a point on the tire tread move in one complete turn? (*Source:* Michelin.) *Bonus question:* How many revolutions does the tire make per mile? Round to the nearest whole number.

24. In August 2011, Hurricane Irene moved along the East Coast of the United States, from North Carolina to New York City. Some 65 million people live along this coast, the greatest number of people threatened by a hurricane in American history. The huge circular storm had a diameter of 500 miles. By contrast, Hurricane Katrina, which flooded New Orleans in 2005, had a diameter of 210 miles. What was the area of each hurricane, to the nearest hundred square miles? (*Source:* Associated Press).

Circular hurricane as seen from a weather satellite.

For Exercises 25–30, first draw a circle and label the radius or diameter. Then solve the problem. Use 3.14 as the approximate value of π and round answers to the nearest tenth, money answers to the nearest cent.

25. A radio station can be heard 150 miles in all directions during evening hours. How many square miles are in the station's broadcast area? (*Hint:* Use the drawing of the broadcast area to help you.)

The radius is 150 miles.

26. An earthquake was felt by people 900 km away in all directions from the epicenter (the source of the earthquake). How much area was affected by the quake?

27. The diameter of Diana Hestwood's wristwatch is 1 in. and the radius of the clock face on her kitchen wall is 3 in. Find the circumference and the area of the watch face and the clock face.

28. The diameter of the largest known ball of twine is 12 ft 9 in. The sign posted near the ball says it has a circumference of 40 ft. Is the sign correct? *Hint:* First change 9 in. to feet and add it to 12 ft. (*Source: Guinness World Records.*)

29. Blaine Fenstad wants to buy a pair of two-way radios. Some models have a range of 2 miles under ideal conditions. More expensive models have a range of 5 miles. What is the difference in the area covered by the 2-mile and 5-mile models? (*Source:* Best Buy.)

30. The National Audubon Society holds an end-of-year bird count. Volunteers count all the birds they see in a circular area during a 24-hour period. Each circle has a diameter of 15 miles. About 1700 circular areas are counted across the United States each December. What is the total area covered by the count?

31. To find the age of a living tree, a forestry specialist measures the circumference of the tree. (See photo.)

   (a) If the circumference of a tree is 144 cm, what is the diameter?

   (b) Explain how you solved part (a).

32. In Atlanta, Interstate 285 circles the city and is known as the "perimeter." If the circumference of the circle made by the highway is 62.8 miles, find

   (a) the diameter of the circle.

   (b) the area inside the circle.

   (*Source: Greater Atlanta Newcomer's Guide.*)

33. Find the cost of sod, at $0.49 per square foot, for this playing field that has a semicircle on each end.

34. Find the area of this skating rink.

*Use the information about prefixes in **Example 7** to answer Exercises 35 and 36.*

35. Explain how you could use the information about prefixes to remember the difference between radius, diameter, and circumference.

36. Explain how you could use the information about prefixes to avoid confusion between parallel and perpendicular lines.

Relating Concepts (Exercises 37–42) For Individual or Group Work

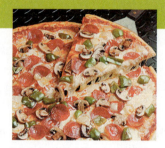

*Use the table below to **work Exercises 37–42 in order.***

Find the best buy for each type of pizza. The best buy is the lowest cost per square inch of pizza. All the pizzas are circular, and the measurement given on the menu board is the diameter of the pizza in inches. Use 3.14 as the approximate value of π. Round the area to the nearest tenth. Round cost per square inch to the nearest thousandth.

Pizza Menu	Small 10"	Medium 12"	Large 14"
Cheese only	$ 7	$ 9	$11
"The Works"	$12	$14	$16
Pepperoni	$ 8	$10	$12

Source: Papa John's.

37. Find the area of a small pizza.

38. Find the area of a medium pizza.

39. Find the area of a large pizza.

40. What is the cost per square inch for each size of cheese pizza? Which size is the best buy?

41. What is the cost per square inch for each size of "The Works" pizza? Which size is the best buy?

42. You use a coupon for $2 off any small pizza. What is the cost per square inch for each size of pepperoni pizza? Which size is the best buy?

Summary Exercises *Perimeter, Circumference, and Area*

1. Draw a sketch of each of these shapes: **(a)** square, **(b)** rectangle, **(c)** parallelogram. On each sketch, indicate 90° angles and show which sides are the same length.

2. (a) Draw a sketch of a circle and show the radius.

(b) Draw another circle and show the diameter.

(c) Describe the relationship between the radius and diameter of a circle.

3. Describe how you can find the perimeter of any flat shape with straight sides.

4. In your own words, describe the difference between finding the perimeter of a shape and finding the area of the shape.

5. Match each shape to its corresponding area formula.

Shapes	Area Formulas
parallelogram ____	**(a)** $A = \dfrac{1}{2} \cdot b \cdot h$
square ____	**(b)** $A = \pi \cdot r^2$
trapezoid ____	**(c)** $A = l \cdot w$
circle ____	**(d)** $A = \dfrac{1}{2} \cdot h \cdot (b + B)$
rectangle ____	**(e)** $A = s^2$
triangle ____	**(f)** $A = b \cdot h$

6. (a) If you know the *radius* of a circle, which formula do you use to find its circumference?

(b) If you know the *diameter* of a circle, which formula do you use to find its circumference?

(c) If you know the *diameter* of a circle, what must you do *before* using the formula for finding the area?

7. Find the perimeter and area of each triangle, to the nearest tenth.

8. CONCEPT CHECK   Some answers from a student's test paper are listed below. The *number* part of each answer is correct, but the *units* are not. Rewrite each answer with the correct units.

(a) $A = 12$ cm

(b) $P = 6\dfrac{1}{2}$ ft^2

(c) $C \approx 28.5$ m^2

(d) $A = 307$ in.

Name each figure and find its perimeter and area. Round answers to the nearest tenth.

9.

10 $\frac{1}{2}$ in.

3 in.

10.

12 yd

12 yd   12 yd

11.

1.9 m

1.7 m

1.3 m

12.

8.5 mm

6 mm

8 mm

9 mm

10 mm

*For each circle, find (**a**) the diameter or radius, (**b**) the circumference, and (**c**) the area.*
Use 3.14 as the approximate value of π. Round answers to the nearest tenth.

13.

6 cm

14.

30 mi

15.

9 ft

Find the area of each shaded region. Note that Exercise 18 contains semicircles.
Use 3.14 as the approximate value of π. Round answers to the nearest tenth.

16.

1.6 m

1.6 m   1.6 m

1 m   1.2 m

0.5 m   0.5 m

3.8 m

17.

|← 15 yd →|

13 yd

21 yd

15 yd

18.

20 in.

20 in.

Solve each application problem. Use 3.14 as the approximate value of π.
Round answers to the nearest tenth.

19. The Mormons traveled west to Utah by covered wagon in 1847. They tied a rag to a wagon wheel to keep track of the distance they traveled. The radius of the wheel was 2.33 ft. How far did the rag travel each time the wheel made a complete revolution? (*Source: Trail of Hope.*) *Bonus question:* How many wheel revolutions equaled one mile?

20. The rectangular front door for a new log home is 0.9 m wide and 2 m high. How much will it cost for weather strip material to go around all edges of the door, if weather strip costs $0.88 per meter?

2 m

|← 0.9 m →|

8.7 Volume

OBJECTIVE ▶ ① **Find the volume of a rectangular solid.** A shoe box and a cereal box are examples of three-dimensional (or solid) figures. The three dimensions are length, width, and height. (A rectangle or square is a two-dimensional figure. The two dimensions are length and width.)

If you want to know how much a shoe box will hold, you find its *volume*. We measure volume by seeing how many cubes of a certain size will fill the space inside the box. Three sizes of *cubic units* are shown below. Notice that all the edges of a cube have the same length and all the sides meet at right angles.

OBJECTIVES

Find the volume of a
① rectangular solid;
② sphere;
③ cylinder;
④ cone and pyramid.

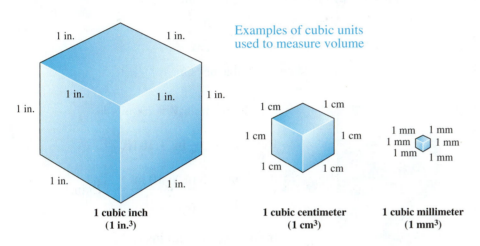

Examples of cubic units used to measure volume

1 cubic inch
(1 in.³)

1 cubic centimeter
(1 cm³)

1 cubic millimeter
(1 mm³)

Some other sizes of cubes that are used to measure volume are 1 cubic foot (1 ft³), 1 cubic yard (1 yd³), and 1 cubic meter (1 m³).

> **CAUTION**
>
> The raised 3 in 4³ means that you multiply 4 • 4 • 4 to get 64. The raised 3 in cm³ or ft³ is a short way to write the word *cubic*. When you see 5 cm³, say "five *cubic* centimeters." Do *not* multiply 5 • 5 • 5. The exponent applies to cm, *not* to the number.

> **Volume**
>
> **Volume** is a measure of the space inside a solid shape. The volume of a solid is how many cubic units it takes to fill the solid.

Use the formula below to find the *volume* of *rectangular solids* (box-like shapes).

> **Finding the Volume of Rectangular Solids**
>
> Volume of a rectangular solid = length • width • height
>
> $$V = l \bullet w \bullet h$$
>
> Remember to use **cubic units** when measuring volume.

① Find the volume of each box. Round your answers to the nearest tenth if necessary.

GS **(a)**

$V = l \cdot w \cdot h$

$V = 3 \text{ m} \cdot 8 \text{ m} \cdot \underline{\hspace{1cm}} \text{ m}$

$V = \underline{\hspace{1.5cm}}$

(b)

(c) Length $6\frac{1}{4}$ ft,

width $3\frac{1}{2}$ ft, height 2 ft

EXAMPLE 1 **Finding the Volume of Rectangular Solids**

Find the volume of each box.

(a)

Each cube that fits in the box is 1 cubic centimeter (1 cm³). To find the volume, you can count the number of cubes.

Bottom layer has 12 cubes. ⎫ total of
Top layer has 12 cubes. ⎭ 24 cubes (24 cm³)

Or, you can use the formula for rectangular solids.

$$V = l \cdot w \cdot h$$
$$V = 4 \text{ cm} \cdot 3 \text{ cm} \cdot 2 \text{ cm}$$
$$V = 24 \text{ cm}^3 \quad \text{Cubic units for volume}$$

(b)

Use the formula $V = l \cdot w \cdot h$.

$$V = 7 \text{ in.} \cdot 2\frac{1}{2} \text{ in.} \cdot 10 \text{ in.}$$

$$V = \frac{7 \text{ in.}}{1} \cdot \frac{5 \text{ in.}}{2} \cdot \frac{\overset{5}{10} \text{ in.}}{1} = 175 \text{ in.}^3$$

If you like, use 2.5 instead of $2\frac{1}{2}$; they are equivalent.

Write **in.³** for volume.

$$V = 7 \text{ in.} \cdot 2.5 \text{ in.} \cdot 10 \text{ in.} = 175 \text{ in.}^3$$

◀ **Work Problem ①** at the Side.

OBJECTIVE **②** **Find the volume of a sphere.** A *sphere* is shown below. Examples of spheres include baseballs, oranges, and Earth. (They aren't perfect spheres, but they're close.)

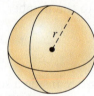

As with circles, the *radius* of a sphere is the distance from the center to the edge of the sphere. Use this formula to find the *volume* of a *sphere*.

Finding the Volume of a Sphere

$$\text{Volume of a sphere} = \frac{4}{3} \cdot \pi \cdot r \cdot r \cdot r$$

$$V = \frac{4}{3} \cdot \pi \cdot r^3 \quad \text{or} \quad \frac{4 \cdot \pi \cdot r^3}{3}$$

Remember to use **cubic units** when measuring volume.

Answers

1. **(a)** 3 m; $V = 72 \text{ m}^3$
   (b) $V \approx 18{,}602.1 \text{ cm}^3$ (rounded)
   (c) $V = 43\frac{3}{4} \text{ ft}^3$ (exactly) or
   $V \approx 43.8 \text{ ft}^3$ (rounded)

EXAMPLE 2   **Finding the Volume of Spheres**

Find the volume of each sphere with the help of a calculator. Use 3.14 as the approximate value of π. Round your answers to the nearest tenth.

(a)

9 m

$V = \dfrac{4}{3} \cdot \pi \cdot r^3$ ⟵ r^3 means $r \cdot r \cdot r$

$V \approx \dfrac{4 \cdot 3.14 \cdot 9\,m \cdot 9\,m \cdot 9\,m}{3}$

$V \approx 3052.08\ m^3$ Now round to tenths.

$V \approx 3052.1\ m^3$ Cubic units for volume

(b)

4.2 ft

$V \approx \dfrac{4 \cdot 3.14 \cdot 4.2\,ft \cdot 4.2\,ft \cdot 4.2\,ft}{3}$

$V \approx 310.18176\ ft^3$ Now round to tenths.

$V \approx 310.2\ ft^3$ Cubic units for volume

▦ Calculator Tip

You can find the volume of the sphere in **Example 2(b)** above on your calculator. The first method works on both scientific and standard calculators:

4 ⊗ 3.14 ⊗ 4.2 ⊗ 4.2 ⊗ 4.2 ⊘ 3 =   Answer is 310.18176

Round the answer to 310.2 ft³

On a *scientific* calculator you can use the ⟨yˣ⟩ key (or the ⟨⌃⟩ key on some models) to calculate r^3 (to multiply the radius times itself three times).

4 ⊗ 3.14 ⊗ 4.2 ⟨yˣ⟩ 3 ⊘ 3 =   Answer is 310.18176
            ‾‾‾‾‾
              r^3

Recall that we are using 3.14 as the approximate value for π instead of using the ⟨π⟩ key.

You can also use the ⟨yˣ⟩ key with other exponents. For example:

To find 2^5, press 2 ⟨yˣ⟩ 5 =   Answer is 32.
To find 6^4, press 6 ⟨yˣ⟩ 4 =   Answer is 1296.

······· **Work Problem ❷ at the Side.** ▶

Half a sphere is called a *hemisphere*. The volume of a hemisphere is *half* the volume of a sphere. Use the following formula to find the *volume* of a hemisphere.

Finding the Volume of a Hemisphere

Volume of a hemisphere $= \dfrac{1}{\underset{1}{2}} \cdot \dfrac{\overset{2}{4}}{3} \cdot \pi \cdot r^3$

$V = \dfrac{2}{3} \cdot \pi \cdot r^3$ or $\dfrac{2 \cdot \pi \cdot r^3}{3}$

Remember to use **cubic units** when measuring volume.

❷ Find the volume of each sphere. ▦ Use 3.14 for π. Round your answers to the nearest tenth.

GS (a)

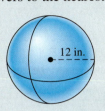

12 in.

$V = \dfrac{4}{3} \cdot \pi \cdot r^3$

$V \approx \dfrac{4 \cdot 3.14 \cdot 12\,in. \cdot 12\,in. \cdot 12\,in.}{3}$

$V \approx$ _____ Round to nearest tenth.

(b)

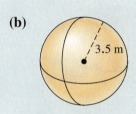

3.5 m

(c) Sphere with a radius of 2.7 cm

Answers

2. (a) $V \approx 7234.6$ in.³ (rounded)
   (b) $V \approx 179.5$ m³ (rounded)
   (c) $V \approx 82.4$ cm³ (rounded)

3 Find the volume of each hemisphere. Use 3.14 for π. Round answers to the nearest tenth.

(a)

$$V = \frac{2 \cdot \pi \cdot r \cdot r \cdot r}{3}$$

(b)

4 Find the volume of each cylinder. Use 3.14 for π. Round your answers to the nearest tenth.

GS (a)

$V = \pi \cdot r^2 \cdot h$

$V \approx 3.14 \cdot 4 \text{ ft} \cdot 4 \text{ ft} \cdot 12 \text{ ft}$

$V \approx \underline{\quad\quad}$

(b)

Hint: Find the **radius** first.

Answers

3. **(a)** $V \approx 7065 \text{ ft}^3$
   (b) $V \approx 452.2 \text{ cm}^3$ (rounded)
4. **(a)** $V \approx 602.9 \text{ ft}^3$ (rounded)
   (b) $V \approx 230.8 \text{ cm}^3$ (rounded)

EXAMPLE 3 **Finding the Volume of a Hemisphere**

Find the volume of the hemisphere with the help of a calculator. Use 3.14 for π. Round your answer to the nearest tenth.

$$V = \frac{2 \cdot \pi \cdot r^3}{3}$$

$$V \approx \frac{2 \cdot 3.14 \cdot 7 \text{ m} \cdot 7 \text{ m} \cdot 7 \text{ m}}{3}$$

Write **m³** for volume. → $V \approx 718.0 \text{ m}^3$ Rounded to nearest tenth

◀ **Work Problem 3 at the Side.**

OBJECTIVE 3 **Find the volume of a cylinder.** Here are several *cylinders.*

The height must be perpendicular to the circular top and bottom of the cylinder.

These are called *right circular cylinders* because the top and bottom are circles, and the side makes a right angle with the top and bottom. Examples of cylinders are a soup can, a home water heater, and a piece of pipe.

Use the formula below to find the *volume* of a *cylinder.* Notice that the first part of the formula, $\pi \cdot r^2$, is the area of the circular base.

> **Finding the Volume of a Cylinder**
>
> Volume of a cylinder $= \pi \cdot r \cdot r \cdot h$
>
> $$V = \pi \cdot r^2 \cdot h$$
>
> Remember to use **cubic units** when measuring volume.

EXAMPLE 4 **Finding the Volume of Cylinders**

Find the volume of each cylinder. Use 3.14 as the approximate value of π. Round your answers to the nearest tenth if necessary.

(a)

The diameter is 20 m so the radius is $\frac{20 \text{ m}}{2} = 10 \text{ m}$. The height is 9 m. Use the formula to find the volume.

$V = \pi \cdot r^2 \cdot h$

$V \approx 3.14 \cdot 10 \text{ m} \cdot 10 \text{ m} \cdot 9 \text{ m}$

$V \approx 2826 \text{ m}^3$ Cubic units for volume

(b)

$V \approx 3.14 \cdot 6.2 \text{ cm} \cdot 6.2 \text{ cm} \cdot 38.4 \text{ cm}$

$V \approx 4634.94144$ Now round to tenths.

$V \approx 4634.9 \text{ cm}^3$ Cubic units for volume

◀ **Work Problem 4 at the Side.**

OBJECTIVE **4** **Find the volume of a cone and a pyramid.** A cone and a pyramid are shown below. Notice that the height line is perpendicular to the base in both solids.

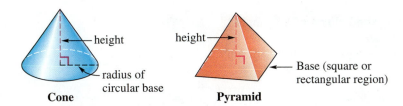

Cone   **Pyramid**

Use the formula below to find the *volume* of a cone.

Finding the Volume of a Cone

$$\text{Volume of a cone} = \frac{1}{3} \cdot B \cdot h$$

$$\text{or} \quad V = \frac{B \cdot h}{3}$$

where B is the area of the circular base of the cone and h is the height of the cone.

Remember to use *cubic units* when measuring volume.

EXAMPLE 5 **Finding the Volume of a Cone**

Find the volume of the cone. Use 3.14 for π. Round your answer to the nearest tenth.

9 cm

4 cm

First find the value of B in the formula, which is the *area of the circular base*. Recall that the formula for the area of a circle is $\pi \cdot r^2$.

$$B = \pi \cdot r \cdot r$$

$$B \approx 3.14 \cdot 4 \text{ cm} \cdot 4 \text{ cm}$$

$$B \approx 50.24 \text{ cm}^2 \quad \leftarrow \text{Do } \textbf{not} \text{ round to tenths yet.}$$

Now find the volume of the cone. The height is 9 cm.

$$V = \frac{B \cdot h}{3}$$

$$V \approx \frac{50.24 \text{ cm}^2 \cdot 9 \text{ cm}}{3}$$

$$V \approx 150.72 \text{ cm}^3 \quad \text{Now round to tenths.}$$

This is **volume**, so write **cm³** in the answer.

$$V \approx 150.7 \text{ cm}^3 \quad \text{Cubic units for volume}$$

Work Problem **5** **at the Side.** ▶

---

5 Find the volume of a cone with **GS** base radius 2 ft and height 11 ft. Use 3.14 for π. Round your answer to the nearest tenth.

First find the area of the circular base.

$$B = \pi \cdot r \cdot r$$

$$B \approx 3.14 \cdot 2 \text{ ft} \cdot 2 \text{ ft}$$

$$B \approx \rule{1.5cm}{0.4pt} \quad \text{Do } \textbf{not} \text{ round yet.}$$

Now find the volume of the cone.

$$V = \frac{B \cdot h}{3}$$

$$V \approx \frac{\rule{1.5cm}{0.4pt} \cdot 11 \text{ ft}}{3}$$

$$V \approx \rule{1.5cm}{0.4pt} \quad \text{Round to nearest tenth.}$$

Answer

5. $B \approx 12.56 \text{ ft}^2$

$$V \approx \frac{12.56 \text{ ft}^2 \cdot 11 \text{ ft}}{3}$$

$$V \approx 46.1 \text{ ft}^3 \text{ (rounded)}$$

6 Find the volume of a pyramid
GS with a square base 10 m by 10 m.
The height of the pyramid is 8 m.
Round your answer to the
nearest tenth.

First find the area of the square
base.

$B = 10 \text{ m} \cdot 10 \text{ m}$

$B = \underline{\hspace{1.5cm}}$

Now find the volume of the
pyramid.

$V = \dfrac{B \cdot h}{3}$

$V = \dfrac{\underline{\hspace{1cm}} \cdot \textbf{8 m}}{3}$

$V \approx \underline{\hspace{1.5cm}}$ Round to
nearest tenth.

Use the same formula to find the *volume* of a *pyramid* as you did to find
the *volume* of a *cone*.

Finding the Volume of a Pyramid

$$\text{Volume of a pyramid} = \frac{1}{3} \cdot B \cdot h$$

$$\text{or} \quad V = \frac{B \cdot h}{3}$$

where B is the area of the square or rectangular base of the pyramid and
h is the height of the pyramid.
Remember to use **cubic units** when measuring volume.

Note

In this book, we will work only with pyramids that have a base with four
sides (square or rectangle). In later math courses you may work with
pyramids that have a base with three sides (triangle), five sides (pentagon),
six sides (hexagon), and so on.

EXAMPLE 6   Finding the Volume of a Pyramid

Find the volume of this pyramid with a rectangular base. Round your answer
to the nearest tenth.

First find the value of B in the formula, which is the *area of the
rectangular base*. Recall that the area of a rectangle is found by multiplying
length times width.

$$B = 5 \text{ cm} \cdot 4 \text{ cm}$$

$$B = 20 \text{ cm}^2 \quad \boxed{\text{First find the } \textit{area} \text{ of the rectangular base.}}$$

Now find the volume of the pyramid.

$$V = \frac{B \cdot h}{3}$$

$$V = \frac{20 \text{ cm}^2 \cdot 11 \text{ cm}}{3}$$

$\boxed{\text{This is } \textbf{volume,} \text{ so write } \textbf{cm}^3 \text{ in the answer.}}$ $V \approx 73.3 \text{ cm}^3$ Rounded to nearest tenth

◄ **Work Problem 6 at the Side.**

Answer

6. $B = 100 \text{ m}^2$

$V = \dfrac{100 \text{ m}^2 \cdot 8 \text{ m}}{3}$

$V \approx 266.7 \text{ m}^3$ (rounded)

8.7 Exercises

FOR EXTRA HELP

Download the MyDashBoard App

▶ MyMathLab®

1. CONCEPT CHECK Look at the solid shape in **Exercise 3** below. Explain why this solid is **not** a cube. When you find the volume of this solid, what will be the units in your answer?

2. CONCEPT CHECK Give two examples of everyday objects that are rectangular solids; two objects that are cylinders; and two objects that are spheres.

Name each solid and find its volume. Use 3.14 as the approximate value of π. Round your answers to the nearest tenth if necessary. **See Examples 1–6.**

3.
12.5 cm
11 cm
4 cm

4.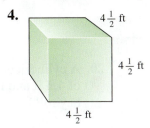
$4\frac{1}{2}$ ft
$4\frac{1}{2}$ ft
$4\frac{1}{2}$ ft

5.
22 m

Hint: This solid is a **sphere**.

6.
1.53 m

Hint: The units for this answer will be **m³**.

7.
12 in.

8.
7.4 in.

9.
5 ft
6 ft

Hint: The units for this answer will be **ft³**.

10.
12 in.
21 in.

Hint: This solid is a **cylinder**.

11.
16 m
5 m

12.
28 cm
40 cm

13.
20 cm
15 cm
8 cm

14.
15 m
5 m
7 m

Solve each application problem. Use 3.14 as the approximate value of π. Round your final answers to the nearest tenth if necessary.

15. A pencil box measures 3 in. by 8 in. by $\frac{3}{4}$ in. high. Find the volume of the box. (*Source:* Faber Castell.)

Hint: A pencil box is a rectangular solid.

16. A train is being loaded with shipping crates. Each one is 6 m long, 3.4 m wide, and 2 m high. How much space will each crate take?

Hint: A shipping crate is a rectangular solid.

17. An oil candle globe made of hand-blown glass has a diameter of 16.8 cm. What is the volume of the globe?

Hint: A globe is another word for a sphere. First find the **radius** of the globe.

18. A metal sphere used as part of a fountain has a diameter of $6\frac{1}{2}$ ft. Find its volume.

Hint: First find the **radius** of the sphere.

19. One of the ancient stone pyramids in Egypt has a square base that measures 145 m on each side. The height is 93 m. What is the volume of the pyramid? (*Source: The Columbia Encyclopedia.*)

20. A cylindrical roll of hay in a farm field has a diameter of 4.5 ft and is 3.5 ft long. What is the volume of the hay roll?

21. A city sewer pipe has a diameter of 5 ft and a length of 200 ft. Find the volume of the pipe.

22. An ice cream cone has a diameter of 2 in. and a height of 4 in. Find its volume.

23. CONCEPT CHECK Explain the *two* errors made by a student in finding the volume of a cylinder with a diameter of 7 cm and a height of 5 cm. Find the correct answer.

$$V \approx 3.14 \cdot 7 \cdot 7 \cdot 5$$

$$V \approx 769.3 \text{ cm}^2$$

24. CONCEPT CHECK Compare the steps in finding the volume of a cylinder and a cone. How are they similar? Suppose you know the volume of a cylinder. How can you find the volume of a cone with the same radius and height by doing just a one-step calculation?

8.8 Pythagorean Theorem

Earlier in **Chapter 8** you used this formula for the area of a square, $A = s^2$. The blue square below has an area of 25 cm² because 5 cm • 5 cm = 25 cm².

5 cm

5 cm

Area = 25 cm²
Area = 5 cm · 5 cm

side = ? cm

Area = 49 cm²
Area = ? cm · ? cm

The red square above has an area of 49 cm². To find the length of a side, ask yourself, "What number can be multiplied by itself to give 49?" Because 7 • 7 = 49, the length of each side is 7 cm.

Also, because 7 • 7 = 49, we say that 7 is the *square root* of 49, or $\sqrt{49} = 7$. Also, $\sqrt{81} = 9$, because 9 • 9 = 81. (See **Chapter 1** for further review.)

Work Problem ❶ at the Side. ▶

A number that has a whole number as its square root is called a *perfect square*. For example, 9 is a perfect square because $\sqrt{9} = 3$, and 3 is a whole number.

Some of the perfect squares are listed below.

Some of the Perfect Squares			
$\sqrt{1} = 1$	$\sqrt{16} = 4$	$\sqrt{49} = 7$	$\sqrt{100} = 10$
$\sqrt{4} = 2$	$\sqrt{25} = 5$	$\sqrt{64} = 8$	$\sqrt{121} = 11$
$\sqrt{9} = 3$	$\sqrt{36} = 6$	$\sqrt{81} = 9$	$\sqrt{144} = 12$

OBJECTIVE ▶ ❶ Find square roots using the square root key on a calculator.
If a number is *not* a perfect square, then you can find its *approximate* square root by using a calculator with a square root key.

> 🖩 **Calculator Tip**
>
> To find a square root, use the ⌐√ key, or, on some calculators, the ⌐√x̄ key. Try these.
>
> To find $\sqrt{16}$ press: 16 ⌐√x̄   Answer is 4.
>
> To find $\sqrt{7}$ press: 7 ⌐√x̄   Answer is 2.645751311
>
> On some scientific calculators, you press the √ key first, then enter the number. (Check your model's operating instructions.)
>
> For $\sqrt{7}$, your calculator shows 2.645751311, an *approximate* answer. (Some calculators show more or fewer digits.) We will be rounding to the nearest thousandth, so $\sqrt{7} \approx 2.646$. To check, multiply 2.646 times 2.646. Do you get 7 as the result? No, you get 7.001316, which is very close to 7. The difference is due to rounding.

OBJECTIVES

❶ Find square roots using the square root key on a calculator.

❷ Find the unknown length in a right triangle.

❸ Solve application problems involving right triangles.

❶ Find each square root.

GS (a) $\sqrt{36}$

What number, multiplied by itself, gives 36? ____

So, $\sqrt{36} =$ ____

(b) $\sqrt{25}$

(c) $\sqrt{9}$

(d) $\sqrt{100}$

(e) $\sqrt{121}$

Answers

1. **(a)** 6; 6 **(b)** 5 **(c)** 3 **(d)** 10 **(e)** 11

2 Use a calculator with a square root key to find each square root. Round to the nearest thousandth if necessary.

(a) $\sqrt{11}$

(b) $\sqrt{40}$

(c) $\sqrt{56}$

(d) $\sqrt{196}$

(e) $\sqrt{147}$

Answers

2. **(a)** $\sqrt{11} \approx 3.317$ **(b)** $\sqrt{40} \approx 6.325$
   (c) $\sqrt{56} \approx 7.483$ **(d)** $\sqrt{196} = 14$
   (e) $\sqrt{147} \approx 12.124$

EXAMPLE 1   **Finding the Square Root of Numbers**

Use a calculator to find each square root. Round answers to the nearest thousandth.

Your calculator may show more or fewer digits.

(a) $\sqrt{35}$ Calculator shows 5.916079783; round to 5.916

(b) $\sqrt{124}$ Calculator shows 11.13552873; round to 11.136

(c) $\sqrt{200}$ Calculator shows 14.14213562; round to 14.142

◀ Work Problem **2** at the Side.

OBJECTIVE **2** **Find the unknown length in a right triangle.** One place you will use square roots is when working with the *Pythagorean Theorem*. This theorem applies only to *right* triangles (triangles with a 90° angle). The longest side of a right triangle is called the **hypotenuse.** It is opposite the right angle. The other two sides are called *legs*. The legs form the right angle.

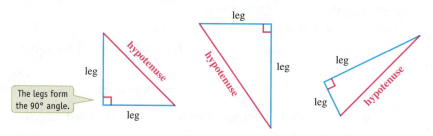

Examples of right triangles

Pythagorean Theorem

$$(\text{hypotenuse})^2 = (\text{leg})^2 + (\text{leg})^2$$

In other words, square the length of each side. After you have squared all the sides, the sum of the squares of the two legs will equal the square of the hypotenuse. An example is shown below.

$$(\text{hypotenuse})^2 = (\text{leg})^2 + (\text{leg})^2$$
$$5^2 = 4^2 + 3^2$$
$$25 = 16 + 9$$
$$25 = 25$$

The theorem is named after Pythagoras, a Greek mathematician who lived about 2500 years ago. He and his followers may have used triangular floor tiles to prove the theorem, as shown below.

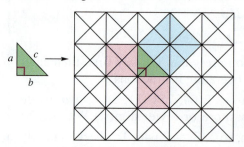

The green right triangle in the center of the floor tiles has sides a, b, and c. The pink square drawn on side a contains four triangular tiles. The pink square on side b contains four tiles. The blue square on side c contains eight tiles. The number of tiles in the square on side c equals the sum of the number of tiles in the squares on sides a and b, that is, 8 tiles = 4 tiles + 4 tiles. As a result, you often see the Pythagorean Theorem written as $c^2 = a^2 + b^2$.

If you know the lengths of any two sides in a right triangle, you can use the Pythagorean Theorem to find the length of the third side.

> ### Formulas Based on the Pythagorean Theorem
> To find the hypotenuse, use this formula:
> $$\text{hypotenuse} = \sqrt{(\text{leg})^2 + (\text{leg})^2}$$
>
> To find a leg, use this formula:
> $$\text{leg} = \sqrt{(\text{hypotenuse})^2 - (\text{leg})^2}$$

> **CAUTION**
>
> *Remember:* A small square drawn in one angle of a triangle indicates a right angle. You can use the Pythagorean Theorem *only* on triangles that have a right angle.

EXAMPLE 2   Finding the Unknown Length in Right Triangles

Find the unknown length in each right triangle. Round answers to the nearest tenth if necessary.

(a)

3 ft

4 ft

The **unknown length** is the side opposite the right angle, which is the **hypotenuse.** Use the formula for finding the hypotenuse.

$\text{hypotenuse} = \sqrt{(\text{leg})^2 + (\text{leg})^2}$ Find the hypotenuse.

$\text{hypotenuse} = \sqrt{(3)^2 + (4)^2}$ Legs are 3 and 4.

$= \sqrt{9 + 16}$ $3 \cdot 3$ is 9   and $4 \cdot 4$ is 16.

$= \sqrt{25}$

$= 5$ This is the length of a side, so write **ft** in the answer (**not** ft²).

The hypotenuse is 5 ft long.

(b)

15 cm

7 cm

We *do* know the length of the hypotenuse (15 cm), so it is **the length of one of the legs that is unknown.** Use the formula for finding a leg.

$\text{leg} = \sqrt{(\text{hypotenuse})^2 - (\text{leg})^2}$ Find a leg.

$\text{leg} = \sqrt{(15)^2 - (7)^2}$ Hypotenuse is 15; one leg is 7.

$= \sqrt{225 - 49}$ $15 \cdot 15$ is 225   and $7 \cdot 7$ is 49.

$= \sqrt{176}$ Use calculator to find $\sqrt{176}$.

≈ 13.3 Round 13.26649916 to 13.3

The length of the leg is approximately 13.3 cm.   Write **cm** in the answer, **not** cm².

·········· **Work Problem ③ at the Side.** ▶

③ Find the unknown length in each right triangle. Round your answers to the nearest tenth if necessary.

GS (a)

5 in.

12 in.

hypotenuse

$\text{hypotenuse} = \sqrt{(5)^2 + (12)^2}$

$= \sqrt{25 + \underline{}}$

$= \sqrt{\underline{}} = \underline{}$

GS (b)

7 cm   90°   25 cm   hypotenuse

$\text{leg} = \sqrt{(25)^2 - (7)^2}$

$= \sqrt{625 - \underline{}}$

$= \sqrt{\underline{}} = \underline{}$

(c)

13 m

17 m

(d)

18 ft   20 ft

Answers

3. **(a)** $\sqrt{25 + 144} = \sqrt{169} = 13$ in.
   (b) $\sqrt{625 - 49} = \sqrt{576} = 24$ cm
   (c) $\sqrt{458} \approx 21.4$ m   **(d)** $\sqrt{76} \approx 8.7$ ft

4 These problems show ladders leaning against buildings. Find the unknown lengths. Round to the nearest tenth of a foot if necessary.

(a)

How far away from the building is the bottom of the ladder? (*Hint:* The ladder is the **hypotenuse.**)

(b)

How long is the ladder?

(c) A 17 ft ladder is leaning against a building. The bottom of the ladder is 10 ft from the building. How high up on the building will the ladder reach? (*Hint:* Start by drawing the building and the ladder.)

Answers

4. (a) leg = $\sqrt{225}$ = 15 ft

(b) hypotenuse = $\sqrt{185}$ ≈ 13.6 ft

(c) leg = $\sqrt{189}$ ≈ 13.7 ft

OBJECTIVE **3** **Solve application problems involving right triangles.**
The next example shows an application of the Pythagorean Theorem.

EXAMPLE 3 **Using the Pythagorean Theorem**

A television antenna is on the roof of a house, as shown below. Find the length of the support wire. Round your answer to the nearest tenth of a meter if necessary.

A right triangle is formed. The total length of the leg on the left is 3 m + 2 m = 5 m.

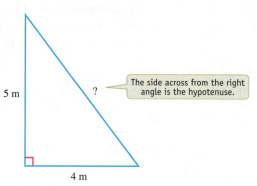

The side across from the right angle is the hypotenuse.

Notice that the support wire is opposite the right angle, so it is the hypotenuse of the right triangle.

hypotenuse = $\sqrt{(\text{leg})^2 + (\text{leg})^2}$ Find the hypotenuse.

hypotenuse = $\sqrt{(5)^2 + (4)^2}$ Legs are 5 and 4.

$= \sqrt{25 + 16}$ 5^2 is 25 and 4^2 is 16.

$= \sqrt{41}$ Use a calculator to find $\sqrt{41}$.

≈ 6.4 Round 6.403124237 to 6.4

The length of the support wire is approximately 6.4 m. — This is **length**, so write **m** in the answer (**not** m²).

CAUTION

You use the Pythagorean Theorem to find the **length** of one side, *not* the area of the triangle. Your answer will be in linear units, such as ft, yd, cm, m, and so on (*not* ft², yd², cm², m²).

◀ **Work Problem** **4** at the Side.

8.8 Exercises

   MyMathLab®

Find each square root. Starting with Exercise 5, use the square root key on a calculator. Round your answers to the nearest thousandth if necessary. **See Example 1.**

1. $\sqrt{16}$

2. $\sqrt{4}$

3. $\sqrt{64}$

4. $\sqrt{81}$

5. $\sqrt{11}$

6. $\sqrt{23}$

7. $\sqrt{5}$

8. $\sqrt{2}$

9. $\sqrt{73}$

10. $\sqrt{80}$

11. $\sqrt{101}$

12. $\sqrt{125}$

13. $\sqrt{190}$

14. $\sqrt{160}$

15. $\sqrt{1000}$

16. $\sqrt{2000}$

17. CONCEPT CHECK You know that $\sqrt{25} = 5$ and $\sqrt{36} = 6$. Using just that information (no calculator), describe how you could *estimate* $\sqrt{30}$. How would you estimate $\sqrt{26}$ or $\sqrt{35}$? Now check your estimates using a calculator.

18. CONCEPT CHECK Explain the relationship between *squaring* a number and finding the *square root* of a number. Include two examples to illustrate your explanation.

Find the unknown length in each right triangle. Use a calculator to find square roots. Round your answers to the nearest tenth if necessary. **See Example 2.**

19.

Hint: The unknown length is the **hypotenuse.**

hypotenuse = $\sqrt{(15)^2 + (36)^2}$

20.

Hint: The unknown length is the **hypotenuse.**

hypotenuse = $\sqrt{(9)^2 + (12)^2}$

21.

22.

23.

Hint: The unknown length is a **leg.**

leg = $\sqrt{(20)^2 - (16)^2}$

24.

Hint: The unknown length is a **leg.**

leg = $\sqrt{(13)^2 - (5)^2}$

25.

3 in.

8 in.

26.

5 cm

11 cm

27.

7 yd

90°

4 yd

28.

7 km

10 km

29.

22 cm

17 cm

30.

16 cm

9 cm

90°

31.

1.3 m

90°

2.5 m

32.

4.2 mi

4.2 mi

33.

11.5 cm

8.2 cm

34.

9.1 mm

10.8 mm

35.

13.2 km

90°

21.6 km

36.

26.5 ft

37.4 ft

37. CONCEPT CHECK Explain the *two* errors made by a student in solving this problem. Also find the correct answer. Round to the nearest tenth.

$$? = \sqrt{(9)^2 + (7)^2}$$
$$= \sqrt{18 + 14}$$
$$= \sqrt{32} \approx 5.657 \text{ in.}$$

9 in.

?

7 in.

38. CONCEPT CHECK Explain the *two* errors made by a student in solving this problem. Also find the correct answer. Round to the nearest tenth.

$$? = \sqrt{(13)^2 + (20)^2}$$
$$= \sqrt{169 + 400}$$
$$= \sqrt{569} \approx 23.9 \text{ m}^2$$

?

13 m

20 m

Solve each application problem. Round your answers to the nearest tenth if necessary.
See Example 3.

39. Find the length of this loading ramp.
Hint: The ramp is the **hypotenuse** in a right triangle.

40. Find the unknown length in this window frame.
Hint: The unknown length is the **hypotenuse** in a right triangle.

41. How high is the airplane above the ground?
Hint: The height of the plane is a **leg** in the right triangle.

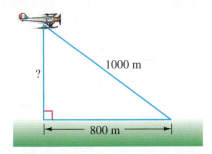

42. Find the height of this farm silo.
Hint: The height of the silo is a **leg** in the right triangle.

43. How long is the diagonal brace on this rectangular gate?

44. Find the height of this rectangular television screen.

45. To reach his ladylove, a knight placed a 12 ft ladder against the castle wall. If the base of the ladder is 3 ft from the building, how high on the castle will the top of the ladder reach? Draw a sketch of the castle and ladder and solve the problem.

46. William drove his car 15 miles north, then made a 90° right turn and drove 7 miles east. How far is he, in a straight line, from his starting point? Draw a sketch to illustrate the problem and solve it.

Relating Concepts (Exercises 47–50) For Individual or Group Work

Use your knowledge of the Pythagorean Theorem to **work Exercises 47–50 in order.** *Round answers to the nearest tenth.*

47. A major league baseball diamond is a square shape measuring 90 ft on each side. If the catcher throws a ball from home plate to second base, how far is he throwing the ball? (*Source:* American League of Professional Baseball Clubs.)

48. A softball diamond is only 60 ft on each side. (*Source:* Amateur Softball Association.)

(a) Draw a sketch of the softball diamond and label the bases and the lengths of the sides.

(b) How far is it to throw a ball from home plate to second base?

49. Look back at your answer to **Exercise 47.** Explain how you can tell the distance from third base to first base without doing any further calculations.

50. Show how you could set up a proportion to answer **Exercise 48** instead of using the Pythagorean Theorem. (You'll need your answer from **Exercise 47.**)

8.9 Similar Triangles

Two triangles with the same *shape* (but not necessarily the same size) are called **similar triangles.** Three pairs of similar triangles are shown below.

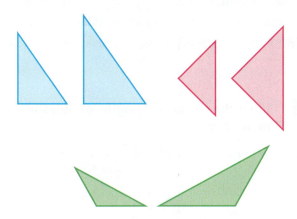

OBJECTIVE **1** **Identify corresponding parts in similar triangles.** The two triangles shown below are different sizes but have the same shape, so they are *similar triangles.* Angles *A* and *P* measure the same number of degrees and are called *corresponding angles.* Angles *B* and *Q* are corresponding angles, as are angles *C* and *R.* The triangles have the same shape because the corresponding angles have the same measure.

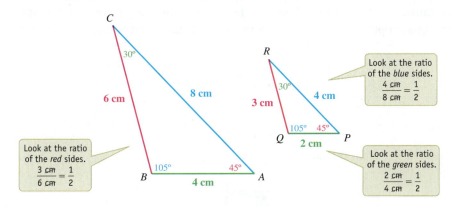

Look at the ratio of the *red* sides.
$$\frac{3\text{ cm}}{6\text{ cm}} = \frac{1}{2}$$

Look at the ratio of the *blue* sides.
$$\frac{4\text{ cm}}{8\text{ cm}} = \frac{1}{2}$$

Look at the ratio of the *green* sides.
$$\frac{2\text{ cm}}{4\text{ cm}} = \frac{1}{2}$$

$\overline{PR}$ and $\overline{AC}$ are called *corresponding sides* because they are *opposite* corresponding angles. Also, $\overline{QR}$ and $\overline{BC}$ are corresponding sides, as are $\overline{PQ}$ and $\overline{AB}$. Although corresponding angles measure the same number of degrees, corresponding sides do *not* need to be the same length. In the triangles here, each side in the smaller triangle is *half* the length of the corresponding side in the larger triangle.

Work Problem 1 at the Side. ▶

OBJECTIVES

1. Identify corresponding parts in similar triangles.

2. Find the unknown lengths of sides in similar triangles.

3. Solve application problems involving similar triangles.

1 Identify corresponding angles and sides in these similar triangles.

(a)

Angles:

1 and _____

2 and _____

3 and _____

Sides:

$\overline{PN}$ and _____

$\overline{PM}$ and _____

$\overline{NM}$ and _____

(b)

Hint: The smaller triangle is "flipped over" compared to the larger triangle.

Angles:

1 and _____

2 and _____

3 and _____

Sides:

$\overline{AB}$ and _____

$\overline{BC}$ and _____

$\overline{AC}$ and _____

Answers

1. **(a)** 4; 5; 6; $\overline{ZX}$; $\overline{ZY}$; $\overline{XY}$
   (b) 6; 4; 5; $\overline{EF}$; $\overline{FG}$; $\overline{EG}$

2 Find the length of $\overline{EF}$ in
Example 1 by setting up and
solving a proportion. Let x
represent the unknown length.

$$\begin{array}{c}EF \to \\ CB \to\end{array} \dfrac{x}{33} = \dfrac{}{} \begin{array}{c}\leftarrow ED \\ \leftarrow CA\end{array}$$

OBJECTIVE **2** **Find the unknown lengths of sides in similar triangles.**
Similar triangles are useful because of the following definition.

> **Similar Triangles**
>
> If two triangles are **similar**, then
>
> 1. Corresponding angles have the same measure, and
> 2. The **ratios** of the lengths of corresponding sides are equal.

EXAMPLE 1 Finding the Unknown Lengths of Sides in Similar Triangles

Find the length of y in the smaller triangle. Assume the triangles are similar.

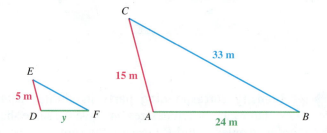

The length you want to find in the smaller triangle is $\overline{DF}$, and it corresponds to $\overline{AB}$ in the larger triangle. Then, notice that $\overline{ED}$ in the smaller triangle corresponds to $\overline{CA}$ in the larger triangle, and you know both of their lengths. Because the *ratios* of the lengths of corresponding sides are equal, you can set up a proportion. (Recall that a proportion states that two ratios are equal.)

$$\text{Corresponding sides} \begin{cases} DF \to \\ AB \to \end{cases} \dfrac{y}{24} = \dfrac{5}{15} \begin{cases} \leftarrow ED \\ \leftarrow CA \end{cases} \text{Corresponding sides}$$

$$\dfrac{y}{24} = \dfrac{1}{3} \qquad \text{Write } \tfrac{5}{15} \text{ in lowest terms as } \tfrac{1}{3}.$$

Find the cross products.

$$\dfrac{y}{24} = \dfrac{1}{3} \qquad \begin{array}{l} 24 \cdot 1 = 24 \\ \\ y \cdot 3 \end{array}$$

$$y \cdot 3 = 24 \qquad \text{Show that the cross products are equal.}$$

$$\dfrac{y \cdot \overset{1}{\cancel{3}}}{\cancel{3}_{1}} = \dfrac{24}{3} \qquad \text{Divide both sides by 3.}$$

Write **m** in the answer.

$$y = 8$$

$\overline{DF}$ has a length of 8 m.

◀ **Work Problem 2** at the Side.

Answer

2. $\dfrac{x}{33} = \dfrac{5}{15}; x = 11$ m

EXAMPLE 2   **Finding an Unknown Length and the Perimeter**

Find the perimeter of the smaller triangle. Assume the triangles are similar.

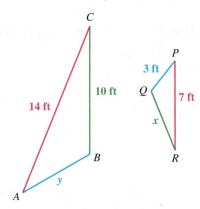

First find x, the length of $\overline{QR}$, then add the lengths of all three sides to find the perimeter.

The smaller triangle is turned "upside down" compared to the larger triangle, so be careful when identifying corresponding sides. $\overline{PR}$ is the longest side in the smaller triangle, and $\overline{AC}$ is the longest side in the larger triangle. So $\overline{PR}$ and $\overline{AC}$ are corresponding sides and you know both of their lengths. $\overline{QR}$, the length you want to find in the smaller triangle, corresponds to $\overline{BC}$ in the larger triangle. The ratios of the lengths of corresponding sides are equal, so you can set up a proportion.

$$\begin{array}{ll} QR \rightarrow \dfrac{x}{10} = \dfrac{7}{14} & \leftarrow PR \\ BC \rightarrow & \leftarrow AC \end{array}$$

$$\dfrac{x}{10} = \dfrac{1}{2} \qquad \text{Write } \tfrac{7}{14} \text{ in lowest terms as } \tfrac{1}{2}.$$

Find the cross products.

$$\dfrac{x}{10} = \dfrac{1}{2} \qquad \begin{array}{l} 10 \cdot 1 = 10 \\[6pt] x \cdot 2 \end{array}$$

$$x \cdot 2 = 10 \qquad \text{Show that the cross products are equal.}$$

$$\dfrac{x \cdot \overset{1}{2}}{\underset{1}{2}} = \dfrac{10}{2} \qquad \text{Divide both sides by 2.}$$

> Write **ft** for the length.

$$x = 5$$

$\overline{QR}$ has a length of 5 ft.

Now add the lengths of all three sides to find the perimeter of the smaller triangle.

$$\text{Perimeter} = 5 \text{ ft} + 3 \text{ ft} + 7 \text{ ft} = 15 \text{ ft}$$

· ▸ **Work Problem ❸ at the Side.** ▸

---

❸ (a) Find the perimeter of
GS triangle ABC in **Example 2** at the left. Let y represent the unknown length of side AB.

$$\begin{array}{ll} PQ \rightarrow \dfrac{3}{y} = \underline{\hphantom{xxx}} & \leftarrow PR \\ AB \rightarrow & \leftarrow AC \end{array}$$

(b) Find the perimeter of each triangle. Assume the triangles are similar.

4 Find the height of each flagpole.

GS **(a)**

$$\underset{\substack{\text{Longer} \\ \text{height} \\ \text{Shorter} \\ \text{height}}}{\longrightarrow} \frac{h}{5} = \underline{\qquad} \underset{\substack{\text{Longer} \\ \text{shadow} \\ \text{Shorter} \\ \text{shadow}}}{\longleftarrow}$$

(b)

The next example shows an application of similar triangles.

EXAMPLE 3 **Using Similar Triangles in an Application**

A flagpole casts a shadow 99 m long at the same time that a pole 10 m tall casts a shadow 18 m long. Find the height of the flagpole.

The triangles shown are similar, so write a proportion to find h.

$$\underset{\text{\textbf{Height} in smaller triangle}}{\overset{\text{\textbf{Height} in larger triangle}}{\longrightarrow}} \frac{h}{10} = \frac{99}{18} \underset{\longleftarrow \text{\textbf{Shadow} in smaller triangle}}{\overset{\longleftarrow \text{\textbf{Shadow} in larger triangle}}{}}$$

Find the cross products and show that they are equal.

$$h \cdot 18 = 10 \cdot 99$$

$$h \cdot 18 = 990$$

$$\frac{h \cdot \overset{1}{\cancel{18}}}{\underset{1}{\cancel{18}}} = \frac{990}{18} \qquad \text{Divide both sides by 18.}$$

$$h = 55$$

The flagpole is 55 m high.

Note

There are several other correct ways to set up the proportion in **Example 3** above. One way is to simply flip the ratios on *both* sides of the equal sign.

$$\frac{10}{h} = \frac{18}{99}$$

But there is another option, shown below.

$$\underset{\text{Shadow in \textbf{larger} triangle}}{\overset{\text{Height in \textbf{larger} triangle}}{\longrightarrow}} \frac{h}{99} = \frac{10}{18} \underset{\longleftarrow \text{Shadow in \textbf{smaller} triangle}}{\overset{\longleftarrow \text{Height in \textbf{smaller} triangle}}{}}$$

Notice that both ratios compare *height* to *shadow* in the same order. The ratio on the left describes the larger triangle, and the ratio on the right describes the smaller triangle.

◀ **Work Problem** **4** at the Side.

Answers

4. (a) $\dfrac{h}{5} = \dfrac{48}{12}$; $h = 20$ ft   (b) $h = 18$ m

8.9 Exercises

FOR EXTRA HELP

MyMathLab®

CONCEPT CHECK *Which pairs of triangles appear to be similar? Write* similar *or* not similar *for each pair.*

1.

2.

3.

4.

5. CONCEPT CHECK In two triangles that are similar,

(a) explain what is special about the angles.

(b) explain what is special about the lengths of the sides.

6. CONCEPT CHECK Look up the word *similar* in a dictionary. What is the nonmathematical definition of this word? Describe two examples of similar objects at home, school, or work.

Name the corresponding angles and the corresponding sides in each pair of similar triangles. **See Margin Problem 1.**

7.

8.

9.

10.

*Write the ratio for each pair of corresponding sides in the similar triangles shown below. Write the ratios as fractions in lowest terms. **See Examples 1 and 2.***

11. $\dfrac{AB}{PQ}; \dfrac{AC}{PR}; \dfrac{BC}{QR}$

12. $\dfrac{AB}{PQ}; \dfrac{AC}{PR}; \dfrac{BC}{QR}$

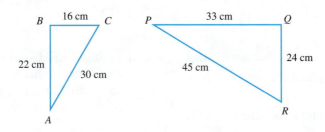

*Find the unknown lengths in each pair of similar triangles. **See Example 1.***

13.

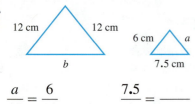

$$\dfrac{a}{12} = \dfrac{6}{12} \qquad \dfrac{7.5}{b} = \underline{\hspace{1cm}}$$

14.

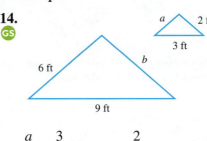

$$\dfrac{a}{6} = \dfrac{3}{} \qquad \dfrac{2}{b} = \underline{\hspace{1cm}}$$

15.

16.

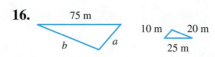

*Find the perimeter of each triangle. Assume the triangles are similar. See **Example 2**.*

17.

18.

19. Triangles *CDE* and *FGH* are similar. Find the perimeter and area of triangle *FGH*. *Note:* The heights of similar triangles have the same ratio as corresponding sides. Round to the nearest tenth when necessary.

20. Triangles *JKL* and *MNO* are similar. Find the perimeter and area of triangle *MNO*.

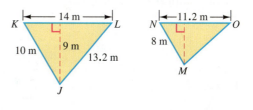

Solve each application problem. See Example 3.

21. The height of the house shown here can be found by comparing its shadow to the shadow cast by a 3 ft stick. Find the height of the house by writing a proportion and solving it.

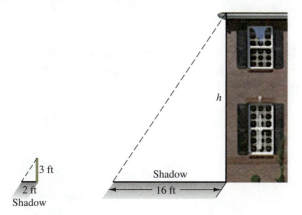

22. A fire lookout tower provides an excellent view of the surrounding countryside. The height of the tower can be found by lining up the top of the tower with the top of a 2-meter stick. Use similar triangles to find the height of the tower.

23. Refer to the building in **Exercise 21.** Later in the day, the same building had a shadow 6 ft long. How long would the stick's shadow be at that time?

24. Refer to the lookout tower in **Exercise 22.**

(a) How far away from the tower was the 2 m stick?

(b) To use a 1-meter stick and have it line up with the same endpoint, how far from the tower would it have to be?

Find the unknown length in Exercises 25–28. Round your answers to the nearest tenth.
Note: When a line is drawn parallel to one side of a triangle, the smaller triangle that
is formed will be similar to the original triangle. In Exercises 25–26, the red segments
are parallel.

25.

Hint: Redraw the two triangles and label the sides.

26.

Hint: Redraw the two triangles and label the sides.

27. Use similar triangles and a proportion to find the length of the lake shown here. (*Hint:* The side 100 m long in the smaller triangle corresponds to the side of 100 m + 120 m = 220 m in the larger triangle.)

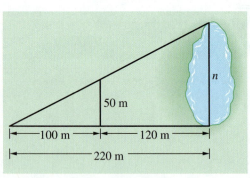

28. To find the height of the tree, find y and then add $5\frac{1}{2}$ ft for the distance from the ground to the person's eye level.

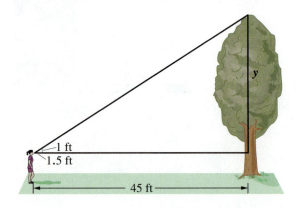

Chapter 8 Summary

Key Terms

8.1

point A point is a location in space. *Example:* Point *P* at the right.

• *P*

line A line is a straight row of points that goes on forever in both directions. *Example:* Line *AB*, written $\overleftrightarrow{AB}$, at the right.

A   *B*

line segment A line segment is a piece of a line with two endpoints. *Example:* Line segment *PQ*, written $\overline{PQ}$, at the right.

P   *Q*

ray A ray is a part of a line that has one endpoint and extends forever in one direction. *Example:* Ray *RS*, written $\overrightarrow{RS}$, at the right.

R   *S*

angle An angle is made up of two rays that have a common endpoint called the vertex. *Example:* Angle 1 at the right.

degrees Degrees are used to measure angles; a complete circle is 360 degrees, written 360°.

right angle A right angle is an angle that measures *exactly* 90°, identified by a small square at the vertex. *Example:* Angle *AOB* at the right.

acute angle An acute angle is an angle that measures less than 90°. *Example:* Angle *E* at the right.

obtuse angle An obtuse angle is an angle that measures more than 90° but less than 180°. *Example:* Angle *F* at the right.

straight angle A straight angle is an angle that measures *exactly* 180°; its sides form a straight line. *Example:* Angle *G* at the right.

intersecting lines Intersecting lines cross. *Example:* $\overleftrightarrow{RQ}$ intersects $\overleftrightarrow{AB}$ at point *P* at the right.

perpendicular lines Perpendicular lines are two lines that intersect to **form a right angle.** *Example:* $\overleftrightarrow{PQ}$ **is perpendicular to** $\overleftrightarrow{RS}$ **at the right.**

parallel lines Parallel lines are two lines in the same plane that never intersect (never cross). *Example:* $\overleftrightarrow{AB}$ **is parallel to** $\overleftrightarrow{ST}$ **at the right.**

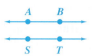

8.2

complementary angles Complementary angles are two angles whose measures add up to 90°.

supplementary angles Supplementary angles are two angles whose measures add up to 180°.

congruent angles Congruent angles are angles that measure the same number of degrees.

vertical angles Vertical angles are two nonadjacent congruent angles formed by two intersecting lines. *Example:* ∠*COA* and ∠*EOF* are vertical angles at the right.

8.3–8.5

perimeter Perimeter is the distance around the outside edges of a flat shape. It is measured in linear units such as ft, yd, cm, m, km, and so on.

8.3–8.6

area Area is the surface inside a two-dimensional (flat) shape. It is measured by determining the number of squares of a certain size needed to cover the surface inside the shape. Some of the commonly used units for measuring area are square inches (in.^{2}), square feet (ft^2), square yards (yd^2), square centimeters (cm^2), and square meters (m^2).

8.3

rectangle A rectangle is a four-sided figure with all sides meeting at 90° angles. The opposite sides are the same length. *Example:* The rectangle measuring 12 cm by 7 cm at the right.

square A square is a rectangle with all four sides the same length. *Example:* The square with a side measurement of 20 inches at the right.

8.4

parallelogram A parallelogram is a four-sided figure with both pairs of opposite sides parallel and equal in length. *Example:* The parallelogram at the right with sides measuring 8 m and 12 m.

trapezoid A trapezoid is a four-sided figure with exactly one pair of parallel sides. *Example:* Trapezoid *PQRS* at the right; $\overline{PQ}$ is parallel to $\overline{SR}$.

8.5

triangle A triangle is a figure with exactly three sides. *Example:* Triangle *ABC* at the right.

8.6

circle A circle is a figure with all points the same distance from a fixed center point. *Example:* See figure at the right.

radius Radius is the distance from the center of a circle to any point on the circle. *Example:* See the red radius in the circle at the right.

diameter Diameter is the distance across a circle, passing through the center. *Example:* See the blue diameter in the circle at the right.

circumference Circumference is the distance around a circle.

π (pi) π is the ratio of the circumference to the diameter of any circle. It is approximately equal to 3.14.

8.7

volume Volume is a measure of the space inside a three-dimensional (solid) shape. Volume is measured in cubic units such as in.^{3}, ft^3, yd^3, mm^3, cm^3, and so on.

8.8

hypotenuse The hypotenuse is the side of a right triangle opposite the 90° angle; it is the longest side. *Example:* See the red side in the triangle at the right.

8.9

similar triangles Similar triangles are triangles with the same shape but not necessarily the same size; corresponding angles measure the same number of degrees, and the *ratios* of the lengths of corresponding sides are equal.

New Symbols

$\overleftrightarrow{AB}$	line *AB*	**Right angle:** (90° angle)	**Square units for measuring area**	in.^{2} ft^2 yd^2 mi^2
		small square in corner		mm^2 cm^2 m^2 km^2
$\overline{EF}$	line segment *EF*			
		⊥ is perpendicular to	**Cubic units for measuring volume**	in.^{3} ft^3 yd^3
$\overrightarrow{RS}$	ray *RS*			mm^3 cm^3 m^3
		≅ is congruent to		
∠*MRN*	angle *MRN*			
		π Greek letter pi; ratio of circumference to diameter of any circle		
1°	one degree			

New Formulas

Perimeter of a rectangle: $P = 2 \cdot l + 2 \cdot w$

Area of a rectangle: $A = l \cdot w$

Perimeter of a square: $P = 4 \cdot s$

Area of a square: $A = s^2$ or $A = s \cdot s$

Area of a parallelogram: $A = b \cdot h$

Area of a trapezoid: $A = \dfrac{1}{2} \cdot h \cdot (b + B)$

or $A = 0.5 \cdot h \cdot (b + B)$

Area of a triangle: $A = \dfrac{1}{2} \cdot b \cdot h$

or $A = 0.5 \cdot b \cdot h$

Diameter of a circle: $d = 2 \cdot r$

Radius of a circle: $r = \dfrac{d}{2}$

Circumference of a circle: $C = \pi \cdot d$

or $C = 2 \cdot \pi \cdot r$

Area of a circle: $A = \pi \cdot r^2$

Area of semicircle: $A = \dfrac{\pi \cdot r^2}{2}$

Volume of rectangular solid: $V = l \cdot w \cdot h$

Volume of a sphere: $V = \dfrac{4}{3} \cdot \pi \cdot r^3$

or $V = \dfrac{4 \cdot \pi \cdot r \cdot r \cdot r}{3}$

Volume of a hemisphere: $V = \dfrac{2}{3} \cdot \pi \cdot r^3$

or $V = \dfrac{2 \cdot \pi \cdot r \cdot r \cdot r}{3}$

Volume of a cylinder: $V = \pi \cdot r^2 \cdot h$

Volume of a cone: $V = \dfrac{1}{3} \cdot B \cdot h$ or $V = \dfrac{B \cdot h}{3}$

Volume of a pyramid: $V = \dfrac{1}{3} \cdot B \cdot h$ or $V = \dfrac{B \cdot h}{3}$

Right triangle: hypotenuse $= \sqrt{(\text{leg})^2 + (\text{leg})^2}$

leg $= \sqrt{(\text{hypotenuse})^2 - (\text{leg})^2}$

Note: Use 3.14 as the approximate value of π.

Test Your Word Power

See how well you have learned the vocabulary in this chapter. The answers are on the next page.

1 Two angles that are **complementary**
   A. have measures that add up to 180°
   B. form a straight angle
   C. have measures that add up to 90°.

2 The **perimeter** of a flat shape is
   A. measured in square units
   B. the distance around the outside edges
   C. measured in cubic units.

3 An **obtuse angle**
   A. is formed by perpendicular lines
   B. measures more than 90° but less than 180°
   C. measures less than 90°.

4 The **hypotenuse** is
   A. the height line in a parallelogram
   B. the longest side in a right triangle
   C. the distance across a circle, passing through the center.

5 π is the ratio of
   A. the diameter to the radius of a circle
   B. the circumference to the diameter of a circle
   C. the diameter to the circumference of a circle.

6 **Perpendicular lines**
   A. intersect to form a right angle
   B. intersect to form an acute angle
   C. never intersect.

7 In a pair of **similar triangles,**
   A. corresponding sides have the same length
   B. all the angles have the same measure
   C. the ratios of the lengths of corresponding sides are equal.

8 The **area of a rectangle** is found by
   A. multiplying length times width
   B. adding the lengths of the sides
   C. using the formula $V = l \cdot w \cdot h$.

Answers to Test Your Word Power

1. C; *Example:* If $\angle 1$ measures 35° and $\angle 2$ measures 55°, the angles are complementary because 35° + 55° = 90°.

2. B; *Example:* If a square measures 5 ft on each side, then the perimeter is 5 ft + 5 ft + 5 ft + 5 ft = 20 ft.

3. B; *Examples:* Angles that measure 91°, 120°, and 175° are all obtuse angles.

4. B; *Example:* In triangle *ABC* at the right, side *AC* is the hypotenuse; sides *AB* and *BC* are the legs.

5. B; *Example:* The ratio of a circumference of 12.57 cm to a diameter of 4 cm is $\frac{12.57}{4} \approx 3.14$ (rounded).

6. A; *Example:* $\overleftrightarrow{EF}$ is perpendicular to $\overleftrightarrow{GH}$, at the right.

7. C; *Example:* Triangle *ABC* is similar to triangle *DEF*, so the ratios of corresponding sides are equal.

$$\frac{AB}{DE} = \frac{3 \text{ m}}{6 \text{ m}} = \frac{1}{2} \qquad \frac{BC}{EF} = \frac{2 \text{ m}}{4 \text{ m}} = \frac{1}{2} \qquad \frac{AC}{DF} = \frac{3.5 \text{ m}}{7 \text{ m}} = \frac{1}{2}$$

8. A; *Example:* In a rectangle with a length of 8 in. and a width of 5 in., Area = 8 in. • 5 in. = 40 in.²

Quick Review

Concepts	Examples

8.1 Lines

A *line* is a straight row of points that goes on forever in both directions.

If a piece of a line has one endpoint, it is a *ray*.

If a piece of a line has two endpoints, it is a *line segment*.

Identify each of the following as a line, line segment, or ray and name it using the appropriate symbol.

Figure **(a)** shows a ray named $\overrightarrow{OS}$.

Figure **(b)** shows a line named $\overleftrightarrow{PQ}$ or $\overleftrightarrow{QP}$.

Figure **(c)** shows a line segment named $\overline{ST}$ or $\overline{TS}$.

If two lines intersect at right angles, they are *perpendicular*.

If two lines in the same plane never intersect, they are *parallel*.

Label each pair of lines as appearing to be parallel or as perpendicular.

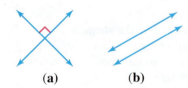

Figure **(a)** shows perpendicular lines (they intersect at 90°).

Figure **(b)** shows lines that appear to be parallel (they never intersect).

8.2 Angles

If the sum of the measures of two angles is 90°, they are *complementary*.

If the sum of the measures of two angles is 180°, they are *supplementary*.

Find the complement and supplement of a 35° angle.

$$90° - 35° = 55° \text{ (the complement)}$$

$$180° - 35° = 145° \text{ (the supplement)}$$

Concepts	Examples

8.2 Angles (*Continued*)

If two angles measure the same number of degrees, the angles are *congruent*. The symbol for congruent is ≅.

Two nonadjacent angles formed by two intersecting lines are called *vertical angles*. Vertical angles are congruent.

Identify the vertical angles in this figure. Which angles are congruent?

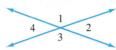

$\angle 1$ and $\angle 3$ are vertical angles.

$\angle 2$ and $\angle 4$ are vertical angles.

Vertical angles are congruent, so $\angle 1 \cong \angle 3$ and $\angle 2 \cong \angle 4$.

8.3 Rectangles and Squares

Use this formula to find the perimeter of a *rectangle*.

$$P = 2 \cdot l + 2 \cdot w$$

Use this formula to find the area of a rectangle.

$$A = l \cdot w$$

Area is measured in **square units.**

Find the perimeter and area of this rectangle.

3 in.
2 in.   2 in.
3 in.

$$P = 2 \cdot l + 2 \cdot w$$
$$= 2 \cdot 3 \text{ in.} + 2 \cdot 2 \text{ in.}$$
$$= 6 \text{ in.} + 4 \text{ in.}$$
$$= 10 \text{ in.}$$
$$A = l \cdot w = 3 \text{ in.} \cdot 2 \text{ in.} = 6 \text{ in.}^2$$

Use these formulas to find the perimeter and area of a *square*.

$$P = 4 \cdot s$$
$$A = s^2$$

Area is measured in **square units.**

Find the perimeter and area of this square.

6 m
6 m   6 m
6 m

$$P = 4 \cdot s = 4 \cdot 6 \text{ m} = 24 \text{ m}$$
$$A = s^2 = s \cdot s = 6 \text{ m} \cdot 6 \text{ m} = 36 \text{ m}^2$$

8.4 Parallelograms

Use these formulas to find the perimeter and area of a *parallelogram*.

$$P = \text{sum of the lengths of the sides}$$
$$A = b \cdot h$$

Area is measured in **square units.**

Find the perimeter and area of this parallelogram.

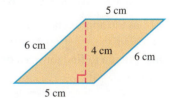

$$P = 5 \text{ cm} + 6 \text{ cm} + 5 \text{ cm} + 6 \text{ cm} = 22 \text{ cm}$$
$$A = 5 \text{ cm} \cdot 4 \text{ cm} = 20 \text{ cm}^2$$

8.4 Trapezoids

Use these formulas to find the perimeter and area of a *trapezoid*.

$$P = \text{sum of the lengths of the sides}$$
$$A = \frac{1}{2} \cdot h \cdot (b + B)$$

where b is the short base and B is the long base.

Area is measured in **square units.**

Find the perimeter and area of this trapezoid.

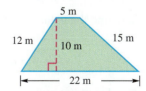

$$P = 5 \text{ m} + 15 \text{ m} + 22 \text{ m} + 12 \text{ m} = 54 \text{ m}$$
$$A = \frac{1}{2} \cdot \overset{5}{\cancel{10}} \text{ m} \cdot (5\text{m} + 22 \text{ m})$$
$$= 5 \text{ m} \cdot (27 \text{ m}) = 135 \text{ m}^2$$

Concepts	**Examples**

8.5 Triangles

Use these formulas to find the perimeter and area of a *triangle*.

$$P = \text{sum of the lengths of the sides}$$

$$A = \frac{1}{2} \cdot b \cdot h$$

$$\downarrow$$

$$\text{or} \quad A = 0.5 \cdot b \cdot h$$

Area is measured in **square units**.

Find the perimeter and area of this triangle.

$$P = 12 \text{ ft} + 10 \text{ ft} + 20 \text{ ft} = 42 \text{ ft}$$

$$A = \frac{1}{2} \cdot b \cdot h$$

$$= \frac{1}{\underset{1}{2}} \cdot \overset{10}{20} \text{ ft} \cdot 5 \text{ ft} = 50 \text{ ft}^2$$

$$\text{or } A = 0.5 \cdot 20 \text{ ft} \cdot 5 \text{ ft} = 50 \text{ ft}^2$$

8.6 Circles

Use this formula to find the *diameter* of a circle when you are given the radius.

$$d = 2 \cdot r$$

Find the diameter of a circle if the radius is 7 yd.

$$d = 2 \cdot r = 2 \cdot 7 \text{ yd} = 14 \text{ yd}$$

Use this formula to find the *radius* of a circle when you are given the diameter.

$$r = \frac{d}{2}$$

Find the radius of a circle if the diameter is 5 cm.

$$r = \frac{d}{2} = \frac{5 \text{ cm}}{2} = 2.5 \text{ cm}$$

Use these formulas to find the *circumference* of a circle.

When you know the radius, use $C = 2 \cdot \pi \cdot r$

When you know the diameter, use $C = \pi \cdot d$

Use 3.14 as the approximate value for π.

Find the circumference of a circle with a radius of 3 cm.

$$C = 2 \cdot \pi \cdot r$$

$$C \approx 2 \cdot 3.14 \cdot 3 \text{ cm} \approx 18.8 \text{ cm} \leftarrow \text{Rounded}$$

Use this formula to find the *area* of a circle.

$$A = \pi \cdot r^2$$

Area is measured in **square units**.

Find the area of this circle.

$$A = \pi \cdot r^2$$

$$A \approx 3.14 \cdot 3 \text{ cm} \cdot 3 \text{ cm}$$

$$A \approx 28.3 \text{ cm}^2 \leftarrow \text{Rounded;}$$
$$\text{square units for area}$$

8.7 Volume of a Rectangular Solid

Use this formula to find the volume of *rectangular solids* (box-like solids).

$$V = l \cdot w \cdot h$$

Volume is measured in **cubic units**.

Find the volume of this box.

$$V = l \cdot w \cdot h$$

$$V = 5 \text{ cm} \cdot 3 \text{ cm} \cdot 6 \text{ cm}$$

$$V = 90 \text{ cm}^3 \quad \text{Cubic units}$$
$$\text{for volume}$$

Concepts	Examples

8.7 Volume of a Sphere

Use this formula to find the volume of a *sphere* (a ball-shaped solid).

$$V = \frac{4}{3} \cdot \pi \cdot r^3$$

$$\text{or} \quad V = \frac{4 \cdot \pi \cdot r \cdot r \cdot r}{3}$$

where *r* is the radius of the sphere.

Volume is measured in **cubic units**.

Find the volume of a sphere with a radius of 5 m.

$$V = \frac{4 \cdot \pi \cdot r^3}{3}$$

$$V \approx \frac{4 \cdot 3.14 \cdot 5 \text{ cm} \cdot 5 \text{ cm} \cdot 5 \text{ cm}}{3}$$

$$V \approx 523.3 \text{ m}^3 \leftarrow \text{Rounded;}$$
cubic units for volume

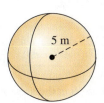

5 m

8.7 Volume of a Hemisphere

Use this formula to find the volume of a *hemisphere* (half of a sphere).

$$V = \frac{2}{3} \cdot \pi \cdot r^3$$

$$\text{or} \quad V = \frac{2 \cdot \pi \cdot r \cdot r \cdot r}{3}$$

where *r* is the radius of the hemisphere.

Volume is measured in **cubic units**.

Find the volume of a hemisphere with a radius of 20 cm.

$$V = \frac{2 \cdot \pi \cdot r^3}{3}$$

$$V \approx \frac{2 \cdot 3.14 \cdot 20 \text{ cm} \cdot 20 \text{ cm} \cdot 20 \text{ cm}}{3}$$

$$V \approx 16{,}746.7 \text{ cm}^3 \leftarrow \text{Rounded;}$$
cubic units for volume

8.7 Volume of a Cylinder

Use this formula to find the volume of a *cylinder*.

$$V = \pi \cdot r^2 \cdot h$$

where *r* is the radius of the circular base and *h* is the height of the cylinder.

Volume is measured in **cubic units**.

Find the volume of this cylinder.

First, find the radius. $r = \dfrac{8 \text{ m}}{2} = 4 \text{ m}$

$$V = \pi \cdot r^2 \cdot h$$

$$V \approx 3.14 \cdot 4 \text{ m} \cdot 4 \text{ m} \cdot 10 \text{ m}$$

$$V \approx 502.4 \text{ m}^3 \leftarrow \text{Cubic units}$$
for volume

8 m

10 m

8.7 Volume of a Cone

Use this formula to find the volume of a *cone*.

$$V = \frac{1}{3} \cdot B \cdot h$$

$$\text{or} \quad V = \frac{B \cdot h}{3}$$

where *B* is the area of the circular base and *h* is the height of the cone.

Volume is measured in **cubic units**.

Find the volume of this cone.

Area of circular **Base** $\approx 3.14 \cdot 4 \text{ in.} \cdot 4 \text{ in.}$

$$B \approx 50.24 \text{ in.}^2$$

$$V = \frac{B \cdot h}{3}$$

$$V \approx \frac{50.24 \text{ in.}^2 \cdot 9 \text{ in.}}{3}$$

$$V \approx 150.7 \text{ in.}^3 \leftarrow \text{Rounded;}$$
cubic units for volume

9 in.

4 in.

Concepts	Examples

8.7 Volume of a Pyramid

Use this formula to find the volume of a *pyramid*.

$$V = \frac{1}{3} \cdot B \cdot h$$

or $\quad V = \dfrac{B \cdot h}{3}$

where B is the area of the square or rectangular base and h is the height of the pyramid.

Volume is measured in **cubic units**.

Find the volume of this pyramid.
Area of square **Base** = 2 cm • 2 cm
$$B = 4 \text{ cm}^2$$

$$V = \frac{B \cdot h}{3}$$

$$V = \frac{4 \text{ cm}^2 \cdot 6 \text{ cm}}{3}$$

$V = 8 \text{ cm}^3 \leftarrow$ Cubic units for volume

8.8 Finding the Square Root of a Number

Use the square root key on a calculator, $\sqrt{}$ or $\sqrt{x}$. Round to the nearest thousandth if necessary.

$\sqrt{64} = 8$ $\qquad$ A perfect square

$\sqrt{43} \approx 6.557$ $\quad$ 6.557438524 is rounded to nearest thousandth.

8.8 Finding the Unknown Length in a Right Triangle

To find the *hypotenuse*, use this formula.

$$\text{hypotenuse} = \sqrt{(\text{leg})^2 + (\text{leg})^2}$$

The hypotenuse is the side opposite the right angle; it is the longest side in a right triangle.

Find the unknown length. Round to the nearest tenth.

6 m

5 m

$$\text{hypotenuse} = \sqrt{(6)^2 + (5)^2}$$
$$= \sqrt{36 + 25}$$
$$= \sqrt{61} \approx 7.8 \text{ m}$$

To find a *leg*, use this formula.

$$\text{leg} = \sqrt{(\text{hypotenuse})^2 - (\text{leg})^2}$$

The legs are the sides that form the right angle.

Find the unknown length. Round to the nearest tenth.

16 cm $\qquad$ 25 cm

$$\text{leg} = \sqrt{(25)^2 - (16)^2}$$
$$= \sqrt{625 - 256}$$
$$= \sqrt{369} \approx 19.2 \text{ cm}$$

8.9 Finding the Unknown Lengths in Similar Triangles

Use the fact that in similar triangles, the *ratios* of the lengths of corresponding sides are equal. Write a proportion. Then find the cross products and show that they are equal. Finish solving for the unknown length.

Find the unknown lengths in this pair of similar triangles.

$$\frac{x}{8} = \frac{5}{10}$$

$$x \cdot 10 = 8 \cdot 5$$

$$\frac{x \cdot \overset{1}{10}}{\underset{1}{10}} = \frac{40}{10}$$

$$x = 4 \text{ m}$$

$$\frac{y}{12} = \frac{5}{10}$$

$$y \cdot 10 = 12 \cdot 5$$

$$\frac{y \cdot \overset{1}{10}}{\underset{1}{10}} = \frac{60}{10}$$

$$y = 6 \text{ m}$$

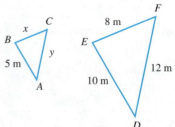

Chapter 8 Review Exercises

8.1 *Identify each figure as a line, line segment, or ray, and name it using the appropriate symbol.*

1.

2.

3.

Label each pair of lines as appearing to be parallel, as perpendicular, or as intersecting.

4.

5.

6.

Label each angle as acute, right, obtuse, or straight. For right and straight angles, indicate the number of degrees in the angle.

7.

8.

9.

10.

8.2 *In each figure you are given the measures of two of the angles. Find the measure of each of the other angles.*

11.

12.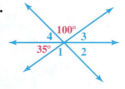

Name the pairs of supplementary angles in each figure.

13.

14.

Find the complement or supplement of each angle.

15. Find the complement of:

   (a) 80°

   (b) 45°

   (c) 7°

16. Find the supplement of:

   (a) 155°

   (b) 90°

   (c) 33°

8.3 *Find the perimeter of each rectangle or square.*

17.

18.

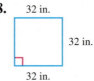

19. A square-shaped pillow measures 38 cm along each side. How much lace is needed to trim all the edges?

20. A rectangular garden plot is $8\frac{1}{2}$ ft wide and 12 ft long. How much fencing is needed to surround the garden?

Find the area of each rectangle or square. Round your answers to the nearest tenth when necessary.

21.

22.

23.

8.4 *Find the perimeter and area of each parallelogram or trapezoid. Round your answers to the nearest tenth when necessary.*

24.

11 cm
10 cm
14 cm
14 cm
11 cm

25.

26 ft
18 ft
18 ft
21.1 ft
37 ft

26.

59.7 m
33.9 m
33.4 m
34.2 m
72.4 m

8.5 *Find the perimeter and area of each triangle.*

27.

153 cm
153 cm
110 cm
212 cm

28.

12.3 m
4.8 m
5 m
9.8 m

29.
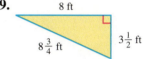
8 ft
$8\frac{3}{4}$ ft
$3\frac{1}{2}$ ft

Find the number of degrees in the third angle of each triangle.

30.

70°
40°

31.

66°

8.6 *Find the unknown length.*

32. The radius of a circular irrigation field is 68.9 m. What is the diameter of the field?

33. The diameter of a juice can is 3 in. What is the radius of the can?

Find the circumference and area of each circle. Use 3.14 as the approximate value for π. Round your answers to the nearest tenth.

34.

1 cm

35.

17.4 m

36.

12 in.

8.3–8.6 *Find each shaded area. Note that Exercises 44 and 45 contain semicircles.*
Use 3.14 as the approximate value for π. Round your answers to the nearest tenth
when necessary.

37.

3.6 m

38.

8 in.
4 in.
4 in.
4 in.
4 in.
4 in.
4 in.
8 in.

39.

12 km
14 km
11 km
24 km
11 km
11 km
12 km
11 km
14 km

40.

15 m
45 m   15 m   10 m
15 m
21 m

41.

15 ft
6 ft
8 ft
7 ft
8 ft
6 ft
15 ft

42.
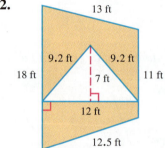
13 ft
9.2 ft   9.2 ft
18 ft   7 ft   11 ft
12 ft
12.5 ft

43.

48 cm   48 cm
74 cm   36 cm   74 cm

44.

21 ft
32 ft   32 ft
21 ft

45.

7 yd
14 yd
21 yd   21 yd
7 yd

8.7 *Name each solid and find its volume. Use 3.14 as the approximate value for π. Round your answers to the nearest tenth when necessary.*

46.

$2\frac{1}{2}$ in.   3 in.   4 in.

47.

4 cm   6 cm   4 cm

48.

75 mm   30 mm   20 mm

49.

4 m

50.

6 ft

51.

7 cm   5 cm

52.

24 m   4 m

53.

10 m   7 m

54.

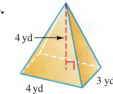

4 yd   4 yd   3 yd

8.8 *Find each square root. Round your answers to the nearest thousandth when necessary.*

55. $\sqrt{49}$ **56.** $\sqrt{8}$ **57.** $\sqrt{3000}$ **58.** $\sqrt{144}$

59. $\sqrt{58}$ **60.** $\sqrt{625}$ **61.** $\sqrt{105}$ **62.** $\sqrt{80}$

Find the unknown length in each right triangle. Use a calculator to find square roots. Round your answers to the nearest tenth when necessary.

63.

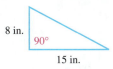

8 in.   90°   15 in.

64.

24 cm   25 cm

65.

15 cm   90°   11 cm

66.

4 in.   90°   6 in.

67.

2.2 m   1.3 m

68.

12 km   8.5 km

8.9 *Find the unknown lengths in each pair of similar triangles. Then find the perimeter of the larger triangle in each pair.*

69.

70.

71.

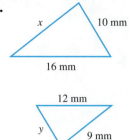

Mixed Review Exercises

Name each figure and find its perimeter (or circumference) and area. Use 3.14 as the approximate value for π. Round your answers to the nearest tenth when necessary.

72.

4 ½ in., 4 ½ in., 4 ½ in., 4 ½ in.

73.

2.3 cm, 2.1 cm, 3.4 cm, 2.5 cm, 2.1 cm

74.

13 m

75.

17 ft, 10 ft, 14 ft, 10 ft, 17 ft

76.

6 ¼ yd, 6 ¼ yd, 5 yd, 7 ½ yd

77.

2.8 km, 0.7 km, 0.7 km, 2.8 km

78.

8.5 m

79.

24 mm, 15 mm, 12 mm, 15 mm, 24 mm

80.

13 mi, 9 mi, 15.8 mi

Label each figure. Choose from these labels: line segment, ray, parallel lines, perpendicular lines, intersecting lines, acute angle, right angle, straight angle, obtuse angle. Indicate the number of degrees in the right angle and the straight angle.

81.

82.

83.

84.

85.

86.

87.

88.

89.

90. What is the complement of an angle measuring 9°?

91. What is the supplement of an angle measuring 42°?

Find the perimeter and area of each figure. In Exercise 92, assume that all angles are 90°.

92.

93.

Name each solid and find its volume. Use 3.14 as the approximate value for π.
Round your answers to the nearest tenth when necessary.

94.

8 ft — 2 ft

95.

$1\frac{1}{2}$ in.

$1\frac{1}{2}$ in.

$1\frac{1}{2}$ in.

96.

0.7 m

3 m

3.5 m

97.

17 cm

9 cm

11 cm

98.

15 cm

9 cm

99.

7 m

Find the unknown angle or side measurement. Round your answers to the
nearest tenth when necessary.

100.

14 km

6 km

x

101.

C

60°

D

48°

E

102. Similar triangles

x

18 mm   16.8 mm

10 mm

15 mm   y

103. Explain how you could use the information about prefixes from earlier in this
chapter to solve a problem that asks, "How many decades are in two centuries?"

Chapter 8 *Test*

The Chapter Test Prep Videos with test solutions are available on DVD, in MyMathLab, and on You Tube—search "LialBasicCollegeMath" and click on "Channels."

Choose the figure that matches each label. For right and straight angles, indicate the number of degrees in the angle.

(a)

(b)

(c)

(d)

(e)

(f)

(g)

1. Acute angle is figure _____.

2. Right angle is figure _____ and its measure is _____.

3. Ray is figure _____.

4. Straight angle is figure _____ and its measure is _____.

5. Write a definition of parallel lines and a definition of perpendicular lines. Make a sketch to illustrate each definition.

6. Find the complement of an 81° angle.

7. Find the supplement of a 20° angle.

8. Find the measure of each unlabeled angle in the figure at the right.

Name each figure and find its perimeter and area.

9.

10.

11.

12.

Find the perimeter and area of each triangle.

13.

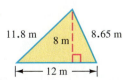

14.

9 yd

13 yd   15$\frac{4}{5}$ yd

15. A triangle has angles that measure 90° and 35°. What does the third angle measure?

In Problems 16–22, use 3.14 as the approximate value for π. Round your answers to the nearest tenth when necessary.

16. Find the radius.

25 in.

17. Find the circumference.

0.9 km

 Find the area of each figure.

18.

16.2 cm

19.

5 m

Name each solid and find its volume.

20.

12 m   30 m
18 m

21.

2 ft

22.

18 ft
5 ft

Find the unknown lengths. Round your answers to the nearest tenth when necessary.

23.

?
6 cm
7 cm

24. Similar triangles

18 cm   15 cm
9 cm
y   10 cm
z

25. Explain the difference between cm, cm², and cm³. In what types of geometry problems might you use each of these units?

Chapters 1–8 *Cumulative Review Exercises*

Simplify. Write answers to fraction problems in lowest terms and as whole or mixed numbers when possible.

1. $3\dfrac{3}{5} \div 8$

2. $1 - 0.0868$

3. $(0.006)(0.013)$

4. $0.7 \div 0.036$ Round answer to the nearest hundredth.

5. $6\dfrac{1}{6} - 1\dfrac{3}{4}$

6. $16 - (10 - 2) \div 2\,(3) + 5$

7. Write 0.0208 in words.

8. Arrange in order from least to greatest.

    2.55   2.505   2.055   2.5005

Solve each proportion or percent problem. Round your answers to hundredths if necessary.

9. $\dfrac{5}{13} = \dfrac{x}{91}$

10. $\dfrac{4.5}{x} = \dfrac{6.7}{3}$

11. 72 patients is what percent of 45 patients?

12. $18 is 3% of what number of dollars?

Convert each measurement.

13. $2\dfrac{1}{4}$ hours to minutes

14. 40 oz to pounds

15. 8 cm to meters

16. 1.8 L to milliliters

Write the most reasonable metric unit in each blank. Choose from km, m, cm, mm, L, mL, kg, g, and mg.

17. Her wristwatch strap is 15 ____ wide.

18. Jon added 2 ____ of oil to his car.

19. The child weighs 15 ____.

20. The bookcase is 90 ____ high.

Name each figure and find its perimeter (or circumference) and area. Use 3.14 as the approximate value of π.

21.

$3\frac{1}{2}$ in.

2 in.   2 in.

$3\frac{1}{2}$ in.

22.

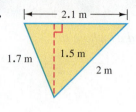

2.1 m

1.7 m   1.5 m

2 m

23.

5 ft

24.

24 cm

14 cm   11 cm

14 cm

24 cm

25. Find the unknown length to the nearest tenth.

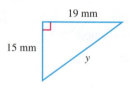

19 mm

15 mm

y

26. Mei Ling must earn 90 credits to receive an associate of arts degree. She has 53 credits. What percent of the necessary credits does she have? Round to the nearest whole percent.

27. A Folger's coffee can has a diameter of 15.6 cm and a height of 16 cm. Find the volume of the can. Use 3.14 for π and round your answer to the nearest tenth. (*Source:* Folger's.)

28. Which bag of chips is the best buy: Brand T is $8\frac{1}{2}$ oz for $3.09, Brand F is $10\frac{1}{2}$ oz for $3, and Brand H is $9\frac{1}{2}$ oz for $3.69. You have a coupon for 50¢ off Brand H and another for 75¢ off Brand T.

29. Steven bought a $4\frac{1}{2}$ yd length of canvas material to repair the tents used by the scout troop. He used $1\frac{2}{3}$ yd on one tent and $1\frac{3}{4}$ yd on another. What length of material is left?

30. The cooks at a homeless shelter used 25 lb of meat, 35 lb of potatoes, and 15 lb of carrots to make stew for 140 people. At that rate, how much of each item is needed for stew to feed 200 people? Round to the nearest tenth.

31. Graciela needs 85 cm of yarn to make a tassel for one corner of a pillow. How many meters of yarn does she need to put a tassel on each corner of a square-shaped pillow?

32. Swimsuits are on sale in August at 65% off the regular price. How much will Lanece pay for a suit that has a regular price of $64?

9

Basic Algebra

In this chapter, you'll see how we need both *positive* and *negative* numbers. For example, we use them to report temperatures, check our bank account balance, and interpret medical test results.

625

9.1 Signed Numbers

OBJECTIVES

1. Write negative numbers.
2. Graph signed numbers on a number line.
3. Use the < and > symbols.
4. Find absolute value.
5. Find the opposite of a number.

All the numbers you have studied so far in this book have been either 0 or greater than 0. Numbers *greater* than 0 are called *positive numbers*. For example, you have worked with these positive numbers.

Salary of $45,000

Temperature of 98.6 °F

Length of $3\frac{1}{2}$ feet

OBJECTIVE ▶ 1 Write negative numbers. Not all numbers are positive. For example, "15 degrees below 0" or "a loss of $500" is expressed with a number *less* than 0. Numbers less than 0 are called **negative numbers**. Zero is neither positive nor negative.

Writing a Negative Number

To write a negative number, put a **negative sign, −** , in front of it.

For example, "15 degrees below 0" is written with a negative sign, as −15°. And "a loss of $500" is written −$500.

◀ **Work Problem ❶ at the Side.**

OBJECTIVE ▶ 2 Graph signed numbers on a number line. In **Chapter 3** you graphed positive numbers on a number line. Negative numbers can also be shown on a number line. Zero separates the positive numbers from the negative numbers on the number line. The number −5 is read "negative five."

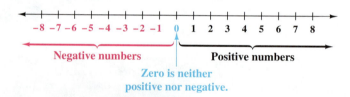
Negative numbers ← | Positive numbers
Zero is neither positive nor negative.

Note

For every *positive* number on a number line, there is a corresponding *negative* number on the *opposite* side of 0.

When you work with both positive and negative numbers (and 0), we say you are working with **signed numbers.**

Writing a Positive Number

A positive number can be written in two ways.

1. Use a " + " sign. For example, +2 is "positive two."

2. Do not write any sign. For example, when you write 3, it is assumed to be "positive three."

◀ **Work Problem ❷ at the Side.**

---

❶ Write each number.

  (a) The temperature at the North Pole is 70 degrees below 0.

  (b) Your bank account is overdrawn by 15 dollars.

  (c) A submarine dived to 284 ft below sea level.

VOCABULARY TIP

Signed numbers refers to the fact that we write a positive sign (+) or negative sign (−) in front of a number to indicate whether it is greater than zero (like +3) or less than zero (like −12). Zero is neither positive nor negative.

❷ Write *positive, negative,* or *neither* for each number.

  (a) −8

  (b) $-\frac{3}{4}$

  (c) 1

  (d) 0

Answers

1. **(a)** −70° **(b)** −$15 **(c)** −284 ft
2. **(a)** negative **(b)** negative
  (c) positive **(d)** neither

The next example shows you how to graph signed numbers on a number line.

> **EXAMPLE 1** Graphing Signed Numbers

Graph these numbers on the number line.

(a) -4 (b) 3 (c) -1 (d) 0 (e) $1\frac{1}{4}$

Place a dot at the correct location for each number.

················ **Work Problem 3** at the Side. ▶

> **OBJECTIVE 3 Use the $<$ and $>$ symbols.** When you look at the number line below, 3 is to the *left* of 5.

3 is to the *left* of 5.

3 is *less than* 5.

Recall the following symbols for comparing two numbers.

$<$ means **"is less than."**

$>$ means **"is greater than."**

Use these symbols to write "3 is less than 5" as shown below.

$$3 \quad < \quad 5$$
$$\downarrow \quad \downarrow \quad \downarrow$$

3 is less than 5.

This example suggests the following.

> **Comparing Numbers on a Number Line**
>
> The *lesser* of two numbers is the one farther to the **left** on a number line.
>
> The *greater* of two numbers is the one farther to the **right** on a number line.

> **EXAMPLE 2** Using the Symbols $<$ and $>$

Use this number line to compare each pair of numbers. Then write $>$ or $<$ to make each statement true.

```
◄─┼─┼─┼─┼─┼─┼─┼─┼─┼─┼─┼─┼─┼─┼─┼─►
 -9 -8 -7 -6 -5 -4 -3 -2 -1 0  1  2  3  4  5  6
```

(a) Compare 2 and 6. $2 < 6$ (read "2 is less than 6") because 2 is to the *left* of 6 on the number line.

(b) Compare -9 and -4. $-9 < -4$ because [Be careful when comparing two *negative* numbers.] -9 is to the *left* of -4.

(c) Compare 2 and -1. $2 > -1$ because 2 is to the *right* of -1.

(d) Compare -4 and 0. $-4 < 0$ because -4 is to the *left* of 0.

················· **Continued on Next Page**

3 Graph each set of numbers.

GS (a) $-1, 1, -3, 3$

```
◄─┼─┼─┼─┼─┼─┼─┼─┼─►
 -4 -3 -2 -1 0 1 2 3 4
```

To graph -1, put a dot at -1 on the number line.

Now put dots in the correct places for $1, -3$, and 3.

(b) $-2, 4, 0, -1, -4$

```
◄─┼─┼─┼─┼─┼─┼─┼─┼─►
 -4 -3 -2 -1 0 1 2 3 4
```

Answers

3. (a)
```
◄─┼─●─┼─●─┼─●─┼─●─┼─►
 -4 -3 -2 -1 0 1 2 3 4
```

(b)
```
◄─●─┼─●─●─┼─●─┼─┼─●─►
 -4 -3 -2 -1 0 1 2 3 4
```

4 Write **<** or **>** in each blank to make a true statement.

GS **(a)** 4 _____ 0

Hint: 0 is the *lesser* of the two numbers because it is *farther to the left* on the number line.

(b) -1 _____ 0

(c) -3 _____ -1

(d) -8 _____ -9

(e) 0 _____ -3

5 Simplify each absolute value expression.

(a) $|5|$

(b) $|-5|$

GS **(c)** $-|-17|$

_____

(d) $-|-9|$

(e) $-|2|$

Note

When using **>** and **<**, the *smaller* pointed end of the symbol points to the *smaller* (lesser) number.

◀ **Work Problem** **4** **at the Side.**

OBJECTIVE **4** **Find absolute value.** In order to graph a number on the number line, you need to answer two questions:

1. Which *direction* is it from 0? It can be in a *positive* direction or a *negative* direction. You can tell the direction by looking for a positive sign or a negative sign (or no sign, which is positive).

2. How *far* is it from 0? The *distance* from 0 is the **absolute value** of a number.

Absolute value is indicated by two vertical bars. For example, $|6|$ is read "the **absolute value** of 6."

Note

Absolute value is never negative because it is the *distance* from 0. A distance is never negative.

EXAMPLE 3 **Finding Absolute Values**

Simplify each absolute value expression.

(a) $|8|$ The *distance* from 0 to 8 is 8 spaces, so $|8| = 8$.

Distance is 8 spaces. Direction is positive.

(b) $|-8|$ The *distance* from 0 to -8 is also 8 spaces, so $|-8| = 8$.

Distance is 8 spaces. Direction is negative.

(c) $|0|$ The distance from 0 to 0 is 0 spaces, so $|0| = 0$.

(d) $-|-3|$ First, $|-3|$ is 3. But there is also a negative sign *outside* the absolute value bars. So, -3 is the simplified expression.   *Watch for a negative sign outside the absolute value bars.*

CAUTION

A negative sign *outside* the absolute value bars is *not* affected by the absolute value bars. Therefore, your final answer is negative, as in **Example 3(d)** above.

◀ **Work Problem** **5** **at the Side.**

OBJECTIVE ▶ **5** **Find the opposite of a number.** Two numbers that are the *same distance* from 0 on a number line but on *opposite sides* of 0 are called **opposites** of each other. As this number line shows, −3 and 3 are opposites of each other.

Distance is 3. Distance is 3.

−3   0   3

To indicate the opposite of a number, write a negative sign in front of the number.

EXAMPLE 4 **Finding Opposites**

Find the opposite of each number.

Number	Opposite
5	$-(5) = -5$
9	$-(9) = -9$
$\dfrac{4}{5}$	$-\left(\dfrac{4}{5}\right) = -\dfrac{4}{5}$
0	$-(0) = 0$

— Write a negative sign.

The opposite of 0 is 0. Zero is neither positive nor negative.

No negative sign

Work Problem 6 at the Side. ▶

Some numbers have two negative signs. An example is shown below.

$$-(-3)$$

The negative sign in front of (-3) means the *opposite* of -3. The opposite of -3 is 3.

$$-(-3) = 3$$

Use the following rule to find the opposite of a negative number.

Double Negative Rule

The opposite of a negative number is positive.

For example, $-(-3) = 3$ and $-(-10) = 10$.

EXAMPLE 5 **Finding Opposites**

Find the opposite of each number.

Number	Opposite
−2	$-(-2) = 2$
−9	$-(-9) = 9$
$-\dfrac{1}{2}$	$-\left(-\dfrac{1}{2}\right) = \dfrac{1}{2}$

By the double negative rule (See the green box above.)

Remember, the *opposite* of any *negative* number is *positive*.

Work Problem 7 at the Side. ▶

---

6 Find the opposite of each number.

(a) 4

(b) 10

(c) 49

(d) $\dfrac{2}{5}$

(e) 0

7 Find the opposite of each number.

GS (a) −4 *Hint:* The opposite of a negative number is positive.

(b) −10

(c) −25

(d) −1.9

(e) −0.85

(f) $-\dfrac{3}{4}$

Answers

6. (a) −4 (b) −10 (c) −49
   (d) $-\dfrac{2}{5}$ (e) 0

7. (a) 4 (b) 10 (c) 25 (d) 1.9
   (e) 0.85 (f) $\dfrac{3}{4}$

9.1 Exercises

FOR EXTRA HELP

MyMathLab®

CONCEPT CHECK   *Write a signed number for each situation.*

1. Water freezes at 32 degrees above 0 on the Fahrenheit temperature scale.

2. She made a profit of $920.

3. The price of the stock fell $12.

4. His bank account is overdrawn by $30.79.

5. Keith lost $6\frac{1}{2}$ lb while he was sick with the flu.

6. The river is 20 ft above flood stage.

7. Mount McKinley, the tallest mountain in the United States, rises 20,320 ft above sea level. (*Source: World Almanac and Book of Facts.*)

8. The bottom of Lake Baykal in central Asia is 5315 ft below the surface of the water. (*Source: World Almanac and Book of Facts.*)

Depth below surface

Graph each set of numbers on the number line. **See Example 1.**

9. $4, -1, 2, 0, -5$

$$-5\ -4\ -3\ -2\ -1\ \ 0\ \ 1\ \ 2\ \ 3\ \ 4\ \ 5$$

10. $-2, 1, -3, 5, 0$

$$-5\ -4\ -3\ -2\ -1\ \ 0\ \ 1\ \ 2\ \ 3\ \ 4\ \ 5$$

11. $-\frac{1}{2}, -3, \frac{7}{4}, -4\frac{1}{2}, 3\frac{1}{4}$

$$-5\ -4\ -3\ -2\ -1\ \ 0\ \ 1\ \ 2\ \ 3\ \ 4\ \ 5$$

12. $-4, -\frac{3}{4}, 1, -1\frac{1}{4}, \frac{5}{2}$

$$-5\ -4\ -3\ -2\ -1\ \ 0\ \ 1\ \ 2\ \ 3\ \ 4\ \ 5$$

CONCEPT CHECK   *Write each statement in words.*

13. (a) $0 < 5$

(b) $-10 > -17$

14. (a) $3 > -7$

(b) $12 < 22$

Write $<$ or $>$ in each blank to make a true statement. **See Example 2.**

15. 9 _____ 14

16. 6 _____ 11

17. 0 _____ -2

18. 0 _____ 2

19. -6 _____ 3

20. -9 _____ 9

21. 1 _____ -1

22. -1 _____ 0

23. -11 _____ -2

24. -5 _____ -1

25. -72 _____ -75

26. -50 _____ -60

*Simplify each absolute value expression. **See Example 3.***

27. $|3|$ 　　　　　　**28.** $|9|$ 　　　　　**29.** $|-10|$ 　　　　　**30.** $|-2|$ 　　　　　**31.** $|0|$

32. $\left|-\dfrac{1}{2}\right|$ 　　　**33.** $-\underbrace{|-18|}$ 　　**34.** $-\underbrace{|-5|}$ 　　　**35.** $-|32|$ 　　　　　**36.** $-|20|$

$-\ \_\_\_$ 　　　　$-\ \_\_\_$

*Find the opposite of each number. **See Examples 4 and 5.***

37. 7 　　　　　　**38.** 1 　　　　　**39.** -14 　　　　　**40.** -5 　　　　　**41.** $\dfrac{2}{3}$

42. 0 　　　　　　**43.** -8.3 　　　**44.** 0.2 　　　　　**45.** $-\dfrac{1}{6}$ 　　　　**46.** $-\dfrac{3}{10}$

Write true *or* false *for each statement.*

47. $|-5| > 0$ 　　　　　　**48.** $|-12| > |-15|$ 　　　　　**49.** $0 < -(-6)$

50. $-9 < -(-9)$ 　　　　　**51.** $-|-4| < -|-7|$ 　　　　　**52.** $-|-0| > 0$

Relating Concepts (Exercises 53–56) For Individual or Group Work

A special X-ray test (DEXA) is a quick way to screen patients for osteoporosis (brittle bone disease). Use the information on the Patient Report Form to **work Exercises 53–56 in order.**

Patient Report Form

Your T score measures your bone density compared to that of a young, healthy woman when peak bone mass is achieved.

T Score	Interpretation
Above 0	Above normal
0 to –1.0	Normal
Below –1	You may be at risk; further discussion with your health provider is recommended.

Source: Health Partners, Inc.

53. Here are the T scores for four patients. Draw a number line and graph the four scores.

Patient A:　-1.2
Patient B:　0.6
Patient C:　-0.5
Patient D:　0

54. List the patients' scores in order from lowest to highest.

55. What is the interpretation of each patient's score?

Taking a DEXA test of a patient's spine.

56. (a) Explain what could happen if patient A did not understand the importance of a negative sign.

(b) For which patient does the sign of the score make no difference? Explain your answer.

9.2 Adding and Subtracting Signed Numbers

OBJECTIVES

1 Add signed numbers by using a number line.

2 Add signed numbers without using a number line.

3 Find the additive inverse of a number.

4 Subtract signed numbers.

5 Add or subtract a series of signed numbers.

1 Complete each arrow so it represents the indicated number of units.

(a)

(b)

(c)

(d)

Answers

1. (a)

(b)

(c)

(d)

You can show a *positive* number on a number line by drawing an arrow pointing to the *right*. On the number lines below, both arrows represent positive 4 units.

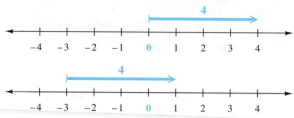

Draw arrows pointing to the *left* to show *negative* numbers. Both of the arrows below represent −3 units.

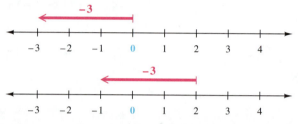

◀ **Work Problem 1 at the Side.**

OBJECTIVE ▶ 1   Add signed numbers by using a number line. You can use a number line to add signed numbers. For example, this number line shows how to add 2 and 3.

Add 2 and 3 by starting at 0 and drawing an arrow 2 units to the right. From the end of this arrow, draw another arrow 3 units to the right. This second arrow ends at 5, showing that the sum of 2 + 3 is 5.

$$2 + 3 = 5$$

EXAMPLE 1   Adding Signed Numbers Using a Number Line

Add using a number line.

(a) $4 + (-1)$

Start at 0 and draw an arrow 4 units to the right. From the end of this arrow, draw an arrow 1 unit to the *left*. (Remember to go to the left for a negative number.) This second arrow ends at 3, so

$$4 + (-1) = 3$$

CAUTION

Always start at 0 when adding on the number line.

········· **Continued on Next Page**

(b) $-6 + 2$

Draw an arrow from 0 going 6 units to the left. From the end of this arrow, draw an arrow 2 units to the right. This second arrow ends at -4, so

$$-6 + 2 = -4$$

(c) $-3 + (-5)$

As the arrows along the number line show, $-3 + (-5) = -8$.

> **Note**
>
> When a negative number *follows* an operation sign, we usually write parentheses around the negative number to prevent confusion. You can see this in **Example 1(c)** above.
>
> $$-3 + (-5) = -8$$
>
> Operation sign ⟶ ⟵ Parentheses around the negative number that follows

················· **Work Problem ② at the Side.** ▶

OBJECTIVE ② Add signed numbers without using a number line. After working with number lines for a while, you will see ways to add signed numbers without drawing arrows. You already know how to add two positive numbers (from **Chapter 1**). Here are the steps for adding two negative numbers.

> **Adding Two Negative Numbers**
>
> **Step 1**   Add the absolute values of the numbers.
>
> **Step 2**   Write a negative sign in front of the sum.

EXAMPLE 2   Adding Two Negative Numbers

Add (without using number lines).

(a) $-4 + (-12)$

The absolute value of -4 is **4.**
The absolute value of -12 is **12.**

Add the absolute values.

$$4 + 12 = 16$$

Write a negative sign in front of the sum.

Write a negative sign in front of 16.

$$-4 + (-12) = -16$$

(b) $-5 + (-25) = -30$

Sum of the absolute values, with a negative sign written in front

(c) $-11 + (-46) = -57$

················· **Continued on Next Page** ▶

---

② Draw arrows to find each sum.

(GS) (a) $3 + (-2) = $ ____

(b) $-4 + 1 = $ ____

(c) $-3 + 7 = $ ____

(d) $-1 + (-4) = $ ____

Answers

2. (a) $3 + (-2) = 1$

(b) $-4 + 1 = -3$

(c) $-3 + 7 = 4$

(d) $-1 + (-4) = -5$

3 Add.

(a) $-4 + (-4)$

(b) $-3 + (-20)$

(c) $-31 + (-5)$

(d) $-10 + (-8)$

GS (e) $-\dfrac{9}{10} + \left(-\dfrac{3}{5}\right)$

Absolute value of $-\dfrac{9}{10}$ is _____.

Absolute value of $-\dfrac{3}{5}$ is _____.

$\dfrac{9}{10} + \dfrac{3}{5} = \dfrac{9}{10} + \dfrac{\rule{1cm}{0.4pt}}{10} = \dfrac{\rule{1cm}{0.4pt}}{10}$

Write $\dfrac{15}{10}$ in lowest terms:

$\dfrac{15 \div \rule{0.6cm}{0.4pt}}{10 \div \rule{0.6cm}{0.4pt}} = \rule{0.8cm}{0.4pt}$

Write a negative sign in front of the final answer.

$-\dfrac{9}{10} + \left(-\dfrac{3}{5}\right) =$

Answers

3. (a) -8 (b) -23 (c) -36 (d) -18

(e) $\dfrac{9}{10}; \dfrac{3}{5}; \dfrac{9}{10} + \dfrac{6}{10} = \dfrac{15}{10}; \dfrac{15 \div 5}{10 \div 5} = \dfrac{3}{2}; -\dfrac{3}{2}$

(d) $-\dfrac{3}{4} + \left(-\dfrac{1}{2}\right)$

The absolute value of $-\dfrac{3}{4}$ is $\dfrac{3}{4}$, and the absolute value of $-\dfrac{1}{2}$ is $\dfrac{1}{2}$. Add the absolute values. Check that the sum is in lowest terms.

> The common denominator for 4 and 2 is 4.

$$\dfrac{3}{4} + \dfrac{1}{2} = \dfrac{3}{4} + \dfrac{2}{4} = \dfrac{5}{4}$$

> Already in lowest terms; do **not** rewrite it as a mixed number.

Write a negative sign in front of the sum.

$$-\dfrac{3}{4} + \left(-\dfrac{1}{2}\right) = -\dfrac{5}{4} \quad \text{Write a negative sign.}$$

Note

In algebra, we always write fractions in lowest terms, but usually do *not* change improper fractions to mixed numbers, because improper fractions are easier to work with. In **Example 2(d)** above, we checked that $-\dfrac{5}{4}$ was in lowest terms but did *not* rewrite it as $-1\dfrac{1}{4}$.

◀ Work Problem **3** at the Side.

Use the following steps to add two numbers with *different* signs.

Adding Two Numbers with Different Signs

Step 1 **Subtract** the lesser absolute value from the greater absolute value.

Step 2 Write the sign of the number with the **greater** absolute value in front of the answer.

EXAMPLE 3 Adding Two Numbers with Different Signs

Find each sum.
(a) $8 + (-3)$ First find this sum with a number line.

Because the second arrow ends at 5,

$$8 + (-3) = 5$$

Now find the sum by using the above rule. First, find the absolute value of each number.

$$|8| = 8 \quad \text{and} \quad |-3| = 3$$

Subtract the lesser absolute value from the greater absolute value.

$$8 - 3 = 5$$

The *positive* number 8 has the greater absolute value, so the answer is *positive*.

$$8 + (-3) = 5 \leftarrow \text{Positive answer}$$

················· **Continued on Next Page**

(b) $4 + (-12)$

First, find the absolute values.

$$|4| = 4 \quad \text{and} \quad |-12| = 12$$

Subtract the lesser absolute value from the greater absolute value.

$$12 - 4 = 8$$

The *negative* number -12 has the greater absolute value, so the answer is *negative*.

$$4 + (-12) = -8$$

Write a negative sign in front of the answer because -12 has the greater absolute value.

(c) $-21 + 15 = -6$

Write a negative sign in front of the answer because -21 has the greater absolute value.

(d) $13 + (-9) = 4$ ← Positive answer because the positive number 13 has the greater absolute value.

(e) $-\dfrac{1}{2} + \dfrac{2}{3}$

The absolute value of $-\frac{1}{2}$ is $\frac{1}{2}$, and the absolute value of $\frac{2}{3}$ is $\frac{2}{3}$. Subtract the lesser absolute value from the greater absolute value.

The common denominator for 3 and 2 is 6. →
$$\frac{2}{3} - \frac{1}{2} = \frac{4}{6} - \frac{3}{6} = \frac{1}{6}$$
← Already in lowest terms

Because the *positive* number $\frac{2}{3}$ has the greater absolute value, the answer is *positive*.

$$-\frac{1}{2} + \frac{2}{3} = \frac{1}{6} \quad \text{← Positive answer}$$

········· **Work Problem 4 at the Side.** ▶

OBJECTIVE 3 Find the additive inverse of a number. Recall that the opposite of 9 is -9, and the opposite of -4 is $-(-4)$, or 4. Adding opposites gives the results shown below.

$$9 + (-9) = 0 \quad \text{and} \quad -4 + 4 = 0$$

The sum of a number and its opposite is always 0. For this reason, opposites are also called *additive inverses* of each other.

Additive Inverse

The opposite of a number is called its **additive inverse.**

The sum of a number and its additive inverse is 0.

---

4 Add.

GS (a) $10 + (-2)$

$|10| = $ ____ and $|-2| = $ ____

What is the sign of the number with the greater absolute value? _____

Will the answer be positive or negative? _____

$10 + (-2) = $ ____

(b) $-7 + 8$

(c) $-11 + 11$

(d) $23 + (-32)$

(e) $-\dfrac{7}{8} + \dfrac{1}{4}$

Answers

4. (a) 10; 2; positive; positive; 8   **(b)** 1   **(c)** 0

(d) -9 **(e)** $-\dfrac{5}{8}$

5 Give the additive inverse of each number. Then find the sum of the number and its inverse.

GS (a) 12 Additive inverse is -12.

$$12 + (-12) = \underline{\hspace{1cm}}$$

(b) -9

(c) 3.5

(d) $-\dfrac{7}{10}$

(e) 0

6 Write each example in words.

GS (a) $-7 - 2$

negative seven minus

_____ _____

(b) -10

(c) $3 - (-5)$

(d) 4

(e) $-8 - (-6)$

(f) $2 - 9$

Answers

5. (a) 0   **(b)** 9; $-9 + 9 = 0$
   (c) -3.5; $3.5 + (-3.5) = 0$
   (d) $\dfrac{7}{10}$; $-\dfrac{7}{10} + \dfrac{7}{10} = 0$
   (e) 0; $0 + 0 = 0$

6. (a) negative seven minus positive two
   (b) negative ten
   (c) positive three minus negative five
   (d) positive four
   (e) negative eight minus negative six
   (f) positive two minus positive nine

EXAMPLE 4   Finding Additive Inverses

This table shows you several numbers and the additive inverse of each.

Number	Additive Inverse	Sum of Number and Inverse
6	-6	$6 + (-6) = 0$
-8	$-(-8)$ or 8	$(-8) + 8 = 0$
4	-4	$4 + (-4) = 0$
-3	$-(-3)$ or 3	$(-3) + 3 = 0$
$\dfrac{5}{8}$	$-\dfrac{5}{8}$	$\dfrac{5}{8} + \left(-\dfrac{5}{8}\right) = 0$
0	0	$0 + 0 = 0$

The sum is always 0.

◀ **Work Problem 5 at the Side.**

OBJECTIVE 4 Subtract signed numbers. You may have noticed that negative numbers are often written with parentheses, like (-8). This is especially helpful when subtracting because the $-$ sign is used both to indicate a **negative number** and to indicate **subtraction**.

Example	How to Say It
(-5)	**negative five**
8	positive eight
$-3 - 2$	**negative three** minus positive two
$6 - (-4)$	positive six **minus negative four**
$-7 - (-1)$	**negative seven minus negative one**
$8 - 3$	positive eight **minus** positive three

CAUTION

Be sure you understand when the " $-$ " sign means "subtract," and when it means "negative number."

◀ **Work Problem 6 at the Side.**

When working with signed numbers, it is helpful to write a subtraction problem as an addition problem. For example, you know that $6 - 4 = 2$. But you get the same result by *adding* 6 and the *opposite* of 4, that is, $6 + (-4)$.

$$6 - 4 = 2$$
$$6 + (-4) = 2$$

Same result

This suggests the following definition of subtraction.

Defining Subtraction

The difference of two numbers, a and b, is

$$a - b = a + (-b)$$

In other words, subtract two numbers by adding the first number and the opposite (additive inverse) of the second number.

Subtracting Signed Numbers

To subtract two signed numbers, add the opposite of the second number to the first number. Use these steps.

Step 1 Change the subtraction sign to addition.

Step 2 Change the sign of the second number to its opposite.

Step 3 Proceed as in addition.

Note

The pattern in a subtraction problem is shown below.

$$\begin{array}{ccc} \text{1st} & & \text{2nd} \\ \text{number} & - & \text{number} \end{array} = \begin{array}{ccc} \text{1st} & & \text{opposite of} \\ \text{number} & + & \text{2nd number} \end{array}$$

EXAMPLE 5   Subtracting Signed Numbers

Subtract.

(a) $8 - 11$

The first number, 8, stays the same. Change the subtraction sign to addition. Change the sign of the second number to its opposite.

Positive 8 stays the same.
$$8 - 11$$
$$\downarrow \quad \downarrow$$
$$8 + (-11)$$
Positive 11 is changed to its opposite, -11

Subtraction is changed to addition.

Now add.

$$8 + (-11) = -3$$
$$\text{So} \quad 8 - 11 = -3 \quad \text{also.}$$

(b) $-9 - 15$
$$\downarrow \quad \downarrow$$
$$-9 + (-15) = -24$$
Subtraction is changed to addition.
Positive 15 is changed to its opposite, (-15).

When subtracting, change two things.

(c) $-5 - (-7)$
$$\downarrow \quad \downarrow$$
$$-5 + (+7) = 2$$
Subtraction is changed to addition.
Negative 7 is changed to its opposite, $(+7)$.

(d) $10 - (-6)$
$$\downarrow \quad \downarrow$$
$$10 + (+6) = 16$$
Subtraction is changed to addition.
Negative 6 is changed to its opposite, $(+6)$.

(e) $-25 - (-20)$
$$\downarrow \quad \downarrow$$
$$-25 + (+20) = -5$$
Subtraction is changed to addition.
Negative 20 is changed to its opposite, $(+20)$.

················· Work Problem **7** at the Side. ▶

7 Subtract.

GS **(a)** $-6 - 5$
$$\downarrow \quad \searrow$$
$$-6 + (-5)$$

Now add.

$$-6 + (-5) = \underline{\quad\quad}$$

(b) $3 - (-10)$
$$\downarrow \quad \downarrow$$

(c) $-8 - (-2)$

(d) $4 - 9$

(e) $-7 - (-15)$

(f) $100 - 300$

Answers

7. (a) -11 **(b)** 13 **(c)** -6 **(d)** -5
(e) 8 **(f)** -200

8 Subtract.

(a) $-0.8 - 0.5$

$-0.8 + (-0.5)$

Now add.

$-0.8 + (-0.5) =$ _____

(b) $4 - 12.03$

(c) $-1.65 - (-20.4)$

(d) $\dfrac{1}{2} - \left(-\dfrac{2}{3}\right)$

(e) $\dfrac{1}{10} - \dfrac{4}{5}$

(f) $-\dfrac{7}{9} - \left(-\dfrac{1}{6}\right)$

Answers

8. **(a)** -1.3 **(b)** -8.03 **(c)** 18.75
   (d) $\dfrac{7}{6}$ **(e)** $-\dfrac{7}{10}$ **(f)** $-\dfrac{11}{18}$

EXAMPLE 6 **Subtracting Signed Decimal and Fraction Numbers**

Subtract.

(a) $7.6 - (-8.3)$ Subtraction is changed to addition.
Negative 8.3 is changed to its opposite, $(+8.3)$.

$7.6 + (+8.3) = 15.9$

Once the subtraction is changed to addition, notice that both numbers are positive. You can add them, as you did in **Chapter 4.**

Line up decimal points.

$\begin{array}{r} 7.6 \\ + 8.3 \\ \hline 15.9 \end{array}$

(b) $-0.7 - 16.5$ Subtraction is changed to addition.
Positive 16.5 is changed to its opposite, (-16.5).

$-0.7 + (-16.5) = -17.2$

To add two *negative* numbers, first add the absolute values.

$|-0.7|$ is 0.7
$|-16.5|$ is 16.5

$\begin{array}{r} 16.5 \\ + 0.7 \\ \hline 17.2 \end{array}$

Then write a *negative* sign in front of 17.2

$$-0.7 + -16.5 = -17.2$$

(c) $\dfrac{5}{8} - \dfrac{3}{4}$ Subtraction is changed to addition.
Positive $\frac{3}{4}$ is changed to its opposite, $(-\frac{3}{4})$.

$$\dfrac{5}{8} + \left(-\dfrac{3}{4}\right) = \dfrac{5}{8} + \left(-\dfrac{6}{8}\right) = -\dfrac{1}{8}$$

Once the subtraction is changed to addition, notice that the numbers have *different* signs. So, we subtract the lesser absolute value from the greater absolute value.

$\left|\dfrac{5}{8}\right|$ is $\dfrac{5}{8}$ and $\left|-\dfrac{6}{8}\right|$ is $\dfrac{6}{8}$ Then $\dfrac{6}{8} - \dfrac{5}{8} = \dfrac{1}{8}$

Lesser absolute value — Greater absolute value

The *negative* number $-\frac{6}{8}$ has the greater absolute value, so the answer is *negative*.

(d) $-\dfrac{1}{4} - \left(-\dfrac{5}{6}\right)$ Subtraction is changed to addition.
Negative $\frac{5}{6}$ is changed to its opposite, $(+\frac{5}{6})$.

$$-\dfrac{3}{12} + \left(+\dfrac{5}{6}\right) = -\dfrac{3}{12} + \left(+\dfrac{10}{12}\right) = \dfrac{7}{12}$$

Once the subtraction is changed to addition, notice that the numbers have *different* signs. So, we subtract the lesser absolute value from the greater absolute value.

$\left|-\dfrac{3}{12}\right|$ is $\dfrac{3}{12}$ and $\left|+\dfrac{10}{12}\right|$ is $\dfrac{10}{12}$ Then $\dfrac{10}{12} - \dfrac{3}{12}$ is $\dfrac{7}{12}$

The *positive* number $\frac{10}{12}$ has the greater absolute value so the answer is *positive*.

◄ Work Problem **8** at the Side.

OBJECTIVE ▶ 5 Add or subtract a series of signed numbers. If a problem involves both addition and subtraction, use the order of operations, from **Chapter 1.**

EXAMPLE 7 **Combining Addition and Subtraction of Signed Numbers**

According to the last step in the order of operations, perform addition and subtraction from left to right.

(a) $-6 + (-11) - 5$

$$-17 \quad - \quad 5$$

Change subtraction to addition; change positive 5 to its opposite, (-5).

$$-17 \quad + \quad (-5)$$
$$-22$$

(b) $4 - (-3) + (-9)$ Change subtraction to addition; change -3 to its opposite, $(+3)$.

$$4 + (+3) + (-9)$$
$$7 \quad + (-9)$$
$$-2$$

(c) $14 + (-9) - (-8) + 10$

$$5 \quad - (-8) + 10$$

Change subtraction to addition; change -8 to its opposite, $(+8)$.

$$5 \quad + (+8) + 10$$
$$13 \quad + 10$$
$$23$$

(d) -6.3
-14.9
8.5
-7.4
5.2

Start at the top.

$$\begin{array}{cccc} -6.3 \\ -14.9 \end{array} \!\!\rightarrow -21.2 \\ \quad 8.5 \qquad\qquad 8.5 \!\!\rightarrow -12.7 \\ \quad -7.4 \qquad -7.4 \qquad -7.4 \!\!\rightarrow -20.1 \\ \quad 5.2 \qquad\quad 5.2 \qquad\quad 5.2 \qquad\quad 5.2 \\ \qquad\qquad\qquad\qquad\qquad\qquad\qquad\quad -14.9$$

·········· **Work Problem 9 at the Side.** ▶

---

9 Perform the additions and subtractions in the correct order.

GS (a) $6 - 7 \quad + (-3)$
$$\downarrow \;\downarrow\; \downarrow$$
$$6 + (-7) + (-3)$$
$$-1 \quad + (-3)$$
$$\underline{}$$

GS (b) $-2 + (-3) - (-5)$
$$-5 \qquad - (-5)$$
$$\downarrow \quad \downarrow$$
$$-5 \qquad (\;\;)$$
$$\underline{}$$

(c) $-3 - (-9) - (-5)$

(d) $8 - (-2) + (-6)$

(e) $-1 - 2 + 3 - 4$

(f) $7 - 6 - 5 + (-4)$

(g) $-6 + (-15) - (-19)$

(h) -19.2
-6.7
15.8
17.1
-5.4

Answers

9. **(a)** -4 **(b)** $-5 + (+5); 0$ **(c)** 11 **(d)** 4
(e) -4 **(f)** -8 **(g)** -2 **(h)** 1.6

*Add by using the number line. **See Example 1.***

1. $-5 + (-2)$

2. $-2 + (-2)$

3. $3 + (-4)$

4. $5 + (-1)$

5. CONCEPT CHECK Explain how to add two numbers with different signs and give an example.

6. CONCEPT CHECK Explain how to add two negative numbers and give an example.

*Add. **See Examples 2 and 3.***

7. $-8 + 5$

8. $-3 + 2$

9. $-1 + 8$

10. $-4 + 10$

11. $-2 + (-5)$

12. $-7 + (-3)$

13. $6 + (-5)$

14. $11 + (-3)$

15. $4 + (-12)$

16. $9 + (-10)$

17. $-10 + (-10)$

18. $-5 + (-20)$

Write and solve an addition problem for each situation.

19. The football team gained 13 yd on the first play and lost 17 yd on the second play.

20. At penguin breeding grounds on Antarctic islands, winter temperatures routinely drop to $-15\ °C$; but at the interior of the continent, temperatures may easily drop another $60°$ below that.

21. Nicole's bank account was overdrawn by $52.50. She deposited $50 in the account.

22. $48.40 was stolen from Jay's car. He got $30 of it back.

23. Use the score sheet to find each player's point total after three rounds in a card game.

	Jeff	Terry
Round 1	Lost 20 pts	Won 42 pts
Round 2	Won 75 pts	Lost 15 pts
Round 3	Lost 55 pts	Won 20 pts

24. Use the information in the table on flood water depths to find the new flood level for each river.

	Red River	Mississippi
Monday	Rose 8 ft	Rose 4 ft
Tuesday	Fell 3 ft	Rose 7 ft
Wednesday	Fell 5 ft	Fell 13 ft

Add.

25. $7.8 + (-14.6)$

26. $4.9 + (-8.1)$

27. $-\dfrac{1}{2} + \dfrac{3}{4}$

28. $-\dfrac{2}{3} + \dfrac{5}{6}$

29. $-\dfrac{7}{10} + \dfrac{2}{5}$

30. $-\dfrac{3}{4} + \dfrac{3}{8}$

31. $-\dfrac{7}{3} + \left(-\dfrac{5}{9}\right)$

32. $-\dfrac{8}{5} + \left(-\dfrac{3}{10}\right)$

Give the additive inverse of each number. **See Example 4.**

33. 3

34. 4

35. -9

36. -14

37. $\dfrac{1}{2}$

38. $\dfrac{7}{8}$

39. -6.2

40. -0.5

CONCEPT CHECK *Write each statement in words.*

41. $-7 - 3$

42. $14 + (-12)$

43. $-20 + (-3)$

44. $-8 - (-5)$

Subtract by changing subtraction to addition. **See Examples 5 and 6.**

45. $19 \ - \ 5$
$\downarrow \quad \downarrow$
$\underbrace{19 \ + \ (-5)}$
___

46. $24 \ - \ 11$
$\downarrow \quad \downarrow$
$\underbrace{24 \ + \ (-11)}$
___

47. $10 - 12$

48. $1 - 8$

49. $7 - 19$

50. $2 - 17$

51. $-15 - 10$

52. $-10 - 4$

53. $-9 - 14$

54. $-3 - 11$

55. $-3 - (-8)$

56. $-1 - (-4)$

57. $6 - (-14)$

58. $8 - (-1)$

59. $1 - (-10)$

60. $6 - (-1)$

61. $-30 - 30$

62. $-25 - 25$

63. $-16 - (-16)$

64. $-20 - (-20)$

65. $-\dfrac{7}{10} - \dfrac{4}{5}$

66. $-\dfrac{8}{15} - \dfrac{3}{10}$

67. $\dfrac{1}{2} - \dfrac{9}{10}$

68. $\dfrac{2}{3} - \dfrac{11}{12}$

69. $-8.3 - (-9)$

70. $-2 - (-3.9)$

71. CONCEPT CHECK Explain and correct the mistakes in these subtraction problems.

(a) $-6 - 6$

$\downarrow \downarrow$

$-6 + 6 = 0$

(b) $-9 - 5$

$\downarrow \downarrow$

$9 + (-5) = 4$

72. CONCEPT CHECK Explain the purpose of the "−" sign in each of these examples.

(a) $6 - 9$

(b) (-9)

(c) $-(-2)$

The windchill table below shows how wind increases a person's heat loss in cold weather. For example, suppose the temperature is 15 °F (along the top of the table) and the wind speed is 20 mph (down the left side of the table). The column and row intersect at −2 °F, the "windchill" temperature. The actual temperature is 15 °F but the wind makes it feel like −2 °F. Use the table to find the windchill temperature under each set of conditions in Exercises 73–76. Then write and solve a subtraction problem to calculate actual temperature minus windchill temperature.

WINDCHILL

Temperature (degrees Fahrenheit)

Calm	40	35	30	25	20	15	10	5	0	−5	−10	−15	−20	−25	−30
5	36	31	25	19	13	7	1	−5	−11	−16	−22	−28	−34	−40	−46
10	34	27	21	15	9	3	−4	−10	−16	−22	−28	−35	−41	−47	−53
15	32	25	19	13	6	0	−7	−13	−19	−26	−32	−39	−45	−51	−58
20	30	24	17	11	4	−2	−9	−15	−22	−29	−35	−42	−48	−55	−61
25	29	23	16	9	3	−4	−11	−17	−24	−31	−37	−44	−51	−58	−64
30	28	22	15	8	1	−5	−12	−19	−26	−33	−39	−46	−53	−60	−67
35	28	21	14	7	0	−7	−14	−21	−27	−34	−41	−48	−55	−62	−69
40	27	20	13	6	−1	−8	−15	−22	−29	−36	−43	−50	−57	−64	−71

Wind Speed (miles per hour)

Shaded area: Frostbite occurs in 15 minutes or less.

Source: National Weather Service

73. (a) 30 °F; 10 mph wind

(b) 15 °F; 15 mph wind

74. (a) 40 °F; 20 mph wind

(b) 20 °F; 35 mph wind

75. (a) 5 °F; 25 mph wind

(b) −10 °F; 35 mph wind

76. (a) 10 °F; 15 mph wind

(b) −5 °F; 30 mph wind

Follow the order of operations to work each problem. **See Example 7.**

77. $-2 + (-11) - (-3)$

78. $-5 - (-2) + (-6)$

79. $4 - (-13) + (-5)$

80. $6 - (-1) + (-10)$

81. $-12 - (-3) - (-2)$

82. $-1 - (-7) - (-4)$

83. $4 - (-4) - 3$

84. $5 - (-2) - 8$

85. $\dfrac{1}{2} - \dfrac{2}{3} + \left(-\dfrac{5}{6}\right)$

86. $\dfrac{2}{5} - \dfrac{7}{10} + \left(-\dfrac{3}{2}\right)$

87. $-5.7 - (-9.4) - 8.1$

88. $-6.5 - (-11.2) - 1.4$

89. $-2 + (-11) + |-2|$

90. $|-7 + 2| + (-2) + 4$

Hint: Work inside the absolute value bars first.

91. $-3 - (-2 + 4) + (-5)$

Hint: Work inside parentheses first.

92. $5 - 8 - (6 - 7) + 1$

Find the balance in each bank account.

93. LaVerle had $37 in her bank account. She made a $626 monthly payment on her tuition, deposited a $908 paycheck, and took out $60 at a cash machine.

94. Yvonne had $478 in her bank account. She deposited a $212 tax refund and paid her $89 electric bill and $605 for rent.

95. Rod's bank account was overdrawn by $89.62. He paid $110.70 for his cell phone and $99.68 for utilities before depositing three $100 bills into his account.

96. Dwayne's bank account was overdrawn by $23.77, so the bank charged a $29 fee. He deposited his $583.29 paycheck before withdrawing $50 in cash.

Relating Concepts (Exercises 97–100) For Individual or Group Work

Use your knowledge of addition and subtraction of signed numbers to *work Exercises 97–100 in order.*

97. Work each pair of examples.

(a) $-5 + 3$ = ____ $3 + (-5)$ = ____

(b) $-2 + (-6)$ = ____ $-6 + (-2)$ = ____

(c) $17 + (-7)$ = ____ $-7 + 17$ = ____

Explain why the answers to each pair are the same.

98. Work each pair of examples.

(a) $-3 - 5$ = ____ $5 - (-3)$ = ____

(b) $-4 - (-6)$ = ____ $-6 - (-4)$ = ____

(c) $3 - 10$ = ____ $10 - 3$ = ____

Explain what happens when you try to apply the commutative property to subtraction.

99. Look back at your answers in **Exercise 98.**

(a) Describe how the two answers for each pair are similar, and how they are different.

(b) Write a rule that explains what happens when you switch the order of the numbers in a subtraction problem.

100. Work each set of exercises to see how 0 functions in addition and subtraction.

(a) Describe the pattern in these addition answers.

$-18 + 0$ = ____ $20 + 0 =$ ____

$0 + (-5) =$ ____ $0 + 4 =$ ____

(b) Describe the pattern in these subtraction answers.

$2 - 0$ = ____ $0 - 10$ = ____

$-3 - 0 =$ ____ $0 - (-7) =$ ____

9.3 Multiplying and Dividing Signed Numbers

1 Multiply.

(a) $5 \cdot (-4)$

The numbers have **different** signs so the product is **negative.**

(b) $(-9)(15)$

(c) $12(-1)$

(d) $-6(6)$

(e) $\left(-\dfrac{7}{8}\right)\left(\dfrac{4}{3}\right)$

In mathematics, the rules or patterns must be consistent. We can use this idea to see how to multiply two numbers with *different* signs. Look for a pattern in this list of products.

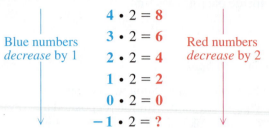

$$4 \cdot 2 = 8$$
$$3 \cdot 2 = 6$$
Blue numbers *decrease* by 1 $2 \cdot 2 = 4$ Red numbers *decrease* by 2
$$1 \cdot 2 = 2$$
$$0 \cdot 2 = 0$$
$$-1 \cdot 2 = \text{?}$$

As the numbers in **blue** decrease by 1, the numbers in **red** decrease by 2. To continue the red pattern, replace **?** with a number 2 *less than* 0, which is -2. We get this result.

$$-1 \cdot 2 = -2$$

OBJECTIVE 1 Multiply or divide two numbers with opposite signs. The pattern above suggests a rule for multiplying two numbers with different signs.

> **Multiplying Two Numbers with Different Signs**
> The product of two numbers with *different* signs is *negative*.

> **EXAMPLE 1 Multiplying Numbers with Different Signs**
>
> Multiply.
>
> **(a)** $-8 \cdot 4 = -32$ Factors have *different* signs, so the product is *negative*.   Negative / Positive
>
> **(b)** $6(-3) = -18$ Factors have *different* signs, so the product is *negative*.   Positive / Negative
>
> **(c)** $(-5)(11) = -55$ **(d)** $12(-7) = -84$

·· **Work Problem 1 at the Side.**

For two numbers with the *same* sign, look at this pattern.

$$4 \cdot (-2) = -8$$
$$3 \cdot (-2) = -6$$
Blue numbers *decrease* by 1 $2 \cdot (-2) = -4$ Red numbers *increase* by 2
$$1 \cdot (-2) = -2$$
$$0 \cdot (-2) = \;\;\;0$$
$$-1 \cdot (-2) = \;\;\;\text{?}$$

This time, as the numbers in **blue** *decrease* by 1, the products *increase* by 2. To continue the **red** pattern, replace **?** with a number 2 *greater than* 0, which is **positive 2.** Therefore, we get this result.

$$-1 \cdot (-2) = 2$$

OBJECTIVE 2 Multiply or divide two numbers with the same sign. In the pattern above, a negative number times a negative number gave a positive result.

Multiplying Two Numbers with the Same Sign

The product of two numbers with the *same* sign is *positive*.

EXAMPLE 2 Multiplying Two Numbers with the Same Sign

Multiply.

(a) $(-9)(-2) = 18$ ← The factors have the **same** sign (both are negative). The product is **positive**.

(b) $-7(-4) = 28$

(c) $(-6)(-2) = 12$ — When signs *match*, the product is *positive*.

(d) $(-10)(-5) = 50$

(e) $7(5) = 35$

············ **Work Problem ❷ at the Side.** ▶

You can use the same sign rules for dividing signed numbers as you use for multiplying signed numbers.

Dividing Signed Numbers

When two nonzero numbers with *different* signs are divided, the result is *negative*.

When two nonzero numbers with the *same* sign are divided, the result is *positive*.

Division involving 0 works the same as it did for whole numbers (see **Chapter 1**). Division by 0 cannot be done; we say it is *undefined*. But 0 divided by any other number is 0.

EXAMPLE 3 Dividing Signed Numbers

Divide.

(a) $\dfrac{-15}{5}$ ← Numbers have **different** signs, so the quotient is **negative**.

$\dfrac{-15}{5} = -3$

(b) $\dfrac{-8}{-4}$ ← Numbers have the **same** sign (both negative), so the quotient is **positive**.

$\dfrac{-8}{-4} = 2$

(c) $\dfrac{-75}{-25} = 3$

(d) $\dfrac{-6}{0}$ is undefined. — Division by 0 *cannot* be done. Write "**undefined**."

(e) $\dfrac{0}{-5} = 0$ — 0 *can* be divided by a nonzero number.

(f) $\dfrac{90}{-9} = -10$

(g) $\dfrac{-\dfrac{2}{3}}{-\dfrac{5}{9}} = \left(-\dfrac{2}{3}\right)\left(-\dfrac{9}{5}\right)$ — Use the reciprocal of $-\frac{5}{9}$, which is $-\frac{9}{5}$.

$= \left(-\dfrac{2}{\overset{}{\underset{1}{3}}}\right)\left(-\dfrac{\overset{3}{9}}{5}\right)$ — Divide out the common factor. Then multiply.

— Numbers have the **same** signs, so the quotient is **positive**.

$= \dfrac{6}{5}$ — Both numbers were **negative**, so the quotient is **positive**.

············ **Work Problem ❸ at the Side.** ▶

---

❷ Multiply.

ɢs (a) $(-5)(-5)$

The numbers have the **same** sign, so the product is **positive**.

(b) $(-14)(-1)$

(c) $-7(-8)$

(d) $3(12)$

(e) $\left(-\dfrac{2}{3}\right)\left(-\dfrac{6}{5}\right)$

❸ Divide.

ɢs (a) $\dfrac{-20}{4}$ **Different** signs; result is **negative**.

ɢs (b) $\dfrac{-50}{-5}$ **Same** signs; result is **positive**.

(c) $\dfrac{44}{2}$

(d) $\dfrac{6}{-6}$

(e) $\dfrac{-15}{-1}$

(f) $\dfrac{-\dfrac{3}{5}}{\dfrac{9}{10}}$

(g) $\dfrac{-35}{0}$

Answers

2. (a) 25 **(b)** 14 **(c)** 56 **(d)** 36 **(e)** $\dfrac{4}{5}$

3. (a) −5 **(b)** 10 **(c)** 22 **(d)** −1 **(e)** 15

  (f) $-\dfrac{2}{3}$ **(g)** undefined

9.3 Exercises

FOR EXTRA HELP

MyMathLab®

1. CONCEPT CHECK Explain how to multiply two negative numbers, and give an example.

2. CONCEPT CHECK Explain how to multiply two numbers with different signs, and give an example.

Multiply. See Examples 1 and 2.

3. $-5 \cdot 7$ **Different** signs; **negative** product.

4. $-10 \cdot 2$ **Different** signs; **negative** product.

5. $(-5)(9)$

6. $(-9)(4)$

7. $3(-6)$

8. $8(-6)$

9. $10(-5)$

10. $5(-11)$

11. $(-1)(40)$

12. $(75)(-1)$

13. $-8(-4)$ **Same** signs; **positive** product.

14. $-3(-9)$ **Same** signs; **positive** product.

15. $11(7)$

16. $4(25)$

17. $-19(-7)$

18. $-21(-3)$

19. $-13(-1)$

20. $-1(-31)$

21. $(0)(-25)$

22. $(-50)(0)$

23. $-\dfrac{1}{2}(-8)$

24. $\dfrac{1}{3}(-15)$

25. $-10\left(\dfrac{2}{5}\right)$

26. $-25\left(-\dfrac{7}{10}\right)$

27. $\left(\dfrac{3}{5}\right)\left(-\dfrac{1}{6}\right)$

28. $\left(-\dfrac{7}{9}\right)\left(-\dfrac{3}{4}\right)$

29. $-\dfrac{7}{5}\left(-\dfrac{10}{3}\right)$

30. $-\dfrac{9}{10}\left(\dfrac{5}{4}\right)$

31. $-\dfrac{7}{15}\cdot\dfrac{25}{14}$

32. $-\dfrac{5}{9}\cdot\dfrac{18}{25}$

33. $-\dfrac{5}{2}\left(-\dfrac{7}{10}\right)$

34. $-\dfrac{8}{5}\left(-\dfrac{15}{16}\right)$

35. $9(-4.7)$

36. $15(-6.3)$

37. $(-0.5)(-12)$

38. $(-3.15)(-5)$

39. $(-6.2)(5.1)$

40. $(-4.3)(9.7)$

41. $-1.25(-3.6)$

42. $6.33(0.2)$

43. $(-8.23)(-1)$

44. $(-1)(-0.69)$

45. $0(-58.6)$

46. $-91.3(0)$

47. CONCEPT CHECK Explain what happens when a signed number is divided by 0, and give an example.

48. CONCEPT CHECK Explain what happens when 0 is divided by a signed number, and give an example.

Divide. **See Example 3.**

49. $\dfrac{-14}{7}$ **Different signs**

50. $\dfrac{-8}{2}$ **Different signs**

51. $\dfrac{30}{-6}$

52. $\dfrac{21}{-7}$

53. $\dfrac{-28}{0}$ **Careful!**

54. $\dfrac{-40}{0}$ **Careful!**

55. $\dfrac{14}{-1}$

56. $\dfrac{25}{-1}$

57. $\dfrac{-20}{-2}$ **Same signs**

58. $\dfrac{-80}{-4}$ **Same signs**

59. $\dfrac{-48}{-12}$

60. $\dfrac{-30}{-15}$

61. $\dfrac{-18}{18}$

62. $\dfrac{50}{-50}$

63. $\dfrac{-573}{-3}$

64. $\dfrac{-580}{-5}$

65. $\dfrac{0}{-9}$

66. $\dfrac{0}{-4}$

67. $\dfrac{-30}{-30}$

68. $\dfrac{-25}{-25}$

69. $\dfrac{-\dfrac{5}{7}}{-\dfrac{15}{14}}$

70. $\dfrac{-\dfrac{3}{4}}{-\dfrac{9}{16}}$

71. $-\dfrac{2}{3} \div (-2)$

72. $-\dfrac{3}{4} \div (-9)$

73. $5 \div \left(-\dfrac{5}{8}\right)$

74. $7 \div \left(-\dfrac{14}{15}\right)$

75. $-\dfrac{7}{5} \div \dfrac{3}{10}$

76. $-\dfrac{4}{9} \div \dfrac{8}{3}$

77. $\dfrac{-18.92}{-4}$

78. $\dfrac{-22.75}{-7}$

79. $\dfrac{-7.05}{1.5}$

80. $\dfrac{-17.02}{7.4}$

81. $\dfrac{45.58}{-8.6}$

82. $\dfrac{6.27}{-0.3}$

Following the order of operations, work from left to right in each exercise.

83. $(-4)(-6)\dfrac{1}{2}$

84. $(-9)(-3)\dfrac{2}{3}$

85. $(-0.6)(-0.2)(-3)$

86. $(-4)(-1.2)(-0.7)$

87. $\left(-\dfrac{1}{2}\right)\left(\dfrac{2}{5}\right)\left(\dfrac{7}{8}\right)$

88. $\left(\dfrac{3}{4}\right)\left(-\dfrac{5}{6}\right)\left(\dfrac{2}{3}\right)$

89. $-36 \div (-2) \div (-3) \div (-3) \div (-1)$

90. $-48 \div (-8) \cdot (-4) \div (-4) \div (-3)$

91. $|-8| \div (-4) \cdot |-5|$

92. $-6 \cdot |-3| \div |9| \cdot (-2)$

Solve each application problem. Be sure to label each answer as a positive or negative number.

93. A new computer-software store had losses of $9950 during each month of its first year. What was the total loss for the year?

94. A college's enrollment dropped by 3245 students over the last 11 years. What was the average drop in enrollment each year?

95. The greatest ocean depth is 36,198 ft below sea level. If an unmanned research sub dives to that depth in 18 equal steps, how far does it dive in each step?

96. Pat ate a dozen crackers as a snack. Each cracker had 17 calories. How many calories did Pat eat? (*Source:* Nabisco, Inc.)

97. Tuition at the state university is $182 per credit for undergraduates. How much tuition will Wei Chen pay for 13 credits?

98. A cell phone company estimates that it is losing 95 customers each week. How many customers will it lose in a year?

99. There is a 3-degree drop in temperature for every thousand feet that an airplane climbs into the sky. If the temperature on the ground is 50 degrees, what will be the temperature when the plane reaches an altitude of 24,000 ft? (*Source:* New Zealand Air.)

100. An unmanned submarine descends to 150 ft below the surface of the ocean. Then it continues to go deeper, taking a water sample every 25 ft. What is its depth when it takes the 15th sample?

Relating Concepts (Exercises 101–102) For Individual or Group Work

Use your knowledge of multiplying and dividing signed numbers and look for patterns as you **work Exercises 101 and 102 in order.**

101. Write three numerical examples for each situation.

    (a) A positive number multiplied by -1

    (b) A negative number multiplied by -1

Now write a rule that explains what happens when you multiply a signed number by -1.

102. Write three numerical examples for each situation.

    (a) A negative number divided by -1

    (b) A positive number divided by -1

Now write a rule that explains what happens when you divide a signed number by -1.

9.4 Order of Operations

OBJECTIVES

1 Use the order of operations.

2 Use the order of operations with exponents.

3 Use the order of operations with fraction bars.

1 Simplify.

(a) $-9 + (-15) + (-3)$

(b) $-8 - (-2) + (-6)$

(c) $3(-4) \div (-6)$

(d) $-18 \div 9(-4)$

2 Use the order of operations to simplify each expression.

GS (a) $10 + \underbrace{8 \div 2}$ Divide first!

$10 + \underline{}$ Now add.

GS (b) $4 - \underbrace{6(-2)}$ Multiply first!

$\underset{\downarrow}{4} - \underset{\downarrow}{(-12)}$ Now change subtraction to addition.

(c) $-3 + (-5) \cdot 2 - 1$

(d) $-6 \div 2 + 3(-2)$

(e) $7 - 6(2) \div (-3)$

Answers

1. (a) -27 (b) -12 (c) 2 (d) 8
2. (a) $10 + 4$; 14 (b) $4 + (+12) = 16$ (c) -14
   (d) -9 (e) 11

In the last two sections, you worked examples that mixed either addition and subtraction or multiplication and division. In those situations, you worked from left to right. Here are two more examples as a review.

Work additions and subtractions from left to right.	Work multiplications and divisions from left to right.
$\underbrace{-8 - (-6)} + (-11)$	$\underbrace{(-15) \div (-3)}(6)$ Divide.
$\underbrace{-2 \quad + (-11)}$	$\underbrace{5(6)}$ Multiply.
-13	30

◀ **Work Problem 1 at the Side.**

OBJECTIVE ▶ **1** **Use the order of operations.** Before working examples that include a mix of operations and/or parentheses, let's review the order of operations from **Chapter 1**.

Order of Operations

1. Do all operations inside **parentheses** or **other grouping symbols**.
2. Simplify any expressions with **exponents** and find any **square roots**.
3. Do the remaining **multiplications and divisions** as they occur from left to right.
4. Do the remaining **additions and subtractions** as they occur from left to right.

EXAMPLE 1   Using the Order of Operations

Use the order of operations to simplify this expression.

$4 - 10 \div 2 + 7$ Check for parentheses: none. Check for exponents and square roots: none.

Move from left to right, checking for multiplying and dividing.

Yes, here is dividing. Use the number on each side of the ÷ sign.

$4 - \underset{\downarrow\downarrow}{\mathbf{10 \div 2}} + 7$ $10 \div 2$ is 5. Bring down the other numbers and signs you haven't used.

$4 - \quad 5 \quad + 7$ Move from left to right, checking for adding and subtracting.

Yes, here is subtracting.

$4 - \quad 5 \quad + 7$ Change subtraction to addition; change 5 to its opposite, (-5).

$\underbrace{4 + (-5)} + 7$ Add $4 + (-5)$ to get -1.

$\underbrace{-1 \quad + 7}$ Add $-1 + 7$ to get 6.

6

◀ **Work Problem 2 at the Side.**

| EXAMPLE 2 | Parentheses and the Order of Operations |

Use the order of operations to simplify each expression.

(a) $-8\,(7-5)-9$ Work inside parentheses first: $7-5$ is 2.
Bring down the other numbers and signs you haven't used yet.

$-8\,(2)\quad-9$ Check for exponents and square roots: none.

$-8\,(2)\quad-9$ Move from left to right, looking for multiplying and dividing. Here is multiplying.

$-16\quad-9$ Move from left to right, doing any adding and subtracting. Change subtraction to addition.

$-16\quad+(-9)$ Add $-16+(-9)$ to get -25.

-25

> $3+2$ is **not** done first.

(b) $3+2\,(6-8)\,(15\div 3)$ Work inside first set of parentheses; change $6-8$ to $6+(-8)$ to get (-2).

$3+\;\;2\,(-2)\,(15\div 3)$ Work inside second set of parentheses; $15\div 3$ is 5.

$3+\;\;2\,(-2)\qquad(5)$ Multiply and divide from left to right. First multiply $2\,(-2)$ to get -4.

$3+\qquad(-4)\qquad(5)$ Then multiply $-4\cdot 5$ to get -20.

$3+\qquad(-20)$ Add last; $3+(-20)$ gives -17.

-17

·········· **Work Problem ❸ at the Side.** ▶

| OBJECTIVE ▶ ❷ | **Use the order of operations with exponents.** Recall

from **Chapter 1** that 2^3 means 2 is used as a factor 3 times.

$$2^3 = 2\cdot 2\cdot 2 = 8$$

The small raised 3 in 2^3 is called an *exponent*. Exponents are also used with signed numbers. Here are three examples.

$$(-3)^2 = (-3)\cdot(-3) = 9$$

$$(-4)^3 = (-4)\cdot(-4)\cdot(-4)$$

> Be careful! Multiply two numbers at a time. Watch the signs.

$$= \quad 16\quad\cdot(-4)$$

$$= \quad -64 \longleftarrow \text{Negative product}$$

$$\left(-\frac{1}{2}\right)^4 = \left(-\frac{1}{2}\right)\cdot\left(-\frac{1}{2}\right)\cdot\left(-\frac{1}{2}\right)\cdot\left(-\frac{1}{2}\right)$$

$$= \quad \frac{1}{4}\quad\cdot\left(-\frac{1}{2}\right)\cdot\left(-\frac{1}{2}\right)$$

$$= \quad -\frac{1}{8}\quad\cdot\left(-\frac{1}{2}\right)$$

$$= \quad \frac{1}{16}$$

Be very careful with exponents and signed numbers. For example,

$$(-3)^2 = (-3)\cdot(-3) = \mathbf{9} \quad\textbf{but}\quad -3^2 = -(3\cdot 3) = \mathbf{-9}$$

---

❸ Simplify.

(a) $2+40\div(-5+3)$ Parentheses

$\quad 2+40\div\quad(-2)$ Divide.

$\quad 2+\quad(\underline{\quad})$ Now add.

$\quad\underline{\quad}$

(b) $5-3\,(2+4)$ Parentheses

$\quad 5-3\quad(6)$ Multiply.

$\quad 5-(\underline{\quad})$ Now subtract.

$\quad\underline{\quad}$

(c) $(-24\div 2)+(15-3)$

(d) $-3\,(2-8)-5\,(4-3)$

(e) $3\,(3)-10\,(3)\div 5$

(f) $6-(2+7)\div(-4+1)$

Answers
3. (a) $2+(-20);\ -18$ (b) $5-(18);\ -13$
   (c) 0   (d) 13   (e) 3   (f) 9

④ Simplify.

⑤ (a) $2^3 - 3^2$ 2^3 is $2 \cdot 2 \cdot 2$

$8 - \underline{}$ 3^2 is $3 \cdot 3$

$\underline{}$

(b) $4^2 - 3^2(5 - 2)$

(c) $-18 \div (-3)(2^3)$

(d) $(-3)^3 + (3 - 8)^2$

(e) $\dfrac{3}{8} + \left(-\dfrac{1}{2}\right)^2 \div \dfrac{1}{4}$

> **CAUTION**
>
> $(-3)^2$ is **not** the same as -3^2.
>
> $$(-3)^2 = (-3) \cdot (-3) = 9 \quad but \quad -3^2 = -(3 \cdot 3) = -9$$
>
> You will need this information as you take more algebra classes.

EXAMPLE 3   **Exponents and the Order of Operations**

Simplify.

(a) $4^2 - (-3)^2$ There are parentheses around (-3), but no work can be done inside these parentheses.

Apply the exponents: $4^2 = 4 \cdot 4 = 16$ and   [Careful! $4^2 = 4 \cdot 4 = 16$ not $4 \cdot 2$]

$(-3)^2 = (-3) \cdot (-3) = 9$

$16 - 9$ Subtract: $16 - 9 = 7$

7

(b) $(-5)^2 - (4 - 6)^2(-3)$ Work inside parentheses first.

$(-5)^2 - (-2)^2(-3)$ Apply the exponents next.   [Careful! $(-5)^2 = (-5) \cdot (-5) = 25$]

$25 - 4(-3)$ Multiply.

$25 - (-12)$ Change subtraction to addition.

$25 + (+12)$ Add.

37

(c) $\left(\dfrac{2}{3} - \dfrac{1}{6}\right)^2 \div \left(-\dfrac{3}{8}\right)$ Inside parentheses: $\dfrac{2}{3} - \dfrac{1}{6} = \dfrac{4}{6} - \dfrac{1}{6} = \dfrac{3}{6} = \dfrac{1}{2}$

$\left(\dfrac{1}{2}\right)^2 \div \left(-\dfrac{3}{8}\right)$ Apply the exponent: $\left(\dfrac{1}{2}\right)^2 = \dfrac{1}{2} \cdot \dfrac{1}{2} = \dfrac{1}{4}$

$\dfrac{1}{4} \div \left(-\dfrac{3}{8}\right)$ Divide by using the reciprocal of the divisor: $-\dfrac{3}{8}$ becomes $-\dfrac{8}{3}$

$\dfrac{1}{4} \cdot \left(-\dfrac{8}{3}\right)$ Divide out common factors, then

multiply: $\dfrac{1}{\overset{}{4}} \cdot -\dfrac{\overset{2}{8}}{3} = -\dfrac{2}{3}$

$-\dfrac{2}{3}$

◀ Work Problem ④ at the Side.

> **Note**
>
> Parentheses can be used in several different ways.
>
> To indicate multiplication: $(4)(-3) = -12$
>
> To separate a negative number $8 - (-2) = 10$
> from a minus sign:
>
> To indicate which operation $35 + (6 - 2)$
> to do first:
>
> $35 + 4$
>
> 39

Answers

4. **(a)** $8 - 9; -1$ **(b)** -11 **(c)** 48 **(d)** -2

(e) $\dfrac{11}{8}$

OBJECTIVE **3** **Use the order of operations with fraction bars.** A fraction bar indicates division, as in $\frac{-6}{2}$, which means $-6 \div 2$. In an expression such as

$$\frac{-5 + 3^2}{16 - 7(2)}$$

the fraction bar also acts as a grouping symbol, like parentheses. It tells us to do the work in the numerator, and then the work in the denominator. The last step is to divide the results.

$$\frac{-5 + 3^2}{16 - 7(2)} \rightarrow \frac{-5 + 9}{16 - 14} \rightarrow \frac{4}{2} \rightarrow \text{Now divide.} \quad 4 \div 2 = 2$$

The final simplified result is 2.

EXAMPLE 4 **Fraction Bars and the Order of Operations**

Simplify.

$$\frac{-8 + 5(4 - 6)}{4 - 4^2 \div 8}$$

First do the work in the numerator.

$-8 + 5(4 - 6)$	Work inside parentheses.
$-8 + 5(-2)$	Multiply.
$-8 + (-10)$	Add.

Numerator $\rightarrow -18$

> When *adding* two negative numbers, the sum is also negative.

Now do the work in the denominator.

$4 - 4^2 \div 8$	No parentheses; apply the exponent.
$4 - 16 \div 8$	Divide.
$4 - 2$	Subtract.

Denominator $\rightarrow 2$

The last step is the division.

$$\frac{\text{Numerator} \rightarrow -18}{\text{Denominator} \rightarrow 2} = -9$$

> This is *division*. Signs are **different**, so the quotient is **negative**.

Work Problem 5 at the Side. ▶

5 Simplify.

(a) $\dfrac{-3(2^3)}{-10 - 6 + 8}$

Hint: First work in the numerator. Then work in the denominator. Last, divide the results.

(b) $\dfrac{(-10)(-5)}{-6 \div 3(5)}$

(c) $\dfrac{6 + 18 \div (-2)}{(1 - 10) \div 3}$

(d) $\dfrac{6^2 - 3^2(4)}{5 + (3 - 7)^2}$

Answers

5. (a) $\dfrac{-24}{-8} = 3$ (b) $\dfrac{50}{-10} = -5$

(c) $\dfrac{-3}{-3} = 1$ (d) $\dfrac{0}{21} = 0$

9.4 Exercises

  MyMathLab®

CONCEPT CHECK   *Simplify each expression following the order of operations. Show the steps you take and explain what you did in each step.*

1. $(-6)^2 - 8 \div 2(-3)$

2. $7 - 2(9 - 11)(-2)^3$

Simplify. See Examples 1–3.

3. $6 + 3(-4)$ **Multiply** first.

$6 + (\underline{\quad})$ Now add.

$\underline{\quad}$

4. $10 - 30 \div 2$ **Divide** first.

$10 - \underline{\quad}$ Change subtraction to addition.

$10 + (\underline{\quad})$

$\underline{\quad}$

5. $-1 + 15 + 2(-7)$

6. $9 + (-5) + 2(-2)$

7. $6^2 + 4^2$ 6^2 is $6 \cdot 6$.

$36 + \underline{\quad}$ 4^2 is $4 \cdot 4$.

$\underline{\quad}$

8. $3^2 + 8^2$ 3^2 is $3 \cdot 3$.

$9 + \underline{\quad}$ 8^2 is $8 \cdot 8$.

$\underline{\quad}$

9. $10 - 7^2$

10. $5 - 5^2$

11. $(-2)^5 + 2$

12. $(-2)^4 - 7$

13. $4^2 + 3^2 + (-8)$

14. $5^2 + 2^2 + (-12)$

15. $2 - (-5) + 3^2$

16. $6 - (-9) + 2^3$

17. $(-4)^2 + (-3)^2 + 5$

18. $(-5)^2 + (-6)^2 + 12$

19. $3 + 5(6 - 2)$

20. $4 + 3(8 - 3)$

21. $-7 + 6(8 - 14)$

22. $-3 + 5(9 - 12)$

23. $-6 + (-5)(9 - 14)$

24. $-5 + (-3)(6 - 7)$

25. $(-5)(7 - 13) \div (-10)$

26. $(-4)(9 - 17) \div (-8)$

27. $9 \div (-3)^2 + (-1)$

28. $-48 \div (-4)^2 + 3$

29. $2 - (-5)(-3)^2$

30. $1 - (-10)(-2)^3$

31. $(-2)(-7) + 3(9)$

32. $4(-2) + (-3)(-5)$

33. $30 \div (-5) - 36 \div (-9)$

34. $8 \div (-4) - 42 \div (-7)$

35. $2(5) - 3(4) + 5(3)$

36. $9(3) - 6(4) + 3(7)$

37. $4(3^2) + 7(3 + 9) - (-6)$

38. $5(4^2) - 6(1 + 4) - (-3)$

Simplify. See Example 4.

39. $\dfrac{-1 + 5^2 - (-3)}{-6 - 9 + 12}$

40. $\dfrac{-6 + 3^2 - (-7)}{7 - 9 - 3}$

41. $\dfrac{-2\,(4^2) - 4\,(6 - 2)}{-4\,(8 - 13) \div (-5)}$

42. $\dfrac{3\,(3^2) - 5\,(9 - 2)}{8\,(6 - 9) \div (-3)}$

43. $\dfrac{2^3(-2 - 5) + 4(-1)}{4 + 5(-6 \cdot 2) + (5 \cdot 11)}$

44. $\dfrac{3^3 + 4\,(-1 - 2) - 25}{-4 + 4\,(3 \cdot 5) + (-6 \cdot 9)}$

Simplify each expression.

45. $(-4)^2(7 - 9)^2 \div 2^3$

46. $(-5)^2(9 - 17)^2 \div (-10)^2$

47. $(-0.3)^2 + (-0.5)^2 + 0.9$

48. $(0.2)^3 - (-0.4)^2 + 3.02$

49. $(-0.75)\,(3.6 - 5)^2$

50. $(-0.3)\,(4 - 6.8)^2$

51. $(0.5)^2(-8) - (0.31)$

52. $(0.3)^3(-5) - (-2.8)$

53. $\dfrac{2}{3} \div \left(-\dfrac{5}{6}\right) - \dfrac{1}{2}$

54. $\dfrac{5}{8} \div \left(-\dfrac{10}{3}\right) - \dfrac{3}{4}$

55. $\left(-\dfrac{1}{2}\right)^2 - \left(\dfrac{3}{4} - \dfrac{7}{4}\right)$

56. $\left(-\dfrac{2}{3}\right)^2 - \left(\dfrac{1}{6} - \dfrac{11}{6}\right)$

57. $\dfrac{3}{5}\left(-\dfrac{7}{6}\right) - \left(\dfrac{1}{6} - \dfrac{5}{3}\right)$

58. $\dfrac{2}{7}\left(-\dfrac{14}{5}\right) - \left(\dfrac{4}{3} - \dfrac{13}{9}\right)$

59. $5^2 (9 - 11)(-3)(-2)^3$

60. $4^2 (13 - 17)(-2)(-3)^2$

61. $1.6(-0.8) \div (-0.32) \div 2^2$

62. $6.5(-4.8) \div (-0.3) \div (-2)^3$

Simplify.

63. $\dfrac{-9 + 18 \div (-3)(-6)}{5 - 4(12) \div 3(2)}$

64. $\dfrac{-20 - 15(-4) - (-40)}{4 + 27 \div 3(-2) - 6}$

65. $-7\left(6 - \dfrac{5}{8} \cdot 24 + 3 \cdot \dfrac{8}{3}\right)$

66. $(-0.3)^2(-5)(3) + (6 \div 2)(0.4)$

67. $\left|-12\right| \div 4 + 2(3^2) \div 6$

68. $6 - (2 - 12) + 4^2 \div (-2)\left(\dfrac{5}{2}\right) + (2)^2$

Summary Exercises *Operations with Signed Numbers*

Simplify each expression.

1. $2 - 8$

2. $(-16)(0)$

3. $-14 - (-7)$

4. $\dfrac{-42}{6}$

5. $-9(-7)$

6. $\dfrac{-12}{12}$

7. $(1)(-56)$

8. $1 + (-23)$

9. $5 - (-7)$

10. $-\dfrac{8}{3} \div \left(-\dfrac{4}{9}\right)$

11. $-18 + 5$

12. $\dfrac{0}{-10}$

13. $-40 - 40$

14. $-17 + 0$

15. $8(-6)$

16. $-\dfrac{1}{10} - \dfrac{9}{10}$

17. $\left(-\dfrac{5}{6}\right)\left(\dfrac{1}{5}\right)$

18. $\dfrac{30}{0}$

19. $0 - 14.6$

20. $\dfrac{1.8}{-3}$

21. $-4(-6)(2)$

22. $-2 + (-12) + (-5)$

23. $-60 \div 10 \div (-3)$

24. $-8 - 4 - 8$

25. $64(0) \div (-8)$

26. $2 - (-5) + 3^2$

27. $-9 + 8 + (-2)$

28. $(-6)(-2)(-3)$

29. $8 + 6 + (-8)$

30. $-72 \div (-9) \div (-4)$

31. $-7 + 28 + (-56) + 3$

32. $9 - 6 - 3 - 5$

33. $-6(-8) \div (-5 - 7)$

34. $-1(9732)(-1)(-1)$

35. $-80 \div 4(-5)$

36. $-10 - 4 + 0 + 18$

37. $-7 \cdot |7| \cdot |-7|$

38. $5 - |-3| + 3$

39. $-2(-3)(7) \div (-7)$

40. $-3 - (-2 + 4) - 5$

41. $0 - |-7 + 2|$

42. $(-4)^2(7 - 9)^2 \div 2^3$

43. $12 \div 4 + 2(-2)^2 \div (-4)$

44. $\dfrac{-9 + 24 \div (-4)(-6)}{32 - 4(12) \div 3(2)}$

45. $\dfrac{5 - |2 - 4 \cdot 4| + (-5)^2 \div 5^2}{-9 \div 3(2 - 2) - (-8)}$

Solve each application problem. Be sure to label your final answer as a positive or negative number.

46. When Ashwini discovered that her checking account was overdrawn by $238, she quickly transferred $450 from her savings to her checking account. What is the balance in her checking account?

47. A discount store found that 174 items were lost to shoplifting last month. The total loss was $4176. What was the average loss on each item?

48. In a laboratory experiment, a mixture started at a temperature of −102 degrees. First the temperature was raised 37 degrees and then raised 52 degrees. What was the final temperature?

49. A plane ascended an average of 730 ft each minute during a 37-minute takeoff. How far did the plane ascend during the takeoff?

50. The Tigers gained 9 yd on the first play of the football game. On the second play they lost 15 yd. Then they gained 13 yd on the third play and lost 19 yd on the fourth play. How many yards did they gain or lose during the four plays?

51. A scuba diver was photographing fish at 65 ft below the surface of the ocean. She swam up 24 ft and then swam down 49 ft. What was her final depth?

Elena Sanchez opened a shop that does alterations and designs custom clothing. Use the table of her income and expenses to answer Exercises 52–55.

Month	Income	Expenses	Profit or Loss
January	$2400	−$3100	
February	$1900	−$2000	
March	$2500	−$1800	
April	$2300	−$1400	
May	$1600	−$1600	
June	$1900	−$1200	

52. Complete the table by finding Elena's profit or loss for each month.

53. Which month had the greatest loss? Which month had the greatest profit?

54. What was Elena's average monthly income?

55. What was the average monthly amount of expenses?

56. Explain in your own words how to add two numbers with different signs. Include two examples in your explanation, one that has a positive answer and one that has a negative answer.

57. Explain what is different and what is similar between multiplying and dividing signed numbers.

9.5 Evaluating Expressions and Formulas

OBJECTIVES

1. Define variable and expression.

2. Evaluate an expression when values of the variables are given.

In formulas, you have seen that numbers can be represented by letters. For example, you used this formula for finding simple interest in **Chapter 6.**

$$I = p \bullet r \bullet t$$

In this formula, p (principal) represents the amount of money borrowed, r is the rate of interest, and t is the time in years. In algebra, we often write multiplication without the multiplication dots. If there is no operation sign written between two letters, or between a letter and a number, you assume it is multiplication.

1 Evaluate $5x - 3y$, if:

GS **(a)** x is **5**, y is **2**

$$5x \quad - \quad 3y$$
$$\downarrow \qquad\quad \downarrow$$
$$5\,(5) - 3\,(2) \quad \text{Multiply first.}$$
$$\underline{\quad} - \underline{\quad} \qquad \text{Subtract.}$$
$$\underline{\quad} \quad \leftarrow \text{Final result}$$

(b) x is -3, y is **4**

Showing Multiplication in Algebra

If there is no operation sign, it is understood to be multiplication. Here are some examples.

$I = p \bullet r \bullet t$	is written	$I = prt$
$2 \bullet r$	is written	$2r$
$3 \bullet x + 4 \bullet y$	is written	$3x + 4y$

OBJECTIVE **1** **Define variable and expression.** Letters (such as the I, p, r, and t used above) that represent numbers are called **variables.** A combination of operations on letters and numbers is an **expression.** Three examples of expressions are shown here.

$$9 + p \qquad\qquad 8r \qquad\qquad 7k - 2m$$

OBJECTIVE **2** **Evaluate an expression when values of the variables are given.** The value of an expression changes depending on the value of each variable. To **evaluate an expression**, replace the variables with their values. It is helpful to write each value inside parentheses when multiplication is involved. Then use the order of operations to do all the calculations.

(c) x is **0**, y is **6**

EXAMPLE 1 **Evaluate an Expression**

Evaluate $5x - 3y$, if x is 2 and y is 7.

Replace x with **2**. Replace y with **7**.

Using the order of operations, do all multiplications first. Then subtract.

$$5 \; x \; - 3 \; y$$
$$\downarrow\downarrow \qquad \downarrow\downarrow$$
$$5\,(2) - 3\,(7)$$
$$\underline{10 \; - \; 21} \quad \left[\begin{array}{l}10 - 21 \\ \text{becomes} \\ 10 + (-21).\end{array}\right]$$
$$-11$$

◀ **Work Problem** **1** **at the Side.**

EXAMPLE 2 **Evaluate an Expression**

Evaluate $\dfrac{6k + 2r}{5s}$, if k is -2, r is 5, and s is -1.

Continued on Next Page

Answers

1. **(a)** $25 - 6$; 19 **(b)** -27 **(c)** -18

Replace k with -2, r with 5, and s with -1.

$$\frac{6k + 2r}{5s} = \frac{6(-2) + 2(5)}{5(-1)}$$ Do all the multiplications first.

$$= \frac{-12 + 10}{-5}$$ Add in the numerator.

$$= \frac{-2}{-5}$$ Dividing two numbers with the **same** sign gives a **positive** answer.

$$= \frac{2}{5}$$ ◁ Check that fraction answers are in lowest terms.

·············· **Work Problem 2 at the Side. ▶**

EXAMPLE 3 **Evaluating an Expression with Negative Signs and Negative Values**

Evaluate $-c - 5b$ when c is -2 and b is -3.

Replace c with -2.
Replace b with -3.

$$-\quad c \quad - \quad 5 \quad b$$

To prevent errors, write parentheses around negative numbers.

$$-(-2) - 5(-3)$$ Multiply $5 \cdot (-3)$.

$-(-2)$ is the opposite of (-2), which is $(+2)$.

$$+2 \quad - (-15)$$ Change subtraction to addition; change (-15) to its opposite, $(+15)$.

$$2 \quad + \quad (+15)$$

$$17$$

CAUTION

Watch the signs carefully when there are negative signs in the expression and the value of a variable is also negative.

·············· **Work Problem 3 at the Side. ▶**

EXAMPLE 4 **Evaluating a Formula**

The formula you used in **Chapter 8** for the area of a triangle can now be written without the multiplication dots.

$$A = \frac{1}{2} \cdot b \cdot h \quad \text{is written} \quad A = \frac{1}{2}bh$$

In this formula, b is the length of the base of the triangle and h is the height of the triangle. What is the area if b is 9 cm and h is 24 cm?

$$A = \frac{1}{2} \quad b \quad h$$

$$A = \frac{1}{2}(9 \text{ cm})(24 \text{ cm})$$ Evaluate the formula by replacing b with 9 cm and h with 24 cm.

$$A = \frac{1}{2}(9 \text{ cm})(\overset{12}{24} \text{ cm})$$ Divide out any common factors.

$$A = 108 \text{ cm}^2$$ ◁ Remember, *area* is measured in *square units*; cm² means *square centimeters*.

The area of the triangle is 108 cm².

·············· **Work Problem 4 at the Side. ▶**

---

2 Evaluate $\dfrac{3k + r}{2s}$ if:

(a) k is 1, r is 1, and s is 2

$$\frac{3k + r}{2s} = \frac{3(1) + (1)}{2(2)}$$

Do all multiplications. $\dfrac{3 + 1}{4}$

Add in the numerator. $\dfrac{}{4}$

Divide numerator by denominator. ____

(b) k is 8, r is -2, and s is -4

3 Evaluate each expression.

(a) $-x - 6y$, if x is -1 and y is -4

$$-x \quad - 6y$$

$$-(-1) - 6(-4) \text{ Multiply.}$$

$$+1 \quad - (-24)$$

$$1 \quad + (+24) \text{ Change subtraction to addition.}$$

____ ← Final result

(b) $-4a - b$, if a is 4, b is -2

4 Evaluate each formula.

(a) $A = \dfrac{1}{2}bh$, when b is 6 yd and h is 12 yd

$$A = \frac{1}{2}(6 \text{ yd})(12 \text{ yd})$$

$$A = \underline{}$$

(b) $P = 2l + 2w$, when l is 10 cm and w is 8 cm

Answers

2. (a) $\dfrac{4}{4} = 1$ (b) $\dfrac{22}{-8} = -\dfrac{11}{4}$

3. (a) 25 (b) -14

4. (a) $A = 36 \text{ yd}^2$ (b) $P = 36 \text{ cm}$

9.5 **Exercises** FOR EXTRA HELP

  MyMathLab®

CONCEPT CHECK *Show how to write each expression without using multiplication dots.*

1. (a) $6 \cdot y$ **(b)** $-2 \cdot x - 7 \cdot w$ **2. (a)** $w \cdot x \cdot y$ **(b)** $-3 \cdot a + 5 \cdot c$

Evaluate the expression $2r + 4s$ for each set of values of r and s. ***See Example 1.***

3. r is 2, s is 6

4. r is 6, s is 1

5. r is 1, s is -3

6. r is 7, s is -2

7. r is -1, s is -7

8. r is -3, s is -5

9. r is 0, s is -2

10. r is -7, s is 0

Using the given values of the variables, evaluate each expression. ***See Examples 1 and 2.***

11. $8x - y$
x is 1, y is 8

12. $a - 5b$
a is 10, b is 2

13. $6k + 2s$
k is 1, s is -2

14. $7p + 7q$
p is -4, q is 1

15. $\dfrac{-m + 5n}{2s + 2}$
m is 4, n is -8, s is 0

16. $\dfrac{2y - z}{x - 2}$
y is 0, z is 5, x is 1

17. $-m - 3n$
m is $\dfrac{1}{2}$, n is $\dfrac{3}{8}$

18. $7k - 3r$
k is $\dfrac{2}{3}$, r is $\dfrac{1}{3}$

CONCEPT CHECK *Find and correct the error in each student's work
in Exercises 19 and 20.*

19. Evaluate $-x - 4y$, when x is -3 and y is -1.

$-3 - 4(-1)$
$-3 - (-4)$
$-3 + (+4)$
1

20. Evaluate $-3b - c - 5$, when b is -2 and c is -10.

$-3(-2) - (-10) - 5$
$6 - (10) - 5$
$6 + (-10) + 5$
$-4 \quad + 5$
1

Be careful when an expression has a negative sign and the value of the variable is also negative. Using the given values, evaluate each expression. **See Example 3.**

21. $-c - 5b$

GS c is -8, b is -4

$$\underset{-(-8)}{-c} \quad \underset{-5(-4)}{-5b}$$

$$+8 \quad -(\underline{})$$

$$\underline{}$$

22. $-c - 5b$

GS c is -1, b is -2

$$\underset{-(-1)}{-c} \quad \underset{-5(-2)}{-5b}$$

$$+1 \quad -(\underline{})$$

$$\underline{}$$

23. $-4x - y$

x is 5, y is -15

24. $-4x - y$

x is 3, y is -8

25. $-k - m - 8n$

k is 6, m is -9, n is 0

26. $-k - m - 8n$

k is 0, m is -7, n is -1

27. $\dfrac{-3s - t - 4}{-s + 6 + t}$

s is -1, t is -13

28. $\dfrac{-3s - t - 4}{-s - 20 - t}$

s is -3, t is -6

Using the given values, evaluate each formula. **See Example 4.**

29. $P = 4s$; s is 7.5 cm   This is **perimeter,** so write **cm** in your answer.

30. $P = 4s$; s is 0.8 km   This is **perimeter,** so write **km** in your answer.

31. $P = 2l + 2w$; l is 9 in., w is 5 in.

32. $P = 2l + 2w$; l is 12 ft, w is 2 ft

33. $A = \pi r^2$; π is approximately 3.14, r is 5 yd
This is **area** so write **yd²** in your answer.

34. $A = \pi r^2$; π is approximately 3.14, r is 10 mm
This is **area** so write **mm²** in your answer.

35. $A = \frac{1}{2}bh$; b is 15 ft, h is 3 ft

36. $A = \frac{1}{2}bh$; b is 5 yd, h is 11 yd

37. $V = \frac{1}{3}Bh$; B is 30 mm², h is 60 mm
This is **volume** so write **mm³** in your answer.

38. $V = \frac{1}{3}Bh$; B is 105 cm², h is 5 cm
This is **volume** so write **cm³** in your answer.

39. $V = lwh$; l is 12 in., w is 9 in., h is 10 in.

40. $V = lwh$; l is 16 ft, w is 4 ft, h is 6 ft

41. $C = 2\pi r$; π is approximately 3.14, r is 4 km
This is **circumference**, so write **km** in your answer.

42. $C = 2\pi r$; π is approximately 3.14, r is 18 in.
This is **circumference**, so write **in.** for your answer.

Solve each application problem.

43. The expression for finding the perimeter of a triangle with sides of equal length is $3s$, where s is the length of one side. Evaluate the expression when

  (a) the length of one side is 11 in.

  (b) the length of one side is 3 ft.

44. The expression for finding the perimeter of a pentagon with sides of equal length is $5s$, where s is the length of one side. Evaluate the expression when

  (a) the length of one side is 25 cm.

  (b) the length of one side is 8 in.

45. The expression for figuring a student's average test score is $\frac{p}{t}$, where p is the total points earned on all the tests and t is the number of tests. Evaluate the expression when

  (a) 332 points were earned on 4 tests.

  (b) there were 7 tests and 637 points were earned.

46. The expression for deciding how many buses are needed for a group trip is $\frac{p}{b}$, where p is the total number of people and b is the number of people that one bus will hold. Evaluate the expression when

  (a) 176 people are going on a trip and one bus holds 44 people.

  (b) a bus holds 36 people and 72 people are going on a trip.

9.6 Solving Equations

An **equation** is a statement that says two expressions are equal. Examples of equations are shown here.

$$x + 1 = 9 \qquad 20 = 5k \qquad 6r - 1 = 17$$

The **equal sign** in an equation divides the equation into two parts, the *left side* and the *right side*. In $6r - 1 = 17$, the left side is $6r - 1$, and the right side is 17. The equal sign tells us that the two sides have the same value.

$$6r - 1 = 17$$

Left side = Right side

You solve an equation by finding all numbers that can be substituted for the variable to make the equation true. These numbers are called **solutions** of the equation.

OBJECTIVE ▶ 1 Determine whether a number is a solution of an equation.
To decide whether a number is a solution of an equation, substitute the number in the equation to see whether the result is true.

EXAMPLE 1 **Determining Whether a Number Is a Solution of an Equation**

Is 7 a solution of either one of these equations?

(a) $12 = x + 5$
   Replace x with 7.

$$12 = x + 5 \qquad \text{Replace } x \text{ with 7.} \quad \boxed{\text{7 is the solution, not 12.}}$$
$$12 = 7 + 5$$
$$12 = 12 \checkmark \qquad \text{True}$$

Because the statement is true, 7 is a solution of the equation $12 = x + 5$.

(b) $2y + 1 = 16$
   Replace y with 7.

$$2y + 1 = 16$$
$$2(7) + 1 = 16$$
$$14 + 1 = 16$$
$$15 \neq 16 \qquad \text{False}$$

The *false* statement shows that 7 is *not* a solution of $2y + 1 = 16$.

······································· **Work Problem ① at the Side.** ▶

OBJECTIVE ▶ 2 Solve equations using the addition property of equality.
If the equation $a = b$ is true, and if a number c is added to both a and b, the new equation is also true. This rule, called the **addition property of equality**, means that you can add the *same* number to *both* sides of an equation and still have a true equation.

Addition Property of Equality

If $a = b$, then $a + c = b + c$.

So, you may add the *same* number to *both* sides of an equation.

OBJECTIVES

① Determine whether a number is a solution of an equation.

② Solve equations using the addition property of equality.

③ Solve equations using the multiplication property of equality.

VOCABULARY TIP

Equation is built from the root **equ-**, which you learned in **Chapter 2**. Equ- means "the same." The expression on the left side of the equal sign has the same value as the expression on the right side of the equal sign.

① Decide whether the given number is a solution of the equation.

GS (a) $p + 1 = 10; \quad 9$
   Replace p with 9.

$$p + 1 = 10$$
$$9 + 1 = 10$$
$$\underline{\quad} = 10 \qquad \text{True or false?}$$

(b) $30 = 5r; \quad 6$

(c) $3k - 2 = 4; \quad 3$

(d) $23 = 4y + 3; \quad 5$

Answers

1. **(a)** $10 = 10$; True; solution   **(b)** solution
   (c) not a solution   **(d)** solution

You can use the addition property to solve equations. The idea is to get the variable (the letter) by itself on one side of the equal sign and a number by itself on the other side.

> ### EXAMPLE 2 Solving Equations Using the Addition Property
>
> Solve each equation.

(a) $k - 4 = 6$

To get k by itself on the left side, add 4 to the left side, because $k - 4 + 4$ gives $k + 0$. You must then add 4 to the right side also.

Add 4, because $-4 + 4$ gives 0.		

$$k - 4 = 6 \quad \longleftarrow \text{Original equation}$$
$$k - 4 + 4 = 6 + 4 \quad \text{Add 4 to both sides.} \quad \text{Add the \textbf{same} number to both sides.}$$
$$k + 0 = 10 \quad \text{On the left side, } -4 + 4 \text{ is } 0.$$
$$k = 10 \quad \text{On the left side, } k + 0 \text{ is } k.$$

The solution is 10. To check the solution, replace k with 10 in the original equation.

$$k - 4 = 6 \quad \longleftarrow \text{Original equation}$$
$$10 - 4 = 6 \quad \text{Replace } k \text{ with 10.} \quad \text{10 is the solution, not 6.}$$
$$6 = 6 \checkmark \quad \text{True, so 10 is the correct solution.}$$

This result is true, so **10** is the solution.

> ### CAUTION
>
> When checking the solution in **Example 2(a)** above, we ended up with $6 = 6$. Notice that 6 is **not** the solution. The solution is 10, the number used to replace k in the original equation.

(b) $2 = z + 8$

To get z by itself on the right side, add -8 to both sides.

$$2 = z + 8 \quad \longleftarrow \text{Original equation.}$$
$$2 + (-8) = z + 8 + (-8) \quad \text{Add } (-8) \text{ to both sides, because } 8 + (-8) \text{ gives } 0.$$
$$-6 = z + 0$$
$$-6 = z \quad \longleftarrow \text{The solution is } -6.$$

> ### Note
>
> Notice that we *added* -8 to both sides of the equation to get z by itself. We can accomplish the same thing by *subtracting 8* from both sides. Recall that earlier in this chapter, subtraction was defined in terms of addition. On the left side of the equation above, $2 - 8$ gives the same result as $2 + (-8)$.

Check the solution by replacing z with -6 in the original equation.

$$2 = z + 8 \quad \longleftarrow \text{Original equation}$$
$$2 = -6 + 8 \quad \text{Replace } z \text{ with } -6. \quad -6 \text{ is the solution, not 2.}$$
$$2 = 2 \checkmark \quad \text{True, so } -6 \text{ is the correct solution.}$$

The result is true, so **−6** is the solution (*not* 2).

Continued on Next Page

Here is a summary of the rules for solving equations using the addition property. In these rules, x is the variable and a and b represent numbers.

Solving an Equation Using the Addition Property

Solve $x - a = b$ by adding a to *both* sides.

Solve $x + a = b$ by subtracting a from *both* sides.

·· Work Problem ❷ at the Side. ▶

OBJECTIVE ▸ 3 **Solve equations using the multiplication property of equality.** As long as you do the *same* thing to *both* sides of an equation, it will still be a true equation. So far you have added or subtracted on both sides. Now we will multiply or divide on both sides.

Multiplication Property of Equality

If $a = b$ and c does not equal 0, then we can say the following:

$$a \cdot c = b \cdot c \quad \text{and} \quad \frac{a}{c} = \frac{b}{c}$$

In other words, you may multiply or divide *both* sides of an equation by the *same* number. (The only exception is you *cannot* divide by 0.)

EXAMPLE 3   **Solving Equations Using the Multiplication Property**

Solve each equation.

(a) $9p = 63$

You want to get the variable, p, by itself on the left side. The expression $9p$ means $9 \cdot p$. To undo the multiplication and get p by itself, *divide* both sides by 9.

$$9p = 63$$

| On the left side, divide out 9. |

$$\frac{\overset{1}{9} \cdot p}{\underset{1}{9}} = \frac{63}{9} \qquad \text{Divide \textbf{both} sides by 9.}$$

$$p = 7 \qquad \leftarrow \text{The solution is 7.}$$

CHECK

$$9p = 63 \qquad \leftarrow \text{Original equation}$$

$$9 \cdot 7 = 63 \qquad \text{Replace } p \text{ with 7.} \qquad \boxed{\text{7 is the solution, \textbf{not} 63.}}$$

$$63 = 63 \checkmark \qquad \text{True}$$

The result is true, so **7** is the solution (*not* 63).

·· **Continued on Next Page**

---

❷ Solve each equation. Check each solution.

GS **(a)** $n - 5 = 8$

$$n - \underbrace{5 + 5}_{} = 8 + \underline{\quad}$$

$$n + \underline{\quad} = \underline{\quad}$$

$$n = \underline{\quad}$$

Now check your solution.

(b) $5 = r - 10$

(c) $3 = z + 1$

(d) $k + 9 = 0$

(e) $-2 = y + 9$

(f) $x - 2 = -6$

3 Solve each equation. Check each solution.

(a) $2y = 14$ Divide **both** sides by ____.

$$\frac{\overset{1}{2} \cdot y}{\underset{1}{2}} = \frac{14}{2}$$

$$y = \underline{\hspace{1cm}}$$

(b) $42 = 7p$

(c) $-8a = 32$

(d) $-3r = -15$

(e) $-60 = -6k$

(f) $10x = 0$

Answers

3. **(a)** Divide both sides by 2; $y = 7$
   (b) $p = 6$ **(c)** $a = -4$ **(d)** $r = 5$
   (e) $k = 10$ **(f)** $x = 0$

(b) $-4r = 24$

Divide *both* sides by -4 to get r by itself on the left side.

$$\frac{\overset{1}{-4} \cdot r}{\underset{1}{-4}} = \frac{24}{-4} \qquad \text{Divide \textbf{both} sides by } -4.$$

$$r = -6 \quad \longleftarrow \text{ The solution is } -6.$$

CHECK

$$-4r = 24 \quad \longleftarrow \text{ Original equation}$$

$$-4(-6) = 24 \qquad \text{Replace } r \text{ with } -6. \quad \boxed{\begin{array}{l}-6 \text{ is the}\\ \text{solution,}\\ \textbf{not } 24.\end{array}}$$

$$24 = 24 \ \checkmark \qquad \text{True}$$

The result is true, so -6 is the solution (**not** 24).

(c) $-55 = -11m$

Divide *both* sides by -11 to get m by itself on the right side.

$$\frac{-55}{-11} = \frac{\overset{1}{-11} \cdot m}{\underset{1}{-11}}$$

$$5 = m$$

CHECK

$$-55 = -11m \quad \leftarrow \text{Original equation}$$

$$-55 = -11(5) \qquad \text{Replace } m \text{ with } 5. \quad \boxed{\begin{array}{l}5 \text{ is the}\\ \text{solution,}\\ \textbf{not } -55.\end{array}}$$

$$-55 = -55 \ \checkmark \qquad \text{True}$$

The result is true, so **5** is the solution (**not** -55).

··· ◄ **Work Problem ❸ at the Side.**

EXAMPLE 4   **Solving Equations Using the Multiplication Property**

Solve each equation.

(a) $\dfrac{x}{2} = 9$

Rewrite $\frac{x}{2}$ as $\frac{1}{2}x$, because dividing x by 2 is the same as multiplying x by $\frac{1}{2}$. Then, to get x by itself, multiply both sides by the reciprocal of $\frac{1}{2}$, which is $\frac{2}{1}$. (Recall from **Chapter 2** that when two numbers are reciprocals, their product is 1.)

$$\frac{1}{2}x = 9$$

$$\boxed{\begin{array}{l}\text{Multiply by the}\\ \text{reciprocal of } \frac{1}{2},\\ \text{which is } \frac{2}{1}.\end{array}} \quad \frac{\overset{1}{2}}{1} \cdot \frac{1}{\underset{1}{2}}x = 2 \cdot 9 \qquad \begin{array}{l}\text{Multiply \textbf{both} sides by } \frac{2}{1};\\ (\frac{2}{1} \text{ equals } 2).\end{array}$$

$$1x = 18$$

$$x = 18 \quad \longleftarrow \text{ The solution is 18.}$$

Continued on Next Page

CHECK

$$\frac{x}{2} = 9 \longleftarrow \text{Original equation}$$

$$\frac{18}{2} = 9 \qquad \text{Replace } x \text{ with 18.} \qquad \boxed{\text{18 is the solution, not 9.}}$$

$$9 = 9 \checkmark \qquad \text{True, so 18 is the correct solution.}$$

18 is the correct solution (***not*** 9).

(b) $-\dfrac{2}{3}r = 4$

Multiply both sides by the reciprocal of $-\frac{2}{3}$, which is $-\frac{3}{2}$.

$$-\frac{2}{3}r = 4$$

$$\boxed{\text{Multiply by the reciprocal of } -\frac{2}{3} \text{ which is } -\frac{3}{2}.} \qquad -\frac{\overset{1}{\cancel{3}}}{\underset{1}{\cancel{2}}} \cdot \left(-\frac{\overset{1}{\cancel{2}}}{\underset{1}{\cancel{3}}}r\right) = -\frac{3}{\underset{1}{\cancel{2}}} \cdot \frac{\overset{2}{\cancel{4}}}{1} \qquad \text{Multiply both sides by } -\frac{3}{2}.$$

$$r = -6 \longleftarrow \text{The solution is } -6.$$

Check by replacing r with -6 in the original equation. Write -6 as $\frac{-6}{1}$.

$$-\frac{2}{3}r = 4 \longleftarrow \text{Original equation}$$

$$-\frac{2}{\underset{1}{\cancel{3}}} \cdot \frac{\overset{-2}{\cancel{-6}}}{1} = 4 \qquad \text{Replace } r \text{ with } -6. \qquad \boxed{-6 \text{ is the solution, not 4.}}$$

$$4 = 4 \checkmark \qquad \text{True, so } -6 \text{ is the correct solution.}$$

−6 is the correct solution (***not*** 4).

Here is a summary of the rules for using the multiplication property. In these rules, x is the variable and a, b, and c represent numbers. (Note that a and b are not equal to 0.)

Solving Equations Using the Multiplication Property

Solve the equation $\quad ax = b \quad$ by dividing *both* sides by a.

Solve the equation $\quad \dfrac{a}{b}x = c \quad$ by multiplying *both* sides by $\dfrac{b}{a}$.

$\cdots\cdots$ **Work Problem** ❹ **at the Side.** ▶

---

❹ Solve each equation. Check each solution.

GS (a) $\dfrac{x}{4} = 2 \quad$ Rewrite $\dfrac{x}{4}$ as $\dfrac{1}{4}x$.

$\dfrac{1}{4}x = 2 \quad$ Multiply **both** sides by $\frac{4}{1}$ $\left(\text{the reciprocal of } \frac{1}{4}\right)$.

$\dfrac{\overset{1}{\cancel{4}}}{1} \cdot \dfrac{1}{\underset{1}{\cancel{4}}}x = 2 \cdot 4$

$x = \underline{\qquad}$

(b) $\dfrac{y}{7} = -3$

(c) $-8 = \dfrac{k}{6}$

(d) $8 = -\dfrac{4}{5}z$

(e) $-\dfrac{5}{8}p = -10$

Answers

4. (a) $x = 8$ (b) $y = -21$ (c) $k = -48$
(d) $z = -10$ (e) $p = 16$

9.6 Exercises

Determine whether the given number is a solution of the equation.
See Example 1.

1. $x + 7 = 11$;   4

2. $k - 2 = 7$;   9

3. $4y = 28$;   7

4. $5p = 30$;   6

5. $2z - 1 = -15$; -8

6. $6r - 3 = -14$; -2

CONCEPT CHECK   *To solve each equation, state the number you would add to both sides of the equation. Then explain why you picked that number. Do not finish the solution.*

7. (a) $m - 8 = 1$

(b) $-7 = w + 5$

8. (a) $n + 2 = -9$

(b) $10 = b - 6$

Solve each equation by using the addition property. Check each solution. **See Example 2.**

9. $p + 5 = 9$

$p + 5 - \underline{\quad} = 9 - \underline{\quad}$

$p + \underline{\quad} = \underline{\quad}$

$p = \underline{\quad}$

10. $a + 3 = 12$

$a + 3 - \underline{\quad} = 12 - \underline{\quad}$

$a + \underline{\quad} = \underline{\quad}$

$a = \underline{\quad}$

11. $k + 15 = 0$

12. $y + 6 = 0$

13. $z - 5 = 3$

14. $x \quad 9 = 4$

15. $8 = r - 2$

16. $3 = b - 5$

17. $-5 = n + 3$

18. $-1 = a + 8$

19. $7 = r + 13$

20. $12 = z + 7$

21. $-4 + k = 14$

22. $-9 + y = 7$

23. $-12 + x = -1$

24. $-3 + m = -9$

25. $-5 = -2 + r$

26. $-1 = -10 + y$

27. $d + \dfrac{2}{3} = 3$

28. $x + \dfrac{1}{2} = 4$

29. $z - \dfrac{7}{8} = 10$

30. $m - \dfrac{3}{4} = 6$

31. $\dfrac{1}{2} = k - 2$

32. $\dfrac{3}{5} = t - 1$

33. $m - \dfrac{7}{5} = \dfrac{11}{4}$

34. $z - \dfrac{7}{3} = \dfrac{32}{9}$

35. $x - 0.8 = 5.07$

36. $a - 3.82 = 7.9$

37. $3.25 = 4.76 + r$

38. $8.9 = 10.5 + b$

CONCEPT CHECK   *To solve each equation, you divide both sides by what number?*
Explain why you picked that number. Do not finish the solution.

39. (a) $-10x = 40$

40. (a) $6t = 0$

(b) $-25 = 5z$

(b) $-60 = -12w$

Solve each equation. Check each solution. See Example 3.

41. $6z = 12$

$\dfrac{\overset{1}{\cancel{6}} \cdot z}{\underset{1}{\cancel{6}}} = \dfrac{12}{6}$ Divide both sides by 6.

$z = \underline{\quad}$

42. $8k = 24$

$\dfrac{\overset{1}{\cancel{8}} \cdot k}{\underset{1}{\cancel{8}}} = \dfrac{24}{8}$ Divide both sides by 8.

$k = \underline{\quad}$

43. $48 = 12r$

44. $99 = 11m$

45. $3y = 0$

46. $5a = 0$

47. $-6k = 36$

48. $-7y = 70$

49. $-36 = -4p$

50. $-54 = -9r$

51. $-1.2m = 8.4$

52. $-5.4z = 27$

53. $-8.4p = -9.24$

54. $-3.2y = -16.64$

CONCEPT CHECK   *To solve each equation, you multiply both sides by what number? Explain why you picked that number. Do not finish the solution.*

55. (a) $\dfrac{x}{8} = -4$

(b) $-6 - -\dfrac{2}{3}k$

56. (a) $4 = \dfrac{m}{7}$

(b) $-15 = \dfrac{3}{5}r$

Solve each equation. Check each solution. See Example 4.

57. $\dfrac{k}{2} = 17$

Hint: Rewrite $\dfrac{k}{2}$ as $\dfrac{1}{2}k$.

58. $\dfrac{y}{3} = 5$

Hint: Rewrite $\dfrac{y}{3}$ as $\dfrac{1}{3}y$.

59. $11 = \dfrac{a}{6}$

60. $5 = \dfrac{m}{8}$

61. $\dfrac{r}{3} = -12$

62. $\dfrac{z}{9} = -3$

63. $-\dfrac{2}{5}p = 8$ ▶

64. $-\dfrac{5}{6}k = 15$

65. $-\dfrac{3}{4}m = -3$

66. $-\dfrac{9}{10}b = -18$

67. $6 = \dfrac{3}{8}x$

68. $4 = \dfrac{2}{3}a$

69. $\dfrac{y}{2.6} = 0.5$

70. $\dfrac{k}{0.7} = 3.2$

71. $\dfrac{z}{-3.8} = 1.3$

72. $\dfrac{m}{-5.2} = 2.1$

Solve each equation.

73. $x - 17 = 5 - 3$

74. $y + 4 = 10 - 9$

75. $3 = x + 9 - 15$

76. $-1 = y + 7 - 9$

77. $\dfrac{7}{2}x = \dfrac{4}{3}$

78. $\dfrac{3}{4}x = \dfrac{5}{3}$

79. $\dfrac{1}{2} - \dfrac{3}{4} = \dfrac{a}{5}$

80. $\dfrac{2}{3} - \dfrac{8}{9} = \dfrac{c}{6}$

81. $m - 2 + 18 = |-3 - 4| + 5$

82. $10 - |0 - 8| = n + 1 - 4$

9.7 Solving Equations with Several Steps

OBJECTIVES

1. Solve equations with several steps.
2. Use the distributive property.
3. Combine like terms.
4. Solve more difficult equations.

1 Solve each equation. Check each solution.

GS **(a)** $2r + 7 = 13$

$2r + 7 - 7 = 13 - 7$

$2r = 6$

Now divide both sides by ____.

(b) $20 = 6y - 4$

(c) $7m + 9 = 9$

(d) $-2 = 4p + 10$

(e) $-10z - 9 = 11$

OBJECTIVE ▶ 1 **Solve equations with several steps.** You cannot solve the equation $5m + 1 = 16$ by just adding the same number to both sides; just dividing both sides by the same number doesn't work either. Instead, you use a combination of operations. Here are the steps.

Solving an Equation Using the Addition and Multiplication Properties

Step 1 Add or subtract the *same* amount on *both* sides of the equation so that the variable term ends up by itself on one side.

Step 2 Multiply or divide *both* sides by the *same* number to find the solution.

Step 3 Check the solution in the original equation.

EXAMPLE 1 Solving an Equation with Several Steps

Solve $5m + 1 = 16$.

Step 1 Subtract 1 from *both* sides so that $5m$ will be by itself on the left side.

$$5m + 1 - 1 = 16 - 1$$
$$5m = 15$$

Step 2 Divide *both* sides by 5.

$$\frac{\overset{1}{\cancel{5}} \cdot m}{\underset{1}{\cancel{5}}} = \frac{15}{5}$$

$$m = 3 \leftarrow \text{The solution is 3.}$$

Step 3 Check the solution.

$$5m + 1 = 16 \longleftarrow \text{Original equation}$$
$$5(3) + 1 = 16 \qquad \text{Replace } m \text{ with 3.} \leftarrow \boxed{\text{3 is the solution, not 16.}}$$
$$15 + 1 = 16$$
$$16 = 16 \checkmark \quad \text{True, so 3 is the correct solution.}$$

3 is the correct solution (*not* 16).

◀ **Work Problem** **1** **at the Side.**

OBJECTIVE ▶ 2 **Use the distributive property.** We can use the order of operations to simplify these two expressions.

$$
\begin{array}{ccc}
2(6 + 8) & \text{and} & 2 \cdot 6 + 2 \cdot 8 \\
2(14) & & 12 \quad + \quad 16 \\
28 & & 28
\end{array}
$$

Because both answers are the same, the two expressions are equivalent.

$$2(6 + 8) = 2 \cdot 6 + 2 \cdot 8$$

This is an example of the **distributive property**.

Distributive Property

$$a(b + c) = ab + ac$$

Answers

1. **(a)** Divide both sides by 2; $r = 3$
(b) $y = 4$ **(c)** $m = 0$
(d) $p = -3$ **(e)** $z = -2$

EXAMPLE 2 Using the Distributive Property

Simplify each expression by using the distributive property.

(a) $9(4 + 2) = 9 \cdot 4 + 9 \cdot 2 = 36 + 18 = 54$

Multiplying is done before adding.

The 9 outside the parentheses is *distributed* over the 4 and the 2 inside the parentheses. That means that *every* number inside the parentheses is multiplied by 9.

(b) $-3(k + 9) = -3 \cdot k + (-3) \cdot 9 = -3k + (-27) = -3k - 27$

CAUTION

To understand the final step in **Example 2(b)** above, work backwards from the answer.

Rewrite as addition. $\quad -3k - 27$
$\quad\quad\quad\quad\quad\updownarrow\quad\updownarrow\quad$ **Rewrite as subtraction.**
$\quad\quad\quad\quad -3k + (-27)$

Although the two expressions are equivalent, we write the final answer as $-3k - 27$ because it is simpler and shorter.

(c) $6(y - 5) = 6 \cdot y - 6 \cdot 5 = 6y - 30$

(d) $-2(x - 3) = -2 \cdot x - (-2) \cdot 3 = -2x - (-6) = -2x + 6$

Work Problem ② at the Side. ▶

OBJECTIVE ③ Combine like terms. A single letter or number, or the product of a variable and a number, makes up a *term*. Here are six examples of terms.

$$3y \quad 5 \quad -9 \quad 8r \quad 10r^2 \quad x$$

Terms with exactly the same variable and the same exponent are called **like terms.**

$5x$	and	$3x$	are like terms
$5x$	and	$3m$	are *not* like terms; variables are different
$5x^2$	and	$5x^3$	are *not* like terms; exponents are different
$5x^4$	and	$3x^4$	are like terms

The distributive property can be used to simplify a sum of like terms. We *add* the like terms, which is called *combining like terms.*

$$6r + 3r = (6 + 3)r = 9r$$

EXAMPLE 3 Combining Like Terms

Use the distributive property to combine like terms.

(a) $5k + 11k = (5 + 11)k = 16k$

Think; $10 - 14 + 2$ is $10 + (-14) + 2$, which gives -2.

(b) $10m - 14m + 2m = (10 - 14 + 2)m = -2m$

(c) $-5x + x$ can be written $-5x + 1x = (-5 + 1)x = -4x$

x is equivalent to $1 \cdot x$ or $1x$.

Work Problem ③ at the Side. ▶

---

② Use the distributive property.

GS (a) $3(2 + 6)$

$\quad\quad \underbrace{3 \cdot 2} + \underbrace{3 \cdot 6} \quad$ Do the multiplying; then add.

(b) $8(k - 3)$

(c) $-6(r + 5)$

(d) $-9(s - 8)$

③ Combine like terms.

GS (a) $5y + 3y$

$\quad \underbrace{(5 + 3)y}$

$\quad \underline{\quad\quad} y$

(b) $10a - 28a$

(c) $3x + 3x - 9x$

(d) $k + k$

(e) $6b - b - 7b$

Answers

2. **(a)** $6 + 18 = 24$
   (b) $8 \cdot k - 8 \cdot 3 = 8k - 24$
   (c) $-6 \cdot r + (-6) \cdot 5 = -6r + (-30) = -6r - 30$
   (d) $-9 \cdot s - (-9) \cdot 8 = -9s - (-72) = -9s + 72$
3. **(a)** $8y$ **(b)** $-18a$ **(c)** $-3x$ **(d)** $2k$
   (e) $-2b$

4 Solve each equation. Check each solution.

GS **(a)** $3y - 1 = 2y + 7$

$$3y - 1 - 2y = 2y + 7 - 2y$$

$$\underbrace{3y - 2y} - 1 = 2y - 2y + 7$$

$$1y \quad - 1 = 7$$

$$y - 1 + 1 = 7 + \underline{\quad}$$

$$y = \underline{\quad}$$

(b) $5a + 7 = 3a - 9$

(c) $3p - 2 = p - 6$

OBJECTIVE **4** **Solve more difficult equations.** The next examples show you how to solve more difficult equations using the addition, multiplication, and distributive properties.

EXAMPLE 4 **Solving Equations**

Solve each equation. Check each solution.

(a) $6r + 3r = 36$

You can combine $6r$ and $3r$ because they are *like terms*. $6r + 3r$ is $9r$, so the equation becomes

6r + 3r is (6 + 3)r or 9r.

$$9r = 36$$

$$\frac{\overset{1}{9} \cdot r}{\underset{1}{9}} = \frac{36}{9} \qquad \text{Divide both sides by 9.}$$

$$r = 4 \longleftarrow \text{The solution is 4.}$$

CHECK

$$6r + 3r = 36 \longleftarrow \text{Original equation}$$

$$6(4) + 3(4) = 36 \qquad \text{Replace } r \text{ with 4.} \quad \boxed{\text{4 is the solution, not 36.}}$$

$$24 + 12 = 36$$

$$36 = 36 \checkmark \qquad \text{True, so 4 is the correct solution.}$$

4 is the correct solution (***not*** 36).

(b) $2k - 2 = 5k - 11$

One way to get the variable term by itself on one side is to subtract $5k$ from both sides.

$$2k - 2 - 5k = 5k - 11 - 5k$$

2k − 5k is (2 − 5)k or −3k

$$2k - 5k - 2 = 5k - 5k - 11$$

$$-3k - 2 = -11 \qquad \boxed{\text{5k − 5k is (5 − 5)k or 0k, which leaves −11 by itself.}}$$

Next, add 2 to both sides.

$$-3k - 2 + 2 = -11 + 2$$

$$-3k = -9$$

Finally, divide both sides by -3.

$$\frac{\overset{1}{-3} \cdot k}{\underset{1}{-3}} = \frac{-9}{-3} \qquad \boxed{\text{In division, when signs } match \text{, the result is } positive.}$$

$$k = 3 \longleftarrow \text{The solution is 3.}$$

CHECK

$$2k - 2 = 5k - 11 \longleftarrow \text{Original equation}$$

$$2(3) - 2 = 5(3) - 11 \qquad \text{Replace } k \text{ with 3.}$$

$$6 - 2 = 15 - 11$$

$$4 = 4 \checkmark \qquad \text{True, so 3 is the correct solution.}$$

3 is the correct solution (***not*** 4).

◀ **Work Problem** **4** **at the Side.**

Answers

4. **(a)** $1; y = 8$ **(b)** $a = -8$ **(c)** $p = -2$

Now that you know about the distributive property and combining like terms, here is a summary of all the steps you can use to solve an equation.

Solving an Equation

Step 1 If possible, use the *distributive property* to remove parentheses.

Step 2 *Combine any like terms* on the left side of the equation. Combine any like terms on the right side of the equation.

Step 3 *Add or subtract* the same amount on *both* sides of the equation so that the variable term ends up by itself on one side.

Step 4 *Multiply or divide* both sides by the same number to find the solution.

Step 5 *Check* your solution by going back to the *original equation*. Replace the variable with your solution. Follow the order of operations to complete the calculations. If the two sides of the equation are equal, your solution is correct.

EXAMPLE 5 Solving Equations Using the Distributive Property

Solve $-6 = 3(y - 2)$

Step 1 Use the distributive property on the right side of the equation.

$$3(y - 2) \quad \text{becomes} \quad 3 \cdot y - 3 \cdot 2 \quad \text{or} \quad 3y - 6$$

Now the equation looks like this.

$$-6 = 3y - 6$$

Step 2 Combine like terms. Check the left side of the equation. There are no like terms. Check the right side. No like terms there either, so go on to Step 3.

Step 3 Add 6 to both sides to get the variable term by itself on the right side.

$$-6 + 6 = 3y - 6 + 6$$
$$0 = 3y$$

Step 4 Divide both sides by 3.

> You *can* divide 0 by a nonzero number. The result is 0.

$$\frac{0}{3} = \frac{\overset{1}{\cancel{3}} \cdot y}{\underset{1}{\cancel{3}}}$$

> You can rewrite the solution as $y = 0$ because $y = 0$ means the same as $0 = y$.

$$0 = y \leftarrow \text{The solution is 0.}$$

Step 5 Check. Go back to the original equation.

$$-6 = 3\,(y - 2) \leftarrow \text{Original equation}$$
$$-6 = 3\,(0 - 2) \quad \text{Replace } y \text{ with 0.}$$

> 0 is the solution, **not** −6.

$$-6 = 3(-2)$$
$$-6 = -6 \checkmark \quad \text{True, so 0 is the correct solution.}$$

0 is the correct solution (*not* −6).

··········· Work Problem **5** at the Side. ▶

---

5 Solve each equation. Check each solution.

(a) $-12 = 4(y - 1)$

$$-12 = \underbrace{4 \cdot y} - \underbrace{4 \cdot 1}$$
$$-12 = \underline{} - \underline{} \quad \text{Finish the solution.}$$

(b) $5(m + 4) = 20$

(c) $6(t - 2) = 18$

Answers

5. (a)
$$-12 = 4y - 4$$
$$-12 + 4 = 4y - 4 + 4$$
$$-8 = 4y$$
$$\frac{-8}{4} = \frac{4y}{4}$$
$$-2 = y \quad \text{or} \quad y = -2$$

(b) $m = 0$ **(c)** $t = 5$

9.7 Exercises

MyMathLab®

CONCEPT CHECK *Explain the **first** step in solving each equation. Then show how to do that step. Do **not** finish the solution.*

1. $-4y + 10 = 2$

2. $7 = 6x - 5$

Solve each equation. Check each solution. See Example 1.

3. $7p + 5 = 12$

$7p + 5 - 5 = \underbrace{12 - 5}$

$7p = \underline{\quad}$ Finish the solution.

4. $6k + 3 = 15$

$6k + 3 - 3 = \underbrace{15 - 3}$

$6k = \underline{\quad}$ Finish the solution.

5. $2 = 8y - 6$

6. $10 = 11p - 12$

7. $-3m + 1 = 1$

8. $-4k + 5 = 5$

9. $28 = -9a + 10$

10. $5 = -10p + 25$

11. $-5x - 4 = 16$

12. $-12a - 3 = 21$

13. $-\dfrac{1}{2}z + 2 = -1$

14. $-\dfrac{5}{8}r + 4 = -6$

Use the distributive property to simplify. See Example 2.

15. $6(x + 4)$

$\underbrace{6 \cdot x} + \underbrace{6 \cdot 4}$

$\underline{\quad} + \underline{\quad}$

16. $8(k + 5)$

$\underbrace{8 \cdot k} + \underbrace{8 \cdot 5}$

$\underline{\quad} + \underline{\quad}$

17. $7(p - 8)$

18. $9(t - 4)$

19. $-3(m + 6)$

20. $-5(a + 2)$

21. $-2(y - 3)$

22. $-4(r - 7)$

23. $-8(c + 8)$

24. $-6(n + 5)$

25. $-10(w - 9)$

26. $-11(x - 11)$

CONCEPT CHECK *Explain what is wrong with each student's work and correct it.*

27. (a) $5(x - 4)$

$5x - 4$

(b) $-8(y - 2)$

$-8y - 16$

28. (a) $6(y - 3)$

$6y - 3$

(b) $-5(b - 3)$

$-5b - 15$

Combine like terms. **See Example 3.**

29. $11r + 6r$
GS $\underbrace{(11 + 6)}r$
_____

30. $2m + 5m$
GS $\underbrace{(2 + 5)}m$
_____

31. $8z - 7z$

32. $10x - 2x$

33. $y - 3y$

34. $-10a + a$

35. $-4t + t - 4t$

36. $3y - y - 4y$

37. $7p - 9p + 2p$

38. $-6c - c + 7c$

39. $\dfrac{5}{2}b - \dfrac{11}{2}b$

40. $\dfrac{3}{8}d - \dfrac{9}{8}d$

Show the steps you take to solve each equation. Write a sentence next to each step explaining what you did in that step.

41. $3p - 5p = 32$

42. $-48 = m - 5m$

Solve each equation. Check each solution. **See Example 4.**

43. $4k + 6k = 50$
GS $(4 + 6)k = 50$

$10k = 50$

$\dfrac{\overset{1}{\cancel{10}} \cdot k}{\underset{1}{\cancel{10}}} = \dfrac{50}{10}$

$k = \underline{\quad}$

44. $3a + 2a = 15$
GS $(3 + 2)a = 15$

$5a = 15$

$\dfrac{\overset{1}{\cancel{5}} \cdot a}{\underset{1}{\cancel{5}}} = \dfrac{15}{5}$

$a = \underline{\quad}$

45. $54 = 10m - m$

46. $28 = x + 6x$

47. $2b - 6b = 24$

48. $3r - 9r = 18$

49. $-12 = 6y - 18y$

50. $-5 = 10z - 15z$

51. $6p - 2 = 4p + 6$

52. $5y - 5 = 2y + 10$

53. $9 + 7z = 9z + 13$

54. $8 + 4a = 2a + 2$

55. $-2y + 6 = 6y - 10$

56. $5x - 4 = -3x + 4$

57. $b + 3.05 = 2$

58. $t + 0.8 = -1.7$

59. $2.5r + 9 = -1$

60. $0.5x - 6 = 2$

*Solve each equation by using the distributive property. Check each solution. **See Example 5.***

61. $-10 = 2(y + 4)$

62. $-3 = 3(x + 6)$

63. $-4(t + 2) = 12$

64. $-5(k + 3) = 25$

65. $6(x - 5) = -30$

66. $7(r - 5) = -35$

67. Solve $-2t - 10 = 3t + 5$. Show each step you take to solve it. Next to each step, write a sentence that explains what you did in that step. Be sure to tell when you used the addition property and when you used the multiplication property.

68. Here is one student's solution to an equation.

$$3(2x + 5) = -7$$
$$6x + 5 = -7$$
$$6x + 5 - 5 = -7 - 5$$
$$6x = -12$$
$$x = -2$$

Show how to check the solution. If the solution does not check, find and correct the error.

Solve each equation.

69. $30 - 40 = -2x + 7x - 4x$

$30 + (-40) = (-2 + 7 - 4)x$

$-10 = 1x$

$-10 = x$ **or** $x =$ ____

70. $-6 - 5 + 14 = -50a + 51a$

$-6 + (-5) + (14) = (-50 + 51)a$

$3 = 1a$

$3 = a$ **or** $a =$ ____

71. $0 = -2(y - 2)$

72. $0 = -9(b - 1)$

73. $\dfrac{y}{2} - 2 = \dfrac{y}{4} + 3$

74. $\dfrac{z}{3} + 1 = \dfrac{z}{2} - 3$

75. $-3(w - 2) = \left| 0 - 13 \right| + 4w$

76. $-5b - \left| 47 - 7 \right| = -8(b + 8)$

77. $2(a + 0.3) = 1.2(a - 4)$

78. $-0.5(c - 4) = -3(c - 2.5)$

9.8 Using Equations to Solve Application Problems

OBJECTIVES

1. Translate word phrases into expressions with variables.

2. Translate sentences into equations.

3. Solve application problems.

Usually, an application problem is given in words. You need to *translate* these words into an equation that you can solve.

OBJECTIVE 1 Translate word phrases into expressions with variables.
Examples 1 and 2 show you how to translate word phrases into algebraic expressions.

EXAMPLE 1   Translating Word Phrases into Expressions with Variables

Write each word phrase in symbols, using x as the variable.

Words	Algebraic Expression
A number **plus** 2	$x + 2$ **or** $2 + x$
The **sum** of 8 and a number	$8 + x$ **or** $x + 8$
5 **more than** a number	$x + 5$ **or** $5 + x$
-35 **added to** a number	$-35 + x$ **or** $x + (-35)$
A number **increased by** 6	$x + 6$ **or** $6 + x$
9 **less than** a number	$x - 9$
A number **subtracted from** 3	$3 - x$
A number **decreased by** 4	$x - 4$
10 **minus** a number	$10 - x$

CAUTION

Recall that addition can be done in any order, so $x + 2$ gives the same result as $2 + x$. This is **not** true in subtraction, so be careful. $10 - x$ does **not** give the same result as $x - 10$.

◀ **Work Problem 1 at the Side.**

EXAMPLE 2   Translating Word Phrases into Expressions with Variables

Write each word phrase in symbols, using x as the variable.

Words	Algebraic Expression
8 **times** a number	$8x$
The **product** of 12 and a number	$12x$
Double a number (meaning "2 times")	$2x$
Twice a number (meaning "2 times")	$2x$
The **quotient** of 6 and a number	$\dfrac{6}{x}$
A number **divided by** 10	$\dfrac{x}{10}$
One-third of a number	$\dfrac{1}{3}x$ **or** $\dfrac{x}{3}$
The result **is**	$=$

◀ **Work Problem 2 at the Side.**

Side problems

1 Write each word phrase in symbols, using x as the variable.

(a) 15 less than a number

(b) 12 more than a number

(c) A number increased by 13

(d) A number minus 8

(e) -10 plus a number

(f) A number subtracted from 6

2 Write each word phrase in symbols, using x as the variable.

(a) Double a number

(b) The product of -8 and a number

(c) The quotient of 15 and a number

(d) One-half of a number

Answers

1. (a) $x - 15$ (b) $x + 12$ or $12 + x$
   (c) $x + 13$ or $13 + x$ (d) $x - 8$
   (e) $-10 + x$ or $x + (-10)$ (f) $6 - x$

2. (a) $2x$ (b) $-8x$ (c) $\dfrac{15}{x}$ (d) $\dfrac{1}{2}x$ or $\dfrac{x}{2}$

OBJECTIVE **2** **Translate sentences into equations.** The next example shows you how to translate a sentence into an equation that you can solve.

EXAMPLE 3 Translating a Sentence into an Equation

If 5 times a number is added to 11, the result is 26. Find the number.

Let x represent the unknown number.

Use the information in the problem to write an equation.

5 times a number	added to	11	is	26
↓	↓	↓	↓	↓
$5x$	$+$	11	$=$	26

Note

The phrase "the result is" translates to "$=$."

Next, solve the equation.

$$5x + 11 - 11 = 26 - 11 \quad \text{Subtract 11 from both sides.}$$
$$5x = 15$$
$$\frac{\overset{1}{\cancel{5}} \cdot x}{\underset{1}{\cancel{5}}} = \frac{15}{5} \quad \text{Divide both sides by 5.}$$
$$x = 3 \quad \longleftarrow \text{The solution is 3.}$$

The unknown number is 3.

To check the solution, go back to the words of the *original* problem.

If 5 times	a number	is added to	11,	the result is	26
↓ ↓	↓	↓	↓	↓	↓
5 •	3	$+$	11	$=$	26

Does $5 \cdot 3 + 11$ really equal 26? Yes, $15 + 11 = 26$. So **3** is the correct solution because it "works" when you put it back into the original problem.

················· **Work Problem 3 at the Side.** ▶

OBJECTIVE **3** **Solve application problems.** In **Chapter 1,** you learned how to solve application problems using six steps. Now that you know how to use variables and solve equations, we can include those tools in the problem-solving process. Here is a revised list of the six steps we will use.

Solving an Application Problem Using Algebra
Step 1 **Read** the problem carefully to see what it is about.
Step 2 **Assign a variable** by identifying the unknown(s). Write down what the variable represents. If there are several unknowns, let the variable represent the one you know the least about.
Step 3 **Write an equation** using the variable expression(s).
Step 4 **Solve** the equation.
Step 5 **State the answer.**
Step 6 **Check** the solution in the words of the original problem.

Notice that *Steps 1, 5, and 6* are nearly identical to what you have been using. *Steps 2, 3, and 4* introduce the use of variables and equations.

3 Translate each sentence into an equation and solve it. Check your solution by going back to the words in the original problem.

GS **(a)** If 3 times a number is added to 4, the result is 19. Find the number.

Let x represent the unknown number.

3 times a number	added to	4	is	19
↓	↓	↓	↓	↓
$3x$	$+$	4	$=$	19

(b) If -6 times a number is added to 5, the result is -13. Find the number.

(c) If twice a number is subtracted from 65, the result is -21. Find the number.

4 A college bookstore ordered some red pens. The store sold a dozen of the pens last week and two dozen this week. There are 48 red pens left. How many were originally ordered?

Show your work for each of the six problem-solving steps.

Step 1 Asks for the number of red pens ordered.

Step 2 Let x be the number of pens ordered.

EXAMPLE 4 **Solving an Application Problem with One Unknown**

Jennifer put some money aside in an envelope for small household expenses. Yesterday she took out $30 for groceries. Today a friend paid back a $42 loan, and Jennifer put the cash in the envelope. Now she has $51 in the envelope. How much was in the envelope at the start?

Step 1 **Read.** The problem asks for the amount of money in the envelope at the start.

Step 2 **Assign a variable.** There is only *one* unknown: the amount of money in the envelope at the start. Let x represent the money at the start.

Step 3 **Write an equation.**

Money at the start	Took out $30	Put in $42	Ended up with $51
x	$- \$30$	$+ \$42$	$= \$51$

Step 4 **Solve.**

On the left side, $-30 + 42$ is $+12$.

$$x \underbrace{- 30 + 42}_{} = 51$$
$$x \quad + 12 \quad = 51$$
$$x + 12 - 12 = 51 - 12 \qquad \text{Subtract 12 from both sides.}$$
$$x = 39 \longleftarrow \text{The solution is 39.}$$

Step 5 **State the answer.** Jennifer had $39 in the envelope at the start.

Be sure to write a $ in the answer.

Step 6 **Check.** Use the words of the *original* problem.

The solution is $39.

Had	Took out	Put in	Now has
$\$39$	$- \$30$	$+ \$42$	$= \$51$
$\$9$		$+ \$42$	$= \$51$
	$\$51$		$= \$51$ ✓ True

So **$39** is the correct solution because it "works" in the original problem.

◀ **Work Problem** **4** **at the Side.**

Answer

4. ***Step 3*** $x - 12 - 24 = 48$
Step 4 $x - 36 = 48$
$x - 36 + 36 = 48 + 36$
$x = 84$
Step 5 The bookstore ordered 84 red pens.
Step 6 $84 - 12 - 24$ is 48, which matches the number of pens left.

EXAMPLE 5	Solving an Application Problem with One Unknown

Michael has 5 less than three times as many lab experiments completed as David. If Michael has completed 13 experiments, how many lab experiments has David completed?

Step 1 Read. The problem asks for the number of lab experiments completed by David.

Step 2 Assign a variable. There is only *one* unknown: David's number of experiments.
Let x represent David's number of experiments.

Step 3 Write an equation.

The number Michael did is 5 less than 3 times David's number.

$$13 = 3x - 5$$

> Be careful when subtracting! $5 - 3x$ does **not** mean the same thing as $3x - 5$.

Step 4 Solve.

$$13 = 3x - 5$$

$$13 + 5 = 3x - 5 + 5 \qquad \text{Add 5 to both sides.}$$

$$18 = 3x$$

$$\frac{18}{3} = \frac{\overset{1}{\cancel{3}} \cdot x}{\underset{1}{\cancel{3}}} \qquad \text{Divide both sides by 3.}$$

$$6 = x \longleftarrow \text{The solution is 6.}$$

Step 5 State the answer. David completed 6 lab experiments.

Step 6 Check. Use the words of the *original problem*.

The number Michael did is 5 less than 3 times David's number.

$$13 = (3 \cdot 6) - 5$$

> 6 is the solution, **not** 13.

$$13 = 18 - 5$$

$$13 = 13 \checkmark \qquad \text{True}$$

So **6** is the correct solution because it "works" in the original problem.

> **Note**
>
> In *Step 3* above, you may also write the equation as $3x - 5 = 13$, with the two sides switched. The solution will be the same.

·········· **Work Problem ⑤ at the Side.** ▶

⑤ Susan donated $10 more than twice what LuAnn donated. If Susan donated $22, how much did LuAnn donate?
Show your work for each of the six problem-solving steps.

6 Solve each application using the six problem-solving steps.

(a) In a day of work, Keonda made $12 more than her daughter. Together they made $182. Find the amount made by each person. (*Hint:* Which amount do you know the least about, Keonda's or her daughter's? Let x be that amount.)

EXAMPLE 6 **Solving an Application Problem with Two Unknowns**

During the day, Sheila drove 72 km more than Russell. The total distance traveled by them both was 232 km. Find the distance traveled by each person.

Step 1 **Read.** The problem asks how far Sheila drove and how far Russell drove.

Step 2 **Assign a variable.** There are *two* unknowns: Sheila's distance and Russell's distance.

Let x be the distance traveled by Russell, because you know *less* about his distance than Sheila's distance.

Since Sheila drove 72 km *more* than Russell, the distance she traveled is $x + 72$ km, that is, Russell's distance (x) plus 72 km.

Step 3 **Write an equation.**

$$x \quad + \quad x + 72 \quad = \quad 232$$

Step 4 **Solve.**

$$x + x + 72 = 232 \qquad \text{On the left side, } x + x$$
$$\qquad\qquad\qquad\qquad \text{is } 1x + 1x, \text{ which is } 2x.$$
$$2x + 72 = 232$$
$$2x + 72 - 72 = 232 - 72 \qquad \text{Subtract 72 from both sides.}$$
$$2x = 160$$
$$\frac{\overset{1}{2} \cdot x}{\underset{1}{2}} = \frac{160}{2} \qquad \text{Divide both sides by 2.}$$
$$x = 80 \longleftarrow \text{The solution is 80.}$$

(b) A rope is 21 yd long. Marcos cut it into two pieces, so that one piece is 3 yd longer than the other. Find the length of each piece.

Step 5 **State the answer.**
Russell's distance is x, so Russell traveled 80 km.
Sheila's distance is $x + 72$,
so Sheila traveled 80 km + 72 km = 152 km.

> Be sure to write **km** in both answers.

Step 6 **Check.** Use the words of the *original* problem.
"Sheila drove 72 km more than Russell."
Sheila's 152 km is 72 km more than Russell's 80 km, so that checks.

"The total distance traveled by them both was 232 km."
Sheila's 152 km + Russell's 80 km = 232 km, so that checks.

◄ Work Problem **6** at the Side.

Answers

6. (a) daughter made x;
Keonda made $x + 12$;
$x + x + 12 = 182$;
daughter made $85; Keonda made $97.
Check. $97 - $85 = $12
$97 + $85 = $182
(b) shorter piece is x; longer piece is $x + 3$;
$x + x + 3 = 21$; shorter piece is 9 yd;
longer piece is 12 yd
Check. 12 yd − 9 yd = 3 yd
12 yd + 9 yd = 21 yd

| EXAMPLE 7 | **Solving a Geometry Application Problem** |

The length of a rectangle is 2 cm more than the width. The perimeter is 68 cm. Find the length and width.

Step 1 **Read.** The problem asks for the length and the width of a rectangle.

Step 2 **Assign a variable.** There are *two* unknowns, length and width. You know the *least* about the width, so let x represent the **width.**

Since the length is 2 cm *more* than the width, the **length** is $x + 2$. A drawing of the rectangle will help you see these relationships.

width is x
length is $x + 2$
perimeter is 68

You know the least about the width, so let x represent the width.

Step 3 **Write an equation.**

Use the formula for perimeter of a rectangle, $P = 2l + 2w$.

$$P = 2 \cdot l \qquad + 2 \cdot w$$

Step 4 **Solve.**

$68 = \underline{2\,(x + 2)} + \underline{2 \cdot x}$ Use the distributive property.

$68 = 2x + 4 \quad + 2x$ Combine like terms.

$68 = 4x + 4$

$68 - 4 = 4x + 4 - 4$ Subtract 4 from both sides.

$64 = 4x$

$$\frac{64}{4} = \frac{\overset{1}{\cancel{4}} \cdot x}{\underset{1}{\cancel{4}}}$$ Divide both sides by 4.

$16 = x \longleftarrow$ The solution is 16.

Step 5 **State the answer.**

The width is x, so the width is 16 cm.

The length is $x + 2$, so
the length is $16 + 2$ or 18 cm.

*Be sure to write **cm** in both answers.*

The width is 16 cm and the length is 18 cm.

Step 6 **Check.** Use the words of the original problem. It says the length is 2 cm more than the width. 18 cm is 2 cm more than 16 cm, so that part checks.

The original problem also says the perimeter is 68 cm. Use 18 cm and 16 cm to find the perimeter.

$$P = \underline{2 \cdot 18 \text{ cm}} + \underline{2 \cdot 16 \text{ cm}}$$

$$P = \quad 36 \text{ cm} \quad + \quad 32 \text{ cm} \quad = 68 \text{ cm} \longleftarrow \text{Checks}$$

Work Problem �7 at the Side. ▶

⓻ Make a drawing and use the six steps to solve this problem. The length of Ann's rectangular garden plot is 3 yd more than the width. She used 22 yd of fencing around the edge. Find the length and the width of the garden.

Answer

7.

width is x
length is $x + 3$

$22 = 2(x + 3) + 2 \cdot x$
width is 4 yd
length is 7 yd
Check. 7 yd is 3 more than 4 yd.
$P = 2 \cdot 7 \text{ yd} + 2 \cdot 4 \text{ yd}$
$P = 14 \text{ yd} + 8 \text{ yd}$
$P = 22 \text{ yd} \longleftarrow \text{Checks}$

9.8 Exercises

CONCEPT CHECK *Write each word phrase in symbols, using x as the variable.*
See Examples 1 and 2.

1. 14 plus a number

2. The sum of a number and −8

3. −5 added to a number

4. 16 more than a number

5. 20 minus a number

6. A number decreased by 25

7. 9 less than a number

8. A number subtracted from −7

9. Subtract 4 from a number

10. 3 fewer than a number

11. Six times a number

12. The product of −3 and a number

13. Double a number

14. Half a number

15. A number divided by 2

16. 4 divided by a number

17. Twice a number added to 8

18. Five times a number plus 5

19. 10 fewer than seven times a number

20. 12 less than six times a number

21. The sum of twice a number and the number

22. Triple a number subtracted from the number

23. In your own words, write a definition for each of these words. Give three examples to illustrate each definition.

(a) variable

(b) expression

(c) equation

24. "You can use any letter to represent the unknown in an application problem." Is this statement true or false? Explain your answer.

Translate each sentence into an equation and solve it. **See Example 3.**

25. If four times a number is decreased by 2, the result is 26. Find the number.

4 times a number	decreased by	2	is	26
$4n$	$-$	2	$=$	26

26. The sum of 8 and five times a number is 53. Find the number.

sum of

8	5 times a number	is	53
8	$+$ $5n$	$=$	53

27. If twice a number is added to the number, the result is -15. What is the number?

28. If a number is subtracted from three times the number, the result is -8. What is the number?

29. If the product of some number and 5 is increased by 12, the result is seven times the number. Find the number.

30. If eight times a number is subtracted from eleven times the number, the result is -9. Find the number.

31. When three times a number is subtracted from 30, the result is 2 plus the number. What is the number?

32. When twice a number is decreased by 8, the result is the number increased by 7. Find the number.

33. If half a number is added to twice the number, the answer is 50. Find the number.

34. If one-third of a number is added to three times the number, the result is 30. Find the number.

Solve each application problem by using the six problem-solving steps you learned in this section. ***See Examples 4, 5, and 6.***

35. Ricardo gained 15 pounds over the winter. He went on a diet and lost 28 pounds. Then he regained 5 pounds and weighed 177 pounds. How much did he weigh originally?

36. Mr. Chee deposited $80 into his bank account. Then, after paying $43 for gas, and $190 for his son's day care, the balance in his account was $47. How much was in his account before he made the deposit?

37. While shopping for clothes, Consuelo spent $3 less than twice what Brenda spent. Consuelo spent $81. How much did Brenda spend?

38. Dennis weighs 184 lb. His weight is 2 lb less than six times his child's weight. How much does his child weigh?

39. My sister is 9 years younger than I am. The sum of our ages is 51. Find our ages.

40. Ed and Marge were candidates for city council. Marge won, with 93 more votes than Ed. The total number of votes cast in the election was 587. Find the number of votes received by each candidate.

41. Last year, Lien earned $1500 more than her husband. Together they earned $37,500. How much did each of them earn?

42. A $149,000 estate is to be divided between two charities so that one charity receives $18,000 less than the other. How much will each charity receive?

43. Jason paid eight times as much for his notebook computer as he did for his printer. He paid a total of $396 for both items. What did each item cost?

44. The attendance at the Saturday night baseball game was three times the attendance at Sunday's game. In all, 56,000 fans attended the games. How many fans were at each game?

45. A board is 78 cm long. Rosa cut the board into two pieces, with one piece 10 cm longer than the other. Find the length of both pieces. (*Hint:* Make a sketch of the board.)

longer piece   shorter piece

46. A rope is 50 ft long. Juan cut it into two pieces so that one piece is 8 ft longer than the other. Find the length of each piece (*Hint:* Make a sketch of the rope.)

longer piece   shorter piece

47. A wire is cut into two pieces, with one piece 7 ft shorter than the other. The wire was 31 ft long before it was cut. How long is each piece?

48. A 90 cm pipe is cut into two pieces so that one piece is 6 cm shorter than the other. Find the length of each piece.

In Exercises 49–54, use the formula for the perimeter of a rectangle, $P = 2l + 2w$. Make a drawing to help you solve each problem, and use the six problem-solving steps. **See Example 7.**

49. The perimeter of a rectangle is 48 yd. The width is 5 yd. Find the length.

50. The length of a rectangle is 27 cm, and the perimeter is 74 cm. Find the width of the rectangle.

51. A rectangular dog pen is twice as long as it is wide. The perimeter of the pen is 36 ft. Find the length and the width of the pen.

52. A new city park has a rectangular shape. The length is triple the width. It will take 240 yd of fencing to go around the park. Find the length and width of the park.

53. The length of a rectangular jewelry box is 3 inches more than twice the width. The perimeter is 36 inches. Find the length and the width.

54. The perimeter of a rectangular house is 122 ft. The width is 5 ft less than the length. Find the length and the width.

Chapter 9 *Summary*

Key Terms

9.1

negative numbers Negative numbers are numbers that are less than 0.

signed numbers Signed numbers include positive numbers, negative numbers, and 0.

absolute value Absolute value is the distance of a number from 0 on a number line. Absolute value is never negative.

opposite of a number The opposite of a number is a number the same distance from 0 on a number line, but on the opposite side of 0.

9.2

additive inverse The additive inverse is the opposite of a number. The sum of a number and its additive inverse is always 0.

9.5

variables Variables are letters that represent numbers.

expression An expression is a combination of operations on variables and numbers.

evaluate an expression To evaluate an expression, replace the variables with the values you are given. Then do all possible calculations, following the order of operations.

9.6

equation An equation is a statement that says two expressions are equal. An equation contains an equal sign.

solution The solution of an equation is a number that can replace the variable so that the equation is true.

addition property of equality The addition property of equality states that the same number can be added or subtracted on both sides of an equation.

multiplication property of equality The multiplication property of equality states that both sides of an equation can be multiplied or divided by the same number, except division by 0 cannot be done.

9.7

distributive property If a, b, and c are three numbers, the distributive property says that $a(b + c) = ab + ac$.

like terms Like terms are terms with exactly the same variables and the same exponents.

New Symbols

-6 negative sign; read as "negative 6"; indicates the number is *less* than 0.

$|4|$ and $|-2|$ absolute value of 4; absolute value of -2

$a, b, c, d,$ etc.   variables that represent numbers

Test Your Word Power

See how well you have learned the vocabulary in this chapter. The answers are on the next page.

1 An **expression**
A. has an equal sign
B. follows the order of operations
C. is a combination of operations on variables and numbers.

2 An **equation**
A. always has more than one solution
B. is evaluated by adding the same number to both sides
C. contains an equal sign.

3 The **absolute value** of a number is
A. its distance from 0
B. never positive
C. less than 0.

4 **Like terms**
A. are always positive
B. have matching variables and exponents
C. can be multiplied but not added.

5 The **opposite** of a number is
A. never negative
B. called the additive inverse
C. called the absolute value.

6 A **variable** is
A. a solution of an equation
B. a letter representing a number
C. one of several like terms.

Answers to Test Your Word Power

1. C; *Examples:* $3x - 5$; $2y^2$; $4a + b$.

2. C; *Examples:* $2x + 6 = -4$; $-8 + y = 7y - 14$.

3. A; *Example:* $|-3| = 3$ and $|3| = 3$ because both -3 and 3 are 3 spaces away from 0.

4. B; *Examples:* $2y$ and $-7y$ are like terms; $4x^2$ and x^2 are like terms; $3y^2$ and $3y^3$ are not like terms.

5. B; *Example:* The opposite of 5 is -5; it is the additive inverse because $5 + (-5) = 0$.

6. B; *Examples:* x, y, and d are variables.

Quick Review

Concepts	Examples
9.1 Graphing Signed Numbers Place a dot at the correct location on the number line. Positive numbers are to the right of 0; negative numbers are to the left of 0.	Graph -2, 1, 0, and $2\frac{1}{2}$.
9.1 Using the $<$ and $>$ Symbols Place the symbol $<$ (less than) or $>$ (greater than) between two numbers to make the statement true. The smaller, pointed end of the symbol points to the smaller (lesser) number.	Use the symbol $<$ or $>$ to make each statement true. $2 \underline{\ >\ } 1$ $-3 \underline{\ >\ } -5$ $-6 \underline{\ <\ } 2$
9.1 Finding the Absolute Value of a Number Determine the distance from 0 to the given number on the number line. Because the absolute value is a distance, it is never negative.	Simplify each absolute value expression. **(a)** $\|8\|$ **(b)** $\|-7\|$ **(c)** $-\|-5\|$ $\|8\| = 8$ $\|-7\| = 7$ -5
9.1 Finding the Opposite of a Number Find the number that is the same distance from 0 as the given number, but on the opposite side of 0 on the number line.	Find the opposite of each number. **(a)** -6 **(b)** $+9$ $-(-6) = 6$ $-(+9) = -9$
9.2 Adding Two Signed Numbers **Adding Two Positive Numbers** Add the absolute values. The sum is positive. **Adding Two Negative Numbers** Add the absolute values and write a negative sign in front of the sum.	Add. **(a)** $8 + 6 = 14$ **(b)** $-8 + (-6)$ Find absolute values. $\|-8\| = 8$ $\|-6\| = 6$ Add absolute values: $8 + 6 = 14$ Write a negative sign in front of the sum. $$-8 + (-6) = -14$$

(Continued)

Concepts	Examples

9.2 Adding Two Signed Numbers (*Continued*)

Adding Two Numbers with Different Signs
Subtract the lesser absolute value from the greater absolute value. Write the sign of the number with the greater absolute value in front of the answer.

(c) $5 + (-7)$

Find absolute values. $|5| = 5$ $|-7| = 7$

Subtract the lesser absolute value from the greater: $7 - 5 = 2$.

The number with the greater absolute value is -7. Its sign is negative, so write a negative sign in front of the answer.

$$5 + (-7) = -2$$

9.2 Subtracting Two Signed Numbers

To subtract two numbers, add the first number to the opposite of the second number. Follow these steps.

Step 1 Change the subtraction sign to addition.

Step 2 Change the sign of the second number to its opposite.

Step 3 Proceed as in addition.

Subtract.

(a) $-6 - 5$

$-6 + (-5) = -11$

(b) $5 - (-8)$

$5 + (+8) = 13$

9.3 Multiplying Signed Numbers

The product of two numbers with the *same* sign is *positive*.

The product of two numbers with *different* signs is *negative*.

Multiply.

(a) $7 \cdot 3 = 21$

(b) $(-3) \cdot 4 = -12$

(c) $5(-8) = -40$

(d) $(-9)(-6) = 54$

9.3 Dividing Signed Numbers

Use the same sign rules as for multiplying signed numbers. When two numbers have the *same* sign, the quotient is *positive*.
When two numbers have *different* signs, the quotient is *negative*.

Divide.

(a) $\dfrac{8}{4} = 2$

(b) $\dfrac{-20}{5} = -4$

(c) $\dfrac{50}{-5} = -10$

(d) $\dfrac{-12}{-6} = 2$

9.4 Using the Order of Operations to Simplify Numerical Expressions

Use the order of operations to simplify numerical expressions.
1. Do all operations inside *parentheses* or *other grouping symbols*.
2. Simplify any expressions with *exponents* and find any *square roots*.
3. Do the remaining **multiplications and divisions** as they occur from left to right.
4. Do the remaining **additions and subtractions** as they occur from left to right.

Simplify.

(a) $-4 + 6 \div (-2)$

$-4 + (-3)$

-7

(b) $3^2 + 3(8 \div 2)$

$3^2 + 3(4)$

$9 + 3(4)$

$9 + 12$

21

Concepts	Examples

9.5 Evaluating Expressions

Replace each variable in the expression with the specified number.

Use the order of operations to simplify the expression. Remember that when no operation sign is written between a number and a letter, you assume it is multiplication. For example, $7x$ means $7 \cdot x$.

Evaluate $6p - 5s$, if p is -3 and s is -4.

$$
\begin{array}{cc}
6 \quad p & -5 \quad s \\
\downarrow \downarrow & \downarrow \downarrow \\
6(-3) & -5(-4) \\
\underbrace{-18} & \underbrace{-(-20)} \\
\end{array}
$$
$$2$$

9.6 Determining Whether a Number Is a Solution of an Equation

Substitute the number for the variable in the equation.
If the resulting statement is *true,* the number is a solution of the equation.
If the resulting statement is *false,* the number is *not* a solution of the equation.

Is 4 a solution of the equation $3x - 5 = 7$?

Replace x with 4.
$$3(4) - 5 = 7$$
$$12 - 5 = 7$$
$$7 = 7 \quad \checkmark \quad \text{True}$$

The result is true, so **4** is a solution of $3x - 5 = 7$.

9.6 Using the Addition Property of Equality to Solve an Equation

Add or subtract the same number on both sides of the equation so that you get the variable by itself on one side.

Solve each equation.

(a)
$$x - 6 = 9$$
$$x - 6 + 6 = 9 + 6 \qquad \text{Add 6 to both sides.}$$
$$x + 0 = 15$$
$$x = 15 \quad \longleftarrow \text{The solution is 15.}$$

(b)
$$-7 = x + 9$$
$$-7 - 9 = x + 9 - 9 \qquad \text{Subtract 9 from both sides.}$$
$$-16 = x + 0$$
$$-16 = x \quad \longleftarrow \text{The solution is } -16.$$

9.6 Using the Multiplication Property of Equality to Solve an Equation

Multiply or divide both sides of the original equation by the same number so that you get the variable by itself on one side. (Do not divide by 0.)

Solve each equation.

(a) $-54 = 6x$

$$\frac{-54}{6} = \frac{\overset{1}{6} \cdot x}{\underset{1}{6}}$$

$$-9 = x$$

(b) $\dfrac{1}{3}x = 8$

$$\frac{\overset{1}{3}}{1} \cdot \frac{1}{\underset{1}{3}}x = 3 \cdot 8$$

$$1x = 24$$
$$x = 24$$

Concepts	Examples

9.7 Solving Equations with Several Steps

Use the following steps.

Step 1 Add or subtract the same amount on both sides of the equation so that the variable ends up by itself on one side.

Step 2 Multiply or divide both sides by the same number to find the solution.

Step 3 Check the solution by going back to the *original* equation.

Solve: $2p - 3 = 9$.

$2p - 3 + 3 = 9 + 3$ Add 3 to both sides.

$2p = 12$

$$\frac{\overset{1}{2} \cdot p}{\underset{1}{2}} = \frac{12}{2}$$ Divide both sides by 2.

$p = 6$ ⟵ The solution is 6.

CHECK $2p - 3 = 9$ ⟵ Original equation

$2(6) - 3 = 9$ Replace p with 6.

$12 - 3 = 9$

$9 = 9$ ✓   True, so 6 is the solution.

6 is the solution (*not* 9).

9.7 Using the Distributive Property

To simplify expressions, use the distributive property.

$$a(b + c) = ab + ac$$

Simplify: $-2(x + 4)$

$$= -2 \cdot x + (-2) \cdot 4$$
$$= -2x + (-8)$$
$$= -2x - 8$$

9.7 Combining Like Terms

To combine like terms, add or subtract the number parts of the terms. The variable part stays the same.

Combine like terms.

(a) $6p + 7p = (6 + 7)p = 13p$

(b) $8m - 11m = (8 - 11)m = -3m$

9.8 Translating Word Phrases into Expressions with Variables

Use x (or any other letter) as a variable and use symbols to show the operations described by the words.

Write each word phrase in symbols, using x as the variable.

(a) Two more than a number $x + 2$ or $2 + x$

(b) 8 subtracted from a number $x - 8$

(c) The product of a number and 15 $15x$

(d) A number divided by 9 $\dfrac{x}{9}$

Chapter 9 Review Exercises

9.1 *Graph each set of numbers on the number line.*

1. $2, -3, 4, 1, 0, -5$

$-7 \ -6 \ -5 \ -4 \ -3 \ -2 \ -1 \ \ 0 \ \ 1 \ \ 2 \ \ 3 \ \ 4$

2. $-2, 5, -4, -1, 3, -6$

$-6 \ -5 \ -4 \ -3 \ -2 \ -1 \ \ 0 \ \ 1 \ \ 2 \ \ 3 \ \ 4 \ \ 5$

3. $-1\dfrac{1}{4}, -\dfrac{5}{8}, -3\dfrac{3}{4}, 2\dfrac{1}{8}, \dfrac{3}{2}, -2\dfrac{1}{8}$

$-8 \ -7 \ -6 \ -5 \ -4 \ -3 \ -2 \ -1 \ \ 0 \ \ 1 \ \ 2 \ \ 3$

4. $0, -\dfrac{3}{4}, \dfrac{5}{4}, -4\dfrac{1}{2}, \dfrac{7}{8}, -7\dfrac{2}{3}$

$-8 \ -7 \ -6 \ -5 \ -4 \ -3 \ -2 \ -1 \ \ 0 \ \ 1 \ \ 2 \ \ 3$

Write $<$ or $>$ in each blank to make a true statement.

5. $0 \ \rule{2cm}{0.4pt} \ -2$

6. $-5 \ \rule{2cm}{0.4pt} \ 0$

7. $-1 \ \rule{2cm}{0.4pt} \ -4$

8. $-9 \ \rule{2cm}{0.4pt} \ -6$

Simplify each absolute value expression.

9. $|8|$

10. $|-19|$

11. $-|-7|$

12. $-|15|$

9.2 *Add.*

13. $-4 + 6$

14. $-10 + 3$

15. $-11 + (-8)$

16. $-9 + (-24)$

17. $12 + (-11)$

18. $1 + (-20)$

19. $\dfrac{9}{10} + \left(-\dfrac{3}{5}\right)$

20. $-\dfrac{7}{8} + \dfrac{1}{2}$

21. $-6.7 + 1.5$

22. $-0.8 + (-0.7)$

Give the additive inverse (opposite) of each number.

23. 6

24. −14

25. $-\dfrac{5}{8}$

26. 3.75

Subtract.

27. 4 − 10

28. 7 − 15

29. −6 − 1

30. −12 − 5

31. 8 − (−3)

32. 2 − (−9)

33. −1 − (−14)

34. −10 − (−4)

35. −40 − 40

36. −15 − (−15)

37. $\dfrac{1}{3} - \dfrac{5}{6}$

38. 2.8 − (−6.2)

9.3 *Multiply or divide.*

39. −4 (6)

40. 5 • (−4)

41. −3 (−5)

42. −8 (−8)

43. $\dfrac{80}{-10}$

44. $\dfrac{-9}{3}$

45. $\dfrac{-25}{-5}$

46. $\dfrac{-120}{-6}$

47. (−37) (0)

48. (−1) (81)

49. $\dfrac{0}{-10}$

50. $\dfrac{-20}{0}$

51. $\left(\dfrac{2}{3}\right)\left(-\dfrac{6}{7}\right)$

52. $-\dfrac{4}{5} \div \left(-\dfrac{2}{15}\right)$

53. (−0.5) (−2.8)

54. $\dfrac{-5.28}{0.8}$

9.4 *Simplify each expression.*

55. $2 - 11(-5)$

56. $(-4)(-8) - 9$

57. $48 \div (-2)^3 - (-5)$

58. $-36 \div (-3)^2 - (-2)$

59. $5(4) - 7(6) + 3(-4)$

60. $2(8) - 4(9) + 2(-6)$

61. $3^3(-4) - 2(5 - 9)$

62. $6(-4)^2 - 3(7 - 14)$

63. $\dfrac{3 - (5^2 - 4^2)}{14 + 24 \div (-3)}$

64. $(-0.8)^2(0.2) - (-1.2)$

65. $\left(-\dfrac{1}{3}\right)^2 + \dfrac{1}{4}\left(-\dfrac{4}{9}\right)$

66. $\dfrac{12 \div (2 - 5) + 12(-1)}{2^3 - (-4)^2}$

9.5 *Evaluate each expression using the given values of the variables.*

67. $3k + 5m$
    k is 4, m is 3

68. $3k + 5m$
    k is -6, m is 2

69. $2p - q$
    p is -5, q is -10

70. $2p - q$
    p is 6, q is -7

71. $\dfrac{5a - 7y}{2 + m}$
    a is 1, y is 4, m is -3

72. $\dfrac{5a - 7y}{2 + m}$
    a is 2, y is -2, m is -26

Using the given values, evaluate each formula.

73. $P = a + b + c$; a is 9 cm, b is 12 cm, c is 14 cm

74. $A = \dfrac{1}{2}bh$; b is 6 ft, h is 9 ft

9.6–9.7 *Solve each equation. Check each solution.*

75. $y + 3 = 0$

76. $a - 8 = 8$

77. $-5 = z - 6$

78. $-8 = -9 + r$

79. $-\dfrac{3}{4} + x = -2$

80. $12.92 + k = 4.87$

81. $-8r = 56$

82. $3p = 24$

83. $\dfrac{z}{4} = 5$

84. $\dfrac{a}{5} = -11$

85. $20 = 3y - 7$

86. $-5 = 2b + 3$

Use the distributive property to simplify each expression.

87. $6(r - 5)$

88. $11(p + 7)$

89. $-9(z - 3)$

90. $-8(x + 4)$

Combine like terms.

91. $3r + 8r$

92. $10z - 15z$

93. $3p - 12p + p$

94. $-6x - x + 9x$

Solve each equation. Show your work.

95. $-4z + 2z = 18$

96. $-35 = 9k - 2k$

97. $4y - 3 = 7y + 12$

98. $b + 6 = 3b - 8$

99. $-14 = 2(a - 3)$

100. $42 = 7(t + 6)$

9.8 *Write each word phrase in symbols, using x to represent the variable.*

101. 18 plus a number

102. Half a number

103. −5 times a number

104. A number subtracted from 20

Translate each sentence into an equation and solve it.

105. The sum of four times a number and 6 is −14. Find the number.

106. If a number is subtracted from five times the number, the result is 100. Find the number.

Solve each application problem using the six problem-solving steps.

107. A $30,000 scholarship is being divided between two students so that one student receives three times as much as the other. How much will each student receive?

108. The perimeter of a rectangular game board is 48 in. The length is 4 in. more than the width. Find the length and width of the game board. (Draw a sketch of the board.)

Mixed Review Exercises

Add, subtract, multiply, or divide as indicated.

109. $-6 - (-9)$

110. $(-8)(-5)$

111. $-12 + 11$

112. $\dfrac{-70}{10}$

113. $-4(4)$

114. $5 - 14$

115. $\dfrac{-42}{-7}$

116. $16 + (-11)$

117. $-10 - 10$

118. $\dfrac{-5}{0}$

119. $-\dfrac{2}{3} + \dfrac{1}{9}$

120. $0.7(-0.5)$

121. $|-6| + 2 - 3(-8) - 5^2$

122. $9 \div |-3| + 6(-5) + 2^3$

Solve each equation. Show your work.

123. $-45 = -5y$

124. $b - 8 = -12$

125. $6z - 3 = 3z + 9$

126. $-5 = r + 5$

127. $-3x = 33$

128. $2z - 7z = -15$

129. $3(k - 6) = 6 - 12$

130. $6(t + 3) = -2 + 20$

131. $-10 = \dfrac{a}{5} - 2$

132. $4 + 8p = 4p + 16$

Solve each application problem using the six problem-solving steps.

133. The recommended daily intake of vitamin C for an adult female is 8 mg more than four times the recommended amount for a young child. The amount for an adult female is 60 mg. How much should the young child receive? (*Source:* Food and Nutrition Board.)

134. In the U.S. Congress, the number of Representatives is 65 less than five times the number of Senators. There are a total of 535 members of Congress. Find the number of Senators and the number of Representatives. (*Source: World Almanac and Book of Facts.*)

135. A cheetah's sprinting speed is 25 miles per hour (mph) faster than a zebra can run. The sum of their running speeds is 111 mph. How fast can each animal run? (*Source: Grolier Multimedia Encyclopedia.*)

136. A rectangular park has a perimeter of 1320 yd. The length of the park is five times the width. Find the length and width of the park.

Chapter 9 *Test*

The Chapter Test Prep Videos with test solutions are available on DVD, in MyMathLab, and on YouTube—search "LialBasicCollegeMath" and click on "Channels."

Work each problem.

1. Graph $-4, -1, 1\frac{1}{2}, 3,$ and 0 on the number line at the right.

$$-5 \; -4 \; -3 \; -2 \; -1 \quad 0 \quad 1 \quad 2 \quad 3 \quad 4 \quad 5$$

2. Write $<$ or $>$ in each blank to make a true statement.

-3 ____ 0 -4 ____ -8

3. Find each absolute value: $|-7|$ and $|15|$

Add, subtract, multiply, or divide.

4. $-8 + 7$

5. $-11 + (-2)$

6. $6.7 + (-1.4)$

7. $8 - 15$

8. $4 - (-12)$

9. $-\frac{1}{2} - \left(-\frac{3}{4}\right)$

10. $8(-4)$

11. $-7(-12)$

12. $(-16)(0)$

13. $\dfrac{-100}{4}$

14. $\dfrac{-24}{-3}$

15. $-\frac{1}{4} \div \frac{5}{12}$

Simplify each expression.

16. $-5 + 3(-2) - (-12)$

17. $2 - (6 - 8) - (-5)^2$

Evaluate $8k - 3m$, given each set of values.

18. k is -4, m is 2

19. k is 3, m is -1

20. In **Exercises 18 and 19,** you were evaluating an expression. Explain the difference between evaluating an expression and solving an equation.

21. The formula for the area of a triangle is $A = \frac{1}{2}bh$. Find A, if b is 20 ft and h is 11 ft.

Solve each equation. Show your work.

22. $x - 9 = -4$

23. $30 = -1 + r$

24. $3t - 8t = -25$

25. $\dfrac{p}{5} = -3$

26. $-15 = 3(a - 2)$

27. $3m - 5 = 7m - 13$

Solve each application problem using the six problem-solving steps.

28. A board is 118 cm long. Karin cut it into two pieces, with one piece 4 cm longer than the other. Find the length of both pieces.

29. The perimeter of a rectangular building is 420 ft. The length is four times as long as the width. Find the length and the width. Make a drawing to help solve this problem.

30. Marcella and her husband, Tim, spent a total of 19 hours redecorating their living room. Tim spent 3 hours less time than Marcella. How long did each person work on the room?

Chapters 1–9 *Cumulative Review Exercises*

First use front end rounding to round each number and estimate an answer. Then find the exact answer. Write answers to fraction problems in lowest terms and as whole or mixed numbers when possible.

1. *Estimate:*   *Exact:*

$$\overline{)}$$ $78\overline{)39{,}234}$

2. *Exact:*

$$5\frac{5}{6} \cdot \frac{9}{10} = \underline{}$$

Estimate:

$$\underline{} \cdot \underline{} = \underline{}$$

3. *Exact:*

$$4\frac{1}{6} \div 1\frac{2}{3} = \underline{}$$

Estimate:

$$\underline{} \div \underline{} = \underline{}$$

Simplify.

4. $17 - 8.094$

5. $4.06 \div 0.072$
Round to nearest tenth.

6. $\dfrac{30}{-6}$

7. $-3 - (-7)$

8. $3.2 + (-4.5)$

9. $\dfrac{1}{4} - \dfrac{3}{4}$

10. $45 \div \sqrt{25} - 2(3) + (10 \div 5)$

11. $-6 - (4 - 5) + (-3)^2$

Solve each proportion or percent problem.

12. $\dfrac{x}{12} = \dfrac{1.5}{45}$

13. 90 cars is 180% of what number of cars?

14. $5.80 is what percent of $145?

Convert these measurements.

15. $3\frac{1}{2}$ gal to quarts

16. 72 hr to days

17. 3.75 kg to grams

18. 40 cm to meters

Name each figure and find its perimeter (or circumference) and the area. Use 3.14 as the approximate value of π. Round answers to the nearest tenth, if necessary.

19.

20.

21.

22.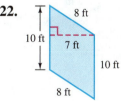

Solve each equation. Show your work.

23. $-2t - 6t = 40$

24. $3x + 5 = 5x - 11$

25. $6(p + 3) = -6$

26. Write an equation and solve it. If 40 is added to four times a number, the result is zero. Find the number.

27. Use the six problem-solving steps. $1000 in prize money is being split between Reggie and Donald. Donald should get $300 more than Reggie. How much will each man receive?

28. Make a drawing to help solve this problem. The length of a photograph is 5 cm more than the width. The perimeter of the photograph is 82 cm. Find the length and the width.

29. Find the unknown length in this figure. Round your answer to the nearest tenth if necessary.

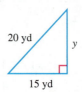

20 yd y 15 yd

Solve each application problem. Round money answers to the nearest cent.

30. Portia bought two CDs at $14.98 each. The sales tax rate is $6\frac{1}{2}\%$. Find the total amount charged to Portia's credit card.

31. Rich spent 25 minutes reading 14 pages in his sociology textbook. At that rate, how long will it take him to read 30 pages? Round to the nearest whole number of minutes.

32. A packing crate measures 2.4 m long, 1.2 m wide, and 1.2 m high. A trucking company wants crates that hold 4 m³. The crate's volume is how much more or less than 4 m³?

1.2 m

1.2 m

2.4 m

33. Jackie drove her car 364 miles on 14.5 gallons of gas. Maya used 16.3 gallons to drive 406 miles. Naomi drove 300 miles on 11.9 gallons. Which car had the highest number of miles per gallon? How many miles per gallon did that car get, rounded to the nearest tenth?

34. The local food pantry received 2480 lb of food this month. Their goal was 2000 lb. What percent of their goal was received?

35. Brian's spaghetti sauce recipe calls for $3\frac{1}{3}$ cups of tomato sauce. He wants to make $2\frac{1}{2}$ times the usual amount. How much tomato sauce does he need?

10 Statistics

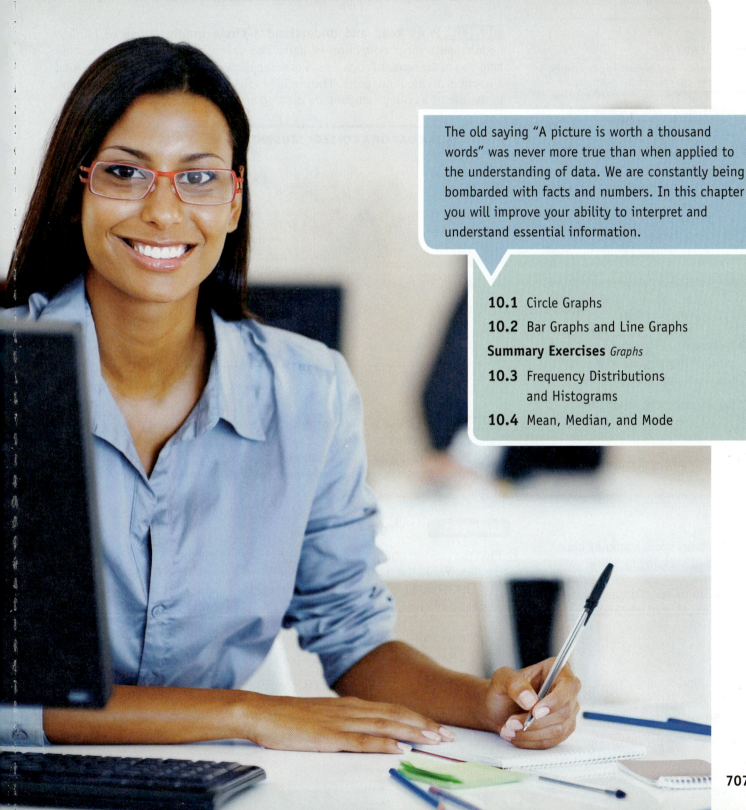

The old saying "A picture is worth a thousand words" was never more true than when applied to the understanding of data. We are constantly being bombarded with facts and numbers. In this chapter you will improve your ability to interpret and understand essential information.

10.1 Circle Graphs

OBJECTIVES

1. Read and understand a circle graph.
2. Use a circle graph.
3. Draw a circle graph.

VOCABULARY TIP

Circle graphs, also known as **pie graphs** or **pie charts,** are the best type of graph for showing the relative size of different items being compared.

1 Use the circle graph to answer each question.

(a) The greatest number of hours is spent on which activity?

(b) How many more hours are spent working than studying?

(c) Find the total number of hours spent studying, working, and attending classes.

2 Use the circle graph to find each ratio. Write the ratios as fractions in lowest terms.

GS **(a)** Hours spent driving to whole day

$$\frac{2 \text{ hr driving}}{\underline{\hspace{1cm}} \text{ hr in whole day}} = \underline{\hspace{1cm}}$$

(b) Hours spent sleeping to whole day

(c) Hours spent attending class and studying to whole day

(d) Hours spent driving and working to whole day

Answers

1. **(a)** sleeping   **(b)** 2 hr   **(c)** 13 hr

2. **(a)** 24; $\frac{1}{12}$ **(b)** $\frac{7}{24}$ **(c)** $\frac{7}{24}$ **(d)** $\frac{1}{3}$

The word *statistics* originally came from words that mean *state numbers.* State numbers refer to numerical information, or *data,* gathered by the government such as the number of births, deaths, or marriages in a population. Today, the word *statistics* has a much broader application; data from the fields of economics, social science, and business can all be organized and studied under the branch of mathematics called *statistics.*

OBJECTIVE **1** **Read and understand a circle graph.** It can be hard to understand a large collection of data. The graphs described in this section help you make sense of such data. For example, a **circle graph** shows how a total amount is divided into parts. The circle graph below shows you how 24 hours in the life of a college student are divided among different activities.

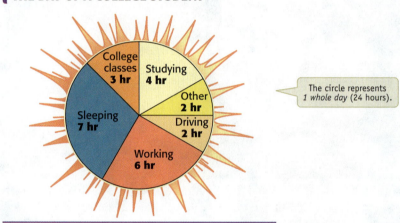

THE DAY OF A COLLEGE STUDENT

The circle represents *1 whole day (24 hours).*

◀ **Work Problem** **1** **at the Side.**

OBJECTIVE **2** **Use a circle graph.** The above circle graph uses pie-shaped pieces called *sectors* to show the amount of time spent on each activity (the total must be one day, which is 24 hours). The circle graph can therefore be used to compare the time spent on any one activity to the total number of hours in the day.

EXAMPLE 1   **Using a Circle Graph**

Find the ratio of hours spent in college classes to the total number of hours in the day. Write the ratio as a fraction in lowest terms. (See **Chapter 5.**)

The circle graph shows that 3 of the 24 hours in a day are spent in class. The ratio of class time to the hours in a day is shown below.

$$\frac{3 \text{ hours (college classes)}}{24 \text{ hours (whole day)}} = \frac{3 \text{ hours}}{24 \text{ hours}} = \frac{3 \div 3}{24 \div 3} = \frac{1}{8} \leftarrow \text{Lowest terms}$$

◀ **Work Problem** **2** **at the Side.**

The circle graph above can also be used to find the ratio of the time spent on one activity to the time spent on any other activity.

| EXAMPLE 2 | Finding a Ratio from a Circle Graph |

Use the circle graph about a student's day to find the ratio of study time to class time. Write the ratio as a fraction in lowest terms.

The circle graph shows 4 hours spent studying and 3 hours spent in class. The ratio of study time to class time is shown below.

The common units (hours) divide out.

$$\frac{4 \text{ hours } (\textbf{study})}{3 \text{ hours } (\textbf{class})} = \frac{4 \text{ hours}}{3 \text{ hours}} = \frac{4}{3} \leftarrow \text{Lowest terms}$$

Work Problem ❸ at the Side. ▶

A circle graph often shows data as percents. For example, suppose that the yearly vending machine snack food sales in the United States were $36 billion. The circle graph below shows how sales were divided among various types of snack foods. The entire circle represents the total $36 billion in sales. Each sector represents the sales of one snack item as a percent of the total sales (the total must be 100%).

YEARLY U.S. SNACK MARKET SALES
($36 BILLION)

32.5% Candy

14.4% Misc. (popcorn, pretzels, granola snacks, etc.)

5.1% Corn chips

The circle represents 100% of the snack sales. (100% of something is all of it.)

5.3% Cakes/Pies

23.8% Cookies/ Crackers

5.9% Nuts

13.0% Potato chips

Source: Natural Choice—USA.

| EXAMPLE 3 | Calculating an Amount Using a Circle Graph |

Use the circle graph above on vending machine snack sales to find the amount spent on candy for the year.

Recall the percent equation.

$$\textbf{part} = \textbf{percent} \cdot \textbf{whole}$$

The total sales are $36 billion, so the whole is $36 billion. The percent is 32.5% or, as a decimal, 0.325. Find the part.

$$\textbf{part} = \textbf{percent} \cdot \textbf{whole}$$

32.5% = 0.325

$$x = (0.325)(36 \text{ billion})$$
$$x = 11.7 \text{ billion}$$

The amount spent on candy was $11.7 billion or $11,700,000,000.

Work Problem ❹ at the Side. ▶

❸ Use the circle graph on a student's day to find the following ratios. Write the ratios as fractions in lowest terms.

GS (a) Hours spent in class to hours spent studying

$$\frac{\_\_\_\_ \text{ hr (class)}}{4 \text{ hr (study)}}$$

GS (b) Hours spent working to hours spent sleeping

$$\frac{6 \text{ hr}}{\_\_\_\_ \text{ hr}}$$

(c) Hours spent driving to hours spent working

(d) Hours spent in class to hours spent for "Other"

❹ Use the circle graph on vending machine snack sales to find the following.

GS (a) Amount spent on corn chips; 5.1% is 0.051 as a decimal

$$x = (0.051)(36 \text{ billion})$$
$$x = _____$$

(b) The amount spent on misc. (popcorn, pretzels, granola snacks, etc.)

(c) The amount spent on cakes/pies

(d) The amount spent on cookies/crackers

Answers

3. (a) $3; \frac{3}{4}$ (b) $7; \frac{6}{7}$ (c) $\frac{1}{3}$ (d) $\frac{3}{2}$

4. (a) $1.836 billion or $1,836,000,000
   (b) $5.184 billion or $5,184,000,000
   (c) $1.908 billion or $1,908,000,000
   (d) $8.568 billion or $8,568,000,000

OBJECTIVE ▶ ③ **Draw a circle graph.** In a recent year, Goodwill Industries donors helped fund programs that let nearly 1 million people take their first steps toward new and better jobs and financial independence. The following table shows those who were served.

A HELPING HAND

Group Served	Percent of Total
People with disabilities	25%
Welfare recipients	15%
Working poor	10%
Ex-offenders	10%
At-risk youth	5%
Unemployed	35%
Total	**100%**

Source: Goodwill Industries.

You can show these percents visually by using a circle graph. The entire circle will represent all of the groups served (all 100%).

EXAMPLE 4 **Drawing a Circle Graph**

Using the data in the table, find the number of degrees in the sector that would represent the "People with disabilities" and begin constructing a circle graph.

Recall that a complete circle has 360°. (See **Chapter 8.**) Because the "People with disabilities" make up 25% of the total number of people, the number of degrees needed for the "People with disabilities" sector of the circle graph is 25% of 360°.

$$(360°)\,(25\%) = (360°)\,(0.25) = 90°$$

Use a tool called a **protractor** to make a circle graph. First, using a straightedge, draw a line from the center of a circle to the left edge. Place the hole in the protractor over the center of the circle, making sure that 0 on the protractor lines up with the line that was drawn. Find 90° and make a mark as shown in the illustration. Then remove the protractor and use the straightedge to draw a line from the center of the circle to the 90° mark at the edge of the circle. This sector is 90° and represents "People with disabilities." Label the sector with the group name and percent.

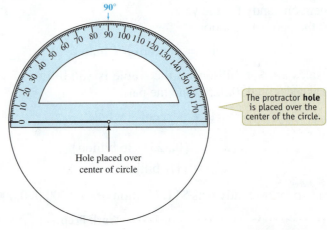

90°

The protractor **hole** is placed over the center of the circle.

Hole placed over center of circle

Continued on Next Page

To draw the "Welfare recipients" sector, begin by finding the number of degrees in the sector. From the table, you see that "Welfare recipients" represent 15% of the total.

$$(360°)(15\%) = (360°)(0.15) = 54°$$

Again, place the hole of the protractor over the center of the circle, but this time align 0 on the second line that was drawn. Make a mark at 54° and draw a line as before. This sector is 54° and represents those who are "Welfare recipients."

Line up 0 with previous 90° mark.

90°

People with
disabilities
25%
90°

54°

Hole placed over
center of circle

CAUTION

You must be certain that the hole in the protractor is placed over the exact center of the circle each time you measure the size of a sector.

· **Work Problem ❺ at the Side.** ▶

Use this circle for **Margin Problem 5** at the side.

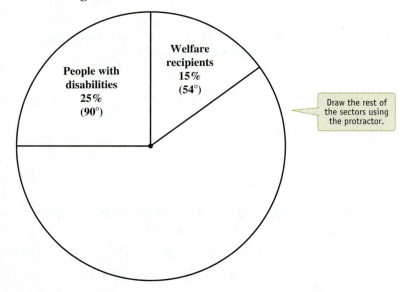

**Welfare
recipients
15%
(54°)**

**People with
disabilities
25%
(90°)**

Draw the rest of
the sectors using
the protractor.

❺ Using the information in the table on the groups served, find the number of degrees needed for each sector. Complete the circle graph at the bottom. Label each sector with the group name and percent.

GS **(a)** Working poor

$$(360°)(10\%)$$
$$= (360°)(\underline{\quad})$$
$$= \underline{\quad}$$

(b) Ex-offenders

(c) At-risk youth

(d) Unemployed

Answers

5. (a) 0.10; 36° **(b)** 36° **(c)** 18° **(d)** 126°

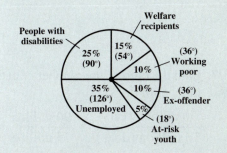

Welfare
recipients

People with
disabilities

25%
(90°)

15%
(54°)

(36°)
Working
poor

10%

35%
(126°)
Unemployed

10%

5%

(36°)
Ex-offender

(18°)
At-risk
youth

10.1 Exercises

CONCEPT CHECK.   *Fill in each blank with the correct response.*

1. A circle graph shows how a total amount is divided into _____.

2. When making a circle graph, a tool called a _____ is used to measure degrees in the circle.

This circle graph shows the number of pets owned in the United States. Use the circle graph to answer Exercises 3–8. Write ratios as fractions in lowest terms. See Examples 1 and 2.

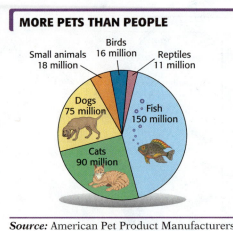

MORE PETS THAN PEOPLE

Small animals 18 million

Birds 16 million

Reptiles 11 million

Dogs 75 million

Fish 150 million

Cats 90 million

Source: American Pet Product Manufacturers Association.

3. Find the number of pets owned in the United States.

4. Which type of pet is owned by the greatest number of people? How many of these pets are owned?

5. Find the ratio of the number of cats owned to the total number of pets.

6. Find the ratio of the number of small animals owned to the total number of pets.

7. Find the ratio of the number of cats owned to the number of dogs.

$$\frac{90 \text{ million (cats)}}{75 \text{ million (dogs)}} = \frac{90 \text{ million}}{75 \text{ million}} = \frac{90 \div 15}{75 \div 15} =$$

8. Find the ratio of the number of fish owned to the number of cats.

$$\frac{150 \text{ million (fish)}}{90 \text{ million (cats)}} = \frac{150 \text{ million}}{90 \text{ million}} = \frac{150 \div 30}{90 \div 30} =$$

This circle graph shows how Americans spend their time in a typical day. Use this circle graph to answer Exercises 9–16. **See Examples 1 and 2.** *(Hint: There are 60 minutes in one hour and 24 hours in one day.)*

ALL IN A DAY

Leisure, TV, and sports
5 hours 10 minutes

Educational/civic/ religious activities
50 minutes

Other
1 hour 16 minutes

Personal care
48 minutes

Eating and drinking
1 hour 16 minutes

Sleeping
8 hours 40 minutes

Working
3 hours 30 minutes

Household duties/ caring for others
2 hours 30 minutes

Source: Bureau of Labor Statistics.

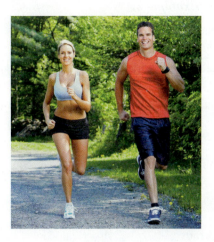

9. Americans spend most of their time doing which activity? How much time do they spend?

10. Americans spend the second most amount of time doing which activity? How much time do they spend?

Answer Exercises 11–16 by writing a ratio as a fraction in lowest terms.

11. Americans spend how much time sleeping compared to the total number of minutes in a day?
(*Hint:* First find the total minutes in a day by multiplying $(24)(60) =$ _____ minutes.)

12. Americans spend how much time on personal care compared to the total number of minutes in a day?

13. Americans spend how much time on educational/ civic/religious activities compared to eating and drinking?

14. Americans spend how much time on personal care compared to sleeping?

15. Americans spend how much time working compared to household duties/caring for others?

16. Americans spend how much time on leisure, TV, and sports compared to working?

*This circle graph shows the favorite hot dog toppings in the United States. Each topping is expressed as a percent of the 3200 people in the survey. Use the graph to find the number of people in the survey who favored each of the toppings in Exercises 17–22. **See Example 3.***

17. Onions

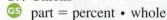 part = percent • whole

$x = (0.05)(3200)$

$x = $ _____

18. Ketchup

19. Sauerkraut

20. Relish

21. Mustard

22. Chili

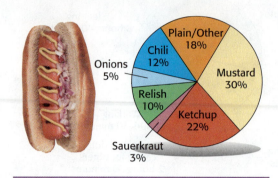

FAVORITE HOT DOG TOPPINGS

Plain/Other 18%
Chili 12%
Onions 5%
Mustard 30%
Relish 10%
Ketchup 22%
Sauerkraut 3%

Source: National Hot Dog and Sausage Council.

*J.D. Power and Associates estimates that by 2016 there will be 64,722 electric cars sold in the United States. The primary benefit to the driver is that an electric car will cost considerably less to operate than a traditional gasoline-powered car. The circle graph shows the number of miles an electric car would need to travel between charges before a person would consider buying one. If 3021 people were surveyed for this study, find the number of people giving each response in Exercises 23–28. Round answers to the nearest whole number. **See Example 3.***

23. Need the car to go 400 miles or more

24. Need the car to go 200 miles

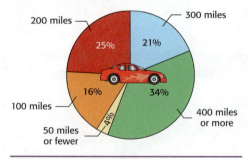

GOING ELECTRIC
How far must it go for you to buy an electric car?

200 miles — 25%
300 miles — 21%
100 miles — 16%
50 miles or fewer — 4%
400 miles or more — 34%

Source: J.D. Power and Associates.

25. Need the car to go 50 miles or fewer

26. Need the car to go 300 miles

27. Need the car to go 100 miles

28. How many people would consider buying an electric car if it will go 300 miles or more between charges?

29. Describe the procedure for determining how large each sector must be to represent each of the items in a circle graph.

30. A protractor is the tool used to draw a circle graph. Give a brief explanation of what the protractor does and how you would use it to measure and draw each sector in the circle graph.

During one month the Orangevale Parks and Recreation District spent $5460 for the activities shown in the following table. Find all numbers missing from the table.

Item	Dollar Amount	Percent of Total	Degrees of a Circle
31. Adult sports	$1365	25%	_____
32. Children's sports	$1092	_____	72°
33. Day camp	$546	_____	_____
34. Senior fitness	$546	10%	_____
35. Annual egg hunt	$819	15%	_____
36. Arts and crafts	$273	_____	_____
37. Mommy and baby exercise	$819	_____	54°

38. Draw a circle graph by using the information from **Exercises 31–37.** Label each sector in your graph. **See Example 4.**

39. White Water Rafting Company divides its annual sales into five categories as follows.

Category	Annual Sales
Adventure classes	$12,500
Grocery and provision sales	$40,000
Equipment rentals	$60,000
Rafting tours	$50,000
Equipment sales	$37,500

(a) Find the total sales for the year.

(b) Find the number of degrees in a circle graph for each item.

(c) Make a circle graph showing the percent for each category. Label each sector in your graph.

40. Online retail sales in the United States are growing rapidly. The top online retail sales categories are 50% for travel; 20% for apparel, accessories, and footwear; 15% for computer hardware and software; 10% for autos and auto parts; and 5% for home furnishings. (*Source:* Forester Research.)

(a) Find the number of degrees in a circle graph for each online retail sales category.

(b) Draw a circle graph showing the percent for each category. Label each sector in your graph.

41. Match.com and Chadwick Martin Bailey surveyed adults who met and married within the last three years. Each person was asked how he or she met his or her spouse. The results are shown in the figure on the right.

(a) Use this information to complete the table and draw a circle graph. Round to the nearest whole percent and to the nearest degree. Label each sector in your graph.

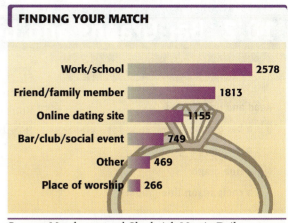

FINDING YOUR MATCH

Work/school 2578
Friend/family member 1813
Online dating site 1155
Bar/club/social event 749
Other 469
Place of worship 266

Source: Match.com and Chadwick Martin Bailey.

How They Met	Number of Adults	Percent of Total	Number of degrees
Work/school			
Place of worship			
Online dating site			
Friend/family member			
Bar/club/social event			
Other			

(b) Add up the percents. Is the total 100%? Explain why or why not.

(c) Add up the degrees. Is the total 360°? Explain why or why not.

10.2 Bar Graphs and Line Graphs

OBJECTIVES

Read and understand

1 a bar graph;

2 a double-bar graph;

3 a line graph;

4 a comparison line graph.

VOCABULARY TIP

Bar Graphs are typically used to display data that fit into different categories. The bars are usually separated by spaces.

1 Use the bar graph in the text to find the number of members in the Fitness Centers in each of these years.

GS **(a)** 2010

Bar rises to _____.

Multiply _____ by 1000.

There were _____ members.

(b) 2009

(c) 2013

(d) 2012

OBJECTIVE **1** **Read and understand a bar graph.** **Bar graphs** are useful when showing comparisons. For example, the bar graph below compares the total number of members in all the Fitness Center locations during each of five years.

Notice that the label says "in thousands."

FITNESS CENTER MEMBERSHIP

EXAMPLE 1 **Using a Bar Graph**

How many members did the Fitness Center have in 2011?

The bar for 2011 rises to 25. Notice the label along the left side of the graph that says "Number of Members (in thousands)." The phrase *in thousands* means you have to multiply 25 by 1000 to get 25,000. So, there were 25,000 (**not** 25) members in the Fitness Center locations in 2011.

◀ **Work Problem** **1** **at the Side.**

OBJECTIVE **2** **Read and understand a double-bar graph.** A **double-bar graph** can be used to compare two sets of data. The graph below shows the number of new E-flix subscribers each quarter for two different years.

NEW E-FLIX SUBSCRIBERS

The color key shows which bars represent each year.

Answers

1. **(a)** 20; 20; 20,000 members
   (b) 25,000 members   **(c)** 35,000 members
   (d) 30,000 members

| **EXAMPLE 2** | **Reading a Double-Bar Graph** |

Use the double-bar graph on the previous page to find the following.

(a) The number of new E-flix subscribers in the second quarter of 2012

There are two bars for the second quarter. The color code in the upper center of the graph tells you that the **red bars** represent 2012. So the **red bar** on the *left* is for the 2nd quarter of 2012. It rises to 600. Multiply 600 by 1000 because the label on the left side of the graph says *in thousands*. So there were 600,000 new E-flix subscribers for the second quarter in 2012.

(b) The number of new E-flix subscribers in the second quarter of 2013

The **green bar** for the second quarter rises to 400, and 400 times 1000 is 400,000. So, in the second quarter of 2013, there were 400,000 new E-flix subscribers.

> **CAUTION**
>
> Use a ruler or straightedge to line up the top of the bar with the number on the left side of the graph.

······························ **Work Problem ②** at the Side. ▶

OBJECTIVE ③   Read and understand a line graph. A **line graph** is often useful for showing a trend. The line graph below shows the number of trout stocked along the Feather River over a 5-month period. Each dot indicates the number of trout stocked during the month directly below that dot.

TROUT STOCKED IN THE FEATHER RIVER

A trend can often be seen from a line graph.

| **EXAMPLE 3** | **Understanding a Line Graph** |

Use the line graph to find the following.

(a) In which month were the least number of trout stocked?

The lowest point on the graph is the dot directly over August, so the least number of trout were stocked in August.

(b) How many trout were stocked in August?

Use a ruler or straightedge to line up the August dot with the numbers along the left edge of the graph. The August dot is halfway between the 2 and the 3. Notice that the label on the left side says *in ten-thousands*. So August is halfway between (2 • 10,000) and (3 • 10,000). It is halfway between 20,000 and 30,000. That means 25,000 trout were stocked in August.

······························ **Work Problem ③** at the Side. ▶

---

② Use the double-bar graph to find the number of new E-flix subscribers in 2012 and 2013 for each quarter.

⑤ (a) 1st quarter

> 2012 is the red bar. Multiply $(400)(1000) = $ _____.
>
> 2013 is the green bar. Multiply $(700)(1000) = $ _____.

(b) 3rd quarter

(c) 4th quarter

(d) Find the greatest number of new subscribers. Identify the quarter and the year in which that occurred.

VOCABULARY TIP

Line Graphs can be used to show how something changes over time. These graphs are most often characterized by peaks and valleys to indicate upward and downward trends.

---

③ Use the line graph in the text to find the number of trout stocked in each month.

(a) June

(b) May

(c) April

(d) July

Answers

2. (a) 400,000; 700,000 subscribers
   (b) 400,000; 500,000 subscribers
   (c) 700,000; 800,000 subscribers
   (d) 800,000 subscribers; 4th quarter 2013
3. (a) 55,000 trout   **(b)** 30,000 trout
   (c) 40,000 trout   **(d)** 60,000 trout

4 Use the comparison line graph in the text to find the following.

GS **(a)** The number of desktop computers sold in 2009, 2011, 2012, and 2013

2009: $(30)(1000) =$ _____

2011: $(40)(1000) =$ _____

2012: $(20)(1000) =$ _____

2013: $(15)(1000) =$ _____

(b) The number of laptop computers sold in 2009, 2010, 2011, and 2012

(c) The first full year in which the number of laptop computers sold was greater than the number of desktop computers sold

OBJECTIVE ▶ **4** **Read and understand a comparison line graph.** Two sets of data can also be compared by drawing two line graphs together as a **comparison line graph.** For example, the line graph below compares the number of desktop computers and the number of laptop computers sold during each of 5 years by a major electronics retailer.

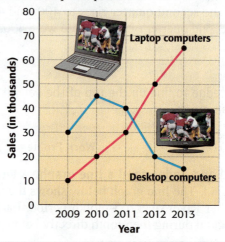

COMPUTE THIS!

Sales of Laptop Computers vs. Sales of Desktop Computers.

Use a ruler to line up each dot with the numbers along the left edge.

EXAMPLE 4 **Interpreting a Comparison Line Graph**

Use the comparison line graph above to find the following.

(a) The number of desktop computers sold in 2010
Find the dot on the **blue line** above 2010. Use a ruler or straightedge to line up the dot with the numbers along the left edge. The dot is halfway between 40 and 50, which is 45. Then, 45 times 1000 is 45,000 desktop computers sold in 2010.

(b) The number of laptop computers sold in 2013
The **red line** on the graph shows that 65 times 1000, or 65,000 laptop computers were sold in 2013.

Note

Both the double-bar graph and the comparison line graph are used to compare two sets of data.

◀ Work Problem **4** at the Side.

Answers

4. (a) 30,000; 40,000; 20,000; 15,000 desktop computers
(b) 10,000; 20,000; 30,000; 50,000 laptop computers **(c)** 2012

10.2 Exercises

FOR EXTRA HELP

Download the MyDashBoard App

MyMathLab®

CONCEPT CHECK   *Label each statement as* true *or* false.

1. A double-bar graph cannot be used to compare two sets of data.

2. A line graph is often useful for showing a trend.

The American Farm Bureau Federation reports that the average adult in the United States will work 40 days (rounded to the nearest day) to earn enough to pay the annual household food bill. This was found by multiplying the average percent of household income spent on food by 365 (the number of days in a year). This bar graph shows the percent of income spent in various countries of the world. Use this graph to answer Exercises 3–8. **See Example 1.**

TAKING A BITE OUT OF HOUSEHOLD INCOME

The average American adult will work 40 days each year to earn enough to pay the household food bill. Percent of household income spent on food in:

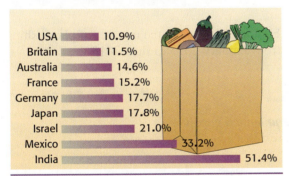

Source: American Farm Bureau Federation.

3. In which country is the highest percent of income spent on food? What percent is this?

4. In which country is the lowest percent of income spent on food? What percent is this?

5. List all countries in the graph in which less than 15% of household income is spent, on average, for food.

6. List all countries in the graph in which more than 20% of household income is spent, on average, for food.

7. How many days each year will the average adult have to work to earn enough to pay for food in Mexico? Round to the nearest day.

33.2% of 365 days

$x = (0.332)(365 \text{ days})$

$x = \underline{\hphantom{xxxx}}$ days (rounded)

8. How many days each year will the average adult have to work to earn enough to pay for food in Israel? Round to the nearest day.

21.0% of 365 days

$x = (0.21)(365)$

$x = \underline{\hphantom{xxxx}}$ days (rounded)

This double-bar graph shows the number of outdoor plants shipped by Capital Growers during the first six months of 2012 and 2013. Use this graph to answer Exercises 9–14. See Example 2.

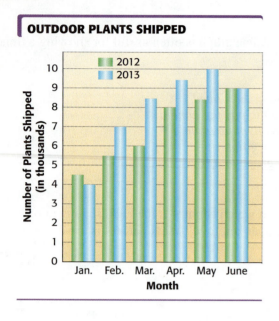

OUTDOOR PLANTS SHIPPED

9. In which month in 2013 were the greatest number of plants shipped? What was the total number of plants shipped in that month?

10. How many plants were shipped in January of 2012?

11. How many more plants were shipped in February of 2013 than in February of 2012?

12. How many fewer plants were shipped in March of 2012 than in March of 2013?

13. Find the increase in the number of plants shipped from February 2012 to April 2013.

14. Find the increase in the number of plants shipped from January 2013 to June 2013.

This double-bar graph shows sales of super unleaded and supreme unleaded gasoline at a service station for each of 5 years. Use this graph to answer Exercises 15–20. See Example 2.

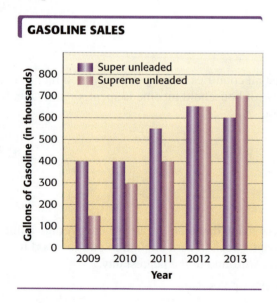

GASOLINE SALES

15. How many gallons of supreme unleaded gasoline were sold in 2009?

16. How many gallons of super unleaded gasoline were sold in 2012?

17. In which year did the greatest difference in sales between super unleaded and supreme unleaded gasoline occur? Find the difference.

18. In which year did the sales of supreme unleaded gasoline surpass the sales of super unleaded gasoline?

19. Find the increase in supreme unleaded gasoline sales from 2009 to 2013.

20. Find the increase in super unleaded gasoline sales from 2009 to 2013.

*This line graph shows how the personal computer (PC) has evolved over three decades of its existence. What began as a technician's dream is now a common tool of business and home life. Use this line graph to answer Exercises 21–26. **See Example 3.***

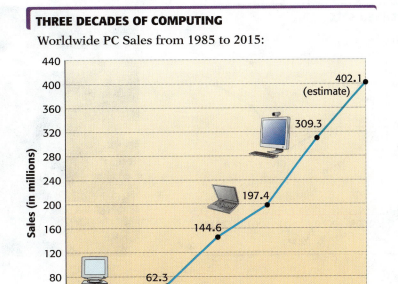

THREE DECADES OF COMPUTING

Worldwide PC Sales from 1985 to 2015:

Source: Gartner Dataquest and ARS Technic.

21. Find the number of PCs sold in 1990.

In 1990: $(24.1)(1,000,000) = $ _____

22. What was the number of PCs sold in 1995?

In 1995: $(62.3)(1,000,000) = $ _____

23. Find the increase in the estimated number of PCs to be sold in 2015 from the number sold in 2005.

$$\underbrace{(402.1)}_{2015}(1,000,000) - \underbrace{(197.4)}_{2005}(1,000,000) = $$

24. How many more PCs were sold in 2000 than in 1990?

$$\underbrace{(144.6)}_{2000}(1,000,000) - \underbrace{(24.1)}_{1990}(1,000,000) = $$

25. Give two possible explanations for the increase in the number of PCs sold.

26. Give two possible conditions that could result in a decrease in PC sales in the future.

This comparison line graph shows the number of MP3 players sold by two different chain stores during each of 5 years. Use this graph to find the annual number of MP3 players sold each year in Exercises 27–34. **See Example 4.**

27. Store A in 2013

28. Store A in 2012

29. Store A in 2011

30. Store B in 2013

31. Store B in 2012

32. Store B in 2011

33. Looking at the comparison line graph above, which store would you like to own? Explain why. Based on the graph, what amount of MP3 player sales would you predict for your store in 2014?

34. In the comparison line graph above, Store B used to have lower sales than Store A. What might have happened to cause this change? Give two possible explanations.

35. Explain in your own words why a bar graph or a line graph (not a double-bar graph or comparison line graph) can be used to show only one set of data.

36. The double-bar graph and the comparison line graph are both useful for comparing two sets of data. Explain how this works and give your own example.

This comparison line graph shows the sales and profits of Refresh Salon and Day Spa for each of 4 years. Use the graph to answer Exercises 37–44. **See Example 4.**

37. Total sales in 2013

38. Total sales in 2012

39. Total sales in 2011

40. Profit in 2013

41. Profit in 2012

42. Profit in 2011

43. Give two possible explanations for the decrease in sales from 2010 to 2011 and two possible explanations for the increase in sales from 2011 to 2013.

44. Based on the graph, what conclusion can you make about the relationship between sales and profits?

Relating Concepts (Exercises 45–50) For Individual or Group Work

The following information includes statistics regarding Life Savers.
Use this information to **work Exercises 45–50 in order.**

> **ROLL WITH IT**
>
> Sweet-toothed fans recently voted to change three
> of the original flavors. The new five-flavor roll
> includes cherry, watermelon, pineapple, raspberry,
> and blackberry. The orange, lemon, and lime
> flavors have been replaced.
>
> **Life Savers by the numbers**
>
> Year first flavor invented: **1912 (Pep-O-Mint)**
>
> Year five-flavor roll invented: **1935**
>
> Total number of flavors today: **25**
>
> Number of candies per roll: **14**
>
> Life Saver candies produced daily: **3 million**
>
> Pounds of sugar used per day: **250,000**
>
> Number of miniature rolls given out at
> Halloween: **88 million**

Source: *USA Today* and Wrigley Company.

45. The first Life Savers flavor was Pep-O-Mint. How
long after the Pep-O-Mint Life Saver was invented
was the five-flavor roll invented?

46. In addition to the five flavors, how many more
flavors of Life Savers are there?

47. Find the number of rolls of Life Savers produced
daily. Round to the nearest whole number.

48. Use your answer from **Exercise 47** to find the
amount of sugar in one roll of Life Savers. Round to
the nearest hundredth of a pound.

49. Check your work in **Exercise 48.** Is the answer
reasonable? Explain why or why not.

50. Name three possible causes of errors in statistics.

Summary Exercises   *Graphs*

This circle graph shows the number of people in a recent survey who preferred each of the five top ice cream flavors. Use the graph to answer Exercises 1–6. Write ratios as fractions in lowest terms.

ICE CREAM—FAVORITE FLAVORS

All other flavors 640 people
Vanilla 660 people
Chocolate 380 people
Strawberry 80 people
Neapolitan 100 people
Nut caramel 140 people

Source: The NPD Group, National Eating Trends Group.

1. Find the total number of people in the favorite ice cream flavor survey.

2. Which ice cream flavor is preferred by the greatest number of people in the survey? How many people picked this flavor?

3. Find the ratio of the number of people preferring chocolate to the total number of people in the survey.

4. Find the ratio of the number of people preferring Neapolitan to the total number of people in the survey.

5. Find the ratio of the number of people preferring all other flavors to those preferring strawberry.

6. Find the ratio of the number of people preferring vanilla to those preferring nut caramel.

The bar graph shows the projected growth in world population. Use the graph to answer Exercises 7–11.

WORLD POPULATION GROWTH

The United Nations projects the world's population to grow as shown (in billions).

2000	6.1
2025	8.0
2050	9.3
2075	9.9
2100	10.1

Source: United Nations.

7. When is the world population expected to reach its highest point and what is the estimated population?

8. Find the increase in world population from 2000 to 2100.

9. Find the increase in population from 2050 to 2075.

10. Find the percent of increase in world population from 2000 to 2050, to the nearest tenth of a percent.

11. In 2075, assume that 11.0% of the world's population lives in China and 4.5% of the world's population lives in the United States. Find the population of each country.

The line graph shows the historical trends in marital status in the United States. Use the graph to answer Exercises 12–16.

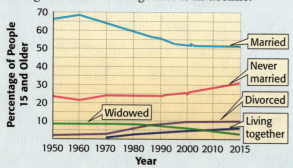

TRENDS IN MARRIAGE

The percentage of people who live together is rising while the marriage rate is in decline.

Source: U.S. Census Bureau.

12. The percent of people who were married was greatest in what year? What percent were married?

13. What percent of the population was divorced in 1950? Has the percent of those divorced more than doubled since 1950?

14. The percent of those who are married has decreased over the years. The percent of what other group has decreased?

15. What percent of the people were living together in 2010?

16. The line graph shows that 0% of the people were living together in 1970. How would you explain this?

10.3 Frequency Distributions and Histograms

OBJECTIVES

1 Understand a frequency distribution.

2 Arrange data in class intervals.

3 Read and understand a histogram.

The owner of Towne Insurance Agency has kept track of her personal phone sales call activity over the past 50 weeks. The number of sales calls made for each of the weeks is given below. Read down the columns, beginning with the left column, for successive weeks of the year.

75	65	40	50	45	30	30	35	45	25
75	70	60	55	30	25	44	30	35	30
75	70	50	30	50	20	30	30	20	25
60	62	45	45	48	40	35	25	20	25
75	45	50	40	35	40	40	30	27	40

OBJECTIVE **1** **Understand a frequency distribution.** A long list of numbers can be confusing. You can make the data easier to read by putting them in a special type of table called a **frequency distribution.**

VOCABULARY TIP

Frequency distribution is a way of organizing data by listing a set of values and their frequency (how many times each one occurs).

EXAMPLE 1   **Preparing a Frequency Distribution**

Using the data above, construct a table that shows each possible number of sales calls. Then go through the original data and place a *tally* mark (|) in the tally column next to each corresponding value. Total the tally marks and place the totals in the third column. The result is a frequency distribution table.

Number of Sales Calls	Tally	Frequency	Number of Sales Calls	Tally	Frequency
20	III	3	48	I	1
25	THL	5	50	IIII	4
27	I	1	55	I	1
30	THL IIII	9	60	II	2
35	IIII	4	62	I	1
40	THL I	6	65	I	1
44	I	1	70	II	2
45	THL	5	75	IIII	4

Work Problem 1 at the Side. ▶

1 Use the frequency distribution table in the text to find the following.

  (a) The least number of sales calls made in a week

  (b) The most common number of sales calls made in a week

  (c) The number of weeks in which 35 calls were made

  (d) The number of weeks in which 45 calls were made

OBJECTIVE **2** **Arrange data in class intervals.** The frequency distribution given in **Example 1** above contains a great deal of information—perhaps too much to digest. It can be simplified by combining the number of sales calls into groups, forming the class intervals shown in the left column below.

GROUPED DATA

Class Intervals (Number of Sales Calls)	Class Frequency (Number of Weeks)
20–29	9
30–39	13
40–49	13
50–59	5
60–69	4
70–79	6

In the table above, look at how many weeks had 20 to 29 sales calls: 3 + 5 + 1 = 9 weeks.

Answers

1. (a) 20 calls **(b)** 30 calls **(c)** 4 weeks
(d) 5 weeks

2 Use the grouped data for the insurance agency on the previous page to answer each question.

GS **(a)** During how many weeks were fewer than 50 calls made?

Number of Calls	Number of Weeks
20–29	9
30–39	13
40–49	13

So, 9 + 13 + ____ = ____ weeks

(b) During how many weeks were 50 or more calls made?

3 Use the histogram at the right to answer each question.

GS **(a)** During how many weeks were 60 or more calls made?

Number of Calls	Number of Weeks
60–69	4
70–79	6

4 + ____ = ____ weeks

(b) During how many weeks were fewer than 60 calls made?

Answers

2. **(a)** 13; 35 weeks **(b)** 15 weeks
3. **(a)** 6; 10 weeks **(b)** 40 weeks

Note

The number of class intervals in the left column of the grouped data table varies. Grouped data usually have between 5 and 15 class intervals.

EXAMPLE 2 **Analyzing a Frequency Distribution**

Use the grouped data for the insurance agency (on the preceding page) to answer the following questions.

(a) During how many weeks were fewer than 30 calls made?
  The first interval in the grouped data table (20–29) is the number of weeks during which fewer than 30 calls were made. Therefore, the owner made fewer than 30 calls during 9 weeks out of the 50 weeks shown.

(b) During how many weeks were 40 or more calls made?
  The last four intervals in the grouped data table are the number of weeks during which 40 or more calls were made.

$$13 + 5 + 4 + 6 = 28 \text{ weeks}$$

◀ **Work Problem** **2** **at the Side.**

OBJECTIVE **3** **Read and understand a histogram.** The results in the grouped data table have been used to draw the special bar graph below, which is called a **histogram.** In a histogram, the width of each bar represents a range of numbers (*class interval*). The height of each bar in a histogram gives the *class frequency,* that is, the number of occurrences in each class interval.

SALES CALL DATA FOR THE PAST 50 WEEKS (GROUPED DATA)

This bar represents the 9 weeks in which 20–29 calls were made.

The histogram bars help you visualize the data.

Number of Weeks (vertical axis)
Number of Sales Calls per Week (horizontal axis)

EXAMPLE 3 **Using a Histogram**

Use the histogram to find the number of weeks in which fewer than 40 calls were made.
  The histogram shows that 20–29 calls were made during 9 of the weeks, and 30–39 calls were made during 13 of the weeks. So, the number of weeks in which fewer than 40 calls were made is 9 + 13 = 22 weeks.

◀ **Work Problem** **3** **at the Side.**

10.3 Exercises

  MyMathLab®

CONCEPT CHECK   *Fill in each blank with the correct response.*

1. A large number of data can be easier to understand if they are in a special type of _____ called a frequency distribution.

2. A histogram is a special type of _____.

Launched in February 2004, Facebook is the most used social networking site in the world. Its mass appeal to people of all demographics has made Facebook an integral part of our lives. The ages of Facebook's users are shown below in the histogram. Use the histogram to answer Exercises 3–8. **See Example 3.**

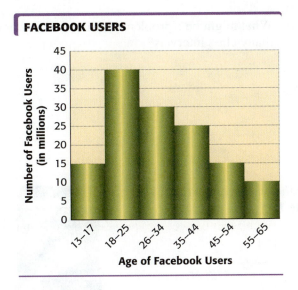

3. The greatest number of users are in which age group? How many users are in that group?

4. The least number of users are in which age group? How many users are in that group?

5. Find the number of users 35 years of age and over.

6. Find the number of users ages 13 to 17.

7. How many users are 35 to 44 years of age?

8. How many users are 26 to 34 years of age?

This histogram shows the annual earnings for the part-time employees of Rally World Amusement Park. Use this histogram to answer Exercises 9–14. **See Example 3.**

9. The greatest number of employees are in which earnings group? How many are in that group?

   The tallest bar for employees is in the $_____ to $_____ group. _____ employees are in that group.

10. The fewest number of employees are in which earnings groups? How many are in each group?

   The shortest bars for employees are in the $_____ to $_____ group and the $_____ to $_____ groups. _____ employees are in each of these groups.

11. Find the number of employees who earn $3100 to $4000.

12. Find the number of employees who earn $1100 to $2000.

13. How many employees earn $5000 or less?

Earning $5000 or less: 6 + 10 + 6 + 11 + 16 = _____ employees.

14. How many employees earn $6100 or more?

Earning $6100 or more: 13 + _____ + 7 = _____ employees.

15. Describe class interval and class frequency. How are they used when preparing a histogram?

16. What might be a problem of using two few or too many class intervals?

This list shows the number of new accounts opened annually by the employees of the Schools Credit Union. Use it to complete the table. **See Example 1.**

186	191	144	198	147	158	174
193	142	155	174	162	151	178
145	151	199	182	147	195	146

	Class Intervals (Number of New Accounts)	Tally	Class Frequency (Number of Employees)
17.	140–149	⟋⟍⟋⟍⟋ I	6
18.	150–159	_____	_____
19.	160–169	I	1
20.	170–179	_____	_____
21.	180–189	_____	_____
22.	190–199	_____	_____

A college professor asked her 30 students how many hours they worked each week.
Use her list of student responses to complete the following table. See Examples 1–3.

14	8	12	28	33	14
6	34	17	20	13	20
25	33	32	4	7	14
0	6	10	8	35	31
25	4	32	18	0	24

	Class Intervals (Number of Hours Worked)	Tally	Class Frequency (Number of Students)
23.	0–5	———	———
24.	6–10	———	———
25.	11–15	———	———
26.	16–20	———	———
27.	21–25	———	———
28.	26–30	———	———
29.	31–35	———	———

30. Construct a histogram by using the data in **Exercises 23–29.**

NUMBER OF HOURS WORKED BY STUDENTS EACH WEEK

Number of Students

Number of Hours Worked Each Week

The popular Los Angeles Lakers were a leading NBA (National Basketball Association) team in the 2011 season. Listed below are the number of points scored by the Lakers during their pre-season and regular season games. Use these numbers to complete the following table. **See Example 1.**

92	120	112	118	92	84	79	108	100	95	92	106	101	139	87
88	105	108	103	96	99	80	99	92	96	113	108	88	112	85
98	112	121	112	99	103	82	101	101	114	75	90	96	102	86
102	114	99	117	113	109	103	109	100	88	89	90	91	110	106
94	107	112	98	115	93	102	112	107	101	99	92	106	96	102
74	124	116	96	87	120	85	115	120	93	104	99	84	90	116

	Class Intervals **(Points Scored)**	**Tally**	**Class Frequency** **(Number of Games)**
31.	70–79	_____	_____
32.	80–89	_____	_____
33.	90–99	_____	_____
34.	100–109	_____	_____
35.	110–119	_____	_____
36.	120–129	_____	_____
37.	130–139	_____	_____

38. Make a histogram showing the results from **Exercises 31–37.**

HOOP IT UP

Number of Games (y-axis: 0, 5, 10, 15, 20, 25, 30)

Number of Points Scored (x-axis: 70–79, 80–89, 90–99, 100–109, 110–119, 120–129, 130–139)

Source: NBA.com

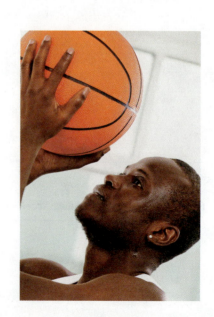

10.4 Mean, Median, and Mode

Businesses, governments, laboratories, colleges, and others working with lists of numbers are often faced with the problem of analyzing a great volume of raw data. The measures discussed in this section are helpful for such analyses.

OBJECTIVES

1. Find the mean of a list of numbers.
2. Find a weighted mean.
3. Find the median.
4. Find the mode.

OBJECTIVE ▶ **1 Find the mean of a list of numbers.** When analyzing data, one of the first things to look for is a *measure of central tendency*— a single number that we can use to represent the entire list of numbers. One such measure is the *average,* or **mean.** The mean can be found with the following formula.

VOCABULARY TIP

Mean The **mean** is the mathematical average.

Finding the Mean (Average)

$$\text{mean} = \frac{\text{sum of all values}}{\text{number of values}}$$

EXAMPLE 1 Finding the Mean

David had test scores of 84, 90, 95, 98, and 88. Find the mean (average) of his scores.

Use the formula for finding the mean. Add up all the test scores and then divide by the number of tests.

$$\text{mean} = \frac{84 + 90 + 95 + 98 + 88}{5} \quad \text{◁ Add up all the test scores.}$$

$$\text{mean} = \frac{455}{5} \quad \text{Divide.} \quad \text{◁ Divide by the number of tests.}$$

$$\text{mean} = 91$$

David has a mean (average) score of 91.

·················· **Work Problem 1 at the Side.** ▶

1 Andrew has math test scores of 96, 98, 84, 88, 82, and 92. Find his mean (average) score.

$$\frac{96 + 98 + 84 + 88 + \underline{\quad} + \underline{\quad}}{6}$$

$$= \frac{\underline{\quad}}{6} = \underline{\quad}$$

2 Find the mean for each list of numbers. Round to the nearest cent if necessary.

(a) Monthly Starbucks purchases:

$50.28, $85.16, $110.50, $78, $120.70, $58.64, $73.80, $86.24, $67.85, $96.56, $138.65, $48.90

EXAMPLE 2 Applying the Average or Mean

Milk sales at a local 7-Eleven for each of the days last week were

$$\$86, \$103, \$118, \$117, \$126, \$158, \text{ and } \$149$$

Find the mean milk sales (rounded to the nearest cent) as shown below.

$$\text{mean} = \frac{\$86 + \$103 + \$118 + \$117 + \$126 + \$158 + \$149}{7}$$

$$\text{mean} = \frac{\$857}{7}$$

$$\text{mean} \approx \$122.43 \text{ (rounded)}$$

The mean daily sales amount for milk was $122.43.

·················· **Work Problem 2 at the Side.** ▶

(b) The attendance at eight major league home games this season:

48,076; 58,595; 37,874; 46,289; 29,235; 59,311; 45,675; 39,721

Answers

1. 82; 92; 540; 90

2. **(a)** $\dfrac{\$1015.28}{12} \approx \$84.61$ (rounded)

   (b) $\dfrac{364{,}776}{8} = 45{,}597$ people

❸ The numbers below show the amount that high school student Matthew Ryan spent for lunch and snacks and the number of days he spent that amount. Find the weighted mean.

Value	Frequency
$ 2	4
$ 4	6
$ 6	5
$ 8	6
$10	12
$12	5
$14	8
$16	4

VOCABULARY TIP

Weighted mean is used when calculating grade point average. See **Example 4.**

OBJECTIVE ❷ Find a weighted mean. Some items in a list might appear more than once. In this case, we find a **weighted mean,** in which each value is "weighted" by multiplying it by the number of times it occurs.

EXAMPLE 3 | **Understanding the Weighted Mean**

The following table shows the family size and the number of families (frequency) who were given groceries in one morning at the Twin Lakes Food Bank. Find the weighted mean for family size.

Family Size	Frequency	
1	7	← 7 families had 1 person.
2	12	
3	8	
4	8	
5	5	
6	9	← 9 families had 6 people.
7	1	
8	6	

The same number of people were in more than one family: for example, there were 2 people in 12 of the families and there were 4 people in 8 of the families. There were 7 people in just 1 of the families. To find the mean, multiply the family size by its frequency. Then add the products. Next, add the numbers in the frequency column to find the total number of families.

Size	Frequency	Product
1	7	$(1 \cdot 7) = 7$
2	12	$(2 \cdot 12) = 24$
3	8	$(3 \cdot 8) = 24$
4	8	$(4 \cdot 8) = 32$
5	5	$(5 \cdot 5) = 25$
6	9	$(6 \cdot 9) = 54$
7	1	$(7 \cdot 1) = 7$
8	6	$(8 \cdot 6) = 48$
Totals	**56**	**221**

Finally, divide the totals. Round to the nearest hundredth.

$$\text{mean} = \frac{221}{56} \approx 3.95 \text{ (rounded)}$$

The mean family size of those using the food bank was 3.95 people.

◀ **Work Problem ❸ at the Side.**

Answer

3. mean $= \dfrac{\$466}{50} = \$9.32$

A common use of the weighted mean is to find a student's *grade point average* (GPA), as shown by the next example.

EXAMPLE 4 **Applying the Weighted Mean**

Find the grade point average for a student earning the following grades. Assume A = 4, B = 3, C = 2, D = 1, and F = 0. The number of credits determines how many times the grade is counted (the frequency).

Course	Credits	Grade	Credits • Grade
Mathematics	4	A (= 4)	4 • 4 = 16
Speech	3	C (= 2)	3 • 2 = 6
English	3	B (= 3)	3 • 3 = 9
Computer science	2	A (= 4)	2 • 4 = 8
Art history	2	D (= 1)	2 • 1 = 2
Totals	**14**		**41**

It is common to round grade point averages to the nearest hundredth. So the grade point average for this student is rounded to 2.93.

$$GPA = \frac{41}{14} \approx 2.93$$

•••••••••••••••••••••••••••• Work Problem **4** at the Side. ▶

OBJECTIVE **3** **Find the median.** Because it can be affected by extremely high or low numbers, the mean is often a poor indicator of central tendency for a list of numbers. In cases like this, another measure of central tendency, called the *median,* can be used. The **median** divides a group of numbers in half; half the numbers lie above the median, and half lie below the median.

Find the median by listing the numbers *in order* from *least* to *greatest.* If the list contains an *odd* number of items, the median is the *middle number.*

EXAMPLE 5 **Finding the Median for an Odd Number of Items**

Find the median for the following list of prices for women's T-shirts.

$9, $23, $15, $8, $18, $12, $24

First arrange the numbers in numerical order from least to greatest.

Least ⟶ 8, 9, 12, 15, 18, 23, 24 ⟵ Greatest

Next, find the middle number in the list.

8, 9, 12, **15**, 18, 23, 24

Three are below. | Three are above.

Middle number

> Remember to list the numbers from least to greatest **before** finding the median.

The median price is $15.

•••••••••••••••••••••••••••• Work Problem **5** at the Side. ▶

---

4 Find the grade point average (GPA) for Laticia Perez, who earned the following grades last semester. Round to the nearest hundredth.

Course	Credits	Grade
Mathematics	3	A (= 4)
P.E.	1	C (= 2)
English	3	C (= 2)
Keyboarding	2	B (= 3)
Biology	4	B (= 3)

VOCABULARY TIP

Median The word **median** actually means "middle." It may help to remember that the grassy area in the middle of two roadways is called a highway median, just as the middle number in a set of data points is called the numerical median.

---

5 Find the median for the following weights of Halloween pumpkins.

14 lb, 18 lb, 10 lb, 17 lb, 15 lb, 19 lb, 20 lb

First list the weights from least to greatest.

10, ___, ___, ___, ___, ___, 20

↑ least   ↑ greatest

Which weight is in the middle of the list? _____

---

6 Find the median for the following list of measurements.

125 m, 87 m, 96 m, 108 m, 136 m, 74 m

VOCABULARY TIP

Mode is the most common or that which appears most often.

7 Find the mode for each list of numbers.

GS **(a)** Number of laps swum:

20, 30, 26, 24, 26
   ↑   ↑

Which number occurs most often? _____

(b) Wait time for Disneyland rides (in minutes):

38, 45, 52, 60, 38, 45

(c) Monthly commissions of salespeople:

$1706, $1289, $1653, $1892, $1301, $1782

If a list contains an *even* number of items, there is no single middle number. In this case, the median is defined as the mean (average) of the **middle two** numbers.

EXAMPLE 6   **Finding the Median for an Even Number of Items**

Find the median for the following list of ages.

74, 7, 15, 13, 25, 28, 47, 59, 32, 68

First arrange the numbers in numerical order from least to greatest. Then find the middle two numbers.

Least $\longrightarrow$ 7, 13, 15, 25, **28, 32**, 47, 59, 68, 74 $\longleftarrow$ Greatest

Middle two numbers

The median age is the mean of the two middle numbers.

$$\text{median} = \frac{28 + 32}{2} = \frac{60}{2} = \textbf{30 years}$$

◄ **Work Problem 6** at the Side.

OBJECTIVE **4** **Find the mode.** Another important statistical measure is the **mode,** the number that occurs *most often* in a list of numbers. For example, if the test scores for 10 students were

To find the mode, you do **not** have to list the numbers from least to greatest.

74, 81, 39, **74**, 82, 80, 100, 92, **74**, and 85,

then the mode is 74. Three students earned a score of 74, so 74 appears more times on the list than any other score. (It is not necessary to place the numbers in numerical order when looking for the mode.)

A list can have two modes; such a list is sometimes called **bimodal.** If no number occurs more frequently than any other number, the list has *no mode*.

EXAMPLE 7   **Finding the Mode**

Find the mode for each list of numbers.

The **mode** is the value that occurs most often.

(a) 51, 32, 49, 73, 49, 90

The number **49** occurs more often than any other number; therefore, 49 is the mode.

(b) 482, 485, 483, 485, 487, 487, 489

Because both **485** and **487** occur twice, each is a mode. This list is *bimodal.*

(c) $10,708; $11,519; $10,972; $12,546; $13,905; $12,182

No number occurs more than once. This list has *no mode.*

◄ **Work Problem 7** at the Side.

Measures of Central Tendency

The **mean** is the sum of all the values divided by the number of values. It is the mathematical average.

The **median** is the middle number (or the average of the middle two numbers) in a group of values that are listed from least to greatest. It divides a group of numbers in half.

The **mode** is the value that occurs most often in a group of values.

Answers

6. $\frac{96 + 108}{2} = \frac{204}{2} = 102\,m$

7. **(a)** 26 laps
   (b) bimodal, 38 minutes and 45 minutes (this list has two modes)
   (c) no mode (no number occurs more than once)

10.4 Exercises

FOR EXTRA HELP   Download the MyDashBoard App   **MyMathLab**®

CONCEPT CHECK   *Label each statement as* true *or* false.

1. Another word for *mean* is *average*.

2. The mode divides a group of numbers in half; half of the numbers lie above the mode, and half below the mode.

CONCEPT CHECK   *Fill in each blank with the correct response.*

3. A common use of the weighted mean is to find a student's _____.

4. The mode is the number that occurs _____ in a list of numbers.

Find the mean for each list of numbers. Round answers to the nearest tenth if necessary.
See Example 1.

5. Hours spent playing video games each week:

   9, 14, 18, 11, 12, 20

6. Minutes of cell phone use each day:

   53, 77, 38, 29, 46, 48, 52

7. Inches of rain per month of 3.1, 1.5, 2.8, 0.8, 4.1

$$\frac{3.1 + 1.5 + 2.8 + 0.8 + 4.1}{\underline{}} =$$

8. Algebra quiz scores of 32, 26, 30, 19, 51, 46, 38, 39

$$\frac{32 + 26 + 30 + 19 + 51 + 46 + 38 + 39}{\underline{}} =$$

9. Annual salaries of $38,500; $39,720; $42,183; $21,982; $43,250

10. Numbers of students enrolled in community colleges: 27,500; 18,250; 17,357; 14,298; 33,110

Solve each application problem. See Example 2.

11. The Sunrise Pharmacy filled prescriptions that cost the following amounts: $18.38, $168.75, $28.63, $72.85, $39.60, $183.74, $15.82, $33.18, $87.45, $98.72, and $50.70. Find the average cost (mean) of the prescriptions sold.

12. In one evening, a waitress collected the following checks from her dinner customers: $30.10, $42.80, $91.60, $51.20, $88.30, $21.90, $43.70, $51.20. Find the average (mean) dinner check amount.

Find the weighted mean. Round answers to the nearest tenth. See Example 3.

13.

Customers Each Hour	Frequency	
8	2	(8 • 2) =
11	12	(11 • 12) =
15	5	(15 • 5) =
26	1	(26 • 1) = _____
		Total →

$$\frac{\text{sum of products}}{\text{total customers}} =$$

14.

Deliveries Each Week	Frequency	
4	1	(4 • 1) =
8	3	(8 • 3) =
16	5	(16 • 5) =
20	1	(20 • 1) = _____
		Total →

$$\frac{\text{sum of products}}{\text{total deliveries}} =$$

15.

Fish per Boat	Frequency
12	4
13	2
15	5
19	3
22	1
23	5

16.

Patients per Clinic	Frequency
25	1
26	2
29	5
30	4
32	3
33	5

*Solve each application problem. **See Example 3.***

17. On a popular television game show, contestants won various amounts of money. The table below shows how many contestants won each amount of money over a two-month period. Find the weighted mean for the amount of money won.

Game Show Winnings	Number of Contestants
$ 5000	5
$ 10,000	10
$ 25,000	7
$ 50,000	5
$ 100,000	12
$ 250,000	5
$ 500,000	5
$1,000,000	1

18. A national health survey provided the information for this table. It shows how often each of the adults in the survey engages in a vigorous, leisure time activity. Find the weighted mean to determine the weekly hours of vigorous activity for the adults surveyed, to the nearest tenth.

Hours per week	Number of Adults
0	224
1	86
2	62
3	45
4	25
5	18
6	24
7	16

*Find the median for each list of numbers. **See Examples 5 and 6.***

19. Number of wedding guests: 125, 100, 150, 135, 114

20. Number of Google searches (in thousands):

140, 85, 122, 114, 98

21. Calories in fast-food menu items:

501, 412, 521, 515, 298, 621, 346, 528

298, 346, 412, $\underbrace{501, 515}$, 521, 528, 621
       middle two numbers

$$\frac{501 + 515}{2} = \underline{\quad} \text{ calories}$$

22. Number of cars in the parking lot each day:

520, 523, 513, 1283, 338, 509, 290, 420, 320, 980

290, 320, 338, 420, $\underbrace{509, 513}$, 520, 523, 980, 1283
       middle two numbers

$$\frac{509 + 513}{2} = \underline{\quad} \text{ cars}$$

*Find the mode or modes for each list of numbers. **See Example 7.***

23. Porosity of soil samples:

21%, 18%, 21%, 28%, 22%, 21%, 25%

24. Daily low temperatures (in degrees Fahrenheit):

21, 32, 46, 32, 49, 32, 49

25. Ages of residents (in years) at Leisure Village:

74, 68, 68, 68, 75, 75, 74, 74, 70

26. Number of pages read:

86, 84, 79, 75, 88, 66, 72, 85, 71

27. When is the median a better measure of central tendency than the mean to describe a set of data? Make up a list of numbers to illustrate your explanation. Calculate both the mean and the median.

28. Suppose you own a hat shop and can order a certain hat in only one size. You look at last year's sales to decide on the size to order. Should you find the mean, median, or mode for these sales? Explain your answer.

Find the grade point average for students earning the following grades. Assume A = 4, B = 3, C = 2, D = 1, and F = 0. Round answers to the nearest hundredth. ***See Example 4.***

29.

Credits	Grade
4	B (= 3)
2	C (= 2)
2	A (= 4)
1	C (= 2)
3	D (= 1)

Credits • Grade

$$4 \cdot 3 = 12$$
$$2 \cdot 2 = 4$$
$$2 \cdot 4 = 8$$
$$1 \cdot 2 = 2$$
$$3 \cdot 1 = \underline{3}$$
$$29$$

$$\text{GPA} = \frac{29}{12} =$$

30.

Credits	Grade
1	C (= 2)
3	A (= 4)
4	B (= 3)
3	C (= 2)
2	A (= 4)

Credits • Grade

$$1 \cdot 2 = 2$$
$$3 \cdot 4 = 12$$
$$4 \cdot 3 = 12$$
$$3 \cdot 2 = 6$$
$$2 \cdot 4 = \underline{8}$$
$$40$$

$$\text{GPA} = \frac{40}{13} =$$

31.

Credits	Grade
4	B
2	A
5	C
1	F
3	B

32.

Credits	Grade
3	A
3	B
4	B
3	C
3	C

33.

Credits	Grade
2	A
3	C
4	A
1	C
4	B

34.

Credits	Grade
3	A
2	A
5	B
4	A
1	A

Relating Concepts (Exercises 35–44)   For Individual or Group Work

*Gluco Industries manufactures and sells glucose monitors and other diabetes-related
products. The number of sales calls made over an 8-week period by two sales
representatives, Scott Samuels and Rob Stricker, is shown below. Use this
information to* **work Exercises 35–44 in order.**

Week	Number of Calls	
	Samuels	Stricker
1	39	21
2	15	22
3	40	20
4	22	23
5	13	19
6	22	24
7	17	25
8	8	22

35. Find the total number of sales calls made by each of the sales representatives.

36. Find the mean number of sales calls made by Samuels and by Stricker.

37. Find the median number of sales calls made by Samuels and by Stricker.

38. Find the mode for the number of sales calls for each of the sales representatives.

39. How do the mean, median, and mode for the two sales representatives compare?

40. Describe how the pattern in the number of weekly sales calls made by Samuels differs from the pattern in Stricker's weekly sales calls.

To show the variation *or* spread *of the number of sales calls made by each of the sales
representatives requires some* **measure of the dispersion,** *or spread of the numbers
around the mean. A common measure of the dispersion is the* **range.** *The range is the*
difference between the greatest value and the least value in the set of numbers.

41. Find the range for the number of sales calls made by Samuels.

42. Find the range for the number of sales calls made by Stricker.

43. Are the performance data for these two sales representatives sufficient to accurately determine which is the best? What else might you want to know?

44. List three possible explanations for the wide variation (range) in the number of sales calls for Samuels.

Chapter 10 *Summary*

Key Terms

10.1

circle graph A circle graph shows how a total amount is divided into parts or sectors. It is based on percents of 360°.

protractor A protractor is a device (usually in the shape of a half-circle) used to measure the number of degrees in angles or parts of a circle.

10.2

bar graph A bar graph uses bars of various heights or lengths to show quantity or frequency.

double-bar graph A double-bar graph compares two sets of data by showing two sets of bars.

line graph A line graph uses dots connected by lines to show trends.

comparison line graph A comparison line graph shows how two sets of data relate to each other by showing a line graph for each set of data.

10.3

frequency distribution A frequency distribution is a table that includes a column showing each possible number in the data collected. The original data are then entered in another column using a tally mark for each corresponding value. The tally marks are counted and the totals are placed in a third column.

histogram A histogram is a bar graph in which the width of each bar represents a range of numbers (class interval) and the height represents the quantity or frequency of items that fall within the interval.

10.4

mean The mean is the sum of all the values divided by the number of values. It is often called the *average*.

weighted mean The weighted mean is a mean calculated so that each value is multiplied by its frequency.

median The median is the middle number in a group of values that are listed from least to greatest. It divides a group of values in half. If there are an even number of values, the median is the mean (average) of the two middle values.

mode The mode is the value that occurs most often in a group of values.

bimodal A list of numbers is bimodal when it has two modes. The two values occur the same number of times.

dispersion The dispersion is the variation, or spread, of the numbers around the mean.

range The range is a common measure of the dispersion of numbers. It is the difference between the greatest value and the least value in the set of numbers.

New Formula

Mean or average: $\text{mean} = \dfrac{\text{sum of all values}}{\text{number of values}}$

Test Your Word Power

See how well you have learned the vocabulary in this chapter.

1 A **circle graph**
A. uses bars of various heights to show quantity or frequency
B. shows how a total amount is divided into parts or sectors
C. uses bars of various widths to represent a range of numbers.

2 A **bar graph**
A. uses bars of various heights to show quantity or frequency
B. shows how a total amount is divided into parts or sectors
C. uses dots connected by lines to show trends.

3 A **histogram** is a graph in which
A. two sets of data are compared using two sets of bars
B. dots are connected by lines to show trends
C. the width of each bar represents a range of numbers and the height represents the frequency of items within that range.

4 A **protractor** is a device used to
A. construct a histogram
B. calculate measures of central tendency
C. measure the number of degrees in angles or parts of a circle.

5 The **mean** is
A. calculated so that each value is multiplied by its frequency
B. the sum of all values divided by the number of values
C. the middle number in a group of values that are listed from least to greatest.

6 The **mode** is
A. calculated so that each value is multiplied by its frequency
B. the middle number in a group of values that are listed from least to greatest
C. the value that occurs most often in a group of values.

Answers to Test Your Word Power

1. B; *Example:*

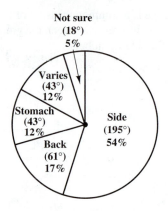

SLEEPING POSITIONS

Not sure (18°) 5%
Varies (43°) 12%
Stomach (43°) 12%
Back (61°) 17%
Side (195°) 54%

2. A; *Example:*

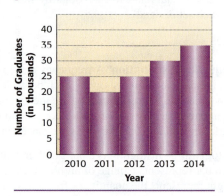

GRADUATION CLASS SIZE

3. C; *Example:*

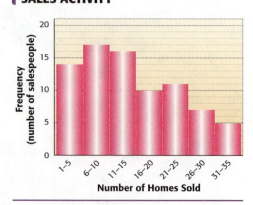

SALES ACTIVITY

4. C; *Example:*

5. B; *Example:* The mean of the values $5, $9, $7, $5, $2, and $8, is
$$\frac{\$5 + \$9 + \$7 + \$5 + \$2 + \$8}{6} = \frac{\$36}{6} = \$6.$$

6. C; *Example:* The mode of the values $5, $9, $7, $5, $2, and $8 is $5 because $5 appears twice in the list.

Quick Review

Concepts	Examples

10.1 Constructing a Circle Graph

Step 1 Determine the percent of the total for each item.

Step 2 Find the number of degrees out of 360° that each percent represents.

Step 3 Use a protractor to measure the number of degrees for each item in the circle.

Item	Amount	Percent of Total	Sector Size
Leather interior	$1600	$\frac{\$1600}{\$8000} = \frac{1}{5} = 20\%$ so 360° • 20% $= 360 • 0.20$	$= 72°$
Wheels/ tires	$2400	$\frac{\$2400}{\$8000} = \frac{3}{10} = 30\%$ so 360° • 30% $= 360 • 0.30$	$= 108°$
Sun roof	$1200	$\frac{\$1200}{\$8000} = \frac{3}{20} = 15\%$ so 360° • 15% $= 360 • 0.15$	$= 54°$
Sport package	$2800	$\frac{\$2800}{\$8000} = \frac{7}{20} = 35\%$ so 360° • 35% $= 360 • 0.35$	$= 126°$

Construct a circle graph from the following table, which lists the costs of options on a new luxury sports car.

Item	Amount
Leather interior	$1600
Wheels/tires	$2400
Sun roof	$1200
Sport package	$2800
Total	$8000

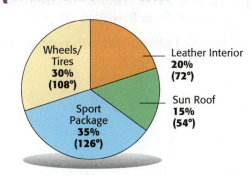

SPORTY CAR OPTION COSTS

Wheels/Tires 30% (108°)
Leather Interior 20% (72°)
Sun Roof 15% (54°)
Sport Package 35% (126°)

10.2 Reading a Bar Graph

The height of the bar is used to show the quantity or frequency (number) in a specific category. Use a ruler or straightedge to line up the top of each bar with the numbers on the left side of the graph.

Use the bar graph below to determine the number of students who earned each letter grade.

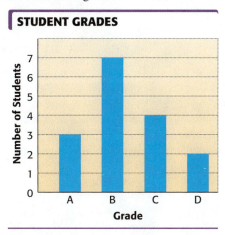

STUDENT GRADES

Grade of A: 3 students; B: 7 students; C: 4 students; D: 2 students.

Concepts	Examples

10.2 Reading a Line Graph

A dot is used to show the number or quantity in a specific class. The dots are connected with lines. This kind of graph is used to show a trend.

The line graph below shows the annual sales for the Fabric Supply Center for each of 4 years.

ANNUAL SALES

Find the sales in 2012.
The dot above 2012 lines up with $500 on the left edge. Then $500 • 1000 = $500,000 in sales.

10.3 Preparing a Frequency Distribution and a Histogram from Raw Data

Step 1   Construct a table listing each value, and the number of times this value occurs.

Draw a histogram for these student quiz scores.

12	15	15	14
13	20	10	12
11	9	10	12
17	20	16	17
14	18	19	13

Quiz Score	Tally	Frequency	
9	I	1	1st
10	II	2	class
11	I	1	interval
12	III	3	2nd
13	II	2	class
14	II	2	interval
15	II	2	3rd
16	I	1	class
17	II	2	interval
18	I	1	4th
19	I	1	class
20	II	2	interval

Concepts

Examples

10.3 **Preparing a Frequency Distribution and a Histogram from Raw Data** (*Continued*)

Step 2   Divide the data into groups, categories, or classes.

Class Interval (Quiz Scores)	Frequency (Number of Students)
9–11	4
12–14	7
15–17	5
18–20	4

Step 3   Draw bars representing these groups to make a histogram.

STUDENT QUIZ SCORES

10.4 **Finding the Mean (Average) of a Set of Numbers**

Step 1   Add all values to obtain a total.

Step 2   Divide the total by the number of values.

Use this formula:

$$\text{mean (average)} = \frac{\text{sum of all values}}{\text{number of values}}$$

The test scores for Kathy West in her algebra course were as follows:

80	92	92	94
76	88	84	93

Find West's mean (average) test score to the nearest tenth.

$$\text{mean} = \frac{80 + 92 + 92 + 94 + 76 + 88 + 84 + 93}{8}$$

$$= \frac{699}{8} \approx 87.4 \text{ (rounded)}$$

West's mean test score is approximately 87.4.

Concepts	Examples

10.4 Finding the Weighted Mean

Step 1 Multiply each value by its frequency.

Step 2 Add all the products from Step 1.

Step 3 Divide the sum in Step 2 by the total number of pieces of data.

This table shows the distribution of the number of school-age children in a survey of 30 families.

Number of School-Age Children	Frequency (Number of Families)
0	12
1	6
2	7
3	3
4	2
Total of 30 families	

Find the mean number of school-age children per family. Round to the nearest hundredth.

Value	Frequency	Product
0	12	$(0 \cdot 12) = 0$
1	6	$(1 \cdot 6) = 6$
2	7	$(2 \cdot 7) = 14$
3	3	$(3 \cdot 3) = 9$
4	2	$(4 \cdot 2) = 8$
Totals	**30**	**37**

$$\text{mean} = \frac{37}{30} \approx 1.23 \text{ (rounded)}$$

The mean number of school-age children per family is approximately 1.23.

10.4 Finding the Median of a Set of Numbers

Step 1 Arrange the data from least to greatest.

Step 2 Select the middle value, or, if there is an even number of values, find the average of the two middle values.

Find the median for Kathy West's test scores from the previous page.

The data arranged from least to greatest is:

76   80   84   88   92   92   93   94
     ↑   ⌣⌣⌣⌣⌣   ↑
   Least   Middle values   Greatest

The middle two values are 88 and 92. The average of these two values is

$$\frac{88 + 92}{2} = 90$$

West's median test score is 90.

10.4 Finding the Mode of a Set of Values

Find the value that appears most often in the list of values. If no value appears more than once, there is no mode. If two different values appear the same number of times, the list is bimodal.

Find the mode for West's test scores shown above.

The most frequently occurring score is 92 (it occurs twice). Therefore, the mode is 92.

Chapter 10 *Review Exercises*

10.1 *The number of girls participating in high school sports in the United States exceeds 3 million. The circle graph shows the number of high school girls' teams in the most popular sports.*

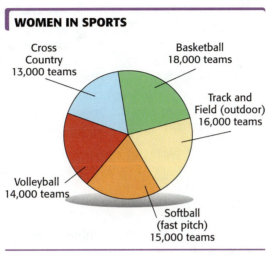

WOMEN IN SPORTS

Cross Country
13,000 teams

Basketball
18,000 teams

Track and Field (outdoor)
16,000 teams

Softball (fast pitch)
15,000 teams

Volleyball
14,000 teams

Source: National Federation of State High School Associations.

Use the circle graph to answer Exercises 1–6. Write ratios as fractions in lowest terms.

1. What women's sport has the greatest number of teams? How many are there?

2. Number of track and field teams to the total number of teams.

3. Number of softball teams to the total number of teams.

4. Number of volleyball teams to the total number of teams.

5. Number of basketball teams to the number of track and field teams.

6. Number of track and field teams to the number of volleyball teams.

10.2 *This bar graph shows the most frequently offered "work perks" and the percent of the responding companies offering them. The survey was conducted online and included 4800 companies ranging in size from 2 to 5000 employees. Use this graph to find the number of companies offering each work perk listed in Exercises 7–10 and to answer Exercises 11 and 12. (Source: Work Perks Survey, Ceridian Employer Services.)*

WORK PERKS

Percent of Companies

- 82% Casual dress
- 60.5% Flexible hours
- 49% Personal development training
- 40% Employee entertainment/Company product discounts
- 36% Free food/Beverages
- 27% Telecommuting
- 15% Fitness centers
- 1% On-site child care

Type of Perk

7. Casual dress

8. Free food/Beverages

9. Fitness centers

10. Flexible hours

11. Which two work perks do companies offer least often? Give one possible explanation why these work perks are not offered.

12. Which two work perks do companies offer most often? Give one possible explanation why these work perks are so popular.

This double-bar graph shows the number of acre-feet of water in Lake Natoma for each of the first six months of 2012 and 2013. Use this graph to answer Exercises 13–18.

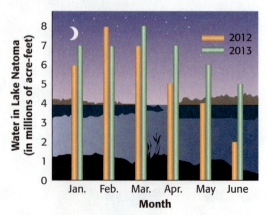

WATER IN LAKE NATOMA

Water in Lake Natoma (in millions of acre-feet)

2012
2013

Jan. Feb. Mar. Apr. May June

Month

13. During which month in 2013 was the greatest amount of water in the lake? How much was there?

14. During which month in 2012 was the least amount of water in the lake? How much was there?

15. How many acre-feet of water were in the lake in June of 2013?

16. How many acre-feet of water were in the lake in May of 2012?

17. Find the decrease in the amount of water in the lake from March 2012 to June 2012.

18. Find the decrease in the amount of water in the lake from April 2013 to June 2013.

This comparison line graph shows the annual floor-covering sales of two different home improvement centers during each of 5 years. Use this graph to find the amount of annual floor-covering sales in each year shown in Exercises 19–22 and to answer Exercises 23 and 24.

ANNUAL FLOOR-COVERING SALES

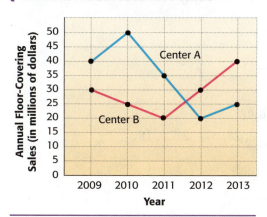

19. Center A in 2010

20. Center A in 2012

21. Center B in 2011

22. Center B in 2013

23. What trend do you see in center A's sales from 2010 to 2013? Why might this have happened?

24. What trend do you see in center B's sales starting in 2011? Why might this have happened?

10.4 *Find the mean for each list of numbers. Round answers to the nearest tenth if necessary.*

25. Digital cameras sold:

18, 12, 15, 24, 9, 42, 54, 87, 21, 3

26. Number of harassment complaints filed:

31, 9, 8, 22, 46, 51, 48, 42, 53, 42

Find the weighted mean for each list. Round to the nearest hundredth if necessary.

27.

Dollar Value	Frequency
$42	3
$47	7
$53	2
$55	3
$59	5

28.

Total Points	Frequency
243	1
247	3
251	5
255	7
263	4
271	2
279	2

Find the median for each list of numbers.

29. The number of accident forms filed:

43, 37, 13, 68, 54, 75, 28, 35, 39

30. Commissions of:

$576, $578, $542, $151, $559, $565, $525, $590

Find the mode or modes for each list of numbers.

31. Running shoes priced at:

$79, $56, $110, $79, $72, $86, $79

32. Boat launchings:

18, 25, 63, 32, 28, 37, 32, 26, 18

Mixed Review Exercises

In her senior year of college Sue Hogan had expenses of $17,920. This amount was spent as shown below. In order to make a circle graph, find all the missing numbers in Exercises 33–37.

Item	Dollar Amount	Percent of Total	Degrees of Circle
33. Books and supplies	$1792	10%	_____
34. Rent	$6272	_____	126°
35. Food	$3584	_____	_____
36. Tuition/fees	$4480	_____	_____
37. Miscellaneous	$1792	_____	_____

38. Draw a circle graph using the information in **Exercises 33–37.**

Find the mean for each list of numbers. Round answers to the nearest tenth if necessary.

39. Number of volunteers for the project:

48, 72, 52, 148, 180

40. Number of flu vaccinations in a day:

122, 135, 146, 159, 128, 147, 168, 139, 158

Find the mode or modes for each list of numbers.

41. Job applicants meeting the qualifications:

48, 43, 46, 47, 48, 48, 43

42. Number of two-bedroom apartments in each building:

26, 31, 31, 37, 43, 51, 31, 43, 43

Find the median for each list of numbers.

43. Hours worked:

4.7, 3.2, 2.9, 5.3, 7.1, 8.2, 9.4, 1.0

44. Number of e-mails each day:

35, 51, 9, 2, 17, 12, 46, 23, 3, 19, 39, 27

Here are the scores of 40 students on a mathematics exam. Complete the table.

78	89	36	59	78	99	92	86
73	78	85	57	99	95	82	76
63	93	53	76	92	79	72	62
74	81	77	76	59	84	76	94
58	37	76	54	80	30	45	38

	Class Intervals (Scores)	Tally	Class Frequency (Number of Students)
45.	30–39	_____	_____
46.	40–49	_____	_____
47.	50–59	_____	_____
48.	60–69	_____	_____
49.	70–79	_____	_____
50.	80–89	_____	_____
51.	90–99	_____	_____

52. Construct a histogram by using the data in **Exercises 45–51.**

Find each weighted mean. Round answers to the nearest tenth if necessary.

53.

Test Score	Frequency
46	4
54	10
62	8
70	12
78	10

54.

Units Sold	Frequency
104	6
112	14
115	21
119	13
123	22
127	6
132	9

Chapter 10 Test CHAPTER **Test Prep** VIDEO

The Chapter Test Prep Videos with test solutions are available on DVD, in MyMathLab, *and on* You Tube—*search "LialBasicCollegeMath" and click on "Channels."*

The circle graph shows the sources of electricity generated in the United States. If the total cost of all electricity generated in one year was $328 billion, find the dollar amount spent on electricity generated by each of these sources. Round to the nearest tenth of a billion.

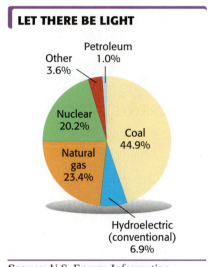

LET THERE BE LIGHT

Source: U.S. Energy Information Administration.

1. Nuclear

2. Hydroelectric (conventional)

3. Petroleum

4. Natural gas

5. Coal

6. Other

During a one-year period, Big 5 Sporting Goods had the following sales in each department. Find all numbers missing from the table.

Item	Dollar Amount	Percent of Total	Degrees of a Circle
7. Team sports	$432,000	30%	_____
8. Golf	$144,000	10%	_____
9. Hunting and fishing	$288,000	20%	_____
10. Athletic shoes	$504,000	35%	_____
11. Water sports	$72,000	_____	18°

12. Draw a circle graph using the information in **Exercises 7–11** and a protractor. Label each sector of the graph.

Here are the profits for each of the past 20 weeks from Alan's Snack Bar vending machines. Complete the table.

$142 $137 $125 $132 $147 $129 $151 $172 $175 $129
$159 $148 $173 $160 $152 $174 $169 $163 $149 $173

Profit	Tally	Number of Weeks
13. $120–129	_____	_____
14. $130–139	_____	_____
15. $140–149	_____	_____
16. $150–159	_____	_____
17. $160–169	_____	_____
18. $170–179	_____	_____

19. Use the information in **Exercises 13–18** to draw a histogram.

Find the mean for each list of numbers. Round answers to the nearest tenth if necessary.

20. Number of miles run each week while training:

52, 61, 68, 69, 73, 75, 79, 84, 91, 98

21. Weight in pounds for the largest bass caught in the lake:

11, 14, 12, 14, 20, 16, 17, 18

22. Airplane speeds in miles per hour:

458, 432, 496, 491, 500, 508, 512, 396, 492, 504

23. Explain why a weighted mean is used to find a student's grade point average. Calculate your own grade point average for last semester or quarter. If you are a new student, make up a grade point average problem of your own and solve it. Round to the nearest hundredth.

24. Explain in your own words the procedure for finding the median when there are an odd number of values in a list. Make up a problem with a list of five numbers and solve for the median.

Find the weighted mean for the following. Round answers to the nearest whole number if necessary.

25.

Cost	Frequency
$12	5
$20	6
$22	8
$28	4
$38	6
$48	2

26.

Points Scored	Frequency
150	15
160	17
170	21
180	28
190	19
200	7

Find the median for each list of numbers.

27. Highest daily temperatures in Alaska in degrees Fahrenheit:

32, 41, 28, 28, 37, 35, 16, 31

28. The lengths of steel beams in meters:

7.6, 11.4, 6.2, 12.5, 31.7, 22.8, 9.1, 10.0, 9.5

Find the mode or modes for each list of numbers.

29. Blood sample amounts in milliliters:

72, 46, 52, 37, 28, 18, 52, 61

30. Hot tub temperatures in degrees Fahrenheit:

96, 104, 103, 104, 103, 104, 91, 74, 103

Chapters 1–10 Cumulative Review Exercises

First use front end rounding to round each number and estimate the answer. Then find the exact answer.

1. *Estimate:*   *Exact:*

$$\begin{array}{r} 875.62 \\ -\ 63.757 \\ \hline \end{array}$$

2. *Estimate:*   *Exact:*

$$\begin{array}{r} 7064 \\ \times\ \ \ \ 635 \\ \hline \end{array}$$

3. *Estimate:*   *Exact:*

$$4.25\overline{)62.56}$$

Simplify. Write answers in lowest terms and as whole or mixed numbers when possible.

4. $4\frac{3}{5} + 5\frac{2}{3}$

5. $6\frac{2}{3} - 4\frac{3}{4}$

6. $\left(9\frac{3}{5}\right)\left(4\frac{5}{8}\right)$

7. $22\left(\frac{2}{5}\right)$

8. $3\frac{1}{3} \div 8\frac{3}{4}$

9. $\frac{2}{3}\left(\frac{7}{8} - \frac{3}{4}\right)$

10. $4 + 10 \div 2 + 7(2)$

11. $\sqrt{81} - 4(2) + 9$

12. $2^2 \cdot 3^3$

Write in order, from least to greatest.

13. $0.218, 0.22, 0.199, 0.207, 0.2215$

14. $0.6319, \frac{5}{8}, 0.608, \frac{13}{20}, 0.58$

Write each ratio in lowest terms. Be sure to make all necessary conversions.

15. $5\frac{1}{2}$ in. to 44 in.

16. 3 hr to 45 min

Find the unknown number in each proportion.

17. $\frac{1}{5} = \frac{x}{30}$

18. $\frac{15}{x} = \frac{390}{156}$

19. $\frac{200}{135} = \frac{24}{x}$

20. $\frac{x}{208} = \frac{6.5}{26}$

Solve each percent problem.

21. Find 5.4% of 6000 homes.

22. $8\frac{1}{2}$% of what number of people is 238 people?

23. What percent of $555 is $1443?

Convert each measurement using unit fractions or the metric conversion line.

24. 28 qt = _____ gal

25. 400 mm to cm

26. 230 g to kg

27. _____ lb = 6 tons

Name each shape and find its area. Use 3.14 as the approximate value of π. Round answers to the nearest tenth.

28.

8.45 m

5.6 m   5.6 m

8.45 m

29.

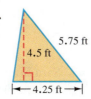

5.75 ft

4.5 ft

4.25 ft

30.

13 cm

In Exercises 31 and 32, name each solid and find its volume. Use 3.14 as the approximate value for π. In Exercise 33, find the unknown length. Round answers to the nearest tenth if necessary.

31.
4.8 cm
7.6 cm

32.
7 m
3 m
9.5 m

33.
30 cm
16 cm
?

Add, subtract, multiply, or divide as indicated.

34. $-12 + (-10)$

35. $-5.7 - (-12.6)$

36. $(-14.6)(-5.7)$

37. $\dfrac{-34.04}{14.8}$

Solve each equation. Show your work.

38. $3x - 5 = 16$

39. $-12 = 3(x + 2)$

40. $3.4x + 6 = 1.4x - 8$

Find the mean, the median, and the mode for each list of numbers. Round to the nearest tenth if necessary.

41. Cable hookups per installer:

16, 37, 27, 31, 19, 25, 15, 38, 43, 19

42. Number of acres plowed each hour:

10.3, 4.3, 1.65, 2.85, 5.3, 5.7, 2.3, 4.35, 2.85

Solve each application problem.

43. In a study of 2082 workers it was found that only 874 of them used all of their paid time-off. What percent used all of their time-off? Round to the nearest tenth of a percent. (*Source:* Hudson Time-Off Survey.)

44. The average increase in residential winter heating bills will be 9.8%. (*Source:* Energy Information Administration.) If the increase amounts to $87.20, find

(a) the average heating bill before the increase,

(b) the average heating bill after the increase.

45. Breathe Right® nasal strips are sold in four package sizes: 12 nasal strips $6.50; 24 nasal strips $7.50; 30 nasal strips $8.95; 38 nasal strips $9.95. You have a $2-off coupon for the 12-strip size and a $1-off coupon for the 30-strip size. Which choice is the best buy?

46. The sketch below shows the plans for a lobby in a large commercial complex. What is the cost of carpeting the lobby, excluding the atrium, if the contractor charges $43.50 per square yard? Use 3.14 for π.

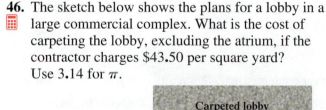
Carpeted lobby
Atrium
5 yd
32 yd
45 yd

47. The Spa Service Center services 140 spas. If each spa needs 125 mL of muriatic acid, how many liters of acid are needed for all the spas?

48. Sara Simon has $11,400 in an 8-month CD (certificate of deposit) account earning $1\frac{3}{4}$% interest. Find the total amount in her account at maturity.

Appendix: Inductive and Deductive Reasoning

Inductive and Deductive Reasoning

OBJECTIVE 1 Use inductive reasoning to analyze patterns. In many scientific experiments, conclusions are drawn from specific outcomes. After many repetitions and similar outcomes, the findings are generalized into statements that appear to be true. When general conclusions are drawn from specific observations, we are using a type of reasoning called **inductive reasoning.** The next examples illustrate this type of reasoning.

EXAMPLE 1 Using Inductive Reasoning

Find the next number in the sequence 3, 7, 11, 15, ...

To discover a pattern, calculate the difference between each pair of successive numbers.

$$7 - 3 = 4$$
$$11 - 7 = 4$$
$$15 - 11 = 4$$

Notice that the difference is always 4. Each number is 4 greater than the previous one. Thus, the next number in the pattern is $15 + 4$, or 19.

···· **Work Problem 1 at the Side.** ▶

EXAMPLE 2 Using Inductive Reasoning

Find the next number in this sequence.

$$7, 11, 8, 12, 9, 13, \ldots$$

The pattern in this example involves addition and subtraction.

$$7 + 4 = 11$$
$$11 - 3 = 8$$
$$8 + 4 = 12$$
$$12 - 3 = 9$$
$$9 + 4 = 13$$

To get the second number, we add 4 to the first number. To get the third number, we subtract 3 from the second number. To obtain subsequent numbers, we continue the pattern. The next number is $13 - 3 = 10$.

···· **Work Problem 2 at the Side.** ▶

OBJECTIVES

1 Use inductive reasoning to analyze patterns.
2 Use deductive reasoning to analyze arguments.
3 Use deductive reasoning to solve problems.

1 Find the next number in the sequence 2, 8, 14, 20, ... Describe the pattern.

2 Find the next number in the sequence 6, 11, 7, 12, 8, 13, ... Describe the pattern.

Answers

1. 26; add 6 each time.
2. 9; add 5, subtract 4.

A-1

3 Find the next number in the sequence 2, 6, 18, 54, ... Describe the pattern.

EXAMPLE 3 **Using Inductive Reasoning**

Find the next number in the sequence 1, 2, 4, 8, 16, ...

Each number after the first is obtained by multiplying the previous number by 2. So the next number would be 16 • **2** = 32.

◀ **Work Problem 3 at the Side.**

EXAMPLE 4 **Using Inductive Reasoning**

(a) Find the next geometric shape in this sequence.

The figures alternate between a blue circle and a red triangle. Also, the number of dots increases by 1 in each subsequent figure. Thus, the next figure should be a blue circle with five dots inside it.

(b) Find the next geometric shape in this sequence.

The first two shapes consist of vertical lines with horizontal lines at the bottom extending first *left* and then *right*. The third shape is a vertical line with a horizontal line at the top extending to the *left*. Therefore, the next shape should be a vertical line with a horizontal line at the top extending to the *right*.

◀ **Work Problem 4 at the Side.**

---

4 Find the next shape in this sequence.

OBJECTIVE ▶ 2 Use deductive reasoning to analyze arguments. In the previous discussion, specific cases were used to find patterns and predict the next event. There is another type of reasoning called **deductive reasoning,** which moves from general cases to specific conclusions.

EXAMPLE 5 **Using Deductive Reasoning**

Does the conclusion follow from the premises in this argument?

All Hondas are automobiles. ← Premise

All automobiles have horns. ← Premise

∴   All Hondas have horns. ← Conclusion

In this example, the first two statements are called *premises* and the third statement (below the line) is called a *conclusion*. The symbol ∴ is a mathematical symbol meaning "**therefore.**" The entire set of statements is called an *argument*.

········ **Continued on Next Page**

Answers

3. 162; multiply by 3.

4.

The focus of deductive reasoning is to determine whether the conclusion follows (is valid) from the premises. A set of circles called **Euler circles** is used to analyze the argument.

In **Example 5,** the statement "All Hondas are automobiles" can be represented by two circles, one for Hondas and one for automobiles. Note that the circle representing Hondas is totally inside the circle representing automobiles because the first premise states that *all* Hondas are automobiles.

Now, a circle is added to represent the second statement, vehicles with horns. This circle must completely surround the circle representing automobiles because the second premise states that *all* automobiles have horns.

To analyze the conclusion, notice that the circle representing Hondas is *completely* inside the circle representing vehicles with horns. Therefore, it must follow that all Hondas have horns. ***The conclusion is valid.***

···· **Work Problem ⑤ at the Side.** ▶

EXAMPLE 6   **Using Deductive Reasoning**

Does the conclusion follow from the premises in this argument?

> All tables are round.
> All glasses are round.
> ∴ All glasses are tables.

Use Euler circles. Draw a circle representing tables *inside* a circle representing round objects, because the first premise states that *all* tables are round.

The second statement requires that a circle representing glasses must now be drawn inside the circle representing round objects, but not necessarily inside the circle representing tables. Therefore, the conclusion does ***not*** follow from the premises. This means that ***the conclusion is invalid.***

···· **Work Problem ⑥ at the Side.** ▶

---

⑤ Does the conclusion follow from the premises in the following argument?

> All cars have four wheels.
> All Fords are cars.
> ∴ All Fords have four wheels.

⑥ Does each conclusion follow from the premises?

(a) All animals are wild.
  All cats are animals.
  ∴ All cats are wild.

(b) All students use math.
  All adults use math.
  ∴ All adults are students.

Answers

5. The conclusion follows from the premises; it is valid.

6. (a) The conclusion follows from the premises; it is valid.

  (b) The conclusion does *not* follow from the premises; it is invalid.

7 In a college class of 100 students, 35 take both math and history, 50 take history, and 40 take math. How many take neither math nor history? Draw a Venn diagram.

8 A Chevy, BMW, Cadillac, and Ford are parked side by side. The known facts are:

(a) The Ford is on the right end.

(b) The BMW is next to the Cadillac.

(c) The Chevy is between the Ford and the Cadillac.

Which car is parked on the left end?

OBJECTIVE ▶ 3 Use deductive reasoning to solve problems. Another type of deductive reasoning problem occurs when a set of facts is given in a problem and a conclusion must be drawn using these facts.

EXAMPLE 7 **Using Deductive Reasoning**

There were 25 students enrolled in a ceramics class. During the class, 10 of the students made a bowl and 8 students made a birdbath. Three students made both a bowl and a birdbath. How many students did not make either a bowl or a birdbath?

This type of problem is best solved by organizing the data using a drawing called a **Venn diagram.** Two overlapping circles are drawn, with each circle representing one item made by the students, as shown below.

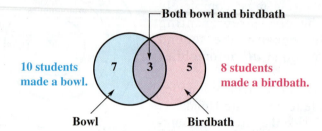

In the region where the circles overlap, write the number of students who made *both* items, namely, 3. In the remaining portion of the birdbath circle, write the number 5, which when added to 3 will give the total number of students who made a birdbath, namely, 8. In a similar manner, write 7 in the remaining portion of the bowl circle, since $7 + 3 = 10$, the total number of students who made a bowl. The total of all three numbers written in the circles is 15. Since there were 25 students in the class, this means $25 - 15$ or 10 students did not make either a birdbath or a bowl.

◀ **Work Problem 7 at the Side.**

EXAMPLE 8 **Using Deductive Reasoning**

Four cars in a race finish first, second, third, and fourth. The following facts are known.

(a) Car A beat Car C.

(b) Car D finished between Cars C and B.

(c) Car C beat Car B.

In which order did the cars finish?

To solve this type of problem, it is helpful to use a line diagram.

1. *Write A before C,* because Car A beat Car C (fact **a**).

$$A \quad C$$

2. *Write B after C,* because Car C beat Car B (fact **c**).

$$A \quad C \quad B$$

3. *Write D between C and B,* because Car D finished between Car C and Car B (fact **b**).

The correct order of finish is shown below.

$$A \quad C \quad D \quad B$$

◀ **Work Problem 8 at the Side.**

Answers

7.

$$15 + 35 + 5 = 55$$
$$100 - 55 = 45$$

45 students take neither math nor history.

8. BMW

Appendix Exercises

  MyMathLab®

Download the
MyDashBoard App

Find the next number in each sequence. Describe the pattern in each sequence.
See Examples 1–3.

1. 2, 9, 16, 23, 30, …

2. 5, 8, 11, 14, 17, …

3. 0, 10, 8, 18, 16, …

4. 3, 9, 7, 13, 11, …

5. 1, 2, 4, 8, …

6. 1, 4, 16, 64, …

7. 1, 3, 9, 27, 81, …

8. 3, 6, 12, 24, 48, …

9. 1, 4, 9, 16, 25, …

10. 6, 7, 9, 12, 16, …

Find the next shape in each sequence. **See Example 4.**

11.

12.

13.

14.

For each argument, draw Euler circles and then state whether or not the conclusion follows from the premises. ***See Examples 5 and 6.***

15. All animals are wild.
All lions are animals.
∴ All lions are wild.

16. All students are hard workers.
All business majors are students.
∴ All business majors are hard workers.

17. All teachers are serious.
All mathematicians are serious.
∴ All mathematicians are teachers.

18. All boys ride bikes.
All Americans ride bikes.
∴ All Americans are boys.

Solve each application problem. ***See Examples 7 and 8.***

19. In a given 30-day period, a husband watched television 20 days and his wife watched television 25 days. If they watched television together 18 days, how many days did neither watch television? Draw a Venn diagram.

20. In a class of 40 students, 21 students take both calculus and physics. If 30 students take calculus and 25 students take physics, how many do not take either calculus or physics? Draw a Venn diagram.

21. Tom, Dick, Mary, and Joan all work for the same company. One is a secretary, one is a computer operator, one is a receptionist, and one is a mail clerk.

   (a) Tom and Joan eat dinner with the computer operator.

   (b) Dick and Mary carpool with the secretary.

   (c) Mary works on the same floor as the computer operator and the mail clerk.

   Who is the computer operator?

22. Four cars—a Ford, a Buick, a Mercedes, and an Audi—are parked in a garage in four spaces.

   (a) The Ford is in the last space.

   (b) The Buick and Mercedes are next to each other.

   (c) The Audi is next to the Ford but not next to the Buick.

   Which car is in the first space?

Answers to Selected Exercises

In this section we provide the answers that we think most students will obtain when they work the exercises using the methods explained in the text. If your answer does not look exactly like the one given here, it is not necessarily wrong. In many cases there are equivalent forms of the answer that are correct. For example, if the answer section shows $\frac{3}{4}$ and your answer is 0.75 you have obtained the right answer but written it in a different (yet equivalent) form. Unless the directions specify otherwise, 0.75 is just as valid an answer as $\frac{3}{4}$.

In general, if your answer does not agree with the one given in the text, see whether it can be transformed into the other form. If it can, then it is the correct answer. If you still have doubts, talk with your instructor.

CHAPTER 1 Whole Numbers

SECTION 1.1 (pages 8–9)

1. c **2.** b **3.** 1; 0 **5.** 8; 2 **7.** 687 **8.** 321 **9.** 3; 561; 435
11. Evidence suggests that this is true. It is common to count using fingers. **12.** No doubt there is a relationship here. One answer might be that people could count using their fingers and toes and, therefore, thought of them as numbers or digits. **13.** false (no "and") **14.** true
15. three hundred forty-six thousand, nine **17.** twenty-five million, seven hundred fifty-six thousand, six hundred sixty-five **19.** 63,163
21. 10,000,223 **23.** 3,200,000 parachute jumps **25.** 50,051,507 cans
27. 54,750,000 Hot Wheels **29.** 800,000,621,020,215 **31.** public transportation; six million, sixty-nine thousand, five hundred eighty-nine workers **33.** seven million, eight hundred ninety-four thousand, nine hundred eleven workers

SECTION 1.2 (pages 18–21)

1. 97 **3.** 89 **5.** 889 **7.** 889 **9.** 7785 **11.** correct **12.** incorrect
13. incorrect **14.** correct **15.** 78,446 **17.** 8928 **19.** 59,224
21. correct **22.** incorrect **23.** correct **24.** incorrect **25.** incorrect
27. 145 **29.** 102 **31.** 1651 **33.** 1154 **35.** 413 **37.** 1771
39. 1410 **41.** 6391 **43.** 11,624 **45.** 17,611 **47.** 15,954 **49.** 10,648
51. 15,594 **53.** 11,557 **55.** 12,078 **57.** 4250 **59.** 12,268 **61.** correct
63. incorrect; should be 769 **65.** correct **67.** incorrect; should be 11,577 **69.** correct **71.** Changing the order in which numbers are added does not change the sum. You can add from bottom to top when checking addition. **72.** Grouping the addition of numbers in any order does not change the sum. You can add numbers in any order. For example, you can add pairs of numbers that add to 10. **73.** 33 miles **75.** 38 miles **77.** $16,342 **79.** 699 people **81.** 20,157 students
83. 970 ft **85.** 72 ft **87.** 9421 **88.** 1249 **89.** 77,762 **90.** 22,267
91. 9,994,433 **92.** 3,334,499 **93.** Write the largest digits on the left, using the smaller digits as you move right. **94.** Write the smallest digits on the left, using the larger digits as you move right.

SECTION 1.3 (pages 28–31)

1. 32 **2.** 4 **3.** 86 **4.** 35 **5.** 17 **7.** 213 **9.** 101 **11.** 7111
13. 3412 **15.** 2111 **17.** 13,160 **19.** 41,110 **21.** correct

23. incorrect; should be 62 **25.** incorrect; should be 121 **27.** correct
29. incorrect; should be 7222 **31.** minuend; subtrahend; below **32.** (b) and (d) **33.** 38 **35.** 45 **37.** 19 **39.** 281 **41.** 519 **43.** 7059
45. 7589 **47.** 7 **49.** 19 **51.** 2833 **53.** 7775 **55.** 503 **57.** 156
59. 2184 **61.** 5687 **63.** 19,038 **65.** 31,556 **67.** 6584 **68.** check
69. subtraction **70.** addition **71.** correct **73.** correct **75.** correct
77. correct **79.** Possible answers are 1. $3 + 2 = 5$ could be changed to $5 - 2 = 3$ or $5 - 3 = 2$ 2. $6 - 4 = 2$ could be changed to $2 + 4 = 6$ or $4 + 2 = 6$. **80.** No, you cannot. Numbers must be subtracted in the order given. The difference found in subtraction is the result of subtracting the subtrahend from the minuend. Changing the order of the minuend and subtrahend does change the answer. **81.** 47 calories **83.** 69 more tornadoes **85.** 246 feet **87.** 3270 jobs eliminated **89.** 9539 flags
91. $263 **93.** 141 miles **95.** 138,888 patients **97.** 310 fewer calories; 26 fewer fat grams **99.** 370 calories; 10 fat grams

SECTION 1.4 (pages 38–41)

1. the same **2.** commutative **3.** zero **4.** zeros; right **5.** 24 **7.** 0
9. 24 **11.** 40 **13.** Factors may be multiplied in any order to get the same answer. They are the same; you may add or multiply numbers in any order. **14.** You may shift the parentheses in a multiplication problem. Just as in addition, the different grouping results in the same answer. **15.** 210
17. 238 **19.** 3210 **21.** 1872 **23.** 8612 **25.** 10,084 **27.** 20,488
29. 258,447 **31.** 86; 172; two; 17,200 **32.** 45; 3510; three; 3,510,000
33. 480 **35.** 2220 **37.** 3600 **39.** 3750 **41.** 65,400 **43.** 270,000
45. 86,000,000 **47.** 48,500 **49.** 720,000 **51.** 1,940,000 **53.** 476
55. 2400 **57.** 3735 **59.** 2378 **61.** 6164 **63.** 15,792 **65.** 21,665
67. 15,730 **69.** 82,320 **71.** 183,996 **73.** 2,468,928 **75.** 66,005
77. 86,028 **79.** 19,422,180 **81.** 2,278,410 **83.** To multiply by 10, 100, or 1000, just attach one, two, or three zeros, respectively, to the number you are multiplying and that's your answer.

84.
$$
\begin{array}{r}
291 \\
\times\ 307 \\
\hline
2037 \\
000 \\
873 \\
\hline
89,337
\end{array}
\qquad
\begin{array}{r}
291 \\
\times\ 307 \\
\hline
2037 \\
8730 \\
\hline
89,337
\end{array}
$$
85. 3000 balls **87.** $2328 **89.** 24,090 gallons

91. $600 **93.** $1560 **95.** $112,888 **97.** 50,568 **99.** 38,250 trees
101. 7,746,712 people **103.** $14,160 **105.** (a) 452 (b) 452
106. commutative **107.** (a) 281 (b) 281 **108.** associative
109. (a) 15,840 (b) 15,840 **110.** commutative
111. (a) 6552 (b) 6552 **112.** associative **113.** No. Some examples are: $7 - 5 = 2$, but $5 - 7$ does not equal 2; $12 - 6 = 6$, but $6 - 12$ does not equal 6; $(8 - 2) - 5 = 1$, but $8 - (2 - 5)$ does not equal 1. **114.** No. Some examples are: $10 \div 2 = 5$, but $2 \div 10$ does not equal 5; $(16 \div 8) \div 2 = 1$, but $16 \div (8 \div 2)$ does not equal 1.

SECTION 1.5 (pages 51–54)

1. $\cdot$; $\times$; ( )( ) **2.** $\div$; $\overline{\smash{)}}$; —— (fraction bar) **3.** $4\overline{)24}$ $\frac{24}{4} = 6$
5. $9\overline{)45}^{5}$ $45 \div 9 = 5$ **7.** $16 \div 2 = 8$ $\frac{16}{2} = 8$ **9.** number **10.** 0

A-7

11. 1 **13.** 7 **15.** undefined **17.** 24 **19.** 0 **21.** undefined
23. 2; 3; 5; 10 **24.** 3; 5 **25.** 2; 3 **26.** 2; 3; 5; 10 **27.** 25 **29.** 18
31. 304 **33.** 627 R1 **35.** 1522 R5 **37.** 309 **39.** 3005 **41.** 5006
43. 811 R1 **45.** 2589 R2 **47.** 2630 **49.** 12,458 R3 **51.** 10,253 R5
53. 18,377 R6 **55.** correct **57.** incorrect; should be 1908 R1
59. incorrect; should be 670 R2 **61.** incorrect; should be 3568 R1
63. correct **65.** correct **67.** incorrect; should be 9628 R3 **69.** correct
71. Multiply the quotient by the divisor and add any remainder. The
result should be the dividend. **72.** Three choices might be: A number is
divisible by 2 if it ends in a 0, 2, 4, 6, or 8; A number is divisible by 5
if it ends in 0 or 5; A number is divisible by 10 if it ends in 0.
73. 328 tables **75.** 9600 each hour **77.** $48,500 **79.** 165 locations
81. $1,137,500 **83.** 862,500 berries **85.** ✓ ✓ ✓ ✓ **87.** ✓ X X X
89. X X ✓ X **91.** X ✓ X X **93.** ✓ ✓ X X **95.** X X X X

SECTION 1.6 (pages 60–62)

1. 53 **3.** 250 **5.** 120 R7 **7.** 1105 R5 **9.** 7134 R12 **11.** 900 R100
13. 73 R5 **15.** 476 R15 **17.** 2407 R1 **19.** 1146 R15 **21.** 3331 R82
23. 850 **25.** incorrect; should be 101 R14 **27.** incorrect; should be 658
29. incorrect; should be 62 **31.** 117 episodes **33.** 56 floor clocks
35. $355 **37.** 43,200 rings **39.** (a) 648 each year (b) less than 2
each day **41.** $0 **42.** 0 **43.** undefined **44.** impossible; if you have
6 cookies, it is not possible to divide them among 0 people. **45.** (a) 14
(b) 17 (c) 38 **46.** Yes. Some examples are $18 \cdot 1 = 18$; $26 \cdot 1 = 26$;
$43 \cdot 1 = 43$. **47.** (a) 3200 (b) 320 (c) 32 **48.** Drop the same
number of zeros that appear in the divisor. The result is the quotient. With
the divisor 10, drop one 0; with 100, drop two zeros; with 1000, drop three
zeros.

SUMMARY EXERCISES Whole Numbers Computation
(pages 63–65)

1. 3; 4 **2.** 6; 0 **3.** 1; 6 **4.** eighty-six thousand, two **5.** four hundred
twenty-five million, two hundred eight thousand, seven hundred thirty-
three **6.** 97 **7.** 905 **8.** 21 **9.** 409 **10.** 17,573 **11.** 82,164
12. 677 **13.** 37,674 **14.** 35,889 **15.** 560 **16.** 5600 **17.** 350,000
18. 252,000 **19.** 1,238,201 **20.** 5,549,375 **21.** 1 **22.** 0
23. undefined **24.** 15 **25.** 56 **26.** 0 **27.** 96 **28.** 304 **29.** 2750 R2
30. 761 R3 **31.** 3380 **32.** 220,545 **33.** 2016 **34.** 1476 **35.** 78
36. 210 **37.** 18,038,816 **38.** 506 R28 **39.** 52 **40.** 1208 R3
41. 573 R3 **42.** 41 **43.** 208,530 **44.** 1,101,744 **45.** 2,154,000,000
46. 40,000,000 **47.** 1520 hours **48.** 7500 tons **49.** Four hundred
twenty-two million, four hundred thousand feet **50.** (a) More than a
mile; 3701 ft more (b) Less than a mile; 1080 ft less

SECTION 1.7 (pages 72–75)

1. ten **2.** ten **3.** hundred **4.** hundred **5.** thousand **6.** thousand
7. ten-thousand **8.** ten-thousand **9.** 860 **11.** 6800 **13.** 28,500
15. 6000 **17.** 16,000 **19.** 8,000,000,000 **21.** 600,000 **23.** 5,000,000
25. 4480; 4500; 4000 **27.** 3370; 3400; 3000 **29.** 6050; 6000; 6000
31. 5340; 5300; 5000 **33.** 19,540; 19,500; 20,000 **35.** 26,290; 26,300;
26,000 **37.** 93,710; 93,700; 94,000 **39.** 1. Locate the place to be
rounded and underline it. 2. Look only at the next digit to the right. If
this digit is 5 or more, increase the underlined digit by 1. 3. Change all
digits to the right of the underlined place to zeros. **40.** 1. Locate the place to

be rounded and underline it. 2. Look only at the next digit to the right. If this
digit is 4 or less do not change the underlined digit. 3. Change all digits to
the right of the underlined place to zeros **41.** 30 60 50 80 220; 219
43. 80 40 40; 35 **45.** 70 30 2100; 2278
47. 900 700 400 800 2800; 2828 **49.** 900 400 500; 435
51. 800 400 320,000; 282,000 **53.** 8000 60 700 4000 12,760; 12,605
55. 700 500 200; 158 **57.** 900 30 27,000; 27,231 **59.** Perhaps the best
explanation is that 3492 is closer to 3500 than 3400, but 3492 is closer to
3000 than 4000. **60.** Rounding numbers usually allows for faster calculation and results in an estimated answer prior to getting an exact answer.
One example is

400	432
− 200	− 209
Estimate: 200	*Exact:* 223

61. 80 million people; 310 million people **63.** 349,000 streets; 350,000
streets **65.** 39,840,000 tickets; 39,800,000 tickets; 40,000,000 tickets
67. 19,266,000 players; 19,270,000 players; 19,300,000 players
69. 71,500 **70.** 72,499 **71.** 7500 **72.** 8499 **73.** 3930; 11,240;
15,970; 17,920; 534,880; 2,788,000 **74.** 4000; 10,000; 20,000; 20,000;
500,000; 3,000,000 **75.** (a) When using front end rounding, all digits
are 0 except the first digit. These numbers are easier to work with when
estimating answers. (b) Sometimes when using front end rounding; the
estimated answer can vary greatly from the exact answer.

SECTION 1.8 (pages 78–81)

1. exponent 2; base 3 **2.** exponent 3; base 2 **3.** exponent 2; base 5
4. exponent 2; base 4 **5.** 2; 8; 64 **7.** 2; 15; 225 **9.** 4 **11.** 8
13. 10 **15.** 12 **17.** false: 5^2 means that 5 is used as a factor 2 times,
$5^2 = 5 \cdot 5 = 25$ **18.** false: $4^2 = 4 \cdot 4 = 16$ **19.** false: 1 raised to any
power is 1. In this example, $1 \cdot 1 \cdot 1 = 1$ **20.** false: a number raisecd to
the first power is the number itself, so $6^1 = 6$ **21.** 36; 36 **23.** 625; 625
25. 10,000; 10,000 **27.** A perfect square is the square of a whole
number. The number 25 is the square of 5 because $5 \cdot 5 = 25$. The number 50
is not a perfect square. There is no whole number that can be squared to get
50. **28.** 1. Do all operations inside parentheses or other grouping symbols. 2. Simplify any expressions with exponents and find any square
roots. 3. Multiply or divide proceeding from left to right. 4. Add or
subtract proceeding from left to right. **29.** true **30.** true
31. false: $6 + 8 \div 2 = 10$. Multiplications and divisions and performed from left to right, then additions and subtractions. **32.** false:
$4 + 5(6 − 4) = 4 + 5 \cdot 2 = 4 + 10 = 14$. A common error is adding
4 to 5 first and then multiplying by 2. Follow the order of operations.
33. 12 **35.** 20 **37.** 45 **39.** 63 **41.** 118 **43.** 22 **45.** 30 **47.** 102
49. 9 **51.** 63 **53.** 33 **55.** 70 **57.** 7 **59.** 17 **61.** 55 **63.** 108
65. 26 **67.** 26 **69.** 27 **71.** 16 **73.** 16 **75.** 21 **77.** 7 **79.** 20
81. 14 **83.** 25 **85.** 16 **87.** 23 **89.** 233

SECTION 1.9 (pages 86–89)

1. 8; 500; $8 \cdot 500 = 4000$ **2.** $8\frac{1}{2}$; 500; $8\frac{1}{2} \cdot 500 = 4250$ **3.** 4500 stores
5. Dollar General; about 5750 stores **7.** 1250 fewer stores **9.** 100
adults **10.** 7 carrer paths **11.** 9 people **13.** (a) Saw ad (b) 25
people **15.** 9 people **17.** 2014; 7000 installations **19.** 4500 installations **21.** Possible answers are 1. shortage of units to install

2. lack of qualified workers   3. poor economy   4. less demand for solar products.   **22.** Possible answers are   1. greater demand   2. more units available   3. many qualified workers   4. tax incentives for solar projects.   **23.** $(7 - 2) \cdot 3 - 6$ **24.** $(4 + 2) \cdot (5 + 1)$ **25.** $36 \div (3 \cdot 3) \cdot 4$ **26.** $56 \div (2 \cdot 2 \cdot 2) + \dfrac{0}{6}$ **27. (a)** 7920 + 1320 + 2640 + (5280 − 1320 − 1320) + 2640 + 1320 + 7920 + 5280   **(b)** $31,680 \times 3 = 95,040$ ft   **(c)** 18 miles

SECTION 1.10 (pages 94–97)

1. (c) There are only 24 hours in a day.   **2. (b)** $100 per hour is not reasonable   **3. (a)** $5 is reasonable   **4. (b)** 25 mpg is reasonable   **5.** *Estimate:* 600 + 900 + 1000 + 800 + 2000 = 5300 sandwiches; *Exact:* 5208 sandwiches   **7.** *Estimate:* $300 − $200 = $100 saved; *Exact:* $104 saved   **9.** *Estimate:* $200 \times 20 = 4000$ kits; *Exact:* 5664 kits   **11.** *Estimate:* $3000 \div 700 \approx 4$ toys; *Exact:* 4 toys   **13.** *Estimate:* $30 \times 5 = $150; *Exact:* $170   **15.** *Estimate:* 100,000 + 300,000 = 400,000 deaths: *Exact:* 360,222 deaths   **17. (a)** *Estimate:* 300,000 − 100,000 = 200,000 deaths; *Exact:* 140,082 deaths   **19.** 100,000 + 300,000 + 90,000 + 200,000 = 690,000 deaths *Exact:* 618,642 deaths   **21.** *Estimate:* $2000 − $700 − $300 − $400 − $200 − $200 = $200; *Exact:* $350   **23.** *Estimate:* 40,000 × 100 = 4,000,000 square feet; *Exact:* 6,011,280 square feet   **25.** *Estimate:* $300 + $200 + $100 + $70 + $400 + $200 = $1270; *Exact:* $1254   **27.** *Estimate:* $300 + $70 + $400 + $200 = $970; $970 − $800 = $170; *Exact:* $185   **29.** *Estimate:* ($1000 × 6) + ($900 × 20) = $24,000; *Exact:* $20,961   **31.** Possible answers are Addition: more; total; gain of. Subtraction: less; loss of; decreased by. Multiplication: twice; of; product. Division: divided by; goes into; per. Equals: is; are.   **32.** 1. Read the problem carefully.   2. Work out a plan. 3. Estimate a reasonable answer.   4. Solve the problem.   5. State the answer.   6. Check your work.   **33.** $20,009   **35.** 2477 pounds   **37.** $378   **39.** $375   **41.** 20 seats

Chapter 1 REVIEW EXERCISES (pages 103–110)

1. 6; 573   **2.** 36; 215   **3.** 105; 724   **4.** 1; 768; 710; 618   **5.** seven hundred twenty-eight   **6.** fifteen thousand, three hundred ten   **7.** three hundred nineteen thousand, two hundred fifteen   **8.** sixty-two million, five hundred thousand, five   **9.** 10,008   **10.** 200,000,455   **11.** 110   **12.** 121   **13.** 5464   **14.** 15,657   **15.** 10,986   **16.** 9845   **17.** 40,602   **18.** 49,855   **19.** 36   **20.** 27   **21.** 189   **22.** 184   **23.** 6849   **24.** 4327   **25.** 224   **26.** 25,866   **27.** 49   **28.** 0   **29.** 32   **30.** 64   **31.** 45   **32.** 42   **33.** 56   **34.** 81   **35.** 40   **36.** 45   **37.** 48   **38.** 8   **39.** 0   **40.** 42   **41.** 48   **42.** 0   **43.** 84   **44.** 368   **45.** 522   **46.** 98   **47.** 5000   **48.** 2992   **49.** 5396   **50.** 45,815   **51.** 14,912   **52.** 20,160   **53.** 465,525   **54.** 174,984   **55.** 875   **56.** 2368   **57.** 1176   **58.** 5100   **59.** 15,576   **60.** 30,184   **61.** 887,169   **62.** 500,856   **63.** $360   **64.** $1064   **65.** $20,352   **66.** $684   **67.** 14,000   **68.** 23,800   **69.** 206,800   **70.** 318,500   **71.** 128,000,000   **72.** 90,300,000   **73.** 5   **74.** 7   **75.** 6   **76.** 2   **77.** 6   **78.** 4   **79.** 7   **80.** 0   **81.** undefined   **82.** 0   **83.** 8   **84.** 9   **85.** 82   **86.** 98   **87.** 4422   **88.** 352   **89.** 150 R4   **90.** 124 R25   **91.** 820   **92.** 15,200   **93.** 21,000   **94.** 70,000   **95.** 3490; 3500; 3000   **96.** 20,070; 20,100; 20,000   **97.** 98,200; 98,200; 98,000   **98.** 352,120; 352,100; 352,000   **99.** 4   **100.** 7   **101.** 12

102. 14   **103.** 3; 7; 343   **104.** 6; 3; 729   **105.** 3; 5; 125   **106.** 5; 4; 1024   **107.** 34   **108.** 26   **109.** 9   **110.** 4   **111.** 9   **112.** 6   **113.** 8 parents   **114.** 5 parents   **115.** Keeping bedroom clean; 25 parents   **116.** Hanging up wet bath towels; 3 parents   **117.** *Estimate:* 40 million × 400 = 16,000 million or 16,000,000,000 checks; *Exact:* 14,600 million or 14,600,000,000 checks   **118.** *Estimate:* 1000 × 60 = 60,000 revolutions; *Exact:* 84,000 revolutions   **119.** *Estimate:* 40,000,000 − 30,000,000 = 10,000,000; *Exact:* 12,073,57 people   **120.** *Estimate:* 700,000 − 600,000 = 100,000; *Exact:* 153,223 people   **121.** *Estimate:* $3000 + (8 × $90) + 200 = $3920; *Exact:* $3583   **122.** *Estimate:* $60 + ($90 × $2) = $240; *Exact:* $233   **123.** *Estimate:* (30 × $2000) + (30 × $900) = $87,000; *Exact:* $74,052   **124.** *Estimate:* (60 × $20) + (20 × $7) = $1340; *Exact:* $1139   **125.** *Estimate:* $600 − $100 = $500; *Exact:* = $513   **126.** *Estimate:* $2000 − $500 − $400 = $1100; *Exact:* $1019   **127.** *Estimate:* 9000 ÷ 200 = 45 pounds; *Exact:* 50 pounds   **128.** *Estimate:* 30,000 ÷ 1000 = 30 hr; *Exact:* 33 hr   **129.** *Estimate:* 30,000 ÷ 600 = 50 acres; *Exact:* 52 acres   **130.** *Estimate:* 6000 ÷ 200 = 30 homes; *Exact:* 32 homes   **131.** 332   **132.** 448   **133.** 253   **134.** 588   **135.** 1041   **136.** 1661   **137.** 32,062   **138.** 24,947   **139.** 3   **140.** 7   **141.** 93,635   **142.** 83,178   **143.** undefined   **144.** 7   **145.** 6900   **146.** 2310   **147.** 1,079,040   **148.** 130,212   **149.** 108   **150.** 207   **151.** three hundred seventy-six thousand, eight hundred fifty-three   **152.** four hundred eight thousand, six hundred ten   **153.** 8700   **154.** 401,000   **155.** 8   **156.** 9   **157.** $5544   **158.** $31,080   **159.** $2288   **160.** $15,782   **161.** 468 cards   **162.** 52,700 pounds   **163.** $280   **164.** $114,635   **165.** $1905   **166.** $12,420   **167.** 1467 ft   **168.** 293 ft   **169. (a)** 11,958 ft   **(b)** greater than two miles by 1398 ft   **170.** more than 4 football fields (a little less than 5)

Chapter 1 TEST (pages 111–112)

1. nine thousand, two hundred five   **2.** twenty-five thousand, sixty-five   **3.** 426,005   **4.** 8530   **5.** 112,630   **6.** 1045   **7.** 6206   **8.** 168   **9.** 171,000   **10.** 1615   **11.** 4,450,743   **12.** 7047   **13.** undefined   **14.** 458 R5   **15.** 160   **16.** 6350   **17.** 76,000   **18.** 41   **19.** 28   **20.** *Estimate:* $500 + $500 + $500 + $400 − $800 = $1100; *Exact:* $1140   **21.** *Estimate:* 90,000 ÷ 400 = 225 acres; *Exact:* 231 acres   **22.** *Estimate:* $2000 − $500 − $200 − $200 = $1100; *Exact:* $948   **23.** *Estimate:* (50 × 60 × 4) + (40 × 60 × 3) = 19,200 chicks; *Exact:* 18,000 chicks   **24.** 1. Locate the place to which you are rounding and underline it. 2. Look only at the next digit to the right. If this digit is a 4 or less, do not change the underlined digit. If the digit is 5 or more, increase the underlined digit by 1. 3. Change all digits to the right of the underlined place to zeros. Each person's rounding example will vary.   **25.** 1. Read the problem carefully. 2. Work out a plan. 3. Estimate a reasonable answer. 4. Solve the problem. 5. State the answer. 6. Check your work.

CHAPTER 2 Multiplying and Dividing Fractions

SECTION 2.1 (pages 116–118)

1. Numerator 4; Denominator 5   2. Numerator 5; Denominator 6

3. Numerator 9; Denominator 8   4. Numerator 7; Denominator 5

5. 3; 8   6. 7; 16   7. 5; 24   8. 24; 32   9. $\frac{3}{4}; \frac{1}{4}$ 11. $\frac{1}{3}; \frac{2}{3}$

13. $\frac{7}{5}; \frac{3}{5}$ 15. $\frac{5}{6}; \frac{4}{6}; \frac{2}{6}$ 17. $\frac{8}{25}$ 19. $\frac{303}{520}$

21. Proper $\frac{1}{3}, \frac{5}{8}, \frac{7}{16}$ Improper $\frac{8}{5}, \frac{6}{6}, \frac{12}{2}$

23. Proper $\frac{3}{4}, \frac{9}{11}, \frac{7}{15}$ Improper $\frac{3}{2}, \frac{5}{5}, \frac{19}{18}$

25. One possibility is

$\frac{3 \leftarrow \text{Numerator}}{4 \leftarrow \text{Denominator}}$

The denominator shows the number of equal parts in the whole and the numerator shows how many of the parts are being considered.

26. An example is $\frac{1}{2}$ as a proper fraction and $\frac{3}{2}$ as an improper fraction. A proper fraction has a numerator smaller than the denominator. An improper fraction has a numerator that is equal to or greater than the denominator. Drawings will vary.

SECTION 2.2 (pages 123–125)

1. true   2. true   3. false: Multiply 7 by 5 and add the numerator. $(7 \cdot 5) + 2 = 37; \frac{37}{5}$ 4. false: Any mixed number can be charged to an improper fraction. The reverse is also true.   5. false: You must add the numerator to the product of the whole number and the denominator. $6\frac{1}{2} = \frac{(6 \cdot 2) + 1}{2} = \frac{13}{2}$ 6. true   7. $\frac{5}{4}$ 9. $\frac{23}{5}$

11. $\frac{17}{2}$ 13. $\frac{81}{8}$ 15. $\frac{43}{4}$ 17. $\frac{29}{5}$ 19. $\frac{43}{5}$ 21. $\frac{54}{11}$ 23. $\frac{131}{4}$

25. $\frac{221}{12}$ 27. $\frac{269}{15}$ 29. $\frac{187}{24}$

31. false: The denominator remains the same. So, $\frac{4}{3} = 1\frac{1}{3}$

32. false: A mixed number always has a value equal to or greater than a whole number.   33. true   34. true   35. $1\frac{1}{3}$ 37. $2\frac{1}{4}$ 39. 9

41. $7\frac{3}{5}$ 43. $15\frac{3}{4}$ 45. $5\frac{2}{9}$ 47. $8\frac{1}{8}$ 49. $16\frac{4}{5}$ 51. 28   53. $26\frac{1}{7}$

55. Multiply the denominator by the whole number and add the numerator. The result becomes the new numerator, which is placed over the original denominator.

$2\frac{1}{2} \quad (2 \cdot 2) + 1 = 5 \quad \frac{5}{2}$

56. Divide the numerator by the denominator. The quotient is the whole number of the mixed number and the remainder is the numerator of the fraction part. The denominator is unchanged.

$\frac{5}{2} \qquad 2\overline{)5} \qquad 2\frac{1}{2} \leftarrow \text{Remainder}$
$\qquad \quad \underline{4}$
$\qquad \quad 1 \leftarrow \text{Remainder}$

57. $\frac{501}{2}$ 59. $\frac{1000}{3}$ 61. $\frac{4179}{8}$ 63. $154\frac{1}{4}$ 65. 171   67. $122\frac{13}{32}$

69. $\frac{2}{3}, \frac{4}{5}, \frac{3}{4}, \frac{7}{10}$ 70. (a) numerator; denominator

(b)

(c) less   71. $\frac{5}{5}, \frac{10}{3}, \frac{6}{5}$ 72. (a) numerator; denominator

(b)

(c) greater   73. $\frac{5}{3} = 1\frac{2}{3}; \frac{7}{7} = 1; \frac{11}{6} = 1\frac{5}{6}$ 74. (a) improper; greater than or equal to

(b)

SECTION 2.3 (pages 130–131)

1. false   2. true   3. true   4. false   5. 1, 2, 3, 4, 6, 8, 12, 16, 24, 48

7. 1, 2, 4, 7, 8, 14, 28, 56   9. 1, 2, 3, 4, 6, 9, 12, 18, 36   11. 1, 2, 4, 5, 8, 10, 20, 40   13. 1, 2, 4, 8, 16, 32, 64   15. 1, 2, 41, 80   17. composite

19. prime   21. composite   23. prime   25. composite   27. prime

29. (b)   30. (c)   31. (a)   33. $2 \cdot 3$ 35. 5^2 37. $2^2 \cdot 17$

39. $2^3 \cdot 3^2$ 41. $2^2 \cdot 11$ 43. $2^2 \cdot 5^2$ 45. 5^3 47. $2^2 \cdot 3^2 \cdot 5$

49. $2^6 \cdot 5$ 51. $2^3 \cdot 3^2 \cdot 5$ 53. A prime number is a whole number that has exactly two *different* factors, itself and 1. Examples include 2, 3, 5, 7, 11. A composite number has a factor(s) other than itself or 1. Examples include 4, 6, 8, 9, 10. The numbers 0 and 1 are neither prime nor composite.

54. No even number other than 2 is prime because all even numbers have 2 as a factor. Many odd numbers are multiples of prime numbers and are not prime. For example, 9, 21, 33, and 45 are all multiples of 3.

55. All the possible factors of 24 are 1, 2, 3, 4, 6, 8, 12, and 24. This list includes both prime numbers and composite numbers. The prime factors of 24 include only prime numbers. The prime factorization of 24 is $2 \cdot 2 \cdot 2 \cdot 3 = 2^3 \cdot 3$.   56. No. The order of division does not matter. As long as you use only prime numbers, your answers will be correct. However, it does seem easier to always start with 2 and then use progressively greater prime numbers. The prime factorization of 36 is $36 = 2 \cdot 2 \cdot 3 \cdot 3 = 2^2 \cdot 3^2$.   57. $2 \cdot 5^2 \cdot 7$ 59. $2^6 \cdot 3 \cdot 5$

61. $2^3 \cdot 3 \cdot 5 \cdot 13$ 63. $2^2 \cdot 3^2 \cdot 5 \cdot 7$ 65. 2, 3, 5, 7, 11, 13, 17, 19, 23, 29, 31, 37, 41, 43, 47   66. A prime number is a whole number that is evenly divisible by itself and 1 only.   67. No. Every other even number is divisible by 2 in addition to being divisible by itself and 1.   68. No. A multiple of a prime number can never be prime because it will always be divisible by the prime number.   69. $2 \cdot 2 \cdot 3 \cdot 5 \cdot 5 \cdot 7$ 70. $2^2 \cdot 3 \cdot 5^2 \cdot 7$

SECTION 2.4 (pages 136–137)

1. even   2. 5; 0   3. 0   4. 3   5. ✓ ✓ ✓ ✓   7. ✓ ✓ ✗ ✗

9. ✓ ✗ ✓ ✓   11. ✓ ✓ ✗ ✗   13. true   14. false   15. false

16. true   17. $\frac{3}{5}$ 19. $\frac{6}{7}$ 21. $\frac{7}{8}$ 23. $\frac{6}{7}$ 25. $\frac{4}{7}$ 27. $\frac{1}{50}$ 29. $\frac{8}{11}$

31. $\frac{5}{9}$ **33.** $\frac{\overset{1}{\cancel{2}} \cdot \overset{1}{\cancel{3}} \cdot 3}{2 \cdot 2 \cdot 2 \cdot \underset{1}{\cancel{3}}} = \frac{3}{4}$ **35.** $\frac{\overset{1}{\cancel{8}} \cdot 7}{2 \cdot 2 \cdot 2 \cdot \underset{1}{\cancel{8}}} = \frac{7}{8}$

37. $\frac{\overset{1}{\cancel{2}} \cdot \overset{1}{\cancel{3}} \cdot \overset{1}{\cancel{3}} \cdot \overset{1}{\cancel{5}}}{2 \cdot 2 \cdot \underset{1}{\cancel{3}} \cdot \underset{1}{\cancel{3}} \cdot \underset{1}{\cancel{5}}} = \frac{1}{2}$ **39.** $\frac{\overset{1}{\cancel{2}} \cdot \overset{1}{\cancel{2}} \cdot \overset{1}{\cancel{3}} \cdot 3}{\underset{1}{\cancel{2}} \cdot \underset{1}{\cancel{2}} \cdot \underset{1}{\cancel{3}}} = 3$

41. $\frac{2 \cdot 2 \cdot 2 \cdot \overset{1}{\cancel{3}} \cdot \overset{1}{\cancel{3}}}{\underset{1}{\cancel{3}} \cdot \underset{1}{\cancel{3}} \cdot 5 \cdot 5} = \frac{8}{25}$ **43.** $\frac{1}{2} = \frac{1}{2}$; equivalent

45. $\frac{5}{12} \neq \frac{2}{5}$; not equivalent   **47.** $\frac{5}{8} \neq \frac{35}{52}$; not equivalent

49. $\frac{7}{8} = \frac{7}{8}$; equivalent   **51.** $8 \neq 9$; not equivalent   **53.** $\frac{5}{6} = \frac{5}{6}$; equivalent

55. A fraction is in lowest terms when the numerator and the denominator have no common factors other than 1. Some examples are $\frac{1}{2}, \frac{3}{8}$, and $\frac{2}{3}$.

56. Two fractions are equivalent when they represent the same portion of a whole. For example, the fractions $\frac{10}{15}$ and $\frac{8}{12}$ are equivalent.

$$\frac{10}{15} = \frac{2 \cdot \overset{1}{\cancel{5}}}{3 \cdot \underset{1}{\cancel{5}}} = \frac{2 \cdot 1}{3 \cdot 1} = \frac{2}{3} \leftarrow$$

$$\text{Equivalent} \left(\frac{2}{3} = \frac{2}{3}\right)$$

$$\frac{8}{12} = \frac{\overset{1}{\cancel{2}} \cdot \overset{1}{\cancel{2}} \cdot 2}{\underset{1}{\cancel{2}} \cdot \underset{1}{\cancel{2}} \cdot 3} = \frac{1 \cdot 1 \cdot 2}{1 \cdot 1 \cdot 3} = \frac{2}{3} \leftarrow$$

57. $\frac{5}{8}$ **59.** $\frac{2}{1} = 2$

SUMMARY EXERCISES   Fraction Basics (pages 140–141)

1. $\frac{5}{6}; \frac{1}{6}$ **2.** $\frac{1}{3}; \frac{2}{3}$ **3.** $\frac{5}{8}; \frac{3}{8}$ **4.** 3; 4   **5.** 8; 5   **6.** Proper $\frac{3}{5}, \frac{4}{25}, \frac{1}{32}$

Improper $\frac{8}{2}, \frac{16}{7}, \frac{8}{8}$ **7.** $\frac{7}{36}$ **8.** $\frac{24}{36}$ **9.** $\frac{7}{18}$ **10.** $\frac{33}{36}$ **11.** $2\frac{1}{2}$ **12.** $1\frac{3}{8}$

13. $1\frac{2}{7}$ **14.** $2\frac{2}{3}$ **15.** 8   **16.** 5   **17.** $7\frac{1}{5}$ **18.** $4\frac{7}{10}$ **19.** $\frac{10}{3}$ **20.** $\frac{43}{8}$

21. $\frac{34}{5}$ **22.** $\frac{53}{5}$ **23.** $\frac{51}{4}$ **24.** $\frac{62}{13}$ **25.** $\frac{71}{6}$ **26.** $\frac{189}{8}$ **27.** $2 \cdot 5$

28. $5 \cdot 11$ **29.** $2^2 \cdot 3^2$ **30.** 3^4 **31.** $2^3 \cdot 5 \cdot 7$ **32.** $2^3 \cdot 3^2 \cdot 5$

33. $\frac{1}{3}$ **34.** $\frac{1}{2}$ **35.** $\frac{1}{4}$ **36.** $\frac{3}{5}$ **37.** $\frac{3}{4}$ **38.** $\frac{5}{6}$ **39.** $\frac{7}{8}$ **40.** $\frac{1}{50}$

41. $\frac{1}{8}$ **42.** $\frac{5}{9}$ **43.** $\frac{4}{7}$ **44.** $\frac{5}{9}$ **45.** $\frac{\overset{1}{\cancel{2}} \cdot \overset{1}{\cancel{2}} \cdot 2 \cdot \overset{1}{\cancel{3}}}{\underset{1}{\cancel{2}} \cdot \underset{1}{\cancel{2}} \cdot 3 \cdot \underset{1}{\cancel{3}}} = \frac{2}{3}$

46. $\frac{\overset{1}{\cancel{2}} \cdot \overset{1}{\cancel{2}} \cdot \overset{1}{\cancel{2}} \cdot \overset{1}{\cancel{2}} \cdot \overset{1}{\cancel{5}}}{\underset{1}{\cancel{2}} \cdot \underset{1}{\cancel{2}} \cdot \underset{1}{\cancel{2}} \cdot \underset{1}{\cancel{2}} \cdot 2 \cdot \underset{1}{\cancel{5}}} = \frac{1}{2}$ **47.** $\frac{\overset{1}{\cancel{2}} \cdot \overset{1}{\cancel{3}} \cdot 3 \cdot \overset{1}{\cancel{7}}}{\underset{1}{\cancel{2}} \cdot \underset{1}{\cancel{3}} \cdot \underset{1}{\cancel{7}}} = 3$

48. $\frac{\overset{1}{\cancel{2}} \cdot \overset{1}{\cancel{2}} \cdot \overset{1}{\cancel{2}} \cdot \overset{1}{\cancel{2}} \cdot 2 \cdot 3}{\underset{1}{\cancel{2}} \cdot \underset{1}{\cancel{2}} \cdot \underset{1}{\cancel{2}} \cdot \underset{1}{\cancel{2}} \cdot 7} = \frac{6}{7}$

SECTION 2.5 (pages 148–151)

1. multiply; denominator   **2.** numerator; denominator   **3.** divide; denominator   **4.** lowest terms   **5.** $\frac{1}{4}$ **7.** $\frac{2}{35}$ **9.** $\frac{3}{4}$ **11.** $\frac{1}{4}$ **13.** $\frac{5}{12}$

15. $\frac{4}{5}$ **17.** $\frac{7}{16}$ **19.** $\frac{7}{48}$ **20.** true   **21.** false: $\frac{4}{5} \cdot 8 = \frac{4}{5} \cdot \frac{8}{1} = \frac{32}{5} = 6\frac{2}{5}$

23. 15   **25.** $13\frac{1}{2}$ **27.** $31\frac{1}{2}$ **29.** 80   **31.** $94\frac{2}{3}$ **33.** 400   **35.** 810

37. $\frac{1}{4}$ square mile   **39.** 9 square meters   **41.** $\frac{1}{4}$ square mile

43. Multiply the numerators and multiply the denominators. An example is

$\frac{3}{4} \cdot \frac{1}{2} = \frac{3 \cdot 1}{4 \cdot 2} = \frac{3}{8}$ **44.** You must divide a numerator and a denominator

by the same number. If you do all possible divisions, your answer will be

in lowest terms. One example is $\frac{3}{4} \cdot \frac{2}{3} = \frac{3 \cdot 2}{4 \cdot 3} = \frac{\overset{1}{\cancel{3}} \cdot \overset{1}{\cancel{2}}}{\underset{2}{\cancel{4}} \cdot \underset{1}{\cancel{3}}} = \frac{1}{2}$.

45. $1\frac{1}{2}$ square yards   **47.** $3\frac{1}{2}$ square miles

49. They are both the same size: $\frac{3}{64}$ square mile

51. $4000 + 2000 + 1000 + 800 + 400 + 200 + 100 + 80 =$ 8580 supermarkets   **52.** 8912 supermarkets

53. *Estimate:* $\frac{4}{5} \cdot 2000 = 1600$; *Exact:* 1762 supermarkets (rounded)

54. *Estimate:* $\frac{3}{8} \cdot 200 = 75$; *Exact:* 61 supermarkets (rounded)

55. $\frac{4}{5} \cdot 2200 = 1760$ supermarkets   **56.** $\frac{3}{8} \cdot 160 = 60$ supermarkets

SECTION 2.6 (pages 155–158)

1. times; triple; of; twice; product; twice as much   **2.** Check your work.

3. area   **4.** square   **5.** $\frac{1}{2}$ square foot   **7.** $\frac{8}{9}$ square foot   **9.** $\frac{3}{10}$ square yard

11. $2568   **13.** $35   **15. (a)** 910 women   **(b)** 650 men   **17.** 4 hours;

153 people   **19.** $\frac{3}{4}$; 765 people   **21.** Because everyone is included and fractions are given for *all* groups, the sum of the fractions must be *1* or *all* of the people.   **22.** Answers will vary. Some possibilities are 1. You made an addition error. 2. The fractions on the circle graph are incorrect. 3. The fraction errors were caused by rounding.

23. $76,000   **25.** $15,200   **27.** $4750

29. The correct solution is $\frac{9}{10} \times \frac{20}{21} = \frac{\overset{3}{\cancel{9}}}{\underset{1}{\cancel{10}}} \times \frac{\overset{2}{\cancel{20}}}{\underset{7}{\cancel{21}}} = \frac{6}{7}$

30. Yes, the statements are true. Since whole numbers are 1 or greater, when you multiply, the product will always be greater than either of the numbers multiplied. But, when you multiply two proper fractions, you are finding a fraction of a fraction, and the product will be smaller than either of the two proper fractions.   **31.** $750   **33.** 2 ft   **35.** 9000 votes

37. $\frac{1}{32}$ of the estate

SECTION 2.7 (pages 164–167)

1. reciprocal   **2.** 1   **3.** invert; multiplication   **4.** lowest   **5.** $\frac{8}{3}$

7. $\frac{6}{5}$ **9.** $\frac{5}{8}$ **11.** $\frac{1}{4}$ **13.** $\frac{2}{3}$ **15.** $2\frac{5}{8}$ **17.** $\frac{9}{20}$ **19.** 4   **21.** 6   **23.** $\frac{13}{16}$

25. $\dfrac{4}{5}$ **27.** 18   **29.** 24   **31.** $\dfrac{1}{14}$ **33.** $\dfrac{2}{9}$ quart   **35.** 15 times

37. 88 dispensers   **39.** 160 fasteners   **41.** You can divide two fractions by multiplying the first fraction by the reciprocal of the divisor (the second fraction).   **42.** Sometimes the answer is less and sometimes it is greater.

$$\frac{1}{4} \div \frac{7}{8} = \frac{1}{\overset{1}{\underset{}{4}}} \cdot \frac{\overset{2}{8}}{7} = \frac{2}{7} \quad \text{(less than } \tfrac{7}{8} \text{ but not less than } \tfrac{1}{4})$$

$$\frac{1}{2} \div \frac{1}{4} = \frac{1}{\underset{1}{2}} \cdot \frac{\overset{2}{4}}{1} = \frac{2}{1} = 2 \text{ (greater)}$$

43. 12 pounds   **45.** 208 homes   **47. (a)** 124 visits   **(b)** 62 visits

49. 2432 towels   **51.** double, twice, times, product, twice as much

52. goes into, divide, per, quotient, divided by   **53.** reciprocal

54. $\dfrac{4}{3}; \dfrac{8}{7}; \dfrac{1}{5}; \dfrac{19}{12}$ **55. (a)** Multiply the length of one side by 3, 4, 5, or 6.

(b) $\dfrac{15}{16} \times 4 = \dfrac{15}{\underset{4}{16}} \times \dfrac{\overset{1}{4}}{1} = \dfrac{15}{4} = 3\dfrac{3}{4}$ in.   **56.** $\dfrac{225}{256}$ square inch; Multiply the length by the width.

SECTION 2.8 (pages 175–179)

1. false: A reciprocal is used when dividing fractions.
2. false: The reciprocal of 10 is $\dfrac{1}{10}$.   **3.** true   **4.** true

5. *Estimate:* $5 \cdot 2 = 10$; *Exact:* $7\dfrac{7}{8}$.   **7.** *Estimate:* $2 \cdot 3 = 6$; *Exact:* $4\dfrac{1}{2}$.
9. *Estimate:* $3 \cdot 1 = 3$; *Exact:* 4.   **11.** *Estimate:* $8 \cdot 6 = 48$; *Exact:* 50.
13. *Estimate:* $5 \cdot 2 \cdot 5 = 50$; *Exact:* $49\dfrac{1}{2}$.   **15.** *Estimate:* $3 \cdot 2 \cdot 3 = 18$;
Exact: 12.   **17.** (d)   **18.** (a)   **19.** (b)   **20.** (c)   **21.** *Estimate:* $1 \div 4 = \dfrac{1}{4}$;
Exact: $\dfrac{1}{3}$.   **23.** *Estimate:* $3 \div 3 = 1$; *Exact:* $\dfrac{5}{6}$.   **25.** *Estimate:* $9 \div 3 = 3$;
Exact: $3\dfrac{3}{5}$.   **27.** *Estimate:* $1 \div 2 = \dfrac{1}{2}$; *Exact:* $\dfrac{5}{12}$.   **29.** *Estimate:*
$2 \div 6 = \dfrac{1}{3}$; *Exact:* $\dfrac{3}{10}$.   **31.** *Estimate:* $6 \div 6 = 1$; *Exact:* $\dfrac{17}{18}$.

33. (a) *Estimate:* $1 \cdot 3 = 3$ cups; *Exact:* $1\dfrac{7}{8}$ cups of applesauce

(b) *Estimate:* $1 \cdot 3 = 3$ teaspoons; *Exact:* $1\dfrac{1}{4}$ teaspoons of salt

(c) *Estimate:* $2 \cdot 3 = 6$ cups; *Exact:* $4\dfrac{3}{8}$ cups of flour

35. (a) *Estimate:* $1 \div 2 = \dfrac{1}{2}$ teaspoon; *Exact:* $\dfrac{1}{4}$ teaspoon vanilla extract

(b) *Estimate:* $1 \div 2 = \dfrac{1}{2}$ cup; *Exact:* $\dfrac{3}{8}$ cup applesauce

(c) *Estimate:* $2 \div 2 = 1$ cup; *Exact:* $\dfrac{7}{8}$ cup of flour

37. *Estimate:* $329 \div 12 \approx 27$ units; *Exact:* 28 units

39. *Estimate:* $45 \cdot 20 = 900$ in.; *Exact:* $877\dfrac{1}{2}$ in.

41. The answer should include: *Step 1* Change mixed numbers to improper fractions. *Step 2* Multiply the fractions. *Step 3* Write the answer in lowest terms, changing to mixed or whole numbers where possible.

42. The additional step is to use the reciprocal of the divisor (the second fraction).   **43.** *Estimate:* $130 \cdot 1 = \$130$ million; *Exact:* 182 million or $\$182{,}000{,}000$ **45.** *Estimate:* $140 \cdot 1 = 140$ trips; *Exact:* 112 trips
47. (a) *Estimate:* $18 \cdot 4 = 72$ in.;   **(b)** No, because 6 ft = 72 in.
49. *Estimate:* $6 \cdot 6 = 36$ miles per hour; *Exact:* $33\dfrac{11}{16}$ miles per hour

Chapter 2 REVIEW EXERCISES (pages 188–191)

1. $\dfrac{1}{3}$ **2.** $\dfrac{5}{8}$ **3.** $\dfrac{2}{4}$ **4.** Proper $\dfrac{1}{8}, \dfrac{3}{4}, \dfrac{2}{3}$; Improper $\dfrac{4}{3}, \dfrac{5}{5}$

5. Proper $\dfrac{15}{16}, \dfrac{1}{8}$; Improper $\dfrac{6}{5}, \dfrac{16}{13}, \dfrac{5}{3}$ **6.** $\dfrac{19}{4}$ **7.** $\dfrac{59}{6}$ **8.** $3\dfrac{3}{8}$ **9.** $12\dfrac{3}{5}$

10. 1, 2, 3, 6   **11.** 1, 2, 3, 4, 6, 8, 12, 24   **12.** 1, 5, 11, 55

13. 1, 2, 3, 5, 6, 9, 10, 15, 18, 30, 45, 90   **14.** 3^3 **15.** $2 \cdot 3 \cdot 5^2$

16. $2^2 \cdot 3 \cdot 5 \cdot 7$ **17.** 25   **18.** 288   **19.** 1728   **20.** 2048   **21.** $\dfrac{24}{24} = 1$

22. $\dfrac{18}{24} = \dfrac{3}{4}$ **23.** $\dfrac{14}{24} = \dfrac{7}{12}$ **24.** $\dfrac{10}{24} = \dfrac{5}{12}$ **25.** $\dfrac{\overset{1}{5} \cdot 5}{2 \cdot 2 \cdot 3 \cdot \underset{1}{5}}; \dfrac{5}{12}$

26. $\dfrac{\overset{1}{2} \cdot \overset{1}{2} \cdot \overset{1}{2} \cdot \overset{1}{2} \cdot \overset{1}{2} \cdot 2 \cdot 2 \cdot \overset{1}{2}}{\underset{1}{2} \cdot \underset{1}{2} \cdot \underset{1}{2} \cdot \underset{1}{2} \cdot \underset{1}{2} \cdot \underset{1}{2}}; 4$ **27.** equivalent   **28.** not equivalent

29. equivalent   **30.** $\dfrac{3}{5}$ **31.** $\dfrac{3}{16}$ **32.** $\dfrac{1}{7}$ **33.** $\dfrac{4}{21}$ **34.** 15   **35.** 625

36. $1\dfrac{1}{3}$ **37.** $1\dfrac{2}{3}$ **38.** $2\dfrac{1}{2}$ **39.** 2   **40.** 8   **41.** 24   **42.** $\dfrac{5}{24}$ **43.** $\dfrac{2}{15}$

44. $\dfrac{4}{13}$ **45.** $1\dfrac{3}{8}$ square feet   **46.** $3\dfrac{15}{16}$ square yards   **47.** 7857 square feet

48. $5\dfrac{1}{2}$ square feet   **49.** *Estimate:* $6 \cdot 1 = 6$; *Exact:* $6\dfrac{7}{8}$.

50. *Estimate:* $2 \cdot 7 \cdot 1 = 14$; *Exact:* $21\dfrac{3}{8}$.   **51.** *Estimate:* $16 \div 3 = 5\dfrac{1}{3}$;
Exact: $5\dfrac{1}{6}$.   **52.** *Estimate:* $5 \div 6 = \dfrac{5}{6}$; *Exact:* $\dfrac{3}{4}$.   **53.** 512 bins

54. $\dfrac{3}{20}$ of the business   **55.** *Estimate:* $158 \div 4 \approx 40$ pull cords;
Exact: 36 pull cords   **56.** *Estimate:* $8 \cdot 2 \cdot 50 = 800$ pounds;
Exact: $833\dfrac{1}{3}$ pounds   **57.** 25 pounds   **58.** \$186   **59.** $\dfrac{7}{48}$ of total raised

60. $\dfrac{4}{25}$ ton   **61.** $\dfrac{3}{8}$ **62.** $\dfrac{2}{5}$ **63.** $31\dfrac{1}{4}$ **64.** $28\dfrac{1}{3}$ **65.** $\dfrac{1}{10}$ **66.** $\dfrac{5}{32}$

67. 30   **68.** $2\dfrac{3}{5}$ **69.** $1\dfrac{3}{5}$ **70.** $38\dfrac{1}{4}$ **71.** $\dfrac{17}{3}$ **72.** $\dfrac{307}{8}$

73. $\dfrac{\overset{1}{2} \cdot \overset{1}{2} \cdot 2}{\underset{1}{2} \cdot \underset{1}{2} \cdot 3} = \dfrac{2}{3}$ **74.** $\dfrac{\overset{1}{2} \cdot 2 \cdot \overset{1}{3} \cdot 3 \cdot 3}{\underset{1}{2} \cdot \underset{1}{3} \cdot 5 \cdot 7} = \dfrac{18}{35}$ **75.** $\dfrac{5}{6}$ **76.** $\dfrac{2}{3}$

77. $\dfrac{2}{5}$ **78.** $\dfrac{1}{3}$ **79.** *Estimate:* $3 \cdot 50 = 150$ ounces; *Exact:* 125 ounces

80. *Estimate:* $7 \cdot 9 = 63$ quarts; *Exact:* $67\dfrac{2}{3}$ quarts

81. $1\dfrac{17}{32}$ square inches   **82.** $1\dfrac{31}{32}$ square yards

Chapter 2 TEST (pages 192–193)

1. $\dfrac{5}{6}$ **2.** $\dfrac{3}{8}$ **3.** $\dfrac{2}{3}, \dfrac{6}{7}, \dfrac{1}{4}, \dfrac{5}{8}$ **4.** $\dfrac{27}{8}$ **5.** $30\dfrac{3}{4}$ **6.** 1, 2, 3, 6, 9, 18

7. $3^2 \cdot 5$ **8.** $2^4 \cdot 3^2$ **9.** $2^2 \cdot 5^3$ **10.** $\dfrac{3}{4}$ **11.** $\dfrac{5}{6}$

12. Write the prime factorization of both numerator and denominator. Divide the numerator and denominator by any common factors. Multiply the remaining factors in the numerator and denominator.

$$\frac{56}{84} = \frac{\overset{1}{\cancel{2}} \cdot \overset{1}{\cancel{2}} \cdot 2 \cdot \overset{1}{\cancel{7}}}{\underset{1}{\cancel{2}} \cdot \underset{1}{\cancel{2}} \cdot 3 \cdot \underset{1}{\cancel{7}}} = \frac{2}{3}$$

13. Answers will vary. One possibility is: Multiply fractions by multiplying the numerators and multiplying the denominators. Divide two fractions by using the reciprocal of the divisor (the second fraction) and then changing division to multiplication.

14. $\frac{1}{3}$ **15.** 36   **16.** $\frac{5}{12}$ square yard   **17.** 5475 seedlings   **18.** $\frac{9}{10}$

19. $15\frac{3}{4}$ **20.** 24 pieces   **21.** *Estimate:* $4 \cdot 4 = 16$; *Exact:* $14\frac{7}{16}$

22. *Estimate:* $2 \cdot 4 = 8$; *Exact:* $7\frac{17}{18}$ **23.** *Estimate:* $10 \div 2 = 5$; *Exact:* $4\frac{4}{15}$ **24.** *Estimate:* $9 \div 2 = 4\frac{1}{2}$; *Exact:* $4\frac{6}{7}$

25. *Estimate:* $3 \cdot 12 = 36$; *Exact:* $30\frac{5}{8}$ grams

Chapters 1–2 CUMULATIVE REVIEW EXERCISES (pages 194–195)

1. hundreds 7; tens 8   **2.** millions 8; ten-thousands 2   **3.** 177
4. 149,199   **5.** 4452   **6.** 3,221,821   **7.** 747   **8.** 72   **9.** 2,168,232
10. 450,400   **11.** 9   **12.** 7581   **13.** 8471 R2   **14.** 22 R26
15. 6580; 6600; 7000   **16.** 76,270; 76,300; 76,000   **17.** 8   **18.** 5
19. 540 gallons   **20.** 16,772 cases   **21.** 47,000 hairs   **22.** 188 hours
23. 256 gallons   **24.** 21 pieces   **25.** proper   **26.** improper
27. proper   **28.** $\frac{27}{8}$ **29.** $\frac{32}{5}$ **30.** 2   **31.** $12\frac{7}{8}$ **32.** $2^3 \cdot 3^2$
33. $2 \cdot 3^2 \cdot 7$ **34.** $2 \cdot 5^2 \cdot 7$ **35.** 64   **36.** 288   **37.** 640   **38.** $\frac{7}{8}$
39. $\frac{2}{3}$ **40.** $\frac{5}{9}$ **41.** $\frac{3}{8}$ **42.** 12   **43.** 25   **44.** $\frac{24}{25}$ **45.** $\frac{7}{12}$ **46.** $2\frac{2}{5}$

CHAPTER 3 Adding and Subtracting Fractions

SECTION 3.1 (pages 201–202)

1. like   **2.** unlike   **3.** unlike   **4.** like   **5.** like   **6.** lowest terms
7. $\frac{5}{8}$ **9.** $\frac{1}{2}$ **11.** $1\frac{1}{2}$ **13.** $\frac{5}{6}$ **15.** $1\frac{1}{2}$ **17.** $\frac{11}{27}$ **19.** $\frac{3}{8}$ **21.** $\frac{6}{11}$
23. $\frac{3}{5}$ **25.** $1\frac{1}{7}$ **27.** $\frac{1}{5}$ **29.** $1\frac{1}{6}$ **31.** $1\frac{1}{10}$

33. Three steps to add like fractions are: 1. Add the numerators of the fractions to find the numerator of the sum (the answer). 2. Use the denominator of the fractions as the denominator of the sum. 3. Write the answer in lowest terms.   **34.** Like fractions have the same denominator. Unlike fractions have denominators that are different. Some examples are:

like: $\frac{3}{8}, \frac{1}{8}, \frac{7}{8}$,   unlike: $\frac{1}{2}, \frac{3}{4}, \frac{5}{8}$

35. $\frac{7}{9}$ of their goal   **37.** $\frac{3}{4}$ completed   **39.** $\frac{1}{2}$ acre

SECTION 3.2 (pages 211–214)

1. true   **2.** false: 25 is not divisible by 6. The LCM is 30.   **3.** true
4. false: 28 is not divisible by 3. The LCM is 21.   **5.** 6   **7.** 15   **9.** 36
11. 48   **13.** 100   **15.** 40   **17.** 45   **19.** 120   **21.** 120   **23.** 216
25. 600   **27.** $\frac{9}{24}$ **29.** $\frac{10}{24}$ **31.** $\frac{21}{24}$ **32.** $\frac{8}{12}$ **33.** $\frac{21}{28}$ **34.** $\frac{36}{16}$
35. $\frac{21}{24}$ **37.** 6   **39.** 28   **41.** 12   **43.** 72   **45.** 84   **47.** 96   **49.** 27
51. It probably depends on how large the numbers are. If the numbers are small, the method using multiples of the largest number seems easiest. If the numbers are larger, or there are more than two numbers, then the factorization method or the alternative method will be better.   **52.** Divide the desired denominator by the given denominator. Next, multiply both the given numerator and denominator by this number. Divide 12 by 4 to get 3. Then multiply.

$$\frac{3}{4} = \frac{3 \cdot 3}{4 \cdot 3} = \frac{9}{12}$$

Check your answer by writing $\frac{9}{12}$ in lowest terms. You should get $\frac{3}{4}$.
53. 3600   **55.** 10,584   **57.** like; unlike   **58.** numerators; denominator; lowest   **59.** least; smallest   **60.** 40 is the least common multiple.
61. 70   **62.** 450   **63.** 240 is a common multiple but twice as large as the least common multiple; 120 is the LCM.   **64.** The least common multiple can be no smaller than the largest number in a group and the number 1760 is a multiple of 55.

SECTION 3.3 (pages 219–222)

1. like   **2.** least common denominator   **3.** $\frac{7}{8}$ **5.** $\frac{8}{9}$ **7.** $\frac{3}{4}$ **9.** $\frac{39}{40}$
11. $\frac{23}{36}$ **13.** $\frac{13}{14}$ **15.** $\frac{23}{24}$ **17.** $\frac{29}{36}$ **19.** $\frac{3}{8}$ **21.** $\frac{23}{48}$ **23.** $\frac{1}{2}$
25. $\frac{1}{2}$ **27.** $\frac{7}{15}$ **29.** $\frac{1}{6}$ **31.** $\frac{19}{45}$ **33.** $\frac{3}{40}$ **35.** $\frac{17}{48}$ **37.** $\frac{1}{4}$ in.
39. $\frac{17}{40}$ for reserved seating   **41.** $\frac{31}{40}$ in.   **43.** $\frac{1}{24}$ of the tank
45. You cannot add or subtract until all the fractional pieces are the same size. For example, halves are larger than fourths, so you cannot add $\frac{1}{2} + \frac{1}{4}$ until you rewrite $\frac{1}{2}$ as $\frac{2}{4}$.
46. **Step 1** Rewrite the unlike fractions as like fractions.
   Step 2 Add or subtract the like fractions.
   Step 3 Simplify the answer.
47. $\frac{1}{3}$ **49.** totally honest; 400 people; $\frac{7}{12}$ of the users   **51.** $\frac{3}{16}$ in.

SECTION 3.4 (pages 227–234)

1. 5   **2.** 6   **3.** 9   **4.** 13   **5.** 15   **6.** 21   **7.** 17   **8.** 3
9. *Estimate:* $6 + 3 = 9$; *Exact:* $8\frac{5}{6}$
11. *Estimate:* $7 + 4 = 11$; *Exact:* $11\frac{1}{2}$
13. *Estimate:* $1 + 4 = 5$; *Exact:* $4\frac{5}{24}$
15. *Estimate:* $25 + 19 = 44$; *Exact:* $43\frac{2}{3}$

17. *Estimate:* $34 + 19 = 53$; *Exact:* $52\frac{1}{10}$

19. *Estimate:* $23 + 15 = 38$; *Exact:* $38\frac{5}{28}$

21. *Estimate:* $13 + 19 + 15 = 47$; *Exact:* $45\frac{5}{6}$

23. *Estimate:* $15 - 12 = 3$; *Exact:* $2\frac{5}{8}$

25. *Estimate:* $13 - 1 = 12$; *Exact:* $11\frac{7}{15}$

27. *Estimate:* $28 - 6 = 22$; *Exact:* $22\frac{7}{30}$

29. *Estimate:* $17 - 7 = 10$; *Exact:* $10\frac{3}{8}$

31. *Estimate:* $19 - 6 = 13$; *Exact:* $12\frac{19}{20}$

33. *Estimate:* $20 - 12 = 8$; *Exact:* $7\frac{11}{12}$

35. $\frac{15}{4}$ **36.** $\frac{63}{8}$ **37.** $2\frac{2}{5}$ **38.** $3\frac{3}{7}$ **39.** $\frac{43}{8}$ **40.** $\frac{92}{15}$ **41.** $18\frac{2}{3}$

42. $3\frac{5}{8}$ **43.** $9\frac{1}{8}$ **45.** $11\frac{1}{2}$ **47.** $3\frac{5}{6}$ **49.** $6\frac{11}{12}$ **51.** $8\frac{1}{8}$ **53.** $\frac{5}{6}$

55. $2\frac{7}{8}$ **57.** $2\frac{7}{12}$ **59.** $5\frac{9}{20}$ **61.** $3\frac{16}{21}$ **63.** Find the least common denominator. Change the fraction parts so that they have the same denominator. Add the fraction parts. Add the whole number parts. Write the answer as a mixed number. Simplify the answer.

64. You need to regroup when the minuend (top number) is a whole number or when the fraction in the minuend is smaller than the fraction in the subtrahend (bottom number). Examples are shown below.

$$10 = 9\frac{3}{3} \qquad 5\frac{1}{4} = 5\frac{1}{4} = 4\frac{5}{4}$$

$$\frac{-6\frac{2}{3} = 6\frac{2}{3}}{3\frac{1}{3}} \qquad \frac{-3\frac{1}{2} = 3\frac{2}{4} = 3\frac{2}{4}}{1\frac{3}{4}}$$

65. *Estimate:* $26 - 15 = 11$ ft; *Exact:* $11\frac{1}{4}$ ft

67. *Estimate:* $3 - 3 = 0$ in.; *Exact:* $\frac{7}{16}$ in.

69. *Estimate:* $3 - 1 = 2$ in.; *Exact:* $2\frac{1}{2}$ in.

71. *Estimate:* $3 - 1 = 2$ in.; *Exact:* $2\frac{3}{16}$ in.

73. *Estimate:* $16 + 19 + 24 + 31 = 90$ ft; *Exact:* $87\frac{8}{4} = 89$ ft

75. *Estimate:* $24 + 35 + 24 + 35 = 118$ in.; *Exact:* $116\frac{1}{2}$ in.

77. *Estimate:* $100 - 10 - 14 - 9 - 19 - 12 - 10 - 14 = 12$ gallons; *Exact:* $12\frac{5}{8}$ gallons

79. *Estimate:* $527 - 108 - 151 - 139 = 129$ ft; *Exact:* 130 ft

81. *Estimate:* $59 + 24 + 17 + 29 + 58 = 187$ tons; *Exact:* $186\frac{13}{24}$ tons

83. $4\frac{11}{16}$ in. **85.** $21\frac{3}{8}$ in. **87.** (a) 30 (b) 28 (c) 25 (d) 264

88. least common denominator **89.** (a) $\frac{23}{24}$ (b) $\frac{8}{15}$ (c) $\frac{43}{48}$ (d) $\frac{4}{21}$

90. fraction parts **91.** improper; large

92. (a) $4\frac{5}{8} + 3\frac{6}{8} = 7\frac{11}{8} = 8\frac{3}{8}$; $\frac{37}{8} + \frac{30}{8} = \frac{67}{8} = 8\frac{3}{8}$

(b) $11\frac{56}{40} - 8\frac{35}{40} = 3\frac{21}{40}$; $\frac{496}{40} - \frac{355}{40} = \frac{141}{40} = 3\frac{21}{40}$

SUMMARY EXERCISES Adding and Subtracting Fractions (pages 235–236)

1. proper **2.** improper **3.** improper **4.** proper **5.** $\frac{5}{6}$ **6.** $\frac{7}{8}$

7. $\frac{3}{7}$ **8.** $\frac{23}{47}$ **9.** $\frac{1}{2}$ **10.** $\frac{3}{8}$ **11.** 35 **12.** $\frac{5}{6}$ **13.** $1\frac{1}{6}$ **14.** 56

15. $1\frac{13}{24}$ **16.** $1\frac{13}{16}$ **17.** $2\frac{1}{12}$ **18.** $\frac{1}{12}$ **19.** $\frac{11}{24}$ **20.** $\frac{2}{15}$

21. *Estimate:* $4 \cdot 2 = 8$; *Exact:* $7\frac{7}{8}$

22. *Estimate:* $5 \cdot 3 = 15$; *Exact:* $17\frac{15}{32}$

23. *Estimate:* $8 \cdot 6 \cdot 2 = 96$; *Exact:* $107\frac{2}{3}$

24. *Estimate:* $4 \div 4 = 1$; *Exact:* $1\frac{1}{6}$

25. *Estimate:* $7 \div 2 = 3\frac{1}{2}$; *Exact:* $3\frac{7}{16}$

26. *Estimate:* $5 \div 1 = 5$; *Exact:* $6\frac{1}{6}$

27. *Estimate:* $6 + 4 = 10$; *Exact:* $9\frac{11}{12}$

28. *Estimate:* $18 + 10 = 28$; *Exact:* $28\frac{1}{6}$

29. *Estimate:* $15 + 11 = 26$; *Exact:* $25\frac{4}{15}$

30. *Estimate:* $9 - 4 = 5$; *Exact:* $4\frac{19}{20}$

31. *Estimate:* $14 - 7 = 7$; *Exact:* $6\frac{5}{8}$

32. *Estimate:* $32 - 23 = 9$; *Exact:* $9\frac{1}{4}$

33. 40 **34.** 60 **35.** 30 **36.** 48 **37.** 72 **38.** 84 **39.** 35 **40.** 12 **41.** 12 **42.** 25 **43.** 15 **44.** 55

SECTION 3.5 (pages 243–546)

1.–12.

2. 1. 10. 4. 3. 12. 7. 5. 6. 11. 9. 8.

13. $>$ **15.** $<$ **17.** $>$ **19.** $>$ **21.** true; $\left(\frac{1}{2}\right)^2 = \frac{1}{2} \cdot \frac{1}{2} = \frac{1}{4}$

22. false; $\left(\frac{3}{8}\right)^2 = \frac{3}{8} \cdot \frac{3}{8} = \frac{9}{64}$ **23.** false; $\left(\frac{2}{5}\right)^3 = \frac{2}{5} \cdot \frac{2}{5} \cdot \frac{2}{5} = \frac{8}{125}$

24. true; $\left(\frac{5}{6}\right)^3 = \frac{5}{6} \cdot \frac{5}{6} \cdot \frac{5}{6} = \frac{125}{216}$ **25.** $\frac{1}{9}$ **27.** $\frac{25}{64}$ **29.** $\frac{9}{16}$ **31.** $\frac{64}{125}$

33. $\frac{81}{16} = 5\frac{1}{16}$ **35.** $\frac{81}{256}$

37. A number line is a horizontal line with a range of equally spaced whole numbers placed on it. The lowest number is on the left and the greatest number is on the right. It can be used to compare the size or value of numbers.

38. 1. Do all operations inside parentheses or other grouping symbols. 2. Simplify any expressions with exponents or square roots.   3. Multiply or divide, proceeding from left to right.   4. Add or subtract, proceeding from left to right.

39. 4   **41.** 10   **43.** 1   **45.** $\frac{3}{16}$ **47.** $\frac{4}{9}$ **49.** $\frac{1}{3}$ **51.** $\frac{1}{2}$ **53.** $\frac{3}{8}$ **55.** $\frac{1}{4}$
57. $1\frac{1}{2}$ **59.** $\frac{1}{12}$ **61.** 3   **63.** $\frac{5}{16}$ **65.** $\frac{1}{4}$ **67.** $\frac{1}{32}$ **69.** $\frac{11}{50}$ in Las Vegas
is greater.   **71.** $<; >$ **72.** **(a)** like; numerators; numerator
(b) Answers will vary.   **73.** $\frac{2}{45}$ **74.** $2\frac{1}{32}$

75. – 80.

Chapter 3 REVIEW EXERCISES (pages 253–258)

1. $\frac{6}{7}$ **2.** $\frac{7}{9}$ **3.** $\frac{3}{4}$ **4.** $\frac{1}{8}$ **5.** $\frac{4}{5}$ **6.** $\frac{1}{6}$ **7.** $\frac{13}{31}$ **8.** $\frac{1}{3}$
9. $\frac{11}{12}$ of his total income   **10.** $\frac{1}{4}$ more of the events   **11.** 10   **12.** 12
13. 60   **14.** 24   **15.** 120   **16.** 180   **17.** 8   **18.** 21   **19.** 10
20. 45   **21.** 32   **22.** 20   **23.** $\frac{5}{6}$ **24.** $\frac{7}{8}$ **25.** $\frac{5}{8}$ **26.** $\frac{5}{12}$ **27.** $\frac{13}{24}$
28. $\frac{17}{36}$ **29.** $\frac{9}{10}$ of the students   **30.** $\frac{23}{24}$ of her budget

31. *Estimate:* $19 + 14 = 33$; *Exact:* $32\frac{3}{8}$

32. *Estimate:* $23 + 15 = 38$; *Exact:* $38\frac{1}{9}$

33. *Estimate:* $13 + 9 + 10 = 32$; *Exact:* $31\frac{43}{80}$

34. *Estimate:* $32 - 15 = 17$; *Exact:* $17\frac{1}{12}$

35. *Estimate:* $34 - 16 = 18$; *Exact:* $18\frac{1}{3}$

36. *Estimate:* $215 - 136 = 79$; *Exact:* $79\frac{7}{16}$

37. $9\frac{1}{10}$ **38.** $10\frac{5}{12}$ **39.** $3\frac{1}{4}$ **40.** $1\frac{2}{3}$ **41.** $5\frac{1}{2}$ **42.** $2\frac{19}{24}$

43. *Estimate:* $19 - 6 - 7 = 6$ miles; *Exact:* $5\frac{19}{24}$ miles

44. *Estimate:* $29 + 25 = 54$ tons; *Exact:* $53\frac{5}{12}$ tons

45. *Estimate:* $8 + 3 + 5 + 3 = 19$ pounds; *Exact:* $18\frac{1}{2}$ pounds

46. *Estimate:* $1535 - 1476 = 59$ pounds; *Exact:* $59\frac{11}{16}$ pounds

47. – 50.

47.   48. 49. 50.

51. $<$ **52.** $<$ **53.** $>$ **54.** $>$ **55.** $<$ **56.** $>$ **57.** $<$ **58.** $>$
59. $\frac{1}{4}$ **60.** $\frac{4}{9}$ **61.** $\frac{27}{1000}$ **62.** $\frac{81}{4096}$ **63.** $\frac{1}{2}$ **64.** $6\frac{3}{4}$ **65.** $\frac{1}{16}$
66. 1   **67.** $\frac{3}{16}$ **68.** $1\frac{25}{64}$ **69.** $\frac{3}{4}$ **70.** $\frac{2}{5}$ **71.** $\frac{19}{32}$ **72.** $\frac{11}{16}$ **73.** $2\frac{1}{6}$
74. $26\frac{1}{4}$ **75.** $5\frac{3}{8}$ **76.** $11\frac{43}{80}$ **77.** $15\frac{5}{12}$ **78.** $\frac{8}{11}$ **79.** $\frac{1}{250}$ **80.** $\frac{1}{2}$
81. $\frac{2}{9}$ **82.** $\frac{11}{27}$ **83.** $>$ **84.** $<$ **85.** $<$ **86.** $>$ **87.** 36   **88.** 120
89. 126   **90.** 18   **91.** 108   **92.** 60
93. *Estimate:* $93 - 14 - 22 = 57$ ft; *Exact:* $56\frac{7}{8}$ ft
94. *Estimate:* $4 \cdot 50 = 200$ pounds of sugar; $200 - 69 - 77 - 33 = 21$ pounds; *Exact:* $21\frac{5}{8}$ pounds

Chapter 3 TEST (pages 259–260)

1. $\frac{3}{4}$ **2.** $\frac{1}{2}$ **3.** $\frac{2}{5}$ **4.** $\frac{1}{6}$ **5.** 12   **6.** 30   **7.** 108   **8.** $\frac{5}{8}$ **9.** $\frac{23}{36}$
10. $\frac{5}{24}$ **11.** $\frac{1}{40}$ **12.** *Estimate:* $8 + 5 = 13$; *Exact:* $12\frac{1}{2}$
13. *Estimate:* $16 - 12 = 4$; *Exact:* $4\frac{11}{15}$
14. *Estimate:* $19 + 9 + 12 = 40$; *Exact:* $40\frac{29}{60}$
15. *Estimate:* $24 - 18 = 6$; *Exact:* $5\frac{5}{8}$
16. Answers will vary. One possibility is: Probably addition and subtraction of fractions are more difficult because you have to find the least common denominator and then change the fractions to the same denominator.
17. Answers will vary. One possibility is: Round mixed numbers to the nearest whole number. Then add or subtract to estimate the answer. The estimate may vary from the exact answer but it lets you know if your answer is reasonable.
18. *Estimate:* $10 + 85 + 37 + 8 = 140$ pounds; *Exact:* $140\frac{1}{24}$ pounds
19. *Estimate:* $148 - 69 - 37 = 42$ gallons; *Exact:* $41\frac{5}{8}$ gallons
20. $>$ **21.** $>$ **22.** 2   **23.** $\frac{13}{48}$ **24.** $1\frac{3}{4}$ **25.** $1\frac{1}{3}$

Chapters 1–3 CUMULATIVE REVIEW EXERCISES (pages 261–262)

1. 5, 3, 9, 2   **2.** 59,800; 59,800; 60,000
3. *Estimate:* $20,000 - 10,000 = 10,000$; *Exact:* 14,389
4. *Estimate:* $100,000 \div 40 = 2500$; *Exact:* 3211
5. 1,255,609   **6.** 2,801,695   **7.** 160   **8.** 369,408   **9.** 135   **10.** 2693 R2
11. *Estimate:* $20 + 9 + 5 + 20 + 9 + 5 = 68$ ft; *Exact:* 64 ft
12. *Estimate:* $20 \cdot 10 = 200$ ft²; *Exact:* 252 ft²
13. *Estimate:* $47 \cdot 9 = 423$ miles; *Exact:* $413\frac{7}{8}$ miles
14. *Estimate:* $70 \cdot 130 = 9100$ in.; *Exact:* $9132\frac{1}{2}$ in.

15. 144 **16.** 9 **17.** 44 **18.** $\dfrac{4}{45}$ **19.** $\dfrac{9}{10}$ **20.** $1\dfrac{1}{16}$ **21.** $\dfrac{1}{2}$

22. $36\dfrac{3}{4}$ **23.** $2\dfrac{3}{16}$ **24.** $13\dfrac{1}{2}$ **25.** *Estimate:* $3 + 5 = 8$; *Exact:* $7\dfrac{7}{8}$

26. *Estimate:* $22 + 4 = 26$; *Exact:* $26\dfrac{7}{24}$

27. *Estimate:* $5 - 2 = 3$; *Exact:* $2\dfrac{5}{8}$

28. – 31.

$\dfrac{1}{9}$ $\dfrac{5}{3}$ $2\dfrac{3}{4}$ $\dfrac{10}{3}$

0 1 2 3 4

29.   **30.** **28. 31.**

32. < **33.** > **34.** <

CHAPTER 4 Decimals

SECTION 4.1 (pages 269–273)

1. 3 **2.** 4 **3.** 8 tenths **4.** 3 thousandths **5.** 7; 0; 4 **7.** 4; 7; 0
9. 1; 8; 9 **11.** 6; 2; 1 **13.** 410.25 **15.** 6.5432 **17.** 5406.045
19. $\dfrac{7}{10}$ **21.** $13\dfrac{2}{5}$ **23.** $\dfrac{1}{4}$ **25.** $\dfrac{33}{50}$ **27.** $10\dfrac{17}{100}$ **29.** $\dfrac{3}{50}$ **31.** $\dfrac{41}{200}$
33. $5\dfrac{1}{500}$ **35.** $\dfrac{343}{500}$ **37.** five tenths **39.** seventy-eight hundredths
41. one hundred five thousandths **43.** twelve and four hundredths
45. one and seventy-five thousandths **47.** 6.7 **49.** 0.32
51. 420.008 **53.** 0.0703 **55.** 75.030 **57.** Anne should not say "and" because that denotes a decimal point. **58.** Jerry used "and" twice; only the first "and" is correct. **59.** ten thousandths inch; $\dfrac{10}{1000} = \dfrac{1}{100}$ inch
61. 12 pounds **63.** 3-C **64.** 4-C **65.** 4-A **66.** 3-B **67.** One and six hundred two thousandths centimeters **68.** One and twenty-six thousandths centimeters. **69.** millionths, ten-millionths, hundred-millionths, billionths; these match the words on the left side of the place value chart in Chapter 1 with "ths" attached. **70.** The first place to the left of the decimal point is ones, so the first place to the right could be one*ths,* like tens and ten*ths.* But anything that is 1 or more is to the *left* of the decimal point. **71.** Seventy-two million four hundred thirty-six thousand nine hundred fifty-five hundred-millionths **72.** six hundred seventy-eight thousand five hundred fifty-four billionths **73.** eight thousand six and five hundred thousand one millionths **74.** twenty thousand sixty and five hundred five millionths
75. 0.0302040 **76.** 9,876,543,210.100200300

SECTION 4.2 (pages 279–280)

1. 0 **2.** 9 **3.** Look only at the 6, which is *5 or more.* So round up by adding one thousandth to 5.709 to get 5.710. **4.** Look only at the 2, which is *4 or less.* So the part you are keeping, 10.0, stays the same. **5.** 16.9
7. 0.956 **9.** 0.80 **11.** 3.661 **13.** 794.0 **15.** 0.0980 **17.** 9.09
19. 82.0002 **21.** $0.82 **23.** $1.22 **25.** $0.70 **27.** $48,650
29. $840 **31.** $500 **33.** $1.00 **35.** $1000 **37. (a)** 253 miles per hour **(b)** 135 miles per hour **39. (a)** 186.0 miles per hour
(b) 763.0 miles per hour **41.** Rounds to $0 (zero dollars) because $0.499 is closer to $0 than to $1. **42.** Round amounts less than $1.00 to the nearest cent instead of the nearest dollar. **43.** Rounds to $0.00 (zero cents) because $0.0015 is closer to $0.00 than to $0.01.

44. Both round to $0.60. Rounding to nearest thousandth (tenth of a cent) would allow you to identify $0.597 as less than $0.601.

SECTION 4.3 (pages 285–288)

1. $\begin{array}{r} 6.420 \\ +\,10.163 \end{array}$ **2.** $\begin{array}{r} 7.000 \\ +\,9.204 \end{array}$ **3.** $\begin{array}{r} 20.0000 \\ -\,9.1263 \end{array}$ **4.** $\begin{array}{r} 137.06 \\ -\,12.00 \end{array}$ **5.** 17.48
7. 7.763 **9.** 77.006 **11.** 20.104 **13.** 0.109 **15.** 330.86895
17. (a) 24.75 in. **(b)** 3.95 in. **19. (a)** 62.27 in. **(b)** 0.39 in.
21. 6 should be written 6.00; sum is 46.22.
22. The two problems are done in a different order: $8 - 2.9$ is not the same as $2.9 - 8$ because subtraction is not commutative. **23.** *Estimate:* $20 - 7 = 13$; *Exact:* $13.16 **25.** *Estimate:* $400 + 1 + 20 = 421$; *Exact:* 414.645 **27.** *Estimate:* $9 - 4 = 5$; *Exact:* 4.849 **29.** *Estimate:* $60 + 500 + 6 = 566$; *Exact:* 608.4363 **31.** 0.275 **33.** 6.507
35. 1.81 **37.** 6056.7202 **39.** *Estimate:* $80 - 30 = 50$ million people; *Exact:* 50.4 million people
41. *Estimate:* $400 + 200 + 100 + 80 + 40 + 30 + 30 = 880$ million people; *Exact:* 940.82 million people **43.** *Estimate:* $2 + 2 + 2 = 6$ meters; *Exact:* 6.19 meters, which is less than the rhino by 0.21 meter
45. *Estimate:* $11 - 10 = 1$ ounce; *Exact:* 0.65 ounce
47. *Estimate:* $20 + 6 + 20 + 6 = 52$ in.; *Exact:* 52.1 in.
49. *Estimate:* $5 - 5 = 0$; *Exact:* $0.30
51. *Estimate:* $19 + 2 + 2 + 10 + 2 = 35$; *Exact:* $35.25
53. $2059.36 **55.** $103.97 **57.** $498.22 **59.** $b = 1.39$ centimeters
61. $q = 7.943$ ft

SECTION 4.4 (pages 291–294)

1. 3 **2.** 5 **3.** 2 **4.** 4 **5.** 0.1344 **7.** 159.10 **9.** 15.5844
11. $34,500.20 **13.** 43.2 **14.** 43.2 **15.** 0.432
16. 0.432 **17.** 0.0432 **18.** 0.0432 **19.** 0.0000312 **21.** 0.000025
23. *Estimate:* $40 \times 5 = 200$; *Exact:* 190.08 **25.** *Estimate:* $40 \times 40 = 1600$; *Exact:* 1558.2 **27.** *Estimate:* $7 \times 5 = 35$; *Exact:* 30.038
29. *Estimate:* $3 \times 7 = 21$; *Exact:* 19.24165 **31.** unreasonable; $289.00
33. reasonable **35.** unreasonable; $4.19 **37.** unreasonable; 9.5 pounds
39. 945.87 (rounded) **41.** $2.45 (rounded) **43.** $77.10 (rounded)
45. $20,265 **47. (a)** Area before 1929 $\approx$ 23.2 in.²; Area today $\approx$ 16.0 in.²
(b) 7.2 in.² **49. (a)** 0.43 in. **(b)** 4.3 in. **51.** $1264.04; $2723.72
53. $4.09 (rounded) **55. (a)** 72.05 **(b)** $27.80
57. 59.6; 32; 4.76; 803.5; 7226; 9. Multiplying by 10, decimal point moves one place to the right; by 100, two places to the right; by 1000, three places to the right. **58.** 5.96; 0.32; 0.0476; 8.035; 6.5; 52.3. Multiplying by 0.1, decimal point moves one place to the left; by 0.01, two places to the left; by 0.001, three places to the left.

SUMMARY EXERCISES Adding, Subtracting, and Multiplying Decimal Numbers (pages 295–296)

1. $\dfrac{4}{5}$ **2.** $6\dfrac{1}{250}$ **3.** $\dfrac{7}{20}$ **4.** ninety-four and five tenths **5.** two and three ten-thousandths **6.** seven hundred six thousandths **7.** 0.05 **8.** 0.0309
9. 10.7 **10.** 6.19 **11.** 1.0 **12.** 0.420 **13.** $0.89 **14.** $3.00
15. $100 **16.** 0.945 **17.** 49.6199 **18.** 1.845 **19.** $93.50
20. 0.00488 **21.** 2.15 **22.** 10.955 **23.** 18.4009 **24.** 4.3043

25. 0.87 **26.** $110.84 **27.** 82.84 (rounded) **28.** $P = 3.1$ in.; $A \approx 0.5$ in.^{2} **29.** $P = 3.25$ in.; $A \approx 0.66$ in.^{2} **30.** (a) Pumpkin is heaviest; strawberry is lightest. (b) 1809.5 pounds **31.** 92.8 pounds (rounded) **32.** (a) 40.5 pounds; (b) 6.0 or 6 pounds **33.** (a) 3.6875 pounds (b) 2.4625 pounds **34.** (a) 4; 7; 17; 2; 69; 1810; 1; 8 (all pounds) (b) *Estimate:* 1918 pounds; *Exact:* 1915.7455 pounds **35.** $91.28

SECTION 4.5 (pages 303–306)

1. c; $5\overline{)25.5}$ **2.** b; $7\overline{)42.3}$ **3.** $2.2\overline{)8.24}$ **4.** $5.1\overline{)10.5}$ **5.** 3.9 **7.** 0.47 **9.** 400.2 **11.** 36 **13.** 0.06 **14.** 0.6 **15.** 6000 **16.** 6 **17.** 25.3 **19.** 516.67 (rounded) **21.** 26.756 (rounded) **23.** 10,082.647 (rounded) **25.** unreasonable; $40 \div 8 = 5$; Correct answer is 4.725. **27.** reasonable; $50 \div 50 = 1$ **29.** unreasonable; $300 \div 5 = 60$; Correct answer is 60.2. **31.** $4.00 (rounded) **33.** $67.08 **35.** $11.92 per hour **37.** 28.0 miles per gallon (rounded) **39.** $0.03 per can (rounded) **41.** 7.37 meters (rounded) **43.** 0.08 meter **45.** 22.49 meters **47.** (a) Work inside parentheses; subtract $9.5 - 3.1$ to get 6.4. (b) Apply the exponent; multiply $(2.2)(2.2)$ to get 4.84. (c) Add 4.84 plus 6.4 to get 11.24. **48.** (a) Work inside parentheses; add $0.4 + 5.07$ to get 5.47 (b) Multiply 5.47 times 3 to get 16.41 (c) Subtract $60.41 - 16.41$ to get 44 **49.** 14.25 **51.** 73.4 **53.** 1.205 **55.** 0.334 **57.** (a) 1,083,333 pieces (rounded) (b) 18,056 pieces (rounded) (c) 301 pieces (rounded). **59.** 100,000 box tops **61.** 2632 box tops (rounded) **63.** 0.377; 0.91; 0.0886; 3.019; 40.65; 662.57. (a) Dividing by 10, decimal point moves one place to the left; by 100, two places to the left; by 1000, three places to the left. (b) The decimal point moved to the *right* when *multiplying* by 10 or 100 or 1000. Here it moves to the *left* when *dividing* by those numbers. **64.** 402; 71; 3.39; 157.7; 460; 8730 (a) Dividing by 0.1, decimal point moves one place to the right; by 0.01, two places to the right; by 0.001, three places to the right (b) The decimal point moved to the *left* when *multiplying* by 0.1 or 0.01 or 0.001. Here the decimal point moves to the *right* when *dividing* by those numbers.

SECTION 4.6 (pages 311–314)

1. (a) not correct, should be $5\overline{)2}$; (b) correct (c) correct **2.** (a) correct (b) not correct, should be $3\overline{)1}$; (c) correct **3.** $4\overline{)3.}^{\,0.}$ To continue dividing, write zeros in the dividend. **4.** $8\overline{)1.}^{\,0.}$ To continue dividing, write zeros in the dividend. **5.** (a) less than (b) greater than (c) less than **6.** (a) less than (b) greater than (c) greater than **7.** 0.5 **9.** 0.75 **11.** 0.3 **13.** 0.9 **15.** 0.6 **17.** 0.875 **19.** 2.25 **21.** 14.7 **23.** 3.625 **25.** 6.333 (rounded) **27.** 0.833 (rounded) **29.** 1.889 (rounded) **31.** $\frac{2}{5}$ **33.** $\frac{5}{8}$ **35.** $\frac{7}{20}$ **37.** 0.35 **39.** $\frac{1}{25}$ **41.** 0.2 **43.** $\frac{9}{100}$ **45.** shorter; 0.72 inch **47.** more; 0.05 inch **49.** too much; 0.005 gram **51.** 0.9991 cm, 1.0007 cm **52.** 3.0 ounces, 2.995 ounces, 3.005 ounces **53.** 0.5399, 0.54, 0.5455 **55.** 5.0079, 5.79, 5.8, 5.804 **57.** 0.6009, 0.609, 0.628, 0.62812 **59.** 2.8902, 3.88, 4.876, 5.8751 **61.** 0.006, 0.043, $\frac{1}{20}$, 0.051 **63.** 0.37, $\frac{3}{8}$, $\frac{2}{5}$, 0.4001 **65.** red box **67.** green box **69.** 1.4 in. (rounded) **71.** 0.3 in. (rounded)

73. 0.4 in. (rounded) **75.** (a) A proper fraction like $\frac{5}{9}$ is less than 1, so it cannot be equivalent to a decimal number that is greater than 1. (b) $\frac{5}{9}$ means $5 \div 9$ or $9\overline{)5}$ so correct answer is 0.556 (rounded). This makes sense because both the fraction and decimal are less than 1. **76.** (a) $2.035 = 2\frac{35}{1000} = 2\frac{7}{200}$, not $2\frac{7}{20}$. (b) Adding the whole number part gives $2 + 0.35$, which is 2.35, not 2.035. To check, $2.35 = 2\frac{35}{100} = 2\frac{7}{20}$. **77.** Just add the whole number part to 0.375. So $1\frac{3}{8} = 1.375$; $3\frac{3}{8} = 3.375$; $295\frac{3}{8} = 295.375$. **78.** It works only when the fraction part has a one-digit numerator and a denominator of 10, a two-digit numerator and a denominator of 100, and so on.

Chapter 4 REVIEW EXERCISES (pages 320–323)

1. 0; 5 **2.** 0; 6 **3.** 8; 9 **4.** 5; 9 **5.** 7; 6 **6.** $\frac{1}{2}$ **7.** $\frac{3}{4}$ **8.** $4\frac{1}{20}$ **9.** $\frac{7}{8}$ **10.** $\frac{27}{1000}$ **11.** $27\frac{4}{5}$ **12.** eight tenths **13.** four hundred and twenty-nine hundredths **14.** twelve and seven thousandths **15.** three hundred six ten-thousandths **16.** 8.3 **17.** 0.205 **18.** 70.0066 **19.** 0.30 **20.** 275.6 **21.** 72.79 **22.** 0.160 **23.** 0.091 **24.** 1.0 **25.** $15.83 **26.** $0.70 **27.** $17,625.79 **28.** $350 **29.** $130 **30.** $100 **31.** $29 **32.** *Estimate:* $6 + 400 + 20 = 426$; *Exact:* 444.86 **33.** *Estimate:* $80 + 1 + 100 + 1 + 30 = 212$; *Exact:* 233.515 **34.** *Estimate:* $300 - 20 = 280$; *Exact:* 290.7 **35.** *Estimate:* $9 - 8 = 1$; *Exact:* 1.2684 **36.** *Estimate:* 90 million $-$ 20 million $=$ 70 million; *Exact:* 77.0 million more cats **37.** *Estimate:* $400 - $300 - $70 = 30; *Exact:* $15.80 **38.** *Estimate:* $2 + $5 + $20 = 27; $30 - $27 = 3; *Exact:* $4.14 **39.** *Estimate:* $2 + 4 + 5 = 11$ kilometers; *Exact:* 11.55 kilometers **40.** *Estimate:* $6 \times 4 = 24$; *Exact:* 22.7106 **41.** *Estimate:* $40 \times 3 = 120$; *Exact:* 141.57 **42.** 0.0112 **43.** 0.000355 **44.** reasonable; $700 \div 10 = 70$ **45.** unreasonable; $30 \div 3 = 10$; Correct answer is 9.5. **46.** 14.467 (rounded) **47.** 1200 **48.** 0.4 **49.** $708 (rounded) **50.** $2.99 (rounded) **51.** 133 shares (rounded) **52.** $3.47 (rounded) **53.** 29.215 **54.** 10.15 **55.** 3.8 **56.** 0.64 **57.** 1.875 **58.** 0.111 (rounded) **59.** 3.6008, 3.68, 3.806 **60.** 0.209, 0.2102, 0.215, 0.22 **61.** $\frac{1}{8}$, $\frac{3}{20}$, 0.159, 0.17 **62.** 404.865 **63.** 254.8 **64.** 3583.261 (rounded) **65.** 29.0898 **66.** 0.03066 **67.** 9.4 **68.** 175.675 **69.** 9.04 **70.** 19.50 **71.** 8.19 **72.** 0.928 **73.** 35 **74.** 0.259 **75.** 0.3 **76.** $3.00 (rounded) **77.** $2.17 (rounded) **78.** $35.96 **79.** $199.71 **80.** $78.50 **81.** (a) baked potato with skin (b) $\frac{1}{2}$ cup green peas (c) 0.59 milligram **82.** (a) 2.08 milligrams (b) more, by 0.08 milligram

Chapter 4 TEST (pages 324–325)

1. $18\frac{2}{5}$ **2.** $\frac{3}{40}$ **3.** sixty and seven thousandths **4.** two hundred eight ten-thousandths **5.** 725.6 **6.** 0.630 **7.** $1.49 **8.** $7860 **9.** *Estimate:* $8 + 80 + 40 = 128$; *Exact:* 129.2028 **10.** *Estimate:* $80 - 4 = 76$; *Exact:* 75.498 **11.** *Estimate:* $6(1) = 6$; *Exact:* 6.948 **12.** *Estimate:* $20 \div 5 = 4$; *Exact:* 4.175 **13.** 839.762

14. 669.004 **15.** 0.0000483 **16.** 480 **17.** 2.625
18. 0.44, $\frac{9}{20}$, 0.4506, 0.451 **19.** 35.49 **20.** $1294.47 **21.** pintails, wigeons, gadwalls **22.** $5.35 (rounded) **23.** 2.8 degrees **24.** $3.79 per foot (rounded) **25.** Answer varies.

Chapters 1–4 CUMULATIVE REVIEW EXERCISES (pages 326–327)

1. 500,000 **2.** 602.49 **3.** $710 **4.** $0.05 **5.** 9.671 **6.** $1\frac{4}{9}$
7. $1\frac{2}{5}$ **8.** 4914 **9.** 93.603 **10.** 404 R3 **11.** $1\frac{17}{24}$ **12.** 233,728
13. 0.03264 **14.** 8 **15.** 45 **16.** $\frac{4}{31}$ **17.** $\frac{2}{3}$ **18.** 0.51 (rounded)
19. 4 **20.** 14 **21.** forty and thirty-five thousandths **22.** 0.0306
23. 7.005, 7.5, 7.5005, 7.505 **24.** 0.8, 0.8015, $\frac{21}{25}$, $\frac{7}{8}$
25. *Estimate:* $60 − $50 − $1 = $9; *Exact:* $11.17
26. *Estimate:* 50 − 47 = 3 in.; *Exact:* $3\frac{3}{8}$ in.
27. *Estimate:* $10 × 20 = $200; *Exact:* $191.90 (rounded)
28. *Estimate:* (8 × 20) + (10 × 30) = 160 + 300 = 460 students;
Exact: 488 students **29.** *Estimate:* 2 + 4 = 6 yards; *Exact:* $6\frac{5}{24}$ yards
30. *Estimate:* $30 + $200 − $40 − $40 = $150; *Exact:* $174.50
31. *Estimate:* $6 ÷ 3 = $2 per pound; *Exact:* $2.29 per pound (rounded)
32. *Estimate:* $80,000 ÷ 100 = $800; *Exact:* $729 (rounded)
33. size M (medium) **34.** XXS $1\frac{1}{2}$ in.; XS $\frac{3}{4}$ in.; S $3\frac{3}{4}$ in.; M $\frac{7}{8}$ in.; L $3\frac{3}{4}$ in.
35. 21.125 − 20.25 = 0.875 in.; 7 ÷ 8 = 0.875 in.
or $\frac{875}{1000} = \frac{875 \div 125}{1000 \div 125} = \frac{7}{8}$ in.
36. Answers will vary. Many people prefer using decimals because you do not need to find a common denominator or rewrite answers in lowest terms.

CHAPTER 5 Ratio and Proportion

SECTION 5.1 (pages 335–338)

1. Answers will vary. One possibility is: A ratio compares two quantities with the same units. Examples will vary. **2.** You can divide out the common (same) units. **3.** 20; $\frac{4}{1}$ **4.** 5; $\frac{4}{15}$ **5.** $\frac{8}{9}$ **7.** $\frac{2}{1}$ **9.** $\frac{1}{3}$ **11.** $\frac{8}{5}$
13. $\frac{3}{8}$ **15.** $\frac{9}{7}$ **17.** $\frac{6}{1}$ **19.** $\frac{5}{6}$ **21.** ounces; takes fewer steps to solve and you do not have to work with a mixed number **22.** hours; takes fewer steps to solve and you do not have to work with a mixed number **23.** $\frac{8}{5}$
25. $\frac{1}{12}$ **27.** $\frac{5}{16}$ **29.** $\frac{4}{1}$ **31.** $\frac{1}{6}$ **33.** $\frac{15}{1}$ **35.** Answers will vary. One possibility is stocking cards of various types in the same ratios as those in the table. **36.** Answers will vary. Examples: a person may send Valentine's Day cards every year but graduation cards only once few years; Valentine's Day may have more advertising. **37.** $\frac{6}{5}$; $\frac{36}{17}$ **39.** *White Christmas* to *It's Now or Never*; *White Christmas* to *I Will Always Love You*; *Candle in the Wind* to *I Want to Hold Your Hand*. **41.** $\frac{7}{5}$ **43.** $\frac{6}{1}$ **45.** $\frac{38}{17}$ **47.** $\frac{1}{4}$ **49.** $\frac{34}{35}$
51. $\frac{1}{1}$; as long as the sides all have the same length, any measurement you choose will maintain the ratio.

52. Answers will vary. Some possibilities are:
$\frac{4}{5} = \frac{8}{10} = \frac{12}{15} = \frac{16}{20} = \frac{20}{25} = \frac{24}{30} = \frac{28}{35}$. **53.** It is not possible. Amelia would have to be older than her mother to have a ratio of 5 to 3.
54. Answers will vary, but a ratio of 3 to 1 means your income is 3 times your friend's income.

SECTION 5.2 (pages 343–346)

1. $\frac{5 \text{ cups}}{3 \text{ people}}$ **3.** $\frac{3 \text{ feet}}{7 \text{ seconds}}$ **5.** $\frac{18 \text{ miles}}{1 \text{ gallon}}$
7. Divide; correct set-up is $\frac{\$5.85}{3 \text{ boxes}}$ or $3\overline{)5.85}$ **8.** 5 rooms for 1 nurse.
9. $12 per hour or $12/hour **11.** 1.25 pounds/person **13.** 325.9; 21.0 (rounded) **15.** 338.6; 20.9 (rounded) **17.** $1.125/oz; $0.913/oz (rounded); $0.913/oz; best buy is 4 oz for $3.65.
19. 14 ounces for $2.89, about $0.206/ounce **21.** 18 ounces for $1.79, about $0.099/ounce **23.** Answers will vary. For example, you might choose Brand B because you like more chicken, so the cost per chicken chunk may actually be the same as or less than Brand A. **24.** Answers will vary. For example, if you use only half of the larger bag, you really pay $0.30 per pound, so the smaller bag is the better buy. **25.** 1.75 pounds/week
27. $12.26/hour **29. (a)** Penny Saver, $0.415; Most Minutes, $0.503 (rounded); USA Card, $0.30 **(b)** Penny Saver, $0.083/min; Most Minutes, $0.101/min (rounded); USA Card, $0.06/min; USA Card is the best buy.
31. Penny Saver total $0.54, $0.018/min; Most Minutes total $0.565, $0.019/min (rounded); USA card total $0.55, $0.018/min (rounded); all three unit rates are very similar. **33.** one battery for $1.79; like getting 3 batteries so $1.79 ÷ 3 ≈ $0.597 per battery **35.** Brand P with the 50¢ coupon is the best buy. ($3.39 − $0.50 = $2.89; $2.89 ÷ 16.5 ounces ≈ $0.175 per ounce) **37.** Plan A: $0.09/min; Plan B: $0.07/min; Plan B is the better buy. **38. (a)** 15 min/day **(b)** 30 min/day
39. (60 extra min) ($0.45) = $27 overage
$39.95 + $27 overage = $66.95
$\frac{\$66.95}{450 \text{ min} + 60 \text{ min}} = \frac{\$66.95}{510 \text{ min}} \approx \$0.13 \text{ min}$
40. (90 extra min) ($0.40) = $36 overage
$59.95 + $36 overage = $95.95
$\frac{\$95.95}{900 \text{ min} + 90 \text{ min}} = \frac{\$95.95}{990 \text{ min}} \approx \$0.10/\text{min}$
41. From Exercise 39: $\frac{\$69.95}{510 \text{ min}} \approx \$0.14 \text{ min}$
From Exercise 40: $\frac{\$69.95}{990 \text{ min}} \approx \$0.07 \text{ min}$

SECTION 5.3 (pages 350–351)

1. $\frac{\$9}{12 \text{ cans}} = \frac{\$18}{24 \text{ cans}}$ **3.** $\frac{200 \text{ adults}}{450 \text{ children}} = \frac{4 \text{ adults}}{9 \text{ children}}$ **5.** $\frac{120}{150} = \frac{8}{10}$
7. $\frac{3}{5} = \frac{3}{5}$; true **9.** $\frac{5}{8} = \frac{5}{8}$; true **11.** $\frac{3}{4} \neq \frac{2}{3}$; false **13.** $\frac{14}{5} = \frac{14}{5}$; true
15. $\frac{16}{9} = \frac{16}{9}$; true **17.** $\frac{7}{6} \neq \frac{9}{8}$; false **19.** Answers may vary. One example: A proportion shows that two ratios are equal. The multiplications on the proportion should show 6 • 45 = 270 and 30 • 9 = 270.
20. Answers will vary. One possibility: If the cross products are equal, the proportion is true. If the cross products are not equal, the proportion is false.
21. 54 = 54; True **23.** 336 ≠ 320; False **25.** 2880 ≠ 2970; False

27. $28 = 28$; True   **29.** $44.8 \neq 45$; False   **31.** $66 = 66$; True

33. $68\frac{1}{4} = 68\frac{1}{4}$; True   **35.** $5\frac{2}{5} \neq 5\frac{1}{3}$; False

37. $2\frac{1}{80} \neq 2\frac{7}{100}$ or $2.0125 \neq 2.07$; False

39. $\dfrac{68 \text{ hits}}{200 \text{ at bats}} = \dfrac{153 \text{ hits}}{450 \text{ at bats}}$ $200 \cdot 153 = 30{,}600$ $68 \cdot 450 = 30{,}600$

Cross products are *equal* so the proportion is *true;* they hit equally well.
40. Left-hand ratio compares hours to cartons, but right-hand ratio compares cartons to hours. Correct proportion is shown.

$\dfrac{3.5 \text{ hours}}{91 \text{ cartons}} = \dfrac{5.25 \text{ hours}}{126 \text{ cartons}}$

$91 \cdot 5.25 = 477.75$ Cross products are *not* equal so the proportion is *false;*
$3.5 \cdot 126 = 441$ the men do not work equally fast.

SECTION 5.4 (pages 356–357)

1. Finding the cross products; $10 \cdot 5 = 50$ and $7 \cdot x$

2. Divide both sides by 7; $\dfrac{\overset{1}{\cancel{7}} \cdot x}{\cancel{7}_1} = \dfrac{50}{7}$ **3.** 12; 12; $x = 4$ **5.** $x = 2$

7. $x = 88$ **9.** $x = 91$ **11.** $x = 5$ **13.** $x = 10$ **15.** $x \approx 24.44$ (rounded)

17. $x = 50.4$ **19.** $x \approx 17.64$ (rounded)   **21.** $x = 1$ **23.** $x = 3\frac{1}{2}$

25. $x = 0.2$ or $x = \frac{1}{5}$ **27.** $x = 0.005$ or $x = \frac{1}{200}$

29. Find cross products: $20 \neq 30$, so the proportion is false.

$\dfrac{6\frac{2}{3}}{4} = \dfrac{5}{3}$ or $\dfrac{10}{6} = \dfrac{5}{3}$ or $\dfrac{10}{4} = \dfrac{7.5}{3}$ or $\dfrac{10}{4} = \dfrac{5}{2}$

30. Find cross products: $192 \neq 180$, so the proportion is false.

$\dfrac{6.4}{8} = \dfrac{24}{30}$ or $\dfrac{6}{7.5} = \dfrac{24}{30}$ or $\dfrac{6}{8} = \dfrac{22.5}{30}$ or $\dfrac{6}{8} = \dfrac{24}{32}$

SUMMARY EXERCISES Ratios, Rates, and Proportions (pages 358–359)

1. $\dfrac{50 \text{ million}}{25 \text{ million}} = \dfrac{2}{1}$ **2.** $\dfrac{15 \text{ million}}{20 \text{ million}} = \dfrac{3}{4}$ **3.** $\dfrac{20 \text{ million} + 25 \text{ million}}{50 \text{ million}} = \dfrac{9}{10}$

4. $\dfrac{380 \text{ million}}{110 \text{ million}} = \dfrac{38}{11}$ **5.** Comparing the violin to piano, guitar, organ, clarinet, and drums gives ratios of $\frac{1}{11}, \frac{1}{10}, \frac{1}{3}, \frac{1}{2}$, and $\frac{2}{3}$ respectively.

6. (a) guitar to clarinet   **(b)** organ to drums, or clarinet to violin
7. 2.1 points/min; 0.5 min/point (both rounded)   **8.** 1.7 points/min; 0.6 min/point (both rounded)   **9.** $16.32/hour; $24.48/hour for overtime   **10.** $0.20/channel; $0.17/channel; $0.18/channel (all answers rounded)   **11.** 12 ounces at $0.60/ounce (rounded)   **12.** Brand P with the $2 coupon is the best buy at $0.57 per pound.   **13.** $\dfrac{4}{3} = \dfrac{4}{3}$

or $924 = 924$; true   **14.** $2.0125 \neq 2.07$; false   **15.** $68\frac{1}{4} = 68\frac{1}{4}$; true   **16.** $x = 28$ **17.** $x = 3.2$ **18.** $x = 182$ **19.** $x \approx 3.64$ (rounded)   **20.** $x \approx 0.93$ (rounded)   **21.** $x = 1.56$ **22.** $x \approx 0.05$ (rounded)   **23.** $x = 1$ **24.** $x = \frac{3}{4}$

SECTION 5.5 (pages 363–366)

1. $\dfrac{6 \text{ potatoes}}{4 \text{ eggs}} = \dfrac{12 \text{ potatoes}}{x \text{ eggs}}$ **2.** $\dfrac{1 \text{ serving}}{15 \text{ chips}} = \dfrac{x \text{ serving}}{70 \text{ chips}}$ **3.** 18; x;

22.5 hours   **5.** $7.20   **7.** 42 pounds   **9.** $403.68   **11.** 10 ounces (rounded)   **13.** 5 quarts   **15.** 14 ft, 10 ft   **17.** 14 ft, 8 ft   **19.** 96 pieces chicken; 33.6 pounds lasagna; 10.8 pounds deli meats; $5\frac{3}{5}$ pounds cheese; 7.2 dozen (about 86) buns; 14.4 pounds of salad   **21.** 44 students is an *unreasonable* answer because there are only 35 students in the class.
22. 2 minutes is an *unreasonable* answer because it is less than the 30 minutes she gets in just one day.   **23.** 2065 students (reasonable); about 4214 students with incorrect set-up (only 2950 students in the group)   **25.** about 83 people (reasonable); about 750 people with incorrect set-up (only 250 people attended)   **27.** 625 stocks   **29.** 4.06 meters (rounded)   **31.** 311 calories (rounded)   **33.** 10.53 meters (rounded)
35. You cannot solve this problem using a proportion because the ratio of age to weight is not constant. As Jim's age increases, his weight may decrease, stay the same, or increase.   **36.** Answers will vary; Exercises 3–34 are all examples of application problems.
37. 5610 students use cream   **39.** 120 calories and 12 grams of fiber
41. $1\frac{3}{4}$ cups water, 3 Tbsp margarine, $\frac{3}{4}$ cup milk, 2 cups flakes
42. $5\frac{1}{4}$ cups water, 9 Tbsp margarine, $2\frac{1}{4}$ cups milk, 6 cups flakes

Chapter 5 REVIEW EXERCISES (pages 372–375)

1. $\frac{3}{4}$ **2.** $\frac{4}{1}$ **3.** great white shark to whale shark; whale shark to blue whale   **4.** $\frac{2}{1}$ **5.** $\frac{2}{3}$ **6.** $\frac{5}{2}$ **7.** $\frac{1}{6}$ **8.** $\frac{3}{1}$ **9.** $\frac{3}{8}$ **10.** $\frac{4}{3}$ **11.** $\frac{1}{9}$
12. $\frac{10}{7}$ **13.** $\frac{7}{5}$ **14.** $\frac{5}{6}$ **15.** $\dfrac{\$11}{1 \text{ dozen}}$ **16.** $\dfrac{12 \text{ children}}{5 \text{ families}}$
17. 0.2 page/minute or $\frac{1}{5}$ page/minute; 5 minutes/page
18. $20/hour; 0.05 hour/dollar or $\frac{1}{20}$ hour/dollar
19. 8 ounces for $4.98, about $0.623/ounce
20. 17.6 pounds for $18.69 − $1 coupon, about $1.005/pound
21. $\frac{3}{5} = \frac{3}{5}$ or $90 = 90$; true   **22.** $\frac{1}{8} \neq \frac{1}{4}$ or $432 \neq 216$; false
23. $\frac{47}{10} \neq \frac{49}{10}$ or $980 \neq 940$; false   **24.** $\frac{16}{9} = \frac{16}{9}$ or $3456 = 3456$; true
25. $4.8 = 4.8$; true   **26.** $14 = 14$; true   **27.** $x = 1575$ **28.** $x = 20$
29. $x = 400$ **30.** $x = 12.5$ **31.** $x \approx 14.67$ (rounded)   **32.** $x \approx 8.17$ (rounded)   **33.** $x = 50.4$ **34.** $x \approx 0.57$ (rounded)   **35.** $x \approx 2.47$ (rounded)   **36.** 27 cats   **37.** 46 hits   **38.** $15.63 (rounded)   **39.** 3299 students (rounded)   **40.** 68 feet   **41.** $27\frac{1}{2}$ hours or 27.5 hours
42. 511 calories (rounded)   **43.** 14.7 milligrams   **44.** $x = 105$
45. $x = 0$ **46.** $x = 128$ **47.** $x \approx 23.08$ (rounded)   **48.** $x = 6.5$
49. $x \approx 117.36$ (rounded)   **50.** $1440 \neq 1485$; False
51. $10.8 \neq 10.864$; False   **52.** $2\frac{1}{3} = 2\frac{1}{3}$; True   **53.** $\frac{8}{5}$ **54.** $\frac{33}{80}$
55. $\frac{15}{4}$ **56.** $\frac{4}{1}$ **57.** $\frac{4}{5}$ **58.** $\frac{37}{7}$ **59.** $\frac{3}{8}$ **60.** $\frac{1}{12}$ **61.** $\frac{45}{13}$
62. 24,900 fans (rounded)   **63.** $\frac{8}{3}$ **64.** 75 ft for $1.99 − $0.50 coupon, about $0.020/ft   **65.** 21 ft long; 15 ft wide   **66.** 7.5 hours or $7\frac{1}{2}$ hours
67. $\frac{1}{2}$ teaspoon or 0.5 teaspoon   **68.** 21 points (rounded)

69. Set up the proportion to compare teaspoons to pounds on both sides.

$$\frac{1.5 \text{ teaspoons}}{24 \text{ pounds}} = \frac{x \text{ teaspoons}}{8 \text{ pounds}}$$

Show that cross products are equal. $(24)(x) = (1.5)(8)$

Divide both sides by 24. $\dfrac{\overset{1}{\cancel{24}}(x)}{\cancel{24}} = \dfrac{12}{24}$ so $x = \dfrac{1}{2}$ teaspoon or 0.5 teaspoon

70. (a) 1400 milligrams   **(b)** 100 milligrams

Chapter 5 TEST (pages 376–377)

1. $\dfrac{4}{5}$ **2.** $\dfrac{20 \text{ miles}}{1 \text{ gallon}}$ **3.** $\dfrac{\$1}{5 \text{ minutes}}$ **4.** $\dfrac{9}{2}$ **5.** $\dfrac{15}{4}$

6. Best buy is 8 in. sub, about $0.861/inch   **7.** 16 ounces for
$1.89 − $0.50 coupon, about $0.087/ounce

8. You earned less this year. An example is: $\begin{array}{l} \text{Last year} \to \\ \text{This year} \to \end{array} \dfrac{\$30,000}{\$20,000} = \dfrac{3}{2}$

9. $\dfrac{3}{7} \neq \dfrac{2}{5}$ or $252 \neq 270$; false   **10.** $5.88 = 5.88$; true   **11.** $x = 25$

12. $x \approx 2.67$ (rounded)   **13.** $x = 325$ **14.** $x = 10\dfrac{1}{2}$ **15.** 24 orders

16. 3.6 ounces   **17.** 87 students (rounded)   **18.** No, 4875 cannot be
correct because there are only 650 students in the whole school.
19. 23.8 grams (rounded)   **20.** 60 ft

Chapters 1–5 CUMULATIVE REVIEW EXERCISES
(pages 378–379)

1. 9900   **2.** 617.1   **3.** $100   **4.** $3.06   **5.** 29.34
6. 610 R27   **7.** 0.0076   **8.** 2312   **9.** 68.381   **10.** 55.6
11. 39   **12.** 18   **13.** 64   **14.** 0.95   **15.** one hundred five
ten-thousandths   **16.** 60.071

17. $\dfrac{1}{5}$ **18.** $\dfrac{4}{1}$ **19.** $1\dfrac{1}{8}$, 1.25, $1\dfrac{3}{8}$, 1.5, $1\dfrac{9}{16}$ (all inches)

20. $\dfrac{1}{16}$ inch or 0.063 inch (rounded)   **21.** $x = 21$ **22.** $x \approx 17.14$
(rounded)   **23.** $x \approx 0.98$ (rounded)   **24.** 250 pounds

25. 26.7 centimeters (rounded)   **26.** $7\dfrac{3}{20}$ miles   **27.** 34 servings for

$3.35 − $0.50 coupon, about $0.084/serving   **28.** 140 residents

29. $1\dfrac{1}{4}$ teaspoons   **30.** $0.018, $0.02, $0.084, $0.099, $0.10, $0.106,
$0.134   **31.** 202 minutes (rounded)   **32.** 189 minutes (rounded)
per $20 card, so buy 2 cards to cover 360 minutes (6 hours).

33. $\dfrac{1000}{200} = \dfrac{5}{1}$ **34.** 89.7 min rounds to 90, but the call would be cut off
after 89 min.

CHAPTER 6 Percent

SECTION 6.1 (pages 387–392)

1. %; 100   **2.** decimal point; left   **3.** 0.12   **5.** 0.70 or 0.7   **7.** 0.25
9. 1.40 or 1.4   **11.** 0.055   **13.** 1.00 or 1   **15.** 0.005   **17.** 0.0035
19. 100; %   **20.** decimal point; right   **21.** 60%   **23.** 1%   **25.** 37.5%
27. 200%   **29.** 370%   **31.** 3.12%   **33.** 416.2%   **35.** 0.28%

37. Answers will vary. Some possibilities are: No common denominators
are needed with percents. The denominator is always 100 with percent,
which makes comparisons easier to understand.   **38.** Answers will vary.

Some answers might be: when using discounts on purchases, calculating
sales tax, figuring interest on loans, examining investments, finding tips in
restaurants, calculating interest on savings, and doing math problems in
this book.   **39.** 0.13   **41.** 21.8%   **43.** 0.151   **45.** 17.7%   **47.** 0.146
49. 8%   **51.** 0.30 or 0.3   **53.** 12 children   **54.** 500 adults
55. 420 employees   **57.** 270 chairs   **59.** $377.50   **61.** 820 commuters
63. 26 plants   **65. (a)** Since 100% means 100 parts out of 100 parts,
100% is all of the number.   **(b)** Answers will vary. For example, 100% of
$72 is $72.   **66. (a)** 50% means 50 parts out of 100 parts. That's half of
the number. A shortcut for finding 50% of a number is to divide the number
by 2.   **(b)** Answers will vary. For example, 50% of $14 is $14 ÷ 2 = $7.
67. (a) Since 200% is two times a number, find 200% of the number by
multiplying the number by 2 (double it).   **(b)** Answers will vary. For example, 200% of $20 is 2 • $20 = $40.   **68. (a)** Since 300% is three times
a number, find 300% of the number by multiplying the number by 3 (triple
it).   **(b)** Answers will vary. For example, 300% of $10 is 3 • $10 − $30.
69. (a) Since 10% means 10 parts out of 100 parts or $\frac{1}{10}$, the shortcut for
finding 10% of a number is to move the decimal point in the number one
place to the left.   **(b)** Answers will vary. For example, 10% of $90 is
$9.   **70. (a)** Since 1% means 1 part out of 100 parts or $\frac{1}{100}$, the shortcut
for finding 1% of a number is to move the decimal point in the number two
places to the left.   **(b)** Answers will vary. For example, 1% of $500 is $5.
71. 12%; 0.12   **73. (a)** witch costume   **(b)** 4.5%; 0.045   **75.** 9%; 0.09
77. (a) speeding   **(b)** 5%; 0.05   **79.** 21%; 0.21   **81. (a)** candy
(b) 10%; 0.10   **83.** 95% shaded; 5% unshaded   **85.** 30% shaded; 70%
unshaded   **87.** 55% shaded; 45% unshaded   **89.** 64% shaded; 36% unshaded

SECTION 6.2 (pages 398–403)

1. false   **2.** true   **3.** true   **4.** false   **5.** $\dfrac{17}{20}$ **7.** $\dfrac{5}{8}$ **9.** $\dfrac{1}{16}$ **11.** $\dfrac{1}{6}$

13. $\dfrac{1}{15}$ **15.** $\dfrac{1}{200}$ **17.** $1\dfrac{4}{5}$ **19.** $3\dfrac{3}{4}$ **21.** true   **22.** false   **23.** false

24. true   **25.** 70%   **27.** 37%   **29.** 62.5%   **31.** 87.5%   **33.** 48%

35. 46%   **37.** 35%   **39.** 83.3% (rounded)   **41.** 55.6% (rounded)

43. 14.3% (rounded)   **45. (b)**   **46. (d)**   **47.** $\dfrac{1}{2}$; 50%   **49.** $\dfrac{7}{8}$; 0.875

51. 0.167 (rounded); 16.7% (rounded)   **53.** $\dfrac{7}{10}$; 70%   **55.** $\dfrac{1}{8}$; 0.125

57. 0.667 (rounded); 66.7% (rounded)   **59.** 0.06; 6%   **61.** 0.08; 8%

63. 0.005; 0.5%   **65.** $2\dfrac{1}{2}$; 250%   **67.** 3.25; 325%   **69.** There are
many possible answers. Examples 2 and 3 show the steps that students
should include in their answers.   **70.** There are many correct answers.
The table of percent equivalents shows some of the possibilities.

71. $\dfrac{9}{50}$; 0.18; 18%   **73.** $\dfrac{13}{100}$; 0.13; 13%   **75.** $\dfrac{1}{5}$; 0.2; 20%

77. (a) $\dfrac{4}{5}$; 0.80; 80%   **(b)** $\dfrac{1}{5}$; 0.20; 20%   **79.** $\dfrac{1}{10}$; 0.1; 10%

81. $\dfrac{2}{25}$; 0.08; 8%   **83.** $\dfrac{13}{50}$; 0.26; 26%   **85.** 100; 100

86. (a) 765 workers   **(b)** 96 letters   **(c)** 21 DVDs

87. 50; 100; half or $\dfrac{1}{2}$ **88.** 10; 100; 1; left   **89.** 1; 100; 2; left

90. (a) 525 homes   **(b)** 37 printers   **(c)** $0.08

77. 500%   **78.** 471%   **79.** 0.086   **80.** 62.1%   **81.** 0.00375

82. 0.06%   **83.** 75%   **84.** $\frac{21}{50}$ **85.** $\frac{7}{8}$ **86.** 37.5% or $37\frac{1}{2}$%

87. $\frac{13}{40}$ **88.** 60%   **89.** $\frac{1}{400}$ **90.** 375%   **91.** $3331.25   **92.** $16,520

93. (a) 96.0% (rounded)   **(b)** 45.9% (rounded)   **94. (a)** $179,894.93

(rounded)   **(b)** $31,894.93 (rounded)   **95.** $6225   **96.** 18.2% (rounded)

97. $848.40 (rounded)   **98.** 1100 boaters   **99.** $13,005   **100.** 33.4%

(rounded)

Chapter 6 TEST (pages 472–473)

1. 0.65   **2.** 80%   **3.** 175%   **4.** 87.5% or $87\frac{1}{2}$%   **5.** 3.00 or 3   **6.** 0.02

7. $\frac{1}{8}$ **8.** $\frac{1}{400}$ **9.** 60%   **10.** 62.5% or $62\frac{1}{2}$%   **11.** 250%

12. 800 sacks   **13.** 20%   **14.** 125,000,000 households   **15.** $3450.60

16. 8%   **17.** 25%   **18.** A possible answer is: Part is the increase in salary. Whole is last year's salary. Percent of increase is unknown.

$$\frac{\text{amount of increase}}{\text{last year's salary}} = \frac{p}{100}$$

19. The interest formula is $I = p \cdot r \cdot t$. If time is in months, it is expressed
as a fraction with 12 as the denominator. If time is expressed in years, it is
placed over 1 or shown as a decimal number.

Problems will vary.

Some possibilities are:

$$I = (1000)(0.05)\left(\frac{9}{12}\right) = \$37.50$$

$$I = (1000)(0.05)(2.5) = \$125$$

20. $11.52; $84.48   **21.** $91; $189   **22.** $378   **23.** $192   **24.** $20,880

25. (a) $10,668 (rounded)   **(b)** $1668

Chapters 1–6 CUMULATIVE REVIEW EXERCISES (pages 474–475)

1. *Estimate:* $80,000 - 50,000 = 30,000$; *Exact:* 28,988

2. *Estimate:* $8 - 4 = 4$; *Exact:* 4.2971

3. *Estimate:* $7000 \times 700 = 4,900,000$; *Exact:* 4,628,904

4. *Estimate:* $70 \times 9 = 630$; *Exact:* 568.11

5. *Estimate:* $40,000 \div 40 = 1000$; *Exact:* 902

6. *Estimate:* $7 \div 1 = 7$; *Exact:* 8.45

7. 18   **8.** 19   **9.** 39   **10.** 7,600,000   **11.** $513   **12.** $362.74

13. $13\frac{3}{8}$ **14.** $3\frac{7}{8}$ **15.** $28\frac{4}{5}$ **16.** 16   **17.** $12\frac{1}{2}$ ft^2 **18.** $22\frac{1}{2}$ hours

19. <   **20.** <   **21.** >   **22.** $\frac{1}{4}$ **23.** 1   **24.** $\frac{1}{4}$ **25.** 0.75   **26.** 0.375

27. 0.583 (rounded)   **28.** 0.55   **29.** $x = 6$ **30.** $x = 4$ **31.** $x = 16$

32. $x = 30$ **33.** 0.03   **34.** 2.00 or 2   **35.** 87%   **36.** 380%   **37.** $\frac{2}{25}$

38. $\frac{5}{8}$ **39.** $1\frac{3}{4}$ **40.** 87.5% or $87\frac{1}{2}$%   **41.** 420%   **42.** 2132 DVDs

43. 1700 miles   **44.** 140%   **45.** 6.5% or $6\frac{1}{2}$%   **46.** $117.18

47. $205.20; $250.80   **48.** $47,341.75   **49.** 45 children

50. 255 ounces   **51.** 12.5%   **52.** 2%   **53.** 5.5% (rounded)

54. 8.1% (rounded)

CHAPTER 7 Measurement

SECTION 7.1 (pages 484–487)

1. (a) inch, foot, yard, mile   **(b)** fluid ounce, cup, pint, quart, gallon

2. (a) ounce, pound, ton   **(b)** second, minute, hour, day, week   **3. (a)** 3;

(b) 12   **5. (a)** 8; **(b)** 2   **7. (a)** 2000; **(b)** 16   **9. (a)** 60; **(b)** 60

11. (a) 2   **(b)** 240   **13. (a)** $\frac{1}{2}$ or 0.5   **(b)** 78   **15.** 14,000 to 16,000 lb

17. $\frac{3 \text{ ft}}{1 \text{ yd}}$; 27   **19.** 112   **21.** 10   **23.** $1\frac{1}{2}$ or 1.5   **25.** $\frac{1}{4}$ or 0.25   **27.** $1\frac{1}{2}$

or 1.5   **29.** $2\frac{1}{2}$ or 2.5   **31.** snowmobile/ATV; person walking   **33.** 5000

35. 17   **37.** 4 to 8 in.   **39.** 216   **41.** 28   **43.** 518,400   **45.** 48,000

47. (a) pound/ounces   **(b)** quarts/pints or pints/cups

(c) minutes/hours or seconds/minutes   **(d)** feet/inches   **(e)** pounds/tons

(f) days/weeks   **48.** Feet and inches both measure length, so you can add
them once you've changed 12 inches into 1 ft. But pounds measure weight
and cannot be added to a length measurement.   **49.** 174,240   **51.** 800

53. 0.75 or $\frac{3}{4}$ **55.** $1.83 (rounded)   **57.** $140   **59. (a)** 1056 sec

(b) 17.6 min   **61. (a)** $12\frac{1}{2}$ qt   **(b)** 4 containers, because you can't buy
part of a container   **63. (a)** 0.8 mi (rounded)   **(b)** 13,031 ft   **(c)** 2.5 mi
(rounded)   **64. (a)** 183 yd (rounded)   **(b)** 6600 in.   **(c)** 0.1 mi (rounded)

(d) 9 ft (rounded)   **65. (a)** $1\frac{1}{2}$ to $2\frac{1}{2}$ ft, or 1.5 to 2.5 ft   **(b)** 8400 months

(c) 140 more years, for a total of 840 years in all   **66. (a)** 283 mi (rounded)

(b) 377 mi (rounded)

SECTION 7.2 (pages 493–494)

1. 1000   **2.** 10   **3.** $\frac{1}{1000}$ or 0.001; $\frac{1}{1000}$ or 0.001

4. $\frac{1}{10}$ or 0.1; $\frac{1}{10}$ or 0.1   **5.** $\frac{1}{100}$ or 0.01; $\frac{1}{100}$ or 0.01   **6.** 100; 100

7. (a) cm   **(b)** mm   **(c)** cm   **9.** Answers will vary; about 8 to 10 cm.

11. Answers will vary; about 20 to 25 mm.   **13.** cm   **15.** m   **17.** km

19. mm   **21.** cm   **23.** m   **25.** 700 cm   **27.** 0.040 m or 0.04 m

29. 9400 m   **31.** 5.09 m   **33.** 40 cm   **35.** 910 mm   **37.** less;
18 cm or 0.18 m   **38.** greater; 22 m or 0.022 km

39.
⤴ 5 mm = 0.5 cm
☐ ←1 mm = 0.1 cm

41. 0.061 km   **43.** 164 cm; 1640 mm   **45.** 0.0000056 km

SECTION 7.3 (pages 501–504)

1. (a) milliliters   **(b)** liters   **(c)** milligrams   **(d)** kilograms

2. (a) grams   **(b)** liters   **(c)** milliliters   **(d)** milligrams   **3.** mL

5. L   **7.** kg   **9.** g   **11.** mL   **13.** mg   **15.** L   **17.** kg

19. unreasonable; too much   **20.** reasonable   **21.** unreasonable; too
much   **22.** unreasonable; too much   **23.** reasonable   **24.** reasonable

25. reasonable   **26.** unreasonable; too little   **27.** Unit for your answer
(g) is in numerator; unit being changed (kg) is in denominator so it will

divide out. The unit fraction is $\frac{1000 \text{ g}}{1 \text{ kg}}$.

28. From mg to g is three places to the left on the metric conversion line,
so move the decimal point three places to the left. $\underset{\smile}{020}$. mg = 0.02 g

29. three, right; 15,000 mL   **31.** 3 L   **33.** 0.925 L   **35.** 0.008 L

37. 4150 mL   **39.** 8 kg   **41.** 5200 g   **43.** 850 mg   **45.** 30 g   **47.** 0.598 g

49. 0.06 L   **51.** 0.003 kg   **53.** 990 mL   **55.** mm   **57.** mL   **59.** cm

61. mg **63.** 0.3 L **65.** 1340 g; 5000 g **67.** 0.9 L **69.** 3 kg to 4 kg
71. greater; 5 mg or 0.005 g **72.** less; 10 mL or 0.01 L **73.** 200 nickels
75. (a) 1,000,000 **(b)** $\dfrac{3.5 \text{ Mm}}{1} \cdot \dfrac{1,000,000 \text{ m}}{1 \text{ Mm}} = 3,500,000 \text{ m}$
76. (a) 1,000,000,000 **(b)** $\dfrac{2500 \text{ m}}{1} \cdot \dfrac{1 \text{ Gm}}{1,000,000,000 \text{ m}} = 0.0000025 \text{ Gm}$
77. (a) 1,000,000,000,000 **(b)** 1000; 1,000,000
78. 1,000,000; 1,000,000,000; $2^{20} = 1,048,576$; $2^{30} = 1,073,741,824$

SUMMARY EXERCISES U.S. and Metric Measurement Units (pages 505–506)

1. Length: inch (in.), foot (ft), yard (yd), mile (mi); Weight: ounce (oz), pound (lb), ton (ton); Capacity: fluid ounce (fl oz), cup (c), pint (pt), quart (qt), gallon (gal) **2.** Length: millimeter (mm), centimeter (cm), meter (m), kilometer (km); Weight: milligram (mg), gram (g), kilogram (kg); Capacity: milliliter (mL), liter (L) **3. (a)** ft **(b)** yd **(c)** 5280
4. (a) min **(b)** 60 **(c)** 24 **5. (a)** 8 **(b)** gal **(c)** 2 **6. (a)** lb
(b) 2000 **(c)** 16 **7.** mL **8.** m **9.** kg **10.** mg **11.** cm
12. mm **13.** L **14.** g **15.** 0.45 m **16.** 45 sec **17.** 600 mL
18. 8000 mg **19.** 30 cm **20.** $3\frac{3}{4}$ or 3.75 ft **21.** 0.050 or 0.05 L
22. $4\frac{1}{2}$ or 4.5 gal **23.** 7280 g **24.** 36 oz **25.** 0.009 kg **26.** 180 in.
27. 272 cm; 2720 mm **28.** 2.64 m; 0.00264 km **29.** 0.04 m; 4 cm; 40 mm **30.** 24 cm; 0.24 m; 240 mm **31.** $5.19 (rounded)
32. 9.3 ft (rounded)

SECTION 7.4 (pages 509–510)

1. c **2.** b **3.** kg; kg; $0.83 (rounded) **5.** 89.5 kg **7.** 71 beats (rounded)
9. 180 cm; $0.02/cm (rounded) **11.** 5.03 m **13.** 4 bottles
15. 1.89 g **17. (a)** 1.54 m **(b)** 308 cm **19.** 215 g; 4.3 g; 4300 mg
20. 330 g; 0.33 g; 330 mg **21.** 1550 g; 1500 g; 3 g **22.** 55 g; 50 g; 0.5 g

SECTION 7.5 (pages 515–518)

1. (a) $\dfrac{1 \text{ kg}}{2.20 \text{ lb}}$; $\dfrac{2.20 \text{ lb}}{1 \text{ kg}}$ **(b)** kg; denominator **2. (a)** $\dfrac{1 \text{ qt}}{0.95 \text{ L}}$; $\dfrac{0.95 \text{ L}}{1 \text{ qt}}$
(b) qt; denominator **3.** 21.8 yd **5.** 262.4 ft **7.** 4.8 m **9.** 5.3 oz
11. 111.6 kg **13.** 30.3 qt **15. (a)** about 0.2 oz **(b)** probably not
17. about 31.8 L **19.** about 1.3 cm **21.** −8 °C **22.** 80 °C
23. 40 °C **24.** 25 °C **25.** 150 °C **26.** 10° C **27.** 16 °C (rounded)
29. 40 °C **31.** 46 °C (rounded) **33.** 95 °F **35.** 58 °C (rounded)
37. (a) pleasant weather, above freezing but not hot **(b)** 75 °F to 39 °F (rounded) **(c)** Answers will vary. In Minnesota, it's 0 °C to −40 °C; in California, 24 °C to 0 °C. **39.** about 1.8 m **40.** about 5.0 kg
41. about 3040.2 km **42.** 38 hr 23 min **43.** about 300 m
44. less than 3790 mL **45.** about 59.4 mL **46.** about 1.6%

Chapter 7 REVIEW EXERCISES (pages 525–528)

1. 16 **2.** 3 **3.** 2000 **4.** 4 **5.** 60 **6.** 8 **7.** 60 **8.** 5280
9. 12 **10.** 48 **11.** 3 **12.** 4 **13.** $\frac{3}{4}$ or 0.75 **14.** $2\frac{1}{2}$ or 2.5
15. 28 **16.** 78 **17.** 112 **18.** 345,600
19. (a) $4153\frac{1}{3}$ yd (exact) or 4153.3 yd (rounded) **(b)** 2.4 mi (rounded)
20. $2465.20 **21.** mm **22.** cm **23.** km **24.** m **25.** cm
26. mm **27.** 500 cm **28.** 8500 m **29.** 8.5 cm **30.** 3.7 m
31. 0.07 km **32.** 930 mm **33.** mL **34.** L **35.** g **36.** kg **37.** L

38. mL **39.** mg **40.** g **41.** 5 L **42.** 8000 mL **43.** 4580 mg
44. 700 g **45.** 0.006 g **46.** 0.035 L **47.** 31.5 L
48. 357 g (rounded) **49.** 87.25 kg **50.** $1.42 (rounded)
51. 6.5 yd (rounded) **52.** 11.7 in. (rounded) **53.** 67.0 mi (rounded)
54. 1288 km **55.** 21.9 L (rounded) **56.** 44.0 qt (rounded) **57.** 0 °C
58. 100 °C **59.** 37 °C **60.** 20 °C **61.** 25 °C **62.** 33 °C (rounded)
63. 43 °F (rounded) **64.** 120 °F (rounded) **65.** L **66.** kg **67.** cm
68. mL **69.** mm **70.** km **71.** g **72.** m **73.** mL **74.** g
75. mg **76.** L **77.** 105 mm **78.** $\frac{3}{4}$ hr or 0.75 hr **79.** $7\frac{1}{2}$ ft or 7.5 ft
80. 130 cm **81.** 77 °F **82.** 14 qt **83.** 0.7 g **84.** 810 mL
85. 80 oz **86.** 60,000 g **87.** 1800 mL **88.** 30 °C **89.** 36 cm
90. 0.055 L **91.** 1.26 m **92.** 18 tons **93.** 113 g; 177 °C (both rounded) **94.** 178.0 lb; 6.0 ft (both rounded) **95.** 13.2 m **96.** 22.0 ft
97. 3.6 m **98.** 39.8 in. **99.** 484,000 lb **100.** 19.0 to 22.7 L
101. (a) 242 tons **(b)** 144 in. **102. (a)** 1.02 m **(b)** 670 cm

Chapter 7 TEST (pages 529–530)

1. 36 qt **2.** 15 yd **3.** 2.25 hr or $2\frac{1}{4}$ hr **4.** 0.75 ft or $\frac{3}{4}$ ft
5. 56 oz **6.** 7200 min **7.** kg **8.** km **9.** mL **10.** g **11.** cm
12. mm **13.** L **14.** cm **15.** 2.5 m **16.** 4600 m **17.** 0.5 cm
18. 0.325 g **19.** 16,000 mL **20.** 400 g **21.** 1055 cm **22.** 0.095 L
23. 3.2 ft (rounded) **24. (a)** 1390 mg **(b)** 200 mg or 0.2 g more
25. 95 °C **26.** 0 °C **27.** 1.8 m **28.** 56.3 kg (rounded) **29.** 13 gal
30. 5.0 mi (rounded) **31.** 23 °C (rounded) **32.** 36 °F (rounded)
33. 6 m ≈ 6.54 yd; $26.03 (rounded) **34.** Possible answers: Use same system as rest of the world; easier system for children to learn; less use of fractional numbers; compete internationally.

Chapters 1–7 CUMULATIVE REVIEW EXERCISES (pages 531–532)

1. *Estimate:* $60 \times 5 = 300$; *Exact:* 265.644
2. *Estimate:* $20,000 \div 40 = 500$; *Exact:* 530
3. *Estimate:* $2 + 4 = 6$; *Exact:* $5\frac{1}{2}$ **4.** $\frac{11}{16}$ **5.** $26\frac{2}{3}$
6. 13.0 (rounded) **7.** $1\frac{5}{24}$ **8.** 7.0793 **9.** $\frac{5}{8}$ **10.** 0.00762 **11.** 8
12. 265 **13.** 0.067, 0.6, 0.6007, 0.67 **14.** 27 diapers for $11.71, about $0.434 per diaper. **15.** $x = 12$ **16.** $x \approx 1.67$ (rounded)
17. 5% **18.** 30 hours **19.** 30 in. **20.** $1\frac{3}{4}$ or 1.75 min
21. 280 cm **22.** 122.76 mi **23.** g **24.** mL **25.** m **26.** kg
27. $18.99 (rounded); $170.95 **28.** $6.15 total; $0.14 per print (both rounded) **29.** $12\frac{3}{4}$ or 12.75 lb **30. (a)** 93.6 km **(b)** 58.0 mi (rounded) **31.** $2821.88 (rounded); $21,571.88 **32.** 88.6% (rounded)
33. $9.74 (rounded) **34.** $\frac{1}{6}$ cup more than the amount needed
35. 120 rows **36.** 2100 students **37.** $0.15 **38.** 37.5% decrease

CHAPTER 8 Geometry

SECTION 8.1 (pages 539–540)

1. Answer wording may vary slightly: A **line** is a row of points continuing in both directions forever; the drawing should have arrows at each end. A **line segment** is a piece of a line; the drawing should have an endpoint (dot) at each end. A **ray** has one endpoint and goes on forever in one direction; the drawing should have one endpoint (dot) and an arrowhead on the other end. **2.** Answer wording may vary slightly: An **acute angle** measures less than 90°; a **right angle** measures exactly 90° and forms a "square corner;" an **obtuse angle** measures more than 90° but less than 180°; a **straight angle** measures exactly 180° and forms a straight line. Drawings of an acute and obtuse angle may vary.   **3.** line named $\overleftrightarrow{CD}$ or $\overleftrightarrow{DC}$ **5.** line segment named $\overline{GF}$ or $\overline{FG}$ **7.** ray named $\overrightarrow{PQ}$ **9.** perpendicular   **11.** parallel   **13.** intersecting   **15.** P; $\angle APS$; $\angle SPA$ **17.** $\angle AQC$ or $\angle CQA$ **19.** right (90°)   **21.** acute   **23.** straight (180°)   **25.** **(a)** The car turned around in a complete circle.   **(b)** The governor took the opposite view, for example, having once opposed taxes but now supporting them.   **26.** There are many possibilities. Some examples are the corner of a room, a street corner, the corners of a window, or the corner of a piece of a paper.   **27.** True, because $\overleftrightarrow{UQ}$ is perpendicular to $\overleftrightarrow{ST}$.   **28.** True, because they form a 90° angle, as indicated by the small red square.   **29.** False; the angles have the same measure (both are 180°).   **30.** False; $\overleftrightarrow{ST}$ and $\overleftrightarrow{PR}$ are parallel.   **31.** False; $\overleftrightarrow{QU}$ and $\overleftrightarrow{TS}$ are perpendicular.   **32.** True; both angles are formed by perpendicular lines, so they both measure 90°.

SECTION 8.2 (pages 544–545)

1. Yes, because you can subtract 180° − 5° to get 175°. The sum of the measures of a 5° angle and a 175° degree angle is 180°, so they are supplements of each other.   **2.** You cannot; two angles are complementary if the sum of their measures is 90°. A 135° angle is "too big" to have a complement.   **3.** $\angle EOD$ and $\angle COD$; $\angle AOB$ and $\angle BOC$ **5.** $\angle HNE$ and $\angle ENF$; $\angle ACB$ and $\angle KOL$ **7.** 50°   **9.** 4°   **11.** 50°   **13.** 90°   **15.** No; obtuse angles are >90°, so their sum would be >180°.   **16.** Yes; acute angles are <90° so their sum could equal 90°.   **17.** $\angle SON \cong \angle TOM$; $\angle TOS \cong \angle MON$ **19.** $\angle GOH$ measures 63°; $\angle EOF$ measures 37°; $\angle AOC$ and $\angle GOF$ both measure 80°.   **21.** $\angle ABF \cong \angle ECD$; both are 138°. $\angle ABC \cong \angle BCD$; both are 42°.

SECTION 8.3 (pages 552–555)

1. Perimeter is the distance around the outside edges; add the lengths of all the sides to find the perimeter. The area is the surface inside the figure; multiply length times width to find the area. The units for perimeter will be **yd**; the units for area will be **yd²**.   **2.** A rectangle is a figure with four sides that meet to form four right angles. Each set of opposite sides is parallel and has the same length. A square has four right angles; all four sides have the same length. Drawings will vary; show your instructor.   **3.** $P = 16$ yd $+ 12$ yd $= 28$ yd; $A = 48$ yd²   **5.** $P = 3.6$ km; $A = 0.81$ km²   **7.** $P = 40$ ft; $A = 100$ ft²   **9.** $P = 196.2$ ft; $A = 1674.2$ ft²   **11.** $P = 12$ mi; $A = 9$ mi²   **13.** $P = 38$ m; $A = 39$ m²   **15.** $P = 98$ m; $A = 492$ m²   **17.** unlabeled side − 12 in., $P − 78$ in., $A = 234$ in.²   **19.** $P = 48$ m; $A = 144$ m²   **21.** 96 in.² (rounded); 77 in.² (rounded); 19 in.² (difference)   **23.** \$94.81   **25.** Area of largest field is 9600 yd²; Area of smallest field is 5000 yd²; difference is 9600 yd² − 5000 yd² = 4600 yd².   **27.** 53 yd   **29.** $A = 6528$ ft²

31.

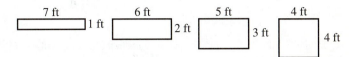

32. (a) 5 ft by 1 ft has area of 5 ft²; 4 ft by 2 ft has area of 8 ft²; 3 ft by 3 ft has area of 9 ft²   **(b)** The square plot 3 ft by 3 ft has greatest area.
33.

34. (a) 7 ft²; 12 ft²; 15 ft²; 16 ft²   **(b)** Square plots have the greatest area.
35. (a) $P = 10$ ft; $A = 6$ ft²

(b) $P = 20$ ft; $A = 24$ ft²

(c) Perimeter is twice the original; area is four times the original.
36. (a) $P = 30$ ft; $A = 54$ ft²

(b) Perimeter is three times the original; area is nine times the original.
(c) Perimeter will be four times the original; area will be 16 times the original.

SECTION 8.4 (pages 560–561)

1. In a parallelogram, opposite sides are parallel and have the same length. Drawings may vary; an example is shown.

2. To find the perimeter of any figure, add the lengths of all the sides. Drawings will vary.   **3.** $P = 46$ m $+ 46$ m $+ 58$ m $+ 58$ m; $P = 208$ m   **5.** $P = 207.2$ m   **7.** $P = 6.06$ km   **9.** $A = 775$ mm²   **11.** $A = 19.25$ ft²   **13.** $A = 3099.6$ cm²   **15.** Height is not part of perimeter; square units are used for area, not perimeter. $P = 2.5$ cm $+ 2.5$ cm $+ 2.5$ cm $+ 2.5$ cm $= 10$ cm   **16.** The bases are the parallel lines, 22 ft and 13 ft; area is measured in square units. $A = 0.5 \cdot 11.6$ ft $\cdot (22$ ft $+ 13$ ft$) = 203$ ft²   **17.** \$1410.75

19. $A = 437.5$ in.^{2}

21. $A = 3.02$ m^2 **23.** $A = 25,344$ ft^2

SECTION 8.5 (pages 566–568)

1. Multiply $\frac{1}{2}$ times 58 m times 66 m; the units in the answers will be m^2.
2. Add 58 m plus 72 m plus 72 m; the units in the answer will be m.
3. $P = 202$ m; $A = 1914$ m^2 **5.** $P = 58.9$ cm; $A = 139.15$ cm^2
7. $P = 26\frac{1}{4}$ yd or 26.25 yd; $A = 30\frac{3}{4}$ yd^2 or 30.75 yd^2 **9.** $P = 85.2$ cm;
$A = 302.46$ cm^2 **11.** Area of triangle $= 0.5 \cdot 12$ m $\cdot$ 9 m $= 54$ m^2
Area of square $= 12$ m $\cdot$ 12 m $= 144$ m^2 Entire area $= 54$ m^2 $+ 144$ m^2 $=$
198 m^2 **13.** $A = 1664$ m^2 **15.** 32°   **17.** 48°   **19.** No. Right angles are
90°, so two right angles are 180°, and the sum of all *three* angles in a triangle
equals 180°.   **20.** Two identical triangles form a parallelogram. The area
of a parallelogram is base times height, so the area of one of the triangles is
$\frac{1}{2} \cdot$ base $\cdot$ height. Sketches will vary.   **21.** $7\frac{7}{8}$ ft^2 or 7.875 ft^2
23. (a) 126.8 m of curb   **(b)** 672 m^2 of sod   **25. (a)** $A = 32$ m^2
(b) $A = 13.41$ m^2 **27. (a)** 175 yd of frontage   **(b)** $A = 8750$ yd^2

SECTION 8.6 (pages 575–578)

1. perimeter; feet   **2.** radius; divide by 2   **3.** $d = 2 \cdot 9$ mm; $d = 18$ mm
5. $r = 0.35$ km   **7.** $C \approx 69.1$ ft; $A \approx 379.9$ ft^2 **9.** $C \approx 8.2$ m;
$A \approx 5.3$ m^2 **11.** $C \approx 47.1$ cm; $A \approx 176.6$ cm^2 **13.** $C \approx 23.6$ ft;
$A \approx 44.2$ ft^2 **15.** $A \approx 76.9$ in.^{2} **17.** $A \approx 57$ cm^2 **19.** π is the ratio
of circumference of a circle to its diameter. If you divide the circumference
of any circle by its diameter, the answer is always a little more than 3. The
approximate value is 3.14, which we call π (pi).   **20.** Test questions will
vary. They should include finding the circumference and/or area of a circle.
21. $A \approx 7850$ yd^2 **23.** $C \approx 91.4$ in.; Bonus: 693 revolutions/mile
(rounded)   **25.** $A \approx 70,650$ mi^2

27.

| Watch | $C \approx 3.1$ in. | Wall Clock | $C \approx 18.8$ in. |
| | $A \approx 0.8$ in.^{2} | | $A \approx 28.3$ in.^{2} |

29.

$A \approx 78.5$ mi^2 $- 12.6$ mi^2;
Difference ≈ 65.9 mi^2

31. (a) $d \approx 45.9$ cm   **(b)** Divide the circumference by π (3.14).
33. $325.83 (rounded).   **35.** The prefix *rad-* tells you that radius is a
ray from the center of the circle. The prefix *dia-* means the diameter goes
through the circle, and the prefix *circum-* means the circumference is the
distance around.   **36.** The prefix *par-* means beside, so parallel lines
are beside each other. (Perpendicular lines cross so that they form a right
angle.)   **37.** $A \approx 78.5$ in.^{2} **38.** $A \approx 113.0$ in.^{2} **39.** $A \approx 153.9$ in.^{2}
40. small: $0.089 (rounded); medium: $0.080 (rounded); large: $0.071
(rounded); Best buy   **41.** small: $0.153 (rounded); medium:

$0.124 (rounded); large: $0.104 (rounded) Best buy;   **42.** Small: $0.076
(rounded) Best buy; medium: $0.088 (rounded); large: $0.078 (rounded)

SUMMARY EXERCISES Perimeter, Circumference, and Area (pages 579–580)

1. (a)   **(b)**

All sides have the same length.   Opposite sides have the same length.

(c)

Opposite sides have the same length.

2. (a)   **(b)**

(c) The diameter is twice the radius, or, the radius is half the diameter.
3. Add up the lengths of all the sides to find the perimeter.   **4.** Perimeter
is the total distance around the outside edges of a shape. Area is the number
of square units needed to cover the space inside the shape.   **5.** f, e, d, b, c, a
6. (a) $C = 2 \cdot \pi \cdot r$ **(b)** $C = \pi \cdot d$ **(c)** Divide the diameter by 2 to
find the radius.   **7.** $P = 63$ ft; $A = 180$ ft^2; $P = 32.4$ cm; $A \approx 45.1$ cm^2
(rounded)   **8. (a)** $A = 12$ cm^2 **(b)** $P = 6\frac{1}{2}$ ft   **(c)** $C \approx 28.5$ m
(d) $A = 307$ in.^{2} **9.** Rectangle; $P = 27$ in.; $A = 31\frac{1}{2}$ in.^{2} or 31.5 in.^{2}
10. Square; $P = 48$ yd; $A = 144$ yd^2 **11.** Parallelogram; $P = 6.4$ m;
$A \approx 2.2$ m^2 (rounded)   **12.** Trapezoid; $P = 33.5$ mm; $A = 64$ mm^2
13. $d = 12$ cm; $C \approx 37.7$ cm (rounded); $A \approx 113.0$ cm^2 (rounded)
14. $r = 15$ mi; $C \approx 94.2$ mi; $A \approx 706.5$ mi^2
15. $r = 4.5$ ft; $C \approx 28.3$ ft (rounded); $A \approx 63.6$ ft^2 (rounded)
16. $A \approx 4.5$ m^2 (rounded)   **17.** $A = 217.5$ yd^2
18. $A \approx 86$ in.^{2} **19.** $C \approx 14.6$ ft; Bonus: About 361.6 revolutions; the
Mormons counted 360 revolutions, which is the answer you get using
$2\frac{1}{3}$ ft instead of 2.33 ft as the radius.   **20.** $5.10 (rounded)

SECTION 8.7 (pages 587–588)

1. It is not a cube because the edges have different measurements; in a
cube, all edges measure the same. The units will be cm^3.   **2.** Examples will
vary. Some possibilities are cereal box and file cabinet for rectangular solid;
water pipe and juice can for cylinder; basketball and moon for sphere.
3. Rectangular solid; $V = 550$ cm^3 **5.** $V \approx 44,579.6$ m^3 (rounded)
7. Hemisphere; $V \approx 3617.3$ in.^{3} (rounded)   **9.** Cylinder; $V \approx 471$ ft^3
11. Cone; $V \approx 418.7$ m^3 (rounded)   **13.** Pyramid; $V = 800$ cm^3
15. $V = 18$ in.^{3} **17.** $V \approx 2481.5$ cm^3 (rounded)   **19.** $V = 651,775$ m^3
21. $V \approx 3925$ ft^3 **23.** Student used diameter of 7 cm; should use radius
of 3.5 cm in formula. Units for volume are cm^3, not cm^2. Correct answer
is $V \approx 192.3$ cm^3.   **24.** Both involve finding the area of a circular base
and multiplying by the height. To find the volume of the cone, divide the
volume of the cylinder by 3.

SECTION 8.8 (pages 593–596)

1. 4 **3.** 8 **5.** 3.317 **7.** 2.236 **9.** 8.544 **11.** 10.050 **13.** 13.784
15. 31.623 **17.** 30 is about halfway between 25 and 36, so $\sqrt{30}$ should
be about halfway between 5 and 6, or about 5.5. Using a calculator,
$\sqrt{30} \approx 5.477$. Similarly, $\sqrt{26}$ should be a little more than $\sqrt{25}$; by
calculator $\sqrt{26} \approx 5.099$. And $\sqrt{35}$ should be a little less than $\sqrt{36}$; by
calculator $\sqrt{35} \approx 5.916$. **18.** Squaring a number is multiplying the number
times itself. Finding the square root is the opposite operation and "undoes"
squaring. Examples will vary; one possibility is $7^2 = 7 \cdot 7 = 49$, so
$\sqrt{49} = 7$. **19.** $\sqrt{1521} = 39$ ft **21.** $\sqrt{289} = 17$ in. **23.** $\sqrt{144} = 12$ mm
25. $\sqrt{73} \approx 8.5$ in. **27.** $\sqrt{65} \approx 8.1$ yd **29.** $\sqrt{195} \approx 14.0$ cm
31. $\sqrt{7.94} \approx 2.8$ m **33.** $\sqrt{65.01} \approx 8.1$ cm **35.** $\sqrt{292.32} \approx 17.1$ km
37. The student did not square the numbers correctly: 9^2 is 81 and 7^2 is 49.
Also, the final answer is rounded to thousandths instead of tenths. Correct
answer is $\sqrt{130} \approx 11.4$ in. **38.** The student used the formula for finding
the hypotenuse, but the unknown side is a leg; so leg $= \sqrt{(20)^2 - (13)^2}$. Also,
the final answer should be m, not m². Correct answer is $\sqrt{231} \approx 15.2$ m.
39. hypotenuse $= \sqrt{65} \approx 8.1$ ft **41.** leg $= \sqrt{360,000} = 600$ m
43. hypotenuse $= \sqrt{32.5} \approx 5.7$ ft **45.** leg $= \sqrt{135} \approx 11.6$ ft

47. $\sqrt{16,200} \approx 127.3$ ft

48. (a)
Second
60 ft
Third ◇ First
60 ft
Home
(b) $\sqrt{7200} \approx 84.9$ ft

49. The distance from third to first is the same as the distance from home
to second because the baseball diamond is a square.

50. One possibility is:

$$\text{major league} \left\{ \frac{90 \text{ ft}}{127.3 \text{ ft}} = \frac{60 \text{ ft}}{x} \right\} \text{softball}$$
$$x \approx 84.9 \text{ ft}$$

SECTION 8.9 (pages 601–604)

1. similar **2.** similar **3.** not similar **4.** not similar
5. (a) Corresponding angles have the same measure. **(b)** The ratios of the
lengths of corresponding sides are equal. **6.** An acceptable dictionary
definition is "resembling, but not identical." Examples of similar objects are sets
of different-size pots or measuring cups; small- and large-size cans of beans.
7. $\angle 1$ and $\angle 4$; $\angle 2$ and $\angle 5$; $\angle 3$ and $\angle 6$; $\overline{AB}$ and $\overline{PQ}$; $\overline{BC}$ and $\overline{QR}$;
$\overline{AC}$ and $\overline{PR}$ **9.** $\angle 1$ and $\angle 6$; $\angle 2$ and $\angle 5$; $\angle 3$ and $\angle 4$; $\overline{MP}$ and $\overline{QS}$;
$\overline{MN}$ and $\overline{QR}$; $\overline{NP}$ and $\overline{RS}$ **11.** $\frac{3}{2}; \frac{3}{2}; \frac{3}{2}$ **13.** $a = 6$ cm: $\frac{7.5}{b} = \frac{6}{12}$;
$b = 15$ cm **15.** $a = 5$ mm; $b = 3$ mm **17.** $x = 24.8$ m; $P = 72.8$ m;
$y = 15$ m; $P = 54.6$ m **19.** $P = 8$ cm $+ 8$ cm $+ 8$ cm $= 24$ cm;
$A \approx 0.5 \cdot 8$ cm $\cdot 6.9$ cm ≈ 27.6 cm² **21.** $h = 24$ ft **23.** $\frac{3}{4}$ ft or 0.75 ft
25. $x = 50$ m **27.** $n = 110$ m

Chapter 8 REVIEW EXERCISES (pages 613–620)

1. line segment named $\overline{AB}$ or $\overline{BA}$ **2.** line named $\overleftrightarrow{CD}$ or $\overleftrightarrow{DC}$ **3.** ray
named $\overrightarrow{OP}$ **4.** parallel **5.** perpendicular **6.** intersecting **7.** acute
8. obtuse **9.** straight; 180° **10.** right; 90° **11.** $\angle 1$ and $\angle 3$ measure
30°; $\angle 2$ measures 90°; $\angle 4$ measures 60°. **12.** $\angle 1$ measures 100°; $\angle 2$ and
$\angle 4$ measure 45°; $\angle 3$ measures 35°. **13.** $\angle AOB$ and $\angle BOC$; $\angle BOC$ and
$\angle COD$; $\angle COD$ and $\angle DOA$; $\angle DOA$ and $\angle AOB$ **14.** $\angle ERH$ and $\angle HRG$;
$\angle HRG$ and $\angle GRF$; $\angle FRG$ and $\angle FRE$; $\angle FRE$ and $\angle ERH$ **15. (a)** 10°
(b) 45° **(c)** 83° **16. (a)** 25° **(b)** 90° **(c)** 147° **17.** $P = 4.84$ m
18. $P = 128$ in. **19.** 152 cm **20.** 41 ft **21.** $A = 486$ mm²
22. $A = 16.5$ ft² or $16\frac{1}{2}$ ft² **23.** $A \approx 39.7$ m² (rounded) **24.** $P = 50$ cm;
$A = 140$ cm² **25.** $P = 102.1$ ft; $A = 567$ ft² **26.** $P = 200.2$ m;
$A \approx 2206.1$ m² (rounded) **27.** $P = 518$ cm; $A = 11,660$ cm²
28. $P = 27.1$ m; $A = 23.52$ m² **29.** $P = 20\frac{1}{4}$ ft (exactly) or
20.3 ft (rounded); $A = 14$ ft² **30.** 70° **31.** 24° **32.** $d = 137.8$ m
33. $r = 1\frac{1}{2}$ in. or 1.5 in. **34.** $C \approx 6.3$ cm (rounded); $A \approx 3.1$ cm²
(rounded) **35.** $C \approx 109.3$ m (rounded); $A \approx 950.7$ m² (rounded)
36. $C \approx 37.7$ in. (rounded); $A \approx 113.0$ in.² (rounded)
37. $A \approx 20.3$ in² (rounded) **38.** $A = 64$ in.² **39.** $A = 673$ km²
40. $A = 1020$ m² **41.** $A = 229$ ft² **42.** $A = 132$ ft² **43.** $A = 5376$ cm²
44. $A \approx 498.9$ ft² (rounded) **45.** $A \approx 447.9$ yd² (rounded)
46. Rectangular solid; $V = 30$ in.³ **47.** Rectangular solid; $V = 96$ cm³
48. Rectangular solid; $V = 45,000$ mm³ **49.** Sphere; $V \approx 267.9$ m³
(rounded) **50.** Hemisphere; $V \approx 452.2$ ft³ (rounded) **51.** Cylinder;
$V \approx 549.5$ cm³ **52.** Cylinder; $V \approx 1808.6$ m³ (rounded) **53.** Cone;
$V \approx 512.9$ m³ (rounded) **54.** Pyramid; $V = 16$ yd³ **55.** 7 **56.** 2.828
(rounded) **57.** 54.772 (rounded) **58.** 12 **59.** 7.616 (rounded)
60. 25 **61.** 10.247 (rounded) **62.** 8.944 (rounded) **63.** $\sqrt{289} = 17$ in.
64. $\sqrt{49} = 7$ cm **65.** $\sqrt{104} \approx 10.2$ cm (rounded) **66.** $\sqrt{52} \approx 7.2$ in.
(rounded) **67.** $\sqrt{6.53} \approx 2.6$ m (rounded) **68.** $\sqrt{71.75} \approx 8.5$ km
(rounded) **69.** $y = 30$ ft; $x = 34$ ft; $P = 104$ ft **70.** $y = 7.5$ m;
$x = 9$ m; $P = 22.5$ m **71.** $x = 12$ mm; $y = 7.5$ mm; $P = 38$ mm
72. Square; $P = 18$ in.; $A \approx 20.3$ in.² (rounded) or
$A = 20\frac{1}{4}$ in.² **73.** Trapezoid; $P = 10.3$ cm; $A \approx 6.2$ cm² (rounded)
74. Circle; $C \approx 40.8$ (rounded); $A \approx 132.7$ m² (rounded)
75. Parallelogram; $P = 54$ ft; $A = 140$ ft² **76.** Triangle;
$P = 20$ yd; $A = 18\frac{3}{4}$ yd² or 18.8 yd² (rounded) **77.** Rectangle;
$P = 7$ km; $A \approx 2.0$ km² (rounded) **78.** Circle; $C \approx 53.4$ m (rounded)
$A \approx 226.9$ m² (rounded) **79.** Parallelogram; $P = 78$ mm; $A = 288$ mm²
80. Triangle; $P = 37.8$ mi; $A = 58.5$ mi² **81.** parallel lines
82. line segment **83.** acute angle **84.** intersecting lines **85.** right
angle; 90° **86.** ray **87.** straight angle; 180° **88.** obtuse angle
89. perpendicular lines **90.** 81° **91.** 138° **92.** $P = 90$ m; $A = 92$ m²
93. $P = 282$ cm; $A = 4190$ cm² **94.** Cylinder; $V \approx 100.5$ ft³ (rounded)
95. Rectangular solid or cube; $V \approx 3.4$ in³ (rounded) or $V = 3\frac{3}{8}$ in.³
(exactly) **96.** Rectangular solid; $V \approx 7.4$ m³ (rounded) **97.** Pyramid;
$V = 561$ cm³ **98.** Cone; $V \approx 1271.7$ cm³ **99.** Sphere; $V \approx 1436.0$ m³
(rounded) **100.** $x \approx 12.6$ km (rounded) **101.** $\angle D = 72°$
102. $x = 12$ mm; $y = 14$ mm **103.** The prefix *dec-* in *dec*ade means 10
and the prefix *cent-* in *cent*ury means 100, so divide 200 (two centuries)
by 10. The answer is 20 decades.

Chapter 8 TEST (Pages 621–622)

1. (e) **2.** (a); 90° **3.** (d) **4.** (g); 180° **5.** Parallel lines are lines in the same plane that never intersect. Perpendicular lines intersect to form a right angle.

Parallel lines   Perpendicular lines

6. 9° **7.** 160° **8.** ∠1 measures 50°; ∠3 measures 90°; ∠2 and ∠4 measure 40° **9.** Rectangle; $P = 23$ ft; $A = 30$ ft^2 **10.** Square; $P = 72$ mm; $A = 324$ mm^2 **11.** Parallelogram; $P = 26.2$ m; $A = 33.12$ m^2 **12.** Trapezoid; $P = 169.4$ cm; $A = 1591$ cm^2 **13.** $P = 32.45$ m; $A = 48$ m^2 **14.** $P = 37.8$ yd or $37\frac{4}{5}$ yd; $A = 58.5$ yd^2 **15.** 55° **16.** $r = 12.5$ in. or $12\frac{1}{2}$ in. **17.** $C \approx 5.7$ km (rounded) **18.** $A \approx 206.0$ cm^2 (rounded) **19.** $A \approx 39.3$ m^2 (rounded) **20.** Rectangular solid; $V = 6480$ m^3 **21.** Sphere; $V \approx 33.5$ ft^3 (rounded) **22.** Cylinder; $V \approx 5086.8$ ft^3 **23.** $\sqrt{85} \approx 9.2$ cm (rounded) **24.** $y = 12$ cm; $z = 6$ cm **25.** Linear units like cm are used to measure perimeter, radius, diameter, and circumference. Area is measured in square units like cm^2 (squares that measure 1 cm on each side). Volume is measured in cubic units like cm^3.

Chapters 1–8 CUMULATIVE REVIEW EXERCISES (pages 623–624)

1. $\frac{9}{20}$ **2.** 0.9132 **3.** 0.000078 **4.** 19.44 (rounded) **5.** $4\frac{5}{12}$ **6.** 9 **7.** two hundred eight ten-thousandths **8.** 2.055; 2.5005; 2.505; 2.55 **9.** $x = 35$ **10.** $x \approx 2.01$ (rounded) **11.** 160% **12.** $600 **13.** 135 min **14.** $2\frac{1}{2}$ or 2.5 lb **15.** 0.08 m **16.** 1800 mL **17.** mm **18.** L **19.** kg **20.** cm **21.** Rectangle; $P = 11$ in.; $A = 7$ in.^{2} **22.** Triangle; $P = 5.8$ m; $A = 1.575$ m^2 **23.** Circle; $C \approx 31.4$ ft; $A \approx 78.5$ ft^2 **24.** Parallelogram; $P = 76$ cm; $A = 264$ cm^2 **25.** $y \approx 24.2$ mm (rounded) **26.** 59% (rounded) **27.** $V \approx 3056.6$ cm^3 (rounded) **28.** Brand T at 8.5 oz for $3.09 − $0.75 coupon, about $0.275/oz **29.** $1\frac{1}{12}$ yd **30.** 35.7 lb meat (rounded); 50 lb potatoes; 21.4 lb carrots (rounded) **31.** 3.4 m **32.** $22.40

CHAPTER 9 Basic Algebra

SECTION 9.1 (pages 630–631)

1. +32 or 32 degrees **2.** +$920 or $920 **3.** −$12 **4.** −$30.79 **5.** $-6\frac{1}{2}$ lb **6.** +20 ft or 20 ft **7.** +20,320 ft or 20,320 ft **8.** −5315 ft

9. ←+|+|+|+◆+|+◆+|+◆+|+◆+|+|→
         −5−4−3−2−1 0 1 2 3 4 5

11. ←|+●|+◆+|+|●|+◆|+◆|●+|→
         −5−4−3−2−1 0 1 2 3 4 5

13. **(a)** zero is less than five, or, zero is less than positive five **(b)** negative ten is greater than negative seventeen **14.** **(a)** three is greater than negative seven, or, positive three is greater than negative seven **(b)** twelve is less than twenty-two, or, positive twelve is less than positive twenty-two.

15. < **17.** > **19.** < **21.** > **23.** < **25.** > **27.** 3 **29.** 10 **31.** 0 **33.** −18 **35.** −32 **37.** −7 **39.** 14 **41.** $-\frac{2}{3}$ **43.** 8.3 **45.** $\frac{1}{6}$ **47.** true **49.** true **51.** false

53.
```
        A  C D  B
←+------●--●●---●---+→
  −2   −1    0    1
```

54. −1.2, −0.5, 0, 0.6 **55.** A: may be at risk; B: above normal; C: normal; D: normal **56.** **(a)** The patient would think the interpretation was "above normal" and wouldn't get treatment. **(b)** Patient D's score of 0; because 0 is neither positive nor negative.

SECTION 9.2 (pages 640–643)

1. −7   −2   −5
```
←+++++++++++++++→
 −7−6−5−4−3−2−1 0 1 2 3
```

3. −1   −4
```
    ←----------+
       3
    +---------→
←+++++++++++++++→
 −7−6−5−4−3−2−1 0 1 2 3
```

5. Subtract the lesser absolute value from the greater absolute value. Write the sign of the number with greater absolute value in front of the answer. Examples will vary. **6.** Add the absolute values of the numbers. Write a negative sign in front of the answer. Examples will vary.

7. −3 **9.** 7 **11.** −7 **13.** 1 **15.** −8 **17.** −20 **19.** $13 + (−17) = −4$ yd **21.** −$52.50 + $50 = −$2.50 **23.** Jeff: −20 + 75 + (−55) = 0 pts; Terry: 42 + (−15) + 20 = 47 pts **25.** −6.8 **27.** $\frac{1}{4}$ **29.** $-\frac{3}{10}$ **31.** $-\frac{26}{9}$ **33.** −3 **35.** 9 **37.** $-\frac{1}{2}$ **39.** 6.2 **41.** negative seven minus three **42.** fourteen plus negative twelve **43.** negative twenty plus negative three **44.** negative eight minus negative five **45.** 14 **47.** −2 **49.** −12 **51.** −25 **53.** −23 **55.** 5 **57.** 20 **59.** 11 **61.** −60 **63.** 0 **65.** $-\frac{3}{2}$ **67.** $-\frac{2}{5}$ **69.** 0.7 **71.** **(a)** Change 6 to its opposite, −6. Correct answer is −6 + (−6) = −12. **(b)** Do *not* change 9 to its opposite. Correct answer is −9 + (−5) = −14. **72.** **(a)** 6 **minus** 9; subtraction **(b)** **negative** 9; a number less than 0 **(c)** the **opposite** of **negative** 2 **73.** **(a)** 21 °F; 30 − 21 = 9 degrees difference **(b)** 0 °F; 15 − 0 = 15 degrees difference **75.** **(a)** −17 °F; 5 − (−17) = 22 degrees difference **(b)** −41 °F; −10 − (−41) = 31 degrees difference **77.** −10 **79.** 12 **81.** −7 **83.** 5 **85.** −1 **87.** −4.4 **89.** −11 **91.** −10 **93.** $259 **95.** $0 **97.** **(a)** −2; −2 **(b)** −8; −8 **(c)** 10; 10. Addition is commutative, so changing the order of the addends does *not* change the sum. **98.** **(a)** −8; 8 **(b)** 2; −2 **(c)** −7; 7. Changing the order of the numbers in subtraction *does change* the result. **99.** **(a)** The answers have the same absolute value but opposite signs. **(b)** When the order of the numbers in a subtraction problem is switched, change the sign on the answer to its opposite. **100.** **(a)** −18; 20; −5; 4. Adding 0 to any number leaves the number unchanged. **(b)** 2; −10; −3; 7. Subtracting 0 from a number leaves the number unchanged, but subtracting a *number from 0* changes the sign of the number.

SECTION 9.3 (pages 646–649)

1. The product of two numbers with the same sign is positive. Examples will vary. One possibility: $(-2)(-4) = 8$ **2.** The product of two numbers with different signs is negative. Examples will vary. One possibility: $(-2)(8) = -16$ **3.** -35 **5.** -45 **7.** -18 **9.** -50 **11.** -40 **13.** 32   **15.** 77   **17.** 133   **19.** 13   **21.** 0   **23.** 4   **25.** -4 **27.** $-\frac{1}{10}$ **29.** $\frac{14}{3}$ **31.** $-\frac{5}{6}$ **33.** $\frac{7}{4}$ **35.** -42.3 **37.** 6   **39.** -31.62 **41.** 4.5   **43.** 8.23   **45.** 0   **47.** Dividing by 0 cannot be done, so it is undefined. Examples will vary. One possibility: $9 \div 0$ is undefined.   **48.** Zero divided by any other number is 0. Examples will vary. One possibility: $0 \div 9 = 0$ **49.** -2 **51.** -5 **53.** undefined   **55.** -14 **57.** 10   **59.** 4   **61.** -1 **63.** 191   **65.** 0   **67.** 1   **69.** $\frac{2}{3}$ **71.** $\frac{1}{3}$ **73.** -8 **75.** $-\frac{14}{3}$ **77.** 4.73   **79.** -4.7 **81.** -5.3 **83.** 12   **85.** -0.36 **87.** $-\frac{7}{40}$ **89.** -2 **91.** -10 **93.** $-\$119,400$ **95.** -2011 ft   **97.** $2366   **99.** -22 degrees   **101. (a)** Examples will vary. Some possibilities: $6(-1) = -6; 2(-1) = -2; 15(-1) = -15$ **(b)** Examples will vary. Some possibilities: $-6(-1) = 6; -2(-1) = 2; -15(-1) = 15$. The result of multiplying any nonzero number times -1 is the number with the opposite sign.   **102. (a)** Examples will vary. Some possibilities: $\frac{-6}{-1} = 6; \frac{-2}{-1} = 2; \frac{-15}{-1} = 15$ **(b)** Some possibilities: $\frac{6}{-1} = -6; \frac{2}{-1} = -2; \frac{15}{-1} = -15$

When dividing by -1, the sign of the number changes to its opposite.

SECTION 9.4 (pages 654–657)

1.

$\underbrace{(-6)^2} - 8 \div 2(-3)$	
$36 - \underbrace{8 \div 2}(-3)$	Apply the exponent. $(-6)(-6)$ is 36
$36 - \underbrace{4(-3)}$	Divide: $8 \div 2$ is 4
$36 - (-12)$	Multiply: $4(-3)$ is -12
	Change subtraction to addition.
$\underbrace{36 + (+12)}$	Add.
48	

2.

$7 - 2\underbrace{(9-11)}(-2)^3$	Work inside parentheses. $9 - 11$ is $9 + (-11) = -2$
$7 - 2(-2)\underbrace{(-2)^3}$	Apply the exponent: $(-2)^3$ is $(-2)(-2)(-2)$ $= -8$
$7 - \underbrace{2(-2)}(-8)$	Multiply from left to right.
$7 - \underbrace{(-4)(-8)}$	
$7 - (+32)$	Change subtraction to addition.
$\underbrace{7 + (-32)}$	Add.
-25	

3. $-12; -6$ **5.** 0   **7.** $16; 52$ **9.** -39 **11.** -30 **13.** 17   **15.** 16   **17.** 30   **19.** 23   **21.** -43 **23.** 19   **25.** -3 **27.** 0   **29.** 47   **31.** 41   **33.** -2 **35.** 13   **37.** 126   **39.** $\frac{27}{-3} = -9$ **41.** $\frac{-48}{-4} = 12$ **43.** $\frac{-60}{-1} = 60$

45. 8   **47.** 1.24   **49.** -1.47 **51.** -2.31 **53.** $-\frac{13}{10}$ **55.** $\frac{5}{4}$ **57.** $\frac{4}{5}$ **59.** -1200 **61.** 1   **63.** $\frac{27}{-27} = -1$ **65.** 7   **67.** 6

SUMMARY EXERCISES Operations with Signed Numbers (pages 658–659)

1. -6 **2.** 0   **3.** -7 **4.** -7 **5.** 63   **6.** -1 **7.** -56 **8.** -22 **9.** 12   **10.** 6   **11.** -13 **12.** 0   **13.** -80 **14.** -17 **15.** -48 **16.** -1 **17.** $-\frac{1}{6}$ **18.** undefined   **19.** -14.6 **20.** -0.6 **21.** 48   **22.** -19 **23.** 2   **24.** -20 **25.** 0   **26.** 16   **27.** -3 **28.** -36 **29.** 6   **30.** -2 **31.** -32 **32.** -5 **33.** -4 **34.** -9732 **35.** 100   **36.** 4   **37.** -343 **38.** 5   **39.** -6 **40.** -10 **41.** -5 **42.** 8   **43.** 1   **44.** $\frac{27}{0}$ is undefined.   **45.** $\frac{-8}{8} = -1$ **46.** $+\$212$ or $\$212$ **47.** $-\$24$ **48.** -13 degrees   **49.** $+27,010$ ft or 27,010 ft   **50.** -12; they lost 12 yd.   **51.** -90 ft   **52.** January: $-\$700$; February: $-\$100$; March: $\$700$; April: $\$900$; May: $\$0$; June: $\$700$ **53.** January; April   **54.** $2100   **55.** $-\$1850$ **56.** Subtract the lesser absolute value from the greater absolute value. The answer has the same sign as the addend with the greater absolute value. Examples: $-6 + 2 = -4$ and $6 + (-2) = 4$ **57.** Similar: If the signs match, the result is positive. If the signs are different, the result is negative. Different: Multiplication is commutative, division is not. You can multiply by 0, but dividing by 0 cannot be done.

SECTION 9.5 (pages 662–664)

1. (a) $6y$ **(b)** $-2x - 7w$ **2. (a)** wxy **(b)** $-3a + 5c$ **3.** 28   **5.** -10 **7.** -30 **9.** -8 **11.** 0   **13.** 2   **15.** -22 **17.** $-\frac{13}{8}$ **19.** The error was made when replacing x with -3; should be $-(-3)$, not -3.

$-x - 4y$	Replace x with (-3) and y with (-1).
$-(-3) - 4(-1)$	Opposite of (-3) is $+3$.
$(+3) - (-4)$	Change subtraction to addition.
$\underbrace{3 + (+4)}$	Add.
7	

20. The error was made when changing subtractions to additions; student did not change the sign of (-10) to 10, and 5 to (-5).

$-3(-2) - (-10) - 5$	Replace b with -2 and c with -10.
$6 - (-10) - 5$	Multiply $-3(-2)$ to get 6.
$6 + (+10) + (-5)$	Change subtractions to additions.
$\underbrace{16} + (-5)$	Add.
11	

21. $-20; 28$ **23.** -5 **25.** 3   **27.** $\frac{12}{-6} = -2$ **29.** $P = 30$ cm   **31.** $P = 28$ in.   **33.** $A \approx 78.5$ yd^2 **35.** $A = \frac{45}{2}$ or $22\frac{1}{2}$ ft^2 **37.** $V = 600$ mm^3 **39.** $V = 1080$ in.^{3} **41.** $C \approx 25.12$ km   **43. (a)** $P = 33$ in.   **(b)** $P = 9$ ft   **45. (a)** average score = 83   **(b)** average score = 91

SECTION 9.6 (pages 670–673)

1. yes **3.** yes **5.** no **7. (a)** Add 8 to both sides because $-8 + 8$ gives 0 on the left side. Then m will be by itself on the left side.
(b) Add -5 to both sides because $5 + (-5)$ gives 0 on the right side. Then w will be by itself on the right side.
8. (a) Add -2 to both sides because $2 + (-2)$ gives 0 on the left side. Then n will be by itself on the left side.
(b) Add 6 to both sides because $-6 + 6$ gives 0 on the right side. Then b will be by itself on the right side.
9. $p + \underline{5 - 5} = 9 - 5$
   $p + \quad 0 \ = 4$
   $\qquad p = 4$

11. $k = -15$ **13.** $z = 8$ **15.** $r = 10$ **17.** $n = -8$ **19.** $r = -6$
21. $k = 18$ **23.** $x = 11$ **25.** $r = -3$ **27.** $d = \dfrac{7}{3}$ or $2\dfrac{1}{3}$
29. $z = \dfrac{87}{8}$ or $10\dfrac{7}{8}$ **31.** $k = \dfrac{5}{2}$ or $2\dfrac{1}{2}$ **33.** $m = \dfrac{83}{20}$ or $4\dfrac{3}{20}$
35. $x = 5.87$ **37.** $r = -1.51$
39. (a) Divide both sides by -10. Then, on the left side, divide out -10 to get x by itself. **(b)** Divide both sides by 5. Then, on the right side, divide out 5 to get z by itself.
40. (a) Divide both sides by 6. Then, on the left side, divide out 6 to get t by itself. **(b)** Divide both sides by -12. Then, on the right side, divide out -12 to get w by itself. **41.** $z = 2$ **43.** $r = 4$ **45.** $y = 0$
47. $k = -6$ **49.** $p = 9$ **51.** $m = -7$ **53.** $p = 1.1$
55. (a) Rewrite $\dfrac{x}{8}$ as $\dfrac{1}{8}x$. Multiply both sides by the reciprocal of $\dfrac{1}{8}$, which is $\dfrac{8}{1}$, in order to get x by itself on the left side.
(b) Multiply both sides by the reciprocal of $-\dfrac{2}{3}$, which is $-\dfrac{3}{2}$, in order to get k by itself on the right side.
56. (a) Rewrite $\dfrac{m}{7}$ as $\dfrac{1}{7}m$. Multiply both sides by the reciprocal of $\dfrac{1}{7}$, which is $\dfrac{7}{1}$, in order to get m by itself on the right side.
(b) Multiply both sides by the reciprocal of $\dfrac{3}{5}$, which is $\dfrac{5}{3}$, in order to get r by itself on the right side.
57. $k = 34$ **59.** $a = 66$ **61.** $r = -36$ **63.** $p = -20$ **65.** $m = 4$
67. $x = 16$ **69.** $y = 1.3$ **71.** $z = -4.94$ **73.** $x = 19$ **75.** $x = 9$
77. $x = \dfrac{8}{21}$ **79.** $a = -\dfrac{5}{4}$ **81.** $m = -4$

SECTION 9.7 (pages 678–681)

1. $-4y + 10 = 2$
Subtract 10 from both sides of the equation.
$-4y + 10 - 10 = 2 - 10$
$\qquad -4y = -8$
Note: You could also add (-10) to both sides as the first step. The result is the same.
2. $7 = 6x - 5$
Add 5 to both sides of the equation.
$7 + 5 = 6x - 5 + 5$
$\quad 12 = 6x$
3. $7p + 5 - 5 = \underline{12 - 5}$
   $\quad 7p = \quad 7$
   $\quad\ p = \quad 1$

5. $y - 1$ **7.** $m = 0$ **9.** $a = 2$ **11.** $x = -4$ **13.** $z - 6$
15. $6x + 24$ **17.** $7p - 56$ **19.** $-3m - 18$ **21.** $-2y + 6$
23. $-8c - 64$ **25.** $-10w + 90$
27. (a) The student did not multiply $5 \cdot 4$. Correct: $5x - 20$
(b) The student has the wrong sign for 16. Correct: $-8y + 16$
28. (a) The student did not multiply $6 \cdot 3$. Correct: $6y - 18$
(b) The student has the wrong sign for 15. Correct: $-5b + 15$
29. $17r$ **31.** $1z$ or z **33.** $-2y$ **35.** $-7t$ **37.** $0p = 0$ **39.** $-3b$
41.

$3p - 5p = 32$	Combine like terms on the left side.
$(3 - 5)p = 32$	$3 - 5$ is $3 + (-5)$, or -2.
$-2p = 32$	
$\dfrac{\overset{1}{-2} \cdot p}{\underset{1}{-2}} = \dfrac{32}{-2}$	Divide both sides by -2 to get p by itself on the left side.
$p = -16$	32 divided by -2 is -16.

42.

$-48 = (1 - 5)m$	Combine like terms on the right side.
$-48 = -4m$	$1 - 5$ is $1 + (-5)$, or -4.
$\dfrac{-48}{-4} = \dfrac{\overset{1}{-4} \cdot m}{\underset{1}{-4}}$	Divide both sides by -4 to get m by itself on the right side.
$12 = m$	-48 divided by -4 is -12.

43. $k = 5$ **45.** $m = 6$ **47.** $b = -6$ **49.** $y = 1$ **51.** $p = 4$
53. $z = -2$ **55.** $y = 2$ **57.** $b = -1.05$ **59.** $r = -4$
61. $y = -9$ **63.** $t = -5$ **65.** $x = 0$
67.

$-2t - 10 = 3t + 5$	
$-2t - 3t - 10 = 3t - 3t + 5$	Subtract $3t$ from both sides (addition property).
$-5t - 10 = 5$	
$-5t - 10 + 10 = 5 + 10$	Add 10 to both sides (addition property).
$-5t = 15$	
$\dfrac{\overset{1}{-5} \cdot t}{\underset{1}{-5}} = \dfrac{15}{-5}$	Divide both sides by -5 (multiplication property).
$t = -3$	

68. Check: $3(-4 + 5) = -7$
   $3(1) = -7$
   $3 = -7$ False
The first step on the left side should have been $3(2x + 5)$ is $6x + 15$. The correct solution is $x = -\dfrac{11}{3}$.
69. $x = -10$ **71.** $y = 2$ **73.** $y = 20$ **75.** $w = -1$ **77.** $a = -6.75$

SECTION 9.8 (pages 688–691)

1. $14 + x$ or $x + 14$ **2.** $x + (-8)$ or $-8 + x$ **3.** $-5 + x$ or $x + (-5)$
4. $x + 16$ or $16 + x$ **5.** $20 - x$ **6.** $x - 25$ **7.** $x - 9$ **8.** $-7 - x$
9. $x - 4$ **10.** $x - 3$ **11.** $6x$ **12.** $-3x$ **13.** $2x$ **14.** $\dfrac{x}{2}$ or $\dfrac{1}{2}x$
15. $\dfrac{x}{2}$ **16.** $\dfrac{4}{x}$ **17.** $8 + 2x$ or $2x + 8$ **18.** $5x + 5$ or $5 + 5x$
19. $7x - 10$ **20.** $6x - 12$ **21.** $2x + x$ or $x + 2x$ **22.** $x - 3x$
23. (a) A variable is a letter that represents an unknown quantity. Examples: x, w, p **(b)** An expression is a combination of operations on variables and numbers. Examples: $6x, w - 5, 2p + 3x$
(c) An equation has an $=$ sign and shows that two expressions are equal. Examples: $2y = 14$; $x + 5 = 2x$; $8p - 10 = 54$ **24.** True. Choose any letter you like, although it may help to choose a letter that reminds you of what it represents, such as w for an unknown width. **25.** $n = 7$
27. $2n + n = -15$; $n = -5$ **29.** $5n + 12 = 7n$; $n = 6$
31. $30 - 3n = 2 + n$; $n = 7$ **33.** $\dfrac{1}{2}n + 2n = 50$; $n = 20$

35. Let x be Ricardo's original weight. $x + 15 - 28 + 5 = 177$. He weighed 185 pounds originally.　**37.** Let x be the amount Brenda spent. $81 = 2x - 3$. Brenda spent \$42.　**39.** Let x be my age; $x - 9$ is my sister's age. $x + x - 9 = 51$. I am 30; my sister is 21.　**41.** Let x be husband's earnings; $x + 1500$ is Lien's earnings. $x + x + 1500 = 37{,}500$. Husband earned \$18,000; Lien earned \$19,500.　**43.** Let x be printer's cost; $8x$ is computer's cost. $x + 8x = 396$. The printer cost \$44; the computer cost \$352.　**45.** Let x be length of shorter piece; $x + 10$ is length of longer piece. $x + x + 10 = 78$. Shorter piece is 34 cm; longer piece is 44 cm.　**47.** Let x be one piece; $x - 7$ is the other piece. $x + x - 7 = 31$. One piece is 19 ft; the other piece is 12 ft.

49. $48 = 2(5) + 2x$; length is 19 yd.

51. $36 = 2(2w) + 2w$; length is 12 ft; width is 6 ft.

53. $36 = 2(2x + 3) + 2x$; length is 13 in.; width is 5 in.

Chapter 9 REVIEW EXERCISES (pages 697–702)

1.

$$-7\;-6\;-5\;-4\;-3\;-2\;-1\;\;0\;\;1\;\;2\;\;3\;\;4$$

2.

$$-6\;-5\;-4\;-3\;-2\;-1\;\;0\;\;1\;\;2\;\;3\;\;4\;\;5$$

3.

$$-7\;-6\;-5\;-4\;-3\;-2\;-1\;\;0\;\;1\;\;2\;\;3$$

4.

$$-8\;-7\;-6\;-5\;-4\;-3\;-2\;-1\;\;0\;\;1\;\;2$$

5. $>$　**6.** $<$　**7.** $>$　**8.** $<$　**9.** 8　**10.** 19　**11.** -7　**12.** -15
13. 2　**14.** -7　**15.** -19　**16.** -33　**17.** 1　**18.** -19　**19.** $\dfrac{3}{10}$
20. $-\dfrac{3}{8}$　**21.** -5.2　**22.** -1.5　**23.** -6　**24.** 14　**25.** $\dfrac{5}{8}$
26. -3.75　**27.** -6　**28.** -8　**29.** -7　**30.** -17　**31.** 11　**32.** 11
33. 13　**34.** -6　**35.** -80　**36.** 0　**37.** $-\dfrac{1}{2}$　**38.** 9　**39.** -24
40. -20　**41.** 15　**42.** 64　**43.** -8　**44.** -3　**45.** 5　**46.** 20
47. 0　**48.** -81　**49.** 0　**50.** undefined　**51.** $-\dfrac{4}{7}$　**52.** 6　**53.** 1.4
54. -6.6　**55.** 57　**56.** 23　**57.** -1　**58.** -2　**59.** -34　**60.** -32
61. -100　**62.** 117　**63.** -1　**64.** 1.328　**65.** 0　**66.** 2　**67.** 27
68. -8　**69.** 0　**70.** 19　**71.** 23　**72.** -1　**73.** $P = 35$ cm
74. $A = 27$ ft^2　**75.** $y = -3$　**76.** $a = 16$　**77.** $z = 1$　**78.** $r = 1$
79. $x = -\dfrac{5}{4}$ or $-1\dfrac{1}{4}$　**80.** $k = -8.05$　**81.** $r = -7$　**82.** $p = 8$

83. $z = 20$　**84.** $a = -55$　**85.** $y = 9$　**86.** $b = -4$　**87.** $6r - 30$
88. $11p + 77$　**89.** $-9z + 27$　**90.** $-8x - 32$　**91.** $11r$　**92.** $-5z$
93. $-8p$　**94.** $2x$　**95.** $z = -9$　**96.** $k = -5$　**97.** $y = -5$
98. $b = 7$　**99.** $a = -4$　**100.** $t = 0$　**101.** $18 + x$　or　$x + 18$
102. $\dfrac{1}{2}x$　or　$\dfrac{x}{2}$　**103.** $-5x$　**104.** $20 - x$　**105.** $4x + 6 = -14; x = -5$
106. $5x - x = 100; x = 25$　**107.** Let x be one student's scholarship; $3x$ is the other student's scholarship. $x + 3x = 30{,}000$. One student receives \$7500; the other student receives \$22,500.
108. Let w be the width; $w + 4$ is the length. $48 = 2(w + 4) + 2w$; width is 10 in.; length is 14 in.

109. 3　**110.** 40　**111.** -1　**112.** -7　**113.** -16　**114.** -9
115. 6　**116.** 5　**117.** -20　**118.** undefined　**119.** $-\dfrac{5}{9}$　**120.** -0.35
121. 7　**122.** -19　**123.** $y = 9$　**124.** $b = -4$　**125.** $z = 4$
126. $r = -10$　**127.** $x = -11$　**128.** $z = 3$　**129.** $k = 4$
130. $t = 0$　**131.** $a = -40$　**132.** $p = 3$
133. Let x be the daily amount for a young child.

$$4x + 8 = 60$$

A young child should receive 13 mg of vitamin C.
134. Let x be the number of Senators; $5x - 65$ is the number of Representatives. $x + 5x - 65 = 535$. 100 Senators; 435 Representatives
135. Let x be the zebra's speed; $x + 25$ is the cheetah's speed. $x + x + 25 = 111$. Zebra's speed is 43 mph; Cheetah's speed is 68 mph.
136. Let w be the width; $5w$ is the length. $1320 = 2(5w) + 2(w)$. Width is 110 yd; length is 550 yd.

Chapter 9 TEST (pages 703–704)

1.

$$-5\;-4\;-3\;-2\;-1\;\;0\;\;1\;\;2\;\;3\;\;4\;\;5$$

2. $<, >$　**3.** 7, 15　**4.** -1　**5.** -13　**6.** 5.3　**7.** -7　**8.** 16
9. $\dfrac{1}{4}$　**10.** -32　**11.** 84　**12.** 0　**13.** -25　**14.** 8　**15.** $-\dfrac{3}{5}$
16. 1　**17.** -21　**18.** -38　**19.** 27　**20.** When evaluating, you are given specific values to replace each variable. When solving an equation, you are not given the value of the variable. You must find a value that "works"; that is, when your solution is substituted for the variable, the two sides of the equation are equal.　**21.** 110 ft^2　**22.** $x = 5$
23. $r = 31$　**24.** $t = 5$　**25.** $p = -15$　**26.** $a = -3$　**27.** $m = 2$
28. Let x be the shorter piece; $x + 4$ is longer piece. $x + x + 4 = 118$. Shorter piece is 57 cm; longer piece is 61 cm.　**29.** Let w be the width; $4w$ is the length. $420 = 2(4w) + 2w$; width is 42 ft; length is 168 ft.

30. Let x be Marcella's time; $x - 3$ is Tim's time. $x + x - 3 = 19$. Marcella spent 11 hr; Tim spent 8 hr.

ANSWERS

Chapters 1–9 CUMULATIVE REVIEW EXERCISES (pages 705–706)

1. *Estimate:* 40,000 ÷ 80 = 500; *Exact:* 503

2. *Exact:* $5\frac{1}{4}$; *Estimate:* 6 • 1 = 6   **3.** *Exact:* $2\frac{1}{2}$; *Estimate:* 4 ÷ 2 = 2

4. 8.906   **5.** 56.4 (rounded)   **6.** −5   **7.** 4   **8.** −1.3

9. $-\frac{1}{2}$ **10.** 5   **11.** 4   **12.** $x = 0.4$ **13.** 50 cars   **14.** 4%

15. 14 qt   **16.** 3 days   **17.** 3750 g   **18.** 0.4 m

19. Square; $P = 9$ ft; $A = 5\frac{1}{16}$ (exactly) or 5.1 ft² (rounded)

20. Circle; $C \approx 28.3$ mm; $A \approx 63.6$ mm² (both rounded)

21. Right triangle; $P = 56$ mi; $A = 84$ mi²   **22.** Parallelogram; $P = 36$ ft; $A = 70$ ft²   **23.** $t = -5$ **24.** $x = 8$ **25.** $p = -4$

26. $4x + 40 = 0$; $x = -10$ **27.** Let m be Reggie's money; $m + 300$ is Donald's money. $m + m + 300 = 1000$; $350 for Reggie; $650 for Donald.   **28.** Let w be the width; $w + 5$ is the length. $82 = 2(w + 5) + 2w$; length is 23 cm; width is 18 cm

29. $y = \sqrt{(20)^2 - (15)^2} = \sqrt{400 - 225} = \sqrt{175}$; $y \approx 13.2$ yd (rounded)

30. $31.91 (rounded)   **31.** 54 min (rounded)   **32.** 0.544 m³ less

33. Naomi's car; 25.2 miles per gallon (rounded).   **34.** 124%   **35.** $8\frac{1}{3}$ cups

CHAPTER 10 Statistics

SECTION 10.1 (pages 712–717)

1. parts   **2.** protractor   **3.** 360 million or 360,000,000 pets

5. $\frac{90}{360} = \frac{1}{4}$ **7.** $\frac{90}{75} = \frac{6}{5}$ **9.** Sleeping; 8 hours 40 minutes

11. $\frac{520}{1440} = \frac{13}{36}$ **13.** $\frac{50}{76} = \frac{25}{38}$ **15.** $\frac{210}{150} = \frac{7}{5}$ **17.** 160 people

19. 96 people   **21.** 960 people   **23.** 1027 people (rounded)

25. 121 people (rounded)   **27.** 483 people (rounded)   **29.** First find the percent of the total that is to be represented by each item. Next, multiply the percent by 360° to find the size of each sector. Finally, use a protractor to draw each sector.   **30.** A protractor is used to measure the number of degrees in a sector. First, you must draw a line from the center of the circle to the left edge. Next, place the hole of the protractor over the center of the circle, making sure that the 0 on the protractor is on the line. Finally, make a mark at the desired number of degrees. This gives you the size of the sector.   **31.** 90°   **33.** 10%; 36°   **35.** 54°   **37.** 15%

39. **(a)** $200,000   **(b)** 22.5°; 72°; 108°; 90°; 67.5°

(c)

41. **(a)** 2578; 37% (rounded); 133°; 266; 4% (rounded); 14°; 1155; 16% (rounded); 58° (rounded); 1813; 26% (rounded); 94° (rounded); 749; 11% (rounded); 40° (rounded); 469; 7% (rounded); 25% (rounded)

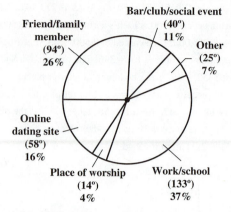

(b) No. The total is 101% due to rounding.   **(c)** No. The total is 364° due to rounding.

SECTION 10.2 (pages 721–726)

1. false   **2.** true   **3.** India; 51.4%   **5.** USA, Britain, and Australia

7. 121 days (rounded)   **9.** May; 10,000 plants   **11.** 1500 plants

13. 4000 plants   **15.** 150,000 gal   **17.** 2009; 250,000 gal

19. 550,000 gal   **21.** 24.1 million or 24,100,000 PCs   **23.** 204.7 million or 204,700,000 PCs   **25.** Answers will vary. Some possibilities are: greater demand as a result of lower price; more uses and applications for the owner; improved technology; multiple computers in each location; more laptop computers sold.   **26.** Answers will vary. Some possibilities are: Fewer people will want a computer because they already own one; new technology will replace the computer with something better.   **27.** 300,000 MP3s

29. 150,000 MP3s   **31.** 350,000 MP3s   **33.** Probably Store B with greater sales. Predicted sales might be 450,000 MP3s to 500,000 MP3s in 2014.   **34.** Some possible answers are that Store B may have started to: do more advertising; keep longer store hours; give better training to their staff; employ more help; give better service than Store A.   **35.** A single bar or a single line must be used for each set of data. To show multiple sets of data, multiple sets of bars or lines must be used.   **36.** You would use a set of bars or a set of lines for each set of data. Examples will vary.   **37.** $400,000   **39.** $250,000   **41.** $50,000   **43.** Answers will vary. Some possibilities are that the decrease in sales may have resulted from poor service or greater competition. The increase in sales may have been a result of more advertising or better service.   **44.** As sales increase or decrease, so do profits.   **45.** 23 years   **46.** 20 additional flavors

47. 214,286 rolls each day (rounded)   **48.** 1.17 lb (rounded)   **49.** The weight of one roll of LifeSavers is much less than 1.17 lb. The answer is "correct" using the information given, but some of the data given must not be accurate. Perhaps 3 million rolls of LifeSavers are produced daily.

50. Answers will vary. Possible answers are: misprints or typographical errors; careless reporting of data; math errors.

SUMMARY EXERCISES Graphs (pages 727–728)

1. 2000 people   **2.** vanilla; 660 people   **3.** $\frac{380}{2000} = \frac{19}{100}$

4. $\frac{100}{2000} = \frac{1}{20}$ **5.** $\frac{640}{80} = \frac{8}{1}$ **6.** $\frac{660}{140} = \frac{33}{7}$ **7.** 2100; 10.1 billion or

10,100,000,000 people   **8.** 4.0 billion or 4,000,000,000 people
9. 0.6 billion or 600,000,000 people   **10.** 52.5% (rounded)
11. 1.089 billion or 1,089,000,000 in China; 0.4455 billion or 445,500,000 in the United States   **12.** 1960; about 68%   **13.** about $2\frac{1}{2}$%; yes
14. widowed   **15.** about 6%   **16.** Answers will vary. Some possibilities are: No one lived together in 1970; this category was very small; no one kept track of this group; no one would admit it.

SECTION 10.3 (pages 731–734)

1. table   **2.** bar graph   **3.** 18–25 years old; 40 million or 40,000,000 users
5. 50 million or 50,000,000 users   **7.** 25 million or 25,000,000 users
9. $4100 to $5000; 16 employees   **11.** 11 employees   **13.** 49 employees
15. Class intervals are the result of combining data into groupings. Class frequency is the number of data items that fit in each class interval. These are used to group data and to have multiple responses (frequency) in a class interval—this makes the data easier to interpret.   **16.** If too few class intervals were used the class frequencies would be high and any differences in the data might not be observable. If too many class intervals were used, interpretation might become impossible because class frequencies would be very low or nonexistent.   **17.** ||||| |; 6   **19.** |; 1   **21.** ||; 2   **23.** ||||; 4
25. ||||| ; 5   **27.** |||; 3   **29.** ||||| ||; 7   **31.** |||; 3   **33.** ||||| ||||| ||||| ||||| ||||| |; 26
35. ||||| ||||| ||||| ||; 17   **37.** |; 1

SECTION 10.4 (pages 739–742)

1. true   **2.** false   **3.** grade point average   **4.** most often   **5.** 14 hours
7. 2.5 in. of rain (rounded)   **9.** $37,127   **11.** $72.53   **13.** 12.5 customers (rounded)   **15.** 17.2 fish (rounded)   **17.** $130,000   **19.** 125 guests
21. 508 calories   **23.** 21%   **25.** 68 and 74 yr (bimodal)   **27.** The median is a better measure of central tendency when the list contains one or more extreme values. Find the mean and the median of the following home values. $182,000; $164,000; $191,000; $115,000; $982,000; mean home value = $326,800; median home value = $182,000   **28.** The size to order is the mode. The mode is the size most worn by customers, and it would be wise to order most hats in this size.   **29.** 2.42 (rounded)   **31.** 2.60
33. 3.14 (rounded)   **35.** Each made 176 sales calls.   **36.** mean for Samuels: 22; mean for Stricker: 22   **37.** median for Samuels: 19.5; median for Stricker: 22   **38.** mode for Samuels: 22; mode for Stricker: 22
39. The mean and mode are identical for both sales representatives, and the medians are close.   **40.** The number of weekly sales calls made by Samuels varies greatly from week to week, while the number of weekly sales calls made by Stricker remains fairly constant.   **41.** range = 40 − 8 = 32   **42.** range = 25 − 19 = 6
43. No, not with any certainty. There probably are additional questions that need to be answered, such as the dollar amount of sales, number of repeat customers, and so on.   **44.** Answers will vary. Some possible answers are: He works hard one week, then takes it easy the next week; the characteristics of the sales territories vary greatly; illness or personal problems may be affecting performance.

Chapter 10 REVIEW EXERCISES (pages 749–754)

1. Basketball; 18,000 teams   **2.** $\frac{16,000}{76,000} = \frac{4}{19}$ **3.** $\frac{15,000}{76,000} = \frac{15}{76}$

4. $\frac{14,000}{76,000} = \frac{7}{38}$ **5.** $\frac{18,000}{16,000} = \frac{9}{8}$ **6.** $\frac{16,000}{14,000} = \frac{8}{7}$ **7.** 3936 companies

8. 1728 companies   **9.** 720 companies   **10.** 2904 companies
11. On-site child care and fitness centers. Answers will vary. Perhaps employers feel that they are not needed or would not be used. Or, it may be that they would be too expensive for the benefit derived.   **12.** Flexible hours and casual dress. Answers will vary. Perhaps employees request them and appreciate them. Or, it may be that neither of them costs the employer anything to offer.   **13.** March; 8,000,000 acre-feet   **14.** June; 2,000,000 acre-feet   **15.** 5,000,000 acre-feet   **16.** 4,000,000 acre-feet
17. 5,000,000 acre-feet   **18.** 2,000,000 acre-feet   **19.** $50,000,000
20. $20,000,000   **21.** $20,000,000   **22.** $40,000,000   **23.** The floor-covering sales decreased for 2 years and then moved up slightly. Answers will vary. Perhaps there is less new construction, remodeling, and home improvement in the area near Center A, or better product selection and service have reversed the decline in sales.   **24.** The floor-covering sales are increasing. Answers will vary. Perhaps new construction and home remodeling have increased in the area near Center B, or greater advertising has attracted more customers.   **25.** 28.5 digital cameras   **26.** 35.2 complaints   **27.** $51.05   **28.** 257.33 points (rounded)   **29.** 39 forms
30. $562   **31.** $79   **32.** 18 and 32 launchings (bimodal)   **33.** 36°
34. 35%   **35.** 20%; 72°   **36.** 25%; 90°   **37.** 10%; 36°
38.

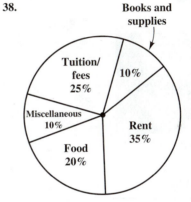

Books and supplies

Tuition/fees 25%   10%

Miscellaneous 10%   Rent 35%

Food 20%

39. 100 volunteers   **40.** 144.7 vaccinations (rounded)   **41.** 48 applicants
42. 31 and 43 two-bedroom apartments (bimodal)   **43.** 5.0 hr
44. 21 e-mails   **45.** ||||; 4   **46.** |; 1   **47.** ||||| |; 6   **48.** ||; 2
49. ||||| ||||| |||; 13   **50.** ||||| ||; 7   **51.** ||||| ||; 7
52.

MATHEMATICS EXAM SCORES

Number of Students (y-axis: 0, 5, 10, 15)
Exam Scores (x-axis: 30–39, 40–49, 50–59, 60–69, 70–79, 80–89, 90–99)

53. 64.5 (rounded)   **54.** 118.8 units (rounded)

Chapter 10 TEST (pages 755–758)

1. $66.3 billion (rounded) **2.** $22.6 billion (rounded) **3.** $3.3 billion (rounded) **4.** $76.8 billion (rounded) **5.** $147.3 billion (rounded) **6.** $11.8 billion (rounded) **7.** 108° **8.** 36° **9.** 72° **10.** 126° **11.** 5%

12.

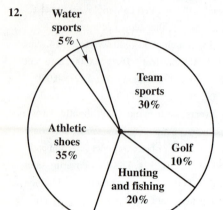

13. |||; 3 **14.** ||; 2 **15.** ||||; 4 **16.** |||; 3 **17.** |||; 3 **18.** ||||; 5

19.

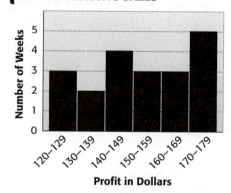

ALAN'S VENDING SALES

20. 75 mi **21.** 15.3 lb (rounded) **22.** 478.9 mph **23.** The weighted mean is used because different classes are worth different numbers of credits. GPA problems will vary; one possibility is shown.

Credits	Grade
3	A 3 • 4 = 12
2	C 2 • 2 = 4
4	B 4 • 3 = 12
9	**Total 28**

28 ÷ 9 ≈ 3.11 (rounded)

24. Arrange the values in order, from least to greatest. When there is an odd number of values in a list, the median is the middle value. Students' problems will vary; one possibility is shown.

8, 17, 23, 32, 64
⌐— Median

25. $26 (rounded) **26.** 174 points (rounded) **27.** 31.5 degrees **28.** 10.0 meters **29.** 52 milliliters **30.** 103 and 104 degrees (bimodal)

Chapters 1–10 CUMULATIVE REVIEW EXERCISES (pages 759–760)

1. *Estimate:* 900 − 60 = 840; *Exact:* 811.863

2. *Estimate:* 7000 × 600 = 4,200,000; *Exact:* 4,485,640

3. *Estimate:* 60 ÷ 4 = 15; *Exact:* 14.72

4. $10\frac{4}{15}$ **5.** $1\frac{11}{12}$ **6.** $44\frac{2}{5}$ **7.** $8\frac{4}{5}$ **8.** $\frac{8}{21}$ **9.** $\frac{1}{12}$

10. 23 **11.** 10 **12.** 108 **13.** 0.199, 0.207, 0.218, 0.22, 0.2215

14. 0.58, 0.608, $\frac{5}{8}$, 0.6319, $\frac{13}{20}$ **15.** $\frac{1}{8}$ **16.** $\frac{4}{1}$ **17.** $x = 6$ **18.** $x = 6$

19. $x = 16.2$ **20.** $x = 52$ **21.** 324 homes **22.** 2800 people

23. 260% **24.** 7 **25.** 40 cm **26.** 0.23 kg **27.** 12,000 lb

28. Rectangle: $A ≈ 47.3 \text{ m}^2$ (rounded) **29.** Triangle: $A ≈ 9.6 \text{ ft}^2$ (rounded) **30.** Circle: $A ≈ 132.7 \text{ cm}^2$ (rounded) **31.** Cylinder: $V ≈ 549.8 \text{ cm}^3$ (rounded) **32.** Rectangular solid: $V = 199.5 \text{ m}^3$

33. 34 cm **34.** −22 **35.** 6.9 **36.** 83.22 **37.** −2.3 **38.** $x = 7$

39. $x = −6$ **40.** $x = −7$ **41.** mean = 27 hookups; median = 26 hookups; mode = 19 hookups **42.** mean = 4.4 acres; median = 4.3 acres; mode = 2.85 acres **43.** 42.0% (rounded) **44.** (a) $889.80 (rounded) (b) $977 **45.** 38 nasal strips for $9.95, about $0.262 per strip **46.** $59,225.25 **47.** 17.5 L **48.** $11,533

Appendix

(pages A-5–A-6)

1. 37; add 7 **3.** 26; add 10, subtract 2 **5.** 16; multiply by 2 **7.** 243; multiply by 3 **9.** 36; add 3, add 5, add 7, etc.; or $1^2, 2^2, 3^2$, etc.

11. ⌐ **13.** ⌐

15. Conclusion follows. **17.** Conclusion does not follow.

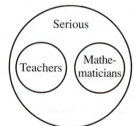

19. Together / Husband / Wife   2 + 18 + 7 = 27   30 − 27 = 3

Neither watched TV on 3 days.

21. Dick is the computer operator.

Solutions to Selected Exercises

CHAPTER 1 Whole Numbers

SECTION 1.1 (pages 8–9)

9. 3,561,435; millions: 3; thousands: 561; ones: 435

15. 346,009 is three hundred forty-six thousand, nine.

31. The least-used method of transportation is public transportation. 6,069,589 in words is six million, sixty-nine thousand, five hundred eighty-nine.

SECTION 1.2 (pages 18–21)

11. Line up the numbers in columns. Then start at the right and add the ones digits. Add the ten digits next, and, finally, the hundreds digits.

$$
\begin{array}{r}
932 \\
44 \\
+\ 613 \\
\hline
1589
\end{array}
$$

49. *Step 1*

Add the digits in the ones column.

Step 2

Add the digits in the tens column, including the regrouped 2.

Step 3

Add the hundreds column, including the regrouped 1.

Step 4

Add the thousands column, including the regrouped 1.

$$
\begin{array}{r}
1\ 12 \\
18 \\
708 \\
9\ 286 \\
+\ 636 \\
\hline
10,648
\end{array}
$$

63. Add up to check addition.

$$
\begin{array}{r}
769 \\
179 \\
214 \\
+\ 376 \\
\hline
759
\end{array}
$$
incorrect; should be 769

77. To find the total amount raised, add the amounts raised at each event.

$$
\begin{array}{rl}
\$3\ 482 & \text{flea market} \\
+\ 12,860 & \text{annual auction} \\
\hline
\$16,342 & \text{total amount raised}
\end{array}
$$

SECTION 1.3 (pages 28–31)

33. In the ones column, 5 is less than 7, so in order to subtract, regroup 1 ten as 10 ones. Then subtract 7 ones from 15 ones in the ones column. Finally, subtract 3 tens from 6 tens in the tens column.

$$
\begin{array}{r}
\overset{6\ 15}{7\cancel{5}} \\
-37 \\
\hline
38
\end{array}
$$

65.
$$
\begin{array}{r}
\overset{9\ \ \ 9}{\underset{5}{6}\cancel{6},\overset{10}{\cancel{0}}\ \overset{10}{\cancel{0}}\ \overset{10}{\cancel{0}}} \\
-34,4\ 4\ 4 \\
\hline
31,5\ 5\ 6
\end{array}
$$

81. To find how many fewer calories a woman burns, subtract the number of calories a woman burns from the number of calories a man burns.

$$
\begin{array}{rl}
187 & \text{calories man burns} \\
-140 & \text{calories woman burns} \\
\hline
47 & \text{fewer calories for a woman}
\end{array}
$$

The woman burned 47 fewer calories.

SECTION 1.4 (pages 38–41)

11.
$$
\underline{(4)(5)}\ (2) \quad \text{or} \quad (4)\underline{(5)(2)}
$$
$$
20(2) = 40 \qquad (4)(10) = 40
$$

37.
$$
\begin{array}{ccc}
600 & 6 & 600 \\
\times\ 6 & \times 6 & \times\ 6 \\
\hline
& 36 & 3600
\end{array}
$$
Attach 00.

73. First multiply 9352 by 4. Then multiply 9352 by 6, making sure to line up the tens. Then multiply 9352 by 2, making sure to line up the hundreds. Then add the partial products.

$$
\begin{array}{r}
9\,3\,5\,2 \\
\times\quad 2\,6\,4 \\
\hline
3\,7\,4\,0\,8 \quad \leftarrow 4 \times 9352 \\
5\,6\,1\,1\,2\quad\ \ \leftarrow 6 \times 9352 \\
1\,8\,7\,0\,4\qquad\ \ \leftarrow 2 \times 9352 \\
\hline
2,4\,6\,8,9\,2\,8
\end{array}
$$

85. *Approach* To find the number of balls purchased, multiply the number of cartons (300) by the number of balls per carton (10).

Solution

$$
\begin{array}{rl}
300 & \text{cartons} \\
\times\ 10 & \text{balls per carton} \\
\hline
3000 & \text{balls}
\end{array}
$$

3000 balls were purchased.

SECTION 1.5 (pages 51–54)

3. $24 \div 4 = 6$: $4\overline{)24}\overset{6}{}$ or $\dfrac{24}{4} = 6$

35. $6\overline{)9\,{}^3 1\,{}^{13}3\,{}^{17}7}\ \overset{1\ 5\ 2\ 2\ \mathbf{R}5}{}$

CHECK $(6 \times 1522) + 5 = 9132 + 5$
$= 9137$

51. $7\overline{)7\,1,{}^{1}7\,{}^{3}7\,{}^{2}6}\ \overset{1\,0,\ 2\ 5\ 3\ \mathbf{R}5}{}$

CHECK $(7 \times 10,253) + 5 = 71,771 + 5$
$= 71,776$

67. $9\overline{)8\,6,6\,5\,5}\ \overset{9\ 6\ 2\ 8\ \mathbf{R}7}{}$

CHECK $(9 \times 9628) + 7 = 86,652 + 7$
$= 86,659$

Does not match dividend, incorrect

Rework:

$9\overline{)8\,6,{}^{5}6\,{}^{2}5\,{}^{7}5}\ \overset{9\ 6\ 2\ 8\ \mathbf{R}3}{}$

CHECK $(9 \times 9628) + 3 = 86,652 + 3$
$= 86,655$

Matches dividend, correct

73. To find the number of tables that can be set, divide the number of napkins (2624) by the number of napkins it takes to set each table (8).

$8\overline{)2\,6\,{}^{2}2\,{}^{6}4}\ \overset{3\ 2\ 8}{}$

328 tables can be set.

SECTION 1.6 (pages 60–62)

13. Use 2 as a trial divisor, since 18 is closer to 20 than to 10.

$$\frac{131}{2} = 6 \text{ with 1 left over}$$

Because $6 \times 18 = 108$ and $131 - 108 = 23$, which is greater than the divisor, use 7 instead. To find the next digit in the quotient, use 2 as a trial divisor again.

$$\frac{5}{2} = 2 \text{ with 1 left over}$$

SECTION 1.4 (pages 38–41) continued content

103. Multiply the number of laptop computers purchased (12) by the price per laptop ($970). Then multiply the number of printers purchased (8) by the price per printer ($315).

$$
\begin{array}{cc}
970 & 315 \\
\times\ 12 & \times\ 8 \\
\hline
11,640 & 2520
\end{array}
$$

The total cost is the sum of these values: $\$11,640 + \$2520 + \$14,160$

Because $2 \times 18 = 36$ and $59 - 36 = 23$, which is greater than the divisor, use 3 instead.

```
      7 3 R5    CHECK        7 3
18)1 3 1 9              ×   1 8
    1 2 6                  5 8 4
      5 9                    7 3
      5 4                  1 3 1 4
        5                 +      5
                          1 3 1 9  ✓ matches
```

25.
```
        1 0 1 R4   CHECK      1 0 1
35)3 5 4 9                ×    3 5
                            5 0 5
                          3 0 3
                          3 5 3 5
                        +      4
                          3 5 3 9    incorrect
```

Rework:
```
        1 0 1 R14  CHECK      1 0 1
35)3 5 4 9                ×    3 5
    3 5                      5 0 5
    0 4 9                  3 0 3
      3 5                  3 5 3 5
      1 4                +    1 4
                          3 5 4 9  ✓ matches
```
The correct answer is 101 **R**14.

39. **(a)** Divide the number of Big Macs eaten by 39 years
```
          6 4 8
39)2 5, 2 7 2
    2 3 4
    1 8 7
    1 5 6
      3 1 2
      3 1 2
          0
```
The average number of Big Macs eaten each year was 648.

(b) 2 Big Macs eaten each day would be $2 \times 365 = 730$. He has eaten less than 2 Big Macs each day.

SECTION 1.7 (pages 72–75)

9. 855 rounded to the nearest ten: 860
855 Underline the 5 in the tens place. Next digit is 5 or more, so add one to the 5 in the tens place. All digits to the right of the underlined place change to zero.

25. **To the nearest ten:** 4476
Underline the 7 in the tens place. Next digit is 5 or more, so add 1 to the 7. All digits to the right of the underlined place are changed to zero.
4480

To the nearest hundred: 4476

Underline the 4 in the hundreds place. Next digit is 5 or more, so add 1 to the 4. All digits to the right of the underlined place are changed to zero.
4500

To the nearest thousand: 4476
Underline the 4 in the thousands place. Next digit is 4 or less, so leave 4 as 4 in the thousands place. All digits to the right of the underlined place are changed to zero.
4000

41. *Estimate:*   *Exact:*

Estimate	Exact
30	25
60	63
50	47
+ 80	+ 84
220	219

47. *Estimate:*   *Exact:*

Estimate	Exact
900	863
700	735
400	438
+ 800	+ 792
2800	2828

53. *Estimate:*   *Exact:*

Estimate	Exact
8000	8 215
60	56
700	729
+ 4000	+ 3 605
12,760	12,605

61. 76,000,000 Next digit is 5 or more, so add 1 to 7. Change all digits to the right of the underlined place to zero. Add 1 to 7.
80 million people

311,000,000 Next digit is 4 or less, so leave 1 as 1. Change all digits to the right of the underlined place to zero.
310 million people

SECTION 1.8 (pages 78–81)

5. 8^2: exponent is 2, base is 8.
$8^2 = 8 \cdot 8 = 64$

21. $6^2 = 36$, so $\sqrt{36} = 6$.

37.
$5 \cdot 3^2 + \dfrac{0}{8}$	Exponent
$5 \cdot 9 + \dfrac{0}{8}$	Multiply.
$45 + \dfrac{0}{8}$	Divide.
$45 + 0 = 45$	Add.

67.
$8 \cdot \sqrt{49} - 6(9 - 4)$	Parentheses
$8 \cdot \sqrt{49} - 6(5)$	Square root
$8 \cdot 7 - 6(5)$	Multiply.
$56 - 6(5)$	Multiply.
$56 - 30 = 26$	Subtract.

85.
$8 \cdot 9 \div \sqrt{36} - 4 \div 2 + (14 - 8)$	Parentheses
$8 \cdot 9 \div \sqrt{36} - 4 \div 2 + 6$	Square root
$8 \cdot 9 \div 6 - 4 \div 2 + 6$	Multiply.
$72 \div 6 - 4 \div 2 + 6$	Divide.
$12 - 4 \div 2 + 6$	Divide.
$12 - 2 + 6$	Subtract.
$10 + 6 = 16$	Add.

SECTION 1.9 (pages 86–89)

7. According to the pictograph, Family Dollar has 4500 stores. Subtract 4500 from 5750 to find the difference.
$5750 - 4500 = 1250$ fewer stores

15. According to the bar graph, $18 - 9 = 9$ more people out of 100 found their career as a result of "Studied in school" than "Luck or chance."

19. According to the line graph the increase in the number of installations from 2113 to 2014 is $7000 - 2500 = 4500$.

23. We want to insert parentheses in $7 - 2 \cdot 3 - 6$ so that it simplifies to 9. There isn't a method to use, so just use trial and error.
$(7 - 2) \cdot 3 - 6 = 5 \cdot 3 - 6$
$= 15 - 6 = 9$

SECTION 1.10 (pages 94–97)

11. *Step 1*
Find the number of toys each child will receive.
Step 2
"Same number of toys to each" indicates division should be used.
Step 3
3000 divided by 700 gives an estimate of about 4 toys.
Step 4
```
          4
657)2 6 2 8
    2 6 2 8
          0
```
Step 5
Each child will receive 4 toys.

Step 6

Exact answer is close to estimate. Also, $4 \times 657 = 2628$ which matches the number given in the problem.

21. *Step 1*

Find her monthly savings.

Step 2

Her monthly take home pay and expenses are given. "Remainder" indicates subtraction may be used.

Step 3

Estimate:

$2000 - $700 - $300 - $400 - $200 - $200 = $200

Step 4

$$
\begin{array}{r}
$695 \\
340 \\
435 \\
240 \\
+ \ 180 \\
\hline
1890
\end{array}
\qquad
\begin{array}{r}
$2240 \\
- \ 1890 \\
\hline
$350
\end{array}
$$

Step 5

Her monthly savings are $350.

Step 6

The exact answer seems reasonable compared to the estimate.

Check: $350 + $1890 = $2240. Re-add the expenses to check the sum of $1890.

29. *Step 1*

The number and cost of wheelchairs and recorder-players are given, and the total cost of all items must be found.

Step 2

Find the cost of all wheelchairs and the cost of all recorder-players. Then add these costs to get the total cost.

Step 3

The cost of the wheelchairs is about $1000 \times 6 = $6000.

The cost of the recorder-players is about $900 \times 20 = $18,000.

Estimate: $6000 + $18,000 = $24,000

Step 4

$$
\text{cost of wheelchairs:} \quad
\begin{array}{r}
$1256 \\
\times \quad 6 \\
\hline
$7536
\end{array}
$$

$$
\text{cost of} \\
\text{recorder-players:} \quad
\begin{array}{r}
$895 \\
\times \quad 15 \\
\hline
4475 \\
895 \\
\hline
$13,425
\end{array}
$$

Step 5

The total cost is $13,425 + $7536 = $20,961.

Step 6

The exact answer is reasonably close to the estimate. Check by repeating Step 4.

41. *Step 1*

The total seating is given along with the information to find the number of seats on the main floor. The number of rows of seats in the balcony is given, and the number of seats in each row of the balcony must be found.

Step 2

First, the number of seats on the main floor must be found. Next, the number of seats on the main floor must be subtracted from the total number of seats to find the number of seats in the balcony. Finally, the number of seats in the balcony must be divided by the number of rows of seats in the balcony.

Step 3

Seats on main floor: $30 \times 25 = 750$

Seats in balcony: $1250 - 750 = 500$

Seats in each row in balcony:
$500 \div 25 = 20$

Since we didn't round, our estimate matches our exact answer.

Step 4

See the calculations in Step 3.

Step 5

The number of seats in each row of the balcony is 20.

Step 6

The answer matches the estimate, as expected.

Check:

Seats in balcony: $20 \times 25 = 500$

Seats on main floor: $30 \times 25 = 750$

Total seats: $500 + 750 = 1250$, which matches the number given in the problem.

CHAPTER 2 Multiplying and Dividing Fractions

SECTION 2.1 (pages 116–118)

13. Each of the two figures is divided into

5 parts and 7 are shaded: $\dfrac{7}{5}$

Three are unshaded: $\dfrac{3}{5}$

21. Proper fractions: numerator *less* than denominator.

$$\frac{1}{3}, \frac{5}{8}, \frac{7}{16}$$

Improper fractions: numerator *greater than or equal to* denominator.

$$\frac{8}{5}, \frac{6}{6}, \frac{12}{2}$$

SECTION 2.2 (pages 123–125)

17. Write $5\dfrac{4}{5}$ as an improper fraction.

$5 \cdot 5 = 25$	Multiply 5 and 5.
$25 + 4 = 29$	Add 4. The numerator is 29.
$5\dfrac{4}{5} = \dfrac{29}{5}$	Use the same denominator.

43. $\dfrac{63}{4}$

Divide 63 by 4.

$$
\begin{array}{r}
15 \quad \longleftarrow \text{ Whole number part} \\
4\overline{)63} \\
4 \\
\hline
23 \\
20 \\
\hline
3 \quad \longleftarrow \text{ Remainder}
\end{array}
$$

The quotient 15 is the whole number part of the mixed number. The remainder 3 is the numerator of the fraction, and the denominator remains as 4.

$$\frac{63}{4} = 15\frac{3}{4}$$

61. Write $522\dfrac{3}{8}$ as an improper fraction.

$522 \cdot 8 = 4176$	
$4176 + 3 = 4179$	Add 3. The numerator is 4179.
$522\dfrac{3}{8} = \dfrac{4179}{8}$	Use the same denominator.

SECTION 2.3 (pages 130–131)

5. Factorizations of 48:

$1 \cdot 48 = 48 \quad 2 \cdot 24 = 48 \quad 3 \cdot 16 = 48$

$4 \cdot 12 = 48 \quad 6 \cdot 8 = 48$

The factors of 48 are 1, 2, 3, 4, 6, 8, 12, 16, 24, and 48.

41. 44

$$
\begin{array}{r}
1 \\
11\overline{)11} \\
\end{array}
\quad \text{Divide 11 by 11.}
$$

$$
\begin{array}{r}
11 \\
2\overline{)22} \\
\end{array}
\quad \text{Divide 22 by 2.}
$$

$$
\begin{array}{r}
22 \\
2\overline{)44} \\
\end{array}
\quad \text{Divide 44 by 2.}
$$

(Continued)

Quotient is 1.

Because all factors (divisors) are
prime, the prime factorization of 44 is
$2 \cdot 2 \cdot 11 = 2^2 \cdot 11$

SECTION 2.4 (pages 136–137)

21. Because the greatest common factor of
56 and 64 is 8, divide both numerator and
denominator by 8.

$$\frac{56}{64} = \frac{56 \div 8}{64 \div 8} = \frac{7}{8}$$

33. Write the prime factorization of both
numerator and denominator. Then divide
both numerator and denominator by any
common factors, and write a 1 by each factor that has been divided. Finally, multiply
the remaining factors in both numerator
and denominator.

$$\frac{18}{24} = \frac{\overset{1}{\cancel{2}} \cdot \overset{1}{\cancel{3}} \cdot 3}{2 \cdot 2 \cdot 2 \cdot \underset{1}{\cancel{3}}} = \frac{1 \cdot 1 \cdot 3}{1 \cdot 2 \cdot 2 \cdot 1} = \frac{3}{4}$$

43. $\dfrac{3}{6}$ and $\dfrac{18}{36}$ Write each fraction in
lowest terms.

$$\frac{3}{6} = \frac{1 \cdot \overset{1}{\cancel{3}}}{2 \cdot \underset{1}{\cancel{3}}} = \frac{1}{2}$$

$$\frac{18}{36} = \frac{\overset{1}{\cancel{2}} \cdot \overset{1}{\cancel{3}} \cdot \overset{1}{\cancel{3}}}{2 \cdot 2 \cdot \underset{1}{\cancel{3}} \cdot \underset{1}{\cancel{3}}} = \frac{1}{2}$$

The fractions are *equivalent* $\left(\dfrac{1}{2} = \dfrac{1}{2} \right)$.

57. Write the prime factorization of both
numerator and denominator. Then divide
both numerator and denominator by any
common factors, and write a 1 by each
factor that has been divided. Finally, multiply the remaining factors in both numerator
and denominator.

$$\frac{160}{256} = \frac{\overset{1}{\cancel{2}} \cdot \overset{1}{\cancel{2}} \cdot \overset{1}{\cancel{2}} \cdot \overset{1}{\cancel{2}} \cdot \overset{1}{\cancel{2}} \cdot 5}{\underset{1}{\cancel{2}} \cdot \underset{1}{\cancel{2}} \cdot \underset{1}{\cancel{2}} \cdot \underset{1}{\cancel{2}} \cdot \underset{1}{\cancel{2}} \cdot 2 \cdot 2 \cdot 2} = \frac{5}{8}$$

SECTION 2.5 (pages 148–151)

11. Divide a numerator and denominator by
the same number to use the multiplication
shortcut.

$$\frac{2}{3} \cdot \frac{7}{12} \cdot \frac{9}{14} = \frac{2}{\underset{1}{\cancel{3}}} \cdot \frac{7}{\underset{6}{\cancel{12}}} \cdot \frac{\overset{1}{\overset{3}{\cancel{9}}}}{\underset{2}{\cancel{14}}}$$

$$= \frac{1 \cdot 1 \cdot 1}{1 \cdot 2 \cdot 2} = \frac{1}{4}$$

25. Write 36 as $\dfrac{36}{1}$ and multiply.

$$36 \cdot \frac{5}{8} \cdot \frac{9}{15} = \frac{\overset{9}{\cancel{36}}}{1} \cdot \frac{\overset{1}{\cancel{5}}}{\underset{2}{\cancel{8}}} \cdot \frac{\overset{3}{\cancel{9}}}{\underset{1}{\underset{3}{\cancel{15}}}} = \frac{27}{2} = 13\frac{1}{2}$$

41. Area = length · width

$$\text{Area} = \frac{5}{6} \cdot \frac{3}{10}$$

$$= \frac{\overset{1}{\cancel{5}}}{\underset{2}{\cancel{6}}} \cdot \frac{\overset{1}{\cancel{3}}}{\underset{2}{\cancel{10}}}$$ Divide 5 and 10 by 5.
    Divide 3 and 6 by 3.

$$= \frac{1}{4} \text{ square mile}$$

45. Area = length · width

$$\text{Area} = 2 \cdot \frac{3}{4}$$

$$= \frac{\overset{1}{\cancel{2}}}{1} \cdot \frac{3}{\underset{2}{\cancel{4}}}$$ Divide numerator and
    denominator by 2.

$$= \frac{3}{2}$$ Write as a mixed number.

$$= 1\frac{1}{2}$$ Square yards

SECTION 2.6 (pages 155–158)

5. *Step 1*
The problem asks for the area of the front
surface of a digital photo frame.

Step 2
To find area, multiply the length of $\dfrac{3}{4}$ foot by
the width of $\dfrac{2}{3}$ foot.

Step 3
An estimate is $\dfrac{6}{12}$ square foot.

$$\frac{3}{4} \cdot \frac{2}{3} = \frac{\overset{1}{\cancel{3}} \cdot \overset{1}{\cancel{2}}}{\underset{2}{\cancel{4}} \cdot \underset{1}{\cancel{3}}} = \frac{1}{2}$$

Step 4
The exact value is $\dfrac{1}{2}$, which is the same as
the estimate since we didn't round.

Step 5
The area of the digital photo frame is
$\dfrac{1}{2}$ square foot.

Step 6
The answer, $\dfrac{1}{2}$ square foot, matches our
estimate.

15. *Step 1*
The problem asks for the number of
runners who are women.

Step 2

$\dfrac{7}{12}$ of the 1560 runners are women, so

multiply $\dfrac{7}{12}$ by 1560 to find the number of

runners who are women.

Step 3
Round the number of runners to 1600;
then, $\dfrac{1}{2}$ of 1600 is 800. Since $\dfrac{7}{12}$ is
more than $\dfrac{1}{2}$, our estimate is that "more
than 800 runners" are women.

Step 4

$$\frac{7}{12} \cdot 1560 = \frac{7 \cdot \overset{130}{\cancel{1560}}}{\underset{1}{\cancel{12}} \cdot 1} = 910$$

Step 5
(a) 910 runners are women.

(b) 1560 total runners − 910 women
    = 650 men

Step 6
The exact answer, 910, fits our estimate of
"more than 800."

25. From Exercise 23 the total income is
$76,000. The circle graph shows that $\dfrac{1}{5}$ of
the income is for rent.

$$\frac{1}{5} \cdot 76,000 = \frac{1}{\underset{1}{\cancel{5}}} \cdot \frac{\overset{15,200}{\cancel{76,000}}}{1} = 15,200$$

The amount of their rent is $15,200.

37. To find the remaining amount of the estate,
subtract $\dfrac{7}{8}$ from 1.

$$1 - \frac{7}{8} = \frac{8}{8} - \frac{7}{8} = \frac{1}{8}$$

Multiply the remaining $\dfrac{1}{8}$ of the estate by
the fraction going to the American Cancer
Society.

$$\frac{1}{4} \cdot \frac{1}{8} = \frac{1}{32}$$

$\dfrac{1}{32}$ of the estate goes to the American
Cancer Society.

SECTION 2.7 (pages 164–167)

5. The reciprocal of $\dfrac{3}{8}$ is $\dfrac{8}{3}$ because

$$\frac{3}{8} \cdot \frac{8}{3} = \frac{24}{24} = 1.$$

SOLUTIONS

29. $\dfrac{18}{\frac{3}{4}} = 18 \div \dfrac{3}{4}$ Rewrite using the ÷ symbol for division.

$\qquad = \dfrac{18}{1} \div \dfrac{3}{4}$ Write 18 as $\frac{18}{1}$.

$\qquad = \dfrac{\overset{6}{\cancel{18}}}{1} \cdot \dfrac{4}{\underset{1}{\cancel{3}}}$ The reciprocal of $\frac{3}{4}$ is $\frac{4}{3}$.

$\qquad\qquad\qquad$ Change "÷" to "•". Divide the numerator and denominator by 3.

$\qquad = \dfrac{6 \cdot 4}{1 \cdot 1}$ Multiply.

$\qquad = \dfrac{24}{1} = 24$

35. Step 1

The problem asks for the number of times a measuring cup needs to be filled.

Step 2

Solve the problem by dividing the total number of cups (5) by the size of the measuring cup $\left(\dfrac{1}{3}\right)$.

Step 3

Because there are three $\dfrac{1}{3}$-cups in one cup, multiply 3 by 5 to get 15. So, our estimate is 15.

Step 4

$5 \div \dfrac{1}{3} = \dfrac{5}{1} \div \dfrac{1}{3} = \dfrac{5}{1} \cdot \dfrac{3}{1} = 15$

Step 5

They need to fill the measuring cup 15 times.

Step 6

The exact answer, 15, is the same as our estimate.

49. Step 1

The problem asks for the number of towels that can be made.

Step 2

Solve the problem by dividing the 912 yards of fabric by the fraction of a yard $\left(\dfrac{3}{8}\right)$ needed for each dish towel.

Step 3

Round 912 yards to 900 yards. Round $\dfrac{3}{8}$ to $\dfrac{1}{2}$.

Multiply 900 by 2, the reciprocal of $\dfrac{1}{2}$, to get the estimate of 1800 towels.

Step 4

$912 \div \dfrac{3}{8} = \dfrac{912}{1} \cdot \dfrac{8}{3} = \dfrac{\overset{304}{\cancel{912}}}{1} \cdot \dfrac{8}{\underset{1}{\cancel{3}}} = 2432$

Step 5

2432 towels can be made.

Step 6

The exact answer, 2432, is close to our estimate, 1800.

SECTION 2.8 (pages 175–179)

5. $4\dfrac{1}{2} \cdot 1\dfrac{3}{4}$

Estimate: $5 \cdot 2 = 10$

To find the exact answer, change each mixed number to an improper fraction and then multiply.

Exact: $4\dfrac{1}{2} \cdot 1\dfrac{3}{4} = \dfrac{9}{2} \cdot \dfrac{7}{4} = \dfrac{63}{8} = 7\dfrac{7}{8}$

21. $1\dfrac{1}{4} \div 3\dfrac{3}{4}$

Estimate: $1 \div 4 = \dfrac{1}{4}$

To find the exact answer, first change each mixed number to an improper fraction. Then, use the reciprocal of the divisor (the second fraction) and multiply.

Exact:

$1\dfrac{1}{4} \div 3\dfrac{3}{4} = \dfrac{5}{4} \div \dfrac{15}{4} = \dfrac{\overset{1}{\cancel{5}}}{\underset{1}{\cancel{4}}} \cdot \dfrac{\overset{1}{\cancel{4}}}{\underset{3}{\cancel{15}}} = \dfrac{1}{3}$

CHAPTER 3 Adding and Subtracting Fractions

SECTION 3.1 (pages 201–202)

15. Add the numerators. Keep the denominator.

$\dfrac{3}{8} + \dfrac{7}{8} + \dfrac{2}{8} = \dfrac{3 + 7 + 2}{8} = \dfrac{12}{8} = 1\dfrac{1}{2}$

29. First, subtract the numerators and keep the denominator. Then write the fraction in lowest terms.

$\dfrac{47}{36} - \dfrac{5}{36} = \dfrac{47 - 5}{36} = \dfrac{42}{36}$

$= \dfrac{42 \div 6}{36 \div 6} = \dfrac{7}{6} = 1\dfrac{1}{6}$

39. First, add the two fractions of land that were purchased.

$\dfrac{9}{10} + \dfrac{3}{10} = \dfrac{9 + 3}{10} = \dfrac{12}{10}$

Next, subtract the fraction of land that was planted in carrots.

$\dfrac{12}{10} - \dfrac{7}{10} = \dfrac{12 - 7}{10} = \dfrac{5}{10} = \dfrac{1}{2}$

$\frac{1}{2}$ acre is planted in squash.

SECTION 3.2 (pages 211–214)

13. 20 and 50

Multiples of 50:

50, <u>100</u>, 150, 200, 250, . . .

100 is the first multiple of 50 that is divisible by 20. $(100 \div 20 = 5)$

The least common multiple of 20 and 50 is 100.

21. Find the prime factorization for each number.

4, 6, 8, 10

$\qquad 4 = 2 \cdot 2$

$\qquad 6 = 2 \cdot ③$

$\qquad 8 = ② \cdot ② \cdot ②$

$\qquad 10 = 2 \cdot ⑤$

LCM $= 2 \cdot 2 \cdot 2 \cdot 3 \cdot 5 = 120$

The LCM of 4, 6, 8, and 10 is 120.

37. $\dfrac{2}{3} = \dfrac{?}{9}$ Divide 9 by 3, to get 3. Now multiply both numerator and denominator of $\dfrac{2}{3}$ by 3.

$\dfrac{2}{3} = \dfrac{2 \cdot 3}{3 \cdot 3} = \dfrac{6}{9}$

41. $\dfrac{3}{16} = \dfrac{?}{64}$ Divide 64 by 16, to get 4. Now multiply both numerator and denominator of $\dfrac{3}{16}$ by 4.

$\dfrac{3}{16} = \dfrac{3 \cdot 4}{16 \cdot 4} = \dfrac{12}{64}$

55. $\dfrac{109}{1512}, \dfrac{23}{392}$

$\begin{array}{r|ll} 2 & 1512 & 392 \\ 2 & 756 & 196 \\ 2 & 378 & 98 \\ 3 & 189 & \cancel{49} \\ 3 & 63 & \cancel{49} \\ 3 & 21 & \cancel{49} \\ 7 & 7 & 49 \\ 7 & \cancel{7} & 7 \\ & 1 & 1 \end{array}$

The LCM of the denominators is

$2 \cdot 2 \cdot 2 \cdot 3 \cdot 3 \cdot 3 \cdot 7 \cdot 7 = 10{,}584.$

61. Find the prime factorization for each number.

5, 7, 14, 10

$\qquad 5 = 5$

$\qquad 7 = ⑦$

$\qquad 14 = ② \cdot 7$

$\qquad 10 = ② \cdot 5$

LCM $= 2 \cdot 5 \cdot 7 = 70$

The LCM of 5, 7, 14, and 10 is 70.

SECTION 3.3 (pages 219–222)

3. The least common multiple of 4 and 8 is 8. Rewrite both fractions as fractions with a least common denominator of 8. Then add the numerators.

$$\frac{3}{4} + \frac{1}{8} = \frac{6}{8} + \frac{1}{8}$$
$$= \frac{6 + 1}{8}$$
$$= \frac{7}{8}$$

29. The least common multiple of 12 and 4 is 12. Rewrite both fractions as fractions with a least common denominator of 12. Then subtract the numerators and write the resulting fraction in lowest terms.

$$\frac{5}{12} - \frac{1}{4} = \frac{5}{12} - \frac{3}{12}$$
$$= \frac{5 - 3}{12}$$
$$= \frac{2}{12} = \frac{1}{6}$$

41. Add the fractions to find the total length of the screw. The least common denominator of the fractions is 40. Rewrite all three fractions with a denominator of 40. Then add the numerators.

$$\frac{1}{8} + \frac{1}{4} + \frac{2}{5} = \frac{5}{40} + \frac{10}{40} + \frac{16}{40}$$
$$= \frac{5 + 10 + 16}{40}$$
$$= \frac{31}{40}$$

The total length of the screw is $\frac{31}{40}$ inch.

49. One way to compare fractions accurately is to rewrite each fraction with a common denominator.

$$\frac{1}{4} = \frac{15}{60}, \frac{1}{3} = \frac{20}{60}, \frac{1}{5} = \frac{12}{60}, \frac{13}{60}$$

The fraction with the largest numerator is the largest fraction. This is $\frac{20}{60}$ or $\frac{1}{3}$, which is "totally honest."

To find the number of hours, multiply.

$$\frac{1}{3} \cdot 1200 = \frac{1}{\cancel{3}} \cdot \frac{\overset{400}{\cancel{1200}}}{1} = \frac{400}{1} = 400$$

400 people responded "totally honest." To find what fraction of the users responding to "totally honest" and "fib a little" add the fractions $\frac{1}{3}$ and $\frac{1}{4}$. The least common denominator is 12. Rewrite $\frac{1}{3}$ and $\frac{1}{4}$ with denominators of 12. Then add the numerators and write in lowest terms.

$$\frac{1}{3} + \frac{1}{4} = \frac{4}{12} + \frac{3}{12} = \frac{4 + 3}{12} = \frac{7}{12}$$

SECTION 3.4 (pages 227–234)

19. *Estimate:*   *Exact:*

$$23 \xleftarrow{\text{Rounds to}} \left\{ 22\frac{3}{4} = 22\frac{21}{28} \right.$$
$$+15 \xleftarrow{\text{Rounds to}} \left\{ +15\frac{3}{7} = 15\frac{12}{28} \right.$$
$$38 \qquad\qquad\qquad 37\frac{33}{28}$$

$$37\frac{33}{28} = 37 + 1\frac{5}{28} = 38\frac{5}{28}$$

29. *Estimate:*   *Exact:*

$$17 \xleftarrow{\text{Rounds to}} \left\{ 17 \right.$$
$$-7 \xleftarrow{\text{Rounds to}} \left\{ -6\frac{5}{8} \right.$$
$$10$$

To subtract $\frac{5}{8}$, first regroup the whole number 17 into $16 + 1$.

$$17 = 16 + 1 = 16 + \frac{8}{8}$$

Now you can subtract.

$$16\frac{8}{8}$$
$$-6\frac{5}{8}$$
$$\overline{10\frac{3}{8}}$$

55.
$$8\frac{3}{4} = \frac{35}{4} = \frac{70}{8}$$
$$-5\frac{7}{8} = \frac{47}{8} = \frac{47}{8}$$
$$\overline{\frac{23}{8} = 2\frac{7}{8}}$$

71. Subtract the size of the smallest hose clamp from the size of the largest hose clamp.

Estimate: $3 - 1 = 2$ in.

Exact:
$$2\frac{3}{4} = \frac{11}{4} = \frac{44}{16}$$
$$-\frac{9}{16} = \frac{9}{16} = \frac{9}{16}$$
$$\overline{\frac{35}{16} = 2\frac{3}{16}}$$

The difference in size is $2\frac{3}{16}$ inches.

81. Add the weights.

Estimate:

$$59 + 24 + 17 + 29 + 58 = 187 \text{ tons}$$

Exact:

$$58\frac{1}{2} = 58\frac{12}{24}$$
$$23\frac{5}{8} = 23\frac{15}{24}$$
$$16\frac{5}{6} = 16\frac{20}{24}$$
$$29\frac{1}{4} = 29\frac{6}{24}$$
$$+58\frac{1}{3} = 58\frac{8}{24}$$
$$\overline{184\frac{61}{24} = 184 + 2\frac{13}{24} = 186\frac{13}{24}}$$

The total weight is $186\frac{13}{24}$ tons.

SECTION 3.5 (pages 243–246)

7.

33. $\left(\dfrac{3}{2}\right)^4 = \dfrac{3}{2} \cdot \dfrac{3}{2} \cdot \dfrac{3}{2} \cdot \dfrac{3}{2} = \dfrac{81}{16} = 5\dfrac{1}{16}$

49. $6\left(\dfrac{2}{3}\right)^2\left(\dfrac{1}{2}\right)^3 = 6\left(\dfrac{2}{3} \cdot \dfrac{2}{3}\right)\left(\dfrac{1}{2} \cdot \dfrac{1}{2} \cdot \dfrac{1}{2}\right)$

Simplify the expressions with exponents.

$$= \frac{\overset{2}{\cancel{6}} \cdot \overset{1}{\cancel{2}} \cdot \overset{1}{\cancel{2}} \cdot 1 \cdot 1 \cdot 1}{\underset{1}{\cancel{3}} \cdot 3 \cdot \underset{1}{\cancel{2}} \cdot \underset{1}{\cancel{2}} \cdot \underset{1}{\cancel{2}}}$$
$$= \frac{1}{3}$$

63. $\left(\dfrac{3}{4}\right)^2 - \left(\dfrac{1}{2} - \dfrac{1}{6}\right) \div \dfrac{4}{3}$

$$= \left(\frac{3}{4}\right)^2 - \left(\frac{3}{6} - \frac{1}{6}\right) \div \frac{4}{3}$$

Work inside parentheses first.

$$= \left(\frac{3}{4}\right)^2 - \left(\frac{2}{6}\right) \cdot \frac{3}{4}$$

Simplify the expression with the exponent.

$$= \frac{3}{4} \cdot \frac{3}{4} - \left(\frac{2}{6}\right) \cdot \frac{3}{4} \qquad \text{Multiply.}$$
$$= \frac{9}{16} - \frac{\overset{1}{\cancel{2}} \cdot \overset{1}{\cancel{3}}}{\underset{2}{\cancel{6}} \cdot \underset{2}{\cancel{4}}}$$
$$= \frac{9}{16} - \frac{1}{4} \qquad \text{The LCD is 16. Rewrite } \frac{1}{4} \text{ as a fraction with a denominator of 16.}$$
$$= \frac{9}{16} - \frac{4}{16} \qquad \text{Subtract.}$$
$$= \frac{5}{16}$$

CHAPTER 4 Decimals

SECTION 4.1 (pages 269–273)

5.
```
tens
 ones
  tenths
7 0 . 4 8 9
```

17. Reorder as 5 thousands, 4 hundreds, 0 tens, 6 ones, 0 tenths, 4 hundredths, 5 thousandths: 5406.045

31. $0.205 = \dfrac{205}{1000} = \dfrac{205 \div 5}{1000 \div 5} = \dfrac{41}{200}$

45. Three decimal places is *thousandths*, so 1.075 is one and seventy-five *thousandths*.

53. Seven hundred three ten-thousandths:
$\dfrac{703}{10{,}000} = 0.0703$

59. 8-pound test line has a diameter of 0.010 inch. 0.010 inch is read "ten thousandths inch."
$0.010 = \dfrac{10}{1000} = \dfrac{10 \div 10}{1000 \div 10} = \dfrac{1}{100}$ inch

SECTION 4.2 (pages 279–280)

13. Draw a cut-off line after the tenths place:
793.9|88
The first digit cut is 8, which is *5 or more*, so round up the tenths place.
```
  793.9
+   0.1
  794.0
```
Answer: 793.988 ≈ 794.0

17. Round 9.0906 to the nearest hundredth.
Draw a cut-off line: 9.09|06
The first digit cut is 0, which is *4 or less*. The part you keep stays the same.
9.0906 ≈ 9.09

35. Round $999.73 to the nearest dollar.
Draw a cut-off line: $999|.73
The first digit cut is 7, which is *5 or more*, so round up.
```
  $999
+    1
 $1000
```
Answer: $999.73 ≈ $1000

SECTION 4.3 (pages 285–288)

13. Subtract 0.291 from 0.4
┌── Line up decimal points.
```
  0.400  ← Write two zeros.
- 0.291
  0.109
```

25.
Estimate:	Exact:
400 ←	392.700
1 ←	0.865
+ 20 ←	+ 21.080
421	414.645

43. First add the three players' heights.

Estimate:	Exact:
2 ←	2.13
2 ←	1.98
+ 2 ←	+ 2.08
6 meters	6.19 meters

Now subtract the players' combined height of 6.19 meters from the rhino's height of 6.4 meters.
$6.4 - 6.19 = 0.21$
So, the NBA stars' combined height is 0.21 meter less than the rhino's height.

49. Subtract the price of the regular fishing line, $4.84, from the price of the fluorescent fishing line, $5.14.

Estimate:	Exact:
$5 ←	$5.14
− 5 ←	− 4.84
$0	$0.30

The fluorescent fishing line costs $0.30 more than the regular fishing line.

SECTION 4.4 (pages 291–296)

15. (7.2) (0.06)
```
   72        7.2  ← 1 decimal place
×   6     × 0.06  ← 2 decimal places
  432      0.432  ← 3 decimal places
```

19. (0.006) (0.0052)
```
    0.0052  ← 4 decimal places
×   0.006   ← 3 decimal places
  0.0000312 ← 7 decimal places
```

43. Multiply the number of gallons that she pumped into her pickup truck by the price per gallon. Use a calculator.
$20.510 \times 3.759 = 77.09709$
Round 77.09709 to the nearest cent.
Michelle paid $77.10 for the gas.

55. (a) Find the cost for three short-sleeved, solid-color shirts
```
  $14.75
×     3
  $44.25
```
Based on this subtotal, shipping is $7.95.
Next, find the cost of the three monograms.
```
  $4.95
×    3
 $14.85
```

Add these amounts, plus $5.00 for a gift box.
```
 $44.25 ←  Shirts
   7.95 ←  Shipping
  14.85 ←  Monograms
+  5.00 ←  Gift box
 $72.05
```
The total cost is $72.05.

(b) Subtract the cost of the shirts to find the difference.
```
  $72.05
- 44.25
  $27.80
```
The monograms, gift box, and shipping added $27.80 to the cost of the gift.

SECTION 4.5 (pages 303–306)

11.
```
          3 6.
    1.5 )5 4.0
          4 5
            9 0
            9 0
              0
```
Move decimal point in divisor and dividend 1 place; write 0 in the dividend.

13. Given: $108 \div 18 = 6$
Find: $0.108 \div 1.8$ by moving the decimal points.
```
        .06
  1.8 )0.1 08
```
So, $0.108 \div 1.8 = 0.06$

19. $\dfrac{3.1}{0.006}$
```
              5 1 6.6 6 6
   0.006 )3.1 0 0 0 0 0
           3 0
             1 0
              6
              4 0
              3 6
                4 0
                3 6
                  4 0
                  3 6
                    4 0
                    3 6
                      4 0
                      3 6
                        4
```
Line up decimal points. Move decimal point in divisor and dividend three places. Write 000 in dividend.

Write 0 in dividend.

Write 0 in dividend.

Write 0 in dividend.

Stop and round answer to the nearest hundredth.

The quotient is 516.67 (rounded).

29. $307.02 \div 5.1 - 6.2$

Estimate: $300 \div 5 = 60$

The answer 6.2 is *unreasonable* because it is so much less than 60.

$$
\begin{array}{r}
6\,0.2 \\
5.1\overline{)3\,0\,7.0\,2} \\
\underline{3\,0\,6} \\
1\ 0 \\
\underline{\quad 0} \\
1\,0\,2 \\
\underline{1\,0\,2} \\
0
\end{array}
$$

The correct answer is 60.2, which is close to the estimate of 60.

55. $33 - 3.2(0.68 + 9) - 1.3^2$ Parentheses

                          Exponent

$33 - 3.2(9.68) - 1.69$ Multiply

$33 - 30.976 - 1.69$ Subtract

$2.024 - 1.69 = 0.334$ Subtract

61. Divide 100,000 box tops (the answer from Exercise 59) by 38 weeks.

$$
\begin{array}{r}
2\,6\,3\,1.5 \\
38\overline{)1\,0\,0,0\,0\,0.0} \\
\underline{7\,6} \\
2\,4\ 0 \\
\underline{2\,2\ 8} \\
1\ 2\,0 \\
\underline{1\ 1\,4} \\
6\,0 \\
\underline{3\,8} \\
2\,2\,0 \\
\underline{1\,9\ 0} \\
3\ 0
\end{array}
$$

2631.5 rounds to 2632.

The school needs to collect 2632 box tops (rounded) during each of the 38 weeks.

SECTION 4.6 (pages 311–314)

23. $3\frac{5}{8} = \frac{29}{8} = 3.625$

$$
\begin{array}{r}
3.6\,2\,5 \\
8\overline{)2\,9.0\,0\,0} \\
\underline{2\,4} \\
5\ 0 \\
\underline{4\ 8} \\
2\ 0 \\
\underline{1\ 6} \\
4\ 0 \\
\underline{4\ 0} \\
0
\end{array}
$$

35. $0.35 = \frac{35}{100} = \frac{35 \div 5}{100 \div 5} = \frac{7}{20}$

---

51. Write zeros so that all the numbers have four decimal places. The acceptable lengths must be greater than 0.9980 cm and less than 1.0020 cm.

$1.0100 > 1.0020$ *unacceptable*

$0.9991 > 0.9980$ and

$0.9991 < 1.0020$ *acceptable*

$1.0007 > 0.9980$ and

$1.0007 < 1.0020$ *acceptable*

$0.9900 < 0.9980$ *unacceptable*

The lengths of 0.9991 cm and 1.0007 cm are acceptable.

63. $\frac{3}{8}, \frac{2}{5}, 0.37, 0.4001$

$\frac{3}{8} = 0.3750$

$\frac{2}{5} = 0.4000$

$0.37 = 0.3700 \leftarrow$ least

$0.4001 = 0.4001 \leftarrow$ greatest

From least to greatest: $0.37, \frac{3}{8}, \frac{2}{5}, 0.4001$

CHAPTER 5 Ratio and Proportion

SECTION 5.1 (pages 335–338)

7. $100 to $50

$\frac{\$100}{\$50} = \frac{100}{50} = \frac{100 \div 50}{50 \div 50} = \frac{2}{1}$

19. $1\frac{1}{4}$ to $1\frac{1}{2}$

$\dfrac{1\frac{1}{4}}{1\frac{1}{2}} = \dfrac{\frac{5}{4}}{\frac{3}{2}} = \frac{5}{4} \div \frac{3}{2} = \frac{5}{\overset{2}{\cancel{4}}} \cdot \frac{\cancel{2}}{3} = \frac{5}{6}$

47. The increase in price is

$12.50 - $10 = $2.50

$\frac{\$2.50}{\$10} = \frac{2.50}{10} = \frac{2.50 \cdot 10}{10 \cdot 10} = \frac{25}{100}$

$= \frac{25 \div 25}{100 \div 25} = \frac{1}{4}$

The ratio of the increase in price to the original price is $\frac{1}{4}$.

SECTION 5.2 (pages 343–346)

15. Miles traveled:

$28,396.7 - 28,058.1 = 338.6$

Miles per gallon:

$\frac{338.6 \text{ miles} \div 16.2}{16.2 \text{ gallons} \div 16.2} \approx 20.90 \text{ or } 20.9$

19.

Size	Cost per Unit
12 ounces	$\frac{\$2.49}{12 \text{ ounces}} \approx \$0.208$
14 ounces	$\frac{\$2.89}{14 \text{ ounces}} \approx \$0.206$
18 ounces	$\frac{\$3.96}{18 \text{ ounces}} = \$0.22$

Because $0.206 is the lowest cost per ounce, the best buy is 14 ounces for $2.89.

33. One battery for $1.79 is like getting 3 batteries, so $1.79 \div 3 \approx \$0.597$ per battery. An eight-pack of AA batteries for $4.99 is $4.99 \div 8 \approx \$0.624$ per battery. Because $0.597 is the lower cost per battery, the better buy is the package with one battery.

SECTION 5.3 (pages 350–351)

11. $\frac{150}{200} = \frac{200}{300}$

$\frac{150 \div 50}{200 \div 50} = \frac{3}{4}$ and $\frac{200 \div 100}{300 \div 100} = \frac{2}{3}$

Because $\frac{3}{4}$ is *not* equal to $\frac{2}{3}$, the proportion is *false.*

33. $\dfrac{2\frac{5}{8}}{3\frac{1}{4}} = \dfrac{21}{26}$

Cross products:

$2\frac{5}{8} \cdot 26 = \frac{21}{\overset{}{\underset{4}{\cancel{8}}}} \cdot \frac{\overset{13}{\cancel{26}}}{1} = \frac{273}{4} = 68\frac{1}{4}$

$3\frac{1}{4} \cdot 21 = \frac{13}{4} \cdot \frac{21}{1} = \frac{273}{4} = 68\frac{1}{4}$

The cross products are *equal,* so the proportion is *true.*

37. $\dfrac{2\frac{3}{10}}{8.05} = \dfrac{\frac{1}{4}}{0.9}$

Cross products:

$2\frac{3}{10}(0.9) = \frac{23}{10} \cdot \frac{9}{10} = \frac{207}{100} = 2\frac{7}{100}$

or $(2.3)(0.9) = 2.07$

$(8.05)\frac{1}{4} = (8.05)(0.25) = 2.0125$

or $8\frac{5}{100} \cdot \frac{1}{4} = \frac{805}{100} \cdot \frac{1}{4} = \frac{805}{400} = 2\frac{1}{80}$

The cross products are *not* equal so the proportion is *false.*

$\left(2\frac{7}{100} \neq 2\frac{1}{80} \text{ or } 2.07 \neq 2.0125\right)$

SECTION 5.4 (pages 356–357)

9. $\dfrac{42}{x} = \dfrac{18}{39}$

$x \cdot 18 = 42 \cdot 39$ Cross products are equal.

$\dfrac{x \cdot \overset{1}{\cancel{18}}}{\cancel{18}_{1}} = \dfrac{1638}{18}$ Divide both sides by 18.

$x = 91$

CHECK

$42 \cdot 39 = 1638$

$91 \cdot 18 = 1638$

15. $\dfrac{99}{55} = \dfrac{44}{x}$ **OR** $\dfrac{9}{5} = \dfrac{44}{x}$

$9 \cdot x = 5 \cdot 44$ Cross products are equal.

$\dfrac{\overset{1}{\cancel{9}} \cdot x}{\cancel{9}_{1}} = \dfrac{220}{9}$ Divide both sides by 9.

$x = \dfrac{220}{9}$

$x \approx 24.44$ (rounded)

CHECK

$55 \cdot 44 = 2420$

$99 \cdot 24.44 = 2419.56$

Slightly different because of rounding.

23. $\dfrac{2\frac{1}{3}}{1\frac{1}{2}} = \dfrac{x}{2\frac{1}{4}}$

$1\frac{1}{2} \cdot x = 2\frac{1}{3} \cdot 2\frac{1}{4}$

$\dfrac{3}{2} \cdot x = \dfrac{7}{\cancel{3}_{1}} \cdot \dfrac{\overset{3}{\cancel{9}}}{4}$

$\dfrac{3}{2} \cdot x = \dfrac{21}{4}$

$\dfrac{\frac{3}{2} \cdot x}{\frac{3}{2}} = \dfrac{\frac{21}{4}}{\frac{3}{2}}$

$x = \dfrac{21}{4} \div \dfrac{3}{2} = \dfrac{\overset{7}{\cancel{21}}}{\cancel{4}_{2}} \cdot \dfrac{\overset{1}{\cancel{2}}}{\cancel{3}_{1}} = \dfrac{7}{2} = 3\frac{1}{2}$

27. $\dfrac{x}{\frac{3}{50}} = \dfrac{0.15}{1\frac{4}{5}}$

Change to decimals:

$\dfrac{3}{50} = 3 \div 50 = 0.06$

$1\frac{4}{5} = \dfrac{9}{5}$ and $9 \div 5 = 1.8$

$\dfrac{x}{0.06} = \dfrac{0.15}{1.8}$

$x \cdot 1.8 = (0.06)(0.15)$

$\dfrac{x \cdot \cancel{1.8}}{\cancel{1.8}} = \dfrac{0.009}{1.8}$

$x = 0.005$

Change to fractions:

$0.15 = \dfrac{15 \div 5}{100 \div 5} = \dfrac{3}{20}$

$\dfrac{x}{\frac{3}{50}} = \dfrac{\frac{3}{20}}{1\frac{4}{5}}$

$1\frac{4}{5} \cdot x = \dfrac{3}{50} \cdot \dfrac{3}{20}$

$\dfrac{1\frac{4}{5} \cdot x}{1\frac{4}{5}} = \dfrac{\frac{9}{1000}}{1\frac{4}{5}}$

$x = \dfrac{9}{1000} \div 1\frac{4}{5} = \dfrac{\cancel{9}}{\cancel{1000}_{200}} \cdot \dfrac{\overset{1}{\cancel{5}}}{\cancel{9}_{1}}$

$= \dfrac{1}{200}$

Compare answers.

$0.005 = \dfrac{5 \div 5}{1000 \div 5} = \dfrac{1}{200} \leftarrow$ Matches

SECTION 5.5 (pages 363–366)

11. $\dfrac{6 \text{ ounces}}{7 \text{ servings}} = \dfrac{x \text{ ounces}}{12 \text{ servings}}$

$7 \cdot x = 6 \cdot 12$

$\dfrac{\overset{1}{\cancel{7}} \cdot x}{\cancel{7}_{1}} = \dfrac{72}{7}$

$x \approx 10.3 \approx 10$ (rounded)

You need about 10 ounces for 12 servings.

17. The length of the dining area is the same as the length of the kitchen, which is 14 feet (from Exercise 15).

Find the width of the dining area.

4.5 inches − 2.5 inches = 2 inches on the floor plan.

$\dfrac{1 \text{ inch}}{4 \text{ feet}} = \dfrac{2 \text{ inches}}{x \text{ feet}}$

$1 \cdot x = 4 \cdot 2$

$x = 8$

The dining area is 8 feet wide.

23. $\dfrac{7 \text{ refresher}}{10 \text{ entering}} = \dfrac{x \text{ refresher}}{2950 \text{ entering}}$

$10 \cdot x = 7 \cdot 2950$

$\dfrac{\overset{1}{\cancel{10}} \cdot x \cdot}{\cancel{10}_{1}} = \dfrac{20,650}{10}$

$x = 2065$

2065 students will probably need a refresher course. This is a reasonable answer because it's more than half the students, but not all the students.

Incorrect set-up

$\dfrac{10 \text{ entering}}{7 \text{ refresher}} = \dfrac{x \text{ refresher}}{2950 \text{ entering}}$

$7 \cdot x = 10 \cdot 2950$

$\dfrac{\overset{1}{\cancel{7}} \cdot x}{\cancel{7}_{1}} = \dfrac{29,500}{7}$

$x \approx 4214$ (rounded)

The *incorrect* set-up gives an *unreasonable* estimate of 4214 entering students; there are only 2950 entering students.

33. Coretta $\left\{\begin{array}{l} \dfrac{1.05 \text{ meters}}{1.68 \text{ meters}} = \dfrac{6.58 \text{ meters}}{x \text{ meters}} \end{array}\right\}$ tree

$1.05 \cdot x = 1.68(6.58)$

$\dfrac{\cancel{1.05} \cdot x}{\cancel{1.05}} = \dfrac{11.0544}{1.05}$

$x = 10.528 \approx 10.53$

The height of the tree is about 10.53 meters.

39. First find the number of calories in a $\frac{1}{2}$-cup serving of bran cereal.

$\dfrac{\frac{1}{3} \text{ cup}}{80 \text{ calories}} = \dfrac{\frac{1}{2} \text{ cup}}{x \text{ calories}}$

$\dfrac{1}{3} \cdot x = 80 \cdot \dfrac{1}{2}$

$\dfrac{\frac{1}{\cancel{3}} \cdot x}{\frac{1}{\cancel{3}}} = \dfrac{40}{\frac{1}{3}}$

$x = \dfrac{40}{1} \cdot \dfrac{3}{1}$

$x = 120$ calories

Then find the number of grams of fiber in a $\frac{1}{2}$-cup serving of bran cereal.

$\dfrac{\frac{1}{3} \text{ cup}}{8 \text{ grams}} = \dfrac{\frac{1}{2} \text{ cup}}{x \text{ grams}}$

$\dfrac{1}{3} \cdot x = 8 \cdot \dfrac{1}{2}$

$\dfrac{\frac{1}{\cancel{3}} \cdot x}{\frac{1}{\cancel{3}}} = \dfrac{4}{\frac{1}{3}}$

$x = \dfrac{4}{1} \cdot \dfrac{3}{1}$

$x = 12$ grams

A $\frac{1}{2}$-cup serving of bran cereal provides 120 calories and 12 grams of fiber.

CHAPTER 6 Percent

SECTION 6.1 (pages 387–392)

15. $0.5\% = 0.005$

Drop the percent symbol. Attach two zeros so the decimal point can be moved two places to the left.

27. $2 = 200\%$

Two zeros are attached so the decimal point can be moved two places to the right. Attach a percent symbol.

53. 100% is all of the children.

So, 100% of 12 children is 12 children.

12 children are present.

79. 21% of the children eat french fries. To write 21% as a decimal, drop the percent symbol and move the decimal point two places to the left, resulting in 0.21. So, 0.21 of the children eat french fries.

SECTION 6.2 (pages 398–403)

9. First write 6.25 over 100. Then to get a whole number in the numerator, multiply the numerator and denominator by 100. Finally, write the fraction in lowest terms.

$$6.25\% = \frac{6.25}{100} = \frac{6.25\,(100)}{100\,(100)}$$

$$= \frac{625}{10{,}000} \overset{\div\,625}{\underset{\div\,625}{=}} \frac{1}{16}$$

41.

$$\frac{5}{9} = \frac{p}{100}$$

$9 \cdot p = 5 \cdot 100$ Find cross products.

$$\frac{9 \cdot p}{9} = \frac{500}{9}$$ Divide both sides by 9.

$$p = \frac{500}{9}$$

$$p \approx 55.5\overline{5}$$

Thus, $\frac{5}{9} = 55.6\%$ (rounded).

49. $87.5\% = 0.875$ decimal

$$= \frac{875 \div 125}{1000 \div 125} = \frac{7}{8}$$ fraction

63. $\frac{1}{200} = 1 \div 200 = 0.005$ decimal

$$= 0.5\%$$ percent

67. $3\frac{1}{4} = \frac{13}{4} = 13 \div 4 = 3.25$ decimal

$$= 325\%$$ percent

---

77. 64 out of 80 employees have cell phones.

(a) $\dfrac{64}{80} = \dfrac{64 \div 16}{80 \div 16} = \dfrac{4}{5}$ fraction

$$\frac{4}{5} = 4 \div 5 = 0.8$$ decimal

$$0.8 = 0.80 = 80\%$$ percent

(b) If $\frac{4}{5}$ (have cell phones)

$$1 - \frac{4}{5} = \frac{1}{5} \text{ do not have cell phones.}$$

$$\frac{1}{5} = 1 \div 5 = 0.2$$ decimal

$$0.2 = 0.20 = 20\%$$ percent

SECTION 6.3 (pages 408–411)

11. part = 15, whole = 60

$$\frac{15}{60} = \frac{x}{100}$$

$$\frac{1}{4} = \frac{x}{100}$$ $\frac{15}{60}$ is $\frac{1}{4}$ in lowest terms.

$4 \cdot x = 1 \cdot 100$ Find cross products.

$$\frac{4 \cdot x}{4} = \frac{100}{4}$$ Divide both sides by 4.

$$x = 25$$

The percent is 25, written as 25%.

21. whole = 5000, part = 20

$$\frac{20}{5000} = \frac{x}{100}$$ Percent proportion

$5000 \cdot x = 20 \cdot 100$ Find cross products.

$$\frac{5000 \cdot x}{5000} = \frac{2000}{5000}$$ Divide both sides by 5000.

$$x = 0.4$$

The percent is 0.4, written as 0.4%.

37. $\underbrace{12 \text{ injections}}_{\text{part}}$ is $\underbrace{20\%}_{\text{percent}}$ $\underbrace{\text{of what number of injections?}}_{\substack{\text{whole} \\ \text{(unknown)}}}$

$$\frac{12}{\text{unknown}} = \frac{20}{100}$$

47. $\underbrace{86}_{\text{part}}$ of $\underbrace{142 \text{ people}}_{\text{whole}}$ is $\underbrace{\text{what percent?}}_{\substack{\text{percent} \\ \text{(unknown)}}}$

$$\frac{86}{142} = \frac{\text{unknown}}{100}$$

SECTION 6.4 (pages 419–424)

15. Write 225% as a decimal, 2.25.

225% of 680 tables

$$(2.25)(680) = 1530$$

$$\text{part} = 1530 \text{ tables}$$

---

29. part is 350; percent is 12.5

$$\frac{\text{part}}{\text{whole}} = \frac{\text{percent}}{100}$$

so $\dfrac{350}{x} = \dfrac{12.5}{100}$

$12.5 \cdot x = 35{,}000$ Cross products

$$\frac{12.5 \cdot x}{12.5} = \frac{35{,}000}{12.5}$$ Divide both sides by 12.5.

$$x = 2800$$

$12\frac{1}{2}\%$ of 2800 is 350.

37. part is 64; whole is 344

$$\frac{\text{part}}{\text{whole}} = \frac{\text{percent}}{100}$$

$$\frac{64}{344} = \frac{x}{100} \text{ OR } \frac{8}{43} = \frac{x}{100}$$

Cross products $43 \cdot x = 8 \cdot 100$

Divide both sides by 43. $\dfrac{43 \cdot x}{43} = \dfrac{800}{43}$

$$x \approx 18.6$$

$64 is 18.6% (rounded) of $344.

43. (a) part is unknown; whole is 240; percent is 22

$$\frac{\text{part}}{\text{whole}} = \frac{\text{percent}}{100}$$

$$\frac{x}{240} = \frac{22}{100} \text{ OR } \frac{x}{240} = \frac{11}{50}$$

Cross products $x \cdot 50 = 240 \cdot 11$

Divide both sides by 50. $\dfrac{x \cdot 50}{50} = \dfrac{2640}{50}$

$$x = 52.8$$

The amount withheld is $52.80.

(b) $240 earnings −$52.80 withheld = $187.20 amount remaining. The amount remaining is $187.20.

55. part is 960; whole is 48,000; percent is unknown

$$\frac{\text{part}}{\text{whole}} = \frac{\text{percent}}{100}$$

$$\frac{960}{48{,}000} = \frac{x}{100} \text{ OR } \frac{1}{50} = \frac{x}{100}$$

$$50 \cdot x = 1 \cdot 100$$

Cross products $\dfrac{50 \cdot x}{50} = \dfrac{100}{50}$

Divide both sides by 50. $x = 2$

There are 2% of these jobs filled by women.

SECTION 6.5 (pages 429–432)

3. Write 25% as the decimal 0.25. The whole is 1080. Let x represent the unknown part.

part = percent • whole

$x = (0.25)(1080)$

$x = 270$

25% of 1080 blood donors is 270 blood donors.

23. The part is 3.75 and the percent is $1\frac{1}{4}\% = 1.25\%$ or 0.0125 as a decimal. The whole is unknown.

part = percent • whole

$3.75 = (0.0125)(x)$

$\dfrac{3.75}{0.0125} = \dfrac{(0.0125)(x)}{0.0125}$

$300 = x$

$1\frac{1}{4}\%$ of 300 gallons is 3.75 gallons.

29. Because 160 follows *of*, the whole is 160. The part is 2.4, and the percent is unknown.

part = percent • whole

$2.4 = x \cdot 160$

$\dfrac{2.4}{160} = \dfrac{x \cdot 160}{160}$

$0.015 = x$

0.015 is 1.5%

1.5% of 160 liters is 2.4 liters.

43. Because 1250 follows *of*, the whole is 1250. The part is 461, and the percent is unknown.

part = percent • whole

$461 = x \cdot 1250$

$\dfrac{461}{1250} = \dfrac{x \cdot 1250}{1250}$

$0.369 \approx x$

0.369 is 36.9%

36.9% (rounded) of these Americans rate their health as excellent.

55. Find the sales tax on the new Polaris unit. The whole is 524. The percent is $7\frac{3}{4}\% = 7.75\%$, which is 0.0775 as a decimal.

part = percent • whole

$x = (0.0775)(524)$

$x = 40.61$

Add the sales tax to the purchase price.

$\$524 + \$40.61 = \$564.61$

Subtract the trade-in.

$\$564.61 - \$125 = \$439.61$

The total cost to the customer is $439.61.

SECTION 6.6 (pages 441–444)

5. Find the sales tax rate using the sales tax formula. The sales tax is $12.75, and the cost of the item is $425.

sales tax = rate of tax • cost of item

$\$12.75 = r \cdot \$425$

$\dfrac{12.75}{425} = \dfrac{r \cdot 425}{425}$ Divide both sides by 425.

$0.03 = r$

0.03 is 3%   Write the decimal as a percent.

The tax rate is 3% and the total cost is $425 + \$12.75 = \$437.75$.

11. The problem asks for the amount of commission. Use the commission formula. The rate of commission is 8%, and the sales are $280.

commission = rate of commission • sales

$= (8\%)(\$280)$

$= (0.08)(\$280)$

$= \$22.40$

The amount of commission is $22.40.

23. The problem asks for the amount of the discount and the sale price. First, find the amount of discount using the discount formula. The rate of discount is 15%, and the original price is $58.40.

amount of discount = rate of discount • original price

$= (15\%)(\$58.40)$

$= (0.15)(\$58.40)$ Write 15% as a decimal.

$= \$8.76$ Amount of discount

Now find the sale price by subtracting the amount of the discount ($8.76) from the original price: $58.40 - \$8.76 = \$49.64$. The amount of discount is $8.76 and the sale price is $49.64.

31. The problem asks for the rate of sales tax. Use the sales tax formula. The sales tax is $99. The cost of the door is $1980. Use r to represent the unknown rate of tax.

sales tax = rate of tax • cost of item

$\$99 = r \cdot \$1980$

$\dfrac{99}{1980} = \dfrac{r \cdot 1980}{1980}$ Divide both sides by 1980.

$0.05 = r$

0.05 is 5%   Write the decimal as a percent.

The rate of sales tax is 5%.

47. The problem asks for the cost of the dictionary. First find the amount of discount using the discount formula. The rate of discount is 6%, and the original price is $18.50.

discount = rate of discount • original price

$= (6\%)(\$18.50)$

$= (0.06)(\$18.50)$ Write 6% as a decimal.

$= \$1.11$ Amount of discount

To find the sale price, subtract the amount of discount ($1.11) from the original price. Sale price

$=$ original price $-$ amount of discount

$= \$18.50 - \$1.11 = \$17.39$ Sale price

To find the sales tax, use the sales tax formula. The rate of tax is 6% and the cost of the dictionary is $17.39.

sales tax = rate of tax • cost of item

$= (6\%)(\$17.39)$

$= (0.06)(\$17.39)$ Write 6% as a decimal.

$\approx \$1.04$ Sales tax

The total cost of the dictionary is $17.39 + \$1.04 = \$18.43$.

57. To find the sales tax, use the sales tax formula.

sales tax = rate of tax • cost of item

$= \left(7\frac{3}{4}\%\right)(\$1248)$

$= (0.0775)(\$1248)$ Write $7\frac{3}{4}\%$ as a decimal.

$= \$96.72$ Sales tax

The total cost is the sum of the ticket, the sales tax, and the excise tax.

$\$1248 + \$96.72 + \$15.40 = \$1360.12$

SECTION 6.7 (pages 447–450)

9. $2300 at $4\frac{1}{2}\%$ for $2\frac{1}{2}$ years

The principal (p) is $2300. The rate ($r$) is $4\frac{1}{2}\%$, or 0.045 as a decimal, and the time (t) is $2\frac{1}{2}$ or 2.5 years.

$I = p \cdot r \cdot t$

$= (2300)(0.045)(2.5)$

$= 258.75$

The interest is $258.75.

19. $940 at 3% for 18 months

The principal is $940. The rate is 3% or

0.03, and the time is $\frac{18}{12}$ of a year.

$$I = p \cdot r \cdot t$$

$$= (940)(0.03)\left(\frac{18}{12}\right)$$

$\qquad$ 18 months $= \frac{18}{12}$ of a year.

$$= (28.2)(1.5) \quad \frac{18}{12} = (1.5)$$

$$= 42.3$$

The interest is $42.30.

33. $16,850 at $7\frac{1}{2}$% for 9 months

First find the interest. The principal is

$16,850. The rate is $7\frac{1}{2}$% or 0.075, and the

time is $\frac{9}{12}$ of a year.

$$I = p \cdot r \cdot t$$

$$= (16,850)(0.075)\left(\frac{9}{12}\right)$$

$$= (1263.75)\left(\frac{3}{4}\right) \quad \text{Write } \frac{9}{12} \text{ in}$$
$$\qquad\qquad\qquad\qquad \text{lowest terms as } \frac{3}{4}.$$

$$= 947.81 \text{ (rounded)}$$

The interest is $947.81.

To find the total amount due, add the

principal and the interest.

$$\text{amount due} = \text{principal} + \text{interest}$$
$$= \$16,850 + \$947.81$$
$$= \$17,797.81$$

The total amount due is $17,797.81.

43. $14,800 at $2\frac{1}{4}$% for 10 months

The principal is $14,800. The rate is $2\frac{1}{4}$%

or 0.0225, and the time is $\frac{10}{12}$ of a year.

$$I = p \cdot r \cdot t$$

$$= (14,800)(0.0225)\left(\frac{10}{12}\right)$$

$$= (333)\left(\frac{5}{6}\right) \quad \text{Write } \frac{10}{12} \text{ in}$$
$$\qquad\qquad\qquad \text{lowest terms as } \frac{5}{6}.$$

$$= 277.50$$

She will earn $277.50 in interest.

SECTION 6.8 (pages 455–458)

3. $500 at 4% for 2 years

Year	Interest	Compound Amount
1	($500)(0.04)(1) = $20	
		$500 + $20 = $520
2	($520)(0.04)(1) = $20.80	
		$520 + $20.80 = $540.80

The compound amount is $540.80.

13. $1400 at 6% for 5 years

$\qquad$ Year 1 Year 2 Year 3 Year 4 Year 5

($1400) (1.06) (1.06)(1.06)(1.06)(1.06) $\approx$ $1873.52

Original deposit $\qquad$ 100% + 6% = 106% = 1.06 $\qquad$ Compound amount

The compound amount is $1873.52 (rounded).

23. $1000 at 4% for 5 years

Look down the column headed 4%, and

across to row 5 (because 5 years = 5 time

periods). At the intersection of the column

and the row, read the compound amount,

1.2167.

$$\text{compound amount} = (\$1000)(1.2167)$$
$$= \$1216.70$$

Find the interest by subtracting the princi-

pal ($1000) from the compound amount.

$$\text{interest} = \$1216.70 - \$1000$$
$$= \$216.70$$

33. (a) 6% column, row 9 from the table is

1.6895.

$$\text{compound amount} = (\$76,000)(1.6895)$$
$$= \$128,402$$

The total amount that should be repaid is

$128,402.

(b) To find the amount of interest earned,

subtract the principal ($76,000) from

the compound amount.

The amount of interest earned is

$128,402 − $76,000 = $52,402

CHAPTER 7 Measurement

SECTION 7.1 (pages 484–487)

25. 3 in. to feet

$$\frac{\overset{1}{\cancel{3}} \text{ in.}}{1} \cdot \frac{1 \text{ ft}}{\underset{4}{\cancel{12}} \text{ in.}} = \frac{1}{4} \text{ ft or } 1 \div 4 = 0.25 \text{ ft}$$

35. $4\frac{1}{4}$ gal to quarts

$$\frac{4\frac{1}{4} \text{ gal}}{1} \cdot \frac{4 \text{ qt}}{1 \text{ gal}} = \frac{17}{\cancel{4}} \cdot \frac{\overset{1}{\cancel{4}}}{1} \text{ qt} = 17 \text{ qt}$$

43. 6 days to seconds

$$\frac{6 \text{ days}}{1} \cdot \frac{24 \text{ hr}}{1 \text{ day}} \cdot \frac{60 \text{ min}}{1 \text{ hr}} \cdot \frac{60 \text{ sec}}{1 \text{ min}}$$

$$= 6 \cdot 24 \cdot 60 \cdot 60 \text{ sec}$$

$$= 518,400 \text{ sec}$$

55. Step 1

The problem asks for the price per pound

of strawberries.

Step 2

Convert ounces to pounds. Then divide the

cost by the number of pounds.

Step 3

To estimate, round $2.29 to $2. Then, there

are 16 oz in a pound, so 20 oz is a little

more than 1 lb. Thus, $2 ÷ 1 = $2 per

pound is our estimate.

Step 4

Use a unit fraction to convert 20 oz to

pounds.

$$\frac{\overset{5}{\cancel{20}} \text{ oz}}{1} \cdot \frac{1 \text{ lb}}{\underset{4}{\cancel{16}} \text{ oz}} = \frac{5}{4} \text{ lb} = 1.25 \text{ lb}$$

$$\begin{array}{l}\text{Cost} \\ \text{per} \\ \text{pound}\end{array} \longrightarrow \frac{\$2.29}{1.25 \text{ lb}} = 1.832 \approx 1.83$$

Step 5

The strawberries are $1.83 per pound

(to the nearest cent).

Step 6

The exact answer, $1.83, is close to our

estimate of $2.

61. (a) Find the total number of cups per

week. Then convert cups to quarts.

$$\frac{2}{3} \cdot 15 \cdot 5 = \frac{2}{\cancel{3}} \cdot \frac{\overset{5}{\cancel{15}}}{1} \cdot \frac{5}{1} = 50 \text{ cups}$$

$$\frac{\overset{25}{\cancel{50}} \text{ cups}}{1} \cdot \frac{1 \text{ pt}}{\underset{1}{\cancel{2}} \text{ cups}} \cdot \frac{1 \text{ qt}}{2 \text{ pt}}$$

$$= \frac{25}{2} \text{ qt} = 12\frac{1}{2} \text{ qt}$$

The center needs $12\frac{1}{2}$ qt of milk per week.

(b) Convert $12\frac{1}{2}$ quarts to gallons.

$$\frac{12\frac{1}{2} \text{ qt}}{1} \cdot \frac{1 \text{ gal}}{4 \text{ qt}}$$

$$= \frac{25}{2} \cdot \frac{1}{4} \text{ gal} = 3.125 \text{ gal}$$

The center should order 4 gallon

containers, because you can't buy

part of a container.

SECTION 7.2 (pages 493–494)

21. A paper clip is about 3 <u>cm</u> long. Both

meters and kilometers would be much too

long, and 3 mm is a very tiny length.

25. 7 m to cm

$$\frac{7 \text{ m}}{1} \cdot \frac{100 \text{ cm}}{1 \text{ m}} = 7 \cdot 100 \text{ cm} = 700 \text{ cm}$$

33. 400 mm to cm

From mm to cm is *one* place to the *left* on the conversion line, so move the decimal point *one* place to the *left* also.

400. mm = 40.0 cm   or   40 cm

37. Convert 82 cm to m

From cm to m is *two* places to the *left* on the conversion line, so move the decimal point *two* places to the *left* also.

82. cm = 0.82 m

0.82 m is less than 1 m, so 82 cm is **less than** 1 m. The difference in length is 1 m − 0.82 m = 0.18 m or 100 cm − 82 cm = 18 cm

45. 5.6 mm to km

$$\frac{5.6 \text{ mm}}{1} \cdot \frac{1 \text{ m}}{1000 \text{ mm}} \cdot \frac{1 \text{ km}}{1000 \text{ m}}$$

$$= \frac{5.6}{1,000,000} \text{ km}$$

$$= 0.0000056 \text{ km}$$

SECTION 7.3 (pages 501–504)

9. Lori caught a small sunfish weighing 150 g. Weight is measured in mg, g, or kg; 150 mg is too light; 150 kg is too heavy.

35. 8 mL to L

$$\frac{8 \text{ mL}}{1} \cdot \frac{1 \text{ L}}{1000 \text{ mL}} = \frac{8}{1000} \text{ L} = 0.008 \text{ L}$$

41. 5.2 kg to g

From kg to g is *three* places to the *right* on the conversion line.

5.200 kg = 5200. g   or   5200 g

67. Convert 900 mL to L.

$$\frac{900 \text{ mL}}{1} \cdot \frac{1 \text{ L}}{1000 \text{ mL}} = \frac{900}{1000} \text{ L} = 0.9 \text{ L}$$

On average, we breathe in and out roughly 0.9 L of air every 10 seconds.

73. Convert 1 kg to g.

$$\frac{1 \text{ kg}}{1} \cdot \frac{1000 \text{ g}}{1 \text{ kg}} = 1000 \text{ g}$$

Divide the 1000 g by the weight of 1 nickel. $\frac{1000 \text{ g}}{5 \text{ g}} = 200$

There are 200 nickels in 1 kg of nickels.

SECTION 7.4 (pages 509–510)

9. Multiply to find the total length of pencil lead in mm, then convert mm to cm.

60 mm • 30 = 1800 mm of pencil lead

$$\frac{\overset{180}{\cancel{1800}} \text{ mm}}{1} \cdot \frac{1 \text{ cm}}{\underset{1}{\cancel{10}} \text{ mm}} = 180 \text{ cm}$$

The total length of pencil lead is 180 cm. Divide to find the cost per cm.

$$\frac{\$3.29}{180 \text{ cm}} \approx \$0.0183/\text{cm} \approx \$0.02/\text{cm}$$

The cost is $0.02/cm (rounded).

15. Three cups a day for one week is 3 • 7 = 21 cups in one week.

$$\frac{21 \text{ cups}}{1} \cdot \frac{90 \text{ mg}}{1 \text{ cup}} = 1890 \text{ mg}$$

Convert mg to g.

$$\frac{1890 \text{ mg}}{1} \cdot \frac{1 \text{ g}}{1000 \text{ mg}} = 1.89 \text{ g}$$

Agnete consumes 1.89 g of caffeine in one week.

SECTION 7.5 (pages 515–518)

5. 80 m to feet

$$\frac{80 \text{ m}}{1} \cdot \frac{3.28 \text{ ft}}{1 \text{ m}} \approx 262.4 \text{ ft}$$

19. Write $\frac{1}{2}$ in. as 0.5 in. Then convert 0.5 in. to centimeters.

$$\frac{0.5 \text{ in.}}{1} \cdot \frac{2.54 \text{ cm}}{1 \text{ in.}} = 1.27 \text{ cm} \approx 1.3 \text{ cm}$$

The dwarf gobie is about 1.3 cm long.

27. Change 60 °F to Celsius.

$$C = \frac{5(60 - 32)}{9} = \frac{5(28)}{9}$$

$$= \frac{140}{9} \approx 15.6 \approx 16$$

60 °F ≈ 16 °C

37. (a) Since the comfort range of the boots is from 24 °C to 4 °C, you would wear these boots in pleasant weather—above freezing, but not hot.

(b) Change 24 °C to Fahrenheit.

$$F = \frac{9 \cdot C}{5} + 32 = \frac{9 \cdot 24}{5} + 32$$

$$= \frac{216}{5} + 32$$

$$= 43.2 + 32 = 75.2$$

Thus, 24 °C ≈ 75 °F.

Change 4 °C to Fahrenheit.

$$F = \frac{9 \cdot C}{5} + 32 = \frac{9 \cdot 4}{5} + 32$$

$$= \frac{36}{5} + 32$$

$$= 7.2 + 32 = 39.2$$

Thus, 4 °C ≈ 39 °F.

The boots are designed for Fahrenheit temperatures of about 75 °F to about 39 °F.

(c) The range of metric temperatures in January would depend on where you live. In Minnesota, it's 0 °C to −40 °C, and in California it's 24 °C to 0 °C.

45. Fuel left after landing

$$\frac{2 \text{ fl oz}}{1} \cdot \frac{1 \text{ qt}}{32 \text{ fl oz}} \cdot \frac{0.95 \text{ L}}{1 \text{ qt}} \cdot \frac{1000 \text{ mL}}{1 \text{ L}}$$

$$= \frac{950}{16} \text{ mL} = 59.375 \text{ mL} \approx 59.4 \text{ mL}$$

CHAPTER 8 Geometry

SECTION 8.1 (pages 539–540)

7. This is a *ray* named $\overrightarrow{PQ}$. A ray is a part of a line that has only one endpoint and goes on forever in one direction.

9. The lines are *perpendicular* because they intersect at right angles; the small red square indicates a right angle.

23. Two rays in a straight line pointing in opposite directions measure 180°. An angle that measures 180° is called a *straight angle*.

SECTION 8.2 (pages 544–545)

9. Find the complement of 86° by subtracting. 90° − 86° = 4°, so the complement is 4°.

13. Find the supplement of 90° by subtracting. 180° − 90° = 90°, so the supplement is 90°.

17. $\angle SON \cong \angle TOM$ because the angles are vertical angles.

$\angle TOS \cong \angle MON$ because the angles are vertical angles.

19. Because $\angle COE$ and $\angle GOH$ are vertical angles, they are also congruent. This means they have the same measure. $\angle COE$ measures 63°, so $\angle GOH$ measures 63°. The sum of the measures of $\angle COE$, $\angle AOC$, and $\angle AOH$ equals 180° because $\angle EOH$ is a straight angle. Therefore, $\angle AOC$ measures 180° − (63° + 37°) = 180° − 100° = 80°.

$\angle AOC$ and $\angle GOF$ are vertical angles, so they are congruent and both measure 80°.

$\angle AOH$ and $\angle EOF$ are vertical angles, so they are congruent and both measure 37°.

SOLUTIONS

SECTION 8.3 (pages 552–555)

5. $P = 4 \cdot s$

$= 4 \cdot 0.9$ km

$= 3.6$ km

$A = s \cdot s$

$= 0.9$ km $\cdot$ 0.9 km

$= 0.81$ km^2 ← Write km^2 for area (**not** km).

9. A storage building that is 76.1 ft by 22 ft

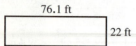

76.1 ft

22 ft

The length of this rectangle is 76.1 ft and the width is 22 ft. Use the formula $P = 2 \cdot l + 2 \cdot w$ to find the perimeter.

$P = 2 \cdot 76.1$ ft $+ 2 \cdot 22$ ft

$= 152.2$ ft $+ 44$ ft

$= 196.2$ ft

Use the formula $A = l \cdot w$ to find the area.

$A = 76.1$ ft $\cdot$ 22 ft

$= 1674.2$ ft^2

17. The unlabeled side is 6 in. + 6 in. or 12 in.

Add the lengths of the sides to find the perimeter.

$P = 15$ in. $+ 6$ in. $+ 6$ in. $+ 6$ in. $+ 6$ in.

$\quad + 6$ in. $+ 9$ in. $+ 12$ in. $+ 6$ in. $+ 6$ in.

$= 78$ in.

Draw two horizontal lines to break up the figure into two smaller squares and one larger rectangle. Add the areas of these squares and rectangle to find the total area.

$A = (6$ in. $\cdot$ 6 in.$) + (18$ in. $\cdot$ 9 in.$)$

$\quad + (6$ in. $\cdot$ 6 in.$)$

$= 36$ in.$^2 + 162$ in.$^2 + 36$ in.^{2}

$= 234$ in.^{2}

You could also draw two vertical lines to break up the figure into three rectangles.

$A = (15$ in. $\cdot$ 6 in.$) + (15$ in. $\cdot$ 6 in.$)$

$\quad + (9$ in. $\cdot$ 6 in.$)$

$= 90$ in.$^2 + 90$ in.$^2 + 54$ in.^{2}

$= 234$ in.^{2} ← Same result

23. Find the perimeter of the room.

$P = 2 \cdot 4.4$ m $+ 2 \cdot 5.1$ m

$= 8.8$ m $+ 10.2$ m $= 19$ m

She will need 19 m of the strip. Multiply by the cost per meter to find the total cost.

$$\text{Cost} = \frac{19 \text{ m}}{1} \cdot \frac{\$4.99}{1 \text{ m}} = \$94.81$$

Tyra will have to spend \$94.81 for the strip.

27.

$A = \text{length} \cdot \text{width}$

5300 yd$^2 = 100$ yd $\cdot$ width

$$\frac{5300 \text{ yd} \cdot \text{yd}}{100 \text{ yd}} = \frac{100 \text{ yd} \cdot \text{width}}{100 \text{ yd}}$$

53 yd $=$ width

The width is 53 yd.

SECTION 8.4 (pages 560–561)

13. The figure is a trapezoid. The height (h) is 42 cm, the short base (b) is 61.4 cm, and the long base (B) is 86.2 cm.

$A = \dfrac{1}{2} \cdot h \cdot (b + B)$

$= 0.5 \cdot 42$ cm $\cdot (61.4$ cm $+ 86.2$ cm$)$

$= 0.5 \cdot 42$ cm $\cdot (147.6$ cm$)$

$= 3099.6$ cm^2

19. Area of one parallelogram:

$A = b \cdot h$

$= 5$ in. $\cdot$ 3.5 in.

$= 17.5$ in.^{2}

To find the total area of the 25 parallelogram pieces, multiply the area of one parallelogram (17.5 in.^{2}) by the number of pieces (25).

Total area $= 25 \cdot 17.5$ in.$^2 = 437.5$ in.^{2}

21. Break the figure into two pieces, a rectangle on the left side and a parallelogram. Find the area of each piece, and then add the areas.

Area of rectangle:

$A = \text{length} \cdot \text{width}$

$= 1.3$ m $\cdot$ 0.8 m

$= 1.04$ m^2

Area of parallelogram:

$A = \text{base} \cdot \text{height}$

$= 1.8$ m $\cdot$ 1.1 m

$= 1.98$ m^2

Total area $= 1.04$ m$^2 + 1.98$ m^2

$= 3.02$ m^2

SECTION 8.5 (pages 566–568)

9. To find the perimeter, add the lengths of the three sides.

$P = 35.5$ cm $+ 21.3$ cm $+ 28.4$ cm

$= 85.2$ cm

When finding the area of this right triangle, the base and height are the perpendicular sides. So the base is 28.4 cm and the height is 21.3 cm.

$A = \dfrac{1}{2} \cdot b \cdot h$

$= 0.5 \cdot 28.4$ cm $\cdot$ 21.3 cm

$= 302.46$ cm^2

13. First, find the area of the entire figure, which is a rectangle.

$A = l \cdot w$

$= 52$ m $\cdot$ 37 m

$= 1924$ m^2

Find the area of the unshaded triangle.

$A = \dfrac{1}{2} \cdot b \cdot h$

$= \dfrac{1}{2} \cdot 52$ m $\cdot$ 10 m

$= 260$ m^2

Subtract the unshaded area from the area of the entire figure to find the shaded area.

Shaded area $= 1924$ m$^2 - 260$ m^2

$= 1664$ m^2

15. *Step 1* Add the two angles given.

$90° + 58° = 148°$

Step 2 Subtract the sum from 180°.

$180° - 148° = 32°$

The third angle measures 32°.

23. **(a)** To find the amount of curbing needed to go around the triangular space, find the perimeter.

$P = 42$ m $+ 32$ m $+ 52.8$ m

$= 126.8$ m

126.8 m of curbing will be needed.

(b) To find the amount of sod needed to cover the space, find the area. It is a right triangle, so the perpendicular sides are the base and height.

$A = \dfrac{1}{2} \cdot 32$ m $\cdot$ 42 m $= 672$ m^2

672 m^2 of sod will be needed.

SECTION 8.6 (pages 575–578)

13. $d = 7\dfrac{1}{2}$ ft or 7.5 ft

$C = \pi \cdot d$

$\approx 3.14 \cdot 7.5$ ft

≈ 23.6 ft

To find the area, first find the radius.

$r = \dfrac{7.5 \text{ ft}}{2} = 3.75$ ft

$A = \pi \cdot r \cdot r$

$\approx 3.14 \cdot 3.75$ ft $\cdot$ 3.75 ft

≈ 44.2 ft^2

17. Find the area of a whole circle with a radius of 10 cm:

$A = \pi \cdot r \cdot r$

$\approx 3.14 \cdot 10$ cm $\cdot$ 10 cm

$= 314$ cm^2

Divide the area of the whole circle by 2 to find the area of the semicircle:

$$\frac{314 \text{ cm}^2}{2} = 157 \text{ cm}^2$$

Area of the triangle:

$$A = \frac{1}{2} \cdot b \cdot h$$

$$= \frac{1}{2} \cdot 20 \text{ cm} \cdot 10 \text{ cm}$$

$$= 100 \text{ cm}^2$$

Subtract to find the shaded area.

$$157 \text{ cm}^2 - 100 \text{ cm}^2 = 57 \text{ cm}^2$$

23. A point on the tire tread moves the length of the circumference in one complete turn.

$$C = \pi \cdot d$$

$$\approx 3.14 \cdot 29.10 \text{ in.}$$

$$\approx 91.4 \text{ in.}$$

Bonus question:

$$\frac{1 \text{ revolution}}{91.4 \text{ inches}} \cdot \frac{12 \text{ inches}}{1 \text{ foot}} \cdot \frac{5280 \text{ feet}}{1 \text{ mile}}$$

$$\approx 693 \text{ revolutions/mile}$$

25.

Find the area of a circle with a radius of 150 miles.

$$A = \pi \cdot r \cdot r$$

$$\approx 3.14 \cdot 150 \text{ mi} \cdot 150 \text{ mi}$$

$$= 70{,}650 \text{ mi}^2$$

There are about 70,650 mi^2 in the broadcast area.

31. (a) $C = 144$ cm

$$C = \pi \cdot d$$

$$144 \text{ cm} = \pi \cdot d$$

$$144 \text{ cm} \approx 3.14 \cdot d$$

$$\frac{144 \text{ cm}}{3.14} \approx \frac{\cancel{3.14} \cdot d}{\cancel{3.14}}$$

$$45.9 \text{ cm} \approx d$$

The diameter is about 45.9 cm.

(b) Divide the circumference by π (3.14).

SECTION 8.7 (pages 587–588)

7. The figure is a hemisphere.

$$V = \frac{2}{3} \cdot \pi \cdot r^3 \quad \text{or} \quad \frac{2 \cdot \pi \cdot r^3}{3}$$

$$\approx \frac{2 \cdot 3.14 \cdot 12 \text{ in.} \cdot 12 \text{ in.} \cdot 12 \text{ in.}}{3}$$

$$\approx 3617.3 \text{ in.}^3 \;\leftarrow \text{Round to tenths.}$$

---

11. The figure is a cone.

First find B, the area of the circular base.

$$B = \pi \cdot r \cdot r$$

$$\approx 3.14 \cdot 5 \text{ m} \cdot 5 \text{ m}$$

$$\approx 78.5 \text{ m}^2$$

Now find the volume of the cone.

$$V = \frac{B \cdot h}{3}$$

$$\approx \frac{78.5 \text{ m}^2 \cdot 16 \text{ m}}{3}$$

$$\approx 418.7 \text{ m}^3 \;\leftarrow \text{Round to tenths.}$$

19. Use the formula for the volume of a pyramid. First find B, the area of the square base.

$$B = s \cdot s$$

$$= 145 \text{ m} \cdot 145 \text{ m}$$

$$= 21{,}025 \text{ m}^2$$

Now find the volume of the pyramid.

$$V = \frac{B \cdot h}{3}$$

$$= \frac{21{,}025 \text{ m}^2 \cdot 93 \text{ m}}{3}$$

$$= 651{,}775 \text{ m}^3$$

The volume of the ancient stone pyramid is 651,775 m^3.

21. Use the formula for the volume of a cylinder. First find the radius of the pipe.

$$\text{radius} = \frac{5 \text{ ft}}{2} = 2.5 \text{ ft}$$

Now find the volume.

$$V = \pi \cdot r^2 \cdot h$$

$$\approx 3.14 \cdot 2.5 \text{ ft} \cdot 2.5 \text{ ft} \cdot 200 \text{ ft}$$

$$\approx 3925 \text{ ft}^3$$

The volume of the city sewer pipe is about 3925 ft^3.

SECTION 8.8 (pages 593–596)

25. The unknown length is the side opposite the right angle, which is the hypotenuse. The lengths of the legs are 8 in. and 3 in.

$$\text{hypotenuse} = \sqrt{(\text{leg})^2 + (\text{leg})^2}$$

$$= \sqrt{(8)^2 + (3)^2}$$

$$= \sqrt{64 + 9}$$

$$= \sqrt{73}$$

$$\approx 8.5 \text{ in.}$$

35. The length of the hypotenuse is 21.6 km. The length of one of the legs is 13.2 km.

$$\text{leg} = \sqrt{(\text{hypotenuse})^2 - (\text{leg})^2}$$

$$= \sqrt{(21.6)^2 - (13.2)^2}$$

$$= \sqrt{466.56 - 174.24}$$

$$= \sqrt{292.32}$$

$$\approx 17.1 \text{ km}$$

---

43. The diagonal brace is the hypotenuse. The lengths of the legs are 4.5 ft and 3.5 ft.

$$\text{hypotenuse} = \sqrt{(\text{leg})^2 + (\text{leg})^2}$$

$$= \sqrt{(4.5)^2 + (3.5)^2}$$

$$= \sqrt{20.25 + 12.25}$$

$$= \sqrt{32.5}$$

$$\approx 5.7 \text{ ft}$$

The diagonal brace is about 5.7 ft long.

45.

The ladder is opposite the right angle so it is the hypotenuse. Thus, the length of the hypotenuse is 12 ft and the length of one of the legs is 3 ft.

$$\text{leg} = \sqrt{(\text{hypotenuse})^2 - (\text{leg})^2}$$

$$= \sqrt{(12)^2 - (3)^2}$$

$$= \sqrt{144 - 9}$$

$$= \sqrt{135}$$

$$\approx 11.6 \text{ ft}$$

The ladder will reach about 11.6 ft high on the building.

SECTION 8.9 (pages 601–604)

15. Set up a ratio of corresponding sides, using the longest side in each triangle.

$$\frac{6 \text{ mm}}{12 \text{ mm}} = \frac{6}{12} = \frac{1}{2}$$

Write a proportion to find a.

$$\frac{a}{10} = \frac{1}{2}$$

$$a \cdot 2 = 10 \cdot 1$$

$$\frac{a \cdot \cancel{2}^{\,1}}{\cancel{2}_{\,1}} = \frac{10}{2}$$

$$a = 5 \text{ mm}$$

Write a proportion to find b.

$$\frac{b}{6} = \frac{1}{2}$$

$$b \cdot 2 = 6 \cdot 1$$

$$\frac{b \cdot \cancel{2}^{\,1}}{\cancel{2}_{\,1}} = \frac{6}{2}$$

$$b = 3 \text{ mm}$$

19. First, find the lengths of $\overline{FH}$ and $\overline{FG}$ in triangle FGH. Set up a ratio of corresponding sides where x is the length of side FH.

$$\frac{DE}{GH} = \frac{CE}{FH}$$
$$\frac{12}{8} = \frac{12}{x}$$
$$12 \cdot x = 8 \cdot 12$$
$$\frac{\overset{1}{\cancel{12}} \cdot x}{\underset{1}{\cancel{12}}} = \frac{96}{12}$$
$$x = 8 \text{ cm}$$

Note that when you set up a proportion to find the length of $\overline{FG}$, it is the same as that used to find the length of $\overline{FH}$. Therefore the length of $\overline{FG}$ is also 8 cm.

Perimeter of triangle FGH.

$= 8 \text{ cm} + 8 \text{ cm} + 8 \text{ cm}$

$= 24 \text{ cm}$

Set up a ratio of corresponding parts to find the height h of triangle FGH.

$$\frac{10.4}{12} = \frac{h}{8}$$
$$12 \cdot h = 10.4 \cdot 8$$
$$\frac{\overset{1}{\cancel{12}} \cdot h}{\underset{1}{\cancel{12}}} = \frac{83.2}{12}$$
$$h \approx 6.9 \text{ cm}$$

Area of triangle FGH

$= 0.5 \cdot b \cdot h$

$\approx 0.5 \cdot 8 \text{ cm} \cdot 6.9 \text{ cm}$

$\approx 27.6 \text{ cm}^2$

23. Let s be the length of the stick's shadow, and h is the height of the house (24 ft) from Exercise 21. Write a proportion to find s.

$$\frac{h}{3} = \frac{6}{s}$$
$$\frac{24}{3} = \frac{6}{s}$$
$$\frac{8}{1} = \frac{6}{s} \qquad \text{Write } \tfrac{24}{3} \text{ in lowest terms as } \tfrac{8}{1}.$$
$$8 \cdot s = 1 \cdot 6$$
$$\frac{\overset{1}{\cancel{8}} \cdot s}{\underset{1}{\cancel{8}}} = \frac{6}{8}$$
$$s = \frac{3}{4} \text{ ft or } 0.75 \text{ ft}$$

27. Write a proportion to find n.

$$\frac{50}{n} = \frac{100}{100 + 120} \quad \text{or} \quad \frac{50}{n} = \frac{100}{220}$$
$$100 \cdot n = 50 \cdot 220$$
$$\frac{\overset{1}{\cancel{100}} \cdot n}{\underset{1}{\cancel{100}}} = \frac{11,000}{100}$$
$$n = 110 \text{ m}$$

The length of the lake is 110 m.

CHAPTER 9 Basic Algebra

SECTION 9.1 (pages 630–631)

21. Compare 1 and -1. Because 1 is to the *right* of -1 on a number line, $1 > -1$.

23. Compare -11 and -2. Because -11 is to the *left* of -2 on a number line, $-11 < -2$.

33. $-|-18|$

First, $|-18| = 18$. But there is also a negative sign *outside* the absolute value bars. So, -18 is the simplified expression.

43. The opposite of -8.3 is $-(-8.3) = 8.3$ because the opposite of any negative number is positive.

51. Because $|-4| = 4$, then $-|-4| = -4$. Because $|-7| = 7$, then $-|-7| = -7$. Since -4 is to the right of -7 on the number line, $-4 > -7$; therefore $-|-4| < -|-7|$ is false.

SECTION 9.2 (pages 640–643)

25. $7.8 + (-14.6)$

The signs are different, so subtract the absolute values.

First, find the absolute values.

$|7.8| = 7.8$ and $|-14.6| = 14.6$

Subtract the lesser absolute value from the greater absolute value.

$14.6 - 7.8 = 6.8$

The *negative* number, -14.6, has the greater absolute value, so the answer is *negative*.

$7.8 + (-14.6) = -6.8$

29. $-\dfrac{7}{10} + \dfrac{2}{5}$

The signs are different, so subtract the absolute values.

First, find the absolute values.

$$\left|-\frac{7}{10}\right| = \frac{7}{10} \quad \text{and} \quad \left|\frac{2}{5}\right| = \frac{2}{5}$$

Subtract the lesser absolute value from the greater absolute value.

$$\frac{7}{10} - \frac{2}{5} = \frac{7}{10} - \frac{4}{10} = \frac{3}{10}$$

The *negative* number, $-\dfrac{7}{10}$, has the greater absolute value, so the answer is *negative*.

$$-\frac{7}{10} + \frac{2}{5} = -\frac{3}{10}$$

49. The first number, 7, stays the same. Change subtraction to addition. Change the sign of the second number to its opposite.

$$7 - 19 = 7 + (-19)$$

Now add.

$$7 + (-19) = -12$$

So $7 - 19 = -12$ also.

55. The first number, -3, stays the same. Change subtraction to addition. Change the sign of the second number to its opposite.

$$-3 - (-8) = -3 + (+8)$$

Now add.

$$-3 + 8 = 5$$

So $-3 - (-8) = 5$ also.

61. The first number, -30, stays the same. Change subtraction to addition. Change the sign of the second number to its opposite.

$$-30 - 30 = -30 + (-30)$$

Now add.

$$-30 + (-30) = -60$$

So $-30 - 30 = -60$ also.

67. The common denominator is 10.

Rewrite $\dfrac{1}{2}$ as $\dfrac{5}{10}$. Change subtraction to addition. Positive $\dfrac{9}{10}$ is changed to its opposite, $\left(-\dfrac{9}{10}\right)$.

$$\frac{1}{2} - \frac{9}{10} = \frac{5}{10} + \left(-\frac{9}{10}\right)$$
$$= -\frac{4}{10} = -\frac{2}{5}$$

Once the subtraction is changed to addition, notice that the numbers have different signs. So, subtract the lesser absolute value from the greater absolute value.

$$\left|\frac{5}{10}\right| \text{ is } \frac{5}{10} \quad \text{and} \quad \left|-\frac{9}{10}\right| = \frac{9}{10}.$$

Then $\dfrac{9}{10} - \dfrac{5}{10} = \dfrac{4}{10} = \dfrac{2}{5}$.

The *negative* number $-\dfrac{9}{10}$ has the greater absolute value so the answer is *negative*.

79. Change subtraction to addition. Change -13 to its opposite, $(+13)$.

$$4 - (-13) + (-5) = 4 + (+13) + (-5)$$
$$= 17 + (-5)$$
$$= 12$$

85. Change subtraction to addition.

Change $\frac{2}{3}$ to its opposite, $\left(-\frac{2}{3}\right)$.

$$\frac{1}{2}-\frac{2}{3}+\left(-\frac{5}{6}\right)=\frac{1}{2}+\left(-\frac{2}{3}\right)+\left(-\frac{5}{6}\right)$$

Common denominator is 6.

$$=\frac{3}{6}+\left(-\frac{4}{6}\right)+\left(-\frac{5}{6}\right)$$
$$=-\frac{1}{6}+\left(-\frac{5}{6}\right)$$
$$=-\frac{6}{6}=-1$$

91. $-3-(-2+4)+(-5)$

Work inside the parentheses: $(-2+4)$ is 2. Next, change subtraction to addition and change 2 to its opposite, (-2).

$$-3-(+2)+(-5)$$
$$=-3+(-2)+(-5)$$
$$=-5+(-5)$$
$$=-10$$

SECTION 9.3 (pages 646–649)

19. $-13(-1)=13$

The numbers have the *same* sign, so the product is *positive*.

23. $-\frac{1}{2}(-8)=-\frac{1}{2}\left(-\frac{\overset{4}{\cancel{8}}}{1}\right)=\frac{4}{1}=4$

The numbers have the *same* sign, so the product is *positive*.

31. $-\frac{7}{15}\cdot\frac{25}{14}=-\frac{\overset{1}{\cancel{7}}}{\underset{3}{\cancel{15}}}\cdot\frac{\overset{5}{\cancel{25}}}{\underset{2}{\cancel{14}}}=-\frac{5}{6}$

The numbers have *different* signs, so the product is *negative*.

41. $-1.25(-3.6)=4.5$

The numbers have the *same* sign, so the product is *positive*.

69. Rewrite the division as multiplication that uses the reciprocal of $-\frac{15}{14}$, which is $-\frac{14}{15}$. Divide out all the common factors. Then multiply.

$$\frac{-\frac{5}{7}}{-\frac{15}{14}}=-\frac{\overset{1}{\cancel{5}}}{\cancel{7}}\cdot\left(-\frac{\overset{2}{\cancel{14}}}{\underset{3}{\cancel{15}}}\right)=\frac{2}{3}$$

The numbers have the *same* sign, so the quotient is *positive*.

71. Rewrite division as multiplication using the reciprocal of -2, which is $-\frac{1}{2}$. Divide out the common factor. Then multiply.

$$-\frac{2}{3}\div(-2)=-\frac{\overset{1}{\cancel{2}}}{3}\cdot\left(-\frac{1}{\underset{1}{\cancel{2}}}\right)=\frac{1}{3}$$

The numbers have the *same* sign, so the quotient is *positive*.

81. $\dfrac{45.58}{-8.6}=-5.3$

The numbers have *different* signs, so the quotient is *negative*.

$$\begin{array}{r}5.3\\8.6\,\overline{\smash)4\,5.5\,8}\\\underline{4\,3\,0}\\2\,5\,8\\\underline{2\,5\,8}\\0\end{array}$$

85. $(-0.6)(-0.2)(-3)=(0.12)(-3)$
$$=-0.36$$

91. $|-8|\div(-4)\cdot|-5|$
$$=8\div(-4)\cdot(5)$$
$$=-2\cdot 5$$
$$=-10$$

99. First, divide 24,000 by 1000.

$$\frac{24,000}{1000}=24$$

Then, multiply 24 by the number of degrees the temperature changes per 1000 feet. The temperature will change $24(-3)=-72$ degrees in 24,000 ft. Finally, add the ground temperature and the change in temperature. The temperature will be $50+(-72)=-22$ degrees.

SECTION 9.4 (pages 654–657)

21. $-7+6(8-14)$ — Work inside parentheses.
$$=-7+6(-6)\quad\text{Multiply }6(-6).$$
$$=-7+(-36)\quad\text{Add.}$$
$$=-43$$

33. $30\div(-5)-36\div(-9)$ — Divide 30 by -5.
$$=-6-36\div(-9)\quad\text{Divide 36 by }-9.$$
$$=-6-(-4)$$
Change subtraction to adding the opposite.
$$=-6+4$$
$$=-2$$

37. $4(3^2)+7(3+9)-(-6)$ — Work inside parentheses.
$$=4\cdot 3^2+7(12)-(-6)\quad\text{Apply exponent: }3^2\text{ is 9.}$$
$$=4\cdot 9+7\cdot 12-(-6)\quad\text{Multiply }4\cdot 9.$$
$$=36+7\cdot 12-(-6)\quad\text{Multiply }7\cdot 12.$$
$$=36+84-(-6)$$
Add $36+84$ to get 120. Change subtraction to adding the opposite.
$$=120+(+6)$$
$$=126$$

43. $\dfrac{2^3(-2-5)+4(-1)}{4+5(-6\cdot 2)+(5\cdot 11)}$

First do the work in the numerator.

$2^3(-2-5)+4(-1)$ — Parentheses
$=2^3(-7)+4(-1)$ — Exponent: 2^3 is 8.
$=8(-7)+4(-1)$ — Multiply $8(-7)$.
$=-56+4(-1)$ — Multiply $4(-1)$.
$=-56+(-4)$ — Add.
$=-60$

Then do the work in the denominator.

$4+5(-6\cdot 2)+(5\cdot 11)$ — First set of parentheses
$=4+5(-12)+(5\cdot 11)$ — Second set of parentheses
$=4+5(-12)+55$ — Multiply $5(-12)$.
$=4+(-60)+55$ — Add $4+(-60)$.
$=-56+55$ — Add.
$=-1$

The last step is the division.

$$\frac{-60}{-1}=60$$

57. $\dfrac{3}{5}\left(-\dfrac{7}{6}\right)-\left(\dfrac{1}{6}-\dfrac{5}{3}\right)$ — Rewrite $\frac{5}{3}$ as $\frac{10}{6}$.

$$=\frac{3}{5}\left(-\frac{7}{6}\right)-\left(\frac{1}{6}-\frac{10}{6}\right)\quad\text{Parentheses}$$
$$=\frac{3}{5}\left(-\frac{7}{6}\right)-\left(-\frac{9}{6}\right)\quad\text{Divide out the common factor.}$$
$$=\frac{\overset{1}{\cancel{3}}}{5}\cdot\left(-\frac{7}{\underset{2}{\cancel{6}}}\right)-\left(-\frac{9}{6}\right)\quad\text{Multiply.}$$
$$=-\frac{7}{10}-\left(-\frac{9}{6}\right)\quad\text{Change subtraction to addition and }-\frac{9}{6}\text{ to }\frac{3}{2}.$$

(Continued)

$$= -\frac{7}{10} + \left(+\frac{3}{2} \right) \qquad \text{Rewrite } \frac{3}{2} \text{ as } \frac{15}{10}.$$

$$= -\frac{7}{10} + \frac{15}{10} \qquad \text{Add.}$$

$$= \frac{8}{10} \qquad \text{Simplify.}$$

$$= \frac{4}{5}$$

65. $-7\left(6 - \frac{5}{8} \cdot 24 + 3 \cdot \frac{8}{3} \right)$ Inside parentheses: multiply.

     $\frac{5}{8}$ and 24 to get 15.

$$= -7\left(6 - 15 + 3 \cdot \frac{8}{3} \right) \qquad \begin{array}{l}\text{Multiply} \\ 3 \text{ and } \frac{8}{3} \text{ to} \\ \text{get } 8.\end{array}$$

$$= -7(6 - 15 + 8) \qquad \begin{array}{l}\text{Subtract} \\ 6 - 15.\end{array}$$

$$= -7(-9 + 8) \qquad \text{Add } -9 + 8.$$

$$= -7(-1) \qquad \text{Multiply.}$$

$$= 7$$

SECTION 9.5 (pages 662–664)

17. $-m - 3n$; m is $\frac{1}{2}$, n is $\frac{3}{8}$

Replace m with $\frac{1}{2}$. Replace n with $\frac{3}{8}$.

$-m - 3n$

$$= -\frac{1}{2} - 3\left(\frac{3}{8} \right) \qquad \begin{array}{l}\text{Multiply } 3\left(\frac{3}{8} \right) \\ \text{to get } \frac{9}{8}.\end{array}$$

$$= -\frac{1}{2} - \frac{9}{8} \qquad \begin{array}{l}\text{Change subtraction to} \\ \text{addition and } \frac{9}{8} \text{ to } -\frac{9}{8}.\end{array}$$

$$= -\frac{1}{2} + \left(-\frac{9}{8} \right) \qquad \text{Rewrite } -\frac{1}{2} \text{ as } -\frac{4}{8}.$$

$$= -\frac{4}{8} + \left(-\frac{9}{8} \right) \qquad \text{Add.}$$

$$= -\frac{13}{8}$$

21. $-c - 5b$; c is -8, b is -4

Replace c with -8. Replace b with -4.

$-c - 5b = -(-8) - 5(-4)$

$-(-8)$ is the opposite of -8 or 8.

Multiply $5(-4)$ to get -20.

$$= 8 - (-20)$$

Change subtraction to addition and -20 to 20.

$$= 8 + (+20)$$

$$= 28$$

27. $\dfrac{-3s - t - 4}{-s + 6 + t}$; s is -1, t is -13

Replace s with -1. Replace t with -13.

$$= \frac{-3(-1) - (-13) - 4}{-(-1) + 6 + (-13)}$$

Multiply $-3(-1)$ in the numerator; $-(-1)$ is 1 in the denominator.

$$= \frac{3 + (+13) - 4}{1 + 6 + (-13)} \qquad \begin{array}{l}\text{Add 3 and 13 in the} \\ \text{numerator.}\end{array}$$

$$= \frac{16 - 4}{1 + 6 + (-13)} \qquad \begin{array}{l}\text{Subtract in the} \\ \text{numerator.}\end{array}$$

$$= \frac{12}{1 + 6 + (-13)} \qquad \begin{array}{l}\text{Add } 1 + 6 \text{ in the} \\ \text{denominator.}\end{array}$$

$$= \frac{12}{7 + (-13)} \qquad \begin{array}{l}\text{Add in the} \\ \text{denominator.}\end{array}$$

$$= \frac{12}{-6} \qquad \text{Divide 12 by } -6.$$

$$= -2$$

33. $A = \pi r^2$; π is approximately 3.14, r is 5

Replace π with 3.14. Replace r with 5.

$A \approx 3.14 \cdot 5^2$

$\approx 3.14 \cdot 5 \cdot 5$

$\approx 78.5 \text{ yd}^2$

SECTION 9.6 (pages 670–673)

11. $k + 15 = 0$

Add -15 to both sides.

$k + 15 + (-15) = 0 + (-15)$

$k + 0 = -15$

$k = -15$

Check the solution by replacing k with -15 in the original equation.

$k + 15 = 0$

$-15 + 15 = 0$

$0 = 0$ ✓ True

The solution is -15.

25. $-5 = -2 + r$

Add 2 to both sides.

$-5 + 2 = -2 + 2 + r$

$-3 = 0 + r$

$-3 = r$

Check the solution by replacing r with -3 in the original equation.

$-5 = -2 + r$

$-5 = -2 + (-3)$

$-5 = -5$ ✓ True

The solution is -3.

31. $\dfrac{1}{2} = k - 2$

Add 2 to both sides.

$\dfrac{1}{2} + 2 = k - 2 + 2$

$\dfrac{1}{2} + \dfrac{4}{2} = k + 0$

$\dfrac{5}{2} = k$

Check the solution by replacing k with $\dfrac{5}{2}$ in the original equation.

$\dfrac{1}{2} = k - 2$

$\dfrac{1}{2} = \dfrac{5}{2} - \dfrac{4}{2}$

$\dfrac{1}{2} = \dfrac{1}{2}$ ✓ True

The solution is $\dfrac{5}{2}$ or $2\dfrac{1}{2}$.

45. $3y = 0$

Divide both sides by 3.

$\dfrac{\overset{1}{\cancel{3}} \cdot y}{\underset{1}{\cancel{3}}} = \dfrac{0}{3}$

$y = 0$

Check the solution by replacing y with 0 in the original equation.

$3y = 0$

$3(0) = 0$

$0 = 0$ ✓ True

The solution is 0.

49. $-36 = -4p$

Divide both sides by -4.

$\dfrac{-36}{-4} = \dfrac{\overset{1}{\cancel{-4}} \cdot p}{\underset{1}{\cancel{-4}}}$

$9 = p$

Check the solution by replacing p with 9 in the original equation.

$-36 = -4p$

$-36 = -4(9)$

$-36 = -36$ ✓ True

The solution is 9.

61. $\dfrac{r}{3} = -12$ can be written as $\dfrac{1}{3}r = -12$

Multiply both sides by $\dfrac{3}{1}$, the reciprocal of $\dfrac{1}{3}$.

$\dfrac{\overset{1}{\cancel{3}}}{1} \cdot \dfrac{1}{\underset{1}{\cancel{3}}} r = -12 \cdot 3$

$1r = -36$

$r = -36$

Check the solution by replacing r with -36 in the original equation.

$$\frac{r}{3} = -12$$

$$\frac{-36}{3} = -12$$

$$-12 = -12 \;\checkmark \text{ True}$$

The solution is -36.

65. $-\frac{3}{4}m = -3$

Multiply both sides by $-\frac{4}{3}$, the reciprocal of $-\frac{3}{4}$.

$$-\frac{\overset{1}{\cancel{4}}}{\underset{1}{\cancel{3}}} \cdot \left(-\frac{\overset{1}{\cancel{3}}}{\underset{1}{\cancel{4}}}m\right) = -\overset{1}{\cancel{3}} \cdot \left(-\frac{4}{\underset{1}{\cancel{3}}}\right)$$

$$1m = 4$$

$$m = 4$$

Check the solution by replacing m with 4 in the original equation.

$$-\frac{3}{4}m = -3$$

$$-\frac{3}{4}(4) = -3$$

$$-3 = -3 \;\checkmark \text{ True}$$

The solution is 4.

79. $\frac{1}{2} - \frac{3}{4} = \frac{a}{5}$ $\quad$ Rewrite $\frac{1}{2}$ as $\frac{2}{4}$.

$$\frac{2}{4} - \frac{3}{4} = \frac{a}{5} \quad \text{Subtract } \tfrac{2}{4} - \tfrac{3}{4}.$$

$$-\frac{1}{4} = \frac{a}{5} \quad \begin{array}{l}\text{Multiply both sides by}\\ \tfrac{5}{1}\text{, the reciprocal of }\tfrac{1}{5}.\end{array}$$

$$\frac{5}{1} \cdot \left(-\frac{1}{4}\right) = \frac{a}{\underset{1}{\cancel{5}}} \cdot \frac{\overset{1}{\cancel{5}}}{1}$$

$$-\frac{5}{4} = a$$

The solution is $-\frac{5}{4}$.

81. $\quad m - 2 + 18 = |-3 - 4| + 5$
$\quad\quad\quad m + 16 = |-3 - 4| + 5$
$\quad\quad\quad m + 16 = |-7| + 5$
$\quad\quad\quad m + 16 = 7 + 5$
$\quad\quad\quad m + 16 = 12$
$\quad m + 16 + (-16) = 12 + (-16)$
$\quad\quad\quad\quad\quad m = -4$

The solution is -4.

SECTION 9.7 (pages 678–681)

5. $\quad 2 = 8y - 6$

Add 6 to both sides.

$$2 + 6 = 8y - 6 + 6$$

$$8 = 8y$$

Divide both sides by 8.

$$\frac{8}{8} = \frac{\overset{1}{\cancel{8}}y}{\underset{1}{\cancel{8}}}$$

$$1 = y$$

CHECK Replace y with 1 in the original equation.

$$2 = 8y - 6$$

$$2 = 8(1) - 6$$

$$2 = 2 \;\checkmark \text{ True}$$

The solution is 1.

13. $-\frac{1}{2}z + 2 = -1$

Subtract 2 from both sides.

$$-\frac{1}{2}z + 2 - 2 = -1 - 2$$

$$-\frac{1}{2}z = -3$$

Multiply both sides by $-\frac{2}{1}$, the reciprocal of $-\frac{1}{2}$.

$$-\frac{\overset{1}{\cancel{2}}}{1} \cdot \left(-\frac{1}{\underset{1}{\cancel{2}}}z\right) = -3 \cdot \left(-\frac{2}{1}\right)$$

$$z = 6$$

CHECK Replace z with 6 in the original equation.

$$-\frac{1}{2}z + 2 = -1$$

$$-\frac{1}{2}(6) + 2 = -1$$

$$-3 + 2 = -1$$

$$-1 = -1 \;\checkmark \text{ True}$$

The solution is 6.

19. $-3(m + 6) = -3 \cdot m + (-3) \cdot 6$
$\quad\quad\quad\quad\quad\quad = -3m - 18$

25. $-10(w - 9) = -10 \cdot w - (-10) \cdot 9$
$\quad\quad\quad\quad\quad\quad = -10w - (-90)$
$\quad\quad\quad\quad\quad\quad = -10w + 90$

39. $\frac{5}{2}b - \frac{11}{2}b = \left(\frac{5}{2} - \frac{11}{2}\right)b$

$$= -\frac{6}{2}b = -3b$$

49. $-12 = 6y - 18y$

Use the distributive property to combine $6y$ and $-18y$.

$$-12 = (6 - 18)y$$

$$-12 = -12y$$

Divide both sides by -12.

$$\frac{-12}{-12} = \frac{\cancel{-12}y}{\cancel{-12}}$$

$$1 = y$$

CHECK Replace y with 1 in the original equation.

$$-12 = 6y - 18y$$

$$-12 = 6(1) - 18(1)$$

$$-12 = 6 - 18$$

$$-12 = -12 \;\checkmark \text{ True}$$

The solution is 1.

55. $-2y + 6 = 6y - 10$

Add 10 to both sides.

$$-2y + 6 + 10 = 6y - 10 + 10$$

$$-2y + 16 = 6y$$

Add $2y$ to both sides.

$$-2y + 2y + 16 = 6y + 2y$$

Combine $6y$ and $2y$ because they are like terms.

$$16 = 8y$$

Divide both sides by 8.

$$\frac{16}{8} = \frac{8y}{8}$$

$$2 = y$$

CHECK Replace y with 2 in the original equation.

$$-2y + 6 = 6y - 10$$

$$-2(2) + 6 = 6(2) - 10$$

$$(-4) + 6 = 12 - 10$$

$$2 = 2 \;\checkmark \text{ True}$$

The solution is 2.

63. $-4(t + 2) = 12$

Use the distributive property on the left side of the equation.

$$-4t + (-4) \cdot 2 = 12$$

$$-4t + (-8) = 12$$

Add 8 to both sides.

$$-4t - 8 + 8 = 12 + 8$$

$$-4t = 20$$

Divide both sides by -4.

$$\frac{\cancel{-4}t}{\cancel{-4}} = \frac{20}{-4}$$

$$t = -5$$

CHECK Replace t with -5 in the original equation.

$$-4(t + 2) = 12$$

$$-4(-5 + 2) = 12$$

$$-4(-3) = 12$$

$$12 = 12 \;\checkmark \text{ True}$$

The solution is -5.

71. $0 = -2(y - 2)$

Use the distributive property on the right side of the equation.

$$0 = -2y - (-2) \cdot 2$$

$$0 = -2y + 4$$

Add 2y to both sides.

$$0 + 2y = -2y + 2y + 4$$
$$2y = 4$$

Divide both sides by 2.

$$\frac{2y}{2} = \frac{4}{2}$$
$$y = 2$$

The solution is 2.

73. $\quad \dfrac{y}{2} - 2 = \dfrac{y}{4} + 3$

Add 2 to both sides.

$$\frac{y}{2} - 2 + 2 = \frac{y}{4} + 3 + 2$$
$$\frac{y}{2} = \frac{y}{4} + 5$$
$$\frac{1}{2}y = \frac{1}{4}y + 5$$

Subtract $\dfrac{1}{4}y$ from both sides.

$$\frac{1}{2}y - \frac{1}{4}y = \frac{1}{4}y - \frac{1}{4}y + 5$$
$$\frac{1}{4}y = 5$$

Multiply both sides by $\dfrac{4}{1}$, the reciprocal of $\dfrac{1}{4}$.

$$\frac{\overset{1}{\cancel{4}}}{1} \cdot \frac{1}{\cancel{4}} y = \frac{4}{1} \cdot 5$$
$$y = 20$$

The solution is 20.

77. $\quad 2(a + 0.3) = 1.2(a - 4)$

Use the distributive property on both sides.

$$2 \cdot a + 2(0.3) = 1.2 \cdot a - 1.2 \cdot 4$$
$$2a + 0.6 = 1.2a - 4.8$$

Subtract 0.6 from both sides.

$$2a + 0.6 - 0.6 = 1.2a - 4.8 - 0.6$$
$$2a = 1.2a - 5.4$$

Subtract $1.2a$ from both sides.

$$2a - 1.2a = 1.2a - 1.2a - 5.4$$

Combine like terms.

$$0.8a = -5.4$$

Divide both sides by 0.8.

$$\frac{0.8a}{0.8} = \frac{-5.4}{0.8}$$
$$a = -6.75$$

The solution is -6.75.

SECTION 9.8 (pages 688–691)

31. Let n represent the unknown number.

30	subtract	3 times a number	is	2 plus	the number
↓	↓	↓	↓	↓	↓
30	$-$	$3n$	$= 2$	$+$	n

$$30 - 3n = 2 + n$$

Add $3n$ to both sides.

$$30 - 3n + 3n = 2 + n + 3n$$

Combine like terms on each side.

$$30 = 2 + 4n$$

Subtract 2 from both sides.

$$30 - 2 = 2 + 4n - 2$$
$$28 = 4n$$

Divide both sides by 4.

$$\frac{28}{4} = \frac{4n}{4}$$
$$7 = n$$

The number is 7.

Check: Three times 7 subtracted from 30 is $[30 - (3 \cdot 7)]$ is 9, which does equal 2 plus 7. True

37. *Step 2* Let x be the amount Brenda spent.

Step 3

Twice what Brenda spent	less $3	is	the amount Consuelo spent
↓	↓ ↓	↓	↓
$2x$	$- \quad 3$	$=$	81

Step 4

$$2x - 3 = 81$$

Add 3 to both sides.

$$2x - 3 + 3 = 81 + 3$$
$$2x = 84$$

Divide both sides by 2.

$$\frac{2x}{2} = \frac{84}{2}$$
$$x = 42$$

Step 5 Brenda spent $42.

Step 6

Check: Consuelo spent $3 less than twice $42 or $[(2 \cdot \$42) - \$3]$, which is $81. True

39. *Step 2*

You know the least about my age, so let x be my age. Then $x - 9$ is my sister's age.

Step 3

The sum of the ages	is	51
↓	↓	↓
$x + (x - 9)$	$=$	51

Step 4

$$x + x - 9 = 51$$
$$2x - 9 = 51$$

Add 9 to both sides.

$$2x - 9 + 9 = 51 + 9$$
$$2x = 60$$

Divide both sides by 2.

$$\frac{2x}{2} = \frac{60}{2}$$
$$x = 30$$

Step 5

x is my age so I am 30; my sister is $x - 9$ or $30 - 9$, so she is 21.

Step 6

Check: 21 is 9 less than 30 and the sum of 30 and 21 is 51. True

43. *Step 2*

You know the least about the cost of the printer, so let x be the cost of the printer. Then $8 \cdot x$ or $8x$ is the cost of the computer.

Step 3

The total cost (add)	is	$396.
↓	↓	↓
$x + 8x$	$=$	396

Step 4

$$x + 8x = 396$$
$$9x = 396$$

Divide both sides by 9.

$$\frac{9x}{9} = \frac{396}{9}$$
$$x = 44$$

Step 5

The cost of the printer is x, so the printer cost $44. The computer cost 8 times as much, so $8(\$44) = \$352$ for the computer.

Step 6

Check: Eight times $44 is $352 and the sum of $44 and $352 is $396. True

45. *Step 2*

You know the least about the shorter piece, so let x be the length of the shorter piece. Then $x + 10$ is the length of the longer piece.

Step 3

The length of the shorter piece	and	the length of the longer piece	are	78 cm.
↓	↓	↓	↓	↓
x	$+$	$x + 10$	$=$	78

Step 4

$$x + x + 10 = 78$$

Combine like terms.

$$2x + 10 = 78$$

Subtract 10 from both sides.

$$2x + 10 - 10 = 78 - 10$$
$$2x = 68$$

Divide both sides by 2.

$$\frac{2x}{2} = \frac{68}{2}$$
$$x = 34$$

Step 5

The length of the shorter piece is x, so the shorter piece is 34 cm. The length of the longer piece is $x + 10$, so $34 + 10 = 44$ cm.

Step 6

Check: 44 cm is 10 cm longer than 34 cm and the sum of 34 cm and 44 cm is 78 cm.

True

53. *Step 2*

You know the least about the width, so let x be the width. Then $2x + 3$ is the length.

$$2x + 3$$
$$x \quad \boxed{} \quad x$$
$$2x + 3$$

Step 3

$$P = 2 \cdot l + 2 \cdot w$$
$$36 = 2 \cdot (2x + 3) + 2 \cdot x$$

Step 4

$$36 = 2 \cdot (2x + 3) + 2 \cdot x$$
$$36 = 4x + 6 + 2x$$

Combine like terms.
$$36 = 6x + 6$$

Subtract 6 from both sides.
$$36 - 6 = 6x + 6 - 6$$
$$30 = 6x$$

Divide both sides by 6.
$$\frac{30}{6} = \frac{6x}{6}$$
$$5 = x$$

Step 5

The width is x, so the width is 5 in. The length is $2x + 3$, so $2(5) + 3$ or 13 in. is the length.

Step 6

Check: $P = 2(13) + 2(5)$
$$P = 26 + 10$$
$$P = 36 \leftarrow \text{Matches perimeter given in problem}$$

CHAPTER 10 Statistics

SECTION 10.1 (pages 712–717)

3. The number of pets owned in the United States is 11 million + 150 million + 90 million + 75 million + 18 million + 16 million = 360 million.

---

15. Time working compared to household duties/caring for others.

3 hours 30 minutes = 210 minutes

2 hours 30 minutes = 150 minutes

$$\frac{210}{150} = \frac{210 \div 30}{150 \div 30} = \frac{7}{5}$$

25. Need the car to go 50 miles or fewer:

$$x = 4\% \text{ of } 3021$$
$$= (0.04)(3021)$$
$$= 121 \text{ people (rounded)}$$

33. Percent for day camp $= \dfrac{\$546}{\$5460}$

$$= 0.10 = 10\%$$

Degrees of a circle $= 10\%$ of $360°$
$$= (0.10)(360°)$$
$$= 36°$$

39. **(a)** Total sales $= \$12,500 + \$40,000 + \$60,000 + \$50,000 + \$37,500 = \$200,000$

(b) Adventure classes $= \$12,500$

$$\text{percent of total} = \frac{12,500}{200,000} = 0.0625$$
$$= 6.25\%$$

$$\text{number of degrees} = (0.0625)(360°)$$
$$= 22.5°$$

Grocery and provision sales $= \$40,000$

$$\text{percent of total} = \frac{40,000}{200,000} = 0.2 = 20\%$$

$$\text{number of degrees} = (0.2)(360°) = 72°$$

Equipment rentals $= \$60,000$

$$\text{percent of total} = \frac{60,000}{200,000} = 0.3 = 30\%$$

$$\text{number of degrees} = (0.3)(360°) = 108°$$

Rafting tours $= \$50,000$

$$\text{percent of total} = \frac{50,000}{200,000} = 0.25 = 25\%$$

$$\text{number of degrees} = (0.25)(360°) = 90°$$

Equipment sales $= \$37,500$

$$\text{percent of total} = \frac{37,500}{200,000} = 0.1875$$
$$= 18.75\%$$

$$\text{number of degrees} = (0.1875)(360°)$$
$$= 67.5°$$

(c)

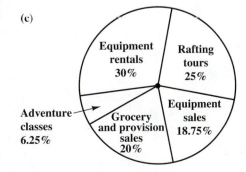

SECTION 10.2 (pages 721–726)

5. The countries in which less than 15% of household income is spent, on average, for food are USA, Britain, and Australia.

11. There are two bars for February. One is for 2012 and one is for 2013. The bar for 2013 rises to 7. Multiply 7 by 1000 because the label on the left side of the graph says *in thousands*. The bar for 2012 rises to halfway between 5 and 6. So multiply 5.5 by 1000.

Plants shipped in February of 2013 = 7000

Plants shipped in February of 2012 = 5500

$$7000 - 5500 = 1500$$

There were 1500 more plants shipped in February of 2013.

21. Above the dot for 1990 is 24.1. Looking at the left edge of the graph, notice that this number is in millions.

The number of PCs sold in 1990 was 24.1 million or 24,100,000.

31. Find the dot above 2012 on the line for store B. Use a ruler or straightedge to line up the dot with the numbers along the left edge. The dot is aligned with 350, so chain store B sold $350 \cdot 1000 = 350,000$ iPods in 2012.

37. Find the dot above 2013 on the line for total sales. Use a ruler or straightedge to line up the dot with the numbers along the left edge. The dot is aligned with 400. Multiply 400 by 1000 because the label on the left side of the graph says *in thousands*. The total sales in 2013 were

$$400 \cdot \$1000 = \$400,000.$$

SECTION 10.3 (pages 731–734)

3. Because the bar associated with 18–25 is the highest, the age group of 18–25 years has the greatest number of members, which is 40,000,000.

11. The bar for 31–40 rises to 11, so there are 11 employees who earn \$3100 to \$4000.

23. 0–5; 4 tally marks and a class frequency of 4

SECTION 10.4 (pages 739–742)

5. Mean = $\dfrac{\text{sum of all values}}{\text{number of values}}$

$= \dfrac{9 + 14 + 18 + 11 + 12 + 20}{6}$

$= \dfrac{84}{6}$

$= 14$ hours

The mean (average) of the time spent playing video games was 14 hours.

13.

Customers Each Hour	Frequency	Product
8	2	$(8 \cdot 2) = 16$
11	12	$(11 \cdot 12) = 132$
15	5	$(15 \cdot 5) = 75$
26	1	$(26 \cdot 1) = 26$
Totals	**20**	**249**

weighted mean

$= \dfrac{\text{sum of products}}{\text{total number of customers}}$

$= \dfrac{249}{20} \approx 12.5$ customers (rounded)

21. Arrange the numbers in numerical order from least to greatest.

298, 346, 412, 501, 515, 521, 528, 621

The list has 8 numbers. The middle numbers are the 4th and 5th numbers, so the median is

$$\dfrac{501 + 515}{2} = \dfrac{1016}{2} = 508 \text{ calories.}$$

25. 74, 68, 68, 68, 75, 75, 74, 74, 70

Because both 68 and 74 occur three times, each is a mode. This list is *bimodal*.

33.

Credits	Grade	Credits • Grade
2	A (= 4)	$2 \cdot 4 = 8$
3	C (= 2)	$3 \cdot 2 = 6$
4	A (= 4)	$4 \cdot 4 = 16$
1	C (= 2)	$1 \cdot 2 = 2$
4	B (= 3)	$4 \cdot 3 = 12$
14		**44**

$\text{GPA} = \dfrac{\text{sum of (Credits} \cdot \text{Grade)}}{\text{total number of credits}}$

$= \dfrac{44}{14} \approx 3.14$ (rounded)

Photo Credits

1 Marcito/Fotolia; 4 Yuri Arcurs/Fotolia; 9 (top left) Nicola Sutton/Life File/Photodisc/ Getty Images, (top right) DKatz/Fotolia, (bottom left) Alexander Crispin/Getty Images, (bottom right) Monart Design/Fotolia; 15 Bew111/Dreamstime; 20 (left) Andres Rodriguez/Fotolia, (right) Emiliau/Fotolia; 30 (top left) Victor Zastol'skiy/Fotolia, (top right) Robhainer/Fotolia, (bottom left) Kenneth Graff/Shutterstock, (bottom right) Stan Salzman; 40 (top) Jonathan Alcorn/ZUMA Press/Newscom, (bottom) John A. Rizzo/ Photodisc/Getty Images; 53 (top left) Anette Linnea Rasmussen/Shutterstock, (top right) Beth Anderson/Pearson Education, Inc., (bottom left) Sueharper/Dreamstime, (bottom right) Stan Salzman; 61 (left) Edwin Remsberg/Alamy, (right) Stan Salzman; 62 (left) Spflaum/Dreamstime, (right) Todd Taulman/Fotolia; 65 Stan Salzman; 84 A.F. Archive/ Alamy; 87 Stan Salzman; 88 Elenathewise/Fotolia; 92 Stan Salzman; 95 Lightpainter/ Dreamstime; 96 Svlumagraphica/Dreamstime; 97 (left) Beth Anderson/Pearson Education, Inc. SUBWAY® is a registered trademark of Doctor's Associates Inc. © Doctor's Associates Inc. All Rights Reserved, (right) Vacclav/Fotolia; 108 (left) Sascha Burkard/ Fotolia, (right) Vladimir Caplinskij/Shutterstock; 109 (left) Jason Stitt/Shutterstock, (right) Lily/Fotolia; 110 Karl Weatherly/Photodisc/Getty Images; 113 (both) Jelly Belly Candy Company. Used with permission; 117 (left) Photodisc/Getty Images, (right) U.S. Mint; 150 (left) Sascha Burkard/Shutterstock, (top right) Pakhnyushchyy/Fotolia, (bottom right) Kotafoty/Fotolia; 151 Iofoto/Fotolia; 155 (left) Antares614/Dreamstime, (right) Mark Breck/Shutterstock; 156 (left) Lynn Watson/Shutterstock, (right) Brodogg1313/ Dreamstime; 157 Stan Salzman; 158 (left) Slocummedia/Fotolia, (right) Arim44/ Dreamstime; 165 (left) V. J. Matthew/Shutterstock, (right) Ulrich Willmünder/ Shutterstock; 166 (left) Stephen Coburn/Fotolia, (right) Oleg Zabielin/Fotolia; 167 Collection of Stan Salzman; 177 Jelly Belly Candy Company. Used with permission; 188 Giuseppe Porzani/Fotolia; 191 (left) Sylvana Rega/Fotolia, (right) Kropic/Fotolia; 196 Olga Nayashkova/Fotolia; 197 ZUMA Press/Newscom; 202 (both) Stan Salzman; 221 (left) Chris Hill/Shutterstock, (right) Stan Salzman; 222 Elenathewise/Fotolia; 230 (left) ZUMA Press/Newscom, (right) Imageegami/Fotolia; 232 (left) Pavel Losevsky/ Fotolia, (right) Stan Salzman; 233 (left) Gudellaphoto/Fotolia, (right) ZUMA Press/ Newscom; 245 (left) Konstantin Sutyagin/Shutterstock, (right) Stan Salzman; 253 (left) Kim Steele/Digital Vision/Getty Images, (right) Stan Salzman; 254 (left) Amy Myers/ Shutterstock, (right) Stan Salzman; 256 (both) Stan Salzman; 258 (left) Alisonhancock/ Fotolia, (right) Lise F. Young/Shutterstock; 262 (left) Stan Salzman, (right) Cosmin Manci/Fotolia; 263 Pressmaster/Fotolia; 272 Pressmaster/Fotolia; 273 Carlosseller/Fotolia; 279 (both) Beth Anderson/Pearson Education, Inc.; 280 (top) Perry Mastrovito/Corbis, (bottom) Charles Smith/Corbis; 296 (top) Stan Salzman, (bottom) JGI/Blend Images/Getty Images; 305 U.K. History/Alamy; 306 (left) Feng Yu/Fotolia, (top right) National Motor Museum/Motoring Picture Library/Alamy, (bottom right) Used with permission of General Mills Marketing, Inc. (GMMI); 312 (left) Diane Macdonald/Stockbyte/Getty Images, (right) Flying Colours Ltd/Photodisc/Getty Images; 313 Neo Edmund/Fotolia; 321 (left) Stockbyte/Getty Images, (right) Beth Anderson/Pearson Education, Inc.; 323 Beth Anderson/Pearson Education, Inc.; 327 Stockbroker/Valueline/PunchStock; 328 Twixx/ Fotolia; 329 Edbockstock/Fotolia; 336 Picturequest/JupiterImages; 338 Alan Pappe/ Photodisc/Getty Images; 345 Graca Victoria/Shutterstock; 346 Edbockstock/Fotolia; 351 Sly/Fotolia; 358 AP Images; 364 Ryan McVay/Photodisc/Getty Images; 366 Jjava/ Fotolia; 374 (left) Doug Menuez/Photodisc/Getty Images, (right) Jack Hollingsworth/ Photodisc/Getty Images; 375 (left) Comstock/Thinkstock, (right) David Buffington/ Photodisc/Getty Images; 378 Tony Campbell/Fotolia; 380 Steve Byland/Fotolia; 381 Pavel Losevsky/Fotolia; 388 (left) Stan Salzman, (right) Digital Vision/Getty Images; 389 (left) Getty Images/Digital Vision, (right) Yuri Arcurs/Fotolia; 390 Annette Shaff/Shutterstock; 401 (left) Stockbyte/Getty Images, (right) Beth Anderson/Pearson Education, Inc.; 402 (left) Regien Paassen/Shutterstock, (right) Siri Stafford/Digital Vision/Getty Images;

410 (top left) Yuri Arcurs/Fotolia, (top right, bottom left) Beth Anderson/Pearson Education, Inc., (bottom right) Lisa F. Young/Fotolia; **411** (left) Beth Anderson/Pearson Education, Inc., (right) Actionpics/Fotolia; **414** Pete Saloutos/Fotolia; **422** (left) Alexander/Fotolia, (right) Stan Salzman; **423** Okea/Fotolia; **424** Givaga/Fotolia; **431** Stan Salzman; **432** (left) Vac/Fotolia, (right) Sean Locke/Photodisc/Getty Images; **433** Evgenyb/Fotolia; **434** (top left) Monkey Business/Fotolia, (top right) Kameel4u/Shutterstock, (bottom left) U.S. Internal Revenue Service, (bottom right) Valueline/PunchStock/Getty Images; **443** (left) PA Saez/Alamy, (top right) Michael Shake/Fotolia, (bottom right) Marc Romanelli/Photodisc/Getty Images; **450** (left) Stan Salzman, (right) Andresr/Shutterstock; **457** (left) Roza/Fotolia, (right) Gabriel Blaj/Fotolia; **467** (left) Peternile/Fotolia, (right) Kratuanoiy/Fotolia; **471** (top left) Jj2theb/Fotolia, (top right) Stan Salzman, (bottom left) Halfdark/PhotoAlto/Getty Images, (bottom right) Jack Hollingsworth/Digital Vision/Getty Images; **476** Michael Flippo/Fotolia; **477** Ricknoll/Fotolia; **484** (left) Paul Banton/Shutterstock, (right) Sharyn Young/Shutterstock; **485** (top left) Denis Pepin/Fotolia, (bottom left) Morey Milbradt/Brand X Pictures/JupiterImages, (right) EPA European Pressphoto Agency B.V./Alamy; **487** (left) Valueline/PunchStock, (top right) Stockbyte/Getty Images, (bottom right) Ricknoll/Fotolia; **494** (left) Sergey Yarochkin/Fotolia, (right) Vladimir Sazonov/Fotolia; **501** (top left) Tinka/Fotolia, (top, middle, and bottom right) Beth Anderson/Pearson Education, Inc., (bottom left) Doug Menuez/Photodisc/Getty Images; **502** (both) Beth Anderson/Pearson Education, Inc.; **503** (top left) Beth Anderson/Pearson Education, Inc., (top right) Comugnero Silvana/Fotolia, (bottom left) Sebastian Kaulitzki/Shutterstock, (bottom right) Max Delson Martins Santos/iStockphoto; **504** (top left) GK Hart/Vikki Hart/Photodisc/Getty Images, (top right) Ryan McVay/Photodisc/Getty Images, (bottom left) Beth Anderson/Pearson Education, Inc., (bottom right) David Frazier/Corbis; **506** (top) AP Images, (bottom) Iofoto/Fotolia; **509** (left) GK Hart/Vikki Hart/Photodisc/Getty Images, (middle) Alan Carey/Spirit/Corbis, (right) Ra'id Khalil/Shutterstock; **510** Beth Anderson/Pearson Education, Inc.; **516** Bongarts/SportsChrome/Newscom; **517** (left) Ryan McVay/Photodisc/Getty Images, (right) Mark Herreid/Fotolia; **526** (top left) Scott Rothstein/Shutterstock, (top right) Rookie72/Fotolia, (middle left) Manchan/Digital Vision/Getty Images, (bottom left) Duncan Smith/Photodisc/Getty Images, (bottom right) Beth Anderson/Pearson Education, Inc.; **528** Olivier Asselin/Alamy; **531** (left) Ryan McVay/Photodisc/Getty Images, (right) John A. Rizzo/Photodisc/Getty Images; **532** Tom Grill/Spirit/Corbis; **533** National Oceanic and Atmospheric Administration (NOAA); **553** (left) Mark Karrass/Spirit/Corbis, (right) Shock/Fotolia; **554** Tom Gannam/AP Images; **576** (left) Beth Anderson/Pearson Education, Inc., (right) Jacques Descloitres, MODIS Rapid Response Team/GSFC/NASA; **577** (left) Stephen Ausmus/U.S. Department of Agriculture, (right) Joe Sohm/Visions of America/Digital Vision/Getty Images; **578** Brian Tolbert/Spirit/Corbis; **588** (left) Neil Beer/Photodisc/Getty Images, (right) Motorlka/Fotolia; **595** Earl Orf; **624** Image100/Corbis; **625** (top) Chris Gloster/Fotolia, (bottom) Monkey Business Images/Shutterstock; **630** J. Luke/PhotoLink/Photodisc/Getty Images; **631** Phanie/SuperStock; **642** Michael Pettigrew/Fotolia; **659** (left) PhotoLink/Photodisc/Getty Images, (right) Ken Usami/Photodisc/Getty Images; **702** (left) Alan Carey/Spirit/Corbis, (right) Beth Anderson/Pearson Education, Inc.; **707** Yuri Arcurs/Fotolia; **710** Beth Anderson/Pearson Education, Inc.; **713** Kurhan/Fotolia; **715** Hwahl3/Fotolia; **716** (left) Karl Weatherly/Photodisc/Getty Images, (right) Sheer Photo, Inc/Stockbyte/Getty Images; **724** Noel Hendrickson/Blend Images/Getty Images; **725** Vibe Images/Fotolia; **726** Beth Anderson/Pearson Education, Inc.; **731** Diego Cervo/Fotolia; **732** Maksym Yemelyanov/Fotolia; **Page 733** Andresr/Shutterstock; **734** Pressmaster/Fotolia; **736** Stan Salzman; **741** Monkey Business Images/Shutterstock; **742** Evgenyb/Fotolia; **745** Michael Shake/Shutterstock; **749** Rena Schild/Shutterstock; **751** Beth Anderson/Pearson Education, Inc.; **753** Robert Kneschke/Fotolia; **754** Corbis; **755** Kristin Smith/Shutterstock; **760** Sean Justice/Corbis

Index

Study Skills

Are you feeling a little uncertain of your ability to learn and remember mathematics? You'll find the help you need on special study skills pages in this book. At the beginning of Chapter 1, read "Your Brain *Can* Learn Mathematics" to find out how your brain learns and remembers best. Then, throughout the rest of Chapters 1–4, watch for special activities to learn brain-friendly ways to use your textbook, take lecture notes, do homework, make study cards, prepare for and take math tests, and much more. It's your ticket to success in mathematics! (See the Table of Contents in the front of the book for study skills titles and locations.)

Formulas

(*See next page for geometry formulas.*)

Percent proportion: $\dfrac{\text{part}}{\text{whole}} = \dfrac{\text{percent}}{100}$

Percent equation: part = percent • whole

Simple interest: Interest = principal • rate • time

Converting from Celsius to Fahrenheit: $F = \dfrac{9 \cdot C}{5} + 32$

Converting from Fahrenheit to Celsius: $C = \dfrac{5(F - 32)}{9}$

$\text{Mean} = \dfrac{\text{sum of all values}}{\text{number of values}}$

Order of Operations

1. Do all operations inside *parentheses* or *other grouping symbols.*
2. Simplify any expressions with *exponents* and find any *square roots.*
3. *Multiply* or *divide,* proceeding from left to right.
4. *Add* or *subtract,* proceeding from left to right.

Example 1

$6^2 - 8 \div 2 + (7 + 3)$ — Work inside parentheses.

$6^2 - 8 \div 2 + 10$ — Simplify exponents.

$36 - 8 \div 2 + 10$ — Divide.

$36 - 4 + 10$ — Add or subtract from left to right.

$32 + 10$

42

Example 2

$4 + 10 \div 2(6) - 3(8 \div 4)$ — Work inside parentheses.

$4 + 10 \div 2(6) - 3(2)$ — Multiply or divide from left to right.

$4 + 5(6) - 3(2)$

$4 + 30 - 3(2)$

$4 + 30 - 6$ — Add or subtract from left to right.

$34 - 6$

28

A phrase to help you remember the order of operations:

Please	**Excuse**	**My**	**Dear**	**Aunt**	**Sally**
Parentheses;	Exponents;	Multiply	or Divide;	Add or	Subtract

Fraction-Decimal-Percent Equivalents

Geometry Formulas

Rectangle
Perimeter: $P = 2 \cdot l + 2 \cdot w$
Area: $A = l \cdot w$

Square
Perimeter: $P = 4 \cdot s$
Area: $A = s^2$, which means $s \cdot s$

Parallelogram
Perimeter: Add the lengths of the sides.
Area: $A = b \cdot h$

Triangle
Perimeter: Add the lengths of the sides.

Area: $A = \dfrac{1}{2} \cdot b \cdot h$

or $A = 0.5 \cdot b \cdot h$

Trapezoid
Perimeter: Add the lengths of the sides.

Area: $A = \dfrac{1}{2} \cdot h \cdot (b + B)$

or $A = 0.5 \cdot h \cdot (b + B)$

Circle
diameter: $d = 2 \cdot r$

radius: $r = \dfrac{d}{2}$

Circumference: $C = \pi \cdot d$
or $C = 2 \cdot \pi \cdot r$

Area: $A = \pi \cdot r^2$, which means $\pi \cdot r \cdot r$

Area of a semicircle: $A = \dfrac{\pi \cdot r^2}{2}$

Use 3.14 as the approximate value of π.

Right Triangle
$\text{hypotenuse} = \sqrt{(\text{leg})^2 + (\text{leg})^2}$

$\text{leg} = \sqrt{(\text{hypotenuse})^2 - (\text{leg})^2}$

Rectangular Solid
Volume: $V = l \cdot w \cdot h$

Cylinder
Volume: $V = \pi \cdot r^2 \cdot h$
or $V = \pi \cdot r \cdot r \cdot h$
Use 3.14 as the approximate value of π.

Cone
Volume: $V = \dfrac{1}{3} \cdot B \cdot h$

or $V = \dfrac{B \cdot h}{3}$

B is area of the circular base or $\pi \cdot r^2$.

Pyramid
Volume: $V = \dfrac{1}{3} \cdot B \cdot h$

or $V = \dfrac{B \cdot h}{3}$

B is area of the base.

Sphere
Volume: $V = \dfrac{4}{3} \cdot \pi \cdot r^3$

or $V = \dfrac{4 \cdot \pi \cdot r \cdot r \cdot r}{3}$

Use 3.14 as the approximate value of π.

Classifying Angles: Four Types of Angles
- Acute angles measure less than 90°.
- Right angles measure *exactly* 90°.
- Obtuse angles measure more than 90° but less than 180°.
- Straight angles measure *exactly* 180°.